AutoCAD 2011
One Step at a Time

Timothy Sean Sykes

Forager Publications
Spring, Texas
ForagerPub.com

Everyone involved in the publication of this text has used his or her best efforts in preparing it. These efforts include the development, research, and testing of the theories and programs to determine their effectiveness. The author and publisher make no warranty of any kind, expressed or implied, with regard to these programs or the documentation contained in this book. The author and publisher shall not be liable in any event for incidental or consequential damages in connection with, or arising out of, the furnishing, performance, or use of these programs.

Trademark info:
AutoCAD® and the AutoCAD® logo are registered trademarks of Autodesk, Inc.
Windows® is a registered trademark of Microsoft.

ISBN 978-0-9819867-4-6 (ebook)
ISBN 978-0-9819867-2-2 (print)

Copyright © 2010; Forager Publications.

Forager Publications
2043 Cherry Laurel
Spring, TX 77386
www.foragerpub.com

AUTHOR'S NOTES

Where to Find the Required Files and Review Questions for Using This Text:

All of the files required to complete the lessons in this text can be downloaded free of charge from this site:

http://foragerpub.com/AcadFiles/2011/2011.htm

Pick on the **Lesson Files** link and save the zip file to your computer. (Windows XP, Vista, and 7 will open a zipped file. If you need a zip utility, however, I suggest the free WinZip download at: http://www.winzip.com/. The evaluation version will do to get your files. You'll have to follow WinZip procedures to unzip the files.)

Once you've downloaded the zip file, follow these instructions to unzip it with Windows XP/Vista/7.

1. Double click on the file. A window will display with the Steps folder inside.
2. Right click on the Steps folder and select **Copy** from the menu that appears.
3. Using Windows Explorer, navigate to the C-drive.
4. Select **Paste** from the Edit pull down menu.

To get the review questions and additional exercises, go to the website indicated at the end of each lesson. There you can select your lesson's link, and a PDF file will open in your Internet browser. These copyrighted files are fully printable, or you can save them to your disk if you wish. The last page of each file contains the answers to the questions in that file.

Where to Find the Software

Autodesk provides tools at their website to purchase online or to help you locate a reseller near you. Go to the Autodesk website (Autodesk.com) and select **Purchase**, then select the option that suits you.

You can also download a free, thirty day trial version of the latest software at the Autodesk website. Again, go to the Autodesk website, select **Products – AutoCAD – Product trial**. The trial version works great for short term training!

Windows Vista and 7 Users

FONTS: The 2011 edition of *AutoCAD: One Step at a Time* falls during a transition period for the Windows operating system – many users have upgraded to the Vista or 7 system while many others prefer to remain with XP. AutoCAD should work with each, but the Vista and 7 systems offered a new font (Calibri) which offers a much finer printed appearance. Unfortunately, it isn't readily available for XP users.

For those who use Vista/7 and who want to take advantage of the Calibri font, use the files referenced in this text but which end with a "- v". For those who prefer the older Times New Roman font and those with XP systems, use the referenced file.

(I created most of the imagery in this text using the Calibri font.)

Contacting the Author/Publisher:

Although we tried awfully hard to avoid errors, typos, and the occasional boo boos, I admit to complete fallibility. Should you find it necessary to let me know of my blunders, or to ask just about anything about the text – or even to make suggestions about how to better the next edition, please feel free to contact me here:

http://foragerpub.com/contact/contact.htm.

I can't promise a fast response (although I'm usually pretty good about responding), but I can promise to read everything that comes my way.

An afterthought for that address: you need to leave me an accurate email address if you want me to respond. I promise that I won't sell, give away, or in any other way distribute your address to anyone else.

Frequently Asked Questions (and responses to less-frequently-asked but more annoying questions):

I went to the web site but got a Page cannot be found *error. What gives?*

This is the most common complaint I get. The fix is simple: make sure you capitalize the "Files" in the address. The web is case-sensitive.

How do I get eBook versions of your texts? What about your other books (I hear you write other things beside AutoCAD textbooks)?

The best place to get my books at a discount is at the Forager Publications site:

http://www.foragerpub.com

But these folks also offer healthy discounts (just search for my *full* name – Timothy Sean Sykes):

Powells.com

Amazon.com

Other sites may also offer discounts. As I learn of them I post them at the end of the specific book's page at the Forager Publications site.

Contents

Lesson 1: In the Beginning...

Let's Get Started 3
   1.1.1 The User Interface and Workspaces 4
   1.1.2 Toolbars and Menus 6
1.1.3 The Ribbon 10
   1.1.4 Tool Palettes 11
   1.1.5 The Graphics Area 12
   1.1.6 The Command Window and Status Bar 13
1.1.7 The View Cube & Navigation Bar 15
1.2 Setting Up a New Drawing 15
   1.2.1 The Groundwork: How AutoCAD Handles Scale, Units, and Paper Size 15
   1.2.2 The Setup 16
1.3 Saving and Leaving a Drawing Session 19
1.4 Opening an Existing Drawing 22
1.5 Navigating Between Drawings with Quick View Drawings 26
1.6 Creating and Using Templates 27
1.7 Extra Steps 29
1.8 What Have We Learned? 30
1.9 Exercises 31

Lesson 2: Drawing Basics - Lines, Circles, & Coordinates

2.1 Lines, Rectangles, and Circles 33
   2.1.1 Lines and Rectangles 33
   2.1.2 Getting Around to Circles 40
2.2 Fixing the Uh-Ohs: *Erase*, *Undo*, and *Redo/MRedo* 42
2.3 Multiple Object Selection Made Easy 45
2.4 The Cartesian Coordinate System 47
   2.4.1 Absolute Coordinates 47
   2.4.2 Relative Coordinates 48
   2.4.3 Polar Coordinates 49
   2.4.4 Practicing with Cartesian Coordinates 51
Extra Steps 52
What Have We Learned? 52
Exercises 53

Lesson 3: Drawing Aids

Simple Stuff: Ortho, Grid, Polar Tracking, and Snap 55
Ortho 55
Grid 56
Polar Tracking 57
Snap (Polar and Grid) 57
Now the Easy Way - *DSettings* 58
Miss the Point with OSNAPs 61
ng OSNAPs 67
nning OSNAPs via Dialog Box 68
nning OSNAPs via Menu 69
lters and Object Snap Tracking 70
t Filters 70
ect Snap Tracking 71

3.6 Isometric Drafting 72
3.7 Extra Steps 75
3.8 What Have We Learned? 76
3.9 Exercises 76

Lesson 4: Display Controls and Basic Annotative Text

4.1 Changing What You See 79
   4.1.1 Zooming 79
   4.1.2 Panning 84
   4.1.3 The Navigation Wheel 85
   4.1.4 The Navigation Bar 88
4.2 Why Find It Twice? – The *View* Command 88
4.2.1 Creating Views 89
   4.2.2 Showing Motion 92
4.3 "Simple" Text 93
4.4 Editing Text 100
4.5 Finding and Replacing Text 100
4.6 Adding Flavor to Text with *Style* 101
4.7 Extra Steps 104
4.8 What Have We Learned? 106
4.9 Exercises 107

Lesson 5: Geometric Shapes (Other Than Lines, Rectangles, and Circles!)

5.1 Ellipses and Isometric Circles 109
5.2 Arcs: The Hard Way! 112
5.3 Polygons 115
5.4 Putting It All Together 118
5.5 Extra Steps 123
5.6 What Have We Learned? 124
5.7 Exercises 124

Lesson 6: Object Properties - Color at Last (and More!)

6.1 Some Preliminaries 127
6.2 Adding Color 128
6.3 Drawing with Transparency 131
6.4 Drawing with Linetypes 132
   6.4.1 Using Linetypes 132
   6.4.2 Modifying Linetype Scale 135
6.5 Using Lineweights 137
6.6 Uh-Ohs, Boo Boos, Ah $%&s: Object Properties 138
   6.6.1 Tell Me About It – The *List* Command 138
   6.6.2 Object Properties and the Quick Properties Panel 139
   6.6.3 Quick Properties Settings 141
   6.6.4 Quick Properties and the Customize User Interface Dialog Box 142
6.7 Extra Steps 145
6.8 What Have We Learned? 146
6.9 Exercises 146

Lesson 7: Colors, Linetypes, and More Made Simple - Layers

7.1 Color, Linetype, and So Much More – Layers and How to Use Them 149
7.2 A Couple Managers 153
   7.2.1 The Layer Properties Manager 153
   7.2.2 The Layer States Manager 162
7.3 The Rest of the (Layer) Story 164
7.4 Sharing Setups: The AutoCAD Design Center and the Layer Translator 169
7.5 Extra Steps 174
7.6 What Have We Learned? 174
7.7 Exercises 175

Lesson 8: Editing Your Drawing - Modification Procedures

8.1 The Location and Number Group 177
   8.1.1 Here to There: The *Move* Command 177
   8.1.2 The *Copy* Command: From One to Many 178
   8.1.3 AddSelected 180
   8.1.4 Okay, Move It – But Then Line It Up: the *Align* Command 180
   8.1.5 Parallels and Concentrics – The *Offset* Cmd 182
   8.1.6 Rows, Columns, and Circles – the *Array* Command 184
   8.1.7 Opposite Copies – the *Mirror* Command 188
8.2 Moving and Copying Objects *between* Drawings 190
8.3 The Change Group 192
   8.3.1 Cutting It Out with the *Trim* Command 192
   8.3.2 Adding to It with the *Extend* Command 194
   8.3.3 Redundancy – Thy Name is AutoCAD: The *Break* Command 195
   8.3.4 Now We Can Round That Corner: The *Fillet* Command 196
   8.3.5 Fillet's Cousin: The *Chamfer* Command 198
8.4 Extra Steps 199
8.5 What Have We Learned? 199
8.6 Exercises 200

Lesson 9: Some More Editing Tools ... & Grips!

9.1 More Commands in the Change Group 203
   9.1.1 Two Ways to Change the Length of Lines and Arcs – The *Lengthen* and *Stretch* Cmds 203
   9.1.2 "Oh, NO! I Drew It Upside Down!" – The *Rotate* Command 206
   9.1.3 "Okay. Give Me Three Just Like It, But Different Sizes." The *Scale* Command 208
9.2 Identifying the Changes – The *RevCloud* Cmd 209
9.3 "A Whole New Ball Game!" Editing with Grips 212
9.4 Putting It All Together 218
9.5 Extra Steps 222
9.6 What Have We Learned? 223
9.7 Exercises 223

Lesson 10: Geometric Constraints & Alternate Selection Methods

10.1 Constraining Relationships: Constraint Symbols, Insertions, and Behaviors 225
10.2 Auto Constraints and Display Settings 230
10.3 Inferred Constraints 232
10.4 Object Selection Tools 234
10.4.1 Selection Modes 234
10.5 Selection Cycling 239
10.6 *SelectSimilar* and the Object Isolation Tools 240
10.7 Extra Steps 242
10.8 What Have We Learned? 242
10.9 Exercises 243

Lesson 11: Some More Cool Tools

11.1 So Where's the *Point*? 245
11.2 Equal or Measured Distances – The *Divide* and *Measure* Commands 246
11.3 From Outlines to Solids – The *Solid*, *Donut*, and *Wipeout* Commands 249
11.4 AutoCAD's Measurement Tools 252
   11.4.1 How Long or How Far – The *Dist* Tool 253
   11.4.2 The Radius Tool 254
   11.4.3 The Angle Tool 255
   11.4.4 Calculating the Area 256
   11.4.5 The Volume Tool 259
   11.4.6 Identifying Any Point with *ID* 260
11.5 More Object Selection Methods 260
11.6 Object Selection Filters – Quick Filters 2
11.7 AutoCAD's Calculator 266
11.8 Extra Steps 270
11.9 What Have We Learned? 270
11.10 Exercises 271

Lesson 12: Polylines and the Actio

12.1 Using Polylines for Wide Lines a Segmented Lines 273
12.2 Editing Polylines – The *PEdit* C
12.3 AutoCAD Macros – The *Actio*
12.4 Extra Steps 291
12.5 What Have We Learned? 29
12.6 Exercises 292

Lesson 13: Guidelines and

13.1 Contour Lines with the
13.2 Changing Splines – Th
13.3 Guidelines 305
13.4 Extra Steps 311
13.5 What Have We Lea
13.6 Exercises 311

Lesson 14: Advanced Lines - Multilines

14.1 Many at Once – AutoCAD's Multilines and the *MLine* Command 313
14.2 Options: The *MLStyle* Command 315
14.3 Editing Multilines: The *MLEdit* Command 321
14.4 The Project 324
14.5 Extra Steps 326
14.6 What Have We Learned? 327
14.7 Exercises 327

Lesson 15: Advanced Text - MText

15.1 AutoCAD's Word Processor: The Multiline Text Editor 329
15.2 Okay I Typed It, but I Don't Know If It's Right! – AutoCAD's *Spell* Command 341
15.3 Find and Replace – without the Multiline Text Editor 344
15.4 Columns 346
15.5 Extra Steps 348
15.6 What Have We Learned? 348
15.7 Exercises 349

Lesson 16: Basic Dimensioning

16.1 First, Some Terminology 351
16.2 Dimension Creation: Dimension Commands 353
  16.2.1 Linear Dimensioning 353
  16.2.2 Dimensioning Angles 355
  16.2.3 Dimensioning Radii and Diameters 357
  16.2.4 Dimension Arc Lengths 358
  16.2.5 Dimension Strings 360
  16.2.6 Aligning Dimensions 362
  16.2.7 Baseline Dimensions & Spacing 363
  16.2.8 Ordinate Dimensions 364
16.3 And Now the Easy Way: Quick Dimensioning (*QDim*) 366
16.4 Dimension Editing: The *DimEdit* and *DimTEdit* Commands 368
  16.4.1 Position the Dimension: The *DimTedit* Command 368
  16.4.2 Changing Value of the Dimension Text: The *DimEdit* Command 369
  16.4.3 Breaking an Extension Line - *DimBreak* 371
16.5 Isometric Dimensioning 372
16.6 Extra Steps 373
16.7 What Have We Learned? 374
16.8 Exercises 375

Lesson 17: Customizing Dimensions and Using Dimensional Constraints

17.1 Using Dimension Styles 377
17.2 Rules of Annotative Dimensioning 390
17.3 391

17.4 Overriding Dimensions 393
17.5 Dimensional Relationships (Constraints) 394
  17.5.1 The Basics 394
  17.5.2 Managing Dimensional Constraints 397
17.6 Extra Steps 400
17.7 What Have We Learned? 401
17.8 Exercises 401

Lesson 18: Leaders

18.1 The *MLeader* Command 405
  18.1.1 Placing Leaders 405
  18.1.2 Aligning and Collecting Leaders 408
  18.1.3 Editing Leaders 411
18.2 Customizing Leaders 413
18.3 Extra Steps 418
18.4 What Have We Learned? 418
18.5 Exercises 419

Lesson 19: Tables and Fields

19.1 Tables 421
  19.1.1 Creating a Table 421
  19.1.2 Editing a Table's Properties 426
  19.1.3 Table Column Tools and Adding Values to Table Cells 430
  19.1.4 Customizing Tables 432
19.2 The Wonders of Automation – Fields 437
  19.2.1 Drawing Properties 437
  19.2.2 Inserting Fields 439
  19.2.3 Calculating Fields 441
19.3 Altogether Now: Tables, Fields, and MS Excel 443
19.4 Extra Steps 446
19.5 What Have We Learned? 447
19.6 Exercises 447

Lesson 20: Hatching and Grouping

20.1 Hatching and Filling 449
20.2 Editing Hatched Areas 455
  20.2.1 Normal Hatch Editing 455
  20.2.2 Non-Associative Hatch Editing 456
20.4 Paper Dolls: The *Group* Command 457
20.5 Extra Steps 464
20.6 What Have We Learned? 465
20.7 Exercises 465

Lesson 21: Many as One - Blocks

21.1 Groups with Backbone – The *Block* Commands 469
  21.1.1 Template Library Creation 470
  21.1.2 Folder Library Creation 473
  21.1.3 Using Blocks in a Drawing – The *Insert* Command 474
21.2 Dynamic Blocks 476

21.3 Getting Blocks from a Folder Library on a Web Site via AutoCAD's iDrop 496
21.4 Extra Steps 498
21.5 What Have We Learned? 498
21.6 Exercises 499

Lesson 22: Advanced Blocks

22.1 Attributes 501
   22.1.1 Creating Attributes 501
   22.1.2 Block Tables 507
22.2 Inserting Attributed Blocks 509
22.3 Editing Attributes 513
   22.3.1 Editing Attribute Values 513
   22.3.2 Editing Attribute Definitions 516
22.4 The Coup de Grace: Using Attribute Information in Bills of Materials, Spreadsheets, or Database Programs 519
22.5 Extra Steps 525
22.6 What Have We Learned? 525
22.7 Exercises 526

Lesson 23: Sharing Your Work with Others

23.1 The Old-Fashioned Way – Putting It on Paper (Plotting) 529
   23.1.1 First Things First – Setting Up Your Printer (or Plotter) 529
   23.1.2 Plot Styles 530
   23.1.3 Setting Up the Page to Be Plotted 530
23.2 Sharing Your Drawing with the *Plot* Command – and *No Paper*! 535
   23.2.1 Quickly Creating a Drawing Web Format (DWF) or Adobe PDF File with the *Export* Command 535
   23.2.2 Viewing a Drawing Web Format (DWF) File 538
   23.2.3 Multiple Plots and Creating a DWF File – with Hyperlinks! 542
   23.2.4 Hyperlinks 544
   23.2.5 AutoCAD can Create Full Web Pages, Too! 547
23.3 Sending the Package over the Internet with *eTransmit* 552
23.4 Extra Steps 558
23.5 What Have We Learned? 558
23.6 Exercises 558

Lesson 24: Space for a New Beginning

24.1 Understanding the Terminology 561
24.2 Using Tiled Viewports 562
24.3 Setting Up a Layout Environment (Paper Space) 568
24.4 Using Floating Viewports 570
   24.4.1 Creating Floating Viewports Using *MView* 571
   24.4.2 The Viewports Panel 573
   24.4.3 Adjusting the Views in Floating Viewports 573
   24.4.4 the Position of the Floating Viewport 576

24.5 And Now the Easy Way – The *LayoutWizard* Command 576
24.6 Extra Steps 580
24.7 What Have We Learned? 580
24.8 Exercises 581

Lesson 25: The New Beginning Continues ...

25.1 Dimensioning and Paper Space 585
   25.1.1 Dimensioning and Paper Space – the Annotative Way 585
   25.1.2 Dimensioning and Paper Space – the Other Way 587
25.2 The Benefits of Layers in Paper Space 588
25.3 Using Text in Paper Space 591
25.4 Plotting the Layout 592
25.5 Tweaking the Layout 593
   25.5.1 Modifying Viewports with the *MVSetup* Command 593
   25.5.2 Changing the Shape of a Viewport with the *VPClip* Command 596
25.6 Putting It All Together 597
25.7 Extra Steps 604
25.8 What Have We Learned? 605
25.9 Exercises 606

Lesson 26: More Sharing Tools - Drawing Sets

26.1 Sheet Sets – A Primer 609
26.2 Using the Sheet Set Manager to Organ' Project 609
   26.2.1 The Sheet List Tab 609
   26.2.2 The Sheet Views Tab 616
   26.2.3 The Model Views Tab 621
26.3 Using Your Sheet Sets to Share Inf
26.4 Extra Steps 628
26.5 What Have We Learned? 628
26.6 Exercises 629

Lesson 27: Externally Referen DWFs, & DGNs

27.1 Working with Externally Re DWFs, and DGNs 6
   27.1.1 Attaching and Detac to Your Drawing 637
   27.1.3 Xrefs and Depend
   27.1.4 Unload, Reload, a
27.2 Editing Xrefs 646
27.3 Using Our Drawing
27.4 Binding an Xref to
27.5 Extra Steps 652
27.6 What Have We L
27.7 Exercises 654

Lesson 28: Other Application Files

28.1 Two Types of Graphics 657
28.2 Working with Raster Images: The Image Mngr 658
   28.2.1 Attaching, Detaching, Loading, and Unloading Image Files 658
   28.2.2 Clipping Image Files 661
   28.2.3 Working with Image Files 662
28.3 Exporting Image Files 665
28.4 Working with Linked Objects – Object Linking and Embedding (OLE) 666
   28.4.1 Inserting Other Application Data into AutoCAD Drawings 666
   28.4.2 Modifying OLE Objects 671
   28.4.3 AutoCAD Data in Other Applications 672
28.5 Extra Steps 673
   28.6 What Have We Learned? 674
28.7 Exercises 675

Appendices

Appendix – A: Initial Setup 678
Appendix – B: Drawing Scales 679
Appendix – C: Function Keys and Their Uses 680
Appendix – D: MText Keystrokes 681
Appendix – E: Dimension Variables 682
Appendix – F: Hotkeys 685
Appendix – G: Actions & Parameters Chart for Dynamic Blocks 686

Lesson 1

Following this lesson, you will:

- ✓ Know how to create a new drawing
- ✓ Know how to open an existing AutoCAD drawing
- ✓ Know how to save and close an AutoCAD drawing
- ✓ Recognize the various parts of the AutoCAD User Interface
- ✓ Be familiar with the AutoCAD Info Center

In the Beginning ...

How do you begin a new drafting project?

The number of answers I've heard to this question over the years might surprise you. Some say you begin with the layout. Others say you begin by positioning the drawing on the page. The most experienced drafters generally come up with a quip about coffee and radio settings.

This doesn't mean that I was questioning bad drafters, but simply that experience has engrained the basics so deeply that good drafters don't think about these simple questions anymore.

So, you might say, how do you begin a drafting project?

Well, first you decide on the scale and units (engineering, architectural, etc.) that you'll use. Then you decide the size of the page on which you'll draw (even if your final drawing will be electronic, you'll need to set up for a page size).

This first lesson will introduce you to the AutoCAD User Interface – the AUI, or the screen on which you'll draw – and show you how to set up things like units and sheet size.

It's probably best to jump right in. So as the heading says ...

1.1 Let's Get Started

AutoCAD works much like any other windows program; that is, you start with the shortcut on the desktop (or the Start menu: **Start – All Programs – Autodesk –** and so forth). You can save a drawing using the Windows' CTRL+S method or the standard **File – Save** approach (although you many not find the usual pull down menus available at first). Finally, you can close a file using the **Exit** button ⊠ at the right end of the title bar.

We'll look at each of these procedures in more detail throughout this lesson. But first, let's open AutoCAD and take a look at it.

> If you haven't opened AutoCAD before, you may discover an initial setup dialog box. Pick the **Skip** button [Skip] in the lower right corner for now. For more information on this dialog box, refer to **Appendix A: Initial Setup.**

Do This: 1.1A	Starting a Drawing Session

I. Follow these steps.

1.1A: STARTING A DRAWING SESSION

1. When you loaded AutoCAD, a shortcut that looks like this appeared on your desktop. Double click this with your left mouse button to launch AutoCAD. It will open (it may take a moment or two), and you'll see something like the AutoCAD User Interface (the AUI) shown in Figure 1.001 (without the annotations)[*]. Don't worry if it doesn't look exactly like this just yet; we'll fix it in the next section.

---

[*] AutoCAD's default background color is actually dark gray. We'll use a white background for clarity throught this text.

1.1A: STARTING A DRAWING SESSION

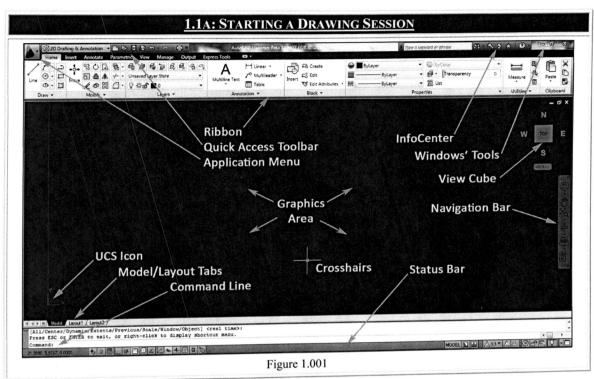

Figure 1.001

Let's take a few minutes to familiarize ourselves with what we're seeing.

> **IMPORTANT IMPORTANT IMPORTANT**
>
> As you can see, AutoCAD's initial work area is fairly dark. To conserve ink (and save on printing costs), we use a lighter background in the instructional and working files for this text. Follow this simple procedure to set your work area to the lighter background.
>
> 1. Enter the **Options** command – that is, type **options**. You'll see that your text appears in the command line (see Figure 1.001).
> 2. When you hit the ENTER key, AutoCAD will display an Options dialog box.
> 3. Select the **Profiles** tab at the right end of the dialog box.
> 4. Pick the **Import** button. AutoCAD displays a standard Windows Open dialog box.
> 5. Navigate to the C:\Steps\Lesson01 folder (or the folder in which you placed the working files for this text), and select the *One Step at a Time Profile.arg* file.
> 6. Pick the **Open** button. AutoCAD presents an Import Profile dialog box.
> 7. Pick the **Apply & Close** button.
> 8. Select *One Step at a Time Profile* in the **Available profiles** list box and pick the **Set Current** button.
> 9. Pick the **OK** button to close the Options dialog box.
>
> Now you're set up just as I am as I begin the text so we're all "on the same page" (so to speak ☺).
> Note: To reset to AutoCAD's defaults, simply repeat Steps 1 – 3 and pick the **Reset** button.)

1.1.1 The User Interface and Workspaces

Where to Find It:

d Line:	Workspace
:	Workspace Switching
	Quick Access – Workspace

AutoCAD comes to you with a few user-interfaces – called *Workspaces* – to help satisfy 2-dimensional users, 3-dimensional users, and your initial setup choices. You'll find these workspaces listed in the **Workspaces** popup menu (Figure 1.002), presented when you pick the **Workspace Switching** button ⚙ on the right end of the status bar (Figure 1.001) or on the left end of the Quick Access Toolbar (Figure 1.001). These include:

- **2D Drafting & Annotation** offers a simple 2D graphics interface (see Figure 1.001). This is the one we'll be using. It mimics the Microsoft Office 7 look (with a ribbon), so it should be an easy transition for its users, and it groups its ribbon tools into tabs and panels for easy access.
- **3D Basics** provides the basic (simpler) tools needed to create and visualize a 3D model. Good for beginners, I can't recommend it for more serious users. (Consider it *3D-Lite!*)
- **3D Modeling** presents all the ribbon tools designed to facilitate three-dimensional work. You'll use this interface when you graduate to the *3D AutoCAD 2011: One Step at a Time* text.
- **AutoCAD Classic** presents the "olde" look with toolbars and no ribbon. I recommend ignoring this one; toolbars won't be with us much longer!

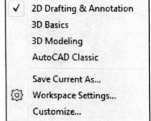

Figure 1.002

Do the following exercise to use the **2D Drafting & Annotation** workspace (the one we'll use throughout this text, and the default shown in Figure 1.001).

Do This: 1.1.1 A	Selecting a Workspace

I. Be sure AutoCAD is open, and follow these steps.

1.1.1A: SELECTING A WORKSPACE

1. Pick the **Workspace** control ⚙ on the left end of the Quick Access toolbar.
2. Select **2D Drafting & Annotation** as shown in Figure 1.002.

Nothing to it!

Of course, you're not limited to just the predefined workspaces. You can open or close palett ribbon panels and other items as you wish, move them around the interface, dock or undock and perform a host of other customizations. Then, once you're pleased with your user-inte *workspace*) – you have all your *gizwitchies* where you want them – you can save the layor case someone else comes along and changes things. (Alternately, you can have a variety setups to use for your different projects.)

Select **Save Current As...** from the menu (Figure 1.002), or use the *WSSave* comman workspace. It looks like this (on the command line* – see Figure 1.001, p.4):

Command: *wssave*

[AutoCAD presents a dialog box asking for the name of the workspace. Ent name and pick the **Save** button.]

That's all there is to saving your workspace so you can return to it at any time. Us *WSCurrent* command to restore your workspace. The command looks like this:

Command: *wscurrent*

Enter new value for WSCURRENT <" ">: *MyWorkspace*

You'll also see options (Figure 1.002) for **Workspace settings**, which presents you control what the popup menu shows, and for **Customiz**(ing) workspaces. presents the Custom User Interface (CUI) dialog box. Customization is beyc so let's move on and take a look at some of the items you can include in you of the other tools AutoCAD makes available.

---

* Commands may also appear on the dynamic input display next to the cursor. V 1.1.6, p.11.

> You can also access the tools available on the popup menu via the command line. Just use the ***Workspace*** command.

1.1.2 Toolbars and Menus

Let's start with what's similar to other Windows' applications. We find the title bar at the top, with a three button toolbar (the Windows tools shown in Figure 1.001, p.4) on the right. From left to right, these buttons and their uses are:

- This reduces the software to a pick on the taskbar (the bar across the bottom of the Windows screen).
- This makes the size of the AutoCAD window adjustable.
- This exits the software.

Okay, you knew those buttons. Let's look at the other end of the title bar – at the Quick Access toolbar and the Application Menu.

The Quick Access Toolbar

Figure 1.003

Figure 1.004

The Quick Access toolbar (Figure 1.003) contains another way to change your workspace (Section 1.1.1, p.4) and some common, often-used buttons. From the left, these include: **New** (or *QNew*), **Open**, **Save**, **SaveAs**, **Undo**, **Redo**, and **Plot**. To determine the function of these or other buttons on the toolbar, place you cursor over the button and wait a second. You'll see a written description (a *tooltip*) appear in a tooltip box. To see an explanation of how the tool works, use the F1 key on your keyboard.

You can easily add or remove items to the Quick Access toolbar by selecting an option from the menu (Figure 1.004) that appears when you select the **More** button ▾ at the right end of the toolbar. Use the **More Commands** option if the command you wish to add doesn't appear in the list, but try not to overdo it; you'll soon find other ways to access commands quickly.

Perhaps more important that the buttons it contains, the Quick Access toolbar also provides access (via the same menu – Figure 1.004) to some other tools. You can use the Quick Access toolbar to move the toolbar **Below the Ribbon** and even **Show** the familiar **Menu bar** (Figure 1.005)! If you're accustomed to using drop-down menus, you'll really appreciate this option.

Figure 1.005

Take a minute to explore these menus. You'll notice that the AutoCAD crosshairs become a cursor when moved out of the graphics area. Place the cursor on one of the menus and click with the left mouse button. A menu will appear from that location showing you various options related to the selection. Picking any of these options will activate an AutoCAD macro.

> A *macro* is a series of commands or events responding to a single input. Generally, you can't tell this from a single command.

You can also make changes to the Quick Access toolbar via a cursor menu (Figure 1.006). (To access a cursor menu, simply select a tool or an object with the right mouse button.)

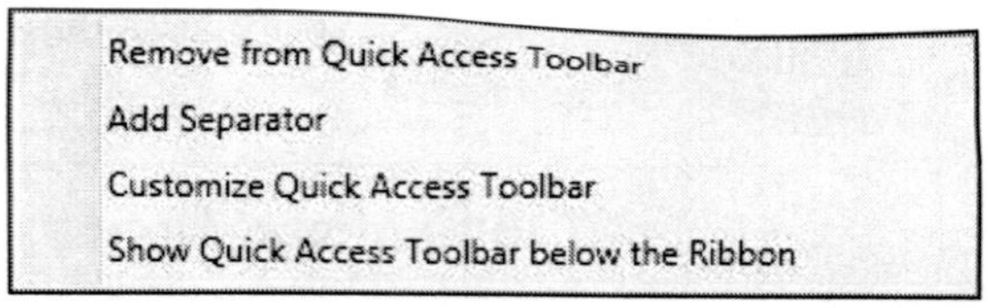

Using this menu, you can **Remove** a tool from the toolbar, **Add** [a] **Separator** between tools or groups of tools, move the **Toolbar below the Ribbon**, or **Customize** [the] **Quick Access Toolbar**. (I don't recommend customizing anything just yet. Let's wait until you're a bit more familiar with the application.)

Figure 1.006

For more information on cursor menus, see **The Cursor Menus** section (p.9).

Adding and Removing Other Toolbars

Although AutoCAD has moved to a ribbon-based tool system, it still provides several toolbars to assist you.

To add a toolbar, right click 🖱 on an existing toolbar and select the desired toolbar from the menu that appears. Alternately, you can use the pull down menus to select a toolbar to add. (See p. 8 for information on how to show the pull down menus.) Follow this path: **Tools – Toolbars – AutoCAD – [select toolbar]**. To remove a toolbar from the screen, undock it (as detailed in the following paragraphs), and then pick the "X" found in the upper right corner of the undocked bar.

A docked toolbar appears "attached" to a side, top, or bottom of the graphics window. An undocked toolbar appears to float over the screen. Drawing objects may exist behind an undocked toolbar, but a docked toolbar forces the drawing to move over.

To undock a toolbar, place your cursor on the two lines found on the left end (or top) of the toolbar and drag it free from its dock. (To drag an object, select it with the left mouse button, but don't release the button until you have relocated the object.)

To dock a toolbar, place your cursor in one of its wings (the gray areas on the sides) and drag it to a docking site (one of the borders of the graphics area).

Once you've docked or undocked your toolbars, avoid accidentally moving them by picking the **Lock Toolbars** icon 🔒 at the right end of the status bar. This will produce a popup menu (Figure 1.007) where you can lock one type of toolbar or all the toolbars on the screen. (Override the lock by holding down the CTRL key while dragging the toolbar.)

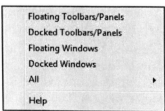

Figure 1.007

I'll have you display an occasional toolbar as we proceed.

The Application Menu

Use the **Application Menu** button on the left end of the title bar to open the Application Menu (Figure 1.008, p.8). Notice the menu list down the left side of the browser? Most of these tools also exist

on the Quick Access toolbar, but you'll find a few more tools here – and more options for the Quick Access tools.

Let's take a look.

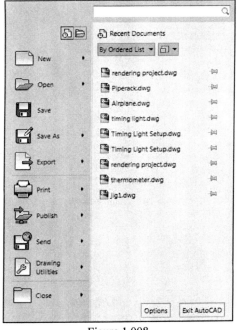
Figure 1.008

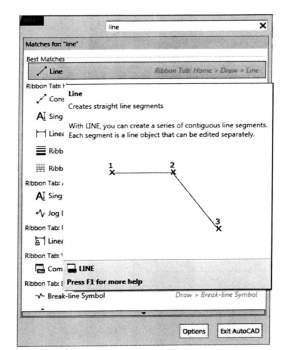
Figure 1.009

- First, there's a *Search* box across the top. Enter a command (or the first letter or two of a command) and AutoCAD will help you find the command.* Then when you hover the cursor over it, AutoCAD will produce an enhanced tooltip (Figure 1.009).
  Pretty cool, huh?
- Below the search box, the menu divides into left and right columns. What you select on the left determines what you see on the right. Let's start with the right.
  o By default, the right side lists either recently accessed drawings or currently open drawings. You'll control which with a pair of buttons found atop the left side list: the **Recent Documents** button and the [currently] **Open Documents** button.
  o When viewing the default menu, you'll find a pair of list buttons (atop the right column) to help you control how AutoCAD displays the right column.
    - The first `By Ordered List` – available when you're viewing recent documents – displays a short menu to help you order the list. You can order by **Ordered List** (most recently opened first), **Access Date**, **Size**, or **Type** of file.
    - The second allows you to change the size of the icon used in the list. This can be handy as the larger images give a thumbnail of the drawing.
  o One of the real benefits of the application menu appears when you hover your cursor over a drawing in the **Recent Documents** or **Open Documents** list (right column). AutoCAD presents an image of the drawing (Figure 1.010, p.9) and some additional information (path, date modified, etc.) to help you select the right one! Pick with left mouse button (that is, *select*) on the drawing you wish to open.
- The left column of the application menu (Figure 1.008) contains a list of function groups. Each selection presents several commands in the right column.

---

* AutoCAD searches the Quick Access toolbar, Application Menu, and current ribbon. If you can't find what you're looking for in these, use the Help menu (F1). We'll discuss Help in Lesson 9.

- **New** allows you to begin a new drawing or a new sheet set (Lesson 26, p.608). We'll look at creating a new drawing shortly.
- **Open** presents tools for opening an existing drawing or sheet set (Lesson 26). You'll also find a tool for importing a DGN file.
- **Save**, the only option which doesn't present a group of tools, presents a standard Windows Save dialog box. (**Saveas** opens a Save Drawing As dialog box.) We'll look at saving your drawing in Section 1.3, p.19.
- **Export** is something like **Saveas** except that you can export your drawing as a DWF (Autodesk's drawing web format), PDF (Adobe Acrobat file), or even a DGN (Microstation drawing file). We'll look at exporting files in Lesson 28, p.656.

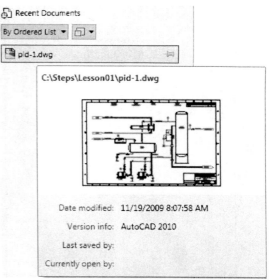
Figure 1.010

AutoCAD also provides a pair of rather sophisticated tools called *PartialOpen* and *PartialLoad*. These rarely used items might serve best to confuse at this point so we won't discuss them. You can, however, pull the supplement *Partial.pdf* from the website (http://www.uneedcad.com/Files) for future reference if you intend to work with extremely large drawing files.

- Autodesk provides three methods of sharing your drawing with others – **Print**, **Publish**, and **Send**. We'll explore these in some detail in Lesson 23.
- **Close** offers options for closing the current drawing or all open drawings.
- **Drawing Utilities** offers several maintenance tools to help you set up the drawing and keep it in good working order.

We'll look at these tools individually as we proceed.

- You'll also find an **Exit AutoCAD** button and an **Options** button at the bottom of the Menu Browser. You can use the **Exit AutoCAD** button to close the application or you can just use the **X** button on the other end of the title bar. Either way, AutoCAD will prompt you to save any unsaved work. The **Options** button opens AutoCAD's Options dialog box. I don't recommend doing anything with this until you're much more familiar with AutoCAD.

That's a busy menu. But we're not finished with menus yet!

The Cursor Menus

One of a less apparent set of menus will appear anytime you right click (pick with the right mouse button) in your drawing. These *cursor menus* (or pop-up screen menus) help you watch the screen more and spend less time searching for the right thing to press on the keyboard!

We have to give the programmers credit for accomplishing a minor miracle with the right mouse button. They've provided one of AutoCAD's most dynamic tools. Which cursor menu appears when the right button is picked depends upon: whether or not there's a command in progress, if there are items selected on the screen, and where the cursor/crosshairs are located when the button is picked!

AutoCAD provides five basic modes (or types) of cursor menus: *default*, *edit-mode*, *dialog-mode*, *command-mode*, and *other* menus.

- The *default* menu appears when you right click anywhere in the graphics area if no command is active and no selection set is available (that is, nothing has been selected on the screen with which to work). The default menu contains general, frequently used commands.

- The *edit-mode* menu appears when objects have been selected on the screen, but no command has been given. You'll see the best example of this when we discuss the Grips menu in Lesson 9, p.212. Generally, you can expect grip-type modifying tools. But object-specific tools will appear for some objects (such as dimensions).
- When you right click 🖱 in a dialog box, AutoCAD may present a *dialog-mode* menu. The tools presented will depend upon which dialog box you're using and the cursor's location when you click.
- When you right click 🖱 while there's a command in progress, AutoCAD presents a *command-mode* menu with tools specific to that command.
- The "*other* menus" category includes all the other menus that don't fit easily into the first four categories. Right clicking 🖱 in different locations and at different times might present some surprising results. Feel free to experiment while you're in a drawing. Try right clicking 🖱 over the different buttons on the status bar.

We'll spend more time exploring the cursor menus as they apply to specific areas throughout this book. But now, let's take a look at the *Ribbon*.

1.1.3 The Ribbon

The *ribbon*, displayed by default across the top of the graphics area, consists of several *tabs* which contain a myriad of *panels*. Each panel, in turn, contains several tools (command buttons). We'll look at individual commands as we proceed through this text. For now, we'll just look at the ribbon itself, and how it works.

The ribbon appears to dominate the entire graphics area; let's face it, it's big ... probably too big to remain in its default location without some adjustments (especially if you're using a widescreen monitor). Luckily, AutoCAD makes it easy to *minimize* the space occupied by the ribbon. Start by picking the down arrow next to the **Minimize** button ▬▼ on the ribbon. select **Minimize to Panel Titles** from the flyout menu (Figure 1.011). This way, even when you hide the ribbon, the panels will remain visible. The ribbon should disappear. But don't

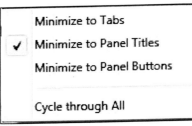

Figure 1.011

worry, it isn't gone for keeps. Move your cursor over one of the panel title bars, and the panel will reappear. But just look at all the screen area you have to work in when the ribbon is hidden!

Alternately, you can opt to **Minimize** [the ribbon] **to Panel Buttons**. This reduces the size of the ribbon as well (although not as much as the **Panel Titles** option). When reduced to buttons, hovering your cursor over a button will cause AutoCAD to display that panel's tools as shown here.

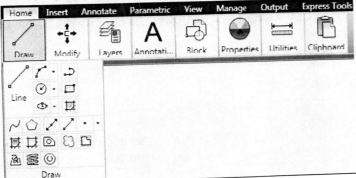

You'll find another menu (Figure 1.012) with several other options for manipulating the size, shape, and location of the ribbon when you right click 🖱 on any dark gray area of the ribbon.

> Select a different tab for a different set of panels; remember, each panel has a different collection of tools. We'll examine each panel in more detail as we proceed.

If you're using a widescreen monitor, you might dock your ribbon to one side (I use the right side as it's consistent with the location of AutoCAD's older dashboard tool which served the same purpose.) Dock it to one side and "auto–hide" it to free up a tremendous about of graphics area where you can work.

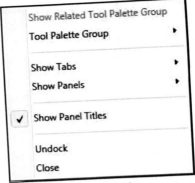

Figure 1.012

When it isn't docked, the ribbon has a titlebar (Figure 1.013) with some useful buttons. From the top, these include:

- The **Exit** button ✕ , which will close the ribbon, thus removing it from the screen and allowing you better access to the drawing (graphics) area. (Restore it with the **Ribbon** command.)

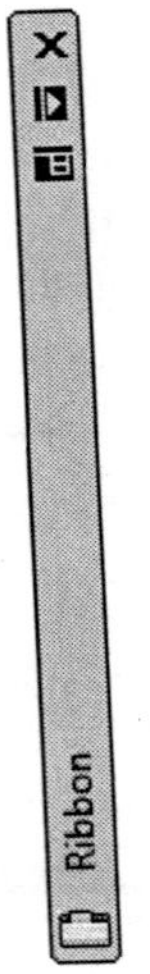

    - An **Auto-hide** button, which looks like this. This is the hide/unhide toggle.
    - A **Properties** button, below the **Auto-hide** button, which calls the menu shown in Figure 1.014. Let's look at the selections available here.
        o When you select **Move**, AutoCAD changes the cursor to a four-sided arrow, which you can use to drag the ribbon to a new location without docking it.
        o The **Size** selection also changes the cursor to a four-sided arrow. This one you can use to resize the ribbon.
        o Of course, **Close** does just that; it closes the ribbon.
        o Selecting **Allow Docking** will either check or uncheck the option. When checked, the ribbon can be docked (just like any toolbar) to the side/top/bottom of the graphics area. When unchecked, it can't be docked.
        o **Anchor Left <** and **Anchor Right >** will dock the ribbon to the selected side window.
    o **Auto-hide** is a menu approach to do the same thing the **Auto-hide** button does.
    o **Show Panel Titles**, when checked, causes a small title bar next to or beneath each pane telling you what the panel contains.
    o **Customize** calls the CUI dialog box. Remember, we don't want to fool around with customization just yet!
    o **Help**, of course, calls AutoCAD's Help dialog box.

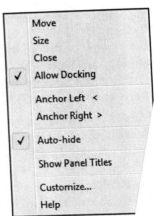

Figure 1.014

Figure 1.013

1.1.4   Tool Palettes

> If your screen doesn't display Tool Palettes (Figure 1.015) use one of the paths shown in the W **Find It** table [aka. the WTFI table]. (The palettes may appear docked – see **Adding and Remo Other Toolbars**, p.7.) You'll find similar tables for most commands discussed in this text.

Where to Find It:	
Command Line:	*ToolPalettes*
Hotkey(s):	*tp*
Ribbon (Tab/Panel):	View – Palettes – **Tool Palettes**
Menu:	Tools – Palettes – **Tool Palettes Window**
Toolbar:	Standard – **Tool Palettes Window**

Tool palettes (see the Tool Palettes Window in Figure 1.014, p.11) provide access to palettes that contain some of the same tools you found on the ribbon, as well as some of the more advanced (and useful) tools you'll discover in your basic studies. We'll go into more detail about these in specific lessons; but let's take a quick look at tool palettes now.

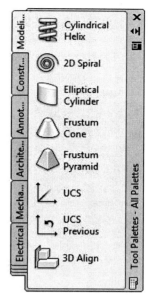

Figure 1.015

Looking at the side of the Tool Palettes window, you'll find several tabs. Selecting one will bring a different palette, with a different set of tools, to the front of the window. Pick on the overlapping tabs at the bottom to produce a list of all the palettes available to you.

The largest part of the palette contains its available tools. We'll discuss specific tools in some detail in later chapters.

The other side of the palette may contain a scrollbar to help access the tools, and a titlebar that works just like the ribbon's titlebar.

AutoCAD's tool palettes will prove useful when we discuss blocks and hatching. For now, however, they may prove a distraction to your basic lessons. You might be better off closing the Tool Palettes window or even docking it to the side of your screen and using **Auto-Hide** until we're ready for it.

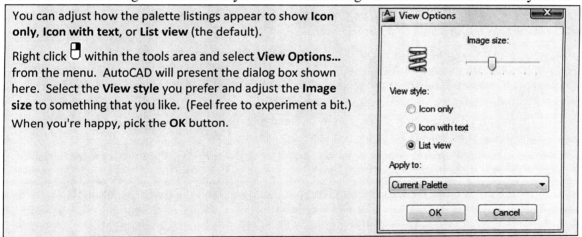

You can adjust how the palette listings appear to show **Icon only**, **Icon with text**, or **List view** (the default).

Right click within the tools area and select **View Options...** from the menu. AutoCAD will present the dialog box shown here. Select the **View style** you prefer and adjust the **Image size** to something that you like. (Feel free to experiment a bit.) When you're happy, pick the **OK** button.

1.1.5 The Graphics Area

AutoCAD dedicates the largest part of the screen – called the *graphics area* – to work space (refer back to Figure 1.001, p.4). You'll do your drawing and editing here.

Until the recent appearance of the ribbon, which we'll remove from the screen (or hide) during most of our work time, the only item allowed to occupy the coveted space on the graphics area was the User Coordinate System (UCS) Icon. The icon serves a very useful purpose in 3-dimensional drafting. For AutoCAD beginners in the 2-dimensional world, however, it serves as a reminder of the X- and Y-planes. (We'll discuss X- and Y-planes in more detail in Lesson 2, p.47, and the UCS Icon in more detail in our 3D AutoCAD text.) You can disable the icon (turn it off) by entering *ucsicon* at the command prompt and responding *off* to the prompt as shown in the following sequence.[*]

Command: *ucsicon*
Enter an option [ON/OFF/All/Noorigin/ORigin/Properties] <ON>: *off*

Repeat the command and enter *on* to restore the UCS Icon. You'll learn more about AutoCAD's prompts throughout this text.

> AutoCAD actually uses two coordinate systems, the UCS and the World Coordinate System (WCS), but that discussion belongs in the 3-dimensional world. The UCS and the WCS are the same for the lessons in this text.

We'll look at those other graphics area tools – the View Cube and the Navigation Bar – in Section 1.1.7, p.15.

1.1.6 The Command Window and Status Bar

Just above the bottom of the screen, you can see a window with several **Command** prompts (or lines of text). Like the ribbon, this window can undock, but I don't recommend it. You can also resize it by placing your cursor on the heavy line above it, picking with the left mouse button, and "dragging" the window up or down. (The cursor will become a double arrow ⇕ when properly located, and you must hold the button down for this procedure.) I usually leave my command window large enough for two or three lines as some prompts require extra lines to display information.

I can't overemphasize the importance of becoming familiar with the **Command** prompt. Look closely. AutoCAD speaks to you on the left side of the colon. You respond on the right. Right now, AutoCAD is "prompting" you, or asking you what you want to do. When you respond, either by keyboard entry or mouse selection, your response will appear to the right of the colon.

Once you enter a command, AutoCAD's prompt may change, asking you for more information or input. Just follow the prompts to complete your task.

> Command line messages also appear on the dynamic input display, which appears to follow your crosshairs. You can remove the command window from your display (and provide more working area) by entering **CommandLineHide** at the prompt. (Redisplay it by entering **CommandLine** ... but be sure the **Dynamic Input** toggle on the status bar is depressed first (it may appear as **DYN** rather than an icon)! (Alternately, you can hold down the CTRL button and press the '9' key on the keyboard.) I don't advise this setup. Some procedures still require the command line.
>
> You can also turn the dynamic input display on and off using the toggle on the status bar. We'll discuss dynamic input in more detail in Lesson 2, p.33.

AutoCAD also provides an enhancement to the command line (and the dynamic input display) that's quite useful. It's called *AutoComplete* functionality. If you're unsure of a command spelling, enter the first few letters and press the TAB key until you find the command you want! You can also use the up and down arrow keys to repeat previously used commands or data entry!

At the bottom of the screen is the *status bar*. On the left end of the status bar, you'll see a small box with three numbers in it. This **Coordinate Display** box shows the X, Y, and Z coordinates of the cursor in your drawing. We'll look at AutoCAD's coordinate system in Lesson 2, p.47.

To the right of the **Coordinate Display** box is a series of toggles (**Snap**, **Grid**, **Ortho**, etc.). We'll look at these in some detail in Lesson 3, p.55.

---

[*] The command sequence may also appear in the dynamic input display. See the first insert in Section 1.1.6, p.11.

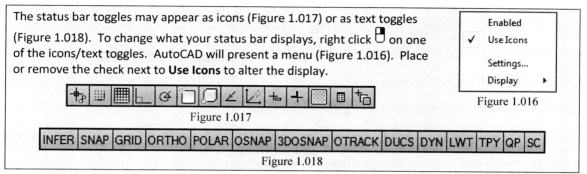

The status bar toggles may appear as icons (Figure 1.017) or as text toggles (Figure 1.018). To change what your status bar displays, right click 🖱 on one of the icons/text toggles. AutoCAD will present a menu (Figure 1.016). Place or remove the check next to **Use Icons** to alter the display.

Figure 1.016

Figure 1.017

Figure 1.018

On the right end of the status bar, you'll see several additional icons (Figure 1.019). These are (from the left):

Figure 1.019

- **Model** and **Layout** toggles – these control what type of "space" your graphics area displays. You'll do most of your actual work in Model space (the default). Layouts (aka paperspace) provide some extremely useful printing tools; we'll discuss these in Lessons 24, p.560, and 25, p.584. The textual **Model** toggle MODEL controls in which space you actually work. Again, this'll become clearer in Lessons 24 and 25.

If your screen displays **Model** and **Layout** tabs \Model / Layout1 / Layout2 /, the toggles may not appear on the status bar. Use the tabs for the same purpose. To show the icons rather than the tabs, right click 🖱 on the tab and select **Hide Layout and Model Tabs** from the menu.

- **Quick View Layouts** and **Quick View Drawings** serve similar functions – although the first works for layouts in the current drawing (more on layouts in Lessons 24/25) and the second for open drawings. We'll examine **Quick View Drawings** in Section 1.5, p.26.

- The next three tools **Annotation Scale**, **Annotation Visibility**, and **Add Scale to Annotative Objects** are related. We'll spend time with these in Lesson 4.

- Next you'll find the **Workspaces** toggle which we've already discussed (Section 1.1.1, p.4), then the **Toolbar/Window Positions Lock** toggle which locks the toolbars in their current positions, a **Hardware Acceleration** toggle to help improve AutoCAD's performance on your system, an **Autodesk TrustedDWG** icon which appears when the current drawing was created in AutoCAD, an **Isolate Objects** button which provides access to the *IsolateObjects* and the *HideObjects* commands (more on these in Lesson 10, p.240), and a down arrow that calls the **Application Status Bar Menu** (Figure 1.020). Use this menu to tell AutoCAD which of the various toggles you'd like to see.

Figure 1.020

Use the **Tray Settings** option in the **Application Status Bar Menu** to open the Tray Settings dialog box (shown here). Use this dialog box to control which icons and notifications AutoCAD displays in the tray on the right end of the status bar.

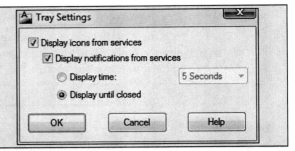

- Finally, you'll find a **Clean Screen** toggle ☐ which clears the monitor's screen of everything except AutoCAD's menu bar, status bar, view cube, navigation bar, and command window. This "expert setting" gives you more work room but limits your use of toolbars, the ribbon, or palettes.

1.1.7 The View Cube & Navigation Bar

The *View Cube* (Figure 1.021) serves no really useful purpose in 2D AutoCAD (although you'll love it in 3D!). To clear it from your 2D screen, follow this command entry:

    Command: *cube*

    Enter an option [ON/OFF/Settings] <ON>: *off*

I know, you had to experiment with it and now your screen is askew. Here's a quick fix:

    Command: *plan*

    Enter an option [Current ucs/Ucs/World] <Current>: *[Enter]*

Nothing to it, right?

(We'll see the *Plan* command again in *3D AutoCAD 2011: One Step at a Time*.)

Quit foolin' around!

Figure 1.021

The Navigation Bar (Figure 1.022) contains several tools that you'll find useful in both 2D and 3D space. These tools prove themselves invaluable when you need to maneuver your way through a large drawing. In Lesson 4, Section 4.1.4, p.88, you'll see how to use the 2D tools and hide the others.

If it distracts you in the meantime, you can control its display with the *NavBar* command:

    Command: *navbar*

    Enter an option [ON/OFF] <OFF>: *off*

Figure 1.022

1.2 Setting Up a New Drawing

1.2.1 The Groundwork: How AutoCAD Handles Scale, Units, and Paper Size

It may surprise you to learn that we won't use a scale when drawing in AutoCAD. In fact, all drawing in AutoCAD is done *full scale*! This simply means that a 3-inch line will actually be drawn 3-inches long. A line 3 miles long will be drawn 3 miles long!

Okay, so if we draw full scale, how can we put a house plan or refinery unit onto a sheet of paper unless the paper is very, very large? Well, we actually *tell* AutoCAD that we have a sheet of paper that's very, very large – a bit larger, in fact, than the house or unit we'll draw. Then later, when we plot (or print) the drawing to an actual sheet of paper, we *plot to scale*.

How do we determine the size of the paper we'll tell AutoCAD to use? I'm glad you asked!

> I've added a discussion of printing/plotting as an independent lesson (Lesson 23, p.528). The idea of making it independent is to allow students/instructors to include it at any point after the first lesson.

We know the standard sizes of paper used in the design world. And we know the standard scales used. To determine what we tell AutoCAD about paper size, we need to determine how many feet or inches (or millimeters) will fit onto each size sheet at each scale. The easiest way to do this is to look at the Drawing Scales Chart in Appendix B. (You'll find a similar chart in the possession of all CAD operators.)

To use the chart, we select the scale at which we'll want the final drawing plotted, and then select the paper size for the plot. Where row and column meet, we find the width x height *limits* of your drawing. (We'll talk more about limits shortly.) There's a place for these numbers in the drawing setup.

1.2.2 The Setup

Actually, you've already started a new drawing – back at the beginning of Section 1.1, p.3. It was just that simple to create a new drawing "from scratch" (using AutoCAD's default settings). You can begin any number of new drawings at any time in the same manner; but to make your drawing time more productive, there are a few things you should set up first – like paper size and units. Of course you could accept AutoCAD's defaults for these and just get started, but let's take a look at the procedures anyway – just in case your boss wants to use company standard settings.

> By default, AutoCAD uses the following settings:
> - Limits (paper size) = 12" x 9" (imperial) or 429mm x 297mm (metric)
> - Units = decimal to 4 places
> - Angles are decimal and measured counterclockwise with 0 degrees East

AutoCAD – never the shirker when it comes to redundancy – provides a couple methods for setting up a new drawing: templates and "from scratch." *

The possibility always exists that you may need to change AutoCAD's basic setup after you've begun a drawing session. (Okay, let's be realistic; you'll more than likely have to change something!) Maybe you need a different sheet size or different units. No sweat! AutoCAD makes it easy for you to adjust these settings.

Where to Find It:	
Command Line:	*Units*
Menu:	Format – **Units**
Application Menu:	Drawing Utilities – **Units**

To change units, enter the ***Units*** command (see the WTFI table). AutoCAD will provide the Drawing Units dialog box (Figure 1.023) where you can identify the new units and unit precision, as well as

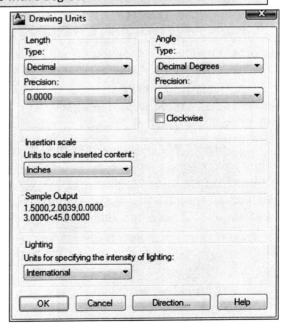

Figure 1.023

---

* We'll mercifully ignore the out-of-date setup wizard!

the angle type, precision, and direction. You can also set the **Insertion scale** for blocks (more on blocks in Lessons 21, p.468, and 22, p.500) and your **Lighting** units (more on those in the 3D text). The **Direction** button calls another dialog box where you can set the compass point from which AutoCAD will begin its angular measures. Let me explain.

Most of us are familiar with the nautical way of determining degrees (a product of a good scout training). Unfortunately, many engineers apparently missed that training. The standard 0°North in the real world has changed to 0°East in the engineering world.

You can change it here if you wish; but be warned, 0°East has become an engineering standard. If another CAD operator discovers that you have changed the angles, you may be in for some grief. Many third-party programs and AutoLISP routines (as well as engineering standards) are based on 0°East.

> Notice that you can also use the Drawing Units dialog box to change the direction in which AutoCAD measures angles. The default – counterclockwise – is also an engineering standard. It's a good idea to check with your CAD supervisor before changing this setting.

That's it for unit setup.

Where to Find It:	
Command Line:	*Limits*
Menu:	Format – **Drawing Limits**

To change the sheet size, enter the ***Limits*** command. AutoCAD will prompt you for the lower left coordinates of your drawing, using 0,0 as the default (see the following insert). Accept the default by hitting ENTER on your keyboard.

> Note that most companies will use 0,0 as the lower left corner of their drawings. Occasionally, however, a company may choose to use *true coordinates*. This is best explained by example.
>
> In the petrochem world, plants are divided into units. Each unit may have one or more drawings specific to that unit. Vessels and other items in the unit are located in the plant by overall *east/west* and *north/south* (*X* and *Y*) coordinates. It's often useful to identify the east/west and north/south coordinates with the X and Y coordinates in AutoCAD. This way, every item in the plant can be quickly located in the drawings simply by doing an ***ID*** command (p.260).
>
> When using true coordinates, the absolute location of the drawing will dictate the lower left as well as the upper right corners of the drawing limits.

AutoCAD will then prompt you for the upper right coordinates. Get these from the Drawing Scales chart (Appendix B) and enter them as ***width,height*** (no spaces). The sequence looks like this:

    Command: *limits*

    Reset Model space limits:

    Specify lower left corner or [ON/OFF] <0.0000,0.0000>: *[ENTER]*

    Specify upper right corner <12.0000,9.0000>: *[Enter the new width and height.]*

While you can make these adjustments at any time during your drawing session, I don't recommend waiting too long. It's always best to groom good habits from the very beginning.

We'll set up several drawings from scratch over the course of our text, so you'll become familiar with the ***Units*** and ***Limits*** commands.

> If you're already in AutoCAD and you want to start a new drawing from scratch, pick the **New** button on the Quick Access toolbar. Alternately, you can select **New** from the File menu or enter *New* at the command prompt. AutoCAD will open a Select Template dialog box. You can use the default *Acad.dwt* template, which is the 2D template, or the *Acad3D.dwt* template for default 3D settings. Alternately, (to open a new drawing from scratch) you can bypass the selection of a template by picking the down arrow next to the **Open** button and selecting **Open with No Template – [Imperial or Metric]** (shown here).

Okay, that's the brief overview, now let's get busy! We'll start by setting up a drawing after starting with the default template. We'll set up a drawing for a ¼"=1'-0" scale on a B-size (11 x 17) sheet of paper.

Do This: 1.2.2A	Setting Up a Drawing

I. If you haven't already, double click the AutoCAD icon on the desktop. If you're still in AutoCAD, enter the *New* command at the command prompt and accept the default template.

II. Pick the dynamic input toggle (✢ or DYN) on the status bar to turn it off. (It should appear unlit.)

III. Follow these steps.

1.2.2A: SETTING UP A DRAWING

1. We'll start by setting up our units. Enter the *Units* command ▩. (Okay, this is your first time, right? Just type "Units" and it'll show up on the command line. Alternately, you can use one of the other methods found in the Units WTFI table, p.16.)

   Command: *units*

AutoCAD presents the Drawing Units dialog box (Figure 1.023, p.16).

2. In the **Length** frame, change the **Type** of units to **Architectural**. Notice that the **Precision** automatically changes to 1/16". You can change that here as well, but we'll accept 1/16" as it's an industry standard.

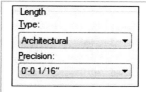

3. Take a moment to explore the other settings in this dialog box, and then accept the rest of the defaults. Pick the **OK** button [OK] to complete your units setup.

4. Now we'll set up the drawing page. Remember, all drawing is full scale so we have to adjust the size of the paper to accommodate the drawing.
Enter the *Limits* command.
   Command: *limits*

5. AutoCAD wants to know what the lower left limits should be. We'll hit ENTER to accept the 0,0 default.
   Specify lower left corner or [ON/OFF] <0'-0",0'-0">: *[ENTER]*

6. Now, check the Drawing Scales chart in Appendix B to determine the limits for a ¼" scale on a 11"x17" sheet of paper. Did you get 44'x68'? Excellent! Enter these at the next prompt, as shown. (Enter the desired page width followed by a comma and the desired page height. Don't use spaces or you'll upset AutoCAD!)
   Specify upper right corner <1'-0",0'-9">: *68',44'*

Don't close your drawing yet; we'll continue with it in our next section.

Congratulations! You've set up your first AutoCAD drawing.

1.3 Saving and Leaving a Drawing Session

Where to Find It:	
Command Line:	*QSave*
Hotkey(s):	CTRL + *s*
Menu:	File – **Save**
Application Menu:	**Save**
Toolbar:	Standard – **Save**
	Quick Access – **Save**

Now that we've created our drawing, we need to save it. We accomplish this by using one of three commands: *Save*, *Saveas*, or *QSave*. Each behaves in a similar manner depending on the status of the drawing. That is, each will present the Save Drawing As dialog box if the drawing has not been previously saved. The *QSave* command, however, will automatically save the drawing without prompting for additional information, provided it has been previously saved and given a name.

Save and *Saveas* are available at the command prompt. *QSave* will occur when you pick the **Save** button on the Quick Access toolbar or one of the menus.

Just as you have several ways to save a drawing, so too, you have several locations to which you can save. These include a local drive, a network drive, or an Internet location. Luckily, the procedures differ only slightly. We'll look at each in the next exercise.

> Saving a drawing to an Internet location is simple enough; it requires that you enter a web address in the **File name** text box of the Save Drawing As dialog box. But because of Internet protocols, you can't transfer information from your computer to an *http* address. You can, however, send your file to an *ftp* address. Don't underestimate the significance of this slight difference. Forgetting to change protocols has driven many an operator into frantic searches for more profound problems, only to wind up feeling foolish when they discovered the oversight.
>
> Once you've entered the correct address and picked the **Save** button, AutoCAD must stop and ask you for a user ID and password. The nature of the Internet requires these to ensure the integrity of the web site. (Imagine opening a file on your web site only to discover that some unknown villain had overwritten your drawing files with ©*Toadal Wisdom* cartoons.)

Where to Find It:	
Command Line:	*Close* or *CloseAll*
Menu:	File – **Close**
Application Menu:	Close – **Current Drawing** (or **All Drawings**)

How you end the drawing depends on whether you wish to end only the current drawing or the entire drawing session. In other words, do you want to close AutoCAD or just close this drawing? To leave the drawing without leaving AutoCAD, use the *Close* command. To close all drawings that are currently open, without leaving AutoCAD, use the *Closeall* command. Otherwise, use the *Quit* command or pick the **X** button on the right end of the title bar.

Let's save and exit our drawing.

Do This: 1.3A	**Saving Your Drawing Changes and Closing the Drawing Session**

I. Continue the previous drawing session.

1.3A: SAVING AND CLOSING YOUR DRAWING

1. Open the Save Drawing As dialog box. (Use one of the methods in the WTFI table, p.19.) AutoCAD presents the Save Drawing As dialog box shown in the following figure.
    Command: *saveas*

Contents

Lesson 1: In the Beginning...

Let's Get Started 3
    1.1.1 The User Interface and Workspaces 4
    1.1.2 Toolbars and Menus 6
    1.1.3 The Ribbon 10
    1.1.4 Tool Palettes 11
    1.1.5 The Graphics Area 12
    1.1.6 The Command Window and Status Bar 13
    1.1.7 The View Cube & Navigation Bar 15
1.2 Setting Up a New Drawing 15
    1.2.1 The Groundwork: How AutoCAD Handles Scale, Units, and Paper Size 15
    1.2.2 The Setup 16
1.3 Saving and Leaving a Drawing Session 19
1.4 Opening an Existing Drawing 22
1.5 Navigating Between Drawings with Quick View Drawings 26
1.6 Creating and Using Templates 27
1.7 Extra Steps 29
1.8 What Have We Learned? 30
1.9 Exercises 31

Lesson 2: Drawing Basics - Lines, Circles, & Coordinates

2.1 Lines, Rectangles, and Circles 33
    2.1.1 Lines and Rectangles 33
    2.1.2 Getting Around to Circles 40
2.2 Fixing the Uh-Ohs: *Erase, Undo,* and *Redo/MRedo* 42
2.3 Multiple Object Selection Made Easy 45
2.4 The Cartesian Coordinate System 47
    2.4.1 Absolute Coordinates 47
    2.4.2 Relative Coordinates 48
    2.4.3 Polar Coordinates 49
    2.4.4 Practicing with Cartesian Coordinates 51
2.5 Extra Steps 52
2.6 What Have We Learned? 52
2.7 Exercises 53

Lesson 3: Drawing Aids

3.1 The Simple Stuff: Ortho, Grid, Polar Tracking, and Snap 55
    3.1.1 Ortho 55
    3.1.2 Grid 56
    3.1.3 Polar Tracking 57
    3.1.4 Snap (Polar and Grid) 57
3.2 And Now the Easy Way - *DSettings* 58
3.3 Never Miss the Point with OSNAPs 61
3.4 Running OSNAPs 67
    3.4.1 Running OSNAPs via Dialog Box 68
    3.4.2 Running OSNAPs via Menu 69
3.5 Point Filters and Object Snap Tracking 70
    3.5.1 Point Filters 70
    3.5.2 Object Snap Tracking 71

3.6 Isometric Drafting 72
3.7 Extra Steps 75
3.8 What Have We Learned? 76
3.9 Exercises 76

Lesson 4: Display Controls and Basic Annotative Text

4.1 Changing What You See 79
    4.1.1 Zooming 79
    4.1.2 Panning 84
    4.1.3 The Navigation Wheel 85
    4.1.4 The Navigation Bar 88
4.2 Why Find It Twice? – The *View* Command 88
4.2.1 Creating Views 89
    4.2.2 Showing Motion 92
4.3 "Simple" Text 93
4.4 Editing Text 100
4.5 Finding and Replacing Text 100
4.6 Adding Flavor to Text with *Style* 101
4.7 Extra Steps 104
4.8 What Have We Learned? 106
4.9 Exercises 107

Lesson 5: Geometric Shapes (Other Than Lines, Rectangles, and Circles!)

5.1 Ellipses and Isometric Circles 109
5.2 Arcs: The Hard Way! 112
5.3 Polygons 115
5.4 Putting It All Together 118
5.5 Extra Steps 123
5.6 What Have We Learned? 124
5.7 Exercises 124

Lesson 6: Object Properties - Color at Last (and More!)

6.1 Some Preliminaries 127
6.2 Adding Color 128
6.3 Drawing with Transparency 131
6.4 Drawing with Linetypes 132
    6.4.1 Using Linetypes 132
    6.4.2 Modifying Linetype Scale 135
6.5 Using Lineweights 137
6.6 Uh-Ohs, Boo Boos, Ah $%&s: Object Properties 138
    6.6.1 Tell Me About It – The *List* Command 138
    6.6.2 Object Properties and the Quick Properties Panel 139
    6.6.3 Quick Properties Settings 141
    6.6.4 Quick Properties and the Customize User Interface Dialog Box 142
6.7 Extra Steps 145
6.8 What Have We Learned? 146
6.9 Exercises 146

Lesson 7: Colors, Linetypes, and More Made Simple - Layers

7.1 Color, Linetype, and So Much More – Layers and How to Use Them 149
7.2 A Couple Managers 153
   7.2.1 The Layer Properties Manager 153
   7.2.2 The Layer States Manager 162
7.3 The Rest of the (Layer) Story 164
7.4 Sharing Setups: The AutoCAD Design Center and the Layer Translator 169
7.5 Extra Steps 174
7.6 What Have We Learned? 174
7.7 Exercises 175

Lesson 8: Editing Your Drawing - Modification Procedures

8.1 The Location and Number Group 177
   8.1.1 Here to There: The *Move* Command 177
   8.1.2 The *Copy* Command: From One to Many 178
   8.1.3 AddSelected 180
   8.1.4 Okay, Move It – But Then Line It Up: the *Align* Command 180
   8.1.5 Parallels and Concentrics – The *Offset* Cmd 182
   8.1.6 Rows, Columns, and Circles – the *Array* Command 184
   8.1.7 Opposite Copies – the *Mirror* Command 188
8.2 Moving and Copying Objects *between* Drawings 190
8.3 The Change Group 192
   8.3.1 Cutting It Out with the *Trim* Command 192
   8.3.2 Adding to It with the *Extend* Command 194
   8.3.3 Redundancy – Thy Name is AutoCAD: The *Break* Command 195
   8.3.4 Now We Can Round That Corner: The *Fillet* Command 196
   8.3.5 Fillet's Cousin: The *Chamfer* Command 198
8.4 Extra Steps 199
8.5 What Have We Learned? 199
8.6 Exercises 200

Lesson 9: Some More Editing Tools ... & Grips!

9.1 More Commands in the Change Group 203
   9.1.1 Two Ways to Change the Length of Lines and Arcs – The *Lengthen* and *Stretch* Cmds 203
   9.1.2 "Oh, NO! I Drew It Upside Down!" – The *Rotate* Command 206
   9.1.3 "Okay. Give Me Three Just Like It, But Different Sizes." The *Scale* Command 208
9.2 Identifying the Changes – The *RevCloud* Cmd 209
9.3 "A Whole New Ball Game!" Editing with Grips 212
9.4 Putting It All Together 218
9.5 Extra Steps 222
9.6 What Have We Learned? 223
9.7 Exercises 223

Lesson 10: Geometric Constraints & Alternate Selection Methods

10.1 Constraining Relationships: Constraint Symbols, Insertions, and Behaviors 225
10.2 Auto Constraints and Display Settings 230
10.3 Inferred Constraints 232
10.4 Object Selection Tools 234
10.4.1 Selection Modes 234
10.5 Selection Cycling 239
10.6 *SelectSimilar* and the Object Isolation Tools 240
10.7 Extra Steps 242
10.8 What Have We Learned? 242
10.9 Exercises 243

Lesson 11: Some More Cool Tools

11.1 So Where's the *Point*? 245
11.2 Equal or Measured Distances – The *Divide* and *Measure* Commands 246
11.3 From Outlines to Solids – The *Solid*, *Donut*, and *Wipeout* Commands 249
11.4 AutoCAD's Measurement Tools 252
   11.4.1 How Long or How Far – The *Dist* Tool 253
   11.4.2 The Radius Tool 254
   11.4.3 The Angle Tool 255
   11.4.4 Calculating the Area 256
   11.4.5 The Volume Tool 259
   11.4.6 Identifying Any Point with *ID* 260
11.5 More Object Selection Methods 260
11.6 Object Selection Filters – Quick Filters 264
11.7 AutoCAD's Calculator 266
11.8 Extra Steps 270
11.9 What Have We Learned? 270
11.10 Exercises 271

Lesson 12: Polylines and the Action Recorder

12.1 Using Polylines for Wide Lines and Multi-Segmented Lines 273
12.2 Editing Polylines – The *PEdit* Command 278
12.3 AutoCAD Macros – The Action Recorder 287
12.4 Extra Steps 291
12.5 What Have We Learned? 292
12.6 Exercises 292

Lesson 13: Guidelines and Splines

13.1 Contour Lines with the *Spline* Command 295
13.2 Changing Splines – The *Splinedit* Command 299
13.3 Guidelines 305
13.4 Extra Steps 311
13.5 What Have We Learned? 311
13.6 Exercises 311

Lesson 14: Advanced Lines - Multilines

14.1 Many at Once – AutoCAD's Multilines and the *MLine* Command 313
14.2 Options: The *MLStyle* Command 315
14.3 Editing Multilines: The *MLEdit* Command 321
14.4 The Project 324
14.5 Extra Steps 326
14.6 What Have We Learned? 327
14.7 Exercises 327

Lesson 15: Advanced Text - MText

15.1 AutoCAD's Word Processor: The Multiline Text Editor 329
15.2 Okay I Typed It, but I Don't Know If It's Right! – AutoCAD's *Spell* Command 341
15.3 Find and Replace – without the Multiline Text Editor 344
15.4 Columns 346
15.5 Extra Steps 348
15.6 What Have We Learned? 348
15.7 Exercises 349

Lesson 16: Basic Dimensioning

16.1 First, Some Terminology 351
16.2 Dimension Creation: Dimension Commands 353
  16.2.1 Linear Dimensioning 353
  16.2.2 Dimensioning Angles 355
  16.2.3 Dimensioning Radii and Diameters 357
  16.2.4 Dimension Arc Lengths 358
  16.2.5 Dimension Strings 360
  16.2.6 Aligning Dimensions 362
  16.2.7 Baseline Dimensions & Spacing 363
  16.2.8 Ordinate Dimensions 364
16.3 And Now the Easy Way: Quick Dimensioning (*QDim*) 366
16.4 Dimension Editing: The *DimEdit* and *DimTEdit* Commands 368
  16.4.1 Position the Dimension: The *DimTedit* Command 368
  16.4.2 Changing Value of the Dimension Text: The *DimEdit* Command 369
  16.4.3 Breaking an Extension Line - *DimBreak* 371
16.5 Isometric Dimensioning 372
16.6 Extra Steps 373
16.7 What Have We Learned? 374
16.8 Exercises 375

Lesson 17: Customizing Dimensions and Using Dimensional Constraints

17.1 Creating Dimension Styles 377
17.2 Miracles of Annotative Dimensioning 390
17.3 Try One 391
17.4 Overriding Dimensions 393
17.5 Dimensional Relationships (Constraints) 394
  17.5.1 The Basics 394
  17.5.2 Managing Dimensional Constraints 397
17.6 Extra Steps 400
17.7 What Have We Learned? 401
17.8 Exercises 401

Lesson 18: Leaders

18.1 The *MLeader* Command 405
  18.1.1 Placing Leaders 405
  18.1.2 Aligning and Collecting Leaders 408
  18.1.3 Editing Leaders 411
18.2 Customizing Leaders 413
18.3 Extra Steps 418
18.4 What Have We Learned? 418
18.5 Exercises 419

Lesson 19: Tables and Fields

19.1 Tables 421
  19.1.1 Creating a Table 421
  19.1.2 Editing a Table's Properties 426
  19.1.3 Table Column Tools and Adding Values to Table Cells 430
  19.1.4 Customizing Tables 432
19.2 The Wonders of Automation – Fields 437
  19.2.1 Drawing Properties 437
  19.2.2 Inserting Fields 439
  19.2.3 Calculating Fields 441
19.3 Altogether Now: Tables, Fields, and MS Excel 443
19.4 Extra Steps 446
19.5 What Have We Learned? 447
19.6 Exercises 447

Lesson 20: Hatching and Grouping

20.1 Hatching and Filling 449
20.2 Editing Hatched Areas 455
  20.2.1 Normal Hatch Editing 455
  20.2.2 Non-Associative Hatch Editing 456
20.4 Paper Dolls: The *Group* Command 457
20.5 Extra Steps 464
20.6 What Have We Learned? 465
20.7 Exercises 465

Lesson 21: Many as One - Blocks

21.1 Groups with Backbone – The *Block* Commands 469
  21.1.1 Template Library Creation 470
  21.1.2 Folder Library Creation 473
  21.1.3 Using Blocks in a Drawing – The *Insert* Command 474
21.2 Dynamic Blocks 476

21.3 Getting Blocks from a Folder Library on a Web Site via AutoCAD's iDrop 496
21.4 Extra Steps 498
21.5 What Have We Learned? 498
21.6 Exercises 499

Lesson 22: Advanced Blocks

22.1 Attributes 501
    22.1.1 Creating Attributes 501
    22.1.2 Block Tables 507
22.2 Inserting Attributed Blocks 509
22.3 Editing Attributes 513
    22.3.1 Editing Attribute Values 513
    22.3.2 Editing Attribute Definitions 516
22.4 The Coup de Grace: Using Attribute Information in Bills of Materials, Spreadsheets, or Database Programs 519
22.5 Extra Steps 525
22.6 What Have We Learned? 525
22.7 Exercises 526

Lesson 23: Sharing Your Work with Others

23.1 The Old-Fashioned Way – Putting It on Paper (Plotting) 529
    23.1.1 First Things First – Setting Up Your Printer (or Plotter) 529
    23.1.2 Plot Styles 530
    23.1.3 Setting Up the Page to Be Plotted 530
23.2 Sharing Your Drawing with the *Plot* Command – and *No Paper*! 535
    23.2.1 Quickly Creating a Drawing Web Format (DWF) or Adobe PDF File with the *Export* Command 535
    23.2.2 Viewing a Drawing Web Format (DWF) File 538
    23.2.3 Multiple Plots and Creating a DWF File – with Hyperlinks! 542
    23.2.4 Hyperlinks 544
    23.2.5 AutoCAD can Create Full Web Pages, Too! 547
23.3 Sending the Package over the Internet with *eTransmit* 552
23.4 Extra Steps 558
23.5 What Have We Learned? 558
23.6 Exercises 558

Lesson 24: Space for a New Beginning

24.1 Understanding the Terminology 561
24.2 Using Tiled Viewports 562
24.3 Setting Up a Layout Environment (Paper Space) 568
24.4 Using Floating Viewports 570
    24.4.1 Creating Floating Viewports Using *MView* 571
    24.4.2 The Viewports Panel 573
    24.4.3 Adjusting the Views in Floating Viewports 573
    24.4.4 the Position of the Floating Viewport 576

24.5 And Now the Easy Way – The *LayoutWizard* Command 576
24.6 Extra Steps 580
24.7 What Have We Learned? 580
24.8 Exercises 581

Lesson 25: The New Beginning Continues ...

25.1 Dimensioning and Paper Space 585
    25.1.1 Dimensioning and Paper Space – the Annotative Way 585
    25.1.2 Dimensioning and Paper Space – the Other Way 587
25.2 The Benefits of Layers in Paper Space 588
25.3 Using Text in Paper Space 591
25.4 Plotting the Layout 592
25.5 Tweaking the Layout 593
    25.5.1 Modifying Viewports with the *MVSetup* Command 593
    25.5.2 Changing the Shape of a Viewport with the *VPClip* Command 596
25.6 Putting It All Together 597
25.7 Extra Steps 604
25.8 What Have We Learned? 605
25.9 Exercises 606

Lesson 26: More Sharing Tools - Drawing Sheet Sets

26.1 Sheet Sets – A Primer 609
26.2 Using the Sheet Set Manager to Organize Your Project 609
    26.2.1 The Sheet List Tab 609
    26.2.2 The Sheet Views Tab 616
    26.2.3 The Model Views Tab 621
26.3 Using Your Sheet Sets to Share Information 623
26.4 Extra Steps 628
26.5 What Have We Learned? 628
26.6 Exercises 629

Lesson 27: Externally Referenced DWGs, PDFs, DWFs, & DGNs

27.1 Working with Externally Referenced DWGs, PDFs, DWFs, and DGNs 631
    27.1.1 Attaching and Detaching External References to Your Drawing 632
    27.1.3 Xrefs and Dependent Symbols 642
    27.1.4 Unload, Reload, and Overlay Xrefs 644
27.2 Editing Xrefs 646
27.3 Using Our Drawing as a Reference 649
27.4 Binding an Xref to Your Drawing 650
27.5 Extra Steps 652
27.6 What Have We Learned? 653
27.7 Exercises 654

Lesson 28: Other Application Files

28.1 Two Types of Graphics 657
28.2 Working with Raster Images: The Image Mngr 658
   28.2.1 Attaching, Detaching, Loading, and Unloading Image Files 658
   28.2.2 Clipping Image Files 661
   28.2.3 Working with Image Files 662
28.3 Exporting Image Files 665
28.4 Working with Linked Objects – Object Linking and Embedding (OLE) 666
   28.4.1 Inserting Other Application Data into AutoCAD Drawings 666
   28.4.2 Modifying OLE Objects 671
   28.4.3 AutoCAD Data in Other Applications 672
28.5 Extra Steps 673
   28.6 What Have We Learned? 674

28.7 Exercises 675

Appendices

Appendix – A: Initial Setup 678
Appendix – B: Drawing Scales 679
Appendix – C: Function Keys and Their Uses 680
Appendix – D: MText Keystrokes 681
Appendix – E: Dimension Variables 682
Appendix – F: Hotkeys 685
Appendix – G: Actions & Parameters Chart for Dynamic Blocks 686

Lesson 1

Following this lesson, you will:

- ✓ Know how to create a new drawing
- ✓ Know how to open an existing AutoCAD drawing
- ✓ Know how to save and close an AutoCAD drawing
- ✓ Recognize the various parts of the AutoCAD User Interface
- ✓ Be familiar with the AutoCAD Info Center

In the Beginning ...

How do you begin a new drafting project?
The number of answers I've heard to this question over the years might surprise you. Some say you begin with the layout. Others say you begin by positioning the drawing on the page. The most experienced drafters generally come up with a quip about coffee and radio settings.
This doesn't mean that I was questioning bad drafters, but simply that experience has engrained the basics so deeply that good drafters don't think about these simple questions anymore.
So, you might say, how do you begin a drafting project?
Well, first you decide on the scale and units (engineering, architectural, etc.) that you'll use. Then you decide the size of the page on which you'll draw (even if your final drawing will be electronic, you'll need to set up for a page size).
This first lesson will introduce you to the AutoCAD User Interface – the AUI, or the screen on which you'll draw – and show you how to set up things like units and sheet size.
It's probably best to jump right in. So as the heading says ...

1.1 Let's Get Started

AutoCAD works much like any other windows program; that is, you start with the shortcut on the desktop (or the Start menu: **Start – All Programs – Autodesk** – and so forth). You can save a drawing using the Windows' CTRL+S method or the standard **File – Save** approach (although you many not find the usual pull down menus available at first). Finally, you can close a file using the **Exit** button at the right end of the title bar.

We'll look at each of these procedures in more detail throughout this lesson. But first, let's open AutoCAD and take a look at it.

> If you haven't opened AutoCAD before, you may discover an initial setup dialog box. Pick the **Skip** button in the lower right corner for now. For more information on this dialog box, refer to **Appendix A: Initial Setup**.

Do This: 1.1A	Starting a Drawing Session

I. Follow these steps.

1.1A: STARTING A DRAWING SESSION

1. When you loaded AutoCAD, a shortcut that looks like this appeared on your desktop. Double click this with your left mouse button to launch AutoCAD. It will open (it may take a moment or two), and you'll see something like the AutoCAD User Interface (the AUI) shown in Figure 1.001 (without the annotations)*. Don't worry if it doesn't look exactly like this just yet; we'll fix it in the next section.

---

* AutoCAD's default background color is actually dark gray. We'll use a white background for clarity throught this text.

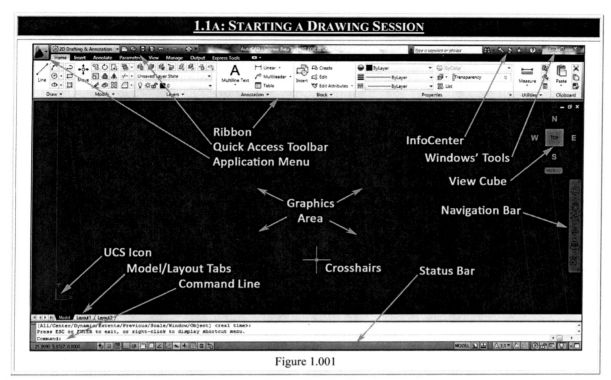

Figure 1.001

Let's take a few minutes to familiarize ourselves with what we're seeing.

> IMPORTANT IMPORTANT IMPORTANT
>
> As you can see, AutoCAD's initial work area is fairly dark. To conserve ink (and save on printing costs), we use a lighter background in the instructional and working files for this text. Follow this simple procedure to set your work area to the lighter background.
>
> 1. Enter the **Options** command – that is, type **options**. You'll see that your text appears in the command line (see Figure 1.001).
> 2. When you hit the ENTER key, AutoCAD will display an Options dialog box.
> 3. Select the **Profiles** tab at the right end of the dialog box.
> 4. Pick the **Import** button. AutoCAD displays a standard Windows Open dialog box.
> 5. Navigate to the C:\Steps\Lesson01 folder (or the folder in which you placed the working files for this text), and select the *One Step at a Time Profile.arg* file.
> 6. Pick the **Open** button. AutoCAD presents an Import Profile dialog box.
> 7. Pick the **Apply & Close** button.
> 8. Select *One Step at a Time Profile* in the **Available profiles** list box and pick the **Set Current** button.
> 9. Pick the **OK** button to close the Options dialog box.
>
> Now you're set up just as I am as I begin the text so we're all "on the same page" (so to speak ☺).
> (Note: To reset to AutoCAD's defaults, simply repeat Steps 1 – 3 and pick the **Reset** button.)

1.1.1 The User Interface and Workspaces

Where to Find It:	
Command Line:	Workspace
Status Bar:	Workspace Switching
Toolbar:	Quick Access – Workspace

AutoCAD comes to you with a few user-interfaces – called *Workspaces* – to help satisfy 2-dimensional users, 3-dimensional users, and your initial setup choices. You'll find these workspaces listed in the **Workspaces** popup menu (Figure 1.002), presented when you pick the **Workspace Switching** button on the right end of the status bar (Figure 1.001) or on the left end of the Quick Access Toolbar (Figure 1.001). These include:

- **2D Drafting & Annotation** offers a simple 2D graphics interface (see Figure 1.001). This is the one we'll be using. It mimics the Microsoft Office 7 look (with a ribbon), so it should be an easy transition for its users, and it groups its ribbon tools into tabs and panels for easy access.
- **3D Basics** provides the basic (simpler) tools needed to create and visualize a 3D model. Good for beginners, I can't recommend it for more serious users. (Consider it *3D-Lite*!)
- **3D Modeling** presents all the ribbon tools designed to facilitate three-dimensional work. You'll use this interface when you graduate to the *3D AutoCAD 2011: One Step at a Time* text.
- **AutoCAD Classic** presents the "olde" look with toolbars and no ribbon. I recommend ignoring this one; toolbars won't be with us much longer!

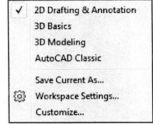

Figure 1.002

Do the following exercise to use the **2D Drafting & Annotation** workspace (the one we'll use throughout this text, and the default shown in Figure 1.001).

Do This: 1.1.1 A	Selecting a Workspace

I. Be sure AutoCAD is open, and follow these steps.

1.1.1A: SELECTING A WORKSPACE
1. Pick the **Workspace** control on the left end of the Quick Access toolbar.
2. Select **2D Drafting & Annotation** as shown in Figure 1.002.

Nothing to it!

Of course, you're not limited to just the predefined workspaces. You can open or close palettes, ribbon panels and other items as you wish, move them around the interface, dock or undock them, and perform a host of other customizations. Then, once you're pleased with your user-interface (your *workspace*) – you have all your *gizwitchies* where you want them – you can save the layout just in case someone else comes along and changes things. (Alternately, you can have a variety of interface setups to use for your different projects.)

Select **Save Current As…** from the menu (Figure 1.002), or use the *WSSave* command to save your workspace. It looks like this (on the command line[*] – see Figure 1.001, p.4):

> **Command:** *wssave*
>
> [AutoCAD presents a dialog box asking for the name of the workspace. Enter an appropriate name and pick the **Save** button.]

That's all there is to saving your workspace so you can return to it at any time. Use the menu or the *WSCurrent* command to restore your workspace. The command looks like this:

> **Command:** *wscurrent*
>
> **Enter new value for WSCURRENT <" ">:** *MyWorkspace*

You'll also see options (Figure 1.002) for **Workspace settings**, which presents a dialog box that lets you control what the popup menu shows, and for **Customiz**(ing) workspaces. The **Customize** option presents the Custom User Interface (CUI) dialog box. Customization is beyond the scope of this text, so let's move on and take a look at some of the items you can include in your workspace … and some of the other tools AutoCAD makes available.

---

[*] Commands may also appear on the dynamic input display next to the cursor. We'll look at this in Section 1.1.6, p.11.

> You can also access the tools available on the popup menu via the command line. Just use the *Workspace* command.

1.1.2 Toolbars and Menus

Let's start with what's similar to other Windows' applications. We find the title bar at the top, with a three button toolbar (the Windows tools shown in Figure 1.001, p.4) on the right. From left to right, these buttons and their uses are:

- This reduces the software to a pick on the taskbar (the bar across the bottom of the Windows screen).
- This makes the size of the AutoCAD window adjustable.
- This exits the software.

Okay, you knew those buttons. Let's look at the other end of the title bar – at the Quick Access toolbar and the Application Menu.

The Quick Access Toolbar

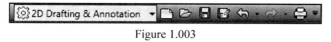
Figure 1.003

Figure 1.004

The Quick Access toolbar (Figure 1.003) contains another way to change your workspace (Section 1.1.1, p.4) and some common, often-used buttons. From the left, these include: **New** (or **QNew**), **Open**, **Save**, **SaveAs**, **Undo**, **Redo**, and **Plot**. To determine the function of these or other buttons on the toolbar, place you cursor over the button and wait a second. You'll see a written description (a *tooltip*) appear in a tooltip box. To see an explanation of how the tool works, use the F1 key on your keyboard.

You can easily add or remove items to the Quick Access toolbar by selecting an option from the menu (Figure 1.004) that appears when you select the **More** button ▼ at the right end of the toolbar. Use the **More Commands** option if the command you wish to add doesn't appear in the list, but try not to overdo it; you'll soon find other ways to access commands quickly.

Perhaps more important that the buttons it contains, the Quick Access toolbar also provides access (via the same menu – Figure 1.004) to some other tools. You can use the Quick Access toolbar to move the toolbar **Below the Ribbon** and even **Show** the familiar **Menu bar** (Figure 1.005)! If you're accustomed to using drop-down menus, you'll really appreciate this option.

Figure 1.005

Take a minute to explore these menus. You'll notice that the AutoCAD crosshairs become a cursor when moved out of the graphics area. Place the cursor on one of the menus and click with the left mouse button. A menu will appear from that location showing you various options related to the selection. Picking any of these options will activate an AutoCAD macro.

> A *macro* is a series of commands or events responding to a single input. Generally, you can't tell this from a single command.

You can also make changes to the Quick Access toolbar via a cursor menu (Figure 1.006). (To access a cursor menu, simply select a tool or an object with the right mouse button.)

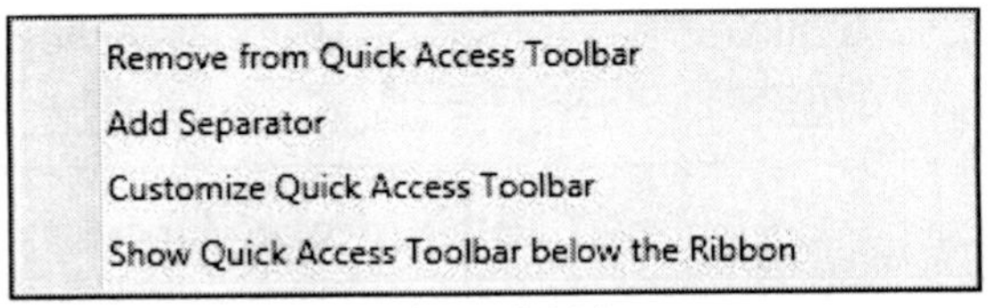

Using this menu, you can **Remove** a tool from the toolbar, **Add** [a] **Separator** between tools or groups of tools, move the **Toolbar below the Ribbon**, or **Customize** [the] **Quick Access Toolbar**. (I don't recommend customizing anything just yet. Let's wait until you're a bit more familiar with the application.)

Figure 1.006

For more information on cursor menus, see **The Cursor Menus** section (p.9).

Adding and Removing Other Toolbars

Although AutoCAD has moved to a ribbon-based tool system, it still provides several toolbars to assist you.

To add a toolbar, right click 🖱 on an existing toolbar and select the desired toolbar from the menu that appears. Alternately, you can use the pull down menus to select a toolbar to add. (See p. 8 for information on how to show the pull down menus.) Follow this path: **Tools – Toolbars – AutoCAD – [select toolbar]**. To remove a toolbar from the screen, undock it (as detailed in the following paragraphs), and then pick the "X" found in the upper right corner of the undocked bar.

A docked toolbar appears "attached" to a side, top, or bottom of the graphics window. An undocked toolbar appears to float over the screen. Drawing objects may exist behind an undocked toolbar, but a docked toolbar forces the drawing to move over.

To undock a toolbar, place your cursor on the two lines found on the left end (or top) of the toolbar and drag it free from its dock. (To drag an object, select it with the left mouse button, but don't release the button until you have relocated the object.)

To dock a toolbar, place your cursor in one of its wings (the gray areas on the sides) and drag it to a docking site (one of the borders of the graphics area).

Once you've docked or undocked your toolbars, avoid accidentally moving them by picking the **Lock Toolbars** icon 🔒 at the right end of the status bar. This will produce a popup menu (Figure 1.007) where you can lock one type of toolbar or all the toolbars on the screen. (Override the lock by holding down the CTRL key while dragging the toolbar.)

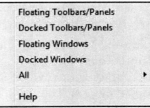

Figure 1.007

I'll have you display an occasional toolbar as we proceed.

The Application Menu

Use the **Application Menu** button 🔺 on the left end of the title bar to open the Application Menu (Figure 1.008, p.8). Notice the menu list down the left side of the browser? Most of these tools also exist

on the Quick Access toolbar, but you'll find a few more tools here – and more options for the Quick Access tools.

Let's take a look.

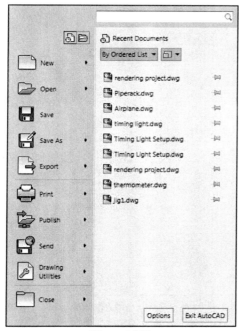

Figure 1.008

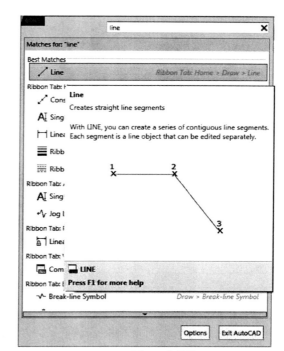

Figure 1.009

- First, there's a *Search* box across the top. Enter a command (or the first letter or two of a command) and AutoCAD will help you find the command.[*] Then when you hover the cursor over it, AutoCAD will produce an enhanced tooltip (Figure 1.009).
  Pretty cool, huh?
- Below the search box, the menu divides into left and right columns. What you select on the left determines what you see on the right. Let's start with the right.
  - By default, the right side lists either recently accessed drawings or currently open drawings. You'll control which with a pair of buttons found atop the left side list: the **Recent Documents** button and the [currently] **Open Documents** button.
  - When viewing the default menu, you'll find a pair of list buttons (atop the right column) to help you control how AutoCAD displays the right column.
    - The first By Ordered List – available when you're viewing recent documents – displays a short menu to help you order the list. You can order by **Ordered List** (most recently opened first), **Access Date**, **Size**, or **Type** of file.
    - The second allows you to change the size of the icon used in the list. This can be handy as the larger images give a thumbnail of the drawing.
  - One of the real benefits of the application menu appears when you hover your cursor over a drawing in the **Recent Documents** or **Open Documents** list (right column). AutoCAD presents an image of the drawing (Figure 1.010, p.9) and some additional information (path, date modified, etc.) to help you select the right one! Pick with left mouse button (that is, *select*) on the drawing you wish to open.
- The left column of the application menu (Figure 1.008) contains a list of function groups. Each selection presents several commands in the right column.

---

[*] AutoCAD searches the Quick Access toolbar, Application Menu, and current ribbon. If you can't find what you're looking for in these, use the Help menu (F1). We'll discuss Help in Lesson 9.

- **New** allows you to begin a new drawing or a new sheet set (Lesson 26, p.608). We'll look at creating a new drawing shortly.
- **Open** presents tools for opening an existing drawing or sheet set (Lesson 26). You'll also find a tool for importing a DGN file.
- **Save**, the only option which doesn't present a group of tools, presents a standard Windows Save dialog box. (**Saveas** opens a Save Drawing As dialog box.) We'll look at saving your drawing in Section 1.3, p.19.
- **Export** is something like **Saveas** except that you can export your drawing as a DWF (Autodesk's drawing web format), PDF (Adobe Acrobat file), or even a DGN (Microstation drawing file). We'll look at exporting files in Lesson 28, p.656.

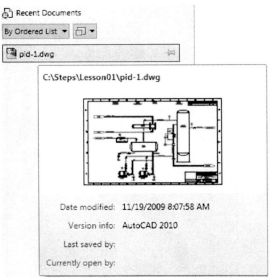

Figure 1.010

AutoCAD also provides a pair of rather sophisticated tools called *PartialOpen* and *PartialLoad*. These rarely used items might serve best to confuse at this point so we won't discuss them. You can, however, pull the supplement *Partial.pdf* from the website (http://www.uneedcad.com/Files) for future reference if you intend to work with extremely large drawing files.

- Autodesk provides three methods of sharing your drawing with others – **Print**, **Publish**, and **Send**. We'll explore these in some detail in Lesson 23.
- **Close** offers options for closing the current drawing or all open drawings.
- **Drawing Utilities** offers several maintenance tools to help you set up the drawing and keep it in good working order.

We'll look at these tools individually as we proceed.

- You'll also find an **Exit AutoCAD** button and an **Options** button at the bottom of the Menu Browser. You can use the **Exit AutoCAD** button to close the application or you can just use the **X** button on the other end of the title bar. Either way, AutoCAD will prompt you to save any unsaved work. The **Options** button opens AutoCAD's Options dialog box. I don't recommend doing anything with this until you're much more familiar with AutoCAD.

That's a busy menu. But we're not finished with menus yet!

The Cursor Menus

One of a less apparent set of menus will appear anytime you right click (pick with the right mouse button) in your drawing. These *cursor menus* (or pop-up screen menus) help you watch the screen more and spend less time searching for the right thing to press on the keyboard!

We have to give the programmers credit for accomplishing a minor miracle with the right mouse button. They've provided one of AutoCAD's most dynamic tools. Which cursor menu appears when the right button is picked depends upon: whether or not there's a command in progress, if there are items selected on the screen, and where the cursor/crosshairs are located when the button is picked!

AutoCAD provides five basic modes (or types) of cursor menus: *default*, *edit-mode*, *dialog-mode*, *command-mode*, and *other* menus.

- The *default* menu appears when you right click anywhere in the graphics area if no command is active and no selection set is available (that is, nothing has been selected on the screen with which to work). The default menu contains general, frequently used commands.

- The *edit-mode* menu appears when objects have been selected on the screen, but no command has been given. You'll see the best example of this when we discuss the Grips menu in Lesson 9, p.212. Generally, you can expect grip-type modifying tools. But object-specific tools will appear for some objects (such as dimensions).
- When you right click 🖱 in a dialog box, AutoCAD may present a *dialog-mode* menu. The tools presented will depend upon which dialog box you're using and the cursor's location when you click.
- When you right click 🖱 while there's a command in progress, AutoCAD presents a *command-mode* menu with tools specific to that command.
- The "*other* menus" category includes all the other menus that don't fit easily into the first four categories. Right clicking 🖱 in different locations and at different times might present some surprising results. Feel free to experiment while you're in a drawing. Try right clicking 🖱 over the different buttons on the status bar.

We'll spend more time exploring the cursor menus as they apply to specific areas throughout this book. But now, let's take a look at the *Ribbon*.

1.1.3	**The Ribbon**

The *ribbon*, displayed by default across the top of the graphics area, consists of several *tabs* which contain a myriad of *panels*. Each panel, in turn, contains several tools (command buttons). We'll look at individual commands as we proceed through this text. For now, we'll just look at the ribbon itself, and how it works.

The ribbon appears to dominate the entire graphics area; let's face it, it's big ... probably too big to remain in its default location without some adjustments (especially if you're using a widescreen monitor). Luckily, AutoCAD makes it easy to *minimize* the space occupied by the ribbon. Start by picking the down arrow next to the **Minimize** button 🔽 on the ribbon. select **Minimize to Panel Titles** from the flyout menu (Figure 1.011). This way, even when you hide the ribbon, the panels will remain visible. The ribbon should disappear. But don't worry, it isn't gone for keeps. Move your cursor over one of the panel title bars, and the panel will reappear. But just look at all the screen area you have to work in when the ribbon is hidden!

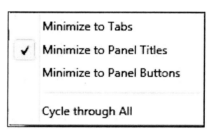
Figure 1.011

Alternately, you can opt to **Minimize** [the ribbon] **to Panel Buttons**. This reduces the size of the ribbon as well (although not as much as the **Panel Titles** option). When reduced to buttons, hovering your cursor over a button will cause AutoCAD to display that panel's tools as shown here.

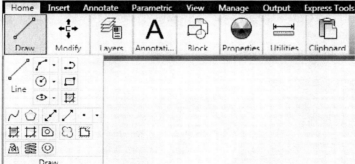

You'll find another menu (Figure 1.012) with several other options for manipulating the size, shape, and location of the ribbon when you right click 🖱 on any dark gray area of the ribbon.

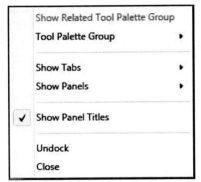

Figure 1.012

Select a different tab for a different set of panels; remember, each panel has a different collection of tools. We'll examine each panel in more detail as we proceed.

If you're using a widescreen monitor, you might dock your ribbon to one side (I use the right side as it's consistent with the location of AutoCAD's older dashboard tool which served the same purpose.) Dock it to one side and "auto–hide" it to free up a tremendous about of graphics area where you can work.

When it isn't docked, the ribbon has a titlebar (Figure 1.013) with some useful buttons. From the top, these include:

- The **Exit** button ✕, which will close the ribbon, thus removing it from the screen and allowing you better access to the drawing (graphics) area. (Restore it with the **Ribbon** command.)

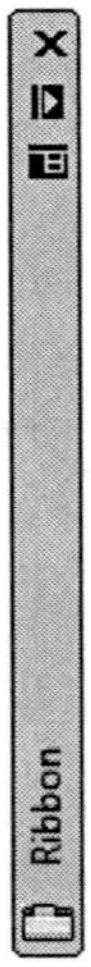

  - An **Auto-hide** button, which looks like this 🔲. This is the hide/unhide toggle.

  - A **Properties** button 🔳, below the **Auto-hide** button, which calls the menu shown in Figure 1.014. Let's look at the selections available here.
    - When you select **Move**, AutoCAD changes the cursor to a four-sided arrow, which you can use to drag the ribbon to a new location without docking it.
    - The **Size** selection also changes the cursor to a four-sided arrow. This one you can use to resize the ribbon.
    - Of course, **Close** does just that; it closes the ribbon.
    - Selecting **Allow Docking** will either check or uncheck the option. When checked, the ribbon can be docked (just like any toolbar) to the side/top/bottom of the graphics area. When unchecked, it can't be docked.
    - **Anchor Left <** and **Anchor Right >** will dock the ribbon to the selected side of the window.
  - **Auto-hide** is a menu approach to do the same thing the **Auto-hide** button does.
  - **Show Panel Titles**, when checked, causes a small title bar next to or beneath each panel telling you what the panel contains.
  - **Customize** calls the CUI dialog box. Remember, we don't want to fool around with customization just yet!
  - **Help**, of course, calls AutoCAD's Help dialog box.

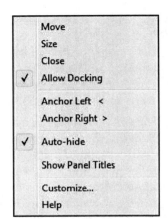

Figure 1.014

Figure 1.013

1.1.4	Tool Palettes

If your screen doesn't display Tool Palettes (Figure 1.015) use one of the paths shown in the **Where to Find It** table [aka. the WTFI table]. (The palettes may appear docked – see **Adding and Removing Other Toolbars**, p.7.) You'll find similar tables for most commands discussed in this text.

Where to Find It:	
Command Line:	*ToolPalettes*
Hotkey(s):	*tp*
Ribbon (Tab/Panel):	View – Palettes – **Tool Palettes**
Menu:	Tools – Palettes – **Tool Palettes Window**
Toolbar:	Standard – **Tool Palettes Window**

Tool palettes (see the Tool Palettes Window in Figure 1.014, p.11) provide access to palettes that contain some of the same tools you found on the ribbon, as well as some of the more advanced (and useful) tools you'll discover in your basic studies. We'll go into more detail about these in specific lessons; but let's take a quick look at tool palettes now.

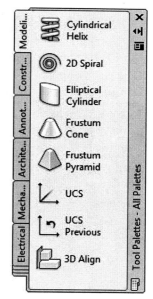

Figure 1.015

Looking at the side of the Tool Palettes window, you'll find several tabs. Selecting one will bring a different palette, with a different set of tools, to the front of the window. Pick on the overlapping tabs at the bottom to produce a list of all the palettes available to you.

The largest part of the palette contains its available tools. We'll discuss specific tools in some detail in later chapters.

The other side of the palette may contain a scrollbar to help access the tools, and a titlebar that works just like the ribbon's titlebar.

AutoCAD's tool palettes will prove useful when we discuss blocks and hatching. For now, however, they may prove a distraction to your basic lessons. You might be better off closing the Tool Palettes window or even docking it to the side of your screen and using **Auto-Hide** until we're ready for it.

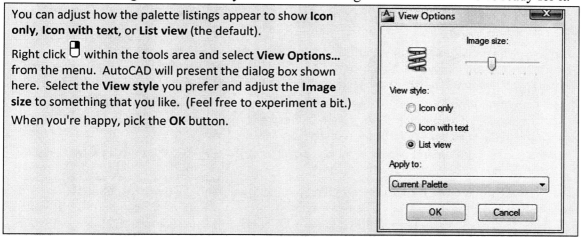

1.1.5   The Graphics Area

AutoCAD dedicates the largest part of the screen – called the *graphics area* – to work space (refer back to Figure 1.001, p.4). You'll do your drawing and editing here.

Until the recent appearance of the ribbon, which we'll remove from the screen (or hide) during most of our work time, the only item allowed to occupy the coveted space on the graphics area was the User Coordinate System (UCS) Icon. The icon serves a very useful purpose in 3-dimensional drafting. For AutoCAD beginners in the 2-dimensional world, however, it serves as a reminder of the X- and Y-planes. (We'll discuss X- and Y-planes in more detail in Lesson 2, p.47, and the UCS Icon in more detail in our 3D AutoCAD text.) You can disable the icon (turn it off) by entering *ucsicon* at the command prompt and responding *off* to the prompt as shown in the following sequence.[*]

    Command: *ucsicon*
    Enter an option [ON/OFF/All/Noorigin/ORigin/Properties] <ON>: *off*

Repeat the command and enter *on* to restore the UCS Icon. You'll learn more about AutoCAD's prompts throughout this text.

> AutoCAD actually uses two coordinate systems, the UCS and the World Coordinate System (WCS), but that discussion belongs in the 3-dimensional world. The UCS and the WCS are the same for the lessons in this text.

We'll look at those other graphics area tools – the View Cube and the Navigation Bar – in Section 1.1.7, p.15.

1.1.6 The Command Window and Status Bar

Just above the bottom of the screen, you can see a window with several **Command** prompts (or lines of text). Like the ribbon, this window can undock, but I don't recommend it. You can also resize it by placing your cursor on the heavy line above it, picking with the left mouse button, and "dragging" the window up or down. (The cursor will become a double arrow ⇕ when properly located, and you must hold the button down for this procedure.) I usually leave my command window large enough for two or three lines as some prompts require extra lines to display information.

I can't overemphasize the importance of becoming familiar with the **Command** prompt. Look closely. AutoCAD speaks to you on the left side of the colon. You respond on the right. Right now, AutoCAD is "prompting" you, or asking you what you want to do. When you respond, either by keyboard entry or mouse selection, your response will appear to the right of the colon.

Once you enter a command, AutoCAD's prompt may change, asking you for more information or input. Just follow the prompts to complete your task.

> Command line messages also appear on the dynamic input display, which appears to follow your crosshairs. You can remove the command window from your display (and provide more working area) by entering *CommandLineHide* at the prompt. (Redisplay it by entering *CommandLine* … but be sure the **Dynamic Input** toggle on the status bar is depressed first (it may appear as **DYN** rather than an icon)! (Alternately, you can hold down the CTRL button and press the '9' key on the keyboard.) I don't advise this setup. Some procedures still require the command line.
>
> You can also turn the dynamic input display on and off using the toggle on the status bar. We'll discuss dynamic input in more detail in Lesson 2, p.33.

AutoCAD also provides an enhancement to the command line (and the dynamic input display) that's quite useful. It's called *AutoComplete* functionality. If you're unsure of a command spelling, enter the first few letters and press the TAB key until you find the command you want! You can also use the up and down arrow keys to repeat previously used commands or data entry!

At the bottom of the screen is the *status bar*. On the left end of the status bar, you'll see a small box with three numbers in it. This **Coordinate Display** box shows the X, Y, and Z coordinates of the cursor in your drawing. We'll look at AutoCAD's coordinate system in Lesson 2, p.47.

To the right of the **Coordinate Display** box is a series of toggles (**Snap, Grid, Ortho**, etc.). We'll look at these in some detail in Lesson 3, p.55.

---

[*] The command sequence may also appear in the dynamic input display. See the first insert in Section 1.1.6, p.11.

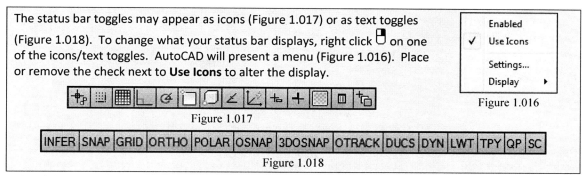

The status bar toggles may appear as icons (Figure 1.017) or as text toggles (Figure 1.018). To change what your status bar displays, right click 🖱 on one of the icons/text toggles. AutoCAD will present a menu (Figure 1.016). Place or remove the check next to **Use Icons** to alter the display.

Figure 1.016

Figure 1.017

Figure 1.018

On the right end of the status bar, you'll see several additional icons (Figure 1.019). These are (from the left):

Figure 1.019

- **Model** 🗔 and **Layout** 🗔 toggles – these control what type of "space" your graphics area displays. You'll do most of your actual work in Model space (the default). Layouts (aka paperspace) provide some extremely useful printing tools; we'll discuss these in Lessons 24, p.560, and 25, p.584. The textual **Model** toggle MODEL controls in which space you actually work. Again, this'll become clearer in Lessons 24 and 25.

If your screen displays **Model** and **Layout** tabs \Model / Layout1 / Layout2 /, the toggles may not appear on the status bar. Use the tabs for the same purpose. To show the icons rather than the tabs, right click 🖱 on the tab and select **Hide Layout and Model Tabs** from the menu.

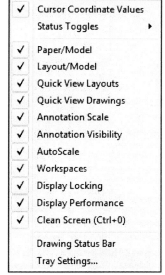

- **Quick View Layouts** 🗔 and **Quick View Drawings** 🗔 serve similar functions – although the first works for layouts in the current drawing (more on layouts in Lessons 24/25) and the second for open drawings. We'll examine **Quick View Drawings** in Section 1.5, p.26.

- The next three tools **Annotation Scale** 🔺 1:1 ▾, **Annotation Visibility** 🗔, and **Add Scale to Annotative Objects** 🗔 are related. We'll spend time with these in Lesson 4.

- Next you'll find the **Workspaces** toggle ⚙ which we've already discussed (Section 1.1.1, p.4), then the **Toolbar/Window Positions Lock** toggle 🗔 which locks the toolbars in their current positions, a **Hardware Acceleration** toggle 🗔 to help improve AutoCAD's performance on your system, an **Autodesk TrustedDWG** icon 🗔 which appears when the current drawing was created in AutoCAD, an **Isolate Objects** button 💡 which provides access to the *IsolateObjects* and the *HideObjects* commands (more on these in Lesson 10, p.240), and a down arrow that calls the **Application Status Bar Menu** (Figure 1.020). Use this menu to tell AutoCAD which of the various toggles you'd like to see.

Figure 1.020

Use the **Tray Settings** option in the **Application Status Bar Menu** to open the Tray Settings dialog box (shown here). Use this dialog box to control which icons and notifications AutoCAD displays in the tray on the right end of the status bar.

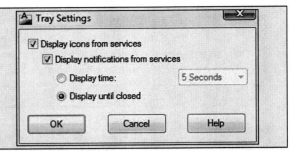

- Finally, you'll find a **Clean Screen** toggle □ which clears the monitor's screen of everything except AutoCAD's menu bar, status bar, view cube, navigation bar, and command window. This "expert setting" gives you more work room but limits your use of toolbars, the ribbon, or palettes.

1.1.7	The View Cube & Navigation Bar

The *View Cube* (Figure 1.021) serves no really useful purpose in 2D AutoCAD (although you'll love it in 3D!). To clear it from your 2D screen, follow this command entry:

   Command: *cube*
   Enter an option [ON/OFF/Settings] <ON>: *off*

I know, you had to experiment with it and now your screen is askew. Here's a quick fix:

   Command: *plan*
   Enter an option [Current ucs/Ucs/World] <Current>: *[Enter]*

Nothing to it, right?

(We'll see the *Plan* command again in *3D AutoCAD 2011: One Step at a Time*.)

Quit foolin' around!

Figure 1.021

The Navigation Bar (Figure 1.022) contains several tools that you'll find useful in both 2D and 3D space. These tools prove themselves invaluable when you need to maneuver your way through a large drawing. In Lesson 4, Section 4.1.4, p.88, you'll see how to use the 2D tools and hide the others.

If it distracts you in the meantime, you can control its display with the *NavBar* command:

   Command: *navbar*
   Enter an option [ON/OFF] <OFF>: *off*

Figure 1.022

1.2	Setting Up a New Drawing
1.2.1	The Groundwork: How AutoCAD Handles Scale, Units, and Paper Size

It may surprise you to learn that we won't use a scale when drawing in AutoCAD. In fact, all drawing in AutoCAD is done *full scale*! This simply means that a 3-inch line will actually be drawn 3-inches long. A line 3 miles long will be drawn 3 miles long!

Okay, so if we draw full scale, how can we put a house plan or refinery unit onto a sheet of paper unless the paper is very, very large? Well, we actually *tell* AutoCAD that we have a sheet of paper that's very, very large – a bit larger, in fact, than the house or unit we'll draw. Then later, when we plot (or print) the drawing to an actual sheet of paper, we *plot to scale*.

How do we determine the size of the paper we'll tell AutoCAD to use? I'm glad you asked!

> I've added a discussion of printing/plotting as an independent lesson (Lesson 23, p.528). The idea of making it independent is to allow students/instructors to include it at any point after the first lesson.

We know the standard sizes of paper used in the design world. And we know the standard scales used. To determine what we tell AutoCAD about paper size, we need to determine how many feet or inches (or millimeters) will fit onto each size sheet at each scale. The easiest way to do this is to look at the Drawing Scales Chart in Appendix B. (You'll find a similar chart in the possession of all CAD operators.)

To use the chart, we select the scale at which we'll want the final drawing plotted, and then select the paper size for the plot. Where row and column meet, we find the width x height *limits* of your drawing. (We'll talk more about limits shortly.) There's a place for these numbers in the drawing setup.

1.2.2 The Setup

Actually, you've already started a new drawing – back at the beginning of Section 1.1, p.3. It was just that simple to create a new drawing "from scratch" (using AutoCAD's default settings). You can begin any number of new drawings at any time in the same manner; but to make your drawing time more productive, there are a few things you should set up first – like paper size and units. Of course you could accept AutoCAD's defaults for these and just get started, but let's take a look at the procedures anyway – just in case your boss wants to use company standard settings.

> By default, AutoCAD uses the following settings:
> - Limits (paper size) = 12" x 9" (imperial) or 429mm x 297mm (metric)
> - Units = decimal to 4 places
> - Angles are decimal and measured counterclockwise with 0 degrees East

AutoCAD – never the shirker when it comes to redundancy – provides a couple methods for setting up a new drawing: templates and "from scratch." *

The possibility always exists that you may need to change AutoCAD's basic setup after you've begun a drawing session. (Okay, let's be realistic; you'll more than likely have to change something!) Maybe you need a different sheet size or different units. No sweat! AutoCAD makes it easy for you to adjust these settings.

Where to Find It:	
Command Line:	*Units*
Menu:	Format – **Units**
Application Menu:	Drawing Utilities – **Units**

To change units, enter the ***Units*** command (see the WTFI table). AutoCAD will provide the Drawing Units dialog box (Figure 1.023) where you can identify the new units and unit precision, as well as

Figure 1.023

---

* We'll mercifully ignore the out-of-date setup wizard!

the angle type, precision, and direction. You can also set the **Insertion scale** for blocks (more on blocks in Lessons 21, p.468, and 22, p.500) and your **Lighting** units (more on those in the 3D text). The **Direction** button calls another dialog box where you can set the compass point from which AutoCAD will begin its angular measures. Let me explain.

Most of us are familiar with the nautical way of determining degrees (a product of a good scout training). Unfortunately, many engineers apparently missed that training. The standard 0°North in the real world has changed to 0°East in the engineering world.

You can change it here if you wish; but be warned, 0°East has become an engineering standard. If another CAD operator discovers that you have changed the angles, you may be in for some grief. Many third-party programs and AutoLISP routines (as well as engineering standards) are based on 0°East.

> Notice that you can also use the Drawing Units dialog box to change the direction in which AutoCAD measures angles. The default – counterclockwise – is also an engineering standard. It's a good idea to check with your CAD supervisor before changing this setting.

That's it for unit setup.

Where to Find It:	
Command Line:	Limits
Menu:	Format – **Drawing Limits**

To change the sheet size, enter the ***Limits*** command. AutoCAD will prompt you for the lower left coordinates of your drawing, using 0,0 as the default (see the following insert). Accept the default by hitting ENTER on your keyboard.

> Note that most companies will use 0,0 as the lower left corner of their drawings. Occasionally, however, a company may choose to use *true coordinates*. This is best explained by example.
>
> In the petrochem world, plants are divided into units. Each unit may have one or more drawings specific to that unit. Vessels and other items in the unit are located in the plant by overall *east/west* and *north/south* (*X* and *Y*) coordinates. It's often useful to identify the east/west and north/south coordinates with the X and Y coordinates in AutoCAD. This way, every item in the plant can be quickly located in the drawings simply by doing an ***ID*** command (p.260).
>
> When using true coordinates, the absolute location of the drawing will dictate the lower left as well as the upper right corners of the drawing limits.

AutoCAD will then prompt you for the upper right coordinates. Get these from the Drawing Scales chart (Appendix B) and enter them as ***width,height*** (no spaces). The sequence looks like this:

    Command: *limits*

    Reset Model space limits:

    Specify lower left corner or [ON/OFF] <0.0000,0.0000>: *[ENTER]*

    Specify upper right corner <12.0000,9.0000>: *[Enter the new width and height.]*

While you can make these adjustments at any time during your drawing session, I don't recommend waiting too long. It's always best to groom good habits from the very beginning.

We'll set up several drawings from scratch over the course of our text, so you'll become familiar with the *Units* and *Limits* commands.

> If you're already in AutoCAD and you want to start a new drawing from scratch, pick the **New** button on the Quick Access toolbar. Alternately, you can select **New** from the File menu or enter *New* at the command prompt. AutoCAD will open a Select Template dialog box. You can use the default *Acad.dwt* template, which is the 2D template, or the *Acad3D.dwt* template for default 3D settings. Alternately, (to open a new drawing from scratch) you can bypass the selection of a template by picking the down arrow next to the **Open** button and selecting **Open with No Template – [Imperial** or **Metric]** (shown here).

Okay, that's the brief overview, now let's get busy! We'll start by setting up a drawing after starting with the default template. We'll set up a drawing for a ¼"=1'-0" scale on a B-size (11 x 17) sheet of paper.

Do This: 1.2.2A	Setting Up a Drawing

I. If you haven't already, double click the AutoCAD icon on the desktop. If you're still in AutoCAD, enter the *New* command at the command prompt and accept the default template.

II. Pick the dynamic input toggle (╬ or DYN) on the status bar to turn it off. (It should appear unlit.)

III. Follow these steps.

1.2.2A: SETTING UP A DRAWING

1. We'll start by setting up our units. Enter the *Units* command. (Okay, this is your first time, right? Just type "Units" and it'll show up on the command line. Alternately, you can use one of the other methods found in the Units WTFI table, p.16.)

   Command: *units*

AutoCAD presents the Drawing Units dialog box (Figure 1.023, p.16).

2. In the **Length** frame, change the **Type** of units to **Architectural**. Notice that the **Precision** automatically changes to 1/16". You can change that here as well, but we'll accept 1/16" as it's an industry standard.

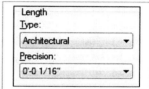

3. Take a moment to explore the other settings in this dialog box, and then accept the rest of the defaults. Pick the **OK** button [OK] to complete your units setup.

4. Now we'll set up the drawing page. Remember, all drawing is full scale so we have to adjust the size of the paper to accommodate the drawing.
Enter the *Limits* command.

   Command: *limits*

5. AutoCAD wants to know what the lower left limits should be. We'll hit ENTER to accept the 0,0 default.

   Specify lower left corner or [ON/OFF] <0'-0",0'-0">: *[ENTER]*

6. Now, check the Drawing Scales chart in Appendix B to determine the limits for a ¼" scale on a 11"x17" sheet of paper. Did you get 44'x68'? Excellent! Enter these at the next prompt, as shown. (Enter the desired page width followed by a comma and the desired page height. Don't use spaces or you'll upset AutoCAD!)

   Specify upper right corner <1'-0",0'-9">: *68',44'*

Don't close your drawing yet; we'll continue with it in our next section.

Congratulations! You've set up your first AutoCAD drawing.

1.3 Saving and Leaving a Drawing Session

Where to Find It:	
Command Line:	*QSave*
Hotkey(s):	CTRL + *s*
Menu:	File – **Save**
Application Menu:	**Save**
Toolbar:	Standard – **Save**
	Quick Access – **Save**

Now that we've created our drawing, we need to save it. We accomplish this by using one of three commands: *Save*, *Saveas*, or *QSave*. Each behaves in a similar manner depending on the status of the drawing. That is, each will present the Save Drawing As dialog box if the drawing has not been previously saved. The *QSave* command, however, will automatically save the drawing without prompting for additional information, provided it has been previously saved and given a name.

Save and *Saveas* are available at the command prompt. *QSave* will occur when you pick the **Save** button on the Quick Access toolbar or one of the menus.

Just as you have several ways to save a drawing, so too, you have several locations to which you can save. These include a local drive, a network drive, or an Internet location. Luckily, the procedures differ only slightly. We'll look at each in the next exercise.

> Saving a drawing to an Internet location is simple enough; it requires that you enter a web address in the **File name** text box of the Save Drawing As dialog box. But because of Internet protocols, you can't transfer information from your computer to an *http* address. You can, however, send your file to an *ftp* address. Don't underestimate the significance of this slight difference. Forgetting to change protocols has driven many an operator into frantic searches for more profound problems, only to wind up feeling foolish when they discovered the oversight.
>
> Once you've entered the correct address and picked the **Save** button, AutoCAD must stop and ask you for a user ID and password. The nature of the Internet requires these to ensure the integrity of the web site. (Imagine opening a file on your web site only to discover that some unknown villain had overwritten your drawing files with ©*Toadal Wisdom* cartoons.)

Where to Find It:	
Command Line:	*Close* or *CloseAll*
Menu:	File – **Close**
Application Menu:	**Close** – **Current Drawing** (or **All Drawings**)

How you end the drawing depends on whether you wish to end only the current drawing or the entire drawing session. In other words, do you want to close AutoCAD or just close this drawing? To leave the drawing without leaving AutoCAD, use the *Close* command. To close all drawings that are currently open, without leaving AutoCAD, use the *Closeall* command. Otherwise, use the *Quit* command or pick the **X** button on the right end of the title bar.

Let's save and exit our drawing.

Do This: 1.3A	Saving Your Drawing Changes and Closing the Drawing Session

I. Continue the previous drawing session.

1.3A: SAVING AND CLOSING YOUR DRAWING

1. Open the Save Drawing As dialog box. (Use one of the methods in the WTFI table, p.19.) AutoCAD presents the Save Drawing As dialog box shown in the following figure.
   Command: *saveas*

1.3A: SAVING AND CLOSING YOUR DRAWING

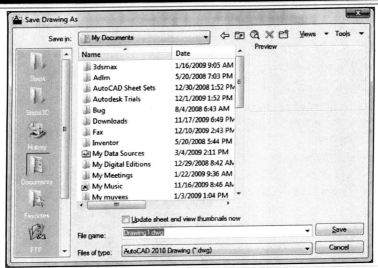

2. Next to the **Save in** control box (where you see *Documents*), you'll find a downward-pointing arrow. Pick on that arrow to see a path showing where you are on your hard drive (as indicated). We're going to save the file to the C:\Steps\Lesson01 folder.

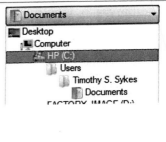

Pick 🖱 on the computer icon next to **(C:)**. The list box below changes to show all the folders on the C-drive. (Use this procedure to save your drawing to a network drive as well. Simply pick the letter that designates the drive to which you wish to save your work.)

3. In the list box, double click on the folder identified as **Steps** 📁 Steps . This folder opens and the list box now shows the contents of the Steps folder.

4. Double click on the folder identified as **Lesson01** 📁 Lesson01 . This folder now opens. Your dialog box should now look like the following figure.

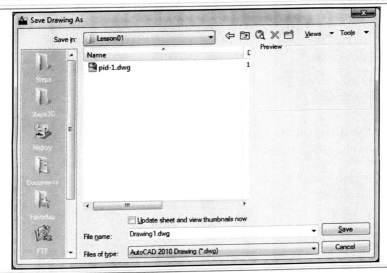

1.3A: SAVING AND CLOSING YOUR DRAWING

5. Type in the name *My First Step* in the box next to the words **File name**. Don't put the extension on the name; AutoCAD will do that for you.	File name: My First Step Files of type: AutoCAD 2010 Drawing (*.dwg)
6. Notice that the type of file you'll save has been identified in the **Files of type** control box. Pick the down arrow to view the different formats available. • The *DWG* file types make it possible to save an AutoCAD drawing so that it may be edited by earlier releases of AutoCAD. • the *DWS* files are used as standard files. • The *DWT* file is an AutoCAD template (we'll look more at this in a few minutes). • Finally, *DXF* files are binary files (computer programming stuff) used to exchange AutoCAD drawings with other programs. Leave the default **AutoCAD 2010 Drawing** selected.	My First Step AutoCAD 2010 Drawing (*.dwg) AutoCAD 2010 Drawing (*.dwg) AutoCAD 2007/LT2007 Drawing (*.dwg) AutoCAD 2004/LT2004 Drawing (*.dwg) AutoCAD 2000/LT2000 Drawing (*.dwg) AutoCAD R14/LT98/LT97 Drawing (*.dwg) AutoCAD Drawing Standards (*.dws) AutoCAD Drawing Template (*.dwt) AutoCAD 2010 DXF (*.dxf) AutoCAD 2007/LT2007 DXF (*.dxf) AutoCAD 2004/LT2004 DXF (*.dxf) AutoCAD 2000/LT2000 DXF (*.dxf) AutoCAD R12/LT2 DXF (*.dxf)

7. Pick the **Save** button [Save] in the lower right corner of the dialog box. AutoCAD saves the drawing and the dialog box closes. Notice that the title and path of the drawing now appear on the title bar.

Normally, you'd just close the drawing now (proceed to Step 13.) But we'll take a moment now to save it to a website, too, just to see how that's done. Be sure your web connection is open before you continue.

* In order to avoid thousands of stray documents finding their way to our website, you can't actually save your document to the Internet in this exercise. Instead, I'll take you to the point where the user ID and password are required, and then we'll cancel the procedure.

---

8. Begin by entering the *Saveas* command.
       Command: *saveas*

9. Enter the ftp address to our Internet site (*ftp://ftp.uneedcad.com/2011/MyDrawing.dwg*[*]). Notice (right) that you'll need to include the extension when saving to an Internet site.	File name: ftp://ftp.uneedcad.com/2011/MyDrawing.dwg Files of type: AutoCAD 2010 Drawing (*.dwg)

10. Pick the **Save** button [Save].

---

[*] I used this site as an example. If you have problems with this one, any ftp site will work.

1.3A: SAVING AND CLOSING YOUR DRAWING

11. AutoCAD asks you for the network password. If you had the proper **User name** and **Password**, you would enter them here and AutoCAD would save the drawing to the appropriate Internet address. Since we can't save without the password, pick the **Cancel** button to exit the command.

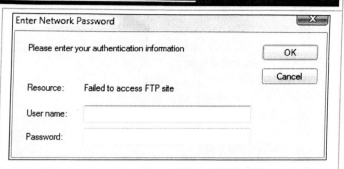

12. Close your Internet connection. (Note: If you have accessed the Internet through a company network, this may not be possible or practical. Move on to Step 13.)

13. Now that your drawing has been saved, you can exit the program. The easiest way to do this is to pick the **X** button on the right end of the title bar, but you can also enter *quit* on the command line or select **Exit** from the File pull-down menu.

   Command: *quit*

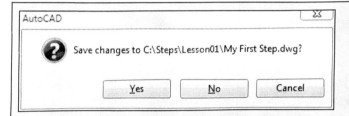

Note that leaving a drawing that has been changed, without first saving it, will cause AutoCAD to present a warning box (shown here). AutoCAD will let you know that you have not saved the drawing and ask if you'd like to save it now. If you haven't saved it before, AutoCAD will take you to the Save Drawing As dialog box that you used in Step 1 of the previous exercise.

1.4 Opening an Existing Drawing

Where to Find It:	
Command Line:	Open
Hotkey(s):	CTRL + o
Menu:	File – Open
Application Menu:	Open – Drawing
Toolbar:	Standard – Open
	Quick Access – Open

How you open a drawing depends in part on where the drawing is located. You won't find opening an AutoCAD file (a drawing) from a local or network location any different from opening a file in most Microsoft applications. In fact, you can open a drawing from an Internet location almost as easily, although AutoCAD does provide a couple methods for this.

- To access a drawing (or Xref, block, or OLE Object – all things that you'll learn about later) via the Internet, simply begin the *Open* command, and then type the file's address in the **File Name** text box. Be sure to type the complete address and include the file's extension (.dwg).

Another approach to accessing a file over the Internet involves AutoCAD's browser (Figure 1.4, p.23). To open it, pick the **Search the Web** button on the *Open* command's Select File dialog box's toolbar (along the top of the Select File dialog box).

AutoCAD's web browser works like any other web browser – except that it makes opening a drawing easier.

Internet Explorer users will find some familiar buttons across the top of the browser. From left to right, these are:

- o **Back** ⇐ – for returning to the last address.
- o **Forward** ⇒ – for moving from an address reached via the **Back** button to the next address in sequence.
- o **Stop** ● – for stopping the browser when a search is taking too long.
- o **Refresh** ↻ – for cleaning up the screen or reloading the current address.
- o **Home** ⌂ – for returning to the home page. By default, AutoCAD uses its own home page here. But you can change the home page with the *InetLocation* command. The command sequence is

    Command: *inetlocation*
    Enter new value for INETLOCATION <"http://www.autodesk.com">: *http://www.foragerpub.com*

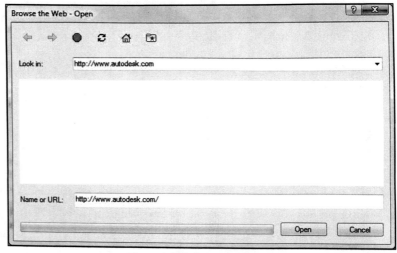

Figure 1.024 (Image may vary)

- **Favorites** ★ – for opening the Windows Favorites folder.
- To go to another web site, type the address into the **Look in** text box (top of the browser). F sure to hit ENTER so AutoCAD will know when you've finished entering the address.
- Once you've reached the web site, you can pick on the file name you wish to open. If the doesn't show the file, enter the file name in the **Name or URL** text box.

But how can so many people open the same internet drawing? Simple – AutoCAD actually the original drawing into a temporary folder on your hard drive. You'll work on it there ur ready to save it. This way, you don't have to maintain an open connection to the Internet

> Although AutoCAD will automatically dial your ISP connection if Windows has been configur you may need to open the connection before trying to access a web site via AutoCAD's bro can do this by executing your dialup connection, or simply by opening your browser as yo would. Many corporate offices or schools connect to the Internet via their own network situation, you probably won't need to do anything since network/Internet access is aut

In our next exercise, we'll open a couple drawings – one that's located on our local our UNeedCAD (ForagerPub) site.

Let's begin.

Do This: 1.4A	Opening an Existing Drawing

I. Begin a new AutoCAD session (double click on the desktop icon).
II. Follow these steps.

1.4A: OPENING A DRAWING
1. Enter the **Open** command 📂 **Command:** *open*
2. Navigate to the C:\Steps\Lesson01 folder just as you did in Ex

23

1.3A: SAVING AND CLOSING YOUR DRAWING

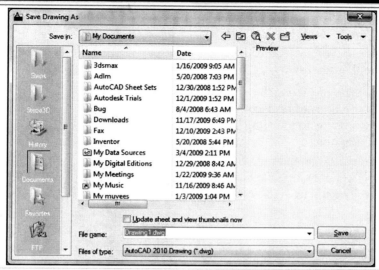

2. Next to the **Save in** control box (where you see *Documents*), you'll find a downward-pointing arrow. Pick on that arrow to see a path showing where you are on your hard drive (as indicated). We're going to save the file to the C:\Steps\Lesson01 folder.

Pick 🖱 on the computer icon next to **(C:)**. The list box below changes to show all the folders on the C-drive. (Use this procedure to save your drawing to a network drive as well. Simply pick the letter that designates the drive to which you wish to save your work.)

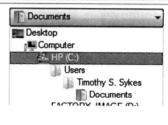

3. In the list box, double click on the folder identified as **Steps** ![Steps]. This folder opens and the list box now shows the contents of the Steps folder.

4. Double click on the folder identified as **Lesson01** ![Lesson01]. This folder now opens. Your dialog box should now look like the following figure.

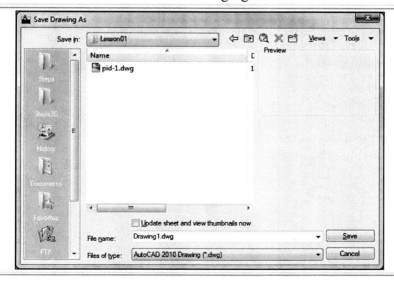

1.3A: SAVING AND CLOSING YOUR DRAWING

5. Type in the name *My First Step* in the box next to the words **File name**. Don't put the extension on the name; AutoCAD will do that for you.	File name: My First Step Files of type: AutoCAD 2010 Drawing (*.dwg)
6. Notice that the type of file you'll save has been identified in the **Files of type** control box. Pick the down arrow to view the different formats available. • The *DWG* file types make it possible to save an AutoCAD drawing so that it may be edited by earlier releases of AutoCAD. • the *DWS* files are used as standard files. • The *DWT* file is an AutoCAD template (we'll look more at this in a few minutes). • Finally, *DXF* files are binary files (computer programming stuff) used to exchange AutoCAD drawings with other programs. Leave the default **AutoCAD 2010 Drawing** selected.	My First Step AutoCAD 2010 Drawing (*.dwg) AutoCAD 2010 Drawing (*.dwg) AutoCAD 2007/LT2007 Drawing (*.dwg) AutoCAD 2004/LT2004 Drawing (*.dwg) AutoCAD 2000/LT2000 Drawing (*.dwg) AutoCAD R14/LT98/LT97 Drawing (*.dwg) AutoCAD Drawing Standards (*.dws) AutoCAD Drawing Template (*.dwt) AutoCAD 2010 DXF (*.dxf) AutoCAD 2007/LT2007 DXF (*.dxf) AutoCAD 2004/LT2004 DXF (*.dxf) AutoCAD 2000/LT2000 DXF (*.dxf) AutoCAD R12/LT2 DXF (*.dxf)
7. Pick the **Save** button [Save] in the lower right corner of the dialog box. AutoCAD saves the drawing and the dialog box closes. Notice that the title and path of the drawing now appear on the title bar.	

Normally, you'd just close the drawing now (proceed to Step 13.) But we'll take a moment now to save it to a website, too, just to see how that's done. Be sure your web connection is open before you continue.

* In order to avoid thousands of stray documents finding their way to our website, you can't actually save your document to the Internet in this exercise. Instead, I'll take you to the point where the user ID and password are required, and then we'll cancel the procedure.

8. Begin by entering the *Saveas* command. **Command:** *saveas*	
9. Enter the ftp address to our Internet site (*ftp://ftp.uneedcad.com/2011/MyDrawing.dwg*[*]). Notice (right) that you'll need to include the extension when saving to an Internet site.	File name: ftp://ftp.uneedcad.com/2011/MyDrawing.dwg Files of type: AutoCAD 2010 Drawing (*.dwg)
10. Pick the **Save** button [Save].	

---

[*] I used this site as an example. If you have problems with this one, any ftp site will work.

1.3A: SAVING AND CLOSING YOUR DRAWING

11. AutoCAD asks you for the network password. If you had the proper **User name** and **Password**, you would enter them here and AutoCAD would save the drawing to the appropriate Internet address. Since we can't save without the password, pick the **Cancel** button [Cancel] to exit the command.

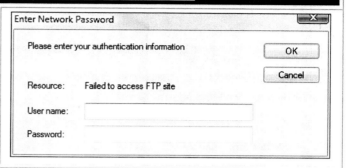

12. Close your Internet connection. (Note: If you have accessed the Internet through a company network, this may not be possible or practical. Move on to Step 13.)

13. Now that your drawing has been saved, you can exit the program. The easiest way to do this is to pick the **X** button [X] on the right end of the title bar, but you can also enter *quit* on the command line or select **Exit** from the File pull-down menu.

Command: *quit*

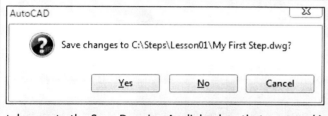

Note that leaving a drawing that has been changed, without first saving it, will cause AutoCAD to present a warning box (shown here). AutoCAD will let you know that you have not saved the drawing and ask if you'd like to save it now. If you haven't saved it before, AutoCAD will take you to the Save Drawing As dialog box that you used in Step 1 of the previous exercise.

1.4 Opening an Existing Drawing

Where to Find It:	
Command Line:	*Open*
Hotkey(s):	CTRL + o
Menu:	File – **Open**
Application Menu:	Open – **Drawing**
Toolbar:	Standard – **Open**
	Quick Access – **Open**

How you open a drawing depends in part on where the drawing is located. You won't find opening an AutoCAD file (a drawing) from a local or network location any different from opening a file in most Microsoft applications. In fact, you can open a drawing from an Internet location almost as easily, although AutoCAD does provide a couple methods for this.

- To access a drawing (or Xref, block, or OLE Object – all things that you'll learn about later) via the Internet, simply begin the ***Open*** command, and then type the file's address in the **File name** text box. Be sure to type the complete address and include the file's extension (.dwg).
- Another approach to accessing a file over the Internet involves AutoCAD's browser (Figure 1.024, p.23). To open it, pick the **Search the Web** button on the ***Open*** command's Select File dialog box's toolbar (along the top of the Select File dialog box).

AutoCAD's web browser works like any other web browser – except that it makes opening a drawing a bit easier.

- Internet Explorer users will find some familiar buttons across the top of the browser. From left to right, these are:

- o **Back** ⇐ – for returning to the last address.
- o **Forward** ⇒ – for moving from an address reached via the **Back** button to the next address in sequence.
- o **Stop** ● – for stopping the browser when a search is taking too long.
- o **Refresh** ↻ – for cleaning up the screen or reloading the current address.
- o **Home** ⌂ – for returning to the home page. By default, AutoCAD uses its own home page here. But you can change the home page with the ***InetLocation*** command. The command sequence is

    Command: *inetlocation*

    Enter new value for INETLOCATION <"http://www.autodesk.com">: *http://www.foragerpub.com*

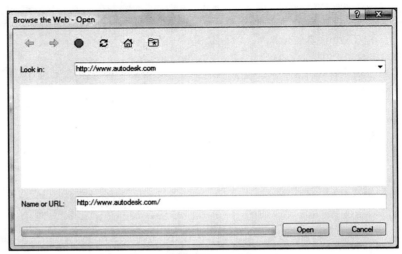

Figure 1.024 (Image may vary)

- **Favorites** – for opening the Windows Favorites folder.
- To go to another web site, type the address into the **Look in** text box (top of the browser). Be sure to hit ENTER so AutoCAD will know when you've finished entering the address.
- Once you've reached the web site, you can pick on the file name you wish to open. If the site doesn't show the file, enter the file name in the **Name or URL** text box.

But how can so many people open the same internet drawing? Simple – AutoCAD actually copies the original drawing into a temporary folder on your hard drive. You'll work on it there until you're ready to save it. This way, you don't have to maintain an open connection to the Internet indefinitely.

> Although AutoCAD will automatically dial your ISP connection if Windows has been configured to do so, you may need to open the connection before trying to access a web site via AutoCAD's browser. You can do this by executing your dialup connection, or simply by opening your browser as you normally would. Many corporate offices or schools connect to the Internet via their own networks. In this situation, you probably won't need to do anything since network/Internet access is automatic.

In our next exercise, we'll open a couple drawings – one that's located on our local drive and one at our UNeedCAD (ForagerPub) site.

Let's begin.

Do This: 1.4A	Opening an Existing Drawing

I. Begin a new AutoCAD session (double click on the desktop icon).
II. Follow these steps.

1.4A: OPENING A DRAWING

1. Enter the **Open** command 📂

    Command: *open*

2. Navigate to the C:\Steps\Lesson01 folder just as you did in Exercise 1.3A, p. 19.

1.4A: OPENING A DRAWING

3. Select the *pid-1* drawing file (following figure). Notice that AutoCAD previews the drawing in the area to the right.

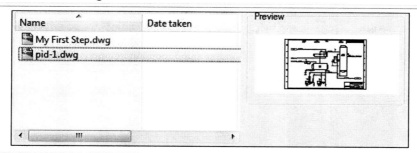

4. Now double click on *pid-1*. AutoCAD will open the drawing for editing. Nothing to it! Now let's look at a couple ways to open a drawing located at an Internet location.

5. Begin as you did in Step 1 📂.

Command: *open*

6. Type the address of the file to open into the **File name** text box. (The address is *http://www.uneedcad.com/Webfiles/web.dwg*. Be sure to include the extension and capitalize "Webfiles".)

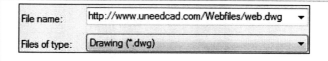

7. Pick the **Open** button [Open] to continue. AutoCAD transfers the file to a local folder and opens it for you. (AutoCAD may present a File Download status box depending on the speed of your connection.) The drawing looks like the figure at right.

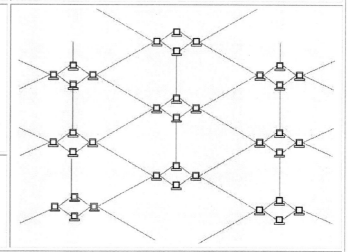

8. Now let's take a look at AutoCAD's browser. Repeat Step 1 📂.

Command: *open*

9. In the Select File dialog box, pick the **Search the Web** button 🔍 (just above the **Preview** frame). (Alternately, you can hold down the ALT key while pressing the 3 key.) AutoCAD presents the browser (Figure 1.024, p.23).

10. First, we'll go to the web site. Enter the address in the **Look in** control box as indicated in the following figure. The address is: *http://www.uneedcad.com/Webfiles*. (Be sure to hit ENTER.)

1.4A: OPENING A DRAWING

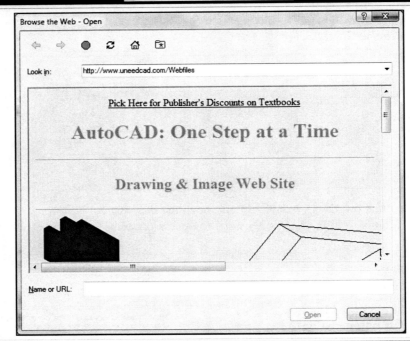

11. Scroll (use the scroll bars) until you see the *Gable* image on the right (following figure).

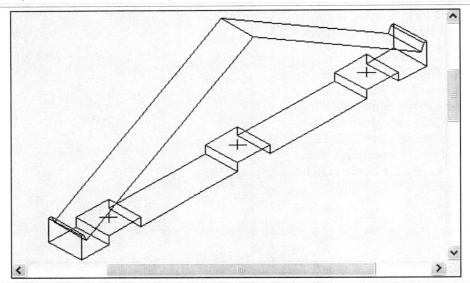

12. Pick on the *Gable* image. AutoCAD places the name and location of the drawing in the **Name or URL** text box (see the following figure).

| Name or URL: | http://www.uneedcad.com/Webfiles/gable.dwg |

13. Pick the **Open** button [Open] to open the drawing. AutoCAD will download the drawing just as it did in Step 7.

14. Leave everything as it is and continue through Section 1.5.

Was that easy? One way to make it even easier is to list the web site in the Favorites folder of your browser. Then you can pick the **Favorites** button on the AutoCAD browser and simply select the site from there. This saves quite a bit of typing.

1.5 Navigating Between Drawings with Quick View Drawings

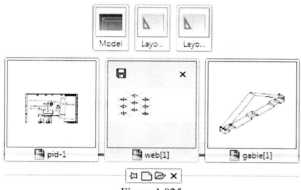

Figure 1.025

Where to Find It:	
Command Line:	*QVDrawing*
Status Bar:	Quick View Drawings

This one deserves a *COOL TOOL* designation! It let's you see all the open drawings at one time – and even select which one you want to use!

Quick View Drawings (*QVDrawing*) will display a thumbnail of all the currently open drawings (Figure 1.025). Select a thumbnail to make that drawing current. Hover your cursor over the smaller thumbnail above to display views of that drawing's layouts or model space. (We'll discuss layouts in Lessons 24, p.560, and 25, p.584.) Select one of the layouts or model space to make that drawing current and opened in the selected layout or model space.

Below the thumbnails, you'll find a smaller toolbar with buttons to: **Pin** the Quick View Drawing display open, start a **New** drawing, **Open** additional drawings, and **Close** the **Quick view drawings** display.

Do This: 1.5A	Quick View Drawings

I. Continue from where we left off in Section 1.4.
II. Follow these steps.

1.5A: QUICK VIEW DRAWINGS

1. Enter the **Quick View Drawings** command.
      Command: *qvdrawing*

AutoCAD presents thumbnails of the open drawings (Figure 1.025, p.26).

2. Pass your cursor over each of the thumbnails and see how the smaller thumbnails above change.

3. Select each thumbnail and see how the current drawing changes. This is the easy way to move from one drawing to another.

4. Right click on the PID-1 drawing file. Notice the cursor menu.

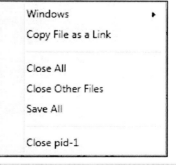

5. Select **Close pid-1**. Notice that this file, along with its thumbnail, closes.

6. Pick anywhere in the drawing. Notice that the thumbnails close.

7. Repeat Step 1.

1.5A: QUICK VIEW DRAWINGS
8. Pick the **Pin** button 📌 on the Quick View Drawings toolbar. Notice that it changes direction.
9. Repeat Step 6. Notice that now, the thumbnails remain open.
10. Finally, move the cursor back and forth between a drawing and one of its layout thumbnails. You can select one of the layout thumbnails and open that drawing in that layout. (Again, we'll look at layouts in some detail in Lessons 24 and 25.)
11. Enter the **Closeall** command to close all the open drawings. (Don't save any changes.)   **Command:** *closeall*

1.6 Creating and Using Templates

We've learned how to set up a new drawing. But let's face it; this is a tedious procedure at best. About midway through the exercise, many students ask if "we have to do this every time." The answer is, "NO!"

You should have to set up a drawing once for each scale and sheet size you'll use. So you may have a ¼"=1'-0" drawing setup for an 11 x 17 sheet of paper, a 3/8"=1'-0" drawing for an 11 x 17 sheet, a ¼"=1'-0" for a 24 x 36, and so forth. But once set up, you shouldn't have to set it up again regardless of how many times you may need it.

Here's how this works. Set up your drawing, but save it as a template. Then, when you need to create a drawing at that scale and with those limits on that size sheet of paper, use the template to create the drawing. Any drawing created with that template will carry the same setup as the template. But the template will never change (unless you overwrite it), so you can use it repeatedly.

Now let's see if we can create a template from the *My First Step* file we created earlier. We'll see how to use the template to start a new drawing, too.

We'll set up our template using some basic stuff – sheet size and scale. But templates can and should also include such things as dimension and text styles, plot styles, layers, and more. You'll learn about these things over the course of this text.
AutoCAD also provides a host of predefined templates that can also be of use with some adjustments for project-specific details.

Do This: 1.6A	Creating a Template

   I. Close any open drawings without saving them, then open the *My First Step* file in the C:\Steps\Lesson01 folder.
  II. Follow these steps.

1.6A: CREATING A TEMPLATE
1. Enter the *Saveas* command.   **Command:** *saveas* AutoCAD presents the Save Drawing As dialog box.

1.6A: CREATING A TEMPLATE

2. Pick the down arrow in the **Files of type** control box and select the **AutoCAD Drawing Template** option (as shown).

3. AutoCAD automatically changes the path to place the template in its own template folder. Change it back to save our template in the C:\Steps\Lesson01 folder.

4. Call it *My First Template* as shown.

5. Pick the **Save** button.

6. AutoCAD presents a Template Options dialog box. Put a bullet next to **Save all layers as reconciled** as indicated (more on layers in Lesson 7, p.148). Notice that you can save the template as either an English or Metric template. Pick the **OK** button to continue.

7. Close the drawing template.
   Command: *close*

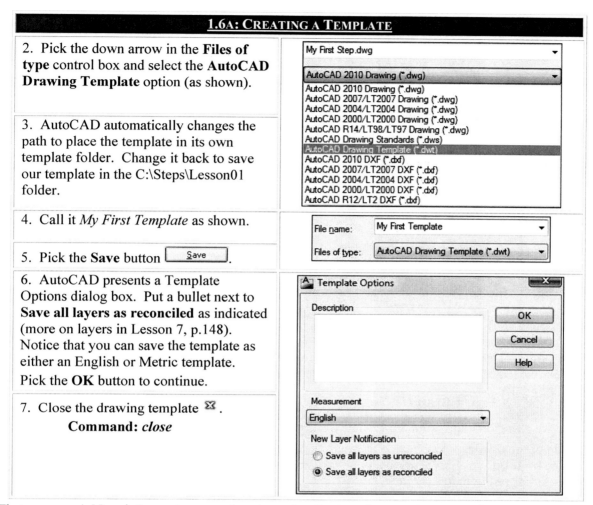

That was easy! Now let's create a new drawing using the template you just created.

Do This: 1.6B — Using a Drawing Template

I. Begin an AutoCAD session.
II. Follow these steps.

1.6B: USING A DRAWING TEMPLATE

1. Enter the *New* command.
   Command: *new*

2. A Select Template dialog box appears (similar to an Open File dialog box). Navigate to the C:\Steps\Lesson01 folder and select the *My First Template* you created in the last exercise (see following figure).

1.6B: USING A DRAWING TEMPLATE

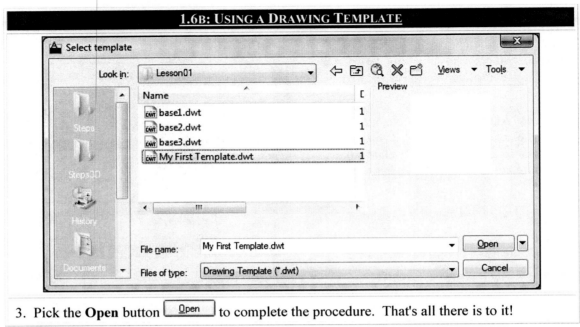

3. Pick the **Open** button to complete the procedure. That's all there is to it!

Remember, you can use the setup in a template file as many times as you wish with this procedure. The drawing you create will contain all the information (setup, drawing, etc.) found in the template, but using the template will in no way affect the template itself.

1.7 Extra Steps

Before we close our exploration of AutoCAD's user interface, let's take a quick look at those tools to the right of the title bar – . From the left, you have the **InfoCenter Input** box, the **Search** button, the **Subscription Center** button, the **Communication Center** button, a **Favorites** shortcut, and a **Help** button.

> For help setting up the InfoCenter or the Communication Center, refer to the supplement – *ComCtrSetup.pdf* – found at our website (http://www.uneedcad.com/Files).

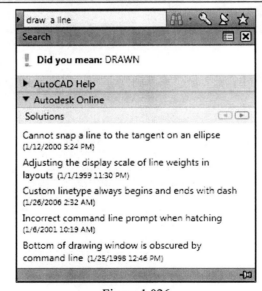

Figure 1.026

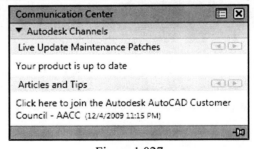

Figure 1.027

- When you have a question about how to do something, enter it (or simply a key word or two) into the **InfoCenter Input** box, then pick the **Search** button. AutoCAD will present a menu of possible solutions to your question (Figure 1.026, p.29) organized by where AutoCAD found the solution. Pick on the listing that will most likely help you, and AutoCAD will open a help window with a detailed explanation.

   It's kinda like having the programmers sitting next to you – pretty cool, huh?

- Next to the **InfoCenter** button, you'll find access to "The Company's" (Autodesk's) way of talking to you – the **Communication Center**. (You won't find many corporations making such an effort to work with their customers!) This button will open the menu in Figure 1.027, p.29.

   Here you can find access to the Communication Center where Autodesk will talk to you.

Obviously, AutoCAD's "Help" tools go far beyond the norm. Have you ever found a company who tried so hard to work with its customers?!

1.8 What Have We Learned?

Items covered in this lesson include:

- *AutoCAD setup*
- *Dynamic Input Display*
- *The InfoCenter*
- *The AutoCAD User Interface:*
    - *Menus*
    - *Quick Access Toolbar*
    - *Ribbon*
    - *Status bar*
- *Commands:*
    - *QSave*
    - *Save*
    - *Saveas*
    - *Close*
    - *Closeall*
    - *Quit*
    - *Open*
    - *Units*
    - *Limits*

- *Workspaces*
- *Templates*

    - *Tool palettes*
    - *Toolbars*
    - *The Navigation Bar*
    - *The View Cube*

    - ***InetLocation***
    - ***UCSIcon***
    - ***CommandLine***
    - ***CommandLineHide***
    - ***WSSave***
    - ***Ribbon***
    - ***QVDrawing***
    - ***NavBar***
    - ***Cube***

Well, now you've gotten your feet wet. How was it?

Any road begins with a first step, and any design begins with some basic decisions; what scale should I use ... what size sheet of paper do I need ... should I listen to country or rock-n-roll while I draw?

We have looked at how to accomplish most of these tasks on a computer, using AutoCAD as our tool. In the next few lessons, you'll see a whole new world of possibilities opening before you as we explore this wonderful drafting tool.

But first, let's get some practice with what we have learned so far.

1.9 Exercises

Create a few templates to use later in this course.

1. Create a template with these parameters:
    - 1.1.1. Use architectural units;
    - 1.1.2. Use a 1=1 scale (this requires no setup; it is the default);
    - 1.1.3. Set up for a sheet size of 8½ x11;
    - 1.1.4. Save this project as *MyBase1.dwt* in the C:\Steps\Lesson01 folder.

2. Create a new template with the following parameters:
    - 2.1.1. Use architectural units;
    - 2.1.2. Use a 1=1 scale;
    - 2.1.3. Set up for a sheet size of 11x17;
    - 2.1.4. Save the project as *MyBase2.dwt* in the C:\Steps\Lesson01 folder.

3. Create a third template with the following parameters:
    - 3.1.1. Use architectural units;
    - 3.1.2. Use a 1=1 scale;
    - 3.1.3. Set up for a sheet size of 17 x 22;
    - 3.1.4. Save the project as *MyBase3.dwt* in the C:\Steps\Lesson01 folder.

1.10 For Web-Based Review Questions and Additional Exercises, visit: http://foragerpub.com/AcadFiles/2011/2011.htm

Don't stop now! We have miles to go before we sleep!

Lesson 2

Following this lesson, you will:

- ✓ *Know how to use the basic draw commands: **Line**, **Rectangle**, and **Circle***
- ✓ *Know how to use the basic modify commands: **Erase**, **Undo**, **Redo** and **MRedo***
- ✓ *Know how to select objects using object selection, window, and crossing window methods*
- ✓ *Know how to use the Direct Distance Option when you draw*
- ✓ *Know how to use AutoCAD's Cartesian Coordinate System*
- ✓ *Know how to use Dynamic Input*

Drawing Basics – Lines, Circles, & Coordinates

We can define drafting as the placing of geometric shapes on paper to represent existing or proposed objects. Of course we're using a computer rather than the traditional paper or vellum medium, but the results are the same.

All geometric shapes can be formed by the constructive use of two simple objects – lines and circles. Did you know that, in a 2-dimensional CAD environment, a drafter spends only about 30% of his time placing lines and circles? He spends the vast majority of his time modifying what he has drawn and placing text on the drawing. Still, without a basic knowledge of placing lines and circles, the drafter will never leave the starting gate.

We'll begin this lesson learning to use the most basic geometric shape with which we can work – the line. Then we'll take a look at circles. Finally, we'll explore the coordinate system AutoCAD uses to identify locations within the drawing.

2.1	Lines, Rectangles, and Circles
2.1.1	Lines and Rectangles

AutoCAD has created commands that are really quite simple to remember. For example, to draw a line, you type *line* at the command prompt. To draw a circle, you type *circle*. To erase an object, what do you think you would type? Did you say *erase*? That's right!

Where to Find It:	
Command Line:	*Line*
Hotkey(s):	*l*
Ribbon (Tab/Panel):	Home – Draw – **Line**
Menu:	Draw – **Line**
Toolbar:	Draw – **Line**
Tool Palette:	Draw – **Line**

But wait; it gets simpler yet! Can't type? Most commands have simple abbreviations, or *hotkeys* (also called *aliases* or *shortcuts*). To draw a line, you can type *l*. To erase, type *e*. (We'll see more on the ***Erase*** command in Section 2.2, p.42.) I'll identify the hotkeys for each command as we progress through the book. Watch the WTFI tables!

Of course, if you have an aversion to the keyboard, there's almost always a ribbon, menu, or palette approach to each command. We'll look at those as well.

The ***Line*** command provides a nice uncluttered approach to seeing the command prompt in action. The sequence involved in drawing a two-point line is as follows:

Command: *line*

Specify first point: *1,1 [Enter the coordinate for a point or pick a point on the screen.]*

Specify next point or [Undo]: *2,2 [Again, enter the coordinate for a point or pick a point on the screen.]*

Specify next point or [Undo]: *[Hit* ENTER *to complete the command, or you can click the right mouse button ᗡ and select* **Enter** *from the cursor menu.]*

If you've toggled dynamic input on (the **DYN** toggle ⊥ on the status bar appears lit), you'll see a lot happening on the screen as you type. Don't let it worry you. You can toggle it off (by picking on its icon on the status bar) if you prefer. We'll look at dynamic input in more detail in a few minutes.

Not too difficult, is it? Try it. Start a new drawing using the *Sample Template 02* template in the Lesson02 folder. (Notice that I've provided a background grid to help guide you through the drawing.) Enter the preceding sequence. Remember to hit the ENTER key on the keyboard after each line of information. Otherwise, AutoCAD won't know you've finished your command. A final ENTER tells AutoCAD that you've finished the command. The numbers represent coordinates on the

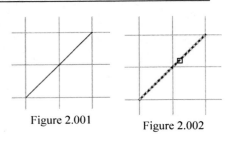

Figure 2.001   Figure 2.002

drawing. We'll discuss them in Section 2.4, p.47.

Does the lower left quadrant of your screen look like Figure 2.001, p.33? Excellent!

Okay, now erase the line. Use this sequence. (We'll spend more time with *Erase* in Section 2.2, p.42.)

> **Command:** *erase* (or *e*)
>
> **Select objects:** *[The crosshairs become a tiny box and the line highlights when you cross over it (Figure 2.002, p.33). Place the box over the line and pick once with the left mouse button 🖰. The line becomes dashed or highlighted.]*
>
> **Select objects:** *[ENTER or click the right mouse button – the line disappears.]*

Now try an exercise.

Try to draw the sample figure shown in the following exercise. Don't worry if you can't get the lines perfect just yet. This is your first time. Besides, we'll soon find much easier and more accurate ways of drawing.

> Your first Hail Mary option: You'll frequently find yourself entering a command too fast or out of order. Don't worry. If you make a mistake, use the ESCAPE key [Esc] in the upper left corner of most keyboards. You'll return to the command prompt where you can start over.

Do This: 2.1.1A	Drawing Lines

I. Start a new drawing using the *Sample Template 02* template file in the C:\Steps\Lesson02 folder.

II. We'll take this opportunity to see two different approaches to locate points in AutoCAD. Begin by turning the Polar, OTrack, and OSNAP tools off (icons for the **POLAR**, **OTRACK**, and **OSNAP** toggles on the status bar should be dark). Then be sure the **DYN** toggle's icon on the status bar is lit.

III. Follow these steps.

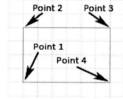

2.1.1A: DRAWING LINES

1. Enter the *Line* command.

    Command: *l*

2. Using the dynamic entry display next to the crosshairs, select a point around grid reference 2,1.

    Specify first point: *[Pick a point near 2,1.]*

3. Now move your cursor to a point around 2,3. (Use the coordinate display at the lower left corner of your screen to help locate the proper coordinate.)

    Ghost dimensions have appeared (right) indicating that you've moved your crosshairs 2" at a 90° angle. The **DYN** button on the status bar toggles this display on or off.

    Pick with the left mouse button 🖰 when you're display looks like ours.

    Specify next point or [Undo]: *[Pick a point near 2,3.]*

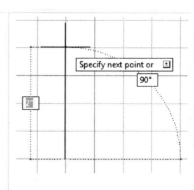

4. Pick the **DYN** toggle on the status bar to turn off the dynamic display.

2.1.1A: DRAWING LINES

5. Pick the **POLAR** toggle ⌕ on the status bar to turn on the polar tracking display.

6. Now move the crosshairs to a point around 5,3.
A tooltip has appeared that reads **Polar: 0'-3"<0°**. This *polar display* tells you that you've moved your cursor 3" over on the X-axis. We'll learn more about coordinate entry methods – including Polar – in Section 2.4, p.47. The **POLAR** toggle ⌕ on the status bar toggles this display on or off.

Additionally, a ghost line has appeared indicating that you're moving in a true 0° direction. (These ghost lines appear for N-S-E-W directions.) The **POLAR** toggle also controls this ghost line.
Pick this point.
   Specify next point or [Undo]: *[Pick a point near 5,3.]*

7. And then select a point around 5,1 (about 2" straight down – or at a 270° angle).
   Specify next point or [Close/Undo]: *[Pick a point near 5,1.]*

8. Finally, type *c* to close the line. Alternately, you can right click 🖱 and pick **Close** 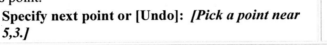 from the cursor menu.
   Specify next point or [Close/Undo]: *c*

Did you notice that last command? When used in response to a *Line* or *Polyline* command prompt, *c* will close the line (that is, it'll draw a line from the last point selected to the first point selected).

Using either the polar display or the dynamic display, you may find it difficult (although not impossible) to draw your lines without some accuracy. But I wouldn't use them at the same time; too much help can be a nuisance! Throughout this text, experiment with both. Then when you go to work, you can use the method you prefer.

> Some things to remember when drawing lines:
> - Hitting ENTER at the **Command** prompt will always repeat the last command regardless of what that command was. (You may also right click in the graphics area and pick the **Repeat** line at the top of the cursor menu.)
> - Hitting ENTER at the **Specify first point** prompt will cause the last point selected to be the first point of the line. In other words, if you leave the line prompt before you actually finish drawing the lines you want, you can hit ENTER to repeat the *Line* command, and then hit ENTER again to continue from where you stopped.
> - *C* will always draw a line from the last point selected to the first point selected during this command sequence ... *but only during this sequence*. In other words, if you return to the **Command** prompt, enter the *Line* command again, and then continue the same line, *C* will draw a line from the last point selected back to the first point of the current sequence. (When entered at the **Command** prompt, *C* will begin the *Circle* command.)
>
> All of this will become clearer as we go.
>
> Try some random lines and experiment with these last few statements. When you're comfortable with them, erase all the lines as explained previously. Then continue.

Let me show you an easier way to draw the same rectangle by using only two picks. We'll use the *Rectang* command this time.

35

The *Rectang* command sequence looks like this:

Command: *rectang*
Specify first corner point or [Chamfer/Elevation/Fillet/Thickness/Width]: *2,1 [Identify the first corner of the rectangle.]*
Specify other corner point or [Area/Dimensions/Rotation]: *5,3 [Identify the opposite corner of the rectangle.]*

⬜ Where to Find It:	
Command Line:	*Rectang*
Hotkey(s):	*rec*
Ribbon (Tab/Panel):	Home – Draw – **Rectangle**
Menu:	Draw – **Rectangle**
Toolbar:	Draw – **Rectangle**
Tool Palette:	Draw – **Rectangle**

Go ahead and try the preceding sequence. Does it look better than the one you drew using the *Line* command? Was it easier to draw?

Now erase it. Did you notice something different here? You only had to select one line and the entire rectangle highlighted. The rectangle is a *polyline*.

> A *polyline* is a multi-segmented line, or a line consisting of one or more than one line segment. We'll cover this more advanced line tool in Lesson 12, p.273.

Did you notice that AutoCAD gave you several options when you drew the rectangle? In the first prompt of the *Rectang* command, you may select to draw a **Chamfered** (mitered – Figure 2.003) or **Filleted** (round-cornered – Figure 2.004) rectangle. Or you can use the **Width** option for a rectangle with heavier lines (Figure 2.004). The other options, **Elevation** and **Thickness**, involve 3-dimensional space. We'll cover those in the 3D text. *The default choice will always precede the bracketed options.* Our response in the preceding sequence accepted the default choice. AutoCAD read *2,1* as the **first corner point** of the rectangle.

The second prompt gave you **Area**, **Dimensions**, and **Rotation** options. After selecting the first corner of the rectangle, you can use one of these prompt to tell AutoCAD how large a rectangle to draw and how to orient it.

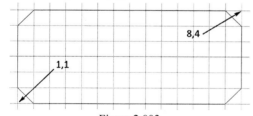

Figure 2.003

Figure 2.004

Let's try some rectangles using some of these options.

Do This: 2.1.1B	**Drawing Rectangles**

I. Start a new drawing using the *Sample Template 02* template file in the C:\Steps\Lesson02 folder.

II. Turn dynamic input off.

III. Follow these steps to draw a simple rectangle.

2.1.1B: DRAWING RECTANGLES

1. Enter the *Rectang* command.
   Command: *rec*

2.1.1B: DRAWING RECTANGLES

2. Type in the lower left coordinate as indicated. (We'll learn more about coordinate entry in Section 2.4, p.47.)

   Specify first corner point or [Chamfer/Elevation/Fillet/Thickness/Width]: *1,1*

3. Type in the upper right coordinate as indicated. Your rectangle will look like the figure that follows.

   Specify other corner point or [Area/Dimensions/Rotation]: *8,4*

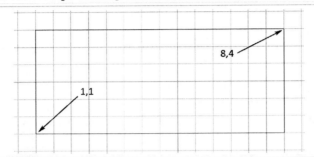

IV. Now draw a ½" chamfered rectangle as follows. (This time, we'll use the **Dimensions** option.)

4. Repeat the **Rectang** command.

   Command: *[ENTER]*

5. You can type *c* for the **Chamfer** option, but this might be a good time to see how useful a cursor menu can be. Right click in the graphics area, and select **Chamfer** from the cursor menu.

   Specify first corner point or [Chamfer/Elevation/Fillet/Thickness/Width]: *c*

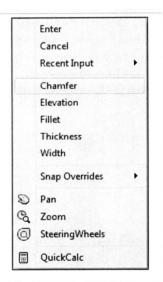

6. Tell AutoCAD what the chamfer distances should be (that is, how far back on each line of the corner to begin the angle).

   Specify first chamfer distance for rectangles <0'-0">: *1/2*

7. AutoCAD automatically sets the second chamfer distance the same distance as the first. You can accept by hitting ENTER, or give a different number. Let's accept.

   Specify second chamfer distance for rectangles <0'-0 1/2">: *[ENTER]*

8. Now tell AutoCAD where to draw. Type in the lower left coordinates.

   Specify first corner point or [Chamfer/Elevation/Fillet/Thickness/Width]: *1,5*

9. When prompted for the other corner, tell AutoCAD you want to use the **Dimensions** option either by typing *d* at the prompt or by selecting **Dimensions** from the right click cursor menu.

   Specify other corner point or [Area/Dimensions/Rotation]: *d*

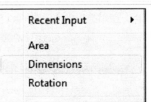

37

2.1.1B: DRAWING RECTANGLES

10. AutoCAD asks you to identify the **length** and **width** of the rectangle. Enter the numbers indicated.

   Specify length for rectangles <0'-10">: 7
   Specify width for rectangles <0'-10">: 3

11. When AutoCAD prompts you again for the other corner, move your cursor around and see how it orients the rectangle according to the cursor's position in the drawing. Pick a point 🖱 above and to the right of the first corner point. AutoCAD creates the rectangle shown in the following figure (without the dimensions).

   Specify other corner point or [Area/Dimensions/Rotation]:

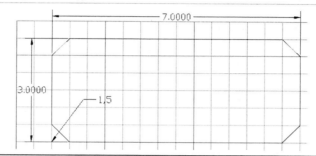

IV. Now draw a ¾" filleted rectangle with 1/16" wide lines and an area of 22 sq. in.

12. Repeat the *Rectang* command ▱.

   Command: *[ENTER]*

13. AutoCAD tells you that you're currently set up to draw a chamfered rectangle. It also tells you the chamfer distances. We'll change that now.

   Type *w* for the **Width** option (or select **Width** from the cursor menu).

   Current rectangle modes: Chamfer= 0'-0 1/2" x 0'-0 1/2"
   Specify first corner point or [Chamfer/Elevation/ Fillet/Thickness/Width]: *W*

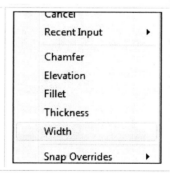

14. Tell AutoCAD how wide to make the lines.

   Specify line width for rectangles <0'-0">: *1/16*

15. Type *f* for the **Fillet** option (or select **Fillet** from the cursor menu).

   Specify first corner point or [Chamfer/Elevation/ Fillet/Thickness/Width]: *F*

16. Tell AutoCAD the radius you want for your fillets.

   Specify fillet radius for rectangles <0'-0 1/2">: *3/4*

17. Tell AutoCAD where to draw. Type in the lower left coordinates.

   Specify first corner point or [Chamfer/Elevation/Fillet/Thickness/Width]: *4,2.5*

2.1.1B: DRAWING RECTANGLES

18. Now tell AutoCAD to base the rectangle on an area that you'll give it.
    Specify other corner point or [Area/Dimensions/Rotation]: *A*

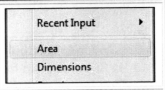

19. And tell it to make the area 22 sq. in.
    Enter area of rectangle in square inches <100.0000>: *22*

20. Use the **Length** option …
    Calculate rectangle dimensions based on [Length/Width] <Length>: *[ENTER]*

21. … and make the length *6"*. Your rectangle will look like the following figure.
    Enter rectangle length <0'-0">: *6*

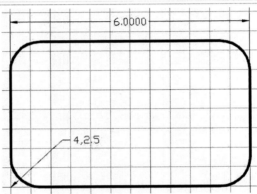

22. Leave the drawing without saving the changes.
    Command: *quit*

The beginning of the sequence for drawing a rectangle by rotation is identical to that for drawing a rectangle by area or dimensions. The rest of the sequence looks like this:

> **Specify other corner point or [Area/Dimensions/Rotation]:** *r [Tell AutoCAD to use the Rotation option.]*
> **Specify rotation angle or [Pick points] <0>:** *[Tell AutoCAD what rotation angle to use.]*
> **Specify other corner point or [Area/Dimensions/Rotation]:** *[Now you can continue by picking points or using the* **Area** *or* **Dimensions** *options; AutoCAD will draw the rectangle at the angle you specified.]*

Have you noticed that you don't need to put inch marks on your numbers? We've set up our drawing using architectural units. The inch is the basic architectural unit. AutoCAD knows this, so we don't have to indicate it. We would have to indicate feet, however, with a prime (').

AutoCAD uses a rather different approach to entering feet and inches in response to a prompt. To enter one-foot-seven-and-one-half-inches, for example, type *1'7-1/2*. Notice the lack of a dash between the feet and inch numbers. Separate these only with the foot mark (the prime, '). Use a dash to separate inches from fractions.

The reason for this is simple: AutoCAD reads a space the same way it reads ENTER. So you can't use a space as a separator. (But you can use the spacebar instead of the ENTER key!)

2.1.2 Getting Around to Circles

Next to lines, circles (and parts of circles) are the most frequent factor in geometric drawings. You might think creating such important objects should be complicated, but AutoCAD has made *Circle* one of the easiest of its commands. We draw lines, as you know, using the *Line* command. In keeping with the simple approach, draw circles by using the *Circle* command.

Here's the command sequence.

Where to Find It:	
Command Line:	*Circle*
Hotkey(s):	*c*
Ribbon (Tab/Panel):	Home – Draw – Circle flyout – **[option]**
Menu:	Draw – Circle – **[option]**
Toolbar:	Draw – **Circle**
Tool Palette:	Draw – **Circle**

    Command: *circle*
    Specify center point for circle or [3P/2P/Ttr (tan tan radius)]: *[Pick or identify a point on the screen.]*
    Specify radius of circle or [Diameter]: *[Drag or type the radius.]*

That seems fairly easy. Open the *circles* file in the C:\Steps\Lesson02 folder and give it a try. Draw a circle in one of the open areas of the screen.

What do you think?

Okay. Let's look at some of the *Circle* command's options. You'll find easy access to these in the menu's and ribbon's **Circle** flyouts (Figure 2.005).

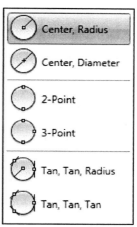

- The default is the **Specify center point** option (**Center, Radius** on the menu). Any point selected on the screen or identified by coordinates will be the center point of the circle.
- **Center, Diameter** (on the menu) works just like the default (**Center, Radius**) but allows you to specify a diameter rather than a radius.
- The **3P** option (**3 Points**) allows you to draw a circle by selecting three points on the circle.
- The **2P** option (**2 Points**) allows you to draw a circle by selecting both ends of an imaginary diameter line.
- **Ttr** (or **Tan, Tan, Radius**) stands for **Tangent-Tangent-Radius** and allows you to draw the circle by selecting two objects to which the circle will be tangent and then entering the required radius.
- Use **Tan, Tan, Tan** (for Tangent, Tangent, Tangent) to create a circle tangent to three lines.

Figure 2.005

Let's try some of these options.

> All of the options are also available via cursor or dynamic input menu once the *Circle* command has begun.

Do This: 2.1.2A	**Circle Practice**

    I. If you haven't yet opened *circles*, open it now. It's in the C:\Steps\Lesson02 folder and looks like the figure at right.

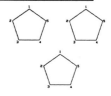

    II. Be sure the **SNAP**, **OSNAP**, and **POLAR** toggles on the status bar are in their off position. (We'll see how these work in Lesson 3, p.55, but turn them off for now so they don't interfere with this exercise.)

    III. Follow these steps.

2.1.2A: CIRCLE PRACTICE

1. Enter the *Circle* command 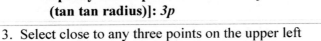.
    Command: *c*

2. Type *3P* (or select it on the cursor menu) for the three-point option. (For clarity, we'll show command line entry throughout this book; but I'd recommend using dynamic entry or cursor menu when possible.)
    Specify center point for circle or [3P/2P/Ttr (tan tan radius)]: *3p*

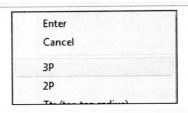

3. Select close to any three points on the upper left polygon. The results should look like the figure shown.
    Specify first point on circle:
    Specify second point on circle:
    Specify third point on circle:

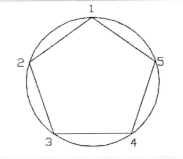

4. Repeat the *Circle* command 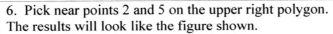.
    Command: *[ENTER]*

5. Use the two-point option.
    Specify center point for circle or [3P/2P/Ttr (tan tan radius)]: *2p*

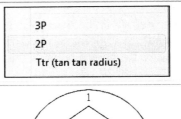

6. Pick near points 2 and 5 on the upper right polygon. The results will look like the figure shown.
    Specify first end point of circle's diameter: *[Select point 2.]*
    Specify second end point of circle's diameter: *[Select point 5.]*

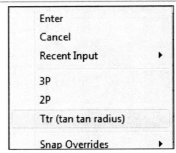

7. Repeat the *Circle* command.
    Command: *[ENTER]*

8. Select the **Ttr** option for tangent-tangent-radius.
    Specify center point for circle or [3P/2P/Ttr (tan tan radius)]: *t*
Notice that AutoCAD has automatically activated the OSNAP toggle. You're going to get a brief peek at one of AutoCAD's cool precision tools; you're going to use a **tangent** OSNAP! You'll see an odd circle symbol where your cursor crosses the lines to help as a guide. (You'll learn more about OSNAPs in Lesson 3, p.61.)

9. Select any point on line **1-2** of the remaining polygon. Notice how the OSNAP symbol for **tangent** appears when you cross the line.
    Specify point on object for first tangent of circle:

41

2.1.2A: CIRCLE PRACTICE

10. Now select any point on line **2-3**.
    Specify point on object for second tangent of circle:

11. Enter a radius of *1*. Your drawing will look like the figure shown. Notice that AutoCAD draws a 1"R circle *tangent* to both the selected lines.
    Specify radius of circle <1.6180>: *1*

12. Now try the **Tan, Tan, Tan** option. Begin the *Circle* command by picking the down arrow next to the **Circle** button on the ribbon ⊘ ▾. (Alternately, you can pick **Circle** on the Draw pull down menu.)

13. Now pick the **Tan, Tan, Tan** option on the flyout menu.

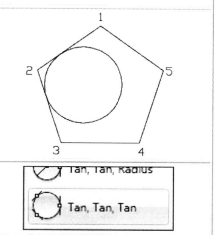

14. Pick line 4-5 on the upper left circle. (Notice the tangent symbol again.)
    Specify center point for circle or [3P/2P/Ttr (tan tan radius)]: _3p Specify first point on circle: _tan to

15. Pick line 2-3 on the upper right circle.
    Specify second point on circle: _tan to

16. Finally, pick line 1-2 on the bottom circle.
    Specify third point on circle: _tan to
    Your circle looks like this – tangent to all three lines!

17. Save 💾 your drawing.
    Command: *qsave*

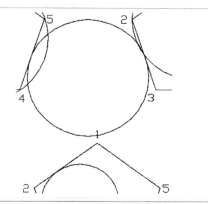

I know. Had you known it was this easy, you'd have learned AutoCAD long ago!

2.2 Fixing the Uh-Ohs: *Erase, Undo*, and *Redo/MRedo*

Where to Find It:	
Command Line:	*Erase*
Hotkey(s):	e
Ribbon (Tab/Panel):	Home – Modify – **Erase**
Menu:	Modify – **Erase**
Toolbar:	Modify – **Erase**
Tool Palette:	Modify – **Erase**

When I first studied drafting – using charcoal from the fire and drawing on the cave walls – I thought the electric eraser was the height of laziness. I mean, how spoiled could a professional be?

Then I got my first drafting job. I was assigned to removing revision clouds from old cloth drawings. I had an electric eraser (and a sore hand) by the second day.

Today, the CAD system makes erasure even easier than that old Bruning did.

You're already familiar with the command sequence for erasing a single object, or a group of objects one at a time:

> **Command:** *erase*
> **Select objects:** *[Select an object.]*
> **Select objects** *[AutoCAD allows you to select more objects if you'd like; otherwise,* **hit ENTER** *to complete the command.]*

We'll take a look at some ways to speed this up a bit in a few minutes. But first, we should look at how we fix mistakes. If, for example, your erasure was a mistake, you can use the *Undo* command to return your drawing to a point before the mistake occurred.

Although the *Undo* command hasn't changed much in the last few releases of AutoCAD, it has become significantly easier to access. Further, its counterpart – the *Redo* command – has been usurped by the newer and more powerful *MRedo* command.

We'll look at *Undo* first. I'll explain the command sequence; but I'll warn you up front that you'll probably never use it. Here's the sequence:

> **Command:** *undo*
> **Current settings: Auto = On, Control = All, Combine = Yes, Layer = Yes**
> **Enter the number of operations to undo or [Auto/Control/BEgin/End/Mark/Back] <1>:** *[ENTER]*

Let's look at the options.

- **Enter the number of operations to undo** (the default) – undoes the specified number of preceding commands. The default number is one, so an ENTER will undo a single command.
- **Auto** – undoes a menu selection as a single command. Remember that menu and toolbar commands cause a macro to run. If **Auto** is set to *Off*, undo will only undo one step of the macro at a time. If it's *On* (the default setting), *Undo* will undo the entire macro at once. (The **Current settings** list before the command line prompt will tell you if **Auto** is on or off.
- **Control** – controls how **Undo** performs. It has five options: **All/None/One/Combine/Layer**.
  - **All** – allows *Undo* to function fully, providing virtually unlimited undos.
  - **None** – turns *Undo* Off.
  - **One** – limits *Undo* to the last command only.
  - **Combine** – controls whether consecutive *Pan* and *Zoom* commands are treated as a single undo or redo operation. (We'll see more on *Pan* and *Zoom* in Lesson 4, p.79.)
  - **Layer** – controls whether AutoCAD will undo all the previous layer changes at once or one at a time. We'll see more on layers in Lesson 7, p.148.
- **Begin** and **End** – **Begin** begins a group. All commands entered after the **Begin** option will be treated as a single command and are undone by a single *Undo*. **End** ends the **Begin** option.
- **Mark** and **Back** – **Mark** places a mark in the command sequence. **Back** undoes back to the mark.

Now, about the *Undo* command ... Most of these options sound good; but as I warned you, you'll never use them. I've rarely, in the years I've worked with CAD, known in advance that I would be undoing a command. How would I know to **Mark** or **Begin** a sequence for later undoing?

I advise people to learn the *U* command discussed next. It'll cover your needs and not require you to memorize the *Undo* options.

The *U* command acts as a macro for the default *Undo* option, and assumes a number of one. If you need to use the keyboard to undo something, you'd be much better off using *U* rather than its more confusing cousin – *Undo*.

Picking the **Undo** arrow on one of the toolbars will

⟵	Where to Find It:
Command Line:	U
Hotkey(s):	CTRL+*u*
Cursor Menu:	Undo
Toolbar:	Quick Access – **Undo**
	Standard – **Undo**

undo the last command – just as it will in most Windows-based software. In this regard, it simply runs the *U* command one time.

Figure 2.006

But if you look to the right of the **Undo** button, you'll find a down arrow that'll list the last several commands you've run (Figure 2.006). You can select to undo from one to all of these in a single stroke!

Perhaps the most important thing to remember about the *Undo* command family is the one thing that you can't undo; you can't undo the undo that you just did! You can, however, *redo* what you undid. (Don't worry ... we'll have an exercise in a minute that'll make this a bit clearer.)

AutoCAD provides two 'redo' tools – ***Redo*** and ***MRedo***. The first – ***Redo*** – simply redoes whatever the last *Undo* undid. *Redo*, however, is only good once, and only after one of the *Undo* commands. *MRedo* is a bit more flexible and allows you to redo several undos at once. Here's the sequence:

Where to Find It:	
Command Line:	*MRedo*
Hotkey(s):	CTRL+y
Cursor Menu:	Redo
Toolbar:	Quick Access – **Redo**
	Standard – **Redo**

Command: *mredo*

Enter number of actions or [All/Last]:

[Tell AutoCAD if you wish to redo only the last undo, or all the previous undos.]

Figure 2.007

The **Redo** button to the right of the **Undo** button on the toolbars (Figure 2.007) actually calls the ***MRedo*** command and sports a down arrow next to it that functions much like the down arrow next to the **Undo** button.

Now you have five of AutoCAD's "Hail Mary" procedures – Escape, ***Erase***, ***Undo***, ***Redo***, and ***MRedo***. You're probably nicely confused. Let's see if an exercise can help clear things up for you.

Do This: 2.2A	Erase, Undo, and Redo

I. Open drawing *erase-samp*, in the C:\Steps\Lesson02 folder.
II. Follow these steps.

2.2A: ERASE, UNDO, AND REDO
1. Enter the ***Erase*** command.   **Command:** *e*
2. Pick one of the lines by placing the selection box on it and clicking the left mouse button as indicated.
3. Confirm that you've finished selecting (hit ENTER or use the right mouse button).   **Select objects:** *[ENTER]*
4. Undo the erasure.   **Command:** *u*   Notice that the erased line returns.
5. Redo the erasure.   **Command:** *redo*

2.2A: ERASE, UNDO, AND REDO

6. Repeat Steps 1, 2, and 3 until you have erased another four lines. (Be sure to repeat all three steps four times.)
   Command: *e*

7. Now pick the down arrow next to the **Undo** button (on the toolbar) and select the last three *Erase* commands.
Notice that the last three lines you erased return to the screen.

8. Pick the down arrow next to the **Redo** button and select two *Erase* commands.
Notice that two of the lines you erased have once again been removed from the drawing. Notice also that they were erased in the order in which they were originally selected.

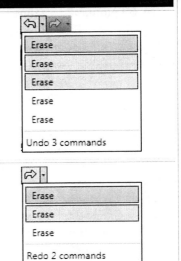

You can also take advantage of the DELETE key in much the same way you can in other Windows programs. Simply select the object to be erased (without entering a command) and then hit the DELETE key. (This works best when you toggle **Quick Properties** off on the status bar. We'll see more on Quick Properties in Lesson 6, p.142.)

2.3 Multiple Object Selection Made Easy

We've seen how easily we can erase a single object. But what if we want to erase 20 or 200 objects? Must we select 20 or 200 times?

The answer, of course, is no! Let's look at a couple of options that'll make multiple object selection easier.

The first option places a window around the objects to be selected (refer to Figure 2.008). What do you suppose it's called? What do you think the shortcut will be? If you said, "window" and *w*, you were quite right. A window selection includes all the objects that are completely within a window.

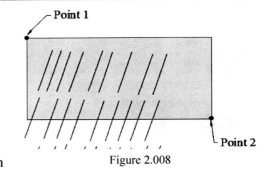

Figure 2.008

The second option places a window around *and across* the objects to be selected. This one is called *crossing* and uses *c* as a shortcut. The difference between this and a standard window is simple. A crossing window will select everything within or *touched by* the window.

Let's give these a try.

Do This: 2.3A	Using Windows to Select Multiple Objects

I. Close and reopen the *erase-samp* drawing you used in the last exercise. Don't save the changes. (It's in the C:\Steps\Lesson02 folder.)
II. Follow these steps.

2.3A: WINDOW SELECTIONS

1. Enter the *Erase* command.
   Command: *e*

2. Tell AutoCAD you want to use a window to make your selections by entering a *w* at the **Select objects** prompt.
   Select objects: *w*

3. Place your first corner near **Point 1** in the drawing (refer to Figure 2.008, p.45).
   Specify first corner:

4. Place the opposite corner near **Point 2**. (Notice that the window has a blue shade to it.)
   Specify opposite corner:

5. Complete the command by hitting ENTER at the **Select objects** prompt.
   Select objects: *[ENTER]*

Notice that only the lines that were completely encircled by the window were erased.

6. Undo the erasure.
   Command: *u*

7. Now let's try a crossing window. Repeat the *Erase* command.
   Command: *e*

8. Tell AutoCAD you want to use a crossing window to make your selections by entering a *c* at the **Select objects** prompt.
   Select objects: *c*

9. Place your first corner near **Point 1** in the drawing.
   Specify first corner:

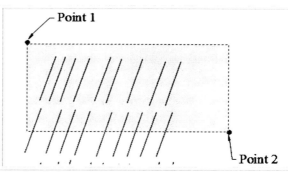

10. Place the opposite corner near **Point 2**.
    Specify opposite corner:

Notice the difference in the way AutoCAD shows the window (this one uses a dashed line and is shaded green).

11. Complete the command by hitting ENTER at the **Select objects** prompt.
    Select objects: *[ENTER]*

Notice that the lines that were completely encircled and the lines that were touched by the crossing window were erased.

12. Undo the erasure.
    Command: *u*

To make windowing even easier, AutoCAD includes *Implied Windowing*. This means that you don't actually have to type *w* or *c* to create a window or a crossing.

When you pick an empty place on your drawing at the **Select objects** prompt, AutoCAD assumes that you want to use Implied Windowing.

To use a window at any **Select objects** prompt, simply pick an empty point to the left of what you want to select, then pick a second point to the right. You'll get a window.

> If you pick the first point to the right and the second point to the left, you'll get a crossing window. Try this. Repeat the last exercise but don't enter **W** or **C**. Simply pick a place near **Point 1** then near **Point 2** for a window. Then try it by selecting a place near **Point 2** and then near **Point 1** for a crossing window.

2.4 The Cartesian Coordinate System

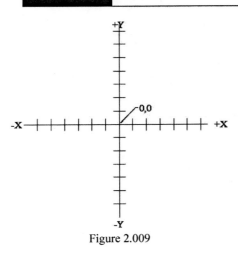

Figure 2.009

Remember suffering through plane geometry back in high school math class? That was the one with the crossing number lines, four quadrants, and probably the one that you insisted to you parents, teachers, and any else who would listen that you would never use. Look at Figure 2.009 – it's your math teacher's revenge!

On the bright side, it's not nearly as difficult as a math teacher might wish. There are three axes – X, Y, and Z. But we're only concerned with two of them in beginning AutoCAD; so the difficulty is already cut by 1/3 (1/2?)!

The *X-axis* runs horizontally (left to right). Everything to the right of a point we call 0 is positive (plus). Everything to the left of the 0 is negative (minus). Thus, we now have +X and –X directions identified. Each number of the X-axis (the axis is the line that runs through the 0) represents an X-*plane*. An X-plane runs infinitely up and down crossing the X-axis.

The *Y-axis* runs vertically (up and down). Everything above the point we call 0 is positive. Everything below 0 is negative. Each number on the Y-axis represents a Y-*plane*. A Y-plane runs infinitely right and left, crossing the Y-axis.

> The X- and Y-planes also run perpendicular to the 3-dimensional Z-axis and might be called the XZ- and YZ-planes. But two dimensions will confuse well enough for now! (What'll we do when Stephen Hawking finds *another* dozen dimensions?!)

Are you still with me?

Okay. Where an X-plane meets a Y-plane in two-dimensional space, we have a *point*. We identify the point by its *coordinate* – or by the number of the X-plane, followed by a comma, and then the number of the Y-plane. Remember this as *X,Y* (no spaces). For example, point 4,3 is found 4 spaces to the right of zero and 3 spaces above, as shown in Figure 2.010.

AutoCAD provides three methods for using these coordinates: *absolute*, *relative*, and *polar*. The use of coordinates is your first step toward a drawing precision that was never possible using conventional drafting tools!

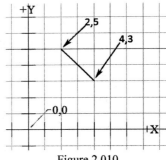

Figure 2.010

2.4.1 Absolute Coordinates

The Absolute System is easiest. You simply enter absolute coordinates as *X,Y* (remember, no spaces) whenever AutoCAD asks for a point.

Give it a try.

| Do This: 2.4.1A | **Using Absolute Coordinates** |

I. Open *ccs* in the C:\Steps\Lesson02 folder. Refer to Figure 2.010.

II. Be sure the dynamic input is off (the status bar icon is dark). By default, you cannot use absolute coordinates with dynamic entry.
III. Turn off polar tracking as well.
IV. Follow these steps.

2.4.1A: USING ABSOLUTE COORDINATES

1. Enter the *Line* command.
    Command: *l*

2. Enter the first absolute coordinate.
    Specify first point: *4,3*

3. Enter the next absolute coordinate.
    Specify next point or [Undo]: *2,5*

4. Hit ENTER to complete the command.
    Specify next point or [Undo]: *[ENTER]*
Your drawing looks like Figure 2.010, p.47.

The Absolute Coordinate System is quite easy to use, but using it means that you must know the exact X and Y value of each point you wish to use. This isn't always possible. (That's why most people don't use it). Let's look, then, at the *Relative System*.

2.4.2 Relative Coordinates

Use a relative coordinate, anytime after identifying the first point, to identify a point *relative to* the last point you identified. (Either select the first point with the mouse, or use an absolute coordinate.) The syntax for relative coordinates is *@X,Y* (read, "at X comma Y").

Let's give the Relative System a try.

Do This: 2.4.2A Using Relative Coordinates

I. Be sure you're still in *ccs* in the C:\Steps\Lesson02 folder. If not, open it now.
II. Erase the line you drew in the last exercise.
III. Follow these steps. (Refer to Figure 2.011.)

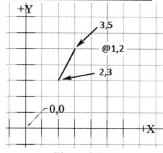

Figure 2.011

2.4.2A: USING RELATIVE COORDINATES

1. Enter the *Line* command.
    Command: *l*

2. Use absolute coordinates to enter the first point.
    Specify first point: *2,3*

3. Now use relative coordinates to draw a line one space in the +X direction and two spaces in the +Y direction.
    Specify next point or [Undo]: *@1,2*

2.4.2A: USING RELATIVE COORDINATES

4. Hit ENTER to complete the command.
   Specify next point or [Undo]: *[ENTER]*

AutoCAD places the second point one unit over on the +X-axis and two units up on the +Y-axis.

5. Undo ⤺ the line.
   Command: *u*

6. Let's look at this procedure using dynamic entry. Turn dynamic entry back on (pick 🖱 dynamic input ⊕ on the status bar).

7. Repeat Steps 1 and 2.
   Command: *l*

8. Rather than using relative coordinates, enter the X and Y coordinate on the keyboard (as X,Y). Notice that the numbers appear in the dynamic display (right).
   Specify next point or [Undo]: *@1,2*
   Notice that when you hit ENTER, AutoCAD places the @ symbol before the numbers on the command line for you. *Dynamic Input uses relative coordinates by default!*

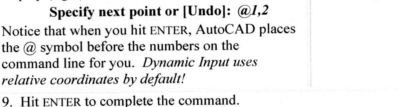

9. Hit ENTER to complete the command.
   Specify next point or [Undo]: *[ENTER]*

Notice that the second point is located 1 unit to the right (a positive 1 on the X-axis) and 2 units up (a positive 2 on the Y-axis), or *at 1X and 2Y* from the last point identified. The point is *relative to* the last identified point.

Using the Relative System means you must know how far (in plus/minus terms) along each axis you want to go from where you are. You'll find this much easier than having to locate each point in absolute terms, especially on larger drawings. But wait – there's an easier way still!

2.4.3 Polar Coordinates

I find polar coordinates the most useful. But I must admit that I'll use the Relative System when needed.

Like relative coordinates, you'll use polar coordinate entry after identifying the first point required. The syntax for polar coordinates is *@dist<angle* (at – *distance* – at an angle of – *angle*). You must know the distance and direction you wish to go from the last selected point. Remember, measure angles counterclockwise beginning at 0°East.

Let's take a look.

Do This: 2.4.3A	Using Polar Coordinates

I. Be sure you're still in *ccs* in the C:\Steps\Lesson02 folder. If not, open it now.
II. Erase the line you drew in the last exercise.
III. Turn off the dynamic input display.
IV. Follow these steps. (Refer to Figure 2.012.)

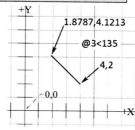

Figure 2.012

2.4.3A: USING POLAR COORDINATES

1. Enter the *Line* command.
   Command: *l*

2. Use absolute coordinates to enter the first point.
   Specify first point: *4,2*

3. Now use polar coordinates to draw a 3" line at an angle of 135°.
   Specify next point or [Undo]: *@3<135*

4. Hit ENTER to complete the command.
   Specify next point or [Undo]: *[ENTER]*

Nothing to it, right? Let's try it with dynamic input.

5. Erase the line you just drew.
   Command: *e*

6. Toggle dynamic input on.

7. Repeat Steps 1 and 2.

8. Enter the distance. Notice that AutoCAD places the distance between ghost dimension lines and shows the current angle in another display box.

9. Hit the TAB key on your keyboard to work in the other display box.

10. Enter the angle for your line (135°). Hit ENTER when you've finished.

11. Complete the command.
    Specify next point or [Undo]: *[ENTER]*

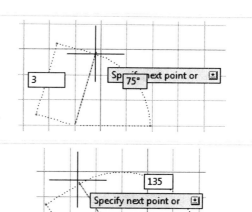

Notice that the coordinate of the second point isn't an integer. This often happens with polar entries. Remember that you're drawing the *hypotenuse* of an angle. This is why I mix my use of polar coordinates with the use of relative coordinates. Sometimes I may know how long a line I need (polar coordinates); other times I know the X and Y distances (relative coordinates).

We've seen several approaches to drawing a simple line with a great deal of precision. It's time now to practice.

2.4.4	Practicing with Cartesian Coordinates
Do This: 2.4.4A	Practice

I. Begin a new drawing using *Sample Template 03* found in the C:\Steps\Lesson02 folder.
II. Toggle the dynamic input display off.
III. Using the chart, draw Figure 2.013. (Note: Grid marks are 1 unit apart.) Draw the figure first using absolute coordinates. Erase it. Draw it using relative coordinates. Erase it. Draw it again using polar coordinates. Erase it.
IV. Now toggle the dynamic input display back on.
V. Draw Figure 2.013 using relative coordinates. Erase it. Draw it again using polar coordinates.

Point	Absolute (X,Y)	Relative (@X,Y)	Polar (@dist<angle)
1	2,15	2,15	2,15
2	5,15	@3,0	@3<0
3	5,9	@0,-6	@6<270
4	6,9	@1,0	@1<0
5	8,11	@2,2	@2.8284<45
6	10,11	@2,0	@2<0
7	12,9	@2,-2	@2.8284<315
8	13,9	@1,0	@1<0
9	13,15	@0,6	@6<90
10	16,15	@3,0	@3<0
11	16,4	@0,-11	@11<270
12	13,4	@-3,0	@3<180
13	10,7	@-3,3	@4.2426<135
14	8,7	@-2,0	@2<180
15	5,4	@-3,-3	@4.2426<225
16	2,4	@-3,0	@3<180
Back to 1	C	C	C

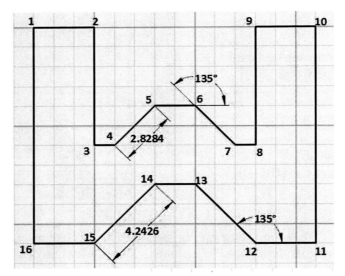

Figure 2.013

| 2.5 | **Extra Steps** |

One of the more useful drawing tools works similarly to the polar coordinate system, but requires less input. This is called the *Direct Distance Option*.

Simply put, this is how it works. At the **Specify next point or [Undo]** prompt, you enter a distance at the keyboard, move the crosshairs in the direction you want the line to go, and then hit ENTER (or pick the right mouse button 🖱 then select **Enter**). Let me demonstrate.

| Do This: 2.5A | **Direct Distance Entry** |

   I. Start a new drawing from scratch. Turn off dynamic input and toggle Ortho on.
   II. Begin to draw a line from any point on the screen.
   III. At the **Specify next point or [Undo]:** prompt, type *3*. Don't hit ENTER yet.
   IV. Move your crosshairs to the right.
   V. Hit ENTER. Notice that AutoCAD has drawn a line 3 units in the direction you moved the crosshairs!

Try drawing the object in Figure 2.013, p.51, using direct distance entry whenever possible. Was it faster?

The marvels never cease!

| 2.6 | **What Have We Learned?** |

Items covered in this lesson include:

- *Object selection*
- *Cartesian Coordinate System*
- *Commands*
    - *Line*
    - *Rectangle*
    - *Circle*
- *Dynamic input display*
- *Direct distance entry*
    - *Erase*
    - *Undo / U*
    - *Redo / MRedo*

We began this lesson by describing drafting as "the placing of geometric shapes on paper to represent existing or proposed structures." We've seen how to use the "backbones" of nearly all geometric structures – lines and circles. We've seen how to draw with precision and accuracy using the "backbone" of AutoCAD precision – the Cartesian Coordinate System.

But if the Cartesian Coordinate System was solely responsible for precision in AutoCAD, the application might never have survived through so many releases. Let's face it, at its best, it can be time-consuming and a bit cumbersome. In Lesson 3, we'll look at some faster ways to draw that, although not replacements for Cartesian coordinates, accent them nicely. Then, in Lesson 5, we'll look at arcs and a few other tools to complete the basic structures of drafting.

| 2.7 | **Exercises** |

1. (Download a blank, printable table from: *http://www.uneedcad.com/Files/table.pdf*.) Using what you've learned, fill in the blanks with the command entries needed to draw the structure in the figure at right. Use each of the Cartesian coordinate methods – absolute, relative, and polar. Then draw the object using each of the methods and the dynamic methods we discussed. (Grid marks are ½" apart.) Save the drawing as *MyM* in the C:\Steps\Lesson02 folder.

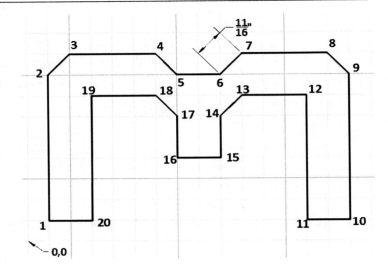

| 2.8 | **For Web-Based Review Questions and Additional Exercises, visit: http://foragerpub.com/AcadFiles/2011/2011.htm** |

Lesson 3

Following this lesson, you will:

- ✓ Know how to use the Ortho, Grid, Snap, and Tracking tools to your advantage
- ✓ Know how to use OSNAPs
- ✓ Know how to use the display controls: **Redraw** and **Regen**
- ✓ Be familiar with the different toggles – keyboard, function keys, and status bar – available to help you
- ✓ Know how to set up an Isometric Drawing

Drawing Aides

Lesson 2 showed us how to draw with accuracy. But without some additional help, we'll need to know exact coordinates or exact distances every time we draw a line or a circle. What if we want to go to a specific point – say, the end of an existing line? Must we break out the calculator every time?

No!

AutoCAD provides many tools and drawing aids that help you minimize the need for calculation. Lesson 2 taught you to draw with accuracy. This lesson will show you how to draw with speed without sacrificing accuracy.

3.1 The Simple Stuff: Ortho, Grid, Polar Tracking, and Snap

Grid, Snap, Polar Tracking, and *Ortho* would not be of much use if we had to stop a drawing or editing command every time we needed to turn one **On** or **Off**. Mercifully, AutoCAD gives us toggles that will work even while we're in the sequence of another command. In other words, we can toggle the *Grid, Snap, Polar Tracking,* or *Ortho* **On** or **Off** while drawing a line (or doing some other command).

Figure 3.001

| INFER | SNAP | GRID | ORTHO | POLAR | OSNAP | 3DOSNAP | OTRACK | DUCS | DYN | LWT | TPY | QP | SC |

Figure 3.002

We'll start this lesson with a quick overview of some of the tools found on the left end of the status bar. By default, you'll find a section of picture icons (Figure 3.001) next to the coordinate display. You can use these or you can replace the picture icons with text icons (Figure 3.002) if you wish. (You'll see how to switch them in a moment.) The picture icons take up less room, but you'll probably find it easier to use the text icons if you're not familiar with the pictures. But whichever display method you choose, when a toggle is "on" or active, it appears to light up; when it's "off" or inactive, it appears gray. (All the icons in Figures 3.001 and 3.002 are inactive.)

Follow the exercise below to change from one to the other.

| Do This: 3.1A | Changing the Status Bar Icons |

I. Begin AutoCAD using the default template.
II. Follow these steps.

3.1A: CHANGING THE STATUS BAR ICONS

1. Right click on any of the icons shown in Figure 3.001 (or 3.002). AutoCAD presents a menu similar to the one shown here. (Different icons produce different menus, but all should have the options shown here. We'll look at the menus in more detail as we look at each tool.)

   ✓ Enabled
   ✓ Use Icons
   Settings...
   Display ▶

2. Pick **Use icons** to add or remove the check.

That's all there is to it!

3.1.1 Ortho

The first drawing aid with which you'll want to familiarize yourself is **Orthomode** (aka. **Ortho**). You may recognize the word as an aberration of the word *Orthographic*. We learned about orthographic projections in the first days of drafting class. These present two-dimensional views necessary to describe an item from all sides. Orthographic views include the following (Figure 3.003): front, back, right side, left side, top, and bottom. The views are placed above or below, left or right of a base view (usually the front) depending on which view it is.

Where to Find It:	
Command Line:	*Ortho*
Hotkey(s):	F8
Status Bar:	⌐ or ORTHO
Temp. Override:	SHIFT

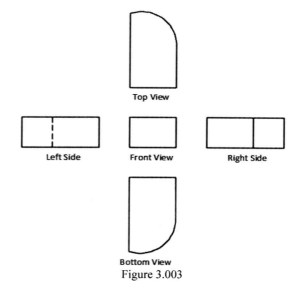

Figure 3.003

Orthomode restricts drawing or editing movement in a drawing to the left/right or up/down directions. When **ORTHO** is toggled on, your lines will be drawn along the X- or Y-plane. (When drawing in the isometric mode, the X- or Y-planes will be located at 30° and 150°.) This will become clearer during our exercises.

The command sequence for **Orthomode** is

> **Command:** *ortho*
> **Enter mode [ON/OFF] <OFF>:** *on*

> You can temporarily toggle **Orthomode** on or off using the SHIFT key. To use this method, begin your command (say, *Line*), pick your start point, hold down the SHIFT key, and pick your second point. If **Orthomode** is on, the SHIFT key will disable it temporarily. Conversely, if **Orthomode** is off, the SHIFT key will temporarily enable it.
> Cool, huh?

3.1.2 Grid

One of the drafting tricks we learned back in the pencil days was to lay a background grid sheet on our drawing board before taping down our drawing sheet. This grid served as a lettering guide and helped in aligning different items.

AutoCAD's grid can be quite useful in aligning things. But as we'll see later, it isn't necessary for sizing text.

The grid toggles on or off quite easily. And controlling the size and shape of the grid, together with some creative use of the ***Snap*** tool, will often dramatically increase your drawing speed. (We'll discuss ***Snap*** in Section 3.1.4, p.57.)

Controlling the grid size is easy. Following is the command sequence:

Where to Find It:	
Command Line:	*Grid*
Hotkey(s):	F7
Status Bar:	▦ or GRID

> **Command:** *grid*
> **Specify grid spacing(X) or [ON/OFF/Snap/Major/aDaptive/Limits/Follow/Aspect] <0.5000>:** *[Enter the desired grid spacing.]*

AutoCAD provides some options for the ***Grid*** command. The default option sets the grid spacing. In this example, the spacing defaulted to **0.5000** drawing units. Other options include:

- **ON/OFF**: Turns the grid on or off.
- **Snap**: Sets the grid spacing equal to the snap increments.

- **Major**: This tool controls the frequency of major (as compared with minor) grid lines. By default, AutoCAD now uses grid lines rather than the points it used in previous releases.
- **aDaptive**: Using this, you can control the density of grid lines when you zoom in or out (more on the *Zoom* command in Lesson 4, p.79).
- **Limits**: Tells AutoCAD to show the grid beyond the limits of the drawing.
- **Follow**: **Follow** tells AutoCAD to force the grid lines to follow the dynamic UCS. "What!?" You ask. Don't worry, it's another 3D thing. (All this 3D stuff gives you something to look forward to in your advanced studies!)
- **Aspect**: Enables you to set a separate spacing for the X- and Y-planes.

3.1.3 Polar Tracking

Where to Find It:	
Hotkey(s):	F10
Status Bar:	or POLAR
Temp. Override:	F10

A remarkable and intuitive tool, polar tracking was designed to assist you by placing a temporary construction line from the last point selected. Additionally, polar tracking provides a tooltip detailing distance and angle from the last point selected (Figure 3.004). This makes it much easier to locate the next point accurately and quickly with a minimal need for absolute, relative, or polar coordinate entry.

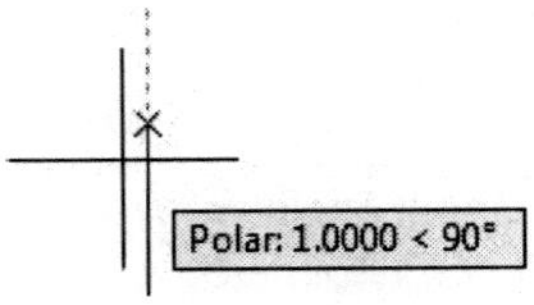

Polar tracking works on the four quadrants (0°, 90°, 180°, and 270°) by default, but you can set it to track at any angle or to override the settings on the fly. Just right click on the toggle and select an angle from the menu (Figure 3.005). (We'll see more on customizing polar tracking in Section 3.2, p.58.)

Figure 3.004

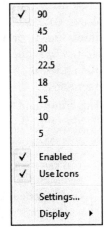

Figure 3.005

When used with **Polar Snap** (See Section 3.1.4), you can snap to any point at any angle with little or no need for the keyboard at all!

3.1.4 Snap (Polar and Grid)

The grid and polar tracking by themselves are only of minimal use. So AutoCAD provides a tool that was not available back in the pencil days. The **Snap** tool actually pulls (or *snaps*) the crosshairs to a grid or polar reference. Controlling the snap while referencing the grid or polar tracking construction lines can provide speed to an otherwise tedious job.

Look at the *Snap* command.

  Command: *snap*
  Specify snap spacing or [ON/OFF/Aspect/Style/Type] <0.5000>:
  [Enter the desired spacing or select an option.]

Where to Find It:	
Command Line:	Snap
Hotkey(s):	sn; F9; CTRL+B
Status Bar:	or SNAP

What's the default spacing? Did you say **0.5000**? Correct. As with other commands, we can see the default inside the <>.

Snap options include:
- **ON/OFF**: For turning the snap on or off.
- **Aspect**: For setting a different snap spacing horizontally and vertically.

- **Style**: For changing from a standard orthographic snap style to an isometric style (we'll look at this in more detail in Section 3.6, p.72).
- **Type**: For setting either **Polar** (to snap to points determined with the polar tracking feature) or **Grid** (to snap to grid referenced points). If you select **Polar** as your snap type, AutoCAD prompts for a **polar snap tracking distance**.

    Enter polar snap tracking distance <>: *[Enter the desired spacing.]*

Conventional wisdom suggests that setting the grid snap to half the grid works best for most drawings. This way, you can snap to each grid, and halfway between each. More than a single snap between grids is difficult to follow. Set the polar snap to the smallest increment you'll be using. You'll find an easier way to toggle between **Polar** or **Grid** snaps in the menu (right) that appears when you right click on the **Snap** toggle.	PolarSnap On Grid Snap On Off ✓ Use Icons Settings... Display ▶

Let's try an exercise using *Grid*, *Snap*, *Polar Tracking*, and *Ortho*.

Do This: 3.1.4A	Grid, Snap, Polar Tracking, and Ortho Practice

Try drawing the "W" in Exercise 2.4.4A, p.51, using only the tools you've just seen.
  I. Start a new drawing using the *SampleTemplate03* template in the C:\Steps\Lesson03 folder.
  II. Set the *Grid* to 1.
  III. Set the *Snap* to ½ and the snap **Type** to **Grid**. Be sure **Grid** and **Snap** are both toggled on.
  IV. Draw the "W" toggling **Orthomode** on or off as needed during the drawing.
  V. Repeat the exercise, but set the **Snap Type** to **Polar**, the **Polar snap tracking distance** to ½, and the **Polar angle** to 45°. (You'll find the angles on the right click **Polar Tracking** menu.)

How much faster did you draw? Usually, at this point, my students begin planning my demise for not having shown them this approach first. I'm really not that cruel. But once these tools are known, it's exceedingly difficult to convince students to learn the Cartesian Coordinate approaches to drawing. As you'll see, both the coordinate system and these drawing tools are quite necessary for speed and accuracy. (Wait until you master OSNAPs!)

Now try drawing the rest of the Lesson 2 exercises using your own settings.

3.2	And Now the Easy Way - *DSettings*

We've seen the command prompt method of setting **Grid** and **Snap**. Any setting that requires several lines of entry, as these do, is a prime candidate for a dialog box, and AutoCAD hasn't disappointed.

There are several ways to display the Drafting Settings dialog box (refer to the WTFI table). AutoCAD will present the Drafting Settings dialog box seen in Figure 3.006, p.59.

This box has five tabs available to help you (refer to the figures that follow).

Where to Find It:	
Command Line:	*DSettings*
Hotkey(s):	*ds*
Menu:	Tools – **Drafting Settings**
Status Bar:	[right click on Snap, Grid, Ortho, Polar Tracking or OSNAP] - **Settings**

- Figure 3.006 shows the **Snap and Grid** tab on top. You'll see this when you select **Settings** from the cursor menu presented when you right click on the SNAP or GRID toggle on the status bar. This tab presents frames where you can set increments for the **Snap** spacing (for the grid snap), the **Polar spacing** (for the polar snap), or the **Grid** spacing. It also provides check boxes for toggling on/off the **Snap** and **Grid**, and radio buttons (see the following insert) for setting the type of snap (**Polar** or **Grid**, **Rectangular** or **Isometric** – more on isometric snap in Section 3.6). The **Grid behavior** frame provides some new and useful tools for 3-dimensional work. You can ignore it for now.

Figure 3.006

A **Radio Button** is a round hole. Selecting a radio button places a black dot (or bullet) inside the round hole. Radio buttons usually come in small groups, but only one button in a group can hold the bullet.

New to 2011, the **Grid style** frame provides check boxes for using the older dot grid (rather than the lined grid you saw in Section 3.1.2, p.56) in **2D model space**, the **Block editor** (Lesson 21, p.476) or in **Sheet/layouts** (Lesson 24, p.560).

- If you access the Drafting Settings dialog box by right clicking on the POLAR toggle and selecting **Settings**, AutoCAD will place the **Polar Tracking** tab on top, as seen in Figure 3.007.

    Here you can toggle **Polar Tracking** on/off using the check box. But more importantly, you can adjust the **Increment angle** settings (the angles at which polar tracking appears) using a drop-down box, or add additional angles (not shown in the drop-down box) by picking the **New** button. **Additional angles** will appear in the list box and be used when you place a check in the check box.

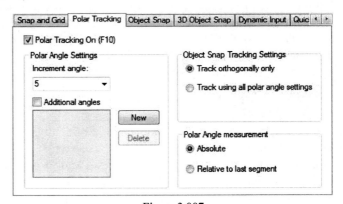

Figure 3.007

    You can also determine whether to use polar tracking **orthogonally only** (at the four compass points: 0°, 90°, 180°, and 270°) or using **all** the **angle settings** identified in the **Polar Angle Settings** frame. The **Object Snap Tracking Settings** frame provides radio buttons for each.

    The **Polar Angle measurement** frame allows you to show polar tracking angles in **absolute** terms (always showing angles as they relate to AutoCAD's compass points) or **relative to** [the] **last segment** (showing angles as they relate to the last line segment drawn). I recommend using the default **Absolute** setting to avoid confusion.

- The **Object Snap** tab (Figure 3.008) will appear on top when the Drafting Settings dialog box is accessed by right clicking on the OSNAP or OTRACK toggles. Here you can set Running Object Snaps (OSNAPs). We'll look at OSNAPs and Running OSNAPs in Sections 3.3, p.61, and 3.4, p.67.

You can also set up Running OSNAPs using a right click menu. We'll see this in Section 3.4.2, p.69.

- We'll discuss the **3D Object Snap** tab in *3D AutoCAD 2011: One Step at a Time*. (It's another of those 3D things.)
- The next tab – **Dynamic Input** (Figure 3.009) controls the appearance and function of AutoCAD's dynamic input (the information you see next to your crosshairs when the DYN toggle on the status bar is lit).

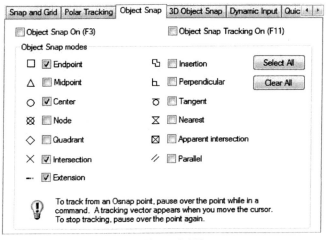

Figure 3.008

  o The two check boxes at the top – **Enable Pointer Input** and **Enable Dimension Input where possible** – control exactly what dynamic input displays. **Enable Pointer Input** controls whether or not coordinates will display as you move your cursor about the screen. **Enable Dimension Input where possible** controls whether or not you'll see a prompt next to the crosshairs when AutoCAD needs a dimension (as when requesting a circle radius or diameter.

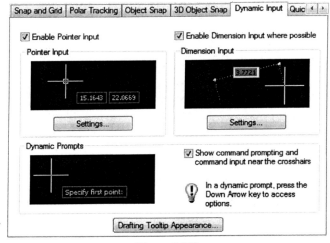

Figure 3.009

  o You can control what the pointer displays with the Pointer Input Settings dialog box (Figure 3.010). Access it by picking the **Settings** button in the **Pointer Input** frame. Here you can change the format from **Relative coordinates** to **Absolute coordinates**, or from a **Polar format** (the second point prompt in a polar-type format, or *X<angle*) to the **Cartesian format** (the second point prompt appears as a Cartesian coordinate).

Options in the **Visibility** frame control when tooltips will be visible.

  o The **Settings** button in the **Dimension Input** frame provides control for stretching with grips. We'll discuss grips in some detail in Lesson 9, p.212.

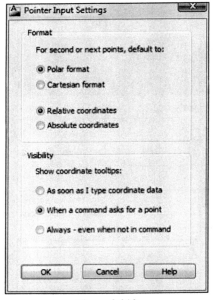

Figure 3.010

- o The **Dynamic Prompts** frame holds a single, handy check box – **Show command prompting and command input near the crosshairs**. Some people prefer to turn this off. Without it, AutoCAD prompts on the command line only (not at the crosshairs). Your screen will still display coordinates but without the additional nuisance of the prompt.
- o The next thing on the **Dynamic Input** tab is the **Drafting Tooltip Appearance** button. This calls a simple dialog box that allows you to adjust the color, size, and transparency of tooltips.
- Finally, you'll see an **Options** button at the bottom of the dialog box. It calls the Options dialog box where a more experienced operator can adjust some of the default settings for AutoCAD.

> We'll discuss this option in more detail in Section 3.4.1, p.68. But for now please consider this carefully:
> *The Options dialog box is not a place for beginners! It contains settings that can render AutoCAD inoperable if set incorrectly, so please, avoid experimentation here until you have some experience.*

- We'll look at the last tab – **Quick Properties** – in some detail in Lesson 6, p.139.

3.3 Never Miss the Point with OSNAPs

We've now seen several ways to draw with precision. What more could we possibly need?

Well, not all of these tools lend themselves easily to large drawing environments like those found in disciplines such as architecture or petrochemical. This leads us back to a need for some tools that free us as much as possible from the Cartesian Coordinate System. Enter Object Snaps – OSNAPs.

These remarkable tools must and will become second nature to the successful CAD operator. After learning OSNAPs, you must engrain this 11th Commandment into your hearts and minds: *Thou shalt not eyeball!* Because after learning OSNAPs, there will never again be a need to guess about the location of a point.

What are OSNAPs? OSNAPs are a means of responding with precision to any prompt directing you to pick a point in your drawing. They provide the means for precisely locating and selecting a point (endpoint, midpoint, center point, etc.) on (or referenced by) any existing object in the drawing – lines, circles, arcs, and so forth.

Study the following chart. The first column shows the OSNAP icons. The second column shows the cheater symbols AutoCAD shows when trying to use an OSNAP in a drawing. The third column shows the equivalent command found when you call up the cursor menu by clicking the right mouse button (with your cursor in the drawing area) while holding down the SHIFT key on the keyboard (SHIFT + 🖱). The last column shows the temporary override keys when they're available.

OSNAP TOOLBAR	SYMBOL	CURSOR MENU	TEMPORARY OVERRIDE (SHIFT +)
		Temporary Track Point	Q or]
		From	
	□	Endpoint	E or P
	△	Midpoint	V or M
	×	Intersection	
	⊠	Apparent Intersection	
	⋯	Extension	
	○	Center	C or ,

OSNAP TOOLBAR	SYMBOL	CURSOR MENU	TEMPORARY OVERRIDE (SHIFT +)
⟡	◇	Quadrant	
⟲	⟲	Tangent	
⊥	⌐	Perpendicular	
∥	∥	Parallel	
⌂	⌂	Insert	
○	⊗	Node	
⋈	⋈	Nearest	
		None	
		Osnap Settings	

Most of these will become obvious once you've used them. Some will require a bit more explanation. The best way to learn how to use object snaps, however, is through a practice exercise. Follow me!

Do This: 3.3A	**OSNAP Practice**

I. Open the *train* file found in the C:\Steps\Lesson03 folder. It'll look like the figure at right.

II. Show the Object Snap toolbar and dock it to the right side of your screen. [Refer to Lesson 1, Section 1.1.2, p.6.]

III. Follow these steps. (For this exercise, I'll use four methods of selecting OSNAPs, but the methods are completely interchangeable.)

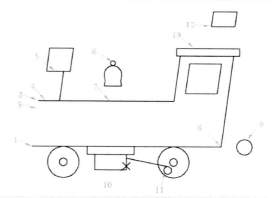

3.3A: OSNAP PRACTICE

1. Check the status bar (following figures) and be sure that all the toggles shown here are off (unlit). If any of the toggles appear lit, pick on it once to turn it off.

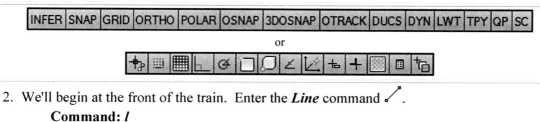

2. We'll begin at the front of the train. Enter the **Line** command.
   Command: *l*

3. Pick the **Endpoint** OSNAP button on the OSNAP toolbar.
   Specify first point: _endp of

3.3A: OSNAP PRACTICE

4. Place the cursor at point 1. Notice how a small square (the symbol for endpoint) appears at the endpoint of the line. Pick here.
Notice that the line begins at the endpoint of the existing line.

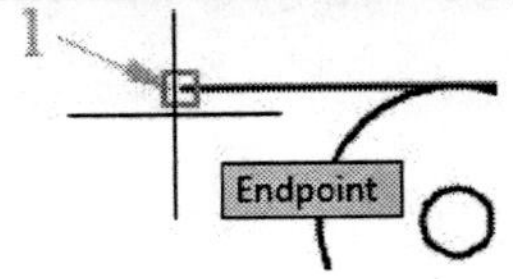

5. Select the **Endpoint** button again.
   Specify next point or [Undo]: _endp of

6. Place the cursor at Point 2. Notice the symbol for endpoint appears again. Pick here. Your line is drawn to the endpoint. Hit ENTER to complete the command.
   Specify next point or [Undo]: *[ENTER]*

7. Next we'll draw a cowcatcher using the *Line* command, the **Extension** OSNAP, and Polar Tracking. First, let's set up polar tracking. Right click on the **Snap** toggle on the status bar, and select **Settings** from the menu.

8. AutoCAD presents the Drafting Settings dialog box with the **Snap and Grid** tab on top, as seen in Figure 3.006, p.59. Place a bullet next to **Polar Snap**, and then set the **Polar distance** at *.25* as shown. Check the **Snap On** check box ☑ Snap On (F9); then pick on the **Polar Tracking** tab.

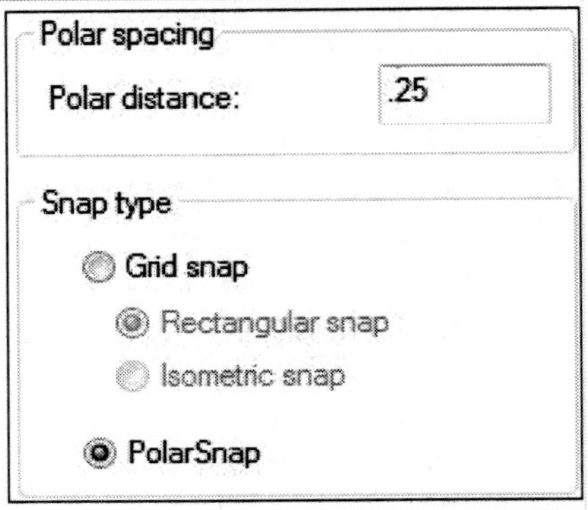

9. We'll want to use Polar Tracking on all angles, so put a bullet in the **Track using all polar angle settings** option of the **Object Snap Tracking Settings** frame as shown.
Check the **Polar Tracking On** check box ☑ Polar Tracking On (F10); then pick the **OK** button to complete the setup.

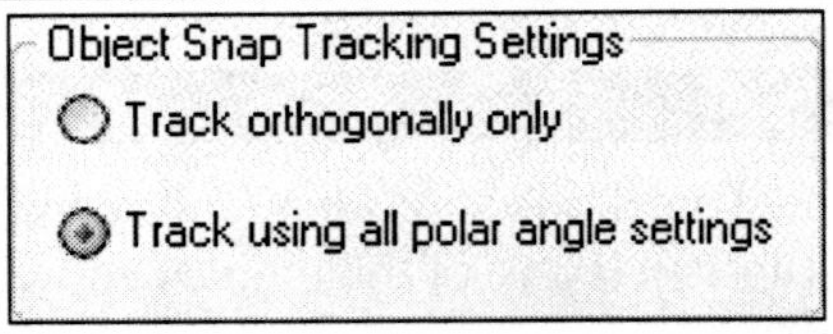

10. Begin the *Line* command.
    Command: *l*

11. Pick the **Extension** OSNAP button.

3.3A: OSNAP PRACTICE

12. Place your crosshairs over Point 1, but *don't pick*. AutoCAD will display a small plus symbol indicating that the object has been located. Move your cursor down and slightly to the left. Notice the tooltip as polar tracking helps you locate a point off of the extension of the line that you created in Steps 2 through 6.
Pick at the point located ¾" at 263° as shown.

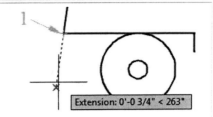

13. Now use polar tracking to create a rectangle ¼" up and 2" to the left.

14. Using the **Endpoint** OSNAP override (hold down the SHIFT and E or P keys), draw a line connecting the upper right corner of the cowcatcher to Point 1, as shown in the following figure.
   Command: *l*

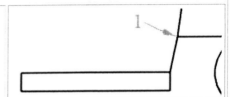

15. Use the **Endpoint** OSNAP to begin a line at the upper left corner of the cowcatcher.
   Command: *l*
   Specify first point: *_endp of*

16. At the **Specify next point** prompt, type *nea* to tell AutoCAD you want the point nearest to where you select.
   Specify next point or [Undo]: *nea*
(Typing "nea" is the keyboard entry method for using the **Nearest** OSNAP.) Select a point near Point 3. Notice the symbol.

17. Complete the command.
   Specify next point or [Undo]: *[ENTER]*

18. Repeat the **Line** command.
   Command: *[ENTER]*
This time we'll select a point where two lines would intersect if they were a bit longer. And we'll use the cursor menu to enter the OSNAPs.

19. Hold down the SHIFT key on the keyboard and right click in the graphics area of the screen. You'll see a cursor menu like the one shown here. Select **Apparent Intersection**.
AutoCAD asks which nonintersecting lines you wish to use. Select the line you drew between the cowcatcher and Point 3; and then select the line at Point 4. Notice the symbols.
   Specify first point: *_appint of* and AutoCAD begins the line.

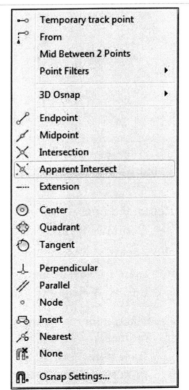

3.3A: OSNAP PRACTICE

20. Now we must identify where to go with the line. Bring up the cursor menu again (as you did in Step 19), and select **Parallel**.

    Specify next point or [Undo]: _par to

Place your cursor over the line between Points 1 and 2 (don't pick). AutoCAD will display a symbol to let you know it has found the line.

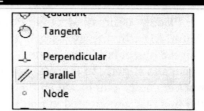

21. Move the cursor to the bottom of the smokestack. Pick when AutoCAD displays the tooltip shown. (Notice that the parallel symbol appears on the line you selected when your line is parallel to it.)

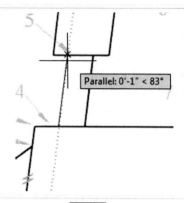

22. Complete the command.

    Specify next point or [Undo]: *[ENTER]*

Your drawing looks like the following figure.

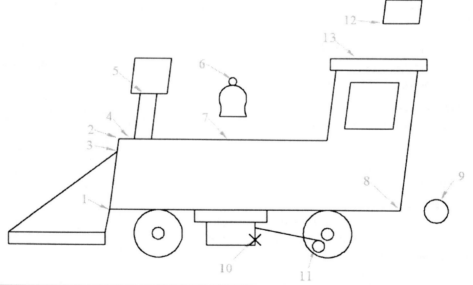

23. Repeat the *Line* command.

    Command: *[ENTER]*

24. We want the bell assembly to start a bit above the top of the bell. There is not a point there to snap, so we'll use the **From** OSNAP.

    Specify first point: *from*

25. We'll start a line ¼" above (**From**) the center of the circle at point 6. We'll need to use polar coordinates.

    Base point: *cen*
    of <Offset>: *@.25<90*

3.3A: OSNAP PRACTICE

26. Perpendicular ⊥ to the line at point 7.
    Specify next point or [Undo]: *per*

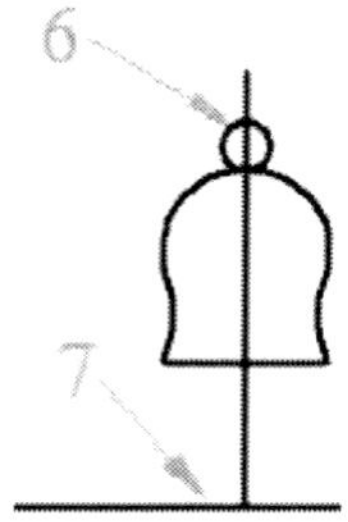

27. Complete the command.
    Specify next point or [Undo]: *[ENTER]*
    The assembly now looks like this.

28. Repeat the *Line* command.
    Command: *[ENTER]*

29. From the **endpoint** at point 8…
    Specify first point: *end*

30. …to the **center** (or SHIFT + C or ,) of the circle at point 9.
    Specify next point or [Undo]: *cen*

31. Complete the command.
    Specify next point or [Undo]: *[ENTER]*

32. Repeat the *Line* command.
    Command: *[ENTER]*

33. From the **node** at point 10 (the node looks like an "X" – more on nodes in Lesson 11, p.245) …
    Specify first point: *nod*

34. …**tangent** to the circle at point 11.
    Specify next point or [Undo]: *tan*

35. Complete the command.
    Specify next point or [Undo]: *[ENTER]*

36. Now the flagpole. Repeat the *Line* command.
    Command: *[ENTER]*

37. From the **intersection** at point 12…
    From point: *int*

38. …to the **midpoint** of the line at point 13. (You can use the override if you wish – hold down the SHIFT & M keys.)
    To point: *mid*

39. Complete the command. Your drawing now looks like the following figure.

3.3A: OSNAP PRACTICE

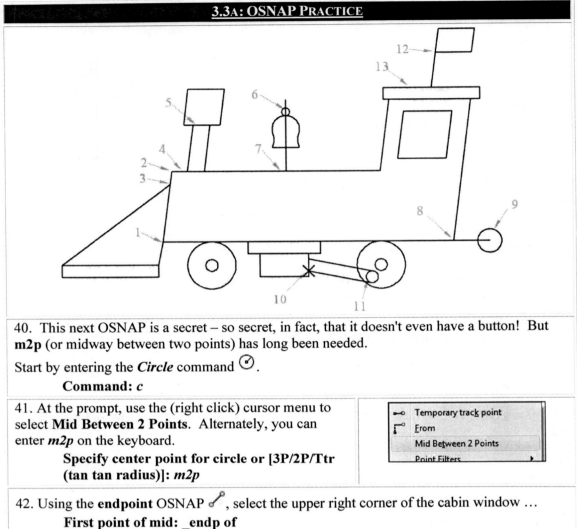

40. This next OSNAP is a secret – so secret, in fact, that it doesn't even have a button! But **m2p** (or midway between two points) has long been needed.

   Start by entering the *Circle* command ⊙.

   Command: *c*

41. At the prompt, use the (right click) cursor menu to select **Mid Between 2 Points**. Alternately, you can enter *m2p* on the keyboard.

   Specify center point for circle or [3P/2P/Ttr (tan tan radius)]: *m2p*

42. Using the **endpoint** OSNAP, select the upper right corner of the cabin window ...

   First point of mid: _endp of

43. ... and then the lower left corner.

   Second point of mid: _endp of

44. Give the circle a ¼" radius. Your drawing looks like the figure at right. (This is by far the easiest method of locating this point, as you will soon see.)

   Specify radius of circle or [Diameter]: *.25*

45. Save the drawing to the C:\Steps\Lesson03\ folder as *MyTrain*.

   Command: *saveas*

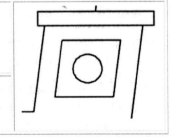

You should now be fairly familiar with four ways to call on OSNAPs – the toolbar, the keyboard, overrides, and the cursor menu.

3.4 Running OSNAPs

Let me show you another way to use OSNAPs. We call this one *Running OSNAPs*.

Before long, you'll discover that having to select an OSNAP every time you want to place a point is a tedious procedure at best (even if it's easier than typing coordinates). Is there not, you might ask, a way to turn OSNAPs on and leave them on? By now, of course, you know that if there weren't, I wouldn't ask.

AutoCAD allows you to set Running OSNAPs in one of two ways – dialog box or menu.

3.4.1 Running OSNAPs via Dialog Box

Where to Find It:	
Command Line:	*OSNAP*
Hotkey(s):	*os*
Toolbar:	Object Snap – **OSNAP Settings**
Status Bar:	[right click **OSNAP** toggle] – **Settings**

Use one of the procedures listed in the WTFI table to display the Drafting Settings dialog box with the **Object Snap** tab on top (Figure 3.008, p.60). Each available OSNAP has a check box beside it. Click in the box to place a check and activate that particular running OSNAP. Look to the left of each box to see a symbol (or marker). We've seen that AutoCAD uses this symbol to indicate that you're selecting a point using this particular OSNAP.

An **Options** button resides at the bottom of the dialog box. This button opens the Options dialog box with the **Drafting** tab on top (Figure 3.011). Here the user has several frames to help control the behavior of object snaps. You can control OSNAP settings from this tab, including:

- the size and color of the symbol (**Marker**);
- whether or not you want a **tooltip** when you hesitate over an OSNAP area;
- whether or not you want the crosshairs to snap to (**Magnet**) the selected OSNAP.

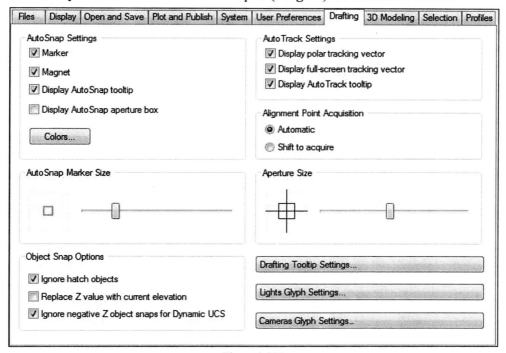

Figure 3.011

But for optimal performance, I suggest leaving all these options at their default settings. (Reminder: DON'T experiment in the Options dialog box!)

3.4.2 Running OSNAPs via Menu

For setting one or two Running OSNAPs, this method works much more quickly and easily than the dialog box, even if it doesn't provide quite the number of options.

Begin by turning on OSNAPs (pick the toggle on the status bar so that it's lit). Now right click on the toggle. Notice the menu (Figure 3.012). Place a check next to the OSNAP you wish to "run" – clear the others. That's all there is to it!

(Now wasn't that easier than calling a dialog box?!)

Okay, before you proceed there are a couple drawbacks:

- You can only check/uncheck one OSNAP at a time – then the menu disappears and you have to call it again. This can get tedious if you have several to set.
- The menu doesn't show the extremely useful **Clear All** button you find on the dialog box.

Well, nothing is perfect … yet!

Let's try using running OSNAPs to redraw the train.

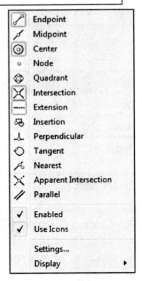

Figure 3.012

You may notice that it's difficult to select the **center** or the **quadrant** of a circle, as AutoCAD doesn't know which you want. If you move your crosshairs over the circle and the center symbol appears instead of the quadrant symbol (or vice versa), hold the mouse steady and press the TAB key on the keyboard. AutoCAD will toggle through the various possibilities until it finds the one you like. This procedure is quite useful for busy drawings in which AutoCAD must choose between several OSNAP possibilities.

Do This: 3.4.2A	More OSNAP Practice

I. Open the *train2* file found in the C:\Steps\Lesson03 folder. This is a pristine copy of the drawing you used in the last exercise.

II. Follow these steps.

3.4.2A: MORE OSNAP PRACTICE

1. Set the running OSNAPs indicated.

2. Now follow the instructions in Exercise 3.3A, p.62, to draw the train, but don't use the OSNAPs menu. AutoCAD will automatically use OSNAP endpoints, midpoints, and so forth. It won't, however, automatically use the **Apparent Intersection** or the **Nearest** OSNAP (you didn't set those). You'll have to select those manually.

3. Exit the drawing without saving.
   Command: *quit*

> Other ways to activate/deactivate Running OSNAPs include clicking on the OSNAP toggle on the status bar, using the F3 function key on the keyboard, or holding down the CTRL key on the keyboard while typing *F*. You can also use the overrides (SHIFT+A, or SHIFT+/) to temporarily toggle OSNAPs on or off.

Was that faster? Easier?

The biggest problem my students have with running OSNAPs is that they forget to deactivate them. They can't understand why their lines or circles keep jumping to an endpoint or intersection. If this happens to you, deactivate Running OSNAPs.

3.5 Point Filters and Object Snap Tracking

Often in drafting, we find it necessary to align objects according to the location of other objects. Parallel bars, triangles, and 4H-lead guidelines made this easy on the board. But what does AutoCAD use to substitute for these proven tools?

Actually, there are two things we can use: *point filters* and *object snap tracking*.

3.5.1 Point Filters

Use point filters to tell AutoCAD to locate an object using the X, Y, and/or Z coordinate of an existing object. Let's see how they work. We'll redraw the stick figure in our train's cabin.

Do This: 3.5.1A — Introducing Point Filters

I. Open the *MyTrain* file found in the C:\Steps\Lesson03 folder.
II. Erase the figure in the cabin window.
III. Follow these steps.

3.5.1A: POINT FILTERS

1. Begin with the ***Circle*** command.
    Command: *c*

2. At the circle prompt, hold down the SHIFT key and right click to open the OSNAP menu.

3. Select *.X* from the **Point Filters** option.* This will tell AutoCAD to use the X coordinate ...
    Specify center point for circle or [3P/2P/Ttr (tan tan radius)]: *.x*
    ... of the midpoint of the lower horizontal line of the cab.
    of _mid of

4. AutoCAD tells you that it needs the Y and Z coordinates. Tell it to use the YZ coordinate of the midpoint of the vertical line.
    (need YZ): _mid of

---

* Alternately, you can simply enter *.x* at the Specify center point... prompt.

3.5.1A: POINT FILTERS

5. AutoCAD locates the center of the circle and asks you for a radius. Tell it to use a radius of ¼".

Specify radius of circle or [Diameter]: *.25*

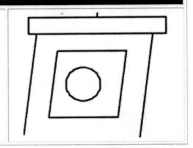

It's not as easy as **m2p**, but this is one way of acquiring coordinates from existing geometry in a drawing. But let's look at yet another, easier method.

3.5.2	Object Snap Tracking

AutoCAD produced the tracking feature in response to complaints that most OSNAPs require a direct contact with an existing object, and point filters were too tedious. However, like point filters, tracking allows you to draw in *relation to* an existing object without actually touching it. Let's see how it works. We'll redraw the stick figure in our train's cabin.

Do This: 3.5.2A	Object Tracking

I. Be sure you're still in the *MyTrain* in the C:\Steps\Lesson03 folder. If not, open it now.
II. Erase the circle you drew in the last exercise.
III. Follow these steps.

3.5.2A: OBJECT TRACKING

1. Turn on the Object Tracking (**OTRACK**) toggle ∠ on the status bar. (You can use the F11 function key on your keyboard if you wish.)

2. Clear all Running OSNAPs except **Midpoint** (see the figure at right). Be sure to turn on running OSNAPs.

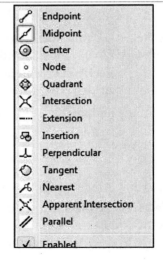

3. Enter the *Circle* command ⊘.

Command: *c*

4. At the **Specify center point** prompt, place your cursor over the midpoint of the lower horizontal line in the cab window. Hesitate for a moment (don't pick anything), and then move your crosshairs upward. Notice the tracking line.

Specify center point for circle or [3P/2P/Ttr (tan tan radius)]:

5. Repeat Step 4, this time placing your crosshairs over the left vertical line of the cab window and moving inward.

Specify center point for circle or [3P/2P/Ttr (tan tan radius)]:

3.5.2A: OBJECT TRACKING

6. Move the crosshairs toward the center of the window. Pick with the left mouse button when Tracking tells you that you're located at 0° from the last point and 90° from the first point (as shown).

7. You'll notice that the center of the circle has been placed at a point near the center of the window.

   Use polar tracking to draw a circle with a ¼" radius (as shown).

8. Draw a line ...
   Command: *l*

9. ... from the lower **quadrant** of the circle you just drew ...
   Specify first point: _qua of

10. ... perpendicular to the bottom of the window.
    Specify next point or [Undo]: _per to

11. Complete the command.
    Specify next point or [Undo]: *[ENTER]*

Your drawing now looks like the following figure.

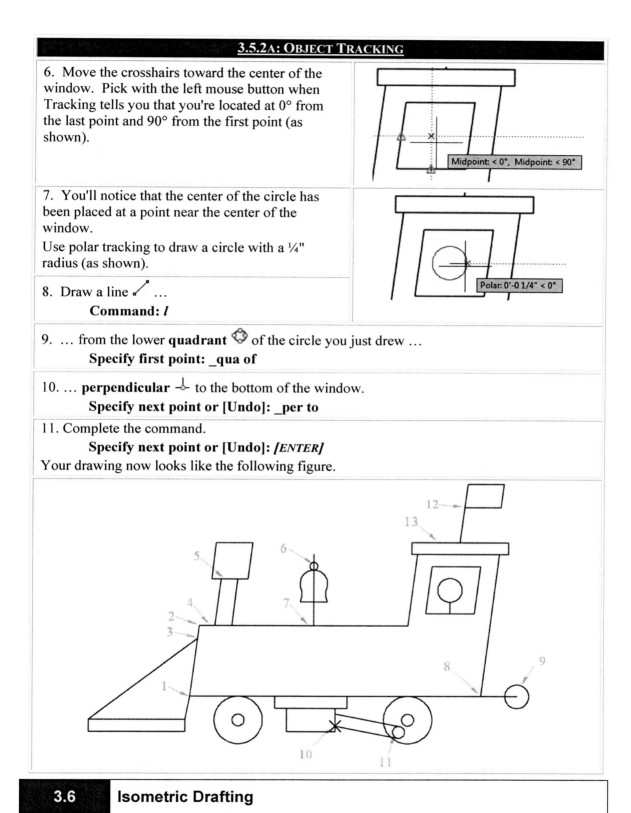

| 3.6 | **Isometric Drafting** |

After having thoroughly confused us with orthographic projections, my old drafting instructor threw *isometric* drawings at us. (At this point, I started thinking about other ways to make a living.) I'll try to make it a bit easier than he did.

Orthographic drawings (projections) show an object from one aspect – left / right / top / bottom / front / back. An isometric drawing shows three aspects at once – left or right side / top or bottom / front or back. Standard isometrics are drawn so that the faces are seen on a plane running 30° above or below the X-axis, as shown in Figure 3.013. Of course, there are variations, but the 30° format is standard.

To draw isometrics in AutoCAD requires adjusting the grid and snap to an isometric format. This is easier than it sounds. Remember the options AutoCAD provided for the snap tool? This is where we make the switch from orthographic layout to isometric. Here's how:

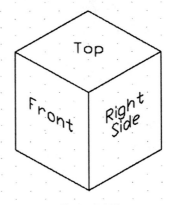

Figure 3.013

    Command: *snap*
    Specify snap spacing or [ON/OFF/Aspect/Style/Type] <0.5000>: *s*
    Enter snap grid style [Standard/Isometric] <S>: *i*
    Specify vertical spacing <0.5000>: *[ENTER]*

That's all there is to it! Ortho will now work along the 30°/90° planes. You'll notice that even your crosshairs have changed, and you'll need a new toggle– the *isometric plane toggle*. This will adjust your crosshairs and help you draw along the 30°/90° plane, the 150°/90° plane, or the 30°/150° plane. The keyboard toggle is CTRL+E, and the function key is F5. As with the other toggles we've learned, these will work regardless of the command sequence we're running.

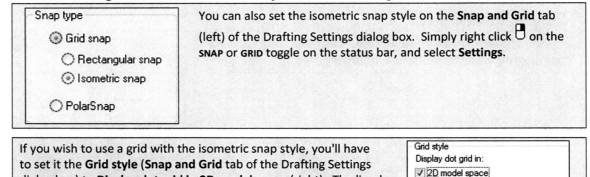

You can also set the isometric snap style on the **Snap and Grid** tab (left) of the Drafting Settings dialog box. Simply right click on the SNAP or GRID toggle on the status bar, and select **Settings**.

If you wish to use a grid with the isometric snap style, you'll have to set it the **Grid style** (**Snap and Grid** tab of the Drafting Settings dialog box) to **Display dot grid in 2D model space** (right). The lined grid doesn't show isometric lines. ☹

Let's try a simple isometric drawing.

Try to draw the isometric shown in Figure 3.013. Don't try the text yet; we'll look at that in Lesson 4, p.93.

Do This: 3.6A	**Isometric Drafting**

    I. Begin a new drawing using the *MyBase1* (or *base2*) template found in the C:\Steps\Lesson01 folder.
    II. Turn on dynamic input ⊞.
    III. Follow these steps.

3.6A: ISOMETRIC DRAFTING

1. Enter the **Snap** command.
    Command: *sn*

3.6A: ISOMETRIC DRAFTING

2. We'll tell AutoCAD to change the snap **Style**, but we'll use a dynamic input trick to do it. Hit the down arrow on the keyboard and select the **Style** option from the dynamic input menu.

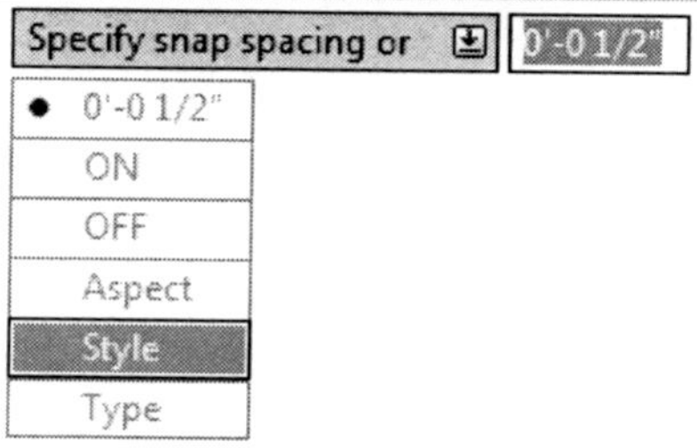

3. Set the **Style** to **Isometric**...
   Enter snap grid style [Standard/Isometric] <S>: *i*

4. ... with a spacing of ¼".
   Specify vertical spacing <0'-0 1/2">: *1/4*

5. Using the **F5** function key, set your crosshairs to a 90°/30° configuration. The crosshairs will look like the figure at right.

6. Turn on the **Grid** and **Ortho** toggles on the status bar.

7. Right click on the **Snap** toggle and open the **Settings** dialog box.

8. Set the **Grid style** to **Display dot grid in 2D model space**, and turn the **Grid snap** on in the **Snap type** frame.

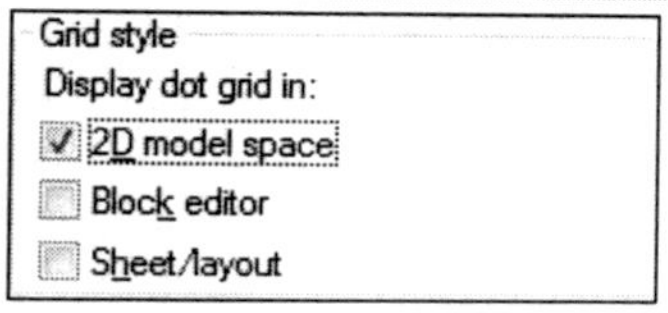

9. Draw the right side of the cube. (Toggle dynamic input on and off as you wish.)
   Command: *l*

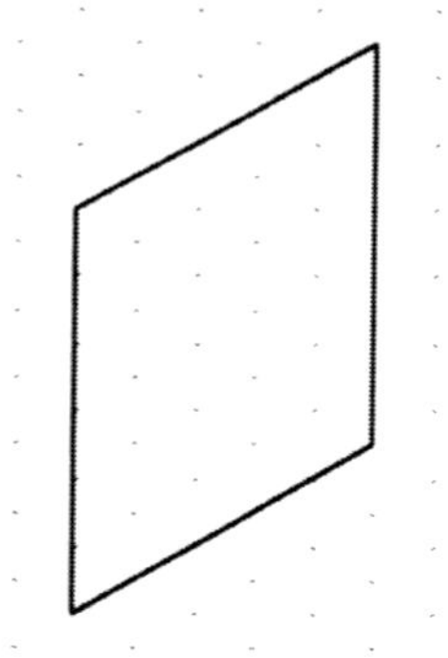

10. Toggle the crosshairs to 90°/150°.

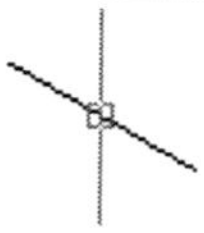

11. Draw the left side of the cube.
    Command: *l*

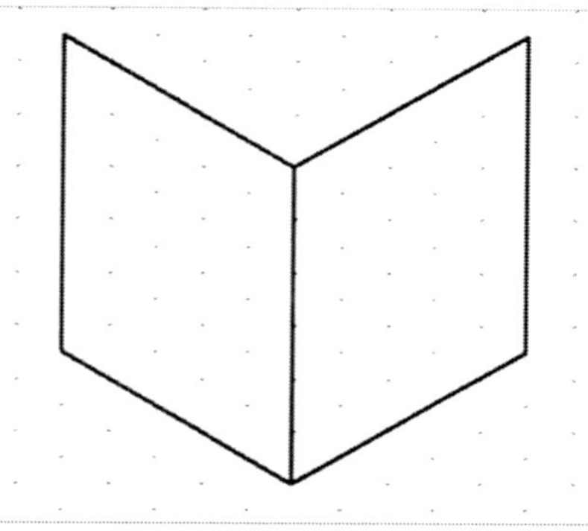

3.6A: ISOMETRIC DRAFTING

12. Toggle the crosshairs [F5] to 30°/150°.

13. Draw the top of the cube.
    Command: *l*

14. Save 💾 your drawing in the C:\Steps\ Lesson03 folder as *MyCube.dwg*.
    Command: *save*

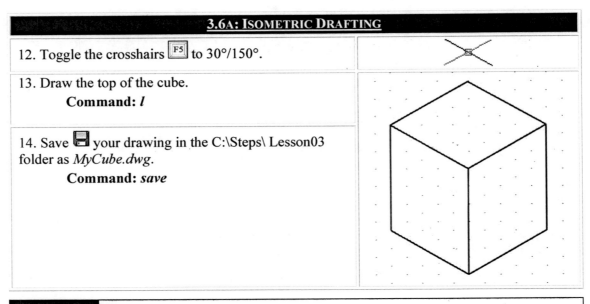

3.7 Extra Steps

With all the drawing and erasing you've done in this chapter, you may have noticed that you'll occasionally "lose" a line (or other object) that you thought was there. Likewise, you may notice that an object may remain highlighted even after you've canceled the command. This problem lies mostly with the video card or monitor you're using. But most programs allow you to *refresh* your screen after doing a great deal of work, to make sure all the screen pixels that should be lit, are lit.

AutoCAD calls its refresh command **Redraw**, but that requires too much typing. When your screen needs refreshing (redrawing), type **R** at the command prompt and hit ENTER. This'll redraw the screen.

> By the way, I indicate commands and hotkeys using capital letters as a matter of convention. AutoCAD isn't case sensitive.

In the event that what you expect to see doesn't materialize after a redraw, you have one other option. Every object within an AutoCAD drawing carries with it quite a bit of information. This information includes details about the type of object, its size, location, layer, and so forth. With AutoCAD's **Regen** command (**Re**), AutoCAD reads and redraws every object in the drawing. You might think this would take a lot of time in a larger drawing, and in fact, it once did. Now, however, with the advent of faster computers and better programming, regens often go unnoticed.

If the object doesn't show after a regen, it doesn't exist as part of the drawing and you'll have to redraw it.

So add these two commands to your "Hail Mary" list, but don't cut it off yet. More will follow!

3.8 What Have We Learned?

Items covered in this lesson include:

- *Drafting Settings: Ortho, Snap, Tracking, Grid, and dynamic input tools*
- *Isometric Drawing Setup*
- *OSNAPs and Running OSNAPs*
- *Commands*
  - *Grid*
  - *Snap*
  - *Ortho*
  - *DSettings*
- *Point Filters*
- *Object Snap Tracking*
- *Polar Tracking*
- *Dynamic Input*
  - *OSNAP*
  - *Redraw*
  - *Regen*

This is pretty cool, but I need some practice!

In this lesson, we've covered the tools that'll make the difference between a computer *doodler* and a CAD *operator*. Anyone can draw lines and circles, but for CAD to be an effective tool in industry, you must have the ability to draw with speed and precision. But don't expect yourself to fly through a drawing yet. First, you must practice the material in this lesson until it becomes second nature – like a draftsman knowing how to begin a drafting project without thinking about it. So take some time to do these exercises. Do them again and again until you're quite comfortable with AutoCAD's drawing aids.

3.9 Exercises

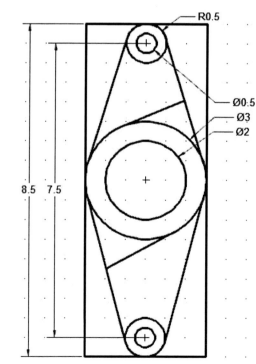

1. Create the drawing at right. Set it up as follows:
   1.1. Grid: ½
   1.2. Snap: ¼
   1.3. Units: use default
   1.4. Limits: use defaults
   1.5. Use OSNAPs whenever possible.
   1.6. Don't try to draw the dimensions yet.
   1.7. Save the file as *OSNAP Practice* in the C:\Steps\Lesson03 folder.
2. Start a new drawing. Set it up as follows:
   2.1. Units: architectural
   2.2. Lower left limits: 0,0
   2.3. Upper right limits: 17,11
   2.4. Grid: ½
   2.5. Snap: ¼
   2.6. Snap style: isometric
   2.7. Save this as a template file called *MyIsoGrid.dwt* to the C:\Steps\Lesson03 folder.

3. For each of the drawings below, start a new drawing using the *MyIsoGrid* template file created in Exercise 2. (If this file is not available, use the *IsoGrid2* template in the same folder.) Draw the figures and save them in the C:\Steps\Lesson03 folder.

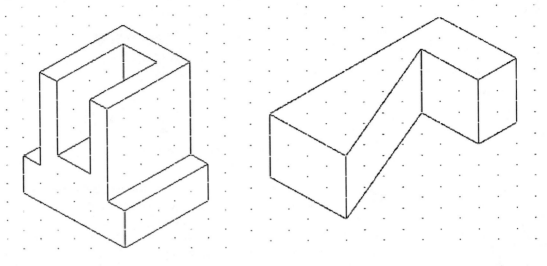

3.10 **For Web-Based Review Questions and Additional Exercises, visit: http://foragerpub.com/AcadFiles/2011/2011.htm**

Lesson 4

Following this lesson, you will:

✓ Know how to manipulate the screen display of your drawing through:
- The **Zoom** command
- The **Pan** command
- The **View** command
- Navigation Wheels
- The Navigation Bar

✓ Know how to use a transparent command

✓ Know how to use Basic Text and Text Editing commands, including:
- **Text/DText**
- **DDEdit**
- **Style**
- **Qtext**
- **Find**

✓ Know how to load a LISP routine

Display Controls and Basic Annotative Text

Our last lesson probably made you feel fairly comfortable with your new ability to create some wildly accurate drawings. But have you tried to use some of these tools in larger drawings? If you have, you might have noticed that grid and snap become difficult (if not impossible) to use as the drawing encompasses more space. You might also have noticed that those wonderful OSNAPs don't help if you can't tell which endpoint you've selected.

We'll begin this lesson by addressing these problems. Then we'll examine some of the tools available to help place text in a drawing.

4.1	**Changing What You See**
4.1.1	**Zooming**

One of the allowances I've had to make as I amassed all this "experience" has been the acquisition of a pair of reading glasses. At first I wore them down on my nose – a look more professorial. Now, however, I wear them closer to my eyes to get a wider field of vision. Ah, ~~age~~ uh, *maturity*!

For this reason, I really appreciate AutoCAD's **Zoom** command. With this, I can enlarge all or part of my drawing as much as I like, without having to search for my "eyes."

To appreciate the demonstrations of the **Zoom** command, we'll need to open the *pid* drawing in the C:\Steps\Lesson04 folder. The drawing is shown in Figure 4.001. Note that the *pid* drawing was created for a D-size sheet of paper, so viewing it on the screen (or on our book-size sheet of paper) makes it difficult to read.

Prof. Otto Kaad

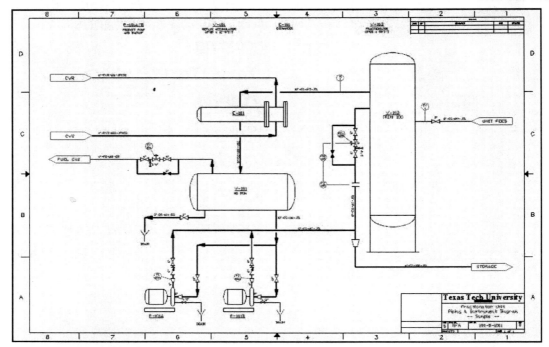

Figure 4.001

The command sequence for the *Zoom* command follows:

> **Command:** *zoom*
> **Specify corner of window, enter a scale factor (nX or nXP), or [All/Center/Dynamic/Extents/Previous/Scale/Window/Object] <real time>:** *[Select a corner of the window.]*
> **Specify opposite corner:** *[Select the opposite corner of the window.]*

Let's look at the various *Zoom* options. Notice that most of the options have an equivalent button in one of the several places indicated in the WTFI table.

	Where to Find It:
Command Line:	*Zoom*
Hotkey(s):	z
Ribbon (Tab/Panel):	View – Navigate – (Extents flyout) [zoom option]
Menu:	View – Zoom – [zoom option]
	Cursor – Zoom (Realtime)
Navigation Bar:	(Extents flyout) [Zoom option]
Toolbar:	Zoom – [zoom option]
	Standard – **Zoom** (**Realtime** or **Window** flyout)

Zoom Option	Alternate Button
Window prompts you to place a window around the area of the drawing you wish to see better. AutoCAD then zooms in to that area.	Zoom Window Button
Realtime appears to be the default, but it's accepted only by hitting the ENTER key. We'll look at realtime zooming shortly. If instead, you select an empty point on the screen, AutoCAD will assume you're placing a window around the objects you wish to view more closely. It'll then prompt you for the other corner. (Remember Implied Windowing – Section 2.3, p.45.)	Zoom Realtime Button
All displays the limits of the drawing, unless something has been drawn outside the limits. In that case, **All** will display the extents, or *all* the objects on the drawing.	Zoom All Button
Center prompts you for the desired center point of the display, and then adjusts the display so that the selected point is in the center of the screen.	Zoom Center Button
When you select the **Object** option of the *Zoom* command, AutoCAD will ask you to select an object(s). It'll then zoom in as close to that object(s) as possible while showing it in its entirety.	Zoom Object Button
One of the most forgotten zoom features, **Dynamic** provides you with the ability to place an adjustable box over that part of the drawing you want to display. When selected, AutoCAD temporarily replaces the screen with a view of the entire drawing. A view box shows what and where your current view area is. A selection box appears that you can manipulate with the mouse as you would the crosshairs. Move this box over the area you wish to view and hit the left mouse button to confirm your selection. AutoCAD redisplays the drawing with the selected area shown (this'll be clearer when we do it in Exercise 4.1.1, p.81).	Zoom Dynamic Button
Extents brings you as close as possible to the drawing while showing *all* the objects in the drawing.	Zoom Extents Button

Zoom Option	Alternate Button
Scale can be a bit confusing. You don't have to type *s*, but you can. If you do, AutoCAD prompts you for the scale you want. Note that this isn't the drawing scale, but the size of the drawing in relation to the graphics area. Hence, a scale of *.5* will cause the drawing to occupy half the graphics area of your screen. If you simply type *.5X* at the *Zoom* prompt (rather than typing *s*), the drawing will appear half its current size. Notice that the *Zoom* prompt suggests the **X** or an **XP** procedure. (Ignore the **XP** for now. We'll cover that in detail in our discussion of layouts – aka. Paper Space – in Lesson 25, p.587.)	Zoom Scale Button
To simplify the **Scale** option, AutoCAD provides two additional tools (not listed at the command prompt) – **Zoom Out** and **Zoom In**. **Zoom In** zooms to a 2X scale and **Zoom Out** zooms to a .5X scale.	Zoom Out Button Zoom In Button
To zoom to a previous view, use **Zoom Previous**. You can also use this one to restore a previous view after using the *View* command (more on that in Section 4.2, p.88).	Zoom Previous Button

In addition to the command line, ribbon panel, toolbars, and menus, all the *Zoom* options can be found on the (right click) cursor and dynamic menus once the *Zoom* command has been entered.

Let's try some of these now.

Do This: 4.1.1A	Practice Zooming

I. If *pid* isn't already open, please open it now. It's in the C:\Steps\Lesson04 folder.
II. Follow these steps.

4.1.1A: PRACTICE ZOOMING

1. We'll begin with the **Window** option of the *Zoom* command.
   Enter the *Zoom* command.

   Command: *z*

2. Select the corners of the window as shown at right.

   Specify corner of window, enter a scale factor (nX or nXP), or [All/ Center/Dynamic/Extents/Previous/ Scale/Window/Object] <real time>: *[Select first window corner.]*

   Specify opposite corner: *[Select other corner.]*

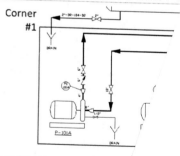

3. Now let's zoom back to where we started by using the **Previous** option on the cursor menu.

   Command: *[Enter]*

   Specify corner of window, enter a scale factor (nX or nXP), or

The command sequence for the *Zoom* command follows:

> **Command:** *zoom*
> **Specify corner of window, enter a scale factor (nX or nXP), or [All/Center/Dynamic/Extents/Previous/Scale/Window/Object] <real time>:** *[Select a corner of the window.]*
> **Specify opposite corner:** *[Select the opposite corner of the window.]*

Let's look at the various *Zoom* options. Notice that most of the options have an equivalent button in one of the several places indicated in the WTFI table.

Where to Find It:	
Command Line:	*Zoom*
Hotkey(s):	*z*
Ribbon (Tab/Panel):	View – Navigate – (Extents flyout) [zoom option]
Menu:	View – Zoom – [zoom option]
	Cursor – Zoom (Realtime)
Navigation Bar:	(Extents flyout) [Zoom option]
Toolbar:	Zoom – [zoom option]
	Standard – **Zoom** (**Realtime** or **Window** flyout)

Zoom Option	Alternate Button
Window prompts you to place a window around the area of the drawing you wish to see better. AutoCAD then zooms in to that area.	Zoom Window Button
Realtime appears to be the default, but it's accepted only by hitting the ENTER key. We'll look at realtime zooming shortly. If instead, you select an empty point on the screen, AutoCAD will assume you're placing a window around the objects you wish to view more closely. It'll then prompt you for the other corner. (Remember Implied Windowing – Section 2.3, p.45.)	Zoom Realtime Button
All displays the limits of the drawing, unless something has been drawn outside the limits. In that case, **All** will display the extents, or *all* the objects on the drawing.	Zoom All Button
Center prompts you for the desired center point of the display, and then adjusts the display so that the selected point is in the center of the screen.	Zoom Center Button
When you select the **Object** option of the *Zoom* command, AutoCAD will ask you to select an object(s). It'll then zoom in as close to that object(s) as possible while showing it in its entirety.	Zoom Object Button
One of the most forgotten zoom features, **Dynamic** provides you with the ability to place an adjustable box over that part of the drawing you want to display. When selected, AutoCAD temporarily replaces the screen with a view of the entire drawing. A view box shows what and where your current view area is. A selection box appears that you can manipulate with the mouse as you would the crosshairs. Move this box over the area you wish to view and hit the right mouse button ⌷ to confirm your selection. AutoCAD redisplays the drawing with the selected area shown (this'll be clearer when we do it in Exercise 4.1.1, p.81).	Zoom Dynamic Button
Extents brings you as close as possible to the drawing while showing *all* the objects in the drawing.	Zoom Extents Button

Zoom Option	Alternate Button
Scale can be a bit confusing. You don't have to type *s*, but you can. If you do, AutoCAD prompts you for the scale you want. Note that this isn't the drawing scale, but the size of the drawing in relation to the graphics area. Hence, a scale of *.5* will cause the drawing to occupy half the graphics area of your screen. If you simply type *.5X* at the *Zoom* prompt (rather than typing *s*), the drawing will appear half its current size. Notice that the *Zoom* prompt suggests the **X** or an **XP** procedure. (Ignore the **XP** for now. We'll cover that in detail in our discussion of layouts – aka. Paper Space – in Lesson 25, p.587.)	Zoom Scale Button
To simplify the **Scale** option, AutoCAD provides two additional tools (not listed at the command prompt) – **Zoom Out** and **Zoom In**. **Zoom In** zooms to a 2X scale and **Zoom Out** zooms to a .5X scale.	Zoom Out Button Zoom In Button
To zoom to a previous view, use **Zoom Previous**. You can also use this one to restore a previous view after using the *View* command (more on that in Section 4.2, p.88).	Zoom Previous Button

In addition to the command line, ribbon panel, toolbars, and menus, all the *Zoom* options can be found on the (right click) cursor and dynamic menus once the *Zoom* command has been entered.

Let's try some of these now.

Do This: 4.1.1A	**Practice Zooming**

I. If *pid* isn't already open, please open it now. It's in the C:\Steps\Lesson04 folder.
II. Follow these steps.

4.1.1A: PRACTICE ZOOMING

1. We'll begin with the **Window** option of the *Zoom* command.
Enter the *Zoom* command.
    Command: *z*

2. Select the corners of the window as shown at right.
    Specify corner of window, enter a scale factor (nX or nXP), or [All/ Center/Dynamic/Extents/Previous/ Scale/Window/Object] <real time>: *[Select first window corner.]*
    Specify opposite corner: *[Select other corner.]*

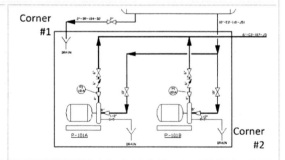

3. Now let's zoom back to where we started by using the **Previous** option on the cursor menu.
    Command: *[Enter]*
    Specify corner of window, enter a scale factor (nX or nXP), or

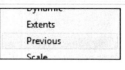

4.1.1A: PRACTICE ZOOMING

[All/Center/Dynamic/Extents/Previous/Scale/Window/
Object] <real time>: *p*

4. Zoom to a .5X scale using the **Zoom Out** option on the menu (**View** – [**Zoom flyout**] – **Out**).

5. Now zoom **All**.
   Command: *[Enter]*
   Specify corner of window, enter a scale factor (nX or nXP), or
   [All/Center/Dynamic/Extents/Previous/Scale/Window/Object] <real time>: *a*

Notice the location of and the spacing around the drawing on the screen. Let's compare that to the **Extents** option.

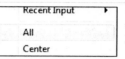

6. Zoom **Extents**.
   Command: *[Enter]*
   Specify corner of window, enter a scale factor (nX or nXP), or
   [All/Center/Dynamic/Extents/Previous/Scale/Window/Object] <real time>: *e*

Notice the location of and spacing around the drawing on the screen now. How does it compare with the results of the zoom **All** option?

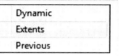

7. Now let's play with the **Dynamic** option. We'll select this one from the dynamic menu – pick (light) the **DYN** option on the status bar. Enter the **Zoom** command and select **Dynamic** from the dynamic input menu 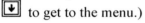. (Remember to use the down arrow to get to the menu.)
   Command: *z*

Notice the change in the display. There is now a box over the display with an "X" in the middle. This is the *location box* (right).

8. Pick with the left mouse button and notice how the box changes. Instead of an "X" in the middle, there's now an arrow pointing toward the right side of the box. This is the *sizing box* (right).

As you move the mouse back and forth, notice how the size of the sizing box changes. When it's just large enough to enclose the exchanger (Step 9), pick again with the left mouse button.

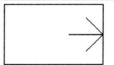

9. Now place the selection box over the exchanger and right click. Your display now looks like the figure at right.

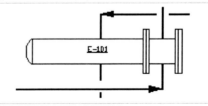

4.1.1A: PRACTICE ZOOMING

10. Now let's center our exchanger. Repeat the *Zoom* command and select the **Center** option [Center]. Alternately, you can pick **Center** on a menu or toolbar.
    Command: *z*
    Specify corner of window, enter a scale factor (nX or nXP), or [All/Center/Dynamic/Extents/Previous/Scale/Window/Object] <real time>: *c*

11. Select a point roughly in the center of the exchanger.
    Specify center point:

12. And hit ENTER to complete the command.
    Enter magnification or height <X.XXXX>: *[ENTER]*

13. Let's take a look at the **Object** option.
    Command: *z*
    Specify corner of window, enter a scale factor (nX or nXP), or [All/Center/Dynamic/Extents/Previous/Scale/Window/Object] <real time>: *o*

14. AutoCAD prompts you to **Select objects**. Select the head of the exchanger (select all of the vertical lines – I used a window). **Select objects:**	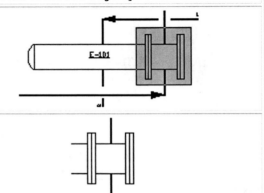
15. Complete the command. Your drawing looks like the figure at right. **Select objects:** *[ENTER]*	

As the commercial says, "Wait! Don't order yet!" I haven't shown you the easy way. I haven't shown you the *Realtime Zoom* feature.

Realtime is just a fancy way to say, "Do it while I watch." In other words, you can judge how far you want to zoom by watching the display change as you move your mouse.

You can find **Zoom Realtime** on the View – Zoom menu or by hitting ENTER at the initial **Zoom** prompt. One you've selected Realtime Zoom, you'll notice the crosshairs change to a cursor that looks like this. Now pick anywhere in the graphics area of the screen with the left mouse button. While holding the button down (dragging), move the cursor up and down. Notice how the display changes. When you're happy with the display, release the left mouse button, click the right button, and pick **Exit** on the cursor menu that appears. (Alternately, you can use the ENTER or ESC keys on your keyboard.)

> AutoCAD will also make use of the wheel located between the two buttons on a wheeled mouse. Rotating the wheel forward or backward works like a **Realtime Zoom**. If you don't have a wheeled mouse, the convenience will more than justify the expense!

> A trick to remember when using realtime zoom is to do a zoom **Center** first. You may have noticed that realtime zoom doesn't allow you to change position during a zoom. In fact, it maintains the same center point on the display, zooming in or out about that point.

4.1.2 Panning

Before we proceed, there is another aspect of display manipulation we should discuss *Pan*. The *Pan* command behaves as though you're putting your hand down on your paper and sliding it across the drawing table (or like panning a camera across your display).

The *Pan* command looks like this:

    Command: *pan*

    Pres ESC or ENTER to exit, or right click to display shortcut menu.

Where to Find It:	
Command Line:	*Pan*
Hotkey(s):	*p*
Ribbon (Tab/Panel):	View – Navigate – **Pan**
Menu:	View – Pan – [Pan option]
Navigation Bar:	**Pan**
Toolbar:	Standard – **Pan Realtime**

Pick anywhere in the drawing with the left mouse button and, holding it down, move the cursor across the screen. AutoCAD moves the page across the screen with your cursor.

The Pan menu (**View – Pan**) lists several options you won't find anywhere else. These include:

- **Point** prompts as follows:

    Specify base point or displacement: *[Pick a base point – the place where you "lay your hand on the paper."]*

    Specify second point: *[Pick the point where your "hand" will be after you've moved the paper.]*

- **Left**, **Right**, **Up**, and **Down** cause your view of the drawing to shift by 20% of the viewed area in the direction you select.

Pan, like *Zoom*, also has a realtime feature; but with *Pan*, it's the default. The **Pan Realtime** feature (called by the **Realtime** option on the **View – Pan** menu, and the *Pan* command) will also change the crosshairs to a cursor that looks like this. It works much like Realtime Zoom. Pick and drag with the left mouse button.

> Here again AutoCAD makes use of the wheeled mouse. Depressing the wheel between the mouse buttons and dragging the cursor across the screen appears as though you are in realtime pan. But no command line or toolbar entries are needed! (This, of course, is the preferred method!)

Exit realtime pan just as you did realtime zoom.

One of the really neat aspects of the realtime features is the cursor menu (Figure 4.002) that displays with a right click. I strongly suggest getting comfortable with this menu (especially if you don't have a wheeled mouse). Expertise with it will enable you to position yourself exactly where you want to be in a drawing, while saving the time you might otherwise spend keyboarding or ribbon panel clicking.

Before continuing, take a few minutes to play with these last few features in the *pid* drawing. Note the increased speed using the realtime features as opposed to the time spent in display manipulation using the methods we learned previously.

Figure 4.002

> AutoCAD's display commands – including the *Zoom* and *Pan* commands are *transparent*. That is, they may be used while at another command's prompt (while running another command). The buttons will work as they always do; however, to enter a transparent command at a prompt other than the command prompt, precede it with an apostrophe. Thus, entering the *Zoom* command while at the *Line* command's prompt will look like this:
>
>   **Specify next point or [Undo]:** *'z*

>>Specify corner of window, enter a scale factor (nX or nXP), or
[All/Center/Dynamic/Extents/Previous/Scale/Window/Object] <real time>:

The double bracket preceding the zoom prompt indicates that it's operating transparently. When you complete the transparent command, AutoCAD returns to the previous command prompt (in this case, the *Line* command's prompt).

4.1.3 The Navigation Wheel

AutoCAD designed its Navigation Wheels for use in a 3-dimensional world (and they are marvels there). But you'll still find them of some use in 2D, and as that is our ultimate goal, we'll go through a brief introduction here.

It's probably not fair to refer to the Navigation Wheel as a single feature – AutoCAD actually provides three *steering* wheels – the Full Navigation Wheel (the default – Figure 4.003), the View Object Wheel, and the Tour Building Wheel. Only one, however, will be of much use in our 2D world (phew!) – the View Object Wheel (Figure 4.004) – and we'll only use part of this one.

Where to Find It:	
Command Line:	**NavSWheel**
Menu:	View – **SteeringWheels**
	Cursor – Steering Wheels
Navigation bar:	(Full Navigation Wheel flyout) [option]

Use one of the methods indicated in the WTFI table to display the default wheel. Notice that it follows your cursor around the screen. You may find this a bit disconcerting at first, but I'll show you some tricks in a moment that'll keep it from becoming too bothersome.

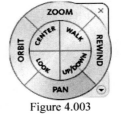
Figure 4.003

Figure 4.004

To change the wheel from the default to the View Object Wheel, right click anywhere on the displayed wheel, select **Basic Wheels** and then **View Object Wheel** as indicated in Figure 4.005. AutoCAD will display the wheel shown in Figure 4.004. Of course, you can also display the Tour Building Wheel (a useful 3D tool for architects) and return to the Full Navigation Wheel from one of the others using this menu as well. (We'll look at the Mini View Object Wheel shortly.)

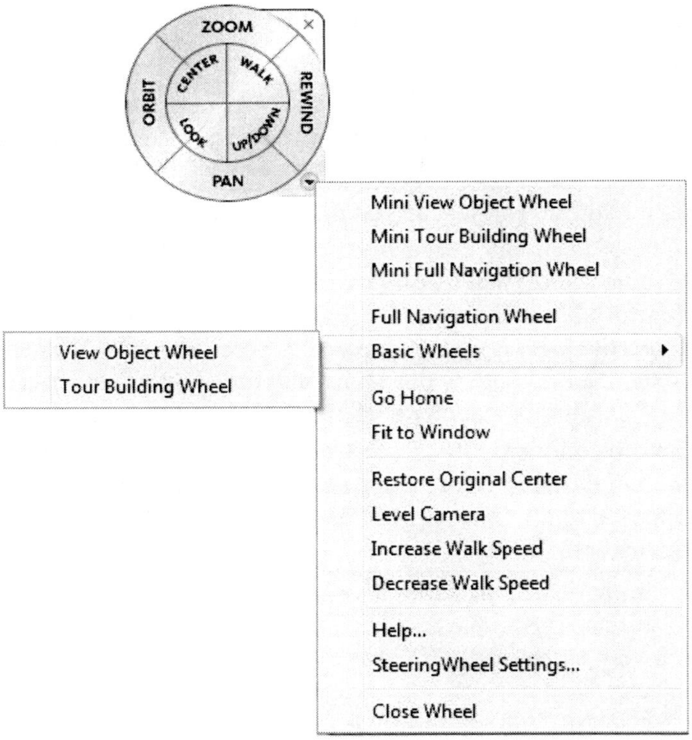

Let's look at the View Object Wheel now.

You have four obvious options:

- **Center** allows you to center a drawing point on your screen. Hold down the **Center** button of the wheel (left mouse button) and move your cursor to an object you wish to center. AutoCAD lets you know when it finds something by change to a green *pivot point* (Figure 4.006). When the location

Figure 4.005

85

satisfies you, release the button and AutoCAD will center that point.

- The **Zoom** button on the wheel manages to cover several of the command line options. Holding the button down allows you to zoom in and out about the pivot point – similar to the way you did with realtime zoom. But you can also click the button to zoom in (also about the pivot point) just as you do with the **Zoom In** option on the Zoom menu.

Figure 4.006

You won't find the **Rewind** option anywhere else – and it's a winner! When you pick (and hold) the **Rewind** button, AutoCAD presents a "reel" of frames from previous views (Figure 4.007). You can move the selection frame (within the corner brackets) back and forth until you've found the display you want. AutoCAD will rewind dynamically – that is, it'll rewind the actual display as you move the selection frame. (Very cool!)

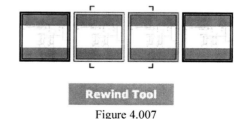
Figure 4.007

- We'll hold our discussion of the **Orbit** button for our 3D text.

> I know, you couldn't help yourself – you had to try the **Orbit** button and now your whole drawing is askew. Just use the *Plan* command you saw in Lesson 1, Section 1.1.7, p.15.

You'll find the wheel's last two items on the corners that extend to the right. The upper corner contains the usual "X" button that closes a window (or in this case, a wheel). The lower right corner contains an arrow that opens the same menu (Figure 4.005, p.85) you found with a right click on the wheel itself. (You can explore the other items on this menu if you wish – most of the remaining options are 3D tools. Just remember the *Plan* command in the previous insert when you're ready to continue.)

Let's look at one more tool before we try an exercise. Notice the **Mini View Object Wheel** option on the wheel's cursor menu (Figure 4.005). This one produces the tool shown in Figure 4.008. The mini wheel contains most of the same tools as its larger cousin, but doesn't overpower the display. Clockwise from the top, the quadrant buttons are: **Zoom, Rewind, Pan,** and **Orbit**. **Zoom** and **Rewind** work as they did on the larger wheel; **Pan** works just like realtime pan.

Figure 4.008

I prefer the mini wheel as it offers the **Pan** button and, as you can see, doesn't take up so much of the display.

Do This: 4.1.3A	**Wheeling Around the Drawing**

I. If *pid* isn't already open, please open it now. It's in the C:\Steps\Lesson04 folder.
II. Zoom extents.
III. Follow these steps.

4.1.3A: WHEELING AROUND THE DRAWING

1. Open the default wheel .

   Command: *navswheel*

   The wheel appears next to your cursor. Notice (Figure 4.003, p.85) that you have the **Zoom, Pan, Rewind,** and **Center** options available on the default wheel (the Full Navigation Wheel). You can use these here or open the View Object Wheel. We'll open the View Object Wheel for this exercise.

4.1.3A: WHEELING AROUND THE DRAWING

2. Right click on the wheel. Select **Basic Wheels – View Object Wheel** as shown. AutoCAD presents the View Object Wheel (Figure 4.004, p.85).	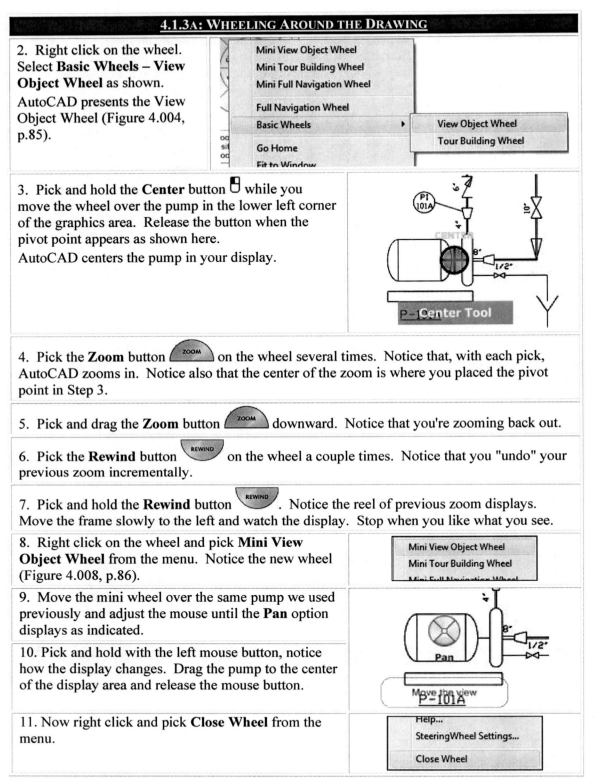
3. Pick and hold the **Center** button 🖱 while you move the wheel over the pump in the lower left corner of the graphics area. Release the button when the pivot point appears as shown here. AutoCAD centers the pump in your display.	
4. Pick the **Zoom** button on the wheel several times. Notice that, with each pick, AutoCAD zooms in. Notice also that the center of the zoom is where you placed the pivot point in Step 3.	
5. Pick and drag the **Zoom** button downward. Notice that you're zooming back out.	
6. Pick the **Rewind** button on the wheel a couple times. Notice that you "undo" your previous zoom incrementally.	
7. Pick and hold the **Rewind** button. Notice the reel of previous zoom displays. Move the frame slowly to the left and watch the display. Stop when you like what you see.	
8. Right click on the wheel and pick **Mini View Object Wheel** from the menu. Notice the new wheel (Figure 4.008, p.86).	
9. Move the mini wheel over the same pump we used previously and adjust the mouse until the **Pan** option displays as indicated.	
10. Pick and hold with the left mouse button, notice how the display changes. Drag the pump to the center of the display area and release the mouse button.	
11. Now right click and pick **Close Wheel** from the menu.	

We'll see these useful navigation wheels again in our 3D studies, but take some time between now and then to get comfortable with them. You'll be glad you did.

You can adjust the size and opacity of the wheels with these commands:
- **NavSWheelSizeBig** – size of the full size wheels (0 to 2; default is 1)
- **NavSWheelOpacityBig** – opacity of the full size wheels (25 – 90; default is 50)
- **NavSWheelSizeMini** – size of the mini wheel
- **NavSWheelOpacityMini** – opacity of the mini wheels

4.1.4 The Navigation Bar

Like the steering wheels, the Navigation Bar (Figure 4.009) services the 3D world as well as the 2D. We'll look at the 2D tools here (the first three) and the **Show Motion** button (the last one).

From the top, you'll find:

Where to Find It:	
Command Line:	*NavBar*
Menu:	View – Display – **Navigation Bar**
Ribbon:	View – Windows – User Interface – **Navigation Bar**

- A **Navigation Wheel** button which calls the steering wheels. In fact, this button has a flyout that you can use to call the specific steering wheel you wish to use.
- The **Pan** button begins the *Pan* command (Section 4.1.2, p.84).

Figure 4.009

- The next button – the **Zoom Extents** button – resides atop a flyout that calls a menu with the other zoom tools.
- The next two buttons – **Orbit** and **Object Mode** (aka. **3Dconnexion**) – refer to 3D tools. We'll spend some time with these in *3D AutoCAD 2011: One Step at a Time*.
- Finally, the **Show Motion** button begins the *NavSMotion* command. We'll discuss this one in some detail in Section 4.2.2, p.92.

You can remove the Navigation Bar by picking the "X" in the upper right corner or by entering *OFF* at the *Navbar* command prompt:

    Command: *navbar*
    Enter an option [ON/OFF] <ON>: *off*

(Turn it back on – or redisplay it – by entering *ON* at the same prompt.)

At the bottom right of the Navigation Bar, you'll find a small circle with a dash in it (see Figure 4.009). Pick this to display the options menu (Figure 4.010). Here you can control which tools the Navigation Bar displays by adding or removing the check next to the button name, and its **Docking position**.

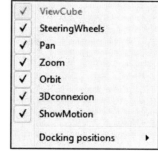

Figure 4.010

You can find most of the tools available on the Navigation Bar somewhere else – toolbars, ribbon panels, and so forth. But you may find it handy to leave it displayed anyway. Its transparency when not in use makes it unobtrusive while keeping its tools available for quick grabs.

4.2 Why Find It Twice? – The *View* Command

One of the things you'll discover after drafting on a computer for a while is that you must return frequently to certain areas of your drawing. A beginning CAD operator will use display tools like *Zoom* and *Pan* because these are simple, easily mastered, and they work. With so much to learn, who can blame them?

But AutoCAD has provided the *View* command to speed past these display controls. With the *View* command, you can create and store certain displays (views) and then restore these views at any time from any position in the drawing. Thus, using our *PID* drawing as an example, we can go directly from a display of the pumps to a display of the exchanger without the need for panning or zooming in and out.

4.2.1 Creating Views

Using *View* requires a bit of setup time at first; however, you can see that it will save quite a bit of time later when you might be panning and zooming all over the place.

View utilizes a dialog box (the View Manager – Figure 4.011) that's easier to use than it looks! Let's see.

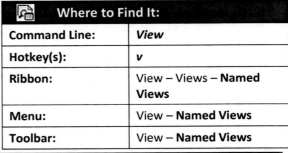

	Where to Find It:
Command Line:	*View*
Hotkey(s):	v
Ribbon:	View – Views – **Named Views**
Menu:	View – **Named Views**
Toolbar:	View – **Named Views**

- The **Views** frame to the left contains a list of four items:
  o **Current** contains only the current view.
  o **Model Views** contains a list of named views and camera views (more on cameras when we get into 3D space). Figure 4.0011 shows several model views available. The selection will have a plus next to it when you define some views. You'll be able to double click (or select and pick the **Set Current** or **Apply** button) to make a view current on your screen. More on that in a few minutes.

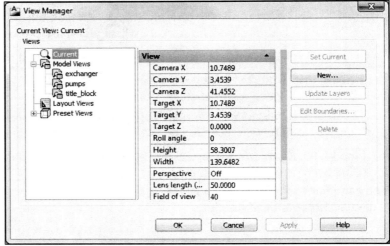

Figure 4.011

  o We'll look at **Layout Views** in depth when we get into layouts in Lessons 24, p.560, and 25, p.584.
  o **Preset Views** contains a list that will prove valuable in 3-dimensional space.
- The center section provides a list of properties of the selected view. These serve as useful references in 3-dimensional space. (You may get tired of me referencing the 3-dimensional tools in AutoCAD, but such is the nature of the beast. Eventually, you'll appreciate those tools as well; but I won't drop you into a three-dimensional ocean until you can walk across the beach.)
- The right side of the View Manager contains several buttons – here's where the work begins.
  o **Set Current** does just that – it sets the currently selected view as the one you'll see on your screen.
  o Use the **Update Layers** button to update the layer settings associated with the selected view to the current settings. (We'll discuss layers in Lesson 7, p.148.)
  o The **Edit Boundaries** button tells AutoCAD to present the drawing with the selected view highlighted. You can select opposite corners to redefine the view's boundaries until you confirm your new boundaries with the ENTER key.

89

- Finally, use the **Delete** button to delete a selected view from the drawing's database.
- The **New** button presents the workhorse of view procedures – the New View / Shot Properties dialog box (Figure 4.012).
    - Here you will name your view (**View name**) and assign it a category (either selected or entered in the **View category** list box).
    - You'll also select the **View type** from a list of choices. These will prove valuable when we create a 3D animated walk–through. But for now, stick with the default **Still View type**.
    - Next you'll find a couple tabs – **View Properties** and **Shot Properties**. To keep it simple, we'll concentrate on the **View Properties** tab and save the **Shot Properties** for our 3D animated walk–through discussion (3D text). (Actually, most of the settings on both tabs relate to 3D work, but we'll discuss the remaining features now.)
        ❖ In the **Boundary** frame, you can tell AutoCAD to use the **Current display**, or you can put a bullet next to the **Define window** option. (Alternately, you can pick the **Define view window** button 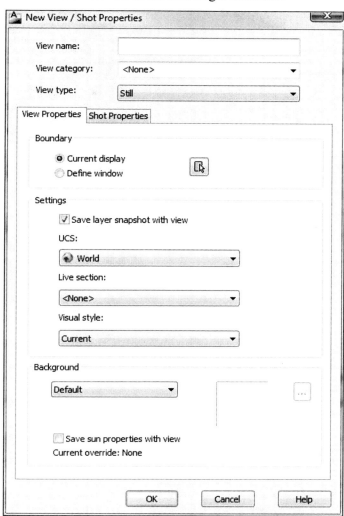 next to the **Define window** option.) When you select the **Define window** option, AutoCAD returns you to the graphics screen and prompts:

Figure 4.012

Specify first corner: Specify opposite corner: *[Place a window around the area to include.]*

Specify first corner (or press ENTER to accept): *[Accept your selection or redefine it.]*
        ❖ The **Settings** frame also contains several options.
            ◊ A check next to **Save layer snapshot with view** tells AutoCAD to save your current layer settings with the view. This can save a lot of work later, but you'll see that when we talk about layers in Lesson 7.
            ◊ The next three selection boxes all refer to 3-dimensional tools. Ignore them for now. (I know, it puts some things off, but it makes life easier for now.)
        ❖ The final frame in the New Views dialog box – **Background** – allows you to change the background color for the newly defined view. This can be useful for a slide show or to insert an image or graphic. Some people will insert the image of a sketch for

tracing into AutoCAD. Once you've traced the image, remove it. This can simplify the process of creating a drawing from a hand-drawn sketch.

Let's try creating and restoring a view using the View dialog box.

Do This: 4.2.1A	Creating a View and Setting It Current

I. If you're not currently in the *pid* file, please open it now. It's in the C:\Steps\Lesson04 folder.

II. **Zoom** All and then follow these steps.

4.2.1A: CREATING A VIEW AND SETTING IT CURRENT

1. Enter the *View* command.

   Command: *v*

2. Pick the **New** button [New...] on the View dialog box. The New View / Shot Properties dialog box will appear (Figure 4.012, p.90).

3. Follow these instructions:
   a. Type **exchanger** in the **View name** text box.
   b. AutoCAD will automatically put the view in the **Model Views** category, so you can ignore the **View category** control box for now.
   c. Leave a check next to **Save layer snapshot with view**.
   d. Place a bullet in the radio button beside the words **Define Window**.

   The dialog boxes will disappear and AutoCAD will prompt you to locate the window.

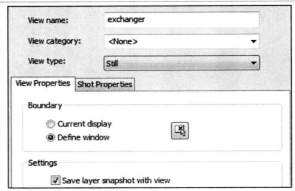

4. Place the window as shown in the following figure. The dialog box will return.

   Specify first corner:
   Specify opposite corner:
   Specify first corner (or press ENTER to accept): *[ENTER]*

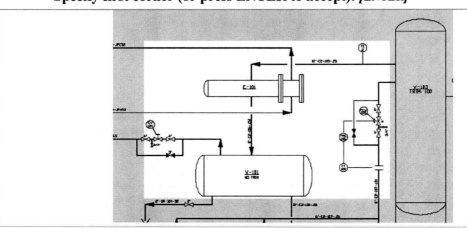

4.2.1A: CREATING A VIEW AND SETTING IT CURRENT

5. Pick the **OK** button [OK] to close the New View dialog box.
The View dialog box returns. You see the exchanger view in the list box.

6. Pick the **OK** button [OK] to close the View Manager.

7. Zoom extents.

8. Now we'll set the exchanger view as current. Reopen the View Manager and double click on **exchanger** in the list box.

9. Pick the **OK** button [OK] to close the View Manager. AutoCAD presents the exchanger view (following image).

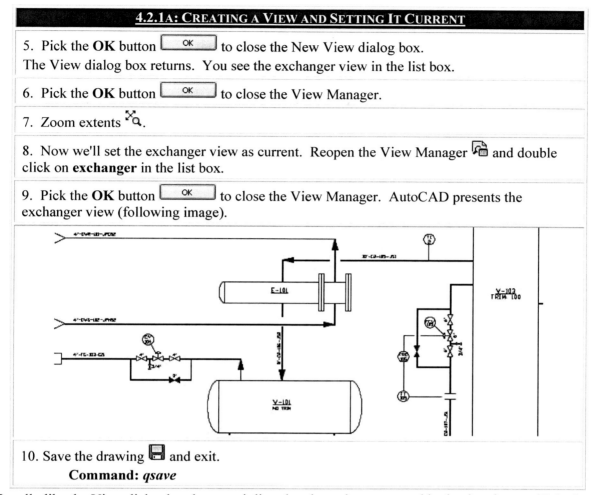

10. Save the drawing and exit.
   Command: *qsave*

I really like the View dialog box because it lists the views that are stored in the drawing, and I don't have to be concerned with remembering their names (or their *spellings!*).

(Note: For a nifty trick, see the **Extra Steps** section of this lesson.)

4.2.2 Showing Motion

I know I promised to show you how to create a 3D walk-through animation in the 3D text, but that doesn't mean you can't see how nice the "projector" works for navigating through the saved views in your drawing. Actually, AutoCAD's projector appears very similar to the Quick View Drawings tool you saw in Lesson 1, p.13.

Where to Find It:	
Command Line:	*NavSMotion*
Menu:	View – Show Motion
Navigation Bar:	Show Motion

It's simple enough to use – pick the **ShowMotion** button on the Navigation Bar and AutoCAD starts the show. That is, it scrolls neatly through each of the views in the drawing!
I'll explain as we go.

Do This: 4.2.2A	Showing Motion

I. Open the *pid2* file in the C:\Steps\Lesson04 folder.

II. **Zoom** All and then follow these steps.

4.2.2A: SHOWING MOTION

1. Begin the command. (Use one of the methods indicated in the WTFI table.)

   Command: *navsmotion*

   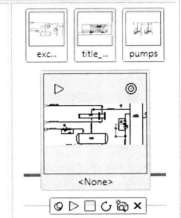

   AutoCAD presents the Show Motion panel (right). Take a moment to examine it.

   The large thumbnail image shows the last view created. The smaller thumbnails above it show the other views in the drawing.

   Below the thumbnails, AutoCAD displays a Show Motion toolbar with the following tools (from the left):

   - **Pin** – for keeping the Show Motion bar open while you work
   - **Play all** – for beginning an animation
   - **Stop** – to stop playing the animation
   - **Loop** – to have the animation repeat until told to stop
   - **New Shot** – opens the New View / Shot Properties dialog box.
   - **Exit** – to close the Show Motion bar

2. Move your cursor over the row of smaller thumbnails. Notice that the smaller and larger thumbnails switch sizes.

3. Pick on each of the larger thumbnails and watch the display. Notice that the selected view becomes current. (You don't need to know the name of the view, although it is shown at the bottom of the thumbnail.)

4. Pick the **Play all** button on the Show Motion toolbar and watch the show!

5. Pick the **Exit** button on the toolbar to close the Show Motion panel.

For now you can use the Show Motion bar to navigate views. It provides a visual interface lacking in the View Manager. But you can see by the cool transitions between views that the walk–though holds quite a bit of promise. This gives you something to look forward to in the 3D text!

4.3 "Simple" Text

Placing text into an AutoCAD drawing appears quite complicated. But once you've mastered the procedures, you'll see that it isn't as difficult as it seems.

The *Text* command allows multiple lines of text to be entered.

The command sequence looks deceptively simple.

> **Command:** *text*
> **Current text style: "Standard" Text height: 0.2000 Annotative: No**
> **Specify start point of text or [Justify/Style]:** *[Pick the starting point of the text.]*
> **Specify height <0.2000>:** *[Enter the desired text height (when using annotative text, this line will prompt for the* **paper height***).]*

	Where to Find It:
Command Line:	*Text*
Hotkey(s):	*dt*
Ribbon (Tab/Panel):	Annotate – Text – Multiline Text flyout – **Single Line**
Menu:	Draw – Text – **Single Line Text**
Toolbar:	Text – **Single Line Text**

Specify rotation angle of text <0>: *[Enter the desired rotation angle.]*
[Here AutoCAD stops prompting and places a text entry box on the screen. Enter your text and hit the ENTER key twice to complete the command.]

> Transparent commands – like *Zoom* or *Pan* – don't work while the *Text* prompt is showing. This way, all the keys are available while entering text.

Let's look at each line.

- The WTFI shows that **DT** (for *Dynamic* Text) is the hotkey for **Text**. Note that **T** is also a hotkey, but it calls the **MText** (MultiLine Text) command – not the **Text** command. (We'll discuss multiline text in Lesson 15, p.328.) Although the **Text** button shown provides a tooltip that reads *Single Line Text*, you can use dynamic text to create several lines of text at once. AutoCAD will treat each line, however, as a single object.
- The second line tells you what style you're using (more on style in Section 4.6, p.101), what the current **Text height** is, and what the **Annotative** setting is. (More on annotative text in a moment.)
- The third line provides access to more **Text** options than initially meet the eye.
  - The default option – **Specify start point** – simply directs you to identify the insertion point of the text. Do this by coordinate input or picking a point on the screen with the mouse.
  - You don't actually have to select the **Justify** option (that is, you don't have to type **J**) to justify your text. Typing J, however, will tell AutoCAD to present the various justification options shown here.
    Enter an option [Align/Fit/Center/Middle/Right/TL/TC/TR/ML/MC/MR/ BL/BC/BR]:

> You can use the *JustifyText* command to reset the justification once the text has been entered. This command won't, however, relocate the text around the insertion point.

Refer to Figure 4.013, to see where each of the options will place the text in relation to the insertion point (the "X"). By default, AutoCAD uses the **Bottom Left** option.

The **Align** and **Fit** justifications behave in much the same manner; however, **Align** will adjust the text height proportionally as it fits the text between the selected points, and **Fit** will maintain the user-defined height.

- The **Style** option enables you to choose among text styles defined within the drawing. More on style in Section 4.6, p.101.

Figure 4.013

- Set the **height** of the text on the next line. We'll talk about that in a moment.
- The next line asks for a **rotation angle**. AutoCAD wants to know if your text will be standard read-from-the-bottom-of-the-page (left-to-right) text or something else. Remember how AutoCAD measures angles! Read-from-the-right-side-of-the-page would be entered at 90°.

- Now AutoCAD starts a text box on your screen. Here you type the desired text, hitting ENTER for a return. When finished, hit ENTER again and the command prompt returns.

Sound simple enough? Take a moment and enter the text in Figure 4.013 just for practice.

How'd you do? Don't worry; we'll have an exercise shortly.

> Did you misspell something? If you did, you may have noticed that AutoCAD placed a dashed red line beneath it just to let you know. This is AutoCAD's automatic spell checker.
>
> Right click on the word above the dashed line. Notice the menu (right).
>
> - The first frame makes several suggestions for possible correct spellings. If the more obvious ones don't satisfy you, try the **More Suggestions** option!
> - Sometimes you may want to use a word unheard of in AutoCAD circles (it happens!), if so, you can either **Add** [it] **to** [the]**Dictionary** or **Ignore** it in the next frame.
> - The next two frames contain some obvious and common tools to assist your spelling edit.
> - **Editor Settings** provides these options:
>   - **Always Display as WYSIWYG** – to display the text at the correct size as you type it
>   - **Opaque Background** – to help you see the text better as you type
>   - **Check Spelling** – to run the automatic spell checker as you type
>   - **Check Spelling Settings** – provides a dialog box to help you set what you want to check
>   - **Dictionaries** – to allow you to change which dictionary AutoCAD will use when it checks your spelling (more on this in Lesson 15, Section X, p.341)
>   - **Text Highlight Color** – to highlight the text
> - We'll look at **Insert Field** when we discuss tables in Lesson 19, p.437.
> - We'll look at **Find and Replace** in Section 4.5 (p.100).
>
> The rest of the options are fairly self–explanatory.

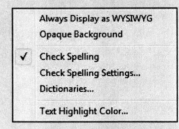

Well, so far so good. Now let's talk about text size and *annotation*.

Remember when we set up our first drawing? I told you that you'll do all your drawing in AutoCAD *full scale* and then scale the full size drawing down to fit on your plotter paper. Well, what happens to text when you downsize the drawing? Of course, it'll downsize, too. Shouldn't you, then, have to adjust your text size to allow for this downsizing? (You must've peaked at the olde way! – see the following insert.)

> As you'll find with each new release, AutoCAD often has an "olde" way of doing things and a "right" way (okay, a "new" way) of doing things. As this is a 2011 textbook, I'll present the current or right way here. But you'll need to know the olde way, too, because many of the files with which you'll work were created before this release became available. Go to the website – *www.uneedcad.com/Files* – and download the *TextSize-theOldeWay.pdf* file to see this section from our '07 text. It'll get you familiar with how we use to do things.

Not anymore! Now there's annotation!

Annotation is a ~~trick~~, ur, tool we use to have AutoCAD automatically size our text for us. We tell AutoCAD the size(s) of our eventual plot(s), set a system variable (**AnnoAutoScale**) to do the work, and then set the text to our desired plotted size. (If you've looked at the olde way, you'll really appreciate this!) And here's the kicker: you can tell AutoCAD to prepare the text for more than one plot size!

Look at these steps.

1. First, "tell AutoCAD the size of our eventual plot." Do this with the annotation tools on the right end of the status bar. Pick the down arrow next to the current scale (1:1) to view a menu of available scales. Simply select the one you want! AutoCAD will automatically size all annotative objects (including text) inserted into the drawing to this scale.

2. Be sure to set the **AnnoAutoScale** system variable to 4, or AutoCAD won't rescale anything. It's a simple procedure; it looks like this:

	Where to Find It:
Command Line:	*AnnoAutoScale*
Status Bar:	**Annotation AutoScale**

    Command: *annoautoscale*
    Enter new value for ANNOAUTOSCALE <-4>: *4*

    Other settings for the **AnnoAutoScale** system variable include:

SETTING	DESCRIPTION	SETTING	DESCRIPTION
-1	This turns **AnnoAutoScale** off, but sets it to 1 when turned on.	-3	This turns **AnnoAutoScale** off, but sets it to 3 when turned on.
1	This adds a newly set annotation scale to all annotative objects that support this scale and are not on layers that are frozen, locked, or off. (More on layers in Lesson 7, p.148.)	3	This adds a newly set annotation scale to all annotative objects that support this scale and are not on locked layers.
-2	This turns **AnnoAutoScale** off, but sets it to 2 when turned on.	-4	This turns **AnnoAutoScale** off, but sets it to 4 when turned on.
2	This adds a newly set annotation scale to all annotative objects that support this scale and are not on layers that are frozen or off.	4	This adds a newly set annotation scale to all annotative objects that support this scale.

    Note: For annotation to work, the text must be created using an annotative text style. More on text styles in Section 4.6, p.101.

3. Finally, at the **Specify height** prompt of the *Text* command, just enter the size at which you wish your text to plot.

No math … no charts … no sweat! Let's try some text.

Do This: 4.3A	Inserting Text

Note: You'll notice that many of the required files for this text are accompanied by another file with the same name except that it includes a "– v" after the title. If you're a Windows Vista or 7 user, use this file; it utilizes a much nicer font (Calibri) unavailable in XP or earlier Windows operating systems.

I. Open the *FlrPln-4* drawing in the C:\Steps\Lesson04 folder. The drawing looks like the figure at right. (I've set up this drawing to use an annotative text style.)

II. Follow these steps.

4.3A: INSERTING TEXT

1. We'll start with the title block, so restore the **TitleBlock** view.
    Command: *v*
   It looks like the figure at right.

2. Tell AutoCAD the eventual size of your plotted drawing. Our drawing has been set up to plot on a ¼"=1'-0" scale. Pick the down arrow next to the current **Annotation Scale** (on the status bar) and select the scale.

   | 1/8" = 1'-0" |
   | 3/16" = 1'-0" |
   | 1/4" = 1'-0" |

3. Set the system variable **AnnoAutoScale** to *4*.
    Command: *annoautoscale*
    Enter new value for ANNOAUTOSCALE <-4>: *4*
   This way, AutoCAD will automatically adjust the size of your text.

4. Now we can enter our text. Begin the *Text* command.
    Command: *dt*

5. Notice that AutoCAD tells you that you're using the **Times** (or **Calibri**) style, a height of **9"**, and that your text is **Annotative**. I set these things up for you, but you'll see how to set them up yourself in Section 4.6 (p.101).
    Current text style: "CALIBRI" Text height: 0'-9" Annotative: Yes
   Tell AutoCAD you want to center justify your text. Remember, you can use the **Justify** option or simply use a *C* at the prompt.
    Specify start point of text or [Justify/Style]: *c*

6. AutoCAD needs to know where to place the text. Pick a point toward the center of the top line in the title block.
    Specify center point of text:

7. Now tell AutoCAD the size at which you'd like the text to plot. We'll use quarter-inch text here.
    Specify paper height <0'-0 3/16">: *1/4*

8. Don't rotate the text.
    Specify rotation angle of text <0>: *[ENTER]*

9. Now enter the name of your favorite school or company. Remember to hit the ENTER key twice to complete the command.
   Your drawing looks something like the figure.

4.3A: INSERTING TEXT

10. Complete the title block. The next line uses 3/16" lettering, and the remaining lines use 1/8". Your drawing looks something like this.

11. Now zoom all.
    Command:

12. Create the rest of the text in the drawing. (The rest of the text is ¼".) When you've finished, your drawing will look like the figure at right.

13. Save the drawing, but don't exit.
    Command: *qsave*

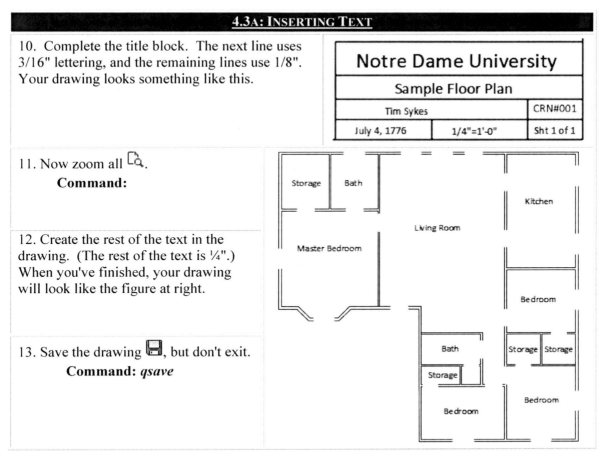

I really like easy; it's so, well, easy!

So, suppose you make a mistake in sizing your text. What're you gonna do? You could erase it and start over ... but that's too much work. Let's check out the *ScaleText* command. It looks like this:

	Where to Find It:
Command Line:	*ScaleText*
Ribbon (Tab/Panel):	Annotate – Text (subpanel) – **Scale**
Menu:	Modify – Object – Text – **Scale**
Toolbar:	Text – **Scale Text**

> **Command:** scaletext
> **Select objects:** *[Select the text object(s) you wish to resize.]*
> **Select objects:** *[Confirm the selection.]*
> **Enter a base point option for scaling [Existing/Left/Center/Middle/Right/TL/TC/TR/ML/MC/MR/BL/BC/BR] <Existing>:** *[Generally, you'll want to use the existing insertion point for your text, but AutoCAD gives you the chance to change it if you wish.]*
> **Specify new model height or [Paper height/Match object/Scale factor] <1/8">:** *[Give AutoCAD the new height for your text or select another option.]*

This works in a fairly straightforward manner, but look at the options in that last line.

- You'll need to specify **Paper height** if you want to resize annotative text. Otherwise, AutoCAD will just ignore you. When you do, AutoCAD will prompt:
  Specify new paper height <0">:
- Use the **Match object** option when you don't know the size you want the text to be, but you have some existing text that already meets the requirements. (The **Match object** option works only for like text – annotative or non-annotative.)

- **Scale factor** matters for non-annotative objects. Go through the *TextSize-theOldeWay.pdf* file mentioned earlier for details on this approach.

Let's resize some text.

Do This: 4.3B	Resizing Text

I. Be sure you're still in the *FlrPln-4* drawing. If not, please open it now. It's in the C:\Steps\Lesson04 folder.

II. Follow these steps.

4.3B: RESIZING TEXT

1. Zoom back in on the title block.
   Command: *z*

2. I think I deserve more credit for my work, so I want to resize my name to the same size as the drawing title. Enter the *ScaleText* command .
   Command: *scaletext*

3. Select the name and accept the existing base point (insertion point).
   Select objects:
   Select objects: *[ENTER]*
   Enter a base point option for scaling [Existing/Left/Center/Middle/Right/ TL/TC/TR/ML/MC/MR/BL/BC/BR] <Existing>: *[ENTER]*

4. Tell AutoCAD you want to enter a new **Paper height** Paper height for the text.
   Specify new model height or [Paper height/Match object/Scale factor] <0">: *p*

5. And change the text to 3/16".
   Specify new paper height <0">: *3/16*

See the difference? (I wonder if that'll make it past the checkers.)

You can use word processing standards to underscore (CTRL+U) or overscore (CTRL+O) a line before and after the text. You can also add some useful symbols using ASCII coding - *%%c* provides a diameter symbol, *%%p* provides a plus/minus symbol, and *%%d* places a degrees symbol in your text. (Note that you can also use most of the ALT+code tricks you use in Microsoft Word!) If you wish to use much beyond this in the way of symbols or text formatting, I suggest you opt for the *MText* command rather than dynamic text. It has nearly full word processor capabilities. (We'll discuss MText in Lesson 15, p.328.)

> One of the ways CAD operators used to save regeneration time was to tell AutoCAD not to regenerate the text in a drawing. As a great portion of the drawing is text, this saved time with larger drawing files. The command they used was **QText** and it looked like this:
>
> **Command:** *qtext*
> **Enter mode [ON/OFF] <OFF>:** *on*
>
> The text in the drawing was replaced with a rectangular locator to help the operator avoid placing geometry on top of the text.
>
> With the advent of faster computers, the **QText** command really isn't necessary anymore. But there'll always be an older operator who likes to **QText** his drawing before passing it on to a freshman operator as a joke. So if you get a drawing with **QText** activated, simply enter the command and turn it off. (Remember to *regen* the drawing afterward.)

4.4 Editing Text

So now you see that creating text isn't that difficult. But suppose you make a mistake or just want to change something. Let's look at AutoCAD's text editor – the ***DDEdit*** command. The command sequence is quite simple:

Command: *ddedit*

Select an annotation object or [Undo]: *[Select the text to edit; despite the prompt, you can edit non-annotative text, as well.]*

Where to Find It:	
Command Line:	*DDEdit*
Hotkey(s):	*ed*
Menu:	Modify – Object – Text – **Edit**
Cursor Menu:	[select text] – **Edit**
Toolbar:	Text – **Edit Text**

AutoCAD highlights the text and allows you to edit it.

Want an easier method? Just double click on the text to edit. AutoCAD automatically opens the editor!

Do This: 4.4A — Editing Text

I. Be sure you're still in the *FlrPln-4* drawing in the C:\Steps\Lesson04 folder. If not, open it now.
II. Be sure the **Quick Properties** icon on the status bar is off.
III. Zoom in on the title block. We'll change the CRN number.
IV. Follow these steps.

4.4A: EDITING TEXT

1. Double click on the line to be edited (*CRN #001*). AutoCAD highlights the text (right) so you can edit it.

    Command: *[Double click on the text without entering a command.]*

2. Type ***CRN #002***, and then hit ENTER twice.

3. Save the drawing.

    Command: *qsave*

That's all there is to editing text.

4.5 Finding and Replacing Text

Where to Find It:	
Command Line:	*Find*
Ribbon (Tab/Panel):	Annotate – Text – **Find text**
Menu:	Edit – **Find**
	Cursor – **Find**
Toolbar:	Text – **Find**

The *Find* command offers the ability to search for (and replace) text in a document. To make it even easier, AutoCAD provides a simple dialog box (Figure 4.014, p.101).

In our example, we're searching for *Notre Dame*. We've entered the text in the **Find what** text box and picked the **Find** button to begin our search.

You can also put the text for which you're searching in the **Find text** box on the ribbon's **Text** panel (**Annotate** tab). But when you pick the **Text Find** button, AutoCAD will open the Find and Replace dialog box.

AutoCAD found one instance of the text and displayed it in the **List results** list box. Now we have some choices of what to do.

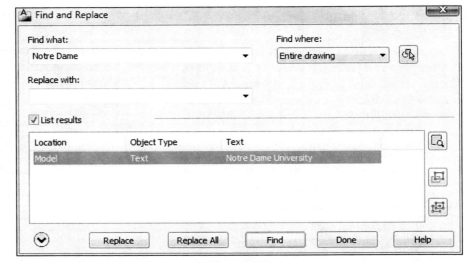

- We can **Replace with**, that is, we can replace the located text with whatever text we place in the **Replace with** box. Once we've added the new text, we'll pick the **Replace** (to

Figure 4.014

replace a single instance) or the **Replace All** button (to replace all instances of the located text).
- We can have AutoCAD zoom to the located text in one of several ways.
   o We can double click on the text in the **List results** box.
   o We can pick the **Zoom to highlighted results** button.
   o We can pick either the **Create Selection Set (Highlighted)** or the **Create Selection Set All** button. These will, respectively, select objects that contain the selected text or select objects that contain all the selected text. AutoCAD will then zoom into the selection and close the dialog box.

Note that we can also search part or all of a drawing using the **Find where** drop-down box.

We're only seeing the tip of the iceberg here; we'll become more familiar with the rest of the Find and Replace options in Lesson 15, p.344.

4.6 Adding Flavor to Text with *Style*

Where to Find It:	
Command Line:	*Style*
Hotkey(s):	*st*
Ribbon (Tab/Panel):	Annotate – Text –Text Style control – **Manage Text Styles**
	Home – Annotation (subpanel) –Text Style control – **Manage Text Styles**
Menu:	Format – **Text Style**
Toolbar:	Text – **Text Style**

Although AutoCAD's default style is called *Standard*, in our next exercise, we'll create text using a style called *Times*. We'll create this style and make it current. We'll use Windows' *Times New Roman* True Type Font instead of AutoCAD's *TXT* font because it shows up better in print.

What exactly is the difference between style *and* font?

Simply put, *font* refers to the physical shape of a letter or number. *Style* refers to all of the characteristics of a letter or number (including font, size, slant, boldness, etc.).

AutoCAD has access to the True Type Fonts used by all the other programs on your Windows computer. So when creating a style, your drawing can be consistent with the other documents in your project.

The Text Style dialog box (Figure 4.015) looks intimidating, but don't let that throw you. Most of the buttons are self-explanatory – **New** to create a new style, **Set Current** to set a selected style current.

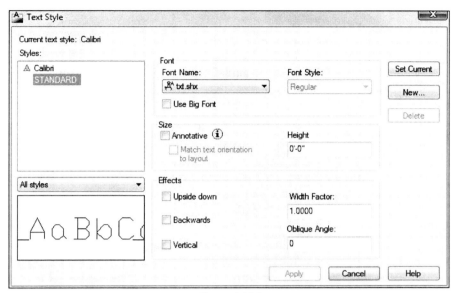

Figure 4.015

Let's create a few styles to see how it's done.

Do This: 4.6A	More Text Editing

I. Start a new drawing from scratch.
II. Set the grid snap to ½". (Be sure to toggle it **On**.)
III. Zoom all.
IV. Follow these steps.

4.6A: MORE TEXT EDITING

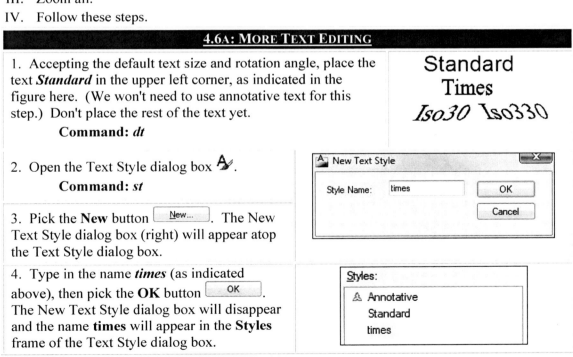

1. Accepting the default text size and rotation angle, place the text *Standard* in the upper left corner, as indicated in the figure here. (We won't need to use annotative text for this step.) Don't place the rest of the text yet.
   Command: dt

2. Open the Text Style dialog box.
   Command: st

3. Pick the **New** button. The New Text Style dialog box (right) will appear atop the Text Style dialog box.

4. Type in the name *times* (as indicated above), then pick the **OK** button. The New Text Style dialog box will disappear and the name **times** will appear in the **Styles** frame of the Text Style dialog box.

4.6A: MORE TEXT EDITING

5. Now let's define the style. Pick the down arrow in the **Font Name** text box (in the **Font** frame). Scroll as necessary to find **Times New Roman**. Select it as shown.	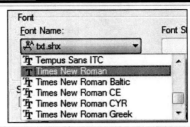
6. Notice that the word **Regular** appears in the **Font Style** text box. Pick the down arrow here to see what your other choices are (as shown), but leave it set to **Regular** for now.	

7. You can set a height for your text in the **Height** text box (**Size** frame). If you do, the text height will be what you have set whenever you use this style, *and AutoCAD won't prompt you for the height when you enter text*. I usually leave the **Height** set to *0* for the flexibility it allows me when I enter the text.

You should also put a check next to **Annotative** here ☑Annotative for your text to be annotative (and behave as it did in Section 4.3, p.93!). Ignore the **Match text orientation to layout** option until we discuss layouts in Lessons 24, p.560, and 25, p.584.

Notice the annotative symbol next to times in the **Styles** list box △ times.

8. You can set additional physical characteristics for the style in the **Effects** frame of the Text Style dialog box. You see the options **Upside down**, **Backwards**, and **Vertical** listed on the left. I've never found a reason for entering text upside down or backward, but the options are available if you find a reason.	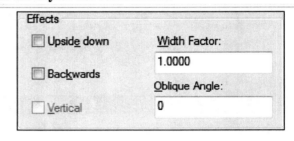
9. The **Width Factor** determines the width of each character in relation to its height (greater than one creates a wider character; less than one creates a narrower character). Most people leave this at *1*, but let's set it to *7/8*. I prefer the narrower characters because it enables me to place more text in a smaller area. The difference is almost imperceptible when plotted.	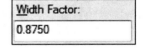

10. You can set the slant of the characters in the **Oblique Angle** text box. *0* is straight text. You'll use the obliquing angle when you set up isometric text on the next page. We'll leave the **times** style at *0*.

11. The **Preview** panel shows you what your settings will look like.	

You'll see a control box above the preview pane. Use this box to filter the list of styles in the **Styles** box. Your choices include **All styles** and **Styles in use**.

4.6A: MORE TEXT EDITING

12. Next pick the **Apply** button [Apply] ...
13. ... and then the **Close** button [Close].

V. Place the word *Times* just below the word **Standard**. (Zoom in as necessary for a better view.) Can you see the difference (refer to the figure in Step 1)?

VI. Now create two more styles with the following settings.

Style Name	Font	Width	Oblique Angle
Iso30	Times New Roman	1	30
Iso330	Times New Roman	1	330

VII. Type the names of these styles below the names of the others. (Hint: Type *S* at the first text prompt to set the style.) Your drawing will look like the figure in Step 1.

VIII. Now let's use the *isotext* in an isometric setting. Follow these steps.

4.6A: MORE TEXT EDITING

14. Set your snap to **isometric**.
    Command: *sn*
    Specify snap spacing or [ON/OFF/Aspect/Rotate/Style/Type] <0.5000>: *s*
    Enter snap grid style [Standard/Isometric] <S>: *i*
    Specify vertical spacing <0.5000>: *[ENTER]*

15. Set the current text style to *Iso30*.
    Command: *dt*
    Current text style: "Standard" Text height: 0.2000 Annotative: Yes
    Specify start point of text or [Justify/Style]: *s*
    Enter style name or [?] <Standard>: *iso30*

16. Pick a point near the other text and enter *30°* as the rotation angle.
    Current text style: "Iso30" Text height: 0.2000 Annotative: No
    Specify start point of text or [Justify/Style]: *[Pick a start point.]*
    Specify height <0.2000>: *[ENTER]*
    Specify rotation angle of text <0>: *30*

17. Enter in the name of the style (*Iso30*).

18. Now repeat the preceding sequence using the **Iso330** style you created and a rotation angle of *330°*. Your text looks like the figure at right.

19. Quit the drawing [X] without saving it.

4.7 Extra Steps

Perhaps the greatest selling point in AutoCAD's favor is its inclusion of AutoLISP as a customizing agent. Certainly, it's too early in our study of AutoCAD to be concerned with customizing it – or at least too early to learn the AutoLISP programming language (although I highly recommend it later). But one thing not included in most textbooks is how to use AutoLISP.

You'll get a chance to create macros using the Action Recorder in Lesson 12. Mastery of this cool tool may relieve you of the chore of learning a programming language like AutoLISP. It'll still help, however, to be able to load the plethora of routines already available!

You can use it to your advantage quite easily, and as there are zillions of lisp routines available in most CAD environments and the Internet (just ask the guy next to you), you should at least be comfortable with loading the programs.

	Where to Find It:
Command Line:	*Appload*
Hotkey(s):	*ap*
Ribbon (Tab/Panel):	Manage – Applications – **Load Application**
Menu:	Tools – **Load Application**

The command sequence is very simple:

   Command: *(load "C:/Steps/lesson04/views")*

That's it! Note that the parentheses are required, as are the quotation marks around the path and file name. Note also that the slashes (normally backslashes) are *front*slashes. (AutoLISP reads backslashes as pauses in its routines.)

But in these days of dialog boxes, there's another way. Use the *Appload* command (see the WTFI table) to open the Load/Unload Applications dialog box (Figure 4.016). Here's how to use it.

1. Use the upper half of the dialog box as you would a typical Windows Open File dialog box.
2. Locate the file you wish to open. In Figure 4.016, I've located the *Views.lsp* file in the C:\Steps\Lesson04 folder. The name of the selected file will appear in the **File name** text box.
3. Pick the **Load** button. The file appears in the **Loaded Applications** list box in the lower half of the dialog box.
4. Pick the **Close** button to finish the procedure.

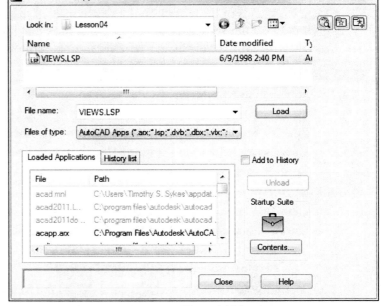

Figure 4.016

You must load each file into the drawing session every time you restart AutoCAD – unless you use the **Startup Suite** to automatically load selected files when AutoCAD begins a new session.

Each file contains one or more *programs* or *routines* intended to shorten or ease your drawing time. Accessing a routine is as easy as typing a command (the command is identified – or programmed – into the routine). Thus, typing *VS* after loading the *Views* routine will enable you to store a view without the dialog box. Typing *VR* will restore a view. Try it! You'll like it!

AutoCAD also provides access to another extremely valuable programming tool – Visual Basic. But an understanding of AutoLISP is necessary to effectively use a VB program. Once you are comfortable with the use of AutoCAD, I highly recommend a study of AutoCAD Customization (including Lisp and VB).

4.8 What Have We Learned?

Items covered in this lesson include:

- *Display commands*
  - *Zoom*
  - *Pan*
  - *NavSWheel*
  - *NavSMotion*
  - *NavBar*
  - *View*
  - *NavSWheelSizeBig*
  - *NavSWheelSizeMini*
  - *NavSWheelOpacityBig*
  - *NavSWheelOpacityMini*
- *Text and Annotative Text Commands*
  - *Text*
  - *DDEdit*
  - *QText*
  - *Style*
  - *ScaleText*
  - *Find*
- *Loading AutoLISP Applications*

Well, this was quite a lesson! We covered AutoCAD's display commands *Zoom*, *Pan*, *NavSWheel*, and *View*. Then we looked at the basic text command *Text*. After that, we covered the *Style* command. We'll see that one again when we cover *MText* in Lesson 15, p.328. Lastly, we took a quick peek at AutoLISP and how to load and use a Lisp routine. Of all the things covered thus far, mastering these few short paragraphs will go further than any other in convincing an employer that you've mastered AutoCAD.

We covered quite a bit of material; but believe it or not, you'll soon be using these tools as second nature.

Don'tcha just love all the texting tools AutoCAD give you?!
(Just wait until Lesson 15!)

4.9 Exercises

1. Start a new drawing with the following parameters:
    1.1. Grid: 1
    1.2. Snap: ½
    1.3. Lower left limits: 0,0
    1.4. Upper right limits: 36,24
    1.5. Text Heights: 3/8", 3/16", 1/4" & 1/8"
    1.6. Create the organizational chart in the figure on the next page. Feel free to substitute names for those used.

(HINT: Most of my students spend an hour or so drawing a number of rectangles only to discover that the text won't fit; then they must redraw them after entering the text. Enter the text *first*.)

    1.7. Save the drawing as: *MyOrg* in the C:\Steps\Lesson04 folder.

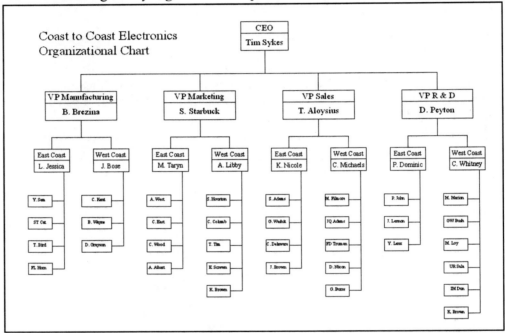

4.10 For Web-Based Review Questions and Additional Exercises, visit: http://foragerpub.com/AcadFiles/2011/2011.htm

Lesson 5

Following this lesson, you will:

✓ *Know how to draw:*
- *Ellipses*
- *Arcs*
- *Polygons*

✓ *Have mastered most of AutoCAD's basic 2-dimensional drawing commands!*

Geometric Shapes (Other Than Lines, Rectangles, and Circles!)

By now, you must have tired of drawing lines, rectangles, and circles. After all, back at the beginning of the second lesson, we discussed lines and circles as the foundation *for drawing geometric shapes. How about building on that foundation?!*

In this lesson, we'll look at drawing arcs and ellipses. We'll also expand our multisided geometry from simple rectangles to include those -gons, collectively known as polygons. Then we'll try our first Putting It All Together *exercise.*

Let's proceed.

5.1 Ellipses and Isometric Circles

I've often been amazed – and frequently aggravated – by the number of incomplete or oddball circles required in drafting. Back in my board days (bored days?), arcs were seldom a problem. I just used my circle template and drew as much as I needed. Ellipses, however, required the purchase of specific templates – often several! There were, of course, templates that tried to provide almost every dimensional ellipse the draftsman might need – from 25° to 80°, and from ¼" to 6". But the ellipse I needed would inevitably fall into an oddball degree or size.

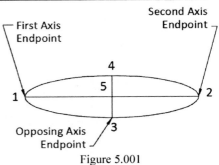

Figure 5.001

Ellipses are one of those things that AutoCAD makes quite a bit easier than plastic templates or clumsy compass attempts to create oddball shapes. Let's look at the command sequence (refer to Figure 5.001).

> **Command:** *ellipse*
> **Specify axis endpoint of ellipse or [Arc/Center]:** *[Select the first axis endpoint.]*
> **Specify other endpoint of axis:** *[Select the opposite axis endpoint.]*
> **Specify distance to other axis or [Rotation]:** *[Select the opposing axis endpoint.]*

Where to Find It:	
Command Line:	*Ellipse*
Hotkey(s):	*el*
Ribbon (Tab/Panel):	Home – Draw – (Ellipse flyout) **[option]**
Menu:	Draw – Ellipse – (Ellipse flyout) **[option]**
Toolbar:	Draw – **Ellipse**
Tool Palette:	Draw – **Ellipse**

The basic ellipse is really very easy to draw regardless of rotation or dimension. But as you can see, we have some options to consider.

- **Specify axis endpoint** is the default. It allows you to draw the ellipse by selecting three axis endpoints, or two axis endpoints and a rotation angle.
- The **Arc** option allows you to draw partial ellipses.
- The **Center** option allows you to create an ellipse using a center point and two axis endpoints (rather than three axis endpoints).

Let's look at each of these options.

> All of the options are also available on the dynamic input or cursor menu once the ***Ellipse*** command has been entered. The menus also contain button selections that automatically enter options for an ellipse – centered ellipse ⊕, the axis-end ellipse (the default), and an arc ellipse ⌒.

Do This: 5.1A	**Drawing Ellipses**

I. Open the *cir-ell* drawing. It's in the C:\Steps\Lesson05 folder and looks like the figure at right.
II. Restore the **ELLIPSE** view.
III. Set the Running OSNAP to **Endpoint**. Clear all other settings.
IV. Follow these steps.

5.1A: DRAWING ELLIPSES

1. Enter the *Ellipse* command . (If you use the ribbon button, be sure to select the **Axis, End** button in the **Ellipse** flyout.)
   Command: *el*

2. Select the endpoint at point 1 of the top set of lines. (The lines aren't necessary; we use them here as guides only.)
   Specify axis endpoint of ellipse or [Arc/Center]:

3. Select the endpoint at point 2.
   Specify other endpoint of axis:

4. Select the endpoint at either point 3 or point 4. Your ellipse looks like the one shown here.
   Specify distance to other axis or [Rotation]:

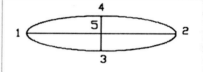

5. Repeat the *Ellipse* command.
   Command: *[ENTER]*

6. Type *c* (or select **Center** on the dynamic or cursor menu) to access the **Center** option.
   Specify axis endpoint of ellipse or [Arc/Center]: *c*

7. When AutoCAD prompts you for the center of the ellipse, select the OSNAP button for intersection, and then select point 5 of the second set of lines.
   Specify center of ellipse: _int of

8. Select the endpoint at either point 1 or point 2.
   Specify endpoint of axis:

9. Let's try the **Rotation** option. Type *R* or select **Rotation** on the menu. (**Rotation** refers to the angle at which you see the circle.)
   Specify distance to other axis or [Rotation]: *r*

10. Type in a rotation angle of 75°. Your ellipse will look like the one in Step 4.
    Specify rotation around major axis: *75*

11. Repeat the *Ellipse* command or pick the **Ellipse Arc** option on the toolbar, ribbon, or one of the menus. (If you use the **Ellipse Arc** button, skip Step 12.)
    Command: *[ENTER]*

5.1A: DRAWING ELLIPSES

12. Type *a* (or select **Arc** [Arc]) to access the **Arc** option.
    Specify axis endpoint of ellipse or [Arc/Center]: *a*

13. Notice that you again have the option to select either an axis endpoint or the center of the ellipse. Let's use the default – **axis endpoint** – and select the endpoint at point 1 on the bottom set of lines.
    Specify axis endpoint of elliptical arc or [Center]:

14. Select the endpoint at point 2.
    Specify other endpoint of axis:

15. Here again your options are repeated. Select point 3.
    Specify distance to other axis or [Rotation]:

16. Now you have some new options. The **Parameter** option uses a complicated formula to determine where the arc will begin and end. We'll use the default – **start angle** to give us more control of our work. Pick the point where you'd like the elliptical arc to begin. (Pick point 1.)
    Specify start angle or [Parameter]:

17. We started our arc at angle 0. An included angle is an angle measured counterclockwise from that point. We can type *I* followed by the angle we wish to use to create our arc. Alternately, we can use the **end angle** approach and simply type the angle we want for the other end of the ellipse. Let's use the **end angle** default and type in *180°*.
    Specify end angle or [Parameter/Included angle]: *180*
    Your ellipse looks like the figure at right.

    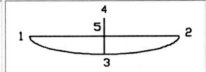

18. Save the drawing 💾, but don't exit.
    Command: *qsave*

By default, AutoCAD draws true ellipses. This enables you to find the center with an OSNAP.

The system variable **PEllipse** allows you to change the way AutoCAD draws ellipses. When set to the default *0*, ellipses work the way we've seen. But set it to *1* and ellipses are drawn as polylines (just as rectangles are drawn). Both ellipses look the same, but you can give the latter width using the *PEdit* command. You can't, however, easily find the center of the ellipse.

We'll learn about polylines in Lesson 12, p.272.

You can now see that drawing ellipses isn't difficult – although mastering the various approaches may take some time.

The ellipses we've drawn thus far have all existed in a true 2-dimensional plane. But you can also use ellipses to draw *isometric circles*. The procedure is simple but requires that you be in *isometric mode*. Otherwise, the *Ellipse* command won't provide the **Isocircle** option.

Let's look at this.

Do This: 5.1B	**Drawing Isometric Circles**

I. Be sure you're still in the *cir-ell* drawing. If not, open it now. It's in the C:\Steps\Lesson05 folder.

II. Restore the **ISO-ELLIPSE** view.
III. Follow these steps.

5.1B: DRAWING ISOMETRIC CIRCLES

1. First, set your snap to the isometric *style*. (Accept the defaults.) Your crosshairs and the grid will change.
   Command: *sn*

2. Begin the *Ellipse* command.
   Command: *el*

3. Select the **Isocircle** option.
   Specify axis endpoint of ellipse or [Arc/Center/Isocircle]: *i*

4. Use the **Node** OSNAP, and then select the node in the center of the isometric rectangle (the node looks like an X).
   Specify center of isocircle: _nod of

5. See how the ellipse drags with the cursor in the current isometric plane? Toggle the plane using the **F5** key to see how the ellipse changes. Stop toggling when the crosshairs return to the 90°/30° position.

6. Notice that you can specify a radius or diameter. Let's use the default (**radius**) and enter *1*. Your drawing looks like the figure shown.
   Specify radius of isocircle or [Diameter]: *1*

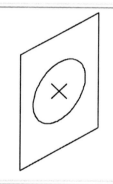

7. Save and close the drawing.
   Command: *qsave*

We've looked at circles and "squished" circles (ellipses), ellipse arcs, and isometric circles. Let's move on then to partial circles – arcs.

5.2 Arcs: The Hard Way!

How's that for a section title to make you want to skip past this part of the lesson?! But wait! "There's gold in them thar pages!"

Although it'll frequently be quicker and easier to create a circle and trim away the part you don't want, there are times when there is simply no substitute for the *Arc* command. (Besides, we haven't learned the *Trim* command yet!)

Drawing an arc isn't difficult – providing you know which of the eleven available procedures to use. The command sequence is

Where to Find It:	
Command Line:	*Arc*
Hotkey(s):	*a*
Ribbon (Tab/Panel):	Home – Draw – (Arc flyout) **[option]**
Menu:	Draw – Arc – (Arc flyout) **[option]**
Toolbar:	Draw – **Arc**
Tool Palette:	Draw – **Arc**

Command: *arc*
   Specify start point of arc or [Center]: *[Select or identify the starting point.]*

Specify second point of arc or [Center/End]: *[Select or identify a point on the arc.]*
Specify end point of arc: *[Select or identify the endpoint.]*

The best way to see each of the options is through an exercise. So fire up the computer and full speed ahead! (The next exercise uses a slightly different format, but it's much easier to follow.)

> The various options of the *Arc* command have their own ribbon and menu picks.
> You can also find the various options in the cursor or dynamic menu once the *Arc* command has been entered.

Do This: 5.2A	Arcs, Arcs, Arcs

I. Open the *arcs* drawing in the C:\Steps\Lesson05 folder.
II. Set the Running OSNAP to **Node** and clear any other settings. Turn off the grid.
III. You'll notice twelve squares in the drawing. Use these as a guide. First, restore the **TOP** view and regenerate the drawing. When you've finished with the first six squares, restore the **BOTTOM** view and continue.
IV. Begin in each square with the *Arc* command.
V. Follow the command/response sequences given in the chart below. You can use keyboard entry, cursor menu, or dynamic entry when making your responses. Alternately, you can use the ribbon button or menu option indicated in the first column.

SQUARE/ PROCEDURE	PROMPT/RESPONSE	RESULTS
1 Three Point	**Command:** *a* **Specify start point of arc or [Center]:** *[Pick point a.]* **Specify second point of arc or [Center/End]:** *[Pick point b.]* **Specify end point of arc:** *[pick point c.]*	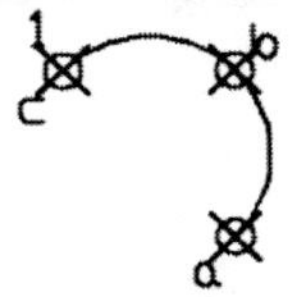
2 Start – Center – End (The most common approach.)	**Command:** *a* **Specify start point of arc or [Center]:** *[Pick point a.]* **Specify second point of arc or [Center/End]:** *c [Select the* **Center** *option.]* **Specify center point of arc:** *[Pick point b.]* **Specify end point of arc or [Angle/chord Length]:** *[Pick point c.]*	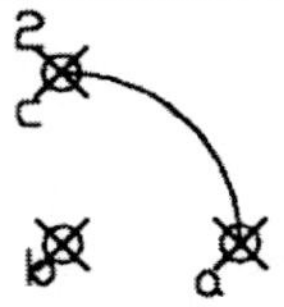
3 Start – Center – Angle	**Command:** *a* **Specify start point of arc or [Center]:** *[Pick point a.]* **Specify second point of arc or [Center/End]:** *c [Select the* **Center** *option.]* **Specify center point of arc:** *[Pick point b.]* **Specify end point of arc or [Angle/chord Length]:** *a [Select the* **Angle** *option.]* **Specify included angle:** *45 [Enter an angle.]*	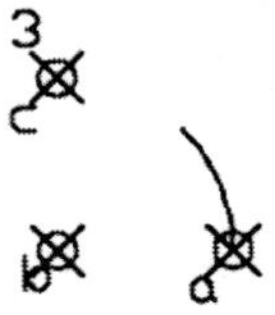

Square/Procedure	Prompt/Response	Results
4 Start – Center – Length (This procedure enables you to control the true length of the arc.)	**Command:** *a* **Specify start point of arc or [Center]:** *[Pick point a.]* **Specify second point of arc or [Center/End]:** *c [Select the Center option.]* **Specify center point of arc:** *[Pick point b.]* **Specify end point of arc or [Angle/chord Length]:** *l [Select the chord Length option.]* **Specify length of chord:** *1.25 [Enter the length of the arc.]*	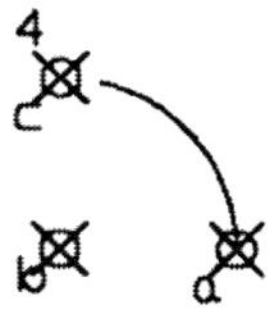
5 Start – End – Angle	**Command:** *a* **Specify start point of arc or [Center]:** *[Pick point a.]* **Specify second point of arc or [Center/End]:** *e [Select the End option.]* **Specify end point of arc:** *[Pick point b.]* **Specify center point of arc or [Angle/Direction/Radius]:** *a [Select the Angle option.]* **Specify included angle:** *45 [Enter an angle for your arc.]*	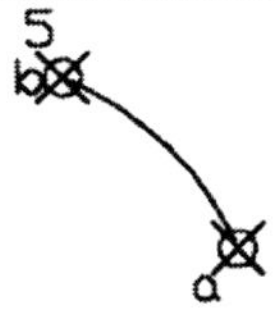
6 Start – End – Direction (This procedure enables you to control the direction of your arc.)	**Command:** *a* **Specify start point of arc or [Center]:** *[Pick point a.]* **Specify second point of arc or [Center/End]:** *e [Select the End option.]* **Specify end point of arc:** *[Pick point c.]* **Specify center point of arc or [Angle/Direction/Radius]:** *d [Select the Direction option.]* **Specify tangent direction for the start point of arc:** *[Pick point b.]*	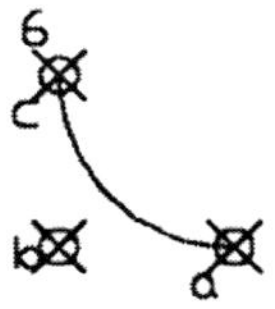
VI.	Set the **BOTTOM** view current.	
7 Start – End – Radius	**Command:** *a* **Specify start point of arc or [Center]:** *[Pick point a.]* **Specify second point of arc or [Center/End]:** *e [select the End option.]* **Specify end point of arc:** *[Pick point b..]* **Specify center point of arc or [Angle/Direction/Radius]:** *r [Select the Radius option.]* **Specify radius of arc:** *1 [Tell AutoCAD what radius to use.]*	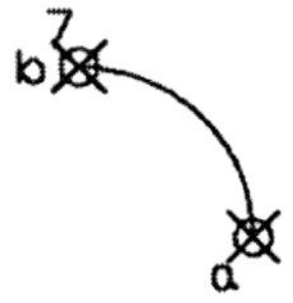

SQUARE/ PROCEDURE	PROMPT/RESPONSE	RESULTS
8 Center – Start – End	**Command:** *a* **Specify start point of arc or [Center]:** *c [Select the Center option.]* **Specify center point of arc:** *[Pick point a.]* **Specify start point of arc:** *[Pick point b.]* **Specify end point of arc or [Angle/chord Length]:** *[Pick point c.]*	
9 Center – Start – Angle	**Command:** *a* **Specify start point of arc or [Center]:** *c [Select the Center option.]* **Specify center point of arc:** *[Pick point a.]* **Specify start point of arc:** *[Pick point b.]* **Specify end point of arc or [Angle/chord Length]:** *a [Select the Angle option.]* **Specify included angle:** *45 [Enter an angle for your arc.]*	
10 Center – Start –Length	**Command:** *a* **Specify start point of arc or [Center]:** *c [Select the Center option.]* **Specify center point of arc:** *[Pick point a.]* **Specify start point of arc:** *[Pick point b.]* **Specify end point of arc or [Angle/chord Length]:** *l [Select the chord Length option.]* **Specify length of chord:** *1.25 [Enter the length of the arc.]*	
11 Continue Arc	*[Create an arc using the Start – Center – End method at points a – b – c.]* *[Repeat the command, and then at the first arc prompt, hit* ENTER *to continue the arc. Pick point d.]*	

Wow! That was a chore! But as you can see, drawing arcs isn't difficult if you just know which method to use.

5.3 Polygons

The polygon has long been one of the foundations on which our world is designed. Everything from the simple triangles of the pyramids to the hex head bolts that hold our automobiles together relies on the mathematics associated with multisided objects. Fortunately, I can leave the mathematics to those better qualified to confuse. All I must do is explain the three simple methods for drawing polygons.

Where to Find It:	
Command Line:	*Polygon*
Hotkey(s):	*pol*
Ribbon (Tab/Panel):	Home – Draw (subpanel) – **Polygon**
Menu:	Draw – **Polygon**
Toolbar:	Draw – **Polygon**
Tool Palette:	Draw – **Polygon**

Consider the chart in Figure 5.002. It's really as simple to draw polygons as this chart suggests. Let's look at the command sequence:

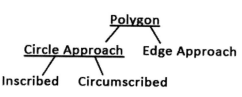

Figure 5.002

Command: *polygon*

Enter number of sides <4>: *[Enter the desired number of equal sides – from 3 to 1024.]*

Specify center of polygon or [Edge]: *[Either identify the location of the center of the polygon (the Circle Approach), or type E to use the Edge Approach.]*

Enter an option [Inscribed in circle/Circumscribed about circle] <I>: *[Let AutoCAD know if you'll draw your polygon* inside *(inscribed) or* outside *(circumscribed) an imaginary circle.]*

Specify radius of circle: *[Tell AutoCAD the radius of the imaginary circle in which (or around which) you want to draw the polygon.]*

Polygon command options are also available on dynamic input or cursor menus once the command has been entered.

Do This: 5.3A	Drawing Polygons

I. Open the *polygons* drawing in the C:\Steps\Lesson05 folder. It looks like the figure at right. (Note: The circles aren't necessary for drawing the polygons. We just use them for demonstration.)

II. Set the **Endpoint** and **Center** Running OSNAPs. Clear all others.

III. Be sure the **DYN** and **OSNAP** buttons on the status bar are lit, and that the other toggles are off.

IV. Follow these steps.

5.3A: DRAWING POLYGONS
1. Enter the *Polygon* command.   **Command:** *pol*
2. AutoCAD asks for the number of sides needed to create the polygon. Enter **6**.   **Enter number of sides <4>:** *6*
3. Now AutoCAD needs to know how to draw the polygon. We'll accept the **center of polygon** default option. Select the center of the larger upper circle. (Let your Running OSNAPs located it.)   **Specify center of polygon or [Edge]:**
4. We'll draw our polygon *inside* the circle (**Inscribed**). Accept the default.   **Enter an option [Inscribed in circle/ Circumscribed about circle] <I>:** *[ENTER]*

5.3A: DRAWING POLYGONS

5. Enter *1.5* as the radius of the circle.
   Specify radius of circle: *1.5*
   Your drawing should look like the figure shown at right.

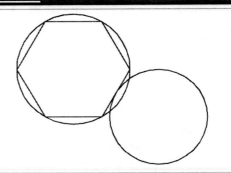

6. Repeat the *Polygon* command ⬠.
   Command: *[ENTER]*

7. This time AutoCAD defaults to six sides, as that was the last number used. Hit ENTER.
   Enter number of sides <6>: *[ENTER]*

8. Select the center of the smaller circle.
   Specify center of polygon or [Edge]:

9. We'll circumscribe this polygon about the outside of the circle.
   Enter an option [Inscribed in circle/ Circumscribed about circle] <I>: *c*

10. When prompted, use the OSNAP to select the bottom quadrant ⊕ of the circle.
    Specify radius of circle: _qua of
    Your drawing looks like the figure shown.

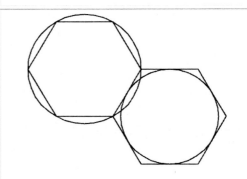

11. Now let's look at the **Edge** option. Repeat the *Polygon* command ⬠.
    Command: *[ENTER]*

12. Accept six as the number of sides.
    Enter number of sides <6>: *[ENTER]*

13. Type *e* (or select **Edge** on the menu) to select the **Edge** option. If the dynamic input menu doesn't show the options, use the down arrow ⬇ on your keyboard to display them (see the figure at right).
    Specify center of polygon or [Edge]: *e*

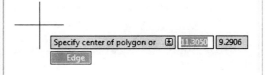

14. Select **Point 1** as indicated.
    Specify first endpoint of edge:

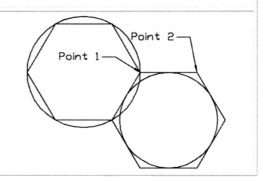

5.3A: DRAWING POLYGONS

15. Select **Point 2**.
    Specify second endpoint of edge:
    Your drawing looks like the figure shown.

16. Save and close the drawing.
    Command: *qsave*

That's it for polygons! Remember, you can draw polygons with anything from 3 to 1024 sides ... and you don't need an actual circle to guide you!

5.4 Putting It All Together

Let's try a project using what we've learned. We'll draw the Ring Stand Base shown in Figure 5.003.

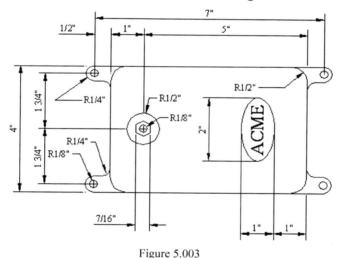

Figure 5.003

Do This: 5.4A — Polygons, Arcs, and Circles – The Project

I. Create a new drawing with the following setup:
   - Lower left limits: 0,0
   - Upper right limits: 17,11
   - Units: Architectural
   - Grid: ½
   - Snap: ¼
   - Textsize: 3/8
   - Font: Times New Roman or Calibri

II. Follow these steps.

5.4A: THE PROJECT

1. **Zoom All** to see the entire drawing.
   Command: *z*

5.4A: THE PROJECT

2. Save the drawing as *MyStand* to the C:\Steps\Lesson05 folder.
   Command: *save*

3. Enter the *Line* command.
   Command: *l*

4. Start at point **4,2½** and draw a 3" line upward. (Feel free to use any method of point entry you prefer – direct distance, dynamic input, coordinate entry, etc.)
   Specify first point: *4,2-1/2*
   Specify next point or [Undo]: *@3<90*
   Specify next point or [Undo]: *[ENTER]*

5. Repeat the *Line* command. Then draw the second line 5" to the east as indicated.
   Command: *[ENTER]*
   Specify first point: *4-1/2,2*
   Specify next point or [Undo]: *@5<0*
   Specify next point or [Undo]: *[ENTER]*

6. Draw a third line 5" to the east …
   Command: *[ENTER]*
   Specify first point: *4-1/2,6*
   Specify next point or [Undo]: *@5<0*
   Specify next point or [Undo]: *[ENTER]*

7. … and a fourth line 3" north.
   Command: *[ENTER]*
   Specify first point: *10,2-1/2*
   Specify next point or [Undo]: *@3<90*
   Specify next point or [Undo]: *[ENTER]*

   Your drawing now looks like the figure at right.

   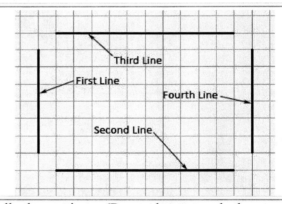

8. Set the Running OSNAP to **Endpoint**. Clear all other settings. (Remember to toggle the Running OSNAP on or off using the F3 function key or the status bar toggle as needed!)
   Command: *os*

9. Enter the *Arc* command.
   Command: *a*

10. Select the endpoint of the second line nearest the fourth line.
    Specify start point of arc or [Center]:

11. Use the **Center** option.
    Specify second point of arc or [Center/End]: *c*

5.4A: THE PROJECT

12. Snap to the grid mark directly above the point you selected in Step 10.
    Specify center point of arc:

13. Select the lower endpoint of the fourth line.
    Specify end point of arc or [Angle/chord Length]:

14. Repeat Steps 9 through 13 to draw the other three arcs. Your drawing now looks like the figure at right.

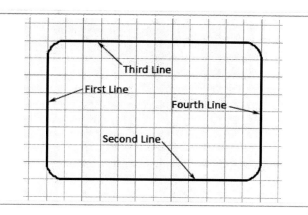

15. Repeat the *Line* command.
    Command: *l*

16. Draw a 1" line from the left endpoint of the second line in the 180° direction.
    Specify first point:
    Specify next point or [Undo]: *@1<180*
    Specify next point or [Undo]: *[ENTER]*

17. Repeat Step 16 at all four corners. Your drawing will look like the figure at right.

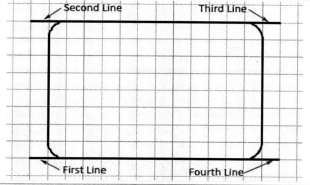

18. Now we'll draw an arc at the end of the first line. Start the command at the point indicated.
    Command: *a*
    Specify start point of arc or [Center]: *3-1/2,2-1/2*

19. Use the **Center** option.
    Specify second point of arc or [Center/End]: *c*

20. Select a point one snap down from the point selected in Step 18.
    Specify center point of arc:

21. Select the endpoint of the first line.
    Specify end point of arc or [Angle/chord Length]:

22. Repeat Steps 18 through 21 at the other four lines. Your drawing will look like the figure at right.

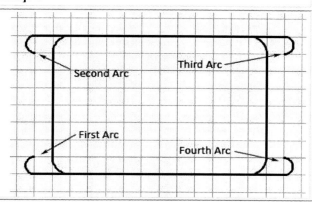

120

5.4A: THE PROJECT

23. Now add a ¼" line at the end of the first arc.
Repeat for each arc.
   Command: *l*
   Specify first point:
   Specify next point or [Undo]: *@1/4<0*
   Specify next point or [Undo]: *[ENTER]*
Your drawing will look like the figure at right.

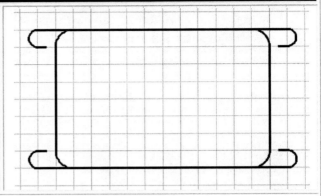

24. Now draw arcs at the end of the new lines (refer to the figure next to Step 29). We'll start at the lower left corner.
   Command: *a*
   Specify start point of arc or [Center]: *[Select the endpoint of the line at the first arc.]*

25. Use the **End** option.
   Specify second point of arc or [Center/ End]: *e*

26. The endpoint is one snap up and one snap to the right.
   Specify end point of arc:

27. Select the **Direction** option.
   Specify center point of arc or [Angle/Direction/Radius]: *d*

28. Select a point to the right of the point selected in Step 24.
   Specify tangent direction for the start point of arc:

29. Repeat Steps 24 through 28 for the other arcs. Your drawing looks like the figure below.

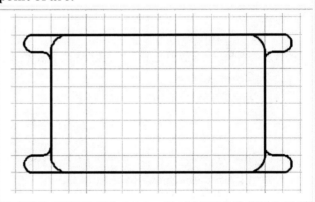

30. It's a good idea to save 💾 your drawing occasionally!
   Command: *qsave*

31. Draw a circle ⊘ in the center of the first anchor leg.
   Command: *c*

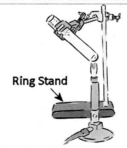
Ring Stand

121

5.4A: THE PROJECT

32. Use the **center** OSNAP to locate the center of the first arc.
    Specify center point for circle or [3P/2P/Ttr (tan tan radius)]: *cen*

33. Use a 1/8" radius for the circle.
    Specify radius of circle or [Diameter]: *1/8*

34. Repeat Steps 31 through 33 for each of the anchor legs. Your drawing will look like the figure at right.

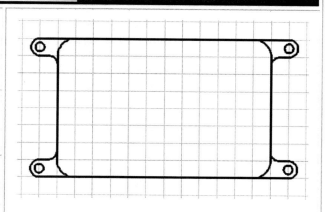

35. Now draw the 1" washer using the *Circle* command.
    Command: *c*
    Specify center point for circle or [3P/2P/Ttr (tan tan radius)]: *5,4*
    Specify radius of circle or [Diameter] <0'-0 1/8">: *1/2*

36. Draw the ¼" bolt center using the *Circle* command.
    Command: *c*
    Specify center point for circle or [3P/2P/Ttr (tan tan radius)]: *5,4*
    Specify radius of circle or [Diameter] <0'-0 1/2">: *1/8*

37. Draw the bolt using the *Polygon* command.
    Command: *pol*

38. Give it six sides.
    Enter number of sides <4>: *6*

39. Place it in the center of the last circle you drew.
    Specify center of polygon or [Edge]: *_cen of*

40. Since we've a dimension on the bolt from flat side to flat side, we'll draw the polygon around a circle with that diameter. Enter c for **Circumscribed**.
    Enter an option [Inscribed in circle/Circumscribed about circle] <I>: *c*

41. Enter *7/32* (half the given diameter).
    Specify radius of circle: *7/32*
    The washer and bolt look like the figure at right.

42. Now let's draw the logo plate using the *Ellipse* command.
    Command: *el*

43. Start at point *8½,3*.
    Specify axis endpoint of ellipse or [Arc/Center]: *8-1/2,3*

44. The ellipse is 2" along the long axis.
    Specify other endpoint of axis: *@2<90*

5.4A: THE PROJECT	
45. It's ½" along the short axis. **Specify distance to other axis or [Rotation]:** *@1/2<0*	
46. Add the *ACME* text 🅰 to finish the project. (Refer to Figure 5.003, p.118.) **Command:** *dt*	

47. We'll middle-justify the text in the center of the ellipse.
   Current text style: "times" Text height: 0'-0 3/8" Annotative: No
   Specify start point of text or [Justify/Style]: *m*
   Specify middle point of text: *8-1/2,4*	

48. If you set the **textsize** to 3/8 during the setup, it will default to that now. Otherwise, set it to *3/8*. **Specify height <0'-0 3/8">:** *[ENTER]*	
49. Set the **rotation angle** so the text may be read from the bottom of the stand. **Specify rotation angle of text <0>:** *90*	
50. Enter the text (*ACME*).	
51. Save 💾 your drawing. It should now look like the sample in Figure 5.003, p.118. **Command:** *qsave*	

5.5 Extra Steps

These may be the most important paragraphs in the text!

AutoCAD provides several system variables (generally called SYSVARS) including one called **Savetime**.

> SYSVARS (system variables) are one of the ways AutoCAD provides for you to configure the software for optimal performance.

The *Savetime* command sequence is

   Command: *savetime*

   Enter new value for SAVETIME <10>:

The default time shown here is 10 minutes. I prefer this setting, but you can set it to whatever makes you feel comfortable. Remember that whatever number you assign to *Savetime* is the amount of drawing time you may lose in case of a system crash.

> You don't have to worry about it overwriting a file if you don't want your changes saved. AutoCAD saves the drawing in a separate folder.
>
> To find where AutoCAD keeps the automatically saved file, do this:
> 1. enter **options** at the command prompt;
> 2. on the **Files** tab, under **Search paths, file names, and file locations**, pick the + next to **Automatic Save File Location**. AutoCAD will show you the path to the saved files.

5.6 What Have We Learned?

Items covered in this lesson include:

- *Isometric circles*
- *Commands*
  - *Arc*
  - *Ellipse*
  - *Polygon*
  - *PEllipse*
  - *Savetime*
  - *PEdit*

Lines and circles (and their various complements) shape our world. With this lesson, we wrap up the basic drawing tools. You're now able to draw quite a few things in the 2-dimensional world. But remember that only about 30% of CAD is drawing. The rest is modifying what you've drawn and entering text. We'll consider many of AutoCAD's modifying tools in Lessons 8, p.176, and 9, p.202, but first, we'll look at adding a bit of flavor to our work in Lessons 6, p.126, and 7, p.148.

Work on the exercises until you're more comfortable with what you've learned so far. Then go on to the next "colorful" lesson.

5.7 Exercises

1. Use the *MyIsoGrid2* template you created in Lesson 3 (or the *IsoGrid2* template in the Lesson03 folder) to create the Isometric Block drawing. Save the drawing as *MyIsoBlockwithEllipses* in the C:\Steps\Lesson05 folder.

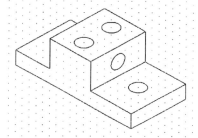

Isometric Block with Isometric Circles

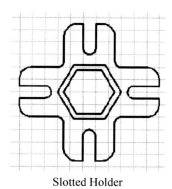

2. Using the *MyBase3* template you created in Lesson 1 (or the *Base3* template in the Lesson01 folder), create the Slotted Holder drawing. Save the drawing as *MyHolder* in the C:\Steps\Lesson05 folder.

Slotted Holder

5.8 For Web-Based Review Questions and Additional Exercises, visit: http://foragerpub.com/AcadFiles/2011/2011.htm

Come on! We're gonna do color next!

We're cruisin' now!

Lesson 6

Following this lesson, you will:

✓ *Know how to add color to a drawing with the **Color** command*

✓ *Know how to use transparency in a drawing*

✓ *Know how to use linetypes in a drawing with the **Linetype** command*

✓ *Know how to use lineweights in a drawing with the **Lineweight** command*

✓ *Know how to modify object properties in a drawing using:*

- ***Properties** (and Quick Properties)*
- ***Matchprop***
- ***List***

Object Properties - Color at Last (and More!)

This is one of my favorite lessons to teach. It may be that I appreciate the respite from new drawing routines. It may be that I like dialog boxes. But more likely, it's just that by this point in the course, I'm really tired of looking at black-and-white drawings!

In the drafting world, we learn to differentiate between objects by well-established uses of linetypes and lineweights (widths). The number and spacing of dashes in a line, the width of a line, or combinations of dashes and width say a lot about what is being represented on the drawing. (For more information about specific representations, look in any basic drafting text.) We'll learn to use these tools in a CAD environment as well. But in the CAD environment, you'll have an additional tool at your disposal – color.

*This lesson will lead you through the first of two different methods of using linetype, lineweight, transparency, and color to differentiate between objects in your drawing; these are the direct approach (using specific commands such as **Linetype**, **LineWeight**, **CETransparency**, and **Color**), and using layers (Lesson 7, p.148). Consider each method exclusive; that is, you shouldn't combine them as the results will no doubt aggravate someone.*

6.1 Some Preliminaries

Let's take a quick look at the **Properties** panel on the ribbon's **Home** tab (Figure 6.001). You see several control boxes (drop down lists).

Figure 6.001

- Use the **Object Color** control box for setting the color with which you'll draw. We'll look at color in more detail in Section 6.2 (p.128).
- Use the **LineWeight** control box to set the lineweight with which you'll draw. We'll look at lineweights in Section 6.5 (p.137).
- Use the **Linetype** control box to set the linetype with which you'll draw. We'll look at linetypes in Section 6.4 (p.132).
- Use the **Plot Style** control box to set the plot style for selected objects. You'll see more on this one in Lesson 23, p.530.
- **Transparency** uses a slider box to control the transparency for the objects you'll draw.
- To the left of the **Transparency** slider box, you'll find a **Transparency** control box with options to
  - create objects using a **Transparency value**,
  - create objects using **Transparency ByLayer** (more on layers in Lesson 7, p.148),
  - create objects using **Transparency ByBlock** (more on blocks in Lessons 21, p.468, and 22, p.500).
- The **List** button starts a command that lists information about an object. We'll look at the *List* command in Section 6.6.1 (p.138).
- You may notice a small arrow in the lower right corner of the panel; this opens the Properties palette. We'll discuss this when we examine the Quick Properties panel in Section 6.6.2 (p.139).

Let's spend some time with each of these.

6.2 Adding Color

"Oh, wow! The colors ... the colors!"
1960s Deep Thinker

Changing the color in which you draw can be as simple as selecting the down arrow in the **Object Color** control box (Figure 6.002, p.128) on the **Home** tab's **Properties** panel, and then selecting your color. Here you'll find the basic options including the basic colors AutoCAD provides. The others are **BYLAYER**, which assigns colors according to the layer setting (more on this in Lesson 7, p.148), and **BYBLOCK**. **ByBlock** means that objects will be drawn in basic black. They'll keep this color until joined together as a block (more on blocks in Lessons 21, p.468, and 22, p.500). When inserted as part of the block, they'll adopt the current color setting in the drawing.

Figure 6.002

If you need more colors from which to choose, use the **Select Colors** option. (Alternately, you can type *Color*, or *Col*, at the command prompt and bypass the **Properties** panel altogether.) AutoCAD will prompt you with the Select Color dialog box (Figure 6.003). This has three tabs that we need to examine more closely.

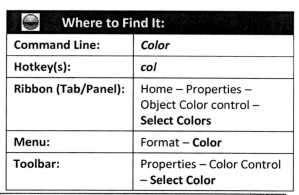

Where to Find It:	
Command Line:	*Color*
Hotkey(s):	*col*
Ribbon (Tab/Panel):	Home – Properties – Object Color control – **Select Colors**
Menu:	Format – **Color**
Toolbar:	Properties – Color Control – **Select Color**

- The first tab – **Index Color** – provides the traditional AutoCAD Color Index (ACI). This index provides 255 colors, shades of black and white, and the **ByLayer/ByBlock** settings we've already discussed.

To select a color from the **Index Color** tab, simply pick on the desired color from the palette. AutoCAD puts a number in the **Color** text box that corresponds to that color (thank goodness you don't have to remember the names of 255 colors!). You can, of course, type in the number yourself if you know it.

- The second tab – **True Color** – allows color settings using 24-bit true color (see Figure 6.004, p.129). Okay, that's computer-eze; what's it mean? Put simply, it means that you have up to 16 million colors from which to choose.

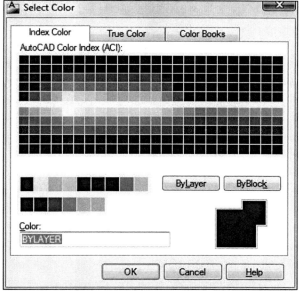

Figure 6.003

You have two color models which you can use on the **True Color** tab – Hue, Saturation, Luminance (**HSL**) or Red, Green, Blue (**RGB**). (Select your preferences from the **Color model** list box in the upper right corner of the tab.) Let's take a look at these.

  o Color can be manipulated by virtue of small changes in three qualities – Hue, Saturation, or Luminance – HSL settings (Figure 6.004). Although these qualities will no doubt be familiar to those with art backgrounds, they won't be of much use to the average CAD operator. Still, a brief discussion is in order.

- *Hue* is a measure of the dominant wavelength of light. To change the hue of your color setting, move the hue box (your cursor) from side to side over the color spectrum area of the tab. Alternately, you can change the number in the **Hue** control box from 0 to 360.
- *Saturation* refers to the purity of a color – that is, the lack of white pollution in the color or how vivid the color appears. To change the saturation of your color setting, move the hue box up and down over the color spectrum area of the tab.

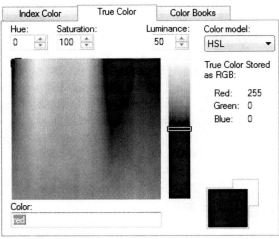

Figure 6.004

Alternately, you can change the number in the **Saturation** control box from 0 to 100.

- *Luminance* refers to how the color reflects light. A value of 0 means that it doesn't reflect light (becomes black); a value of 100 means that it reflects all light (becomes white). The optimal value is 50. To change the luminance, adjust the color slider bar located to the right of the hue box. Alternately, you can enter a value from 0 to 100 in the **Luminance** control box.
  o Another method of adjusting true color is the RGB method (Figure 6.005).

  You can create any color by mixing different amounts of red, green, and blue (just ask the guy who mixes your paint down at Home Depot). That's what you'll do on the **True Color** tab when you select RGB in the **Color model** list box. You can use the slider bars or fill in values (from 0 to 255) in the list boxes. Alternately, you can create a color by entering a series of three numbers (each from 0 to 255) in the **Color** text boxes.

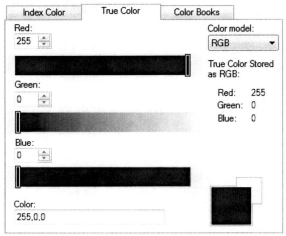

Figure 6.005

- **Color Books** (the final tab – Figure 6.006) provide an interesting addition to AutoCAD's capability. This one may come in handy for the architect or interior designer. Certain companies – by default DIC (Dainippon Nippon Ink & Chemicals), Pantone® (a New Jersey company), and RAL (a German company) – provide color books. These tools are similar to those color cards you use at the

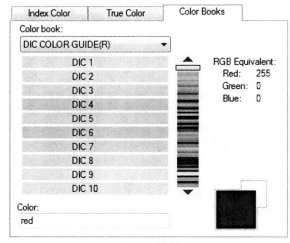

Figure 6.006

129

paint store to decide what color to paint your kitchen. You select which book to use from the **Color book** list box, and then adjust the color bar (to the right of the selection box) to locate a general color. When a general color has been selected, you then select a specific color (or shade of color) from the selection box.

Once you've selected the color, it appears in the **Object Color** control box on the **Properties** panel. This way, you can increase the number of colors and shades of color immediately available for drawing.

Okay, let's try to draw some lines and circles in color!

Do This: 6.2A	**Drawing in Color**

I. Open the *Star* drawing from the C:\Steps\Lesson06 folder. The drawing has a single circle.
II. Follow these steps.

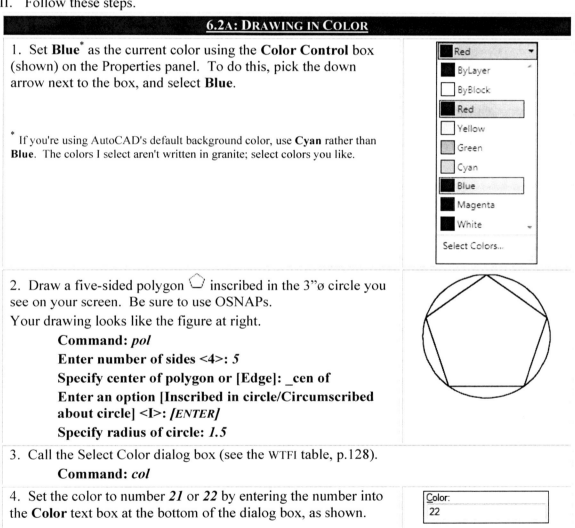

6.2A: DRAWING IN COLOR

1. Set **Blue*** as the current color using the **Color Control** box (shown) on the Properties panel. To do this, pick the down arrow next to the box, and select **Blue**.

    * If you're using AutoCAD's default background color, use **Cyan** rather than **Blue**. The colors I select aren't written in granite; select colors you like.

2. Draw a five-sided polygon inscribed in the 3"ø circle you see on your screen. Be sure to use OSNAPs.
Your drawing looks like the figure at right.
    Command: *pol*
    Enter number of sides <4>: *5*
    Specify center of polygon or [Edge]: _cen of
    Enter an option [Inscribed in circle/Circumscribed about circle] <I>: *[ENTER]*
    Specify radius of circle: *1.5*

3. Call the Select Color dialog box (see the WTFI table, p.128).
    Command: *col*

4. Set the color to number *21* or *22* by entering the number into the **Color** text box at the bottom of the dialog box, as shown.

6.2A: DRAWING IN COLOR

5. Pick the **OK** button.

6. Set the running OSNAP to **Endpoint**, and then draw lines connecting the corners of the polygon to form a star.
   Command: *l*
   Your drawing looks like the figure at right.

7. Now let's use the **RGB** method to set a new color. Call the Select Color dialog box.
   Command: *col*

8. Pick on the **True Color** tab, and select **RGB** from the **Color model** list box.

9. Enter the number *226* into the **Red**, *99* into the **Green**, and *178* into the **Blue** control boxes. Notice that the slider bars reposition themselves in accordance with what you've entered, and that the color indicator (lower right corner) changes to show the color you've set.

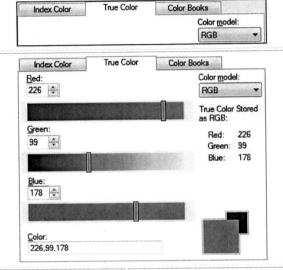

10. Pick the **OK** button to conclude the procedure.
    Notice the change in the **Object Color** control box.

11. Create the text shown here. The text should be middle-justified ... in the center of the circle ... and make it 3/16" high.
    Command: *dt*
    Current text style: "Standard" Text height: 0.2000
    Annotative: No
    Specify start point of text or [Justify/Style]: *m*
    Specify middle point of text: _cen of
    Specify height <0.2000>: *3/16*
    Specify rotation angle of text <0>: *[ENTER]*

12. Save your drawing as *MyStar* in the C:\Steps\Lesson06 folder.
    Command: *saveas*

6.3 Drawing with Transparency

Like **Color**, use **Transparency** to help differentiate between (empahsize or de-emphasize) objects in a drawing. You can even toggle between displaying/not displaying transparency with the status bar toggle.

As with color and the other properties found on the

Where to Find It:	
Command Line:	*CETransparency*
Ribbon (Tab/Panel):	Home – Properties – Transparency slider
Menu:	Format – **Transparency**

Properties panel, you can also use the tools to *change* the properties of selected objects. Give it a try.

Do This: 6.3A	**Drawing with Transparency**

I. Be sure you're still in the *MyStar.dwg* file. If not, please open it now.

II. Pick the **Transparency** toggle on the status bar; be sure it appears lit.

III. Follow these steps.

6.3.1A: DRAWING WITH LINETYPES

1. Without entering a command, select the circle and polygon. They'll look like the figure at right.
   Command: *lt*

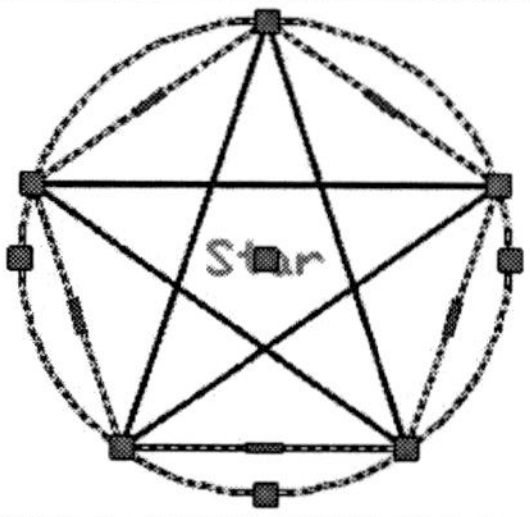

2. On the Properties panel, move the **Transparency** slider until it reads about **75**.

3. Use the ESC key to clear the selection. Notice the difference?

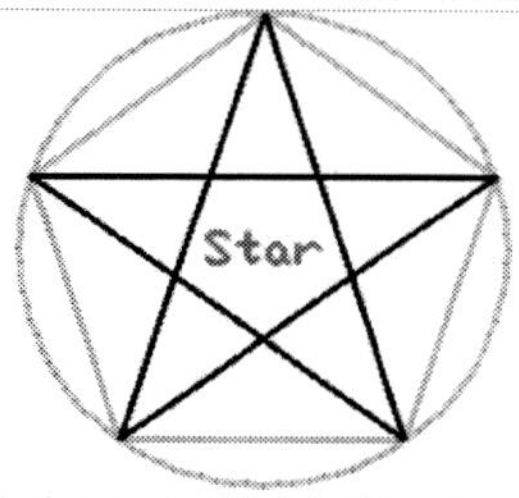

Cool, huh?

6.4	**Drawing with Linetypes**
6.4.1	**Using Linetypes**

What would drafting be without linetypes? *Boring!* And builders wouldn't be able to tell hidden lines from centerlines. I suppose we'd have some pretty interesting buildings out there (not to mention some OSHA nightmares).

Luckily, AutoCAD has continued all the traditional drafting tools – and even added a few. But let's look at linetypes.

Use one of the methods in the WTFI table to bring up the Linetype Manager dialog box (Figure 6.007, p.133).

Where to Find It:	
Command Line:	*Linetype*
Hotkey(s):	*Ltype* or *lt*
Ribbon (Tab/Panel):	Home – Properties – Linetype control - **Other**
Menu:	Format – **Linetype**
Toolbar:	Properties – Linetype Control - **Other**

As you can see in the list box, there are only three options available by default. **ByLayer** and **ByBlock** work in the same way they did with the *Color* command. The only other option is **Continuous**, which provides a solid line. To select one of the options, simply pick on the desired linetype, the **Current** button `Current`, and then the **OK** button `OK`. You'll notice that the **Linetype Control** box on the Properties panel shows your choice as current.

An easier way of making a *loaded* linetype current is simply to pick the down arrow next to the **Linetypes** control box (Figure 6.008) on the Properties panel, and select the linetype of your choice (much as you did with the **Object Color** control box).

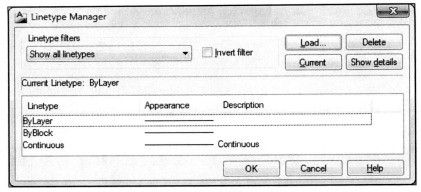

Figure 6.007

So what if the linetype you need isn't shown?

To avoid using a great deal of memory to hold information that may not be used, AutoCAD doesn't automatically load all the available linetypes. But you can load them all or just the ones you want. We'll see how in our next exercise.

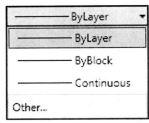

Figure 6.008

> You can still access the command line options of the *Linetype* command by typing a dash in front of it, as in *-linetype*. The command sequence to set a linetype looks like this:
>
> **Command:** *-linetype*
> **Current line type:** "ByLayer"
> **Enter an option [?/Create/Load/Set]:** *s*
> **Specify linetype name or [?] <ByLayer>:** *hidden*
> **Enter an option [?/Create/Load/Set]:** *[ENTER]*
>
> The command line method is only occasionally useful; although *by using this approach, you can set a linetype current without first loading it.* (AutoCAD will load it automatically.)

Do This: 6.4.1A	**Drawing with Linetypes**

I. Open *va* from the C:\Steps\Lesson06 folder. This is a valve attached to a vessel wall (see the figure at right.). We'll load some linetypes, add a centerline to the valve, and add some hidden lines to indicate pipe inside the vessel.

II. Be sure the **Annotation Scale** for the drawing is set to 1:8. (Look at the status bar.)

III. Follow these steps.

6.4.1A: DRAWING WITH LINETYPES

1. Open the Linetype Manager dialog box (Figure 6.007).
    Command: *lt*

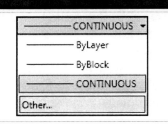

6.4.1A: DRAWING WITH LINETYPES

2. Pick the **Load** button [Load...]. The Load or Reload Linetypes dialog box appears.

Notice the **File** button [File...] at top of this dialog box. AutoCAD stores its linetype definitions in a file called *acad.lin*. There is another file – the *acadiso.lin* file used for metrics – with additional definitions.

Notice also the list of linetype names under the **Linetype** heading in the list box. A description and sample appears to the right of each. Scroll down the list as necessary to find the linetype(s) you want to load.

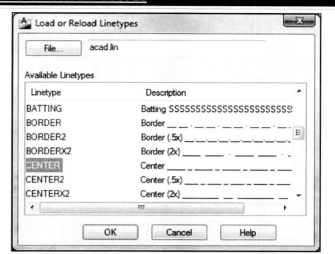

3. For now, scroll down until you see **CENTER**. Select it by picking on the word **CENTER**. Scroll down a bit more until you see **HIDDEN**. Holding down the CTRL key on the keyboard (to enable you to select more than one linetype at a time) select **HIDDEN**.

4. Pick the **OK** button [OK]. AutoCAD closes the dialog box and lists **CENTER** and **HIDDEN** in the Linetype Manager (following figure).

Linetype	Appearance	Description
ByLayer	————	
ByBlock	————	
CENTER	—— — ——	Center _ _ _ _ _ _ _ _
CONTINUOUS	————	Continuous
HIDDEN	– – – – – – –	Hidden _ _ _ _ _ _ _

5. Complete the command [OK].

6. Now let's draw some lines. First, set **CENTER** as the current linetype. Pick the down arrow in the **Linetype** control box on the **Properties** panel, and select the **CENTER** linetype as shown.

7. Draw a line from the node at **Point 1** to **Point 2** (refer to the Step 8 figure). Notice that you've drawn a centerline.
   Command: *l*

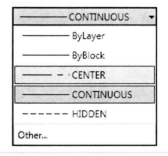

8. Now draw a line from the handwheel – **Point 3** – through the center of the valve to **Point 4** (be sure to use the appropriate OSNAP and Ortho settings).
   Command: *[ENTER]*
Your drawing looks like the figure at right.

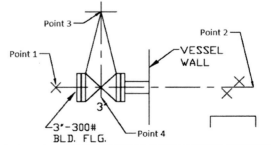

134

6.4.1A: DRAWING WITH LINETYPES

9. Set the current color to **Blue** or **Cyan**.	Cyan / Blue / Magenta
10. Set the linetype to **HIDDEN**.	CONTINUOUS / HIDDEN
11. Now draw the hidden lines to show the pipe inside the vessel (refer to the figure at right). Start at the intersection (use OSNAPs) at **Point 1** and draw to the node at **Point 2**. Continue the line perpendicular to the basin below. Repeat the *Line* command and draw a line from **Point 3** to **Point 4** and down to the basin.	(figure showing vessel wall with Points 1, 2, 3, 4 and 3"-300# BLD. FLG.)
12. Save the drawing, but don't exit.	

6.4.2 Modifying Linetype Scale

You've seen that drawing with linetypes is fairly simple. But there are some other considerations of which you must be aware. The first of these is the length of dashes and spaces in your line – the linetype *scale*.

AutoCAD defines dashed and dotted lines by a code that details how long to make each dash and each space. You can define your own line by learning that code – but that's a topic for a customization guide. At this level, you must know how to adjust the line definitions to appear as dashes or dots on a larger drawing. Otherwise, a dashed line defined as having ¼" dashes separated with 1/8" spaces may appear as a solid line on a scaled drawing.

Of course, the simple approach requires only that you set the **Annotation Scale** for the drawing just as you did when you used it to size your text for you. Try setting the **Annotation Scale** back to 1:1 now and regenerating your drawing; notice the difference (Figure 6.009). Your drawing's lines now have dashes and spaces too tiny to see.

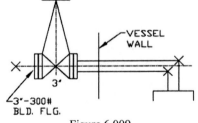

Figure 6.009

The older method of adjusting the dashes and spaces utilized the **LTScale** system variable. Refer to *LTScale.pdf* at our website (www.uneedcad.com/Files) for more details on this approach.

But what if you need the linetypes to use different scales for the dashes and spaces? Well, for that, AutoCAD provides the **Current object scale** option in the Linetype Manager.

Let's look at how this works.

Do This: 6.4.2A	Modifying Linetype Scale

I. If you're not still in the *va* file, open it now. It's in the C:\Steps\Lesson06 folder.

II. Be sure the **Annotation Scale** for the drawing is currently **1:1**. The drawing should look like Figure 6.009 (p.135).
III. Follow these steps.

6.4.2A: MODIFYING LINETYPE SCALE	
1. Erase lines 1 and 2, as indicated. **Command:** *e*	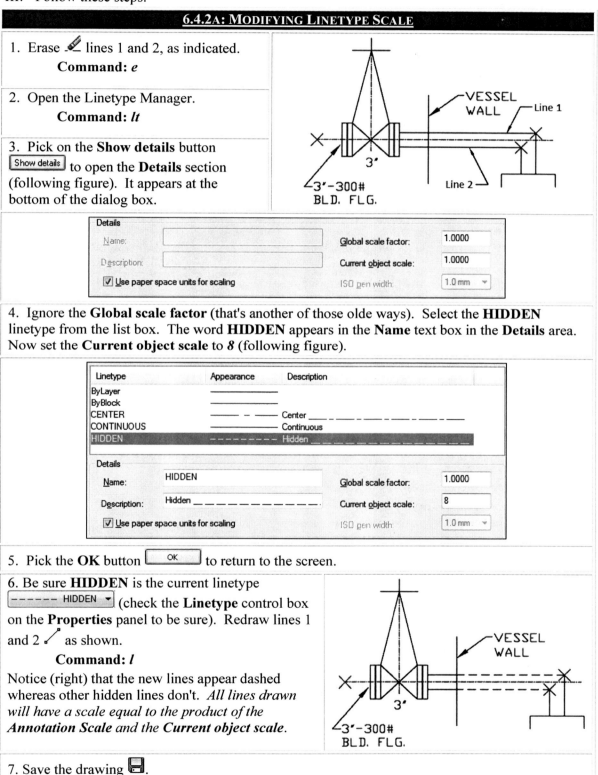
2. Open the Linetype Manager. **Command:** *lt*	
3. Pick on the **Show details** button to open the **Details** section (following figure). It appears at the bottom of the dialog box.	
4. Ignore the **Global scale factor** (that's another of those olde ways). Select the **HIDDEN** linetype from the list box. The word **HIDDEN** appears in the **Name** text box in the **Details** area. Now set the **Current object scale** to *8* (following figure).	
5. Pick the **OK** button to return to the screen.	
6. Be sure **HIDDEN** is the current linetype (check the **Linetype** control box on the **Properties** panel to be sure). Redraw lines 1 and 2 as shown. **Command:** *l* Notice (right) that the new lines appear dashed whereas other hidden lines don't. *All lines drawn will have a scale equal to the product of the Annotation Scale and the Current object scale.*	
7. Save the drawing.	

6.5 Using Lineweights

Exercise caution when using lineweights; they aren't true *WYSIWYG* objects – that is, What You See Is (not necessarily) What You Get. In Model Space, AutoCAD displays lineweight by using a relationship between screen pixels and lineweight. This enables you to tell that an object has weight – even if you can't see how much weight until you plot the drawing.

> Pixels are little lights that make your monitor work. Think of them as tiny flashlights shining through colored lenses in the back of your monitor screen. The number of tiny flashlights depends on the *resolution* of your monitor (see your Windows manual). A standard resolution for AutoCAD would be 1024 x 768 – or 1024 columns of tiny flashlights and 768 rows of tiny flashlights in every square inch of your screen! (They're really *tiny* flashlights!)

Use one of the methods in the WTFI table to call the dialog box for Lineweight Settings (Figure 6.010).

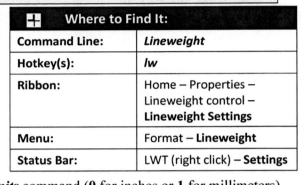

Where to Find It:	
Command Line:	*Lineweight*
Hotkey(s):	*lw*
Ribbon:	Home – Properties – Lineweight control – **Lineweight Settings**
Menu:	Format – **Lineweight**
Status Bar:	LWT (right click) – **Settings**

- The **Lineweights** frame on the left provides a choice of predetermined lineweights from which to select – from **0.00mm** to **2.11mm** (or **0.000"** to **0.083"** – depending on the units selected).
- The **Units for Listing** frame allows you to list the lineweight options in **Inches** or **Millimeters**. You can also set the units system variable at the command line by using the *LWUnits* command (**0** for inches or **1** for millimeters).
- The check in the box next to the line: **Display Lineweight** means that lineweights will be shown in Model Space with weight determined in pixels. You can remove the check (or enter *Off* at the prompt after the *LWDisplay* command, or use the LWT toggle ╋ on the status bar) to prevent seeing lineweights.
- The slider bar in the **Adjust Display Scale** frame allows you to control the pixels-to-lineweight ratio. Slide the bar to the left to minimize or to the right to maximize the number of pixels used to show lineweight.

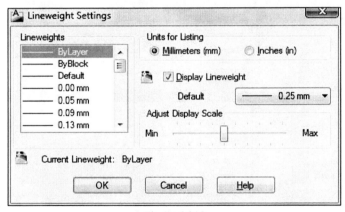

Figure 6.010

Of course, the easiest way you can set lineweight is just as you set color or linetype – using the **Line weight** control box on the ribbon's **Home** tab, **Properties** panel.

Don't forget that you can use the control boxes on the **Properties** panel to *change* (as well as set) lineweight, color, transparency, and linetype! Let's take a look at lineweight.

Do This: 6.5A	Changing the Lineweight

I. If you're not still in the *va* file, open it now. It's in the C:\Steps\Lesson06 folder.
II. Be sure the Quick Properties toggle ▦ on the status bar is off.
III. Follow these steps.

6.5A: CHANGING THE LINEWEIGHT

1. Without entering a command, select the vessel wall. Notice that three tiny squares appear and the line highlights. The squares are grips (more on these wondrous tools in Lesson 9, p.212).

2. Pick the down arrow on the **Lineweight** control box; then scroll down and select **0.30mm** as shown.

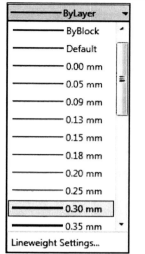

3. Right click in the graphics area and pick **Deselect All** from the cursor menu. This will clear the selection set on the screen. (Alternately, you can simply hit the ESC key on the keyboard.)

Notice the change in the line you selected. (Remember: you can use this technique to change linetypes, colors, transparency, and layers, as well!)

(Note: If you don't see the lineweight, use the status bar's **LWT** toggle to turn it on.)

4. Save the drawing.

Note: Lineweight isn't affected by the Annotation Scale of the drawing.

6.6 Uh-Ohs, Boo Boos, Ah $%&#s: Object Properties

6.6.1 Tell Me About It – The *List* Command

Where to Find It:	
Command Line:	*List*
Hotkey(s):	*ls* or *li*
Ribbon (Tab/Panel):	Home – Properties – **List**
Menu:	Tools – Inquiry – **List**
Toolbar:	Inquiry – **List**

Ever draw something only to find that it didn't look exactly as you expected? For example, you may be trying to draw a red hidden line only to find it comes in solid and green. When something unexpected like this happens, do a *List* (see the WTFI) on the object to see if you can spot the problem. Look at some examples.

Do This: 6.6.1A	Listing an Object's Properties

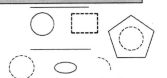

I. Open the *samples* file from the C:\Steps\Lesson06 folder. The drawing looks like the figure at right.
II. Follow these steps.

6.6.1A: LISTING AN OBJECT'S PROPERTIES

1. Enter the *List* command.

   Command: *li*

2. Select the dashed, green circle.

   Select objects:
   Select objects: *[ENTER]*

   AutoCAD switches to the text screen and displays the following information.

   ```
              CIRCLE    Layer: "0"
                        Space: Model space
                Color: 84    Linetype: "HIDDEN"
                Handle = 2b
       center point, X=    1.8116  Y=    2.6301  Z=    0.0000
              radius    1.0352
       circumference    6.5045
                area    3.3668
   ```

   We see the type of object, its color, linetype and layer.

   Note that, had the linetype and color characteristics of this object been assigned by layer, AutoCAD wouldn't have listed these properties. This should indicate which AutoCAD considers the better method for coloring objects and settings linetypes!

3. Repeat the *List* command for the rest of the objects in this drawing. (It's best to list objects one at a time.) Notice that *List* provides slightly different information for each.

Okay. So now you can tell what the object's properties are, how do you change them?

Enter the *Quick Properties* panel (et al). Let's take a look.

6.6.2	Object Properties and the Quick Properties Panel

Make a mistake? Use the wrong color? linetype? Now what do you do?

Fortunately, AutoCAD has a simple and wonderful tool to fix oversights. Behold the Quick Properties panel (Figure 6.011)!

The Quick Properties panel provides an on-screen, editable (and customizable) listing of properties associated with drawing objects. For those of you who are newly computer

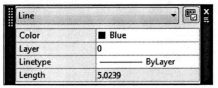

Figure 6.011

literate, this simply means that it's really easy to change things like color, linetype, and lineweight.

The Quick Properties panel opens automatically when you select an object provided you've met a couple conditions:

1. The Quick Properties toggle on the status bar must be lit (on).
2. You must have selected an object defined for Quick Properties in the Customize User Interface dialog box (more on that in Section 6.6.4, p.142) OR you must have told AutoCAD to display the

Quick Properties panel **for any object** in the Drafting Settings dialog box (more on that in Section 6.6.3, p.141).

	Where to Find It:
Command Line:	*Props*
Hotkey(s):	*pr*
Ribbon (Tab/Panel):	View – Palettes – **Properties**
Menu:	Modify – **Properties**
Cursor Menu:	[select object] **Properties**
Toolbar:	Standard – **Properties**

You'll find a most helpful alternative to the Quick Properties panel in the Properties palette, which appears whenever you double click on an object. The palette resembles the panel in many ways without being quite as dynamic. Still, it offers a couple things you won't find on the panel (object filtering – Lesson 11, p.264 – and listing ALL properties by default). Many AutoCAD users opt for this older tool, which operates like any other palette and can be docked and hidden on one edge of the drawing area.

Other ways to open the palette include those found in the WTFI table.

Feel free to experiment with both and use the one you prefer. Operational instructions in this section apply equally to both (although the palette isn't customizable as is the panel).

Let's see how the Quick Properties panel works before we go any further.

Do This: 6.6.2A	Using Quick Properties

I. Be sure you're still in the *va* file in the C:\Steps\Lesson06 folder. If not, please open it now.

II. On the status bar, toggle Quick Properties on ▣ .

III. Follow these steps.

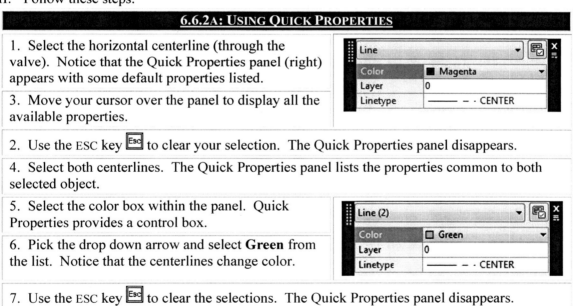

6.6.2A: USING QUICK PROPERTIES

1. Select the horizontal centerline (through the valve). Notice that the Quick Properties panel (right) appears with some default properties listed.
3. Move your cursor over the panel to display all the available properties.
2. Use the ESC key [Esc] to clear your selection. The Quick Properties panel disappears.
4. Select both centerlines. The Quick Properties panel lists the properties common to both selected object.
5. Select the color box within the panel. Quick Properties provides a control box.
6. Pick the drop down arrow and select **Green** from the list. Notice that the centerlines change color.
7. Use the ESC key [Esc] to clear the selections. The Quick Properties panel disappears.

How's that for a quick way to fix a boo boo?!

But wait! AutoCAD makes the Quick Properties panel completely customizable! You have a couple methods you can use: simple **Settings** and the Customize User Interface dialog box (more on that in Section 6.6.4, p.142).

6.6.3 Quick Properties Settings

Use the same technique you learned in Lesson 3, p.58 for opening the Drafting Settings dialog box – either enter **DS** at the command prompt and open the **Quick Properties** tab) or right click on the Quick Properties icon and select settings. AutoCAD will present the three-framed tab shown in Figure 6.012.

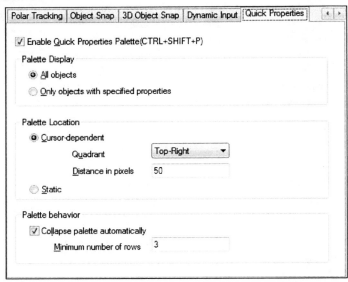

Figure 6.012

- In the **Palette Display** frame, you can tell AutoCAD to display the Quick Properties panel for **All** [the] **objects** you select. Doing so will present a panel with a list of properties that you may or may not be able to change. Or you can tell AutoCAD to display the Quick Properties panel **Only** [for] **objects with specified properties**. We'll see how to specify properties on the Customize User Interface dialog box (Section 6.6.4, p.142).

- The **Palette Location** frame allows you to control where the panel will appear when you select an object.
  o You can located it in relation to the cursor (**Cursor-dependent**) – use the button (showing **Top-Right** in Figure 6.012) to choose a **Quadrant** and the **Distance in Pixels** box to define its distance from the cursor.
  o You can also make the panel **Static**. Doing so means that the panel will float above the graphics area in the same location you left it in the last time you used it.

- In the **Palette behavior** frame, tell AutoCAD whether or not you want the panel to **Auto-collapse**. Use the **Minimum number of rows** box to define the default number of properties to show. Collapsing the panel means that it won't show all of a selected object's properties until you move your cursor over the panel. This saves space but can become a nuisance.

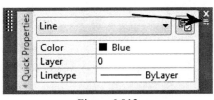

Figure 6.013

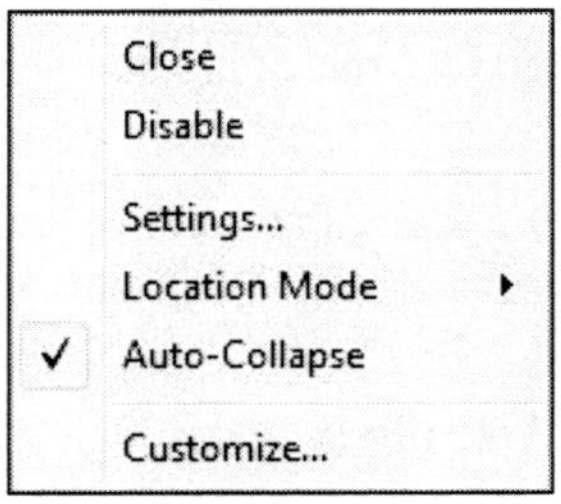

You can reach a short cut menu (Figure 6.014) to some of these settings via the **Options** button on the Quick Properties panel (Figure 6.013). Notice that you can also **Close** the panel or access the Customize User Interface dialog box for the Quick Properties panel (via the **Customize** option). We'll look at that next.

Figure 6.014

6.6.4	Quick Properties and the Customize User Interface Dialog Box

Words like "customize" this early in your training set off screaming alarms in my otherwise serene head.

You'll use the Customize User Interface dialog box to set up the Quick Properties panel the way you like it, but you're probably not ready to experiment here. Wait until you've finished the course (and have time to reinstall the application). You might even consider a customization text/class to help you with the Customize User Interface dialog box and Options dialog box.

Let's do a quick exercise before we proceed. We'll see one of the shortcomings of the Quick Properties panel, then a way to fix (customize) it.

Do This: 6.6.4A	**Limitations of the Quick Properties Panel**

   I. Be sure you're in the *samples* file in the C:\Steps\Lesson06 folder.
   II. On the status bar, toggle Quick Properties on .
   III. Follow these steps.

6.6.4A: LIMITATIONS OF THE QUICK PROPERTIES PANEL

1. Select the ellipse (bottom center object). Notice the Quick Properties panel and the note at the bottom saying that there are **No defined quick properties** and telling you to **click to customize**.

2. Pick the X in the upper right corner of the panel to close it. The Quick Properties panel disappears.

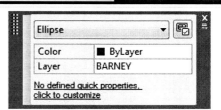

You should have noticed a couple things about the panel you saw in this last exercise.

1. Even though the panel indicated that the selected object (the ellipse) had **no defined quick properties**, it made **Color** and **Layer** available for editing. These **General** properties display by default for any selected object. You'll see more on this in our next exercise.
2. The panel gives you yet another link to the Customize User Interface dialog box (**click to customize**). AutoCAD must really want you to customize this panel!

So, check out the Customize User Interface dialog box. It's a big one (Figure 6.015, p.143)!

A complete discussion of the Customize User Interface dialog box belongs in a customization guide and is beyond the scope of this basic text. We will, however, take a look at how it affects the Quick Properties panel.

Notice that **Quick Properties** has already been selected in the upper left selection box. AutoCAD selected it when you opened the Customize User Interface dialog box via one of the tools on the Quick Properties panel. We'll work now in the right frame.

In the left column of the right frame, under the **Edit Object Type List** button , you'll find a list of objects which will trigger the display of the Quick Properties panel when selected. You can add or remove items from this list by picking the **Edit Object Type List** button; AutoCAD will display a check list of all AutoCAD drawing objects.

In the right column of the right frame, you'll find a check list of properties that the Quick Properties panel will display for the object selected in the left column.

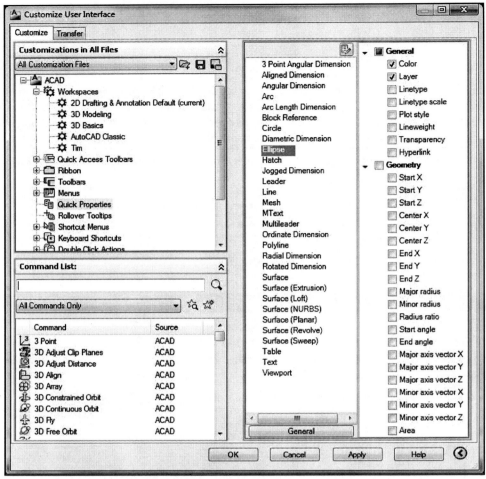

Figure 6.015

Let's use the Customize User Interface dialog box to change what you can modify using the Quick Properties panel.

Do This: 6.6.4B	Customizing the Quick Properties Panel

I. Be sure you're still in the *samples* file in the C:\Steps\Lesson06 folder. If now, please open it now.

II. Be sure the Quick Properties toggle is still on ▦ .

III. Follow these steps.

6.6.4B: CUSTOMIZING THE QUICK PROPERTIES PANEL
1. Open the Customize User Interface dialog box. (Select the ellipse and then use the button 🗗 on the Quick Properties panel.)   **Command:** *cui*
2. Select Quick Properties in the upper left frame (see Figure 6.015).

143

6.6.4B: CUSTOMIZING THE QUICK PROPERTIES PANEL

3. Select **Ellipse** in the left column of the right frame.

Notice that only **Color** and **Layer**, under the **General** category in the right column of the right frame, appear checked. This explains what the Quick Properties panel displayed in our last exercise when we selected the ellipse.

4. Add checks next to **Linetype** under the **General** category, and the first nine X, Y, and Z options under **Geometry**.

The right column now looks like the figure at right.

5. Pick the **Apply** and **OK** buttons to complete the procedure.

6. Select the ellipse again. Notice the difference.

7. Move your cursor over the panel to see the rest of the properties.

Wow! All the properties at once!

8. Now use the **Options** button (Figure 6.013, p.141) on the right to open the menu and remove the check next to **Auto-Collapse**. This displays all the properties without having to move your cursor.

Notice that some of the property values have a gray background while others have a white background. You cannot modify the gray ones; you can modify the white ones.

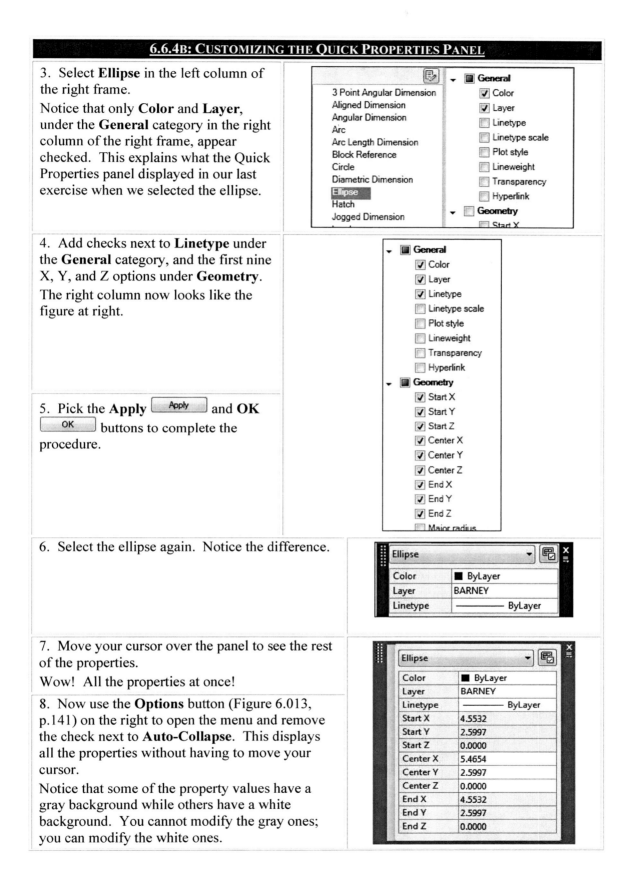

6.6.4B: CUSTOMIZING THE QUICK PROPERTIES PANEL

9. Take a few minutes and change some of the properties. Watch the changes appear on the selected object.

Cool.

6.7 Extra Steps

Let's say you've drawn a chair but used the wrong color by mistake. The chair consists of a rectangle and three lines drawn with an oddball color. You need to know what color you used to draw the chair, but you can't remember. You can check the Quick Properties panel for one of the other chairs just to read the color setting, and then use the Quick Properties approach to fix the chair you just drew … man, that's a lot of work!

Where to Find It:	
Command Line:	*MatchProp* or *Painter*
Hotkey(s):	*ma*
Menu:	Modify – **Match Properties**
Toolbar:	Standard – **Match Properties**

Try the *MatchProp* command. AutoCAD prompts:

    Command: *matchprop*
    Select source object: *[Select an object that was drawn correctly.]*
    Current active settings: Color Layer Ltype Ltscale Lineweight Transparency Thickness PlotStyle Dim Text Hatch Polyline Viewport Table Material Shadow display Multileader
    Select destination object(s) or [Settings]: *[Select the object you want to change.]*
    Select destination object(s) or [Settings]: *[ENTER to complete the command.]*

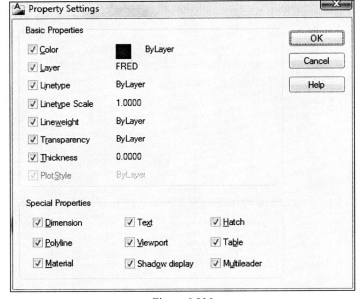

Figure 6.016

Who could ask for anything more?

But wait! There is more! (Will the wonders never cease?!)

Notice that *MatchProp* gives you a **Settings** option. The line above the **Settings** option tells you what the current settings are. AutoCAD will match these properties. Selecting the **Settings** option will produce the Property Settings dialog box (Figure 6.016) in which you can tell AutoCAD the properties you do or don't want to match: **Color, Layer, Linetype, Linetype Scale**, and more!

Personally, I like to leave all the boxes checked. I rarely run into a situation where I might need some properties matched but not others. Still, it's nice to know that I have this option.

6.8 What Have We Learned?

Items covered in this lesson include:
- *Using Properties and Quick Properties*
- *Commands*
    - *Color*
    - *Linetype*
    - *LTScale*
    - *CETransparency*
    - *LineWeight*
    - *LWUnits*
    - *LWDisplay*
    - *List*
    - *Properties*
    - *Matchprop*
    - *Painter*

We've seen how to use colors in our drawing (yea! – no more b&w), and to use linetypes and lineweights to put more into a drawing than we ever thought possible in the world of graphite and paper. (Did you feel like Dorothy walking into Oz?) But we've just scratched the surface of these useful tools. Wait until you master layers in Lesson 7, p.148!

6.9 Exercises

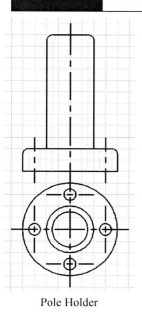

Pole Holder

1. Using the chart and the parameters shown here, create the Pole Holder drawing. Save the drawing as *Colors and Lines* in the C:\Steps\Lesson06 folder.

Object	Color	Linetype
Object	Cyan	Continuous
Hidden Lines	Green	Hidden
Center Lines	Red	Center

    1.1. Lower left limits: 0,0
    1.2. Upper right limits: 12,9
    1.3. Architectural units
    1.4. Grid: ¼"
    1.5. Snap: as needed
    1.6. 1/8"R arcs where shown
    1.7. Hidden line transparency is 65

6.10 For Web-Based Review Questions and Additional Exercises, visit: http://foragerpub.com/AcadFiles/2011/2011.htm

**Jeepers this is work!
(Glad this stuff pays so WELL!)**

Lesson

Following this lesson, you will:

- ✓ Know how to use layers in a drawing
- ✓ Know how to use the Layer States Manager
- ✓ Know how to use the Layer Translator
- ✓ Know how to use the Autodesk Design Center

Colors, Linetypes, and More Made Simple – Layers

*I began our last lesson with an explanation that AutoCAD provided two methods of using linetype, lineweight, transparency, and color to differentiate between objects in your drawing. You're familiar now with the first, direct approach (using specific commands such as **LType**, **LWeight**, and **Color**), but what about that other method – Layers?*

Well, you say, if I'm dedicating an entire lesson just to layers, they must be quite a tool! And you're absolutely right!

Layers take the colored pencils you used in Lesson 6, p.128, and drags them into the computer age. And man-oh-man, will you be glad they do! (Well, you will be once you master these complex tools.)

But remember my admonition in the last lesson: Each method (the direct approach and layers) should be considered exclusive; that is, you shouldn't combine them, as the results will no doubt aggravate someone.

Let's get started.

7.1 Color, Linetype, and So Much More – Layers and How to Use Them

When I was a child (back when we created intricate drafting plans on the cave walls), one of the most coveted possessions of our household was a set of *Encyclopedia Britannica*. I spent hours exploring the world through those books, but my favorite pages were the details of the human body. I wasn't all that interested in anatomy – but I was fascinated by the way the body was shown. There was one page with an outline of the body. Then there were successive pages made of clear plastic overlays with the skeletal, reproductive, digestive, and circulatory systems. As these folded down atop each other, the body took shape. If one system was in the way, all I had to do was fold that sheet back.

This is the idea behind layers in AutoCAD.

You'll assign each layer a specific color, lineweight, linetype, transparency, and plot style. You should also assign a specific name to the layer – like *dim* for dimensions, *txt* for text, or *obj* for objects. All objects referenced by that name are drawn on that layer much as everything related to the skeletal system was found on a single plastic sheet. If something gets in the viewer's way during subsequent drawing sessions or discussions, the appropriate layer can be toggled **Off** or **Frozen** much as I could fold the unwanted sheet back when viewing the body.

Additionally, you can organize the layers into *filtered* groups to make manipulation easier and faster. Using groups, you can change the visibility (**On/Off, Frozen/Thawed**), or even **lock/unlock** several layers simultaneously.

You can access layer tools, like most AutoCAD tools, from a variety of sources – ribbon, palette, command line, etc – but the most common approach (and the easiest) lies in the **Layers** panel (Figure 7.001) on the ribbon. Here are the tools on the primary panel (we'll look at the subpanel tools in Section 7.3, p.164):

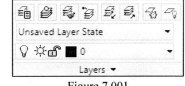

Figure 7.001

You can also find the primary **Layers** panel tools on the Format menu.

- The **Layer Properties** button calls the Layer Properties Manager which we'll see in some detail in Section 7.2 (p.153).
- **Make Object's Layer Current** (*LayMCur*) provides a quick and easy way to set a layer current without having to figure out what layer it is – simply pick an object that already resides on the desired layer. AutoCAD prompts:
    Select object whose layer will become current:

149

- **Match** begins the *LayMch* command which allows you to change the layer of a selected object to a target layer of your choice. You don't need to know the name of either layer. This handy fix saves time when you find you've drawn something on the wrong layer! (We'll see this one in action in our first exercise.)
- **Previous** (*LayerP*) simply sets the previously current layer current again.
- The **Isolate** button initiates the *LayIso* command, which clears a drawing of all the nuisance material by locking and fading all but the selected layer. This way, your drawing is uncluttered while you can still reference all its objects.
- The Layer **UnIsolate** button turns off the affects of the *LayIso* command (with the *LayUnIso* command).
- **Layer Off** (*LayOff*) and **Layer Freeze** (*LayFrz*) appear to do the same thing – they remove (hide) all objects in the drawing that were drawn on (reside on) that layer. The difference between the two meant more in the past than today – objects on a *frozen* layer won't regenerate with the drawing while objects on an *off* layer will. Freezing a layer often saved regen time in earlier releases but the difference doesn't count for much today. Still, you'll freeze layers more often than turn them off … and you'll use this tool quite often!
- The **Layer State** control box lists available layer states. We'll see this tool in detail in Section 7.2.2 (p.162).
- Finally, the **Layer** control box – the workhorse of layers – provides a quick and easy interface for setting layers current, turning them on or off, freezing or thawing them and locking /unlocking them. To accomplish any of this, pick on the icon toggle next to the layer's name (as you'll see in our first exercise).

Let's take a look at layers and how some of these tools work.

Do This: 7.1A	Using Layers

I. Open the *flrpln* file in the C:\Steps\Lesson07 folder. It looks like the figure at right.
II. Follow these steps. We'll start by making a layer current. The preferred method for this lies on the ribbon.

7.1A: USING LAYERS

1. Pick the down arrow on the **Layer** control box and select the **FURNITURE** layer from the list. Notice that **FURNITURE** now appears in the **Layer** control box. It's now current and everything you draw will reside on this layer.
2. Now draw the two 2'-6" x 4' desks missing from the center cubicles. Use the grid and snap as needed.   **Command:** *rec* Notice that the desks assume the color and linetype associated with the **FURNITURE** layer. Your drawing looks like the following figure.

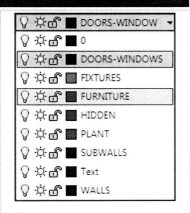

7.1A: USING LAYERS

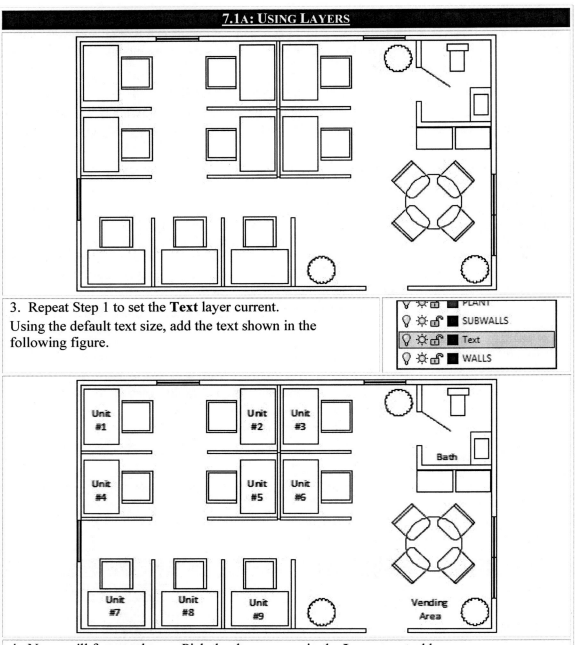

3. Repeat Step 1 to set the **Text** layer current. Using the default text size, add the text shown in the following figure.

4. Now we'll freeze a layer. Pick the down arrow in the **Layer** control box.

5. Pick on the sun icon ☼ next to the **SUBWALLS** layer. Notice that it becomes a snowflake ❄ and that the subwalls in the drawing disappear. (The ones that don't disappear have been drawn on the wrong layer! We'll fix that in a bit.)

Step 5 showed you how to use the **Layer** control box to freeze a layer when you know the name of the layer to freeze. You can also "hide" all objects on the same layer even if you don't know what layer they're on. Use the **Freeze** button ❄ on the **Layers** panel. AutoCAD will prompt you to select an object on the layer to be frozen, as follows:

    Command: _layfrz
    Current settings: Viewports=Vpfreeze, Block nesting level=Block

7.1A: USING LAYERS

> Select an object on the layer to be frozen or [Settings/Undo]: *[Select an object on the layer to be frozen.]*
> Layer "SUBWALLS" has been frozen.
> Select an object on the layer to be frozen or [Settings/Undo]: *[ENTER to complete the command.]*

6. Let's thaw our **SUBWALLS** layer. Pick on the snowflake that now sits next to the **SUBWALLS** layer name. It becomes a sun again and the subwalls reappear.

You'll find a **Thaw All Layers** button on the **Layers** subpanel (pick the arrow on the panel's title bar). Use this with caution as it will thaw *all* the frozen layers in the drawing, not just the one you might want thawed.

7. Let's fix those subwalls that are on the wrong layer. Pick the **Match** button on the **Layers** panel. AutoCAD begins the command.
   Command: *laymch*

8. AutoCAD first asks you to select the **objects to be changed**. Select the three red subwalls along the bottom wall.
   Select objects to be changed:

9. Now AutoCAD needs you to select an object on the **destination** layer. Pick one of the subwalls on the correct layer (one of the purplish subwalls).
   Select object on destination layer or [Name]:
Notice the change!

10. Let's see what happens when we lock a layer. Pick the unlocked icon next to the **WALLS** layer. Notice that it locks.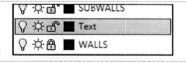

11. Try to erase everything in the upper right corner of the drawing as indicated.
    Command: *e*

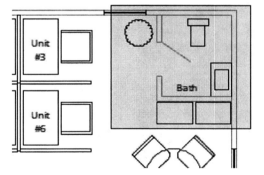

12. When you complete the *Erase* command, notice that AutoCAD removed everything within your selection window except the walls! These reside on the locked **WALLS** layer and can't be modified. (So *cool!*)

13. Undo the erasure.
    Command: *u*

14. Unlock the **WALLS** layer.

15. Save your drawing.
    Command: *qsave*

You've just covered about 80% of what you'll do with layers! (Funny how small a portion of this lesson it took to do that ...) We'll spend the rest of this lesson learning how to manipulate layers in ways that make them one of the most important CAD tools available today ... and one of the most common targets for programming upgrades!

So, let's see how to *manage* the layers behemoth.

> I know; there're other buttons in the **Layers** subpanel. I haven't forgotten them; we'll discuss them in Section 7.3 (p.164).

7.2 A Couple Managers

7.2.1 The Layer Properties Manager

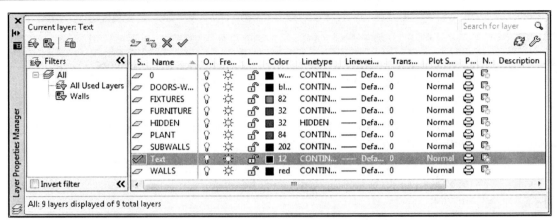

Figure 7.002

Let's look at the Layer Properties Manager (Figure 7.002). Use one of the methods in the WTFI table to open it.

You may notice first that the Layer Properties Manager appears in palette form – just as the Properties palette, the tool palettes, and the undocked ribbon appear. This particular palette takes a bit of set up to optimize it for your convenience.

Where to Find It:	
Command Line:	*Layer*
Hotkey(s):	*la*
Ribbon (Tab/Panel):	Home – Layers – **Layer Properties**
Menu:	Format – **Layer**
Toolbar:	Layers – **Layer Properties Manager**

- First, I recommend docking the Layer Properties Manager palette to one side of your screen and using the **Auto–Hide** feature. You'll use it often enough to justify the minor sacrifice in work space.

> I like to keep some of my palettes open (although hidden with the **Auto–Hide** feature) and docked to the left side of my screen. This makes them available on short notice simply by moving my cursor over them. If you've done this, you'll have noticed that AutoCAD stacks the palettes so you never have more than a single, thin ribbon of palettes to steal your work area.

- Right click on the **Name** row in the right side of the palette (the row that contains the column names). AutoCAD presents the menu in Figure 7.003 (p.154).
  - The check list in the top frame indicates the columns shown in the manager. You can show/hide any columns you consider appropriate. We'll leave them all checked for our work.
  - **Customize** presents a Customize Layers dialog box which allows you to change the order in which the columns appear.
  - The next frame includes options to **Maximize** or **Optimize** the selected (or **all**) columns.
    - A **Maximized** column expands enough to read everything in the column header (title). I recommend maximizing **all columns** as a minimum for operators with smaller monitors or screen resolution.

- An **Optimized** column expands enough to read everything in the column content. **Optimize all columns** works best for those with larger monitors and better resolutions.
  o **Restore all columns to defaults**, of course, does just that.

A right click in the body of the palette presents the menu in Figure 7.004. We'll discuss most of these tools as we proceed (most have a better approach than this menu), but the top frame offers a way to hide the left section of the Layer Properties Manager – the **Filters** area – when you don't use it. Filters provide a terrific method for organizing drawings with a large number of layers, but this section just gets in the way for those that don't. Remove the check next to **Show Filter Tree** if your work falls into the latter category.

Figure 7.003

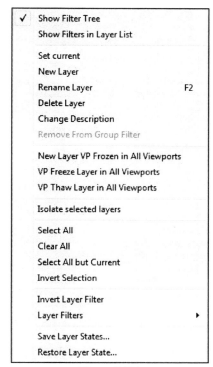

Figure 7.004

If you do use layer filters, I recommend a check next to **Show Filters in Layer List**. This way, AutoCAD will place the name(s) of the filter(s) in the list – and make them accessible in the **Layer** control box on the ribbon's **Layers** panel. Then you can manipulate the filters as easily as you can a single layer.

Okay, now you've set up your Layer Properties Manager, but what's it do?

Well, let's take a look at each column and its command line equivalent as well as buttons available on the **Layers** panel or subpanel.

COLUMN	SYMBOL	COMMAND LINE	BUTTON	DESCRIPTION
Status	Status	(None)	(None)	The **Status** column identifies the currently active layer with a check.
Name	(No Symbol)	(None)	(None)	The **Name** column is just what it implies – the name of the layer.
On	On	*LayOn* *LayOff*	ON/Off Button	The **On** column shows a light bulb that's either lit (yellow) or not lit (gray). If lit, the layer is **On**; if not lit, the layer is **Off** and all objects on the layer will disappear from the screen but remain part of the drawing and be regenerated. The *LayOn* command turns on all layers; *LayOff*

COLUMN	SYMBOL	COMMAND LINE	BUTTON	DESCRIPTION
				allows you to select an object on a layer to turn off.
Freeze	Freeze ☼ ❄	*LayThw* *LayFrz*	Thaw All Button Freeze Button	The **Freeze** column shows an image of the sun when **Thawed** or a snowflake when **Frozen**. When frozen, all objects on the layer will disappear from the screen but remain part of the drawing and *not* be regenerated. Most people prefer **Freeze** to **Off** as it speeds regeneration time. *LayFrz* allows you to select an object on a layer to freeze; *LayThw* thaws all layers.
Lock	Lock 🔓 🔒	*LayUlk* *LayLck*	Unlock Button Lock Button	When a layer is **Locked**, you can't edit or erase any objects on that layer. This is a useful tool when working on a crowded drawing and selecting objects may be difficult, or when you just want to be sure that you don't accidentally move or delete something. The *LayUlk* command unlocks all layers; *LayLck* allows you to select an object on a layer to lock. You can also use the **Locked Layer Fading** slider on the Layers panel to have AutoCAD fade locked layers.
Color	Color ■ white	(None)	(None)	The **Color** column uses a box and a name or number to indicate the current color setting for that layer. (Pick on the color box to open the Color dialog box.)
Linetype	Linetype CONTINUOUS	(None)	(None)	The **Linetype** column lists the specific linetype assigned to that layer. (Pick on the linetype box to open the Linetype dialog box.)
Lineweight	Lineweight ── Default	(None)	(None)	The **Lineweight** column shows the lineweight assigned to the layer. (Pick on the Lineweight box to open the Lineweight dialog box.)
Trans-parency	Transparency 0	*CETrans-parency*	(None)	Indicates the transparency assigned to the layer.
Plot Style	Plot Style Normal	(None)	(None)	The **Plot Style** column shows how this layer will be plotted in the current plot style. (We'll discuss plotting in Lesson 23, p.528.)

COLUMN	SYMBOL	COMMAND LINE	BUTTON	DESCRIPTION
Plot	Plot	(None)	(None)	The **Plot** column allows you to plot the layer or to remove all objects on the layer from the plot. (Note: This doesn't affect the layer's visibility.)
New VP Freeze	New VP Freeze	(None)	(None)	A snowflake shown in this column means that this layer will be automatically frozen in any new viewports you create. (We'll discuss these in Lessons 24, p.560, & 25, p.584.)
Description	No Symbol	(None)	(None)	The **Description** column provides a place for you to enter something a bit more descriptive than the name.

You can select multiple layers for formatting at one time by holding down the CTRL or SHIFT key while making your selections.

You'll notice as you make changes in the Layer Properties Manager that these changes are dynamic – that is, they take place in the drawing as soon as you make them.

Above the Layer Properties Manager list box, you'll find four buttons also associated with the list box. Look at the following chart for an explanation of their function.

BUTTON	DESCRIPTION
	The **New Layer** button enables you to create a new layer. When picked, this button creates a new layer in the list box where you can assign its properties. By default, a new layer will default to the settings associated with the currently selected layer.
	The **New Layer Frozen In All Viewports** button creates a new layer just as the **New Layer** button does, but AutoCAD automatically freezes this layer. You'll learn all about viewports in Lessons 24 and 25.
	Use the **Delete Layer** button to remove a layer. There can be no objects drawn on a layer to be removed. Additionally, you cannot remove layers **0**, **Defpoints**, or the current layer. Alternately, you can use the *LayDel* command to delete a layer along with all the objects that currently reside on it. (As with all deletion commands, use this one with caution!)
	Use the **Current** icon to set the selected layer current. The text box to the right of the **Set Current** button displays the current layer. Of course, using the **Layer** control box on the **Layers** panel makes setting a layer current a lot faster and easier.

Let's explore some of these tools before continuing.

Do This: 7.2.1A	Using the Layer Properties Manager

I. Be sure you're in the *flrpln* file in the C:\Steps\Lesson07 folder. If not, please open it now.
II. Follow these steps.

7.2.1A: USING THE LAYER PROPERTIES MANAGER

1. First, we'll add a new layer. Open the Layer Properties Manager.
   Command: *la*

2. Let's dock it to the left side of the work area. First, we'll have to set the Auto–Hide. Pick the **Auto–Hide** button at the top of the title bar. Be sure it looks like this and that the palette hides.

3. Now pick the **Properties** button on the title bar. AutoCAD presents a properties menu.

4. Select **Anchor Left**.
   AutoCAD moves the Layer Properties Manager to a small ribbon on the left of the screen. Move your cursor over it to display the palette.

5. We want to be able to see all the columns clearly. Right click on one of the column headers. AutoCAD presents the menu in Figure 7.003 (p.154).

6. Pick **Optimize all columns**.
   AutoCAD adjusts all the columns so you can clearly see the content.

7. Okay, we're set up and ready to go. Let's create a new layer.
   Select the **0** layer to use default settings.

8. Pick the **New Layer** button. Notice that AutoCAD creates a new layer and opens a text box so you can name it. (If only they were all that easy!)

9. Call it, *MyLayer*.

10. Pick the **Color** box. AutoCAD presents the Select Color dialog box you saw in our last lesson (Figure 6.003, p.128). Select **Red** and pick the **OK** button to return to the Layer Properties Manager.

11. Repeat Steps 8 – 10 in the **Linetype** column to give your new layer a **HIDDEN** linetype. (The linetype has already been loaded for you.)
    This layer's listing looks like the following figure.

12. Save the drawing but don't exit yet.

That was really too easy! What else can we do to get into trouble? What about filters?

The **Filters** section of the Layer Properties Manager (Figure 7.005) lists all the *groups* of layers existing in the drawing. It also allows you to manipulate the groups by selecting options on a cursor menu. The default filter will show (in the layers list) **All** the layers in the drawing. The only default subgroup is **All Used Layers**. When you select this, the list box will display only those layers which have been used in the drawing.
Let's start at the top.

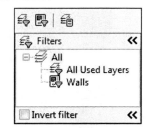

Figure 7.005

- The **New Property Filter** button calls the Layer Filter Properties dialog box (Figure 7.006) that enables you to create a new layer filter based on layer properties.

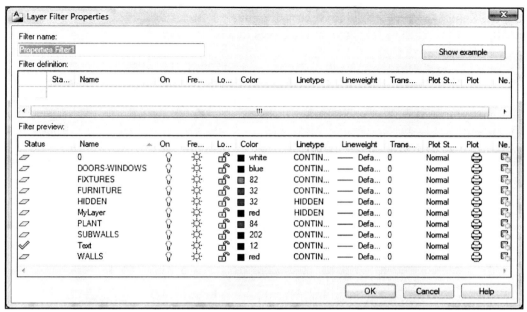

Figure 7.006

Name the new filter in the **Filter name** text box at top. The **Filter definition** box provides a series of lookup columns for you to select or name the filtered value for each property. The list of layers that meet the filter will appear in the **Filter preview** list below.

- You can create additional subgroups picking the **New Group Filter** button above the **Filters** frame. (If you select a group or subgroup before picking the **New Group Filter** button, the new group will be a subgroup of what you've selected.)
  We'll see this in action in our next exercise.

- The **Layer States Manager** button calls the Layer States Manager which we'll discuss in Section 7.2.2 (p.162).

- Below the buttons, the **Filters** title bar contains a pair of arrows that, when picked, will hide the **Filters** section altogether, thus providing more room for the layers list. Once you've set up your layer filters, it might be a good idea to show them in the layers list and hide this section.

- Finally, below the **Filters** list, you'll find a check box that enables you to invert the selected filter – that is, the list box will show everything that does *not* fall into the filtered list.

Right clicking within the **Filters** list area presents a cursor menu (Figure 7.007) that you can use to manipulate all the layers within a group simultaneously. Your options include:

- **Visibility** allows you to turn a filtered group of layers on or off, or freeze/thaw the group.

- **Lock**, of course, allows you to lock/unlock all the layers in a selected group.

- It's best to leave our discussion of viewports to Lesson 24, p.560. This will cover both the **Viewport** and **Isolate Group** options.

- The **New Properties Filter …** and **New Group Filter** options do the same thing as picking the associated button above the tree view.

- **Convert to Group Filter** changes a selected property filter to a group filter.

- **Rename** and **Delete** do just what their names imply.

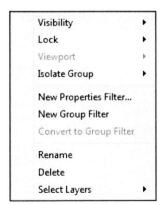

Figure 7.007

- You'll use **Select Layers** (available when you right click on a specific group/subgroup) to manually **Add** layers to or **Replace** all the layers in a selected group.

Let's explore some of these tools.

Do This: 7.2.1B	Using the Layer Properties Manager (continued)

I. Be sure you're in the *flrpln* file in the C:\Steps\Lesson07 folder. If not, please open it now.
II. Follow these steps.

7.2.1B: USING THE LAYER PROPERTIES MANAGER (CONTINUED)

1. Notice that a subgroup for the wall layers already exists in the tree view. Pick on it now. You'll see AutoCAD list (following figure) only those layers already associated with that group.

2. Let's add a layer to the **Walls** group. Be sure **All** is highlighted in the **Filters** box (right). AutoCAD shows all the layers in the list box.

3. Pick and drag the **DOORS-WINDOWS** layer from the list box onto the **Walls** group in the **Filters** box.

4. Repeat Step 1. Notice that the **Walls** filter now includes the **DOORS-WINDOWS** layer.

5. Okay, now pick the **All** filter again, and then make layer **0** current ✓.

6. Now let's freeze all the layers associated with the walls in our drawing. Right click 🖱 on the **Walls** group and select **Visibility – Frozen** from the menu.* Notice in the following figure that all of the walls, subwalls, doors and windows disappear.

Of course; it's all clear to me now!

---

* For another approach to visibility, see the insert following this exercise.

7.2.1B: USING THE LAYER PROPERTIES MANAGER (CONTINUED)

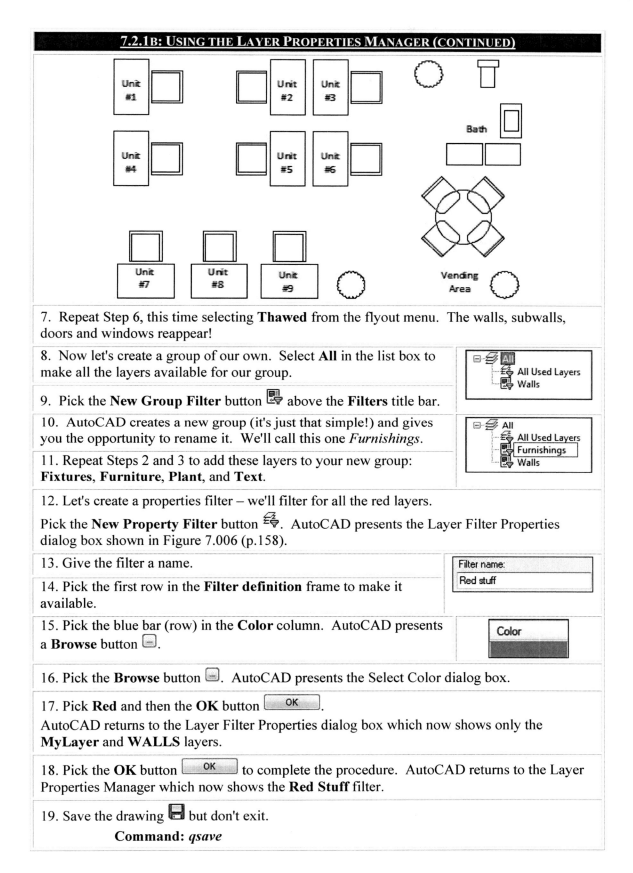

7. Repeat Step 6, this time selecting **Thawed** from the flyout menu. The walls, subwalls, doors and windows reappear!	
8. Now let's create a group of our own. Select **All** in the list box to make all the layers available for our group.	
9. Pick the **New Group Filter** button above the **Filters** title bar.	
10. AutoCAD creates a new group (it's just that simple!) and gives you the opportunity to rename it. We'll call this one *Furnishings*.	
11. Repeat Steps 2 and 3 to add these layers to your new group: **Fixtures, Furniture, Plant,** and **Text**.	
12. Let's create a properties filter – we'll filter for all the red layers. Pick the **New Property Filter** button . AutoCAD presents the Layer Filter Properties dialog box shown in Figure 7.006 (p.158).	
13. Give the filter a name.	Filter name: Red stuff
14. Pick the first row in the **Filter definition** frame to make it available.	
15. Pick the blue bar (row) in the **Color** column. AutoCAD presents a **Browse** button .	Color
16. Pick the **Browse** button . AutoCAD presents the Select Color dialog box.	
17. Pick **Red** and then the **OK** button . AutoCAD returns to the Layer Filter Properties dialog box which now shows only the **MyLayer** and **WALLS** layers.	
18. Pick the **OK** button to complete the procedure. AutoCAD returns to the Layer Properties Manager which now shows the **Red Stuff** filter.	
19. Save the drawing but don't exit. **Command:** *qsave*	

> AutoCAD has another visibility tool that operates independent of layer visibility. Use **Object Visibility** (the **ObjectIsolationMode** system variable) to isolate selected objects. Here's the procedure:
> 1. Select the objects you wish to modify.
> 2. Select **Isolate – Isolate Objects** from the cursor menu.
>
> AutoCAD will hide everything else in the current drawing allowing you to focus on the objects you wish to modify. (Alternately, you can select **Isolate – Hide Objects** to hide the selected objects.) To redisplay the hidden objects, simply select **Isolate – End Object Isolation** from the cursor menu.
> Use this handy tool in a busy drawing to minimize distractions!

One last thing to finish up this section – notice the tools in the upper right corner of the manager (Figure 7.002, p.153). These are:

- The **Search for Layer** box where you can enter the name of a layer that you'd like AutoCAD to display in the list box.
- A **Refresh** button to refresh the palette's display.
- A **Settings** button, which calls the Layer Settings dialog box (Figure 7.008). This dialog box can save you some aggravation.
  - When AutoCAD saves a drawing, it saves a list of layers. It evaluates this list whenever it opens the drawing in the future. (It's a way to prevent drawing corruption.) When you create a new layer, AutoCAD will recognize that it isn't on the list and alert you with a bubble saying that it has found "unreconciled layers" and give you the chance to reconcile them. (The bubble will have a link that opens the Layer Properties Manager with a filtered list of the unreconciled layers. Simply right click on the layers and select **Reconcile** from the menu.)
    This dialog box controls whether or not AutoCAD will evaluate the new layers (for **xrefs** or **all new layers**) and whether or not it will **Notify when new layers are present**.
    If the **Unreconciled Layers** alert bubble bothers you, shut it off by removing the check next to **Notify when new layers are present**.
  - Use the **Isolate Layer Settings** frame to tell AutoCAD whether or not to **Lock and fade** non-isolated layers. When fading, you can control the amount of fading using the **Locked Layer Fading** percentage box.
  - Use the **Apply layer filter to layer toolbar** check box in the **Dialog Settings** frame to have AutoCAD use your current filter settings in the **Layer** control box on the **Layers** ribbon panel.

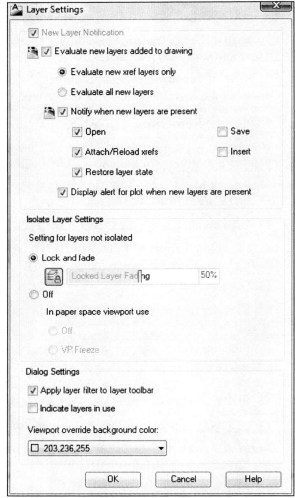

Figure 7.008

Let's take a deeper look now at that Layer States Manager called by the button above the Filters section.

7.2.2 The Layer States Manager

Before we begin, you might want to know just what exactly *Layer States* are.

Well, larger drawings frequently contain dozens of layers (sometimes even more!). That's a lot a of layer settings to adjust when you have a specific job in mind. When you have to work on several jobs simultaneously, you may find yourself having to change the on/off, locked, transparency, and plot settings frequently. That takes a lot of time – and requires that you remember exactly how you had your layers set up for each job!

I don't know about you, but I'm far too lazy for that sort of effort (and this job just doesn't pay *that* well!).

Where to Find It:	
Command Line:	*LayerState*
Hotkey(s):	*las*
Ribbon (Tab/Panel):	Home – Layers – Layer State control – **Manage Layer States**
Menu:	Format – **Layer States Manager**
Toolbar:	Layers – **Layer States Manager**
Palette:	Layer Properties Manager – **Layer States Manager**

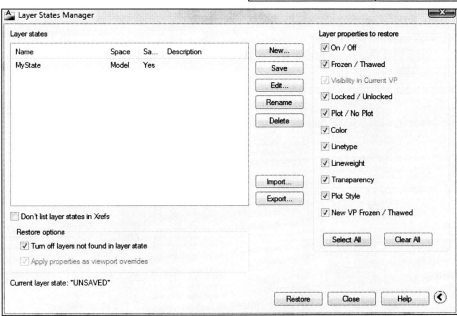

Figure 7.009

Enter Layer States – AutoCAD's ability to remember your layer settings *and return to them on your command*! We'll use the handy Layer States Manager dialog box (Figure 7.009).

- Start with the **Layer properties to restore** list on the right side of the dialog box. (If you don't see this list on your dialog box, pick the right-pointing arrow in the lower right corner.) AutoCAD will remember the checked settings in the layer state you create.

- The list box (the dominant part of the dialog box) has four columns of information: the **Name** of the state, in which **Space** this state occurs (more on Model/Layout spaces in Lessons 24, p.560, & 25, p.584), whether or not the state is the **Same as DWG** (current), and any **Description** you may have given the state when you created it.

- Below the list box, you'll find a **Don't list layer states in Xrefs** check box. We'll look at Xrefs in Lesson 27, p.630.
- Below this, you'll find the **Restore options** frame with two more check boxes:
  - A check next to **Turn off layers not found in layer state** means that, when you restore the selected state, AutoCAD will automatically turn off any layers that you may have added since you created the state.
  - A check next to **Apply properties as viewport overrides** will override viewport layer settings when you restore the layer state. You'll learn about viewports in Lesson 24, p.560.
- The buttons across the bottom of the dialog box do just what they suggest:
  - **Restore** restores the selected state's layer settings.
  - **Close** closes the dialog box.
  - **Help** calls AutoCAD's Help dialog box with information about the Layer States Manager.
- You'll find seven buttons down the center of the Layer States Manager. These are the workhorses of layer states.
  - **New** calls the New Layer State to Save dialog box (Figure 7.010). Here you'll **name** your new state and provide an (optional) **Description**.
  - **Save** saves any changes you made in the selected layer state's settings.
  - The **Rename** and **Delete** buttons do just what their names imply.
  - The last two buttons – **Import** and **Export** – both call select file dialog boxes. Use these to swap layer settings between drawings.

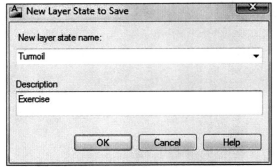

Figure 7.010

  - **Edit** calls the Edit Layer State dialog box (Figure 7.011) where you can make adjustments in your selected layer state's setup. The list box here resembles it's counterpart in the Layer Properties Manager in both form and function, so I'll refer you to that discussion. You will find, however, a couple buttons in the bottom left corner that you'll need to know.

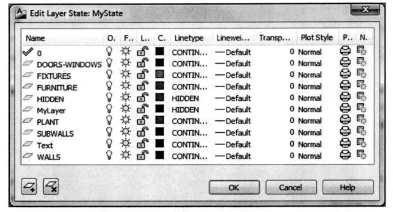

Figure 7.011

    - **Add layer to layer state** presents a dialog box that enables you to add new layers to the state you're editing.
    - **Remove layer from layer state** deletes the selected layer from the state you're editing.

Let's get some experience with Layer States.

Do This: 7.2.2A	Using the Layer States Manager

I. Be sure you're in the *flrpln* file in the C:\Steps\Lesson07 folder. If not, please open it now.

II. Follow these steps.

7.2.2A: USING THE LAYER STATES MANAGER

1. Open the Layer States Manager.
   Command: *las*
   Remember to use the right-pointing arrow in the lower right corner of the manager to see the hidden pane.

2. Pick the **New** button and create a new layer state by putting its name in the New Layer State to Save dialog box (Figure 7.010, p.163). I'll call mine *Confusion* (you know – the state of ... oh, never mind).

3. Pick the **OK** button to complete the procedure. Notice that the name of your new layer state now appears in the list box. That's all there is to creating a layer state based on the current layer settings in the drawing.

4. **Close** the Layer States Manager.

5. Now freeze and lock some of the layers as indicated at right. Use the **Layers** control box in the ribbon panel. Notice the difference in the drawing.

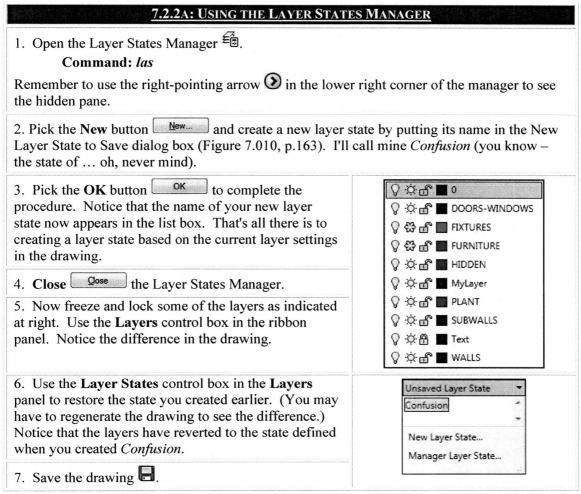

6. Use the **Layer States** control box in the **Layers** panel to restore the state you created earlier. (You may have to regenerate the drawing to see the difference.) Notice that the layers have reverted to the state defined when you created *Confusion*.

7. Save the drawing.

Take a few minutes to play with some of the other Layer State tools we've discussed here. You'll find Layer States will save you tons of time as your drawings become more complex.

7.3 The Rest of the (Layer) Story

Figure 7.012

AutoCAD provides a host of supplementary layer tools – most with buttons on the **Layers** ribbon subpanel (Figure 7.012). (Pick the arrow on the Layers ribbon title bar to expose the subpanel.)

Where to Find It:	
Command Line:	*LayOn* or *LayTHW*
Ribbon (Tab/Panel):	Home – Layers (subpanel) – **Turn All Layers On** or **Thaw All Layers**
Menu:	Format – Layer Tools – **Turn All Layers On** or **Thaw All Layers**

Some of these might save you some time and aggravation. Let's take a look (from the top left).

- The first two buttons – **Turn All Layers On** (*LayOn* command) and **Thaw All Layers** (*LayTHW* command) do just what their names imply but don't require any dialog box interaction (a simple button pick does it)!

- The next two buttons – **Lock** and **Unlock** – launch the *LayLCK* and *LayULK* commands respectively. These will lock/unlock selected layers. AutoCAD prompts:
  Select an object on the layer to be [un]locked: *[Select an object on the layer you wish locked or unlocked.]*

Where to Find It:	
Command Line:	LayLCK or LayULK
Ribbon (Tab/Panel):	Home – Layers (subpanel) – Lock or UnLock
Menu:	Format – Layer Tools – Layer Lock or Layer UnLock
Toolbar:	Layers II – Lock or Unlock

- **Change to Current Layer** (*LayCUR*) moves selected objects from the layer on which they currently reside to the current layer. AutoCAD prompts:
  Select objects to be changed to the current layer: *[Select the object whose layer you wish to change.]*

Where to Find It:	
Command Line:	LayCUR
Ribbon (Tab/Panel):	Home – Layers (subpanel) – Change to Current Layer
Menu:	Format – Layer Tools – Change to Current Layer
Toolbar:	Layers II – Change to Current Layer

- **Copy Objects to New Layer** (*CopyToLayer*) copies selected objects to a user-specified layer and gives you a chance to relocate the new objects. This one's a bit more complex:
  Select objects to copy: *[Select the objects to copy.]*
  Select object on destination layer or [Name] <Name>: *[Either select an object on the target layer or enter the name of the target layer.]*
  1 object(s) copied and placed on layer "WALLS". *[AutoCAD lets you know what it has done.]*
  Specify base point or [Displacement/eXit] <eXit>: *[Now AutoCAD gives you the chance to move the new object – you'll see more on the Move command in Lesson 8.]*
  Specify second point of displacement or <use first point as displacement>:

Where to Find It:	
Command Line:	CopyToLayer
Ribbon (Tab/Panel):	Home – Layers (subpanel) – Copy Objects to New Layer
Menu:	Format – Layer Tools – Copy Objects to New Layer
Toolbar:	Layers II – Copy Objects to New Layer

- A really cool tool – **Layer Walk** (*LayWalk*) – presents a dialog box that allows you to select the layer you want to see. AutoCAD will turn off all but the selected layers allowing you to "walk" through the drawing, one layer at a time. We'll see this one in our next exercise.

Where to Find It:	
Command Line:	LayWalk
Ribbon (Tab/Panel):	Home – Layers (subpanel) – Layer Walk
Menu:	Format – Layer Tools – Layer Walk
Toolbar:	Layers II – Layer Walk

- **Isolate to Current Viewport** (*LayVPI*) freezes a selected layer in all but the current viewport. Again, we'll discuss viewports in Lessons 24, p.560, and 25, p.584.

- **Merge** (*LayMRG*) can also get a bit complicated. It takes all the objects on a selected layer (can't be the current layer), moves them to a second selected layer, and deletes the first layer from the drawing. It looks like this:

	Where to Find It:
Command Line:	*LayMRG*
Ribbon (Tab/Panel):	Home – Layers (subpanel) – Merge
Menu:	Format – Layer Tools – Layer Merge

  Select object on layer to merge or [Name]: *[Select an object on the layer you wish to delete from the drawing.]*
  Selected layers: PLANT. *[AutoCAD confirms which layer you selected.]*
  Select object on layer to merge or [Name/Undo]: *[Continue selecting layers to delete or hit ENTER.]*
  Select object on target layer or [Name]: *[Now pick a layer on which the objects will reside after the initially selected layer is deleted.]*
  ******** WARNING ********
  You are about to merge layer "PLANT" into layer "WALLS".
  Do you wish to continue? [Yes/No] <No>: *[AutoCAD gives you a chance to change your mind.]*
  Deleting layer "PLANT".
  1 layer deleted. *[AutoCAD admits what it has done!]*

- Similar to **Merge**, **Delete** (*LayDEL*) *removes* everything residing on a selected layer and then removes the layer. Be careful with this one!

	Where to Find It:
Command Line:	*LayDEL*
Ribbon (Tab/Panel):	Home – Layers (subpanel) – Delete
Menu:	Format – Layer Tools – Layer Delete

  Select object on layer to delete or [Name]: *[Select an object on the layer you wish to delete.]*
  Selected layers: PLANT. *[AutoCAD confirms which layer you selected]*
  Select object on layer to delete or [Name/Undo]: *[Continue selecting layers to delete or hit ENTER.]*
  ******** WARNING ********
  You are about to delete layer "PLANT" from this drawing.
  Do you wish to continue? [Yes/No] <No>: *[AutoCAD gives you a chance to change your mind.]*
  Deleting layer "PLANT".
  1 layer deleted. *[All finished!]*

- Use the **Locked Layer Fading** toggle with the **Locked Layer Fading** slider bar `Locked layer fading    50%` to control the fading that occurs when you lock a layer. When the toggle is engaged (on), the slider locator appears in the bar and objects residing on the locked layer will fade. Control the amount of fading by moving the slider locator back and forth.

Most of these new tools require little experience to operate, but an exercise might help with others.

Do This: 7.3A	Other Layer Tools

  I. Open the *flrpln-b* file in the C:\Steps\Lesson07 folder. The drawing is a pristine copy of the *flrpln* drawing with which you've been working.
  II. In this exercise, we'll use tools found on the ribbon's **Layers** panel and subpanel (Figure 7.012, p.164). Pick the arrow on the ribbon's **Layers** title bar to expose the subpanel.

III. Follow these steps.

7.3A: OTHER LAYER TOOLS

1. We'll start with the easy one, enter the *LayWalk* command. **Command:** *laywalk* AutoCAD presents the Layer Walk dialog box shown here.	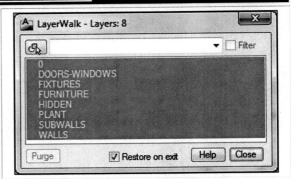
2. Select each of the layers listed in the dialog box and watch the drawing area. AutoCAD turns off all layers except the one selected.	

3. Reselect the entire list (select the top of the list, then hold down the SHIFT key while selecting the bottom). Now pick the **Select Objects** button in the upper left corner of the dialog box.
AutoCAD returns to the screen and prompts you to select objects.

4. Select an object on the screen.
   Select objects:
   Select objects: *[ENTER]*
AutoCAD returns to the dialog box, highlights the layer on which the object resides, and turns the rest of the layers off.

5. Close the dialog box.

6. Enter the *LayLCK* command.
   Command: *laylck*

7. Select an object on the layer you'd like to lock.
   Select an object on the layer to be locked:
   Layer "FURNITURE" has been locked.
Notice that AutoCAD fades all the objects on the selected layer.

8. Use the *LayULK* command to restore the layer.
   Command: *layulk*
   Select an object on the layer to be unlocked:
   Layer "FURNITURE" has been unlocked.

9. Now let's change an object to the current layer. Enter the *LayCUR* command.
   Command: *laycur*

10. Select an object.
    Select objects to be changed to the current layer: *[Select an object.]*
    Select objects to be changed to the current layer: *[ENTER to confirm.]*
    One object changed to layer DOORS-WINDOWS (the current layer).
Notice the difference?

7.3A: OTHER LAYER TOOLS

11. We can use the *LayMCur* command to make an object's layer current. Enter the command.
    Command: *laymcur*

12. Select an object and watch the current layer in the **Layer** control box on the ribbon's **Layers** panel.
    Select object whose layer will become current:

13. Finally, we'll see what happens when we merge a couple layers. Enter the *LayMrg* command.
    Command: *laymrg*

14. AutoCAD asks you to select an object on the layer you wish to merge to another layer. (This layer will be deleted.) Select a desk.
    Select object on layer to merge or [Name]:
    Selected layers: FURNITURE
    Select object on layer to merge or [Name/Undo]: *[ENTER]*

15. Now AutoCAD wants to know onto which layer you'd like to merge the objects on the selected layer. This time, let's use the dialog box. Select the **Name** option by entering *N* or selecting it from the menu.
    Select object on target layer or [Name]: *N*

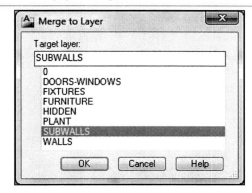

16. AutoCAD presents the Merge to Layer dialog box. Select the **SUBWALLS** layer. Pick the **OK** button to close the dialog box.

17. One last chance to change your mind – select **Yes**.
    AutoCAD takes all the objects on the **FURNITURE** layer and puts them on the **SUBWALLS** layer.

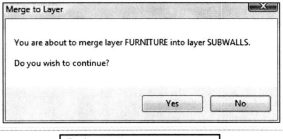

18. Check the **Layers** control box; notice that the **FURNITURE** layer has been removed.

19. Feel free to experiment with the other buttons, then **Close** the drawing without saving the changes.
    Command: *close*

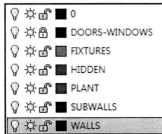

7.4 Sharing Setups: The AutoCAD Design Center and the Layer Translator

You've already seen one way to share drawing information between drawings – templates. These work well; however, AutoCAD provides two considerably more powerful tools – the AutoCAD Design Center and (for layers only) the Layer Translator (more on the Layer Translator on p.171). The first – the AutoCAD Design Center (ADC) – allows you to *mine* another drawing for useful stuff – layers, blocks, text styles, dimension styles, etc. That is, you can dig into another drawing and find and retrieve the pieces you want.

Where to Find It:	
Command Line:	*ADCenter*
Hotkey(s):	*adc*
Ribbon (Tab/Panel):	View – Palettes – **DesignCenter**
Menu:	Tool – Palettes – **DesignCenter**
Toolbar:	Standard – **DesignCenter**

But unlike gold mines, the mined drawing remains unaffected by the task.

AutoCAD presents the ADC floating over the screen (Figure 7.013). You can dock or undock it as you would other palettes.

Let's take a look at the ADC.

- The title bar on the left should be familiar to you after your study of palettes. Its function is exactly the same.
- A toolbar resides along the top of the ADC. The function of each tool follows:

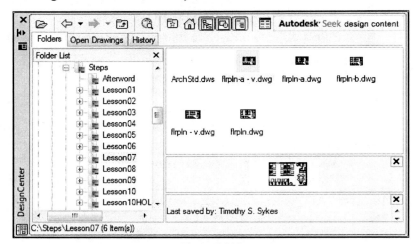

Figure 7.013

TOOL	DESCRIPTION
📂	**Load** provides a standard File ... Open window. With that, you can navigate to the desired folder and open the palette for the desired drawing. The palette, in this case, is simply a list of things available to you – layers, blocks, text styles, etc.)
⬅	The **Back** button returns you to the previous folder. The down arrow next to the **Back** button allows you to select from all the folders you have previously visited.
➡	The **Next** button works conversely to the **Back** button. That is, once you have returned to a previous folder, the **Next** button allows you to navigate back to the folder you occupied prior to using the **Back** button.
🔼	**Up**, another navigation button, changes the display to a step back (up) along the path.
🔍	**Search** presents a dialog box similar to the Windows **Find** program. Using this, you can search for drawings by date modified or included text.
🗂	**Favorites** opens the **Favorites** folder in the folder list.

Tool	Description
🏠	**Home**, of course, returns you to a pre-designated folder. By default, this is the C:\Program Files\AutoCAD 2011\Sample\DesignCenter folder. (To reset the home location, navigate to the desired folder, right click 🖱 on it, and select **Set as Home** from the menu.)
	This is a toggle button for the **Tree** view (the view seen in Figure 7.013, p.169). When depressed, your Design Center shows the folders list in a frame to the left of the window. When raised, the folders list disappears.
	The **Preview** button toggles on or off the drawing preview display just below the content area to the right of the folders list.
	The **Description** button toggles on or off the drawing description display just below the preview display.
	The **Views** button works just like the **Views** button in Windows – it allows you to determine how you will see items in a folder (large icons, small icons, listed, or with details).
Autodesk Seek design content	The **Autodesk Seek** button opens an on-line tool you can use to search the web for available AutoCAD content. You'll often find manufacturers content which you can drag-n-drop into your drawing using iDrop (more in iDrop in Lesson 21, p.496).

- Three tabs appear below the toolbar: **Folders**, **Open Drawings**, and **History**.
  - On the **Folders** tab (Figure 7.013, p.169), you see the **Tree** view. That is, like Windows Explorer, you see the path to a folder on the left (here you see C:\Steps\Lesson07). The right shows the contents of the folder, a preview frame, and a description frame. You can navigate the tree view just as you do Windows Explorer.

    Below the left frame, you see the path that's being shown and the number of items in the current folder.
  - The **Open Drawings** folder shows only those drawings that are currently open.
  - The **History** folder shows the last few drawings that have been opened.

> A useful trick to know is that, once you've navigated to a folder, you can add that folder to your **Favorites** list by selecting **Add to Favorites** on the right click cursor menu.

Let's see just how useful the ADC can be. We'll use it to copy the layers from our *flrpln* drawing file to a new file.

Do This: 7.4A	Using the AutoCAD Design Center

I. Start a new drawing from scratch.
II. Follow these steps.

7.4A: Using the AutoCAD Design Center

1. Open the AutoCAD Design Center 🗔.
   Command: *adc*

2. Navigate to the C:\Steps\Lesson07 folder just as you would using Windows Explorer, and pick on the plus sign next to the *flrpln* drawing.

Pick on **Layers** as indicated in the following figure.

Notice the palette shown in the right window. Since **Layers** has been selected in the left window, the palette shows the layers in the *flrpln* drawing.

7.4A: USING THE AutoCAD DESIGN CENTER

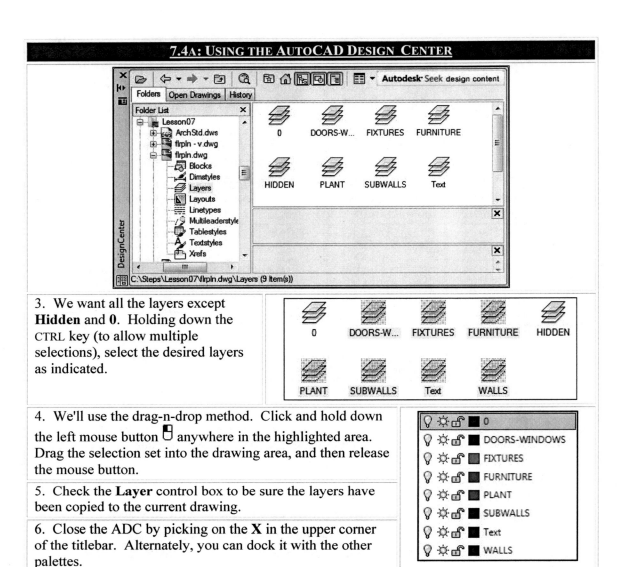

3. We want all the layers except **Hidden** and **0**. Holding down the CTRL key (to allow multiple selections), select the desired layers as indicated.

4. We'll use the drag-n-drop method. Click and hold down the left mouse button anywhere in the highlighted area. Drag the selection set into the drawing area, and then release the mouse button.

5. Check the **Layer** control box to be sure the layers have been copied to the current drawing.

6. Close the ADC by picking on the **X** in the upper corner of the titlebar. Alternately, you can dock it with the other palettes.

The ADC can be used as easily to copy blocks, dimstyles, layouts, leader styles, text styles, table styles, line types, or Xrefs.

Where to Find It:	
Command Line:	*LayTrans*
Ribbon (Tab/Panel):	Manage – CAD Standards – **Layer Translator**
Menu:	Tools – CAD Standards – **Layer Translator**

The second tool AutoCAD provides for sharing layer setups allows you to translate an existing layer or group of layers to meet a set standard. AutoCAD calls this tool the Layer Translator. The standard can be taken from an existing drawing or from a standard drawing (a drawing with a .dws extension) created to help maintain consistency throughout a project.

Let's see how the Layer Translator works.

Do This: 7.4B	**The Layer Translator**

I. Begin in the *flrpln* drawing in the C:\Steps\Lesson07 folder. (Close any other drawings.)

II. Follow these steps.

171

7.4B: THE LAYER TRANSLATOR

1. Call the Layer Translator.
   Command: *laytrans*

   AutoCAD presents the Layer Translator dialog box (see the following figure). Notice that AutoCAD lists the layers in the current drawing in the **Translate From** frame.

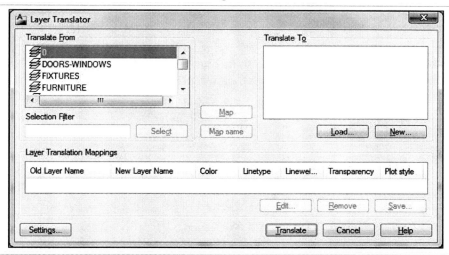

2. First, we must load a *Standards* drawing. Pick the **Load** button below the **Translate To** frame.

3. AutoCAD presents a typical Select ... Files dialog box. First, tell the box to look for *.dws* files as indicated.

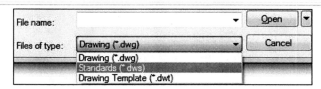

4. Now select the *ArchStd.dws* file in the C:\Steps\Lesson07 folder and pick the **Open** button. Notice now that AutoCAD provides another list of layers – this one in the **Translate To** frame (following figure). These are our standard layers – the ones we wish to use in our current drawing.

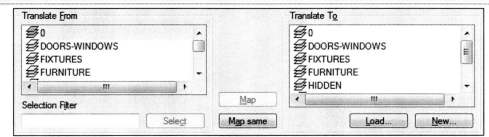

5. When we opened a standards file, AutoCAD activated the **Map same** button (between the frames). This provides a fast and easy way to tell AutoCAD to accept layers in our current drawing that are already standardized (already match the layers in the standards file).

Pick the **Map same** button now. Notice that AutoCAD details the mapped layers in the **Layer Translation Mappings** frame (following figure). Notice also that the **PLANT** layer did not translate (it remains in the **Translate From** frame). The **PLANT** layer isn't a part of the standard setup.

7.4B: THE LAYER TRANSLATOR

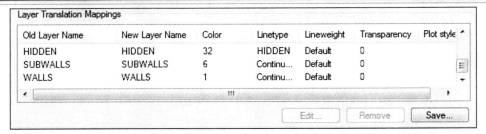

6. Let's translate the **Plant** layer to the standard **Landscaping** layer. Select **PLANT** in the **Translate From** frame. Select **LANDSCAPING** in the **Translate To** frame as shown in the following figure.

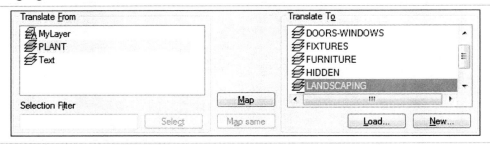

7. Pick the **Map** button to translate objects on the **PLANT** layer to objects on the **LANDSCAPING** layer.

8. Ignore the other layers in the **Translate From** frame – they don't need to be translated. Pick the **Translate** button to complete the procedure. AutoCAD will prompt you with a dialog box (right) giving you the option to **Translate and save mapping information** or to **Translate only**. Pick **Translate only**.

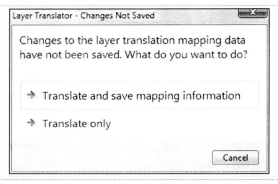

Notice the drawing changes? AutoCAD has translated the layer settings in the *flrpln* drawing file to match those in the project drawing standards file. How simple!

Some additional things to know about the layer translator include:

- You can create a new layer as a target (**Translate To**) layer by picking the **New** button at the bottom of the **Translate To** frame. AutoCAD provides a simple dialog box (Figure 7.014) to help you set up the new layer.
- Use the **Edit** button below the **Layer Translation Mappings** frame to change the settings (color, linetype, plotstyle) on a specific translation layer. Changes here won't affect the settings in the standards file.

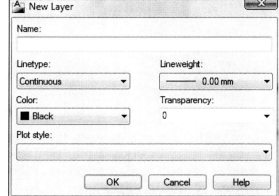

Figure 7.014

173

- The **Remove** button (next to the **Edit** button) will remove a mapped layer translation. The layer that had been mapped will return to the **Translate From** frame for remapping.
- The **Save** button allows you to save the current mappings for later use.
- The **Settings** button in the lower left corner of the Layer Translator calls the Settings dialog box (Figure 7.015). Here you can control exactly what happens during a layer translation.

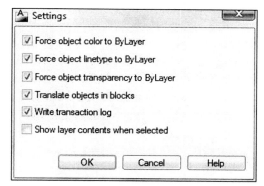

Figure 7.015

7.5 Extra Steps

Have you noticed that you repeat the same commands with some frequency? AutoCAD has provided a way to make those commands available to you in one central location without requiring you to look for on this toolbar or that drop down menu or someplace else. You'll find the tool palettes to be quite easily customized.

Try this.

Let's say you've discovered that you use the *Line*, *Circle*, *Erase*, and *DText* commands quite often. Their buttons are in a variety of places and you can't always quickly remember where to look.

Follow this procedure:

1. If it isn't already open, open the tool palettes. (Lesson 1, Section 1.1.4, p.11.)
2. Pick the **Properties** button on the title bar and select **New Palette** from the menu.
3. AutoCAD will ask you to name the new palette; I'll call mine *My Tools*.
4. Drag a line from the drawing onto the palette (drag the line – not one of the colored boxes). Notice that AutoCAD places a **Line** tool on the palette. (Repeat this step with other objects.)
   Try to draw something from the new palette. Notice that AutoCAD creates the object with the properties (layer, color, linetype, etc.) of the object you used to create the tool.
5. Right click on the tool and select **Properties** from the menu. Use the Tool Properties dialog box to redefine the properties of objects created with this tool.
6. Do the same for DText.

Now, anytime you need one of your more common commands, it'll be right there on your own tool palette!

7.6 What Have We Learned?

Items covered in this lesson include:

- *Using the Layer Properties dialog box*
- *Creating your own tool palette*
- *Using the AutoCAD Design Center*
- *Commands*
    - ***ObjectIsolationMode***
    - ***ADCenter***
    - ***LayCur***
    - ***LayDel***
    - ***LayFrz***
- *Using the Layer translator*
- *An introduction to Autodesk Seek*
- *Working with Layer States*

    - ***LayIso***
    - ***LayUnIso***
    - ***LayLck***
    - ***LayMch***
    - ***LayMCur***

    - ***LayMrg***
    - ***LayOff***
    - ***LayOn***
    - ***LayThw***
    - ***LayUlk***

- *LayWalk*
- *LayLockFadeCTL*
- *CopyToLayer*
- *SetByLayer*
- *Laytrans*

This lesson has challenged you more than any other thus far. We've taken AutoCAD from a simple drawing toy to a real drafting tool. We've seen how to use layers to put more into a drawing than was ever possible in the world of graphite and paper.

Next we'll begin to look at that 70% of CAD work that's *not* drawing – modifying what we've drawn. But try some exercises first and get a bit more comfortable with the material in this chapter.

7.7 Exercises

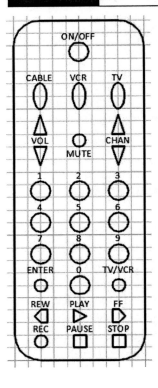

1. Using the layers listed in the chart and the parameters below it, create the drawing shown atop the next page. Save the drawing as *MyRemote* in the C:\Steps\Lesson07 folder.

Layer Name	Color	Linetype
Button	82	continuous
Dim	blue	continuous
Obj	32	continuous
Text	red	continuous
Toggle	blue	continuous
Up-Down	215	continuous

   1.1. Lower left limits: 0,0
   1.2. Upper right limits: 6,10
   1.3. Architectural units
   1.4. Grid: ¼"
   1.5. Snap: as needed
   1.6. Font: Times New Roman
   1.7. Text Height: 1/8

7.8 For Web-Based Review Questions and Additional Exercises, visit: http://foragerpub.com/AcadFiles/2011/2011.htm

Lesson  8

Following this lesson, you will:

✓ Be familiar with several modification procedures, including:
  - **Trim and Extend**
  - **Break**
  - **Fillet** & **Chamfer**
  - **Move & Align**
  - **Copy, Offset, Array, & Mirror**
  - **AddSelected**

✓ Be familiar with Windows' procedures for moving and copying objects between drawings

Editing Your Drawing – Modification Procedures

Although we'll cover some more advanced drawing commands and techniques in later chapters, the first seven lessons have made you comfortable with AutoCAD's basic approach to 2-dimensional drawing. But you probably still feel a bit clumsy with some of your work. This comes partly from the newness of AutoCAD to you, partly from the need to erase and redo an effort because of small mistakes, and partly from the need to draw the same thing over and over (in great Sisyphean efforts) as you did the remote's buttons in our last lesson's exercises. Additionally, you can't create some more complex drawings that might be simple on the drawing board – given templates and an electric eraser.

In this lesson, we'll tackle several commands meant to save drawing time and effort. There are many commands to cover, but they're not difficult.

We'll divide the basic modification routines into two groups: the Change Group, *which will include commands designed to change an object's appearance or basic properties, and the* Location & Number Group, *which will include commands designed to move or duplicate existing objects.*

8.1	The Location and Number Group
8.1.1	Here to There: The *Move* Command

The *Move* command allows you to move one or more objects from one place to another. It has one of the easiest sequences to remember:

Command: *move*

Select objects: *[Select one or more objects.]*

Select objects: *[ENTER to confirm completion of this selection set.]*

Specify base point or [Displacement] <Displacement>: *[Select the point at which you'll "pick up" the object.]*

Specify second point or <use first point as displacement>: *[Pick the target point – the place where you'll put the object "down".]*

Where to Find It:	
Command Line:	*Move*
Hotkey(s):	*m*
Ribbon (Tab/Panel):	Home – Modify – **Move**
Menu:	Modify – **Move**
Toolbar:	Modify – **Move**
Tool Palette:	Modify – **Move**

The only options here involve the **base point** and the **displacement**.

- The easiest way to explain **base point** is this: imagine the object(s) you're moving as a solid object sitting on a table. To move the object, you must first pick it up. The **base point** is the place you grab (a corner, an edge, the middle, etc.). The **second point** is where you put it down – or the point at which you'll place the corner or edge (or whatever) you grabbed.

- Notice the **base point or [Displacement]** prompt. You have two ways to use a displacement method. (Both produce the same results.)
  - First, you can enter an X,Y (or X,Y,Z) distance at the **Specify base point** prompt. Then you'll hit ENTER at the **Specify second point** prompt. AutoCAD will move the selected objects the designated distance along the X- and Y-axes.
  - Alternately, you can hit ENTER at the **Specify base point or [Displacement] <Displacement>** prompt and then enter an X and Y distance. AutoCAD will then move the object as though you'd used the relative coordinate system.

Let's try the *Move* command.

177

Do This: 8.1.1A	Using the *Move* Command

I. Open the *M–C* file in the C:\Steps\Lesson08 folder. The drawing looks like the figure at right.

II. Follow these steps.

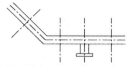

8.1.1A: USING THE MOVE COMMAND

1. Enter the *Move* command.

   Command: *m*

2. Select the nozzle. Then complete the selection set.

   Select objects:
   Select objects: *[ENTER]*

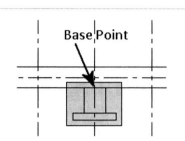

3. Use the intersection 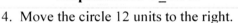 of the nozzles centerline and the pipe as your base point.

   Specify base point or [Displacement] <Displacement>: _int of

4. Move the circle 12 units to the right.

   Specify second point or <use first point as displacement>: *@12<0*

   Notice how the nozzle changes position to the other centerline.

5. Now we'll use the **Displacement** option to move it back. Repeat the command.

   Command: *[ENTER]*

6. Select the nozzle again.

   Select objects:
   Select objects: *[ENTER]*

7. Tell AutoCAD to use the **Displacement** option.

   Specify base point or [Displacement] <Displacement>: *[ENTER]*

8. Return the objects to their original location. Notice that you don't have to enter a Z-distance.

   Specify displacement <0.0000, 0.0000, 0.0000>: *-12,0*

As you can see, there really isn't much to the *Move* command. But you do need to remember things like OSNAPs and Cartesian coordinates to ensure precision in your modifications.

8.1.2	The *Copy* Command: From One to Many

Copy has a command sequence that closely resembles the *Move* command sequence. Compare the following sequence to the *Move* sequence in Section 8.1.1 (p.177).

Command: *copy*
Select objects: *[Select one or more objects.]* 1 found
Select objects: *[ENTER to confirm completion of this selection set.]*

Where to Find It:	
Command Line:	*Copy*
Hotkey(s):	*co* or *cp*
Ribbon (Tab/Panel):	Home – Modify – **Copy**
Menu:	Modify – **Copy**
Toolbar:	Modify – **Copy**
Tool Palette:	Modify – **Copy**

Current settings: Copy mode = Multiple
Specify base point or [Displacement/mOde] <Displacement>: *[Pick a starting (grabbing) point.]*
Specify second point or <use first point as displacement>: *[Pick a target point.]*
Specify second point or [Exit/Undo] <Exit>: *[Continue to pick target points, or ENTER to complete the command.]*

All but one of the options are the same as the *Move* command's – including **Displacement**. But *Copy* provides one big difference. Whereas the *Move* command simply moves an object, the *Copy* command will leave the original and place a copy at the **second point**.

Want more? By default, AutoCAD will continue to place copies until you tell it you have enough (by hitting ENTER). Use the **mOde** option to toggle the default **Multiple** mode off.

Let's give it a try!

Do This: 8.1.2A	Using the *Copy* Command

I. Be sure you're in *M–C* in the C:\Steps\Lesson08 folder.
II. Follow these steps.

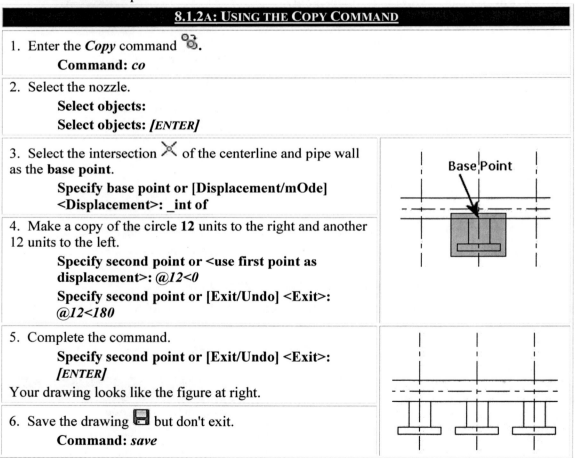

8.1.2A: USING THE COPY COMMAND
1. Enter the *Copy* command.   **Command:** *co*
2. Select the nozzle.   **Select objects:**   **Select objects:** *[ENTER]*
3. Select the intersection of the centerline and pipe wall as the **base point**.   **Specify base point or [Displacement/mOde] <Displacement>:** *_int of*
4. Make a copy of the circle **12** units to the right and another 12 units to the left.   **Specify second point or <use first point as displacement>:** *@12<0*   **Specify second point or [Exit/Undo] <Exit>:** *@12<180*
5. Complete the command.   **Specify second point or [Exit/Undo] <Exit>:** *[ENTER]*   Your drawing looks like the figure at right.
6. Save the drawing but don't exit.   **Command:** *save*

As you can see, *Copy*, like *Move*, requires little effort to master. But the benefits can be wondrous in terms of time and effort saved.

8.1.3 AddSelected

This cool new tool copies the properties (layer, linetype, etc) of a selected object and enables you to create a new object with the same properties.

Where to Find It:	
Command Line:	*AddSelected*
Cursor Menu:	*AddSelected*
Toolbar:	Draw – **AddSelected**

The command sequence begins after you select an object on the screen.

    Command: *addselected*

    [At this point, AutoCAD begins the command sequence associated with the selected object.]

This'll be clearer with an example.

Do This: 8.1.3A	Using AddSelected

I. Open the *star+* drawing in the C:\Steps\Lesson08 folder. The drawing looks like the figure at right.
II. Set the **textsize** system variable to ½".
III. Follow these steps.

8.1.3A: USING ADDSELECTED
1. Select the circle.
2. Select the *AddSelected* command [Add Selected] from the cursor menu 🖰.   **Command:** *addselected*
3. Draw a 3" circle in a clear area of the screen.   **_circle Specify center point for circle or [3P/2P/Ttr (tan tan radius)]:**
4. Select the text within the original circle.
5. Select the *AddSelected* command [Add Selected] from the cursor menu 🖰.   **Command:** *addselected*
6. Add the text *My Star* to the center of the new circle, accepting the defaults at the prompts.
7. Compare the properties of both circles. Which are the same? Which are different?
8. Repeat Step 7 for the text in both circles.
9. Undo the changes ↶.   **Command:** *u*

So you see, *AddSelected* ignores current drawing settings and gives you a quick way to create a new object using the settings used when similar objects were created.

8.1.4 Okay, Move It – But Then Line It Up: the *Align* Command

A variation of the *Move* command, *Align* will move objects and then align them with something else. The command sequence is

    Command: *align*
    Select objects: *[Select the object to be moved/aligned.]*
    Select objects: *[Confirm the selection.]*
    Specify first source point: *[This is the first point where you "grab" the objects.]*

Specify first destination point: *[This is where you put the first grabbed point.]*
Specify second source point: *[This is a second point where you grab the objects – you need at least two points so that you can align the objects.]*
Specify second destination point: *[This is where you put the second grabbed point.]*
Specify third source point or <continue>: *[You can grab the object with three points, but this really isn't necessary in 2D space.]*
Scale objects based on alignment points? [Yes/No] <N>: *[Tell AutoCAD whether or not you want to scale the source object to fit exactly between the points you've selected.]*

	Where to Find It:
Command Line:	*Align*
Hotkey(s):	*al*
Ribbon (Tab/Panel):	Home – Modify (subpanel) – **Align**
Menu:	Modify – 3D Operations – **Align**

Don't worry; it's not as complicated as it looks.

I know, the menu button resides in a 3D location, but don't let that bother you. You can use **Align** in either world.

Do This: 8.1.4A — Using the *Align* Command

I. Be sure you're still in *M–C* in the C:\Steps\Lesson08 folder.
II. Follow these steps.

8.1.4A: USING THE ALIGN COMMAND

1. Turn on the **intersection** and **endpoint** Running OSNAPs. Clear all other OSNAPs.
   Command: *os*

2. Enter the *Align* command.
   Command: *al*

3. Select one of the nozzles.
   Select objects:
   Select objects: *[ENTER]*

4. Using OSNAPs, select the points as indicated.
   Specify first source point: *[Pick point 1a.]*
   Specify first destination point: *[Pick point 2a.]*
   Specify second source point: *[Pick point 1b.]*
   Specify second destination point: *[Pick point 2b.]*

   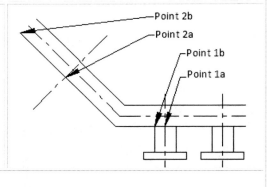

5. Complete the point selection procedure.
   Specify third source point or <continue>: *[ENTER]*

8.1.4A: USING THE ALIGN COMMAND

6. Don't scale the objects.
   Scale objects based on alignment points? [Yes/No] <N>: *[ENTER]*
   Your drawing looks like the figure at right.

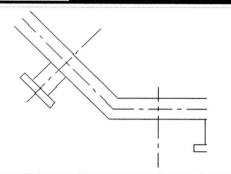

7. Save 💾 and exit your drawing.
   Command: *qsave*

You can see, then, that the *Align* command is related to the *Move* command, although it's a bit more complex. You might say it's also related to the *Rotate* and *Scale* commands (we'll look at these commands in Lesson 9, pp.206 and 208). Wow!

8.1.5 Parallels and Concentrics – The *Offset* Command

Offset gives us another way to create one or many copies of a single object (line, polyline, circle, arc, etc.). The copies will be either parallel (lines and polylines) or concentric (circles & arcs) to the existing object. The command sequence is

Where to Find It:	
Command Line:	*Offset*
Hotkey(s):	*o*
Ribbon (Tab/Panel):	Home – Modify – **Offset**
Menu:	Modify – **Offset**
Toolbar:	Modify – **Offset**
Tool Palette:	Modify – **Offset**

    Command: *offset*
    Current settings: Erase source=No Layer=Source OFFSETGAPTYPE=0
    Specify offset distance or [Through/Erase/Layer] <Through>:
[Either specify an offset distance or hit ENTER to accept the default (Through).]
    Select object to offset or [Exit/Undo] <Exit>: *[Select the object to offset.]*
    Specify through point or [Exit/Multiple/Undo] <Exit>: *[Select a point through which the copy should pass.]*
    Select object to offset or [Exit/Undo] <Exit>: *[Hit ENTER to complete the command.]*

AutoCAD begins by showing you the current *Offset* command settings. Consider the options.

- The first *Offset* prompt asks you for an **Offset distance**. This option allows you to specify *a perpendicular distance away from* the original object to place the copy.
- The **Through** option allows you to select a point in the drawing *through which* the copy will pass. AutoCAD then prompts for the direction in which to place the copy.
- By default, AutoCAD leaves the source object of an offset in the drawing. The **Erase** option allows you to remove the source object once it's been offset. It prompts:
  Erase source object after offsetting? [Yes/No] <No>:
- The **Layer** option lets you choose where you place the new object – either on the current layer or the layer of the source object (the default). It prompts
  Enter layer option for offset objects [Current/Source] <Source>:
- The **Exit** and **Undo** options perform their usual function.
- When offsetting an object, the **Multiple** option (of the **Specify through point** prompt) allows you to make more than a single copy of the source object.

Let's try it.

Do This: 8.1.5A	Using the *Offset* Command

I. Open the *star+* drawing in the C:\Steps\Lesson08 folder.
II. Set the **OBJ1** layer current.
III. Follow these steps.

8.1.5A: USING THE OFFSET COMMAND

1. Enter the *Offset* command.
   Command: *o*
   Current settings: Erase source=No Layer=Source OFFSETGAPTYPE=0

2. Accept the default option.
   Specify offset distance or [Through/Erase/Layer] <Through>: *[ENTER]*

3. Select the circle around the star.
   Select object to offset or [Exit/Undo] <Exit>:

4. Select the endpoint of the upper line.
   Specify through point or [Exit/Multiple/Undo] <Exit>: _endp of

5. Complete the command.
   Select object to offset or [Exit/Undo] <Exit>: *[ENTER]*

Notice (right) that the new circle resides on the same layer as the source object.

6. Repeat the command.
   Command: *[ENTER]*

7. Use the **Layer** option 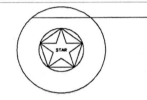 ...
   Specify offset distance or [Through/Erase/Layer] <Through>: *l*

8. ... and tell AutoCAD to place offset copies on the **Current** layer.
   Enter layer option for offset objects [Current/Source] <Source>: *c*

9. AutoCAD returns to the **Specify offset** prompt. Accept the default **Through** option.
   Specify offset distance or [Through/Erase/Layer] <Through>: *[ENTER]*

10. Select the horizontal line above the circle.
    Select object to offset or [Exit/Undo] <Exit>:

11. Select the center of the circle.
    Specify through point or [Exit/Multiple/Undo] <Exit>: _cen of

12. Complete the command.
    Select object to offset or [Exit/Undo] <Exit>: *[ENTER]*

AutoCAD has offset a copy of the source object onto the current layer.

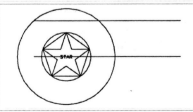

8.1.5A: USING THE OFFSET COMMAND

13. Erase the new line and circle.
    Command: *e*

14. Now let's try a specific distance for our offsets. Repeat the *Offset* command.
    Command: *o*

15. Tell AutoCAD to offset each object ¼ unit.
    Specify offset distance or [Through/Erase/Layer] <Through>: *.25*

16. Select the upper line.
    Select object to offset or [Exit/Undo] <Exit>:

17. Tell AutoCAD to make multiple copies Multiple.
    Specify point on side to offset or [Exit/Multiple/Undo] <Exit>: *m*

18. Pick a point above the line. Continue to pick until you have four lines.
    Specify point on side to offset or [Exit/Undo] <next object>:

19. Tell AutoCAD to move on to the **next object**.
    Specify point on side to offset or [Exit/Undo] <next object>: *[ENTER]*

20. Now select the circle and repeat Step 18 until you have four circles.
    Select object to offset or [Exit/Undo] <Exit>:
    Specify point on side to offset or [Exit/Undo] <next object>: *[ENTER]*

21. Complete the command.
    Specify object to offset or [Exit/Undo] <next object>: *[ENTER]*
    The drawing looks like the figure at right.

22. Erase the new objects. Don't exit.
    Command: *e*

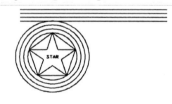

Nothing to it, right?

Let's look at something that can make multiple objects at one time.

8.1.6 Rows, Columns, and Circles – the *Array* Command

Many times we find that not only do we need several copies but also that the copies must be arranged in rows, columns, or even circles. Using a normal *Copy* command, this can evoke groans of tedium from CAD operators. Fortunately, the *Array* command was designed to prevent these groans.

AutoCAD provides a handy dialog box approach to the *Array* command. Let's take a look.

Where to Find It:	
Command Line:	*Array*
Hotkey(s):	*ar*
Ribbon (Tab/Panel):	Home – Modify – **Array**
Menu:	Modify – **Array**
Toolbar:	Modify – **Array**
Tool Palette:	Modify – **Array**

By way of radio buttons across its top, the Array dialog box (Figure 8.001, p.185) provides two distinct types of arrays– the **Rectangular Array** and the **Polar Array**.

- When creating a **Rectangular Array** (rows and columns), you'll need to tell AutoCAD how many **Rows** and **Columns** to create in the corresponding input boxes.

You can enter the distance between rows or columns in the **Row offset** and **Column offset** boxes. Alternately, you can use the pick buttons to indicate a distance by selecting points on the screen. (Notice that AutoCAD provides three buttons – one each to **Pick Row Offset** and **Pick Column Offset**, and a third, larger button that will allow you to **Pick Both Offsets** at one time.)

The dialog box also allows you to define the **Angle of array** by entering the angle in a text box or picking points on the screen to define it.

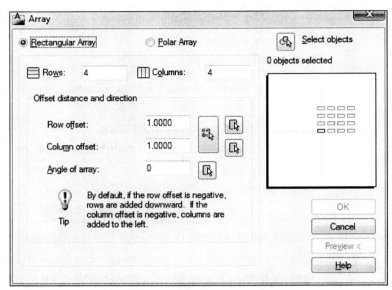

Figure 8.001

Watch the preview box in the right side of the dialog box to see a graphics representation of how the array settings will work.

- The **Polar Array** option presents the choices seen in Figure 8.002.

    Here you can enter the coordinates for the **Center point** of the array in the appropriate text boxes, or you can use the **Pick Center Point** button (to the right of the **X** and **Y** text boxes) to pick the **Center Point** on the screen.

    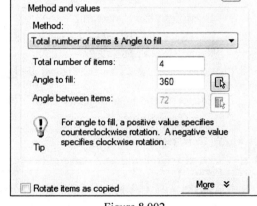

    Figure 8.002

    The Array dialog box allows you to define a polar array in one of three ways using the **Method** control box (**Methods and values** frame):

    o Using the default **Total number of items & Angle to fill** method, you'll define your array by telling AutoCAD how many items you want in the array and what portion of a circle you want to use. Use the **Total number of items** and **Angle to fill** text boxes (or the **Pick Angle to Fill** button).

    o Use the **Total number of items and Angle between items** method to define the array by telling AutoCAD how many items you want in the array and how to space them using angles. Here you'll use the **Total number of items** and the **Angle between items** text boxes (or the **Pick Angle between Items** button).
    This method offers you some flexibility.

    o Use the **Angle to fill & Angle between items** method when you know the area you want to fill with copies of selected objects but aren't sure how many objects it'll take. Here you'll use the **Angle to fill** and the **Angle between items** text boxes (or their associated buttons).

    The **Polar Array** option has a couple other items that require some attention.

    o Just below the **Method and values** frame, you'll find a check box that enables you to **Rotate** [the arrayed] **items as** [they're] **copied**. This valuable tool will come in quite handy!

- o Notice the **More** button 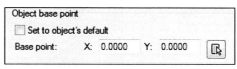 to the right of the **Rotate** check box. This presents the options seen in Figure 8.003. Here you can reset the **base point** of the selected objects for the array. The base point can be critical to the final appearance of your array. This is the point at which you'll 'grab' the objects being arrayed and the point that will define the location of the arrayed objects in relation to the settings on the dialog box.

  Figure 8.003

  Use the **Object base point** frame (revealed by the **More** button) to reset the base point of the selected objects for the array.

- The other options available in the Array dialog box include:
  - o The **Select objects** button to the right of the **Rectangular Array / Polar Array** radio buttons. Use this to return to the screen to select the objects you wish to array.
  - o A **Preview** button (below the preview display box) that enables you to see the results of the array before committing to any changes.

Let's see the Array dialog box in action.

| Do This: 8.1.6A | Using the *Array* Command |

I. Be sure you're still in the *star+* drawing in the C:\Steps\Lesson08 folder.
II. Freeze the **Line** layer.
III. Follow these steps.

8.1.6A: USING THE ARRAY COMMAND

1. Enter the *Array* command.

   Command: *ar*

   AutoCAD presents the Array dialog box (Figure 8.001, p.185).

2. Set up the array as shown:
   - Be sure the **Rectangular Array** option has been selected.
   - Create two rows and four columns.
   - Offset the rows by **3"**.
   - Offset the columns by **3.5"**.

3. Pick the **Select objects** button to return to the graphics area.

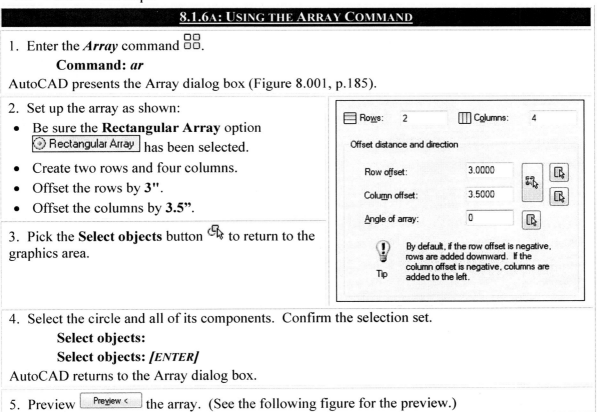

4. Select the circle and all of its components. Confirm the selection set.

   Select objects:

   Select objects: *[ENTER]*

   AutoCAD returns to the Array dialog box.

5. Preview the array. (See the following figure for the preview.)

8.1.6A: USING THE ARRAY COMMAND

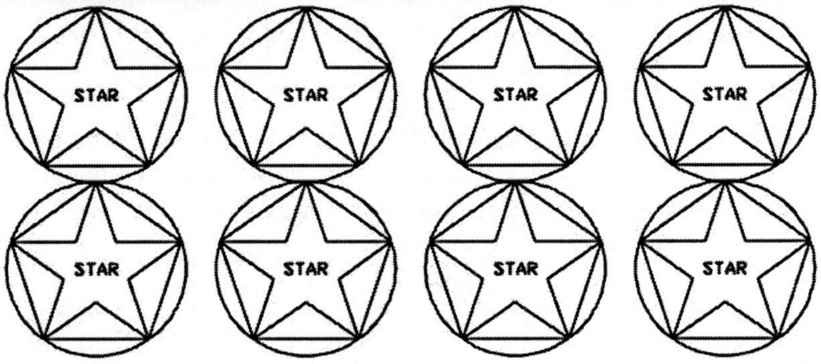

[Figure: Two rows of four circles, each containing a five-pointed star labeled "STAR"]

6. Right click 🖱 to accept the array. AutoCAD creates it and returns to the command line. **Pick or press Esc to return to dialog or <Right-click to accept array>:**
7. Undo the changes ↶. **Command:** *u*
8. Let's try a **Polar** array. Repeat the *Array* command ▦. **Command:** *ar*
9. Set up the array as shown at right: • Be sure the **Polar Array** option ⦿ Polar Array has been selected. • Set the **Center point** to *5,5*. • Use the **Total number of items and Angle to fill** method. • Create five items and fill a 360° angle. • Don't **Rotate the items as** [they're] **copied**.
10. Pick the **More** button [More ⌄]. AutoCAD presents the **Object base point** frame (Figure 8.003, p.186).
11. Be sure the **Set to object's default** check box is clear, and then pick the **Pick Base Point** button 🔲.
12. AutoCAD returns to the graphics screen and prompts you for the base point. Select the center ⊙ of the circle. **Specify the base point of objects: _cen of** AutoCAD returns to the Array dialog box and shows the coordinates of the base point in the **Object base point** frame (right).
13. Pick the **Select objects** button 🗔 to return to the graphics area.
14. Select the circle and all of its components. **Select objects:**

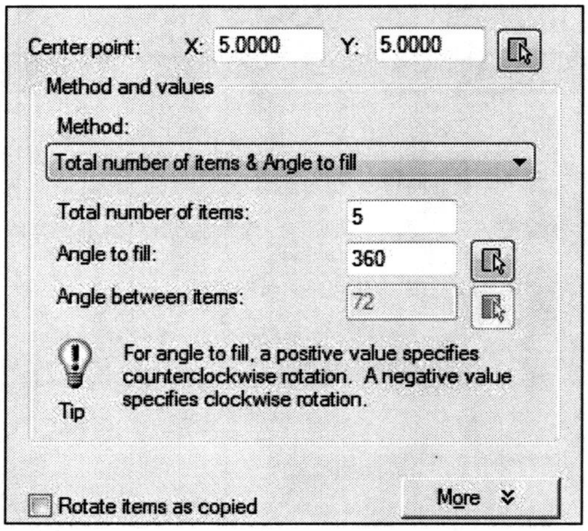

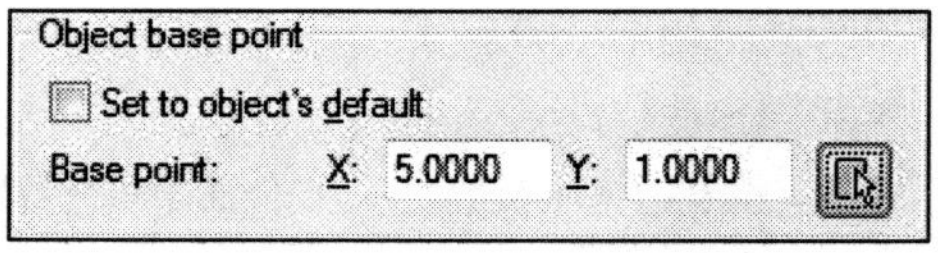

8.1.6A: USING THE ARRAY COMMAND

15. Pick the **OK** button [OK] to complete the procedure. The array appears in the figure shown (left figure following Step 17).

16. Undo the changes ↶.
    Command: *u*

17. Repeat Steps 8 through 15, but this time rotate the copies (Step 9). Notice the difference (following figure – right)?

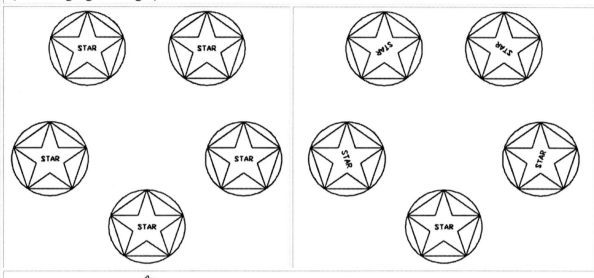

18. Undo the changes ↶.
    Command: *u*

8.1.7 Opposite Copies – the *Mirror* Command

▲	Where to Find It:
Command Line:	*Mirror*
Hotkey(s):	*mi*
Ribbon (Tab/Panel):	Home – Modify – **Mirror**
Menu:	Modify – **Mirror**
Toolbar:	Modify – **Mirror**
Tool Palette:	Modify – **Mirror**

Figure 8.004

Occasionally, you'll run into a situation where you not only need to copy an object, but you need to completely reverse its orientation. This isn't a true rotating of the object(s), which would just stand it on its head. What you see is an opposite or a *mirror* image of the original.

The *Mirror* command is one of AutoCAD's simplest:

>   **Command:** *mirror*
>   **Select objects:** *[Select the object(s) to be mirrored.]*

Select objects: *[Confirm completion of the selection set.]*
Specify first point of mirror line: *[Pick the first point of the mirror line.]*
Specify second point of mirror line: *[Pick the second point of the mirror line.]*
Erase source objects? [Yes/No] <N>: *[Do you want to keep the original object(s)?]*

You may have difficulty understanding the "**mirror line**" in this sequence. Refer to Figure 8.004, p.188. Consider this: To see a mirror image of an object as it lies on a table, you'll lay a mirror alongside the object (at a bit of an angle so as to reflect the object). The mirror *line* is the edge of the mirror where it meets the surface of the table. In AutoCAD, your screen is the table. You must define the *line* (the edge of the mirror) by identifying *two* points on it.

Let's give it a try.

Do This: 8.1.7A	Mirroring an Object

I. Be sure you're still in the *star+* drawing in the C:\Steps\Lesson08 folder.
II. Follow these steps.

8.1.7A: MIRRORING AN OBJECT

1. Enter the *Mirror* command.
   Command: *mi*

2. Select the circle and all of its contents.
   Select objects:
   Select objects: *[ENTER]*

3. Pick the **first point of mirror line** around coordinate *7,3*, and then (with the **Ortho** on) pick a **second point** to the left (or west) of the first point, as shown.
   Specify first point of mirror line: *7,3*
   Specify second point of mirror line:

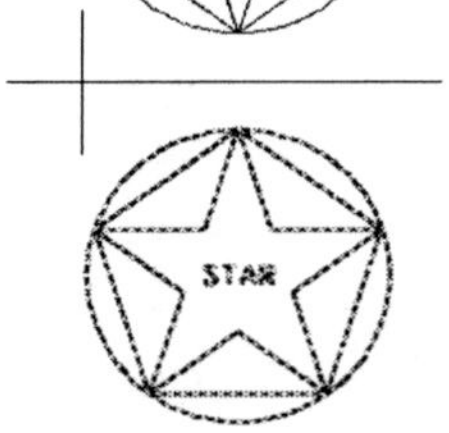

4. AutoCAD asks if you wish to delete the original. Accept the default – **No**.
   Erase source objects? [Yes/No] <N>: *[ENTER]*
   Notice (right) that everything mirrored *except* the text.
   See the following insert for more details on this behavior.

5. Undo the changes.
   Command: *u*

> By default, AutoCAD will mirror all objects *except text*. This default will help you avoid the frustration of having to redo all the text in an object just because you needed a mirror image.
>
> If, however, you want the text mirrored as well, you can change the system variable that controls mirrored text – **Mirrtext**. Here's how:
>
> **Command:** *mirrtext*
>
> **Enter new value for MIRRTEXT <0>:** *1*
>
> Turn on the **Mirrtext** system variable (set it to *1* as indicated), and then redo the previous exercise. Notice the difference? (Be sure to reset **Mirrtext** to *0* when you've finished.)

8.2 Moving and Copying Objects *between* Drawings

AutoCAD has given us the ability to move or copy objects from one drawing to another as well as within a single drawing. This feature makes use of the Windows *Cut & Paste* or *Copy & Paste* commands. It also takes advantage of AutoCAD's Multiple Document Environment (MDE), which enables you to open more than one drawing at a time. (Oh, the wonders that can happen with a little cooperation!)

You've probably seen all the places that you can find the *Copy* and *Move* commands. These, of course, are AutoCAD commands. But did you notice that the *Copy* command is also located in the Edit pull-down and cursor menus? This and the other commands in the Edit menu belong to Windows. They take advantage of the Windows clipboard (Windows' method of copying and moving objects and files within and between Windows' documents). They also make it possible to copy and move objects between AutoCAD drawing files.

The Windows method of copying requires two steps: *Copy* and *Paste*. *Copy* places the item(s) on the clipboard (an *imaginary* clipboard – a location in your computer's memory); *Paste* takes it from the clipboard and puts it into your document.

The Windows method of moving also requires two steps: *Cut* and *Paste*. *Cut* (like *Copy*) places the item(s) on the clipboard. But *Cut* also removes the item(s) from the source location. *Paste* puts the item(s) into your document.

AutoCAD's command line equivalents for Windows commands are:	
WINDOWS	**AUTOCAD**
Copy	*copyclip* or *copybase*
Cut	*cutclip*
Paste	*pasteclip, pasteblock,* or *pasteorig*

Of course, the best way to understand all of this is to see it in action. Let's do an exercise.

Do This: 8.2A	**Copying Objects *between* Drawings**

I. Be sure you're in *M–C* in the C:\Steps\Lesson08 folder. If not, please open it now. (Close any other drawings.)

II. Follow these steps.

8.2A: COPYING BETWEEN DRAWINGS

1. Without closing the *M–C* drawing, start a new drawing from scratch.

   Command: *new*

8.2A: COPYING BETWEEN DRAWINGS

2. The new drawing opens atop the already open *M–C* drawing. We'll use the Quick View Drawing tool to switch back and forth between the two. Open it now.

3. Pick the **Pin** icon so that the panel remains visible. (The **Pin** will turn and point inward as shown.)

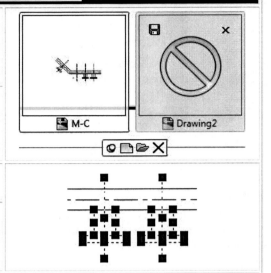

4. Pick the *M–C Model* snapshot to make that drawing active.

5. Without entering a command, place a selection window around the two nozzles on the horizontal pipe (include their centerlines). They now appear highlighted as shown.

6. We could use the *Copy* command in the Edit pull-down menu, but the cursor menu will be more convenient. Right click anywhere in the active document and select **Clipboard – Copy with Base Point** on the menu (the command line equivalent is *Copybase*).
(Had we used the Edit menu's **Copy** command, AutoCAD would have assumed a base point in the lower-left corner of the objects selected.)

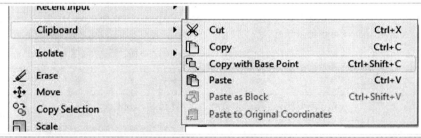

7. Select a convenient base point (I'll use the intersection of one of the centerlines and the pipe wall).
    Specify base point:

8. Use the Quick View Drawing panel to toggle back to the new drawing's model.

9. Right click in an open area of the drawing and pick **Clipboard – Paste** from the cursor menu.

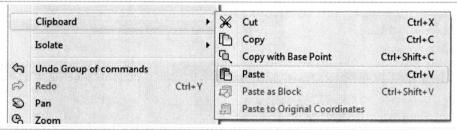

10. AutoCAD prompts for an insertion point. Pick any convenient point in the drawing.
    Specify insertion point:

You may need to zoom out a bit to see the nozzles in the new drawing.

8.2A: COPYING BETWEEN DRAWINGS

11. Exit both drawings without saving.
    Command: *quit*

Some things to remember:

- To move objects, use the **Cut** (*Cutclip* command) option rather than the **Copy** or **Copy with Base Point** option we used in Step 6.
- To clear the highlighting, either hit the ESC key or pick the **Deselect all** option from the cursor menu.
- Using the clipboard requires a small amount of computer memory. Keep this in mind if you use a particularly slow system (not much RAM) or have problems with your computer's memory.
- You can also use the *MatchProp* command from one drawing to another.
- Hitting ENTER to repeat a command will repeat the last command given in the active drawing (i.e., commands are active document-specific).
- *Remember to close a drawing when you've finished with it.* (Type *Close* or select it from the File pull-down menu.)

8.3 The Change Group

8.3.1 Cutting It Out with the *Trim* Command

You'll often find it easier to draw one long line across an area and then cut away the extra pieces than to draw several shorter lines. AutoCAD designed the *Trim* command to remove the "extra" bits of lines and circles. Here's the command sequence:

Where to Find It:	
Command Line:	*Trim*
Hotkey(s):	*tr*
Ribbon (Tab/Panel):	Home – Modify (Trim/ Extend flyout) – **Trim**
Menu:	Modify – **Trim**
Toolbar:	Modify – **Trim**
Tool Palette:	Modify – **Trim**

Command: *trim*
Current settings: Projection=UCS, Edge=None
Select cutting edges ...
Select objects or <select all>: *[Select the cutting edge or hit ENTER to make cutting edges out of all the objects in the drawing.]*
Select objects: *[ENTER or 🖱 to confirm completion of this selection set.]*
Select object to trim or shift-select to extend or [Fence/Crossing/Project/Edge/eRase/Undo]: *[Select the part you wish to remove.]*
Select object to trim or shift-select to extend or [Fence/Crossing/Project/Edge/eRase/Undo]: *[ENTER to confirm completion of this selection set.]*

- The **cutting edges** are usually lines or circles to which you want to trim. In other words, as you select a **cutting edge**, say to yourself, "*I want to cut back to here.*"
- The **object to trim** is the piece of a line or circle you want to remove from the drawing. When selecting the **object to trim**, say to yourself, "*I want to get rid of this.*"
- Notice the subtle **or shift-select to extend** part of the **Select object to trim** prompt. This tells you that you can hold down the SHIFT key on your keyboard and, rather than trimming an object to the cutting edge, you can extend an object to it! This option really cuts down the time required to jump between the *Trim* and *Extend* commands.

- **Fence** allows you to select all objects touched by a *fence*. (A fence behaves much like a single line crossing window.)
- The **Crossing** option forces AutoCAD to use a crossing window regardless of where you pick the corners.
- **Project** refers to a UCS projection – something we'll cover in our 3D text. Ignore it for now.
- **Edge** refers to one of the more useful innovations AutoCAD has provided – the ability to trim an object even if the object doesn't touch your cutting edge.
- **eRase** allows you to erase an object without leaving the *Trim* command. I recommend ignoring this option in favor of its command counterpart. (Just because you can do something doesn't mean that you should!)
- The **Undo** *option*, of course, will undo the last modification within the command. Remember, the *Undo command* will undo the entire *Trim* command modification (all the changes made by the command).

Let's experiment.

Do This: 8.3.1A	Using the *Trim* Command

I. Open *Drill Jig* in the C:\Steps\Lesson08 folder. The drawing looks like the figure at right.
II. Follow these steps. (Work only with the upper figure for now.)

8.3.1A: USING THE TRIM COMMAND	
1. Enter the *Trim* command ⊁. **Command:** *tr*	
2. Select these lines as your cutting edges. **Current settings: Projection=UCS, Edge=None** **Select cutting edges ...** **Select objects or <select all>:**	
3. Confirm that you've completed selecting the **cutting edges**. **Select objects:** *[ENTER]*	
4. Select what you want to trim – all the lines that cross the cutting edges. **Select object to trim or shift-select to extend or** **[Fence/Crossing/Project/Edge/eRase/Undo]:** (Don't complete the command yet.) So far, your drawing should look like the figure at right.	
5. Now pick the top of the second vertical line from the right to trim. Notice that it won't trim. (Don't end the command yet.) **Select object to trim or shift-select to extend or** **[Fence/Crossing/Project/Edge/eRase/Undo]:**	

8.3.1A: USING THE TRIM COMMAND

6. Without leaving the *Trim* command, type *e* to select the **Edge** option. Alternately, you can select **Edge** [Edge] from the cursor or dynamic input menu.

   Select object to trim or shift-select to extend or [Fence/Crossing/Project/Edge/eRase/Undo]: *e*

7. Select the **Extend** option [Extend].

   Enter an implied edge extension mode [Extend/No extend] <No extend>: *e*

8. Now select again the piece that didn't trim in Step 5.

   Select object to trim or shift-select to extend or [Fence/Crossing/Project/Edge/eRase/Undo]:

   Notice the difference? In Steps 6 and 7, you told AutoCAD to extend the cutting edge as an invisible plane in both directions. AutoCAD then used that invisible edge to trim the objects.

9. Now let's use that **shift-select** trick to extend a line. Hold down the SHIFT key and select the right-end of the second horizontal line from the top.

   Select object to trim or shift-select to extend or [Fence/Crossing/Project/Edge/eRase/Undo]:

   Your drawing looks like the figure at right.

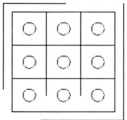

10. Complete the command.

    Select object to trim or shift-select to extend or [Fence/Crossing/Project/Edge/eRase/Undo]: *[ENTER]*

11. Save your drawing 💾 but don't exit.

    Command: *qsave*

The *Trim* command, as useful as it is, is only one side of a coin. The opposite side is the *Extend* command, which is just as useful even if performing the opposite task.

8.3.2   Adding to It with the *Extend* Command

You don't have to enter the *Trim* command to extend an object as you did in our last exercise. It's certainly more logical to use the *Extend* command to extend a line. But considering the two procedures and how frequently they work together, placing them together proves quite a time saver.

To be fair, AutoCAD also placed a **Trim** option in the *Extend* command, which it designed to extend lines that come up short of the mark.

Here is the *Extend* command sequence:

Where to Find It:	
Command Line:	*Extend*
Hotkey(s):	*ex*
Ribbon (Tab/Panel):	Home – Modify (Trim/Extend flyout) – **Extend**
Menu:	Modify – **Extend**
Toolbar:	Modify – **Extend**
Tool Palette:	Modify – **Extend**

   Command: *extend*

   Current settings: Projection=UCS, Edge=Extend

   Select boundary edges ...

   Select objects or <select all>: *[Select the boundary edge or hit ENTER to make boundaries out of all the objects in the drawing.]*

   Select objects: *[ENTER to confirm completion of this selection set.]*

Select object to extend or shift-select to trim or
[Fence/Crossing/Project/Edge/Undo]: *[Select the object to extend.]*
Select object to extend or shift-select to trim or
[Fence/Crossing/Project/Edge/Undo]: *[ENTER to confirm completion of this selection set.]*

Each step corresponds to the same step in the *Trim* command sequence. The only difference is that with the *Extend* command, you'll select a **boundary edge** – the place to which you want to extend a line or arc. (Say to yourself, "I want to extend *to here*.")

Notice that the **Edge**, while defaulting to **None** in the *Trim* command, now defaults to **Extend**. This is because *Trim* and *Extend* share the Edgemode system variable. When you set the **Edge extension mode** to **Extend** in the last exercise, you set it to **Extend** for the *Extend* command as well. This will become clearer as we do the next exercise.

Do This: 8.3.2A	Using the *Extend* Command

I. Be sure you're still in the *Drill Jig* file in the C:\Steps\Lesson08 folder. If not, please open it now.
II. Follow these steps.

8.3.2A: USING THE EXTEND COMMAND	
1. Enter the *Extend* command. **Command:** *ex*	
2. Select the right vertical line and the second horizontal line from the bottom as your **boundary edges**. **Current settings: Projection=UCS, Edge=Extend** **Select boundary edges ...** **Select objects or <select all>:** **Select objects:** *[ENTER]*	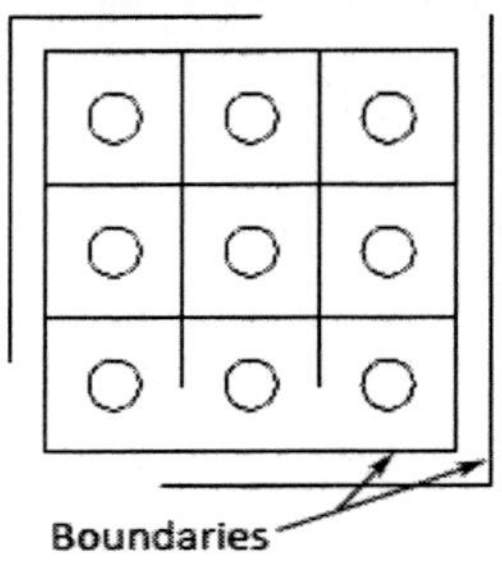 Boundaries
3. Select the top horizontal line and the middle two vertical lines to extend. (Select them toward the ends you wish to extend.) **Select object to extend or shift-select to trim or** **[Fence/Crossing/Project/Edge/Undo]:**	
4. Complete the command. **Select object to extend or shift-select to trim or** **[Fence/Crossing/Project/Edge/Undo]:** *[ENTER]* Your drawing now looks like the figure at right.	
5. Save your drawing but don't exit. **Command:** *qsave*	

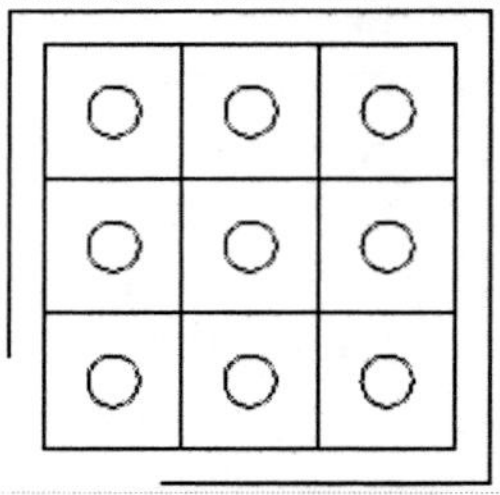

You've probably already recognized the value of these two simple commands. The ability to trim unneeded material or extend a line, circle, or arc will save a lot of redraw time.

8.3.3	Redundancy – Thy Name is AutoCAD: The *Break* Command

AutoCAD provides yet another way to remove parts of lines or circles. But this one is a bit more tedious than the *Trim* command. The *Break* command sequence is

Command: *break*
Select object: *[Select the object to be broken.]*

> Specify second break point or [First point]: *f*
>
> Specify first break point: *[Select the first end of the break.]*
>
> Specify second break point: *[Select the other end of the break.]*

	Where to Find It:
Command Line:	*Break*
Hotkey(s):	*br*
Ribbon (Tab/Panel):	Home – Modify (subpanel) – **Break**
Menu:	Modify – **Break**
Toolbar:	Modify – **Break**
Tool Palette:	Modify – **Break**

The section between the first and second selected points is removed.

There's only one option in this sequence – that of entering *F* for the **First point**. If you enter *F*, AutoCAD prompts for the first end of the break. You can then use OSNAPs for precise selection of both first and second points. If you don't enter *F*, AutoCAD assumes the point at which you selected the line or circle is the first point. As no OSNAP is used to select the line, there is no precision in selecting the first point. So we discover one of those unwritten rules: *always use First point when using the Break command.* (Of course, you know what they say about rules.)

> One of AutoCAD's lesser-known tricks concerns the **Break** command. You can enter an @ symbol at the **Enter second point** prompt to break the object into two pieces without removing anything. The object is broken at the point selected at the **Specify first break point** prompt. (But the easy way is to pick the **Break at Point** button on the **Modify** subpanel.)

8.3.4 Now We Can Round That Corner: The *Fillet* Command

The *Fillet* command provides an easy way to round corners without the need for any of the *Arc* command's routines; or you can use it to *square* corners!

	Where to Find It:
Command Line:	*Fillet*
Hotkey(s):	*f*
Ribbon (Tab/Panel):	Home – Modify (Fillet/Chamfer flyout) – **Fillet**
Menu:	Modify – **Fillet**
Toolbar:	Modify – **Fillet**
Tool Palette:	Modify – **Fillet**

The command sequence is

> Command: *fillet*
>
> Current settings: Mode = TRIM, Radius = 0.0000
>
> Select first object or [Undo/Polyline/Radius/Trim/Multiple]: *[Select the first object – you can fillet lines, arcs, circles, ellipses, polylines, xlines, and splines.]*
>
> Select second object or shift-select to apply corner: *[Select the second object or use shift-select to make a square corner.]*

- The default radius for the *Fillet* command is **0.0**, but you can change that by typing *R* for the **Radius** option (at the **Select first object** prompt). AutoCAD will prompt:

   > Specify fillet radius <0.0000>:

   You can type a different radius, or you can enter *0* for a square corner (but using the SHIFT-select method might be faster).

- The **Polyline** option allows you to fillet all the corners of a polyline at once. (We'll discuss polylines in Lesson 12, p.272.)
- The **Trim** option allows you to decide whether or not to automatically trim away the excess line, circle, arc, etc., when the *Fillet* operation is done.

> The **Trim** option, available for both the **Fillet** and **Chamfer** commands, was added to AutoCAD after users complained that they didn't always want to trim away the excess part of the lines during **Fillet** or **Chamfer** procedures. When selecting the **Trim** option, AutoCAD prompts:
>
> **Enter Trim mode option [Trim/No trim] <Trim>:**
>
> Notice that **Trim** is the default. AutoCAD will automatically trim away the excess. If you want to keep the excess, enter **N**. AutoCAD sets the **Trimmode** (the system variable that controls the **Trim/No trim** option) accordingly. The objects are filleted, but the excess remains.

- Normally, AutoCAD allows one fillet to be completed per command sequence. If you wish to fillet more than one corner, first select the **Multiple** option.
- The second option line provides a useful shortcut. Rather than use the default **Select second object** option (and select a second object), you can hold down the SHIFT key while selecting a second object to create a square corner.
- A final option (one that AutoCAD doesn't show in its options list) allows you to fillet parallel lines. This procedure draws an arc connecting the endpoints of both lines. We'll see how it works in the next exercise.

Let's give it a try.

Do This: 8.3.4A	Using the *Fillet* Command

I. Be sure you're still in the *Drill Jig* file in the C:\Steps\Lesson08 folder. If not, please open it now.

II. Follow these steps.

8.3.4A: USING THE FILLET COMMAND

1. Enter the *Fillet* command ⌐.

   Command: *f*

2. Change the radius [Radius] to ¼".

   Select first object or [Undo/Polyline/Radius/Trim/Multiple]: *r*
   Specify fillet radius <0.0000>: *.25*

3. ... and select the lines indicated here.

   Select first object or
   [Undo/Polyline/Radius/Trim/Multiple]:
   Select second object or shift-select to apply corner:

   Notice that the edges round and the excess lines are automatically trimmed away.

4. Your drawing looks like this. Repeat the command ⌐.

   Command: *[ENTER]*

5. Hold down the SHIFT key while selecting the left and bottom lines.

   Select first object or
   [Undo/Polyline/Radius/Trim/Multiple]:
   Select second object or shift-select to apply corner:

   Notice that AutoCAD squares the corner.

8.3.4A: USING THE FILLET COMMAND

6. Repeat the command.
   Command: *[ENTER]*

7. This time select the left and right vertical lines. Select both toward the top endpoints. The lines are parallel, but AutoCAD will create an arc between them (right).

8. Save the drawing but don't exit.
   Command: *qsave*

8.3.5 Fillet's Cousin: The *Chamfer* Command

Fillet and *Chamfer* are so similar that you might get confused as to which one you want. Look at the pictures on the buttons if you're perplexed.

Where *Fillet* rounds corners, *Chamfer* provides a mitre – a flat edge at a corner – much like a carpenter achieves with a hand plane. You control the size and angle of the edge through responses to the prompts, which look like this:

Command: *chamfer*
(TRIM mode) Current chamfer Dist1 = 0.0000, Dist2 = 0.0000
Select first line or [Undo/Polyline/Distance/Angle/Trim/mEthod/Multiple]: *[Select the first chamfered line.]*
Select second line or shift-select to apply corner: *[Select the second chamfered line.]*

Where to Find It:	
Command Line:	*Chamfer*
Hotkey(s):	*cha*
Ribbon (Tab/Panel):	Home – Modify (Fillet/Chamfer flyout) – **Chamfer**
Menu:	Modify – **Chamfer**
Toolbar:	Modify – **Chamfer**
Tool Palette:	Modify – **Chamfer**

- The default **Method** is **Distance** (shown in the sample sequence). Using the **Distance** option tells AutoCAD to measure a user-defined distance from the corner (real or apparent) on both lines, put a line between these two points, and **Trim** the excess line (or not – depending on the **Trimmode** setting).
- Another method is **Angle**. Here, you define one distance (as in the **Distance** method) and an angle of cut.
- You can switch between methods (**Distance** or **Angle**) using the **mEthod** option.
- The **Polyline** option can be used to chamfer entire polylines, but the results can be surprising. I tend to shy away from chamfering polylines (Lesson 10).
- The **Multiple** option works just is it did in the *Fillet* command.
- Notice that last prompt? That's right; you can square a corner using the *Chamfer* command as easily as you did with the *Fillet* command!

These options will become clearer with an exercise.

Do This: 8.3.5A	Using the *Chamfer* Command

I. Be sure you're still in the *Drill Jig* file in the C:\Steps\Lesson08 folder. If not, please open it now.
II. Follow these steps.

8.3.5A: USING THE CHAMFER COMMAND

1. Enter the *Chamfer* command ⌐.
   Command: *cha*

2. Let's change the **Dist1** and **Dist2** settings so that we can better tell what's happening. Type **D** or pick **Distance** on the menu [Distance].
   (TRIM mode) Current chamfer Dist1 = 0.0000, Dist2 = 0.0000
   Select first line or [Undo/Polyline/Distance/Angle/Trim/mEthod/Multiple]: *d*

3. AutoCAD prompts for each distance. Enter *.25* for the **first chamfer distance** and the **second chamfer distance**.
   Specify first chamfer distance <0.0000>: *.25*
   Specify second chamfer distance <0.2500>: *[ENTER]*

4. Select the top line of the lower object, and then the right vertical line toward the top.
   Select first line or [Undo/Polyline/Distance/Angle/Trim/mEthod/Multiple]:
   Select second line or shift-select to apply corner:

5. Repeat the chamfer on the other side.
   Your drawing looks like the figure at right.

6. Save 💾 and close the drawing.
   Command: *qsave*

8.4 Extra Steps

After completing this lesson, spend some time experimenting with the new Lisp routines mentioned in this and other lessons. (You'll find a pretty cool copy-and-rotate routine – *Coprot.lsp* – in this lesson's folder.) All the lisp routines included with this text are freeware. Play with them, pass them around, alter them as you see fit. Some are quite primitive; others are more advanced. But the value of each has been proved countless times since I wrote them.

Other Lisp routines will be available to you – some with this text, others where you go to work. One thing about the AutoCAD world – there is no shortage of these programs from which to pick and choose. When you get to your job site, ask around. CAD operators love to share!

8.5 What Have We Learned?

Items covered in this lesson include:
- *AutoCAD Modifying Commands*
  - *Trim*
  - *Extend*
  - *Break*
  - *Fillet*
  - *Chamfer*
  - *Move*
  - *Align*
  - *Copy*
  - *Offset*
  - *Array*
  - *Mirror*
  - *Mirrtext*
  - *AddSelected*
- *Windows Commands*
  - *Cut*
  - *Copy*
  - *Paste*

This has been a very good lesson. You've learned the basics of manipulating drawing objects to your advantage. Your drawing time will be noticeably lessened as you become more adept with these tools.

Relax for a moment and think about what you've learned. Sure, it's all still quite new to you. But when you began, did you think you would be able to draw the Bracket (Exercise 2) this soon? Draw it, then pat yourself on the back and go on to Lesson 9, p.202.

8.6 Exercises

1. Open the *Checkers* drawing in the C:\Steps\Lesson08 folder.

   Using the commands you've learned in this lesson, create the drawing in the figure at right from the objects provided. (Be sure to use different layers and colors to mark the different sides.)

   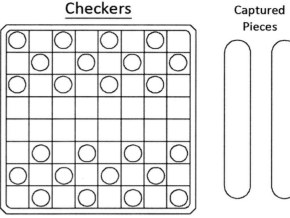

2. Using the *MyBase3* template you created in Lesson 1 (or the *Base3* template in the Lesson01 folder), create the drawing in the figure at right. Reset the grid to ¼" and the snap accordingly. Create your own layers. Don't draw the dimensions.

   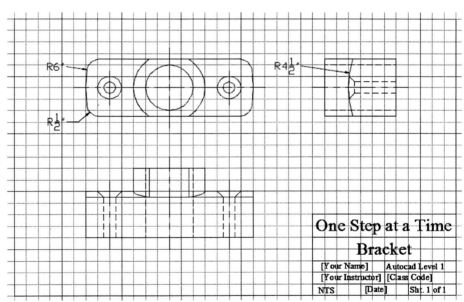

3. Using what you know, create the gear show at right. It should fit on an 11/17 sheet of paper. (Have fun!)

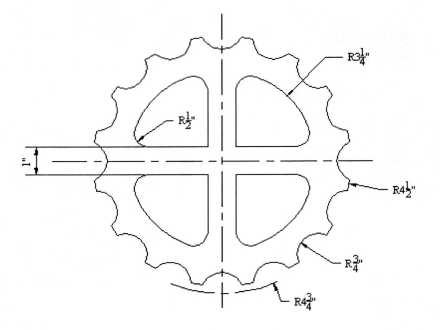

| 8.7 | **For Web-Based Review Questions and Additional Exercises, visit: http://foragerpub.com/AcadFiles/2011/2011.htm** |

This stuff is such a hoot!

Lesson 9

Following this lesson, you will:

- ✓ Know how to use the **Lengthen** command
- ✓ Know how to use the **Stretch** command
- ✓ Know how to use the **Rotate** command
- ✓ Know how to use the **Scale** command
- ✓ Know how to use the **Cloud** command
- ✓ Know how to use Grip commands to modify objects:
    - **Stretch**
    - **Move**
    - **Rotate**
    - **Mirror**
    - **Scale**

Some More Editing Tools... & Grips!

In Lesson 8, p.176, we discussed several modification tools that make computer drafting easier. In this lesson, we'll see some more modifying tools also designed to speed and simplify the drafting process. These include some more advanced tools belonging to the Change *group we began in Lesson 8.*

We'll spend the rest of the lesson discovering grips. These are the culprits responsible for those little squares that dot an object when you select it at the command prompt instead of a **select objects** *prompt. But mastering grips will put a smile on your face that a beginning CAD operator just won't understand!*

Let's begin with the editing tools.

9.1	More Commands in the Change Group
9.1.1	Two Ways to Change the Length of Lines and Arcs – The *Lengthen* and *Stretch* Commands

Prior to the inclusion of the *Lengthen* command, AutoCAD relied on the *Stretch* command to lengthen or shorten lines. The *Stretch* command is still reliable and is really quite simple once you understand the need to use a crossing window to select objects. The command sequence is

Where to Find It:	
Command Line:	*Stretch*
Hotkey(s):	s
Ribbon (Tab/Panel):	Home – Modify – **Stretch**
Menu:	Modify – **Stretch**
Toolbar:	Modify – **Stretch**
Tool Palette:	Modify – **Stretch**

> **Command:** *stretch*
> **Select objects to stretch by crossing-window or crossing-polygon...**
> **Select objects: Specify opposite corner:**
> *[Note the instruction on the previous line – select the object(s) to be stretched using a* **crossing** *window.]*
> **Select objects:** *[ENTER or 🖱 to confirm completion of the selection set.]*
> **Specify base point or [Displacement] <Displacement>:** *[Pick a base point.]*
> **Specify second point or <use first point as displacement>:** *[Pick a target point.]*

The line instructing you to **Select objects to stretch by crossing-window or crossing-polygon...** is easily overlooked as it appears above the actual prompt (and not at all on the dynamic input prompt!). But don't overlook its importance. You must select objects using a crossing window (or crossing-*polygon*, which we'll discuss in Lesson 11, p.260). Unfortunately, AutoCAD doesn't default to a crossing window, so you must use implied windowing (or enter *c*) to create a crossing window for object selection.

Notice that you have the same option provided by the *Copy* and *Move* commands – that is, you can select a **Base point or [Displacement]**. To review how these work (as well as the **Second point of displacement**), see the discussion in Section 8.1.2, p.178.

Let's experiment with the *Stretch* command.

Do This: 9.1.1A	Stretching Objects

I. Open the *star9* drawing in the C:\Steps\Lesson09 folder.
II. Freeze the **Line** layer.
III. Follow these steps.

9.1.1A: STRETCHING OBJECTS

1. Enter the *Stretch* command.
   Command: *s*

2. Place a crossing window around the upper part of the star as shown.
   Select objects to stretch by crossing-window or crossing-polygon...
   Select objects:

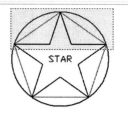

3. Complete the selection set.
   Select objects: *[ENTER]*

4. Use the **displacement** option and enter a displacement of *0,2*.
   Specify base point or [Displacement] <Displacement>: *[ENTER]*
   Specify displacement <0.0000, 0.0000, 0.0000>: *0,2*
   Your drawing looks like the figure at right.

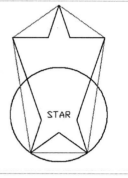

5. Undo your changes.
   Command: *u*

Of course, AutoCAD doesn't limit you to one direction or even one object with the *Stretch* command. So in some ways, it still exceeds the newer *Lengthen* command in desirability. But the *Lengthen* command has its good points, too. Let's look at the command sequence:

	Where to Find It:
Command Line:	*Lengthen*
Hotkey(s):	*len*
Ribbon (Tab/Panel):	Home – Modify (subpanel) – **Lengthen**
Menu:	Modify – **Lengthen**

Command: *lengthen*
Select an object or [DElta/Percent/Total/DYnamic]: *[Tell AutoCAD what you what to do with the object or select a line or arc; subsequent prompts will depend on which option you select.]*
Current length: X.XXXX *[If you select an object, AutoCAD tells you the current length; if you select an arc, it'll tell you the included angle as well.]*
Select an object or [DElta/Percent/Total/DYnamic]: *[ENTER to complete the command.]*

The command includes four fairly simple options.

- **DElta** (the old Greek word for "change") means change. When selected, it prompts:
  Enter delta length or [Angle] <0.0000>: *[Enter the amount of change you wish.]*
  Select an object to change or [Undo]: *[Select the object closest to the end at which you wish to add the length (use a negative number to shorten the line or arc).]*

  If you select the **Angle** choice of the **DElta** option, you can add to an arc by specifying the angle of the arc to add.

- **Percent** allows you to change the length of the selected object by percent. AutoCAD prompts:
  Enter percentage length <100.0000>:
  Note that 100% means no change. More than 100% increases the length of the object and less than 100% decreases the length. Length is added or removed from the end closest to where you select the object.

- **Total** simply means, "How long do you want the object to be?" AutoCAD will add or remove as necessary (again, from the selected end) to make the object as long as you have specified.
- **DYnamic** allows you to manually (or *dynamically*) stretch the object to the desired length. AutoCAD prompts:

   Select an object to change or [Undo]:

   Specify new end point:

   Select the **object to change** and specify a new **end point**.

Let's see what we can do with the *Lengthen* command.

Do This: 9.1.1B	Using the *Lengthen* Command

   I. Be sure you're still in the *star9* drawing in the C:\Steps\Lesson09 folder.
   II. Thaw the **Line** and **Arc** layers.
   III. Zoom all. Notice the line and arc.
   IV. Follow these steps.

9.1.1B: USING THE LENGTHEN COMMAND

1. Enter the *Lengthen* command.
   Command: *len*

2. Select the arc.
   Select an object or [DElta/Percent/Total/DYnamic]:

3. AutoCAD reports the true length of the arc and the included angle.
   Current length: 1.5708, included angle: 90

4. Use the **DElta** option [DElta] to change the length.
   Select an object or [DElta/Percent/Total/DYnamic]: *de*

5. AutoCAD asks if you want to make a change by **Angle** or **length**. Enter a length of **1**.
   Enter delta length or [Angle] <0.0000>: *1*

6. Select the upper end of the arc.
   Select an object to change or [Undo]:
   AutoCAD adds one unit to the upper end (right).

7. Now select the left end of the horizontal line. Notice that AutoCAD adds a piece one unit long to the end.
   Select an object to change or [Undo]:

8. Complete the command.
   Select an object to change or [Undo]: *[ENTER]*

9. Repeat the command.
   Command: *[ENTER]*

10. Let's make an adjustment by the **Percent** method. Select the **Percent** option [Percent].
    Select an object or [DElta/Percent/Total/DYnamic]: *p*

11. Let's cut the objects in half. Enter **50** for the **percentage length**.
    Enter percentage length <100.0000>: *50*

9.1.1B: USING THE LENGTHEN COMMAND

12. Select the lower end of the arc; then select the right end of the line.
    Select an object to change or [Undo]:
    The objects now look like the figure at right.

13. Repeat the command.
    Command: *[ENTER]*

14. This time we'll set the **Total** length of arc and line. Select the **Total** option [Total].
    Select an object or [DElta/Percent/Total/DYnamic]: *t*

15. Enter *4* to set a total length of four units.
    Specify total length or [Angle] <1.0000>: *4*

16. Select the left end of the arc; then select the right end of the line.
    Select an object to change or [Undo]:
    Notice (right) that **Total** added to the arc but subtracted from the line.

17. Repeat the command.
    Command: *[ENTER]*

18. Using the **Dynamic** option [DYnamic], select the line and watch how it changes as you move the cursor back and forth. Then try it with the arc.
    Select an object or [DElta/Percent/Total/DYnamic]: *dy*
    Select an object to change or [Undo]:
    Specify new end point:

19. Close the drawing without saving the changes.
    Command: *quit*

The *Lengthen* and *Stretch* commands have two big differences. First, there's an easy precision allotted by the *Lengthen* command; and second, the *Stretch* command allows you to modify more than one object at a time.

I'm more prone to use the *Stretch* command partly because, more often than not, I must modify multiple objects. But I use it also out of habit. For such a simple command, it's remarkably versatile and quite useful.

9.1.2 "Oh, NO! I Drew It Upside Down!" – The *Rotate* Command

Okay. So this'll probably never happen to you (then again, you might just be surprised). Still, it's not unusual to find a need to rotate text for a better fit or to rotate a piece of equipment or furniture for a more efficient layout. Either way, the *Rotate* command offers a simple solution to problems that, on a drawing board, might cause a redraw.

The *Rotate* command sequence is
    Command: *rotate*

Where to Find It:	
Command Line:	*Rotate*
Hotkey(s):	*ro*
Ribbon (Tab/Panel):	Home – Modify – **Rotate**
Menu:	Modify – **Rotate**
Toolbar:	Modify – **Rotate**
Tool Palette:	Modify – **Rotate**

Current positive angle in UCS: ANGDIR=counterclockwise ANGBASE=0
Select objects: *[Select the object(s) to rotate.]*

Select objects: *[Hit ENTER or 🖱 to confirm the selection set.]*
Specify base point: *[Select a point around which to rotate.]*
Specify rotation angle or [Copy/Reference] <0>: *[How much do you want to rotate the objects?]*

- AutoCAD begins by telling you something about the setup:
  - The **Current positive angle in UCS:** contains two variables:
    - **ANGDIR** simply reminds you that angles are measured counterclockwise (unless you changed it during the setup procedure for the drawing).
    - **ANGBASE** refers to a system variable that allows you to change the angle from which AutoCAD begins to measure. For example, if you tell AutoCAD to use a reference angle of 45°, it'll add 45° to the value of the **Angbase** system variable.
- The default option – **Rotation angle** – simply means, "How much do you want to rotate?" Type in an angle or drag the object on the screen (use Ortho to rotate at 90° increments).
- The next option – **Reference** – prompts again:

    Specify the reference angle <0>: *[Tell AutoCAD what the current rotation is.]*
    Specify the new angle or [Points] <0>: *[Tell AutoCAD what you want the rotation to be either by entering an angle or specifying two points.]*

    Use this option when you know what the current rotation angle is and what you want it to be. If you don't know what it is but do know what you want it to be, use the **Points** option and select two points on a line that represent the desired rotation. AutoCAD will determine the angle from the line you select and rotate to the **new angle**.
- Use the **Copy** option to create a rotated copy of the selected objects.

This may be confusing. Let's try an example.

Do This: 9.1.2A	Rotating Objects

I. Reopen the *star9* drawing in the C:\Steps\Lesson09 folder.
II. Freeze the **Arc** and **Line** layers.
III. Follow these steps.

9.1.2A: ROTATING OBJECTS

1. Enter the *Rotate* command ⟳.
   Command: *ro*

2. Select the circle and all the objects within it.
   Current positive angle in UCS: ANGDIR=counterclockwise ANGBASE=0
   Select objects:
   Select objects: *[ENTER]*

3. Select the center point ⊚ of the circle as the **base point**.
   Specify base point: _cen of

4. Tell AutoCAD to rotate the objects *90°*.
   Specify rotation angle or [Copy/Reference] <0>: *90*
   The star looks like the figure at right.

9.1.2A: ROTATING OBJECTS

5. Let's make a rotated copy. Repeat the *Rotate* command ↻. **Command:** *[ENTER]*	

6. Select the circle and all the objects within it.
    Current positive angle in UCS: ANGDIR=counterclockwise ANGBASE=0
    Select objects:
    Select objects: *[ENTER]*	

7. Specify a point one unit to the left of the object as the base point. **Specify base point:** *3,1*	
8. Tell AutoCAD to make a rotated **Copy**  of the object. **Specify rotation angle or [Copy/Reference] <90>:** *c*	

9. AutoCAD lets you know it'll be rotating a copy of the selected objects.
Now tell it to use the **Reference** option .
    Rotating a copy of the selected objects.
    Specify rotation angle or [Copy/Reference] <90>: *r*	

10. We rotated the objects by 90° in Step 4, so tell AutoCAD that the current angle (or reference angle) is *90*. **Specify the reference angle <0>:** *90*	
11. Tell AutoCAD to rotate all objects to an angle of *270°* (90° + 180°). **Specify the new angle or [Points] <0>:** *270* The stars now look like the figure at right.	
12. Undo ↶ all the changes in this exercise. **Command:** *u*	

If only they were all that easy!

9.1.3 "Okay. Give Me Three Just Like It, But Different Sizes." The *Scale* Command

No, that doesn't mean you have to draw it two more times. You simply need to make two copies and then scale (or resize) the copies to meet the customer's requirements. The command sequence is one of the easy ones:

Where to Find It:	
Command Line:	*Scale*
Hotkey(s):	*sc*
Ribbon (Tab/Panel):	Home – Modify – **Scale**
Menu:	Modify – **Scale**
Toolbar:	Modify – **Scale**
Tool Palette:	Modify – **Scale**

    Command: *scale*
    Select objects: *[Select the object(s) to scale.]*
    Select objects: *[ENTER or 🖱 confirm completion of the selection set.]*
    Specify base point: *[Pick the base point.]*
    Specify scale factor or [Copy/Reference] <1.0000>: *[Enter the scale factor.]*

Copy and Reference work just like they did in the *Rotate* command. Let's see *Scale* in action.

Do This: 9.1.3A	Scaling Objects

I. Be sure you're still in the *star9* drawing in the C:\Steps\Lesson09 folder.
II. Be sure the **Line** and **Arc** layers are frozen.
III. Follow these steps.

9.1.3A: SCALING OBJECTS

1. Use the *Pan* command 🖐 to center the circle in the lower area of the screen.
   Command: *p*

2. Enter the *Scale* command ▫.
   Command: *sc*

3. Select the circle and all the objects within it.
   Select objects:
   Select objects: *[ENTER]*

4. Select the bottom quadrant ◇ of the circle as the **base point**. Notice how the objects change as you move the cursor.
   Specify base point: _qua of

5. Tell AutoCAD you want to scale the objects by a factor of two.
   Specify scale factor or [Copy/Reference] <1.0000>: *2*
   The objects look the same but are twice as large.

6. Repeat Steps 2 through 4, but select only the outer circle.
   Command: *[ENTER]*

7. Tell AutoCAD you want to scale a copy [Copy].
   Specify scale factor or [Copy/Reference] <2.0000>: *c*
   Scaling a copy of the selected objects.

8. Scale the copy to 1.5x the original.
   Specify scale factor or [Copy/Reference] <2.0000>: *1.5*
   Your drawing looks like the figure at right.

10. Quit the drawing. Don't save your changes.
    Command: *quit*

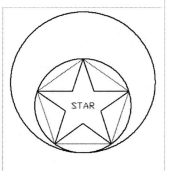

9.2 Identifying the Changes – The *RevCloud* Command

Creating a revision cloud on a paper drawing has always been one of the easiest chores in drafting – simply turn the paper over and scribble a cloudy line around your changes (then add a revision triangle to identify the change). It'll no doubt please you to know that AutoCAD's revision cloud is just as easy as the cloud you put on paper!

Here's the command sequence:

   Command: *revcloud*

	Where to Find It:
Command Line:	*RevCloud*
Ribbon (Tab/Panel):	Home – Draw (subpanel) – **Revision Cloud**
Menu:	Draw – **Revision Cloud**
Toolbar:	Draw – **Revision Cloud**
Tool Palette:	Draw – **Revision Cloud**

Minimum arc length: 0.5000   Maximum arc length: 0.5000   Style: Normal
Specify start point or [Arc length/Object/Style] <Object>: *[Pick the start point for your cloud.]*
Guide crosshairs along cloud path... *[Move your crosshairs along the path you want your cloud to take – no need to pick, AutoCAD will automatically draw the cloud.]*
Revision cloud finished. *[When you move the crosshairs back to the beginning of the cloud, AutoCAD will automatically close it for you!]*

This remarkably easy tool has a couple options you might want to review.

- You have the ability to change the **Arc length** if you wish. The default **0.5000** works quite well, and AutoCAD automatically multiplies it by the **Dimscale** (more on dimension variables in Lesson 16, p.350). Unfortunately, however, it isn't affected by the annotation scale. (Perhaps that's something AutoCAD should consider for future release!)
- The **Object** option allows you to convert any existing closed object (polyline, circle, polygon, or spline) into a cloud.
- The style option gives you two choices:
  Select arc style [Normal/Calligraphy] <Normal>:
  o **Normal** simply draws a polyline cloud.
  o **Calligraphy** makes the cloud fancier. Otherwise, it works just the same as a normal cloud.

You really have to see this tool in action!

Do This: 9.2A	Adding Revision Clouds

I. Reopen the *star9* drawing in the C:\Steps\Lesson09 folder.
II. Make the **OBJ1** layer current and turn ORTHO off.
III. Follow these steps.

9.2A: ADDING REVISION CLOUDS

1. Enter the ***Revcloud*** command.
   Command: *revcloud*

2. AutoCAD tells you what the arc lengths will be and asks you where you'd like to start the cloud. Pick a point above the circle.
   Minimum arc length: 0.5000   Maximum arc length: 0.5000   Style: Normal
   Specify start point or [Arc length/Object/Style] <Object>:

3. Trace the path of the cloud around it.
   Guide crosshairs along cloud path...
   Revision cloud finished.
   Your drawing will look something like the figure at right.

4. Erase the cloud. Notice that, like a polygon, the revcloud is a polyline and a single pick will select the entire cloud. (More on polylines in Lesson 12, p.272.)
   Command: *e*

5. Let's try the **Object** and **Calligraphy** options. Repeat the ***Revcloud*** command.
   Command: *revcloud*

9.2A: ADDING REVISION CLOUDS

6. Select the **Style** option [Style], and then select **Calligraphy** [Calligraphy].
   Specify start point or [Arc length/Object/Style] <Object>: *s*
   Select arc style [Normal/Calligraphy] <Normal>: *c*

7. Now accept the **Object** option.
   Minimum arc length: 0.5000   Maximum arc length: 0.5000
   Arc style: Calligraphy
   Specify start point or [Arc length/Object/Style] <Object>: *[ENTER]*

8. Select the circle.
   Select object:

9. Don't reverse the direction of the cloud this time (we'll do that in a minute).
   Reverse direction [Yes/No] <No>: *[ENTER]*
   Revision cloud finished.
   Your drawing looks like the figure at right.

10. Undo the last command.
    Command: *u*

11. Let's try the **Object** option on a polyline. Repeat the *Revcloud* command.
    Command: *revcloud*

12. Select the **Object** option.
    Minimum arc length: 0.5000   Maximum arc length: 0.5000
    Arc style: Calligraphy
    Specify start point or [Arc length/Object] <Object>: *[ENTER]*

13. Select the polygon just inside the circle.
    Select object:

14. This time, let's reverse the direction of the cloud.
    Reverse direction [Yes/No] <No>: *y*
    Revision cloud finished.
    Your drawing looks like the figure at right.

15. Experiment with the *Revcloud* command on the other objects in the drawing. On which will the command work? On which will it not work? Why/Why not?

As you've seen, *Revcloud's* **Object** option will replace the selected object with a revision cloud. You can prevent the removal of the selected object by changing the **Delobj** system variable from its default of **1** to **0**.

Did you notice what the *Revcloud* command doesn't do? I know, it was so simple that you just hate to hear that it lacks anything … but it's missing one key ingredient for revisions. It doesn't provide a revision *triangle* to allow you to identify the changes you made! You'll have to provide this yourself (use the *Polygon* command)!

> Okay, I couldn't let you go like that. I created a lisp routine (called *Revtriangle.lsp* – the command is *RevT*) and put it in the C:\Steps\Lesson09 folder. You can use it to insert your revision triangles.

9.3 "A Whole New Ball Game!" Editing with Grips

> Before we begin our experiments with grips, we must be sure they're active. Follow this sequence (be sure the *Grips* system variable is set to *2*):
> **Command:** *grips*
> **Enter new value for GRIPS <0>:** *2*

I've always approached grips with reluctance in my basic AutoCAD classes. It's not that they're not remarkable tools; it's just that most students in a basic class are still trying to master the "normal" modifying tools. These are fairly simple and straightforward – if you want to copy something, you type *copy*. AutoCAD then leads you through the necessary steps by dynamic prompts at the cursor menu or by prompts at the command line. Grips, on the other hand, are more intuitive. That is, they require that you *know* how to move from one step to the next, and that you know where the specific desired modification tool is located in the grip prompts.

So, exactly what are grips?

Grips are control points located on all objects, blocks, and groups. We'll look at blocks in Lesson 21, p.468, but we can gain an understanding of grips using simple lines, polylines, and circles.

Grips are assigned to specific locations on specific objects. That is, all lines and arcs will have grips at the endpoints and midpoint. All circles will have grips at the center point and quadrants. And all polylines will have grips at the vertices.

> You can control all aspects of grips display using the Options dialog box. Here, you can set up different grip colors and sizes (not recommended), enable/disable grips in the drawing (the dialog box method of setting the **Grips** system variable), or enable/disable grips within blocks (more on blocks in Lesson 21, p.468).

There are five basic modification tools available using grips: ***Stretch***, ***Move***, ***Rotate***, ***Scale***, and ***Mirror***. ***Copy*** is also available as an option of each of the primary Grip tools or as a default when you hold down the CTRL key while using a grip procedure.

To access the commands, first select the object(s) you wish to modify. (Do this without entering a command.) The blue grips that display for each of the objects are called *unselected* grips. Picking on a grip ▯ will cause it to change to a red, *selected* grip. Your crosshairs will automatically snap to a grip regardless of the **Snap** setting. You must select a grip to modify the object. (To disable [clear] grips, hit the ESC key.)

Note that you can select multiple objects using a window or crossing window just as you've always done. Selecting multiple objects simply means that the grips for several objects will display at once.

To remove an object from a selection set, hold down the SHIFT key and select ▯ the object to remove.

Once a grip has been selected, AutoCAD's command line presents the grip options for the ***Stretch*** command. Toggle through the commands (***Stretch***, ***Move***, ***Rotate***, ***Scale***, ***Mirror***) by hitting the SPACEBAR.

> The cursor menu also presents the Grip commands once you have a red (or *hot*) grip.

The initial grips prompt looks like this:
   ** STRETCH **
   Specify stretch point or [Base point/Copy/Undo/eXit]:

- By default, you'll stretch the object using the opposite endpoint as an anchor (center point for circles).
- Selecting the **Base point** option will allow you to change the point at which you "hold" the object while stretching it. The default base point is the selected grip.

- The **Copy** option provides one of the great benefits of using grips – you can leave the original object as it is and create stretched copies. (You can create multiple copies by default!)

> A nifty trick to using grips involves the CTRL key. Holding it down while locating the "to point" of the stretched line will automatically start the **Copy** option. Continuing to hold it down after the first "to point" is selected will cause grips to function in *offset* mode. That is, AutoCAD will determine an offset distance from the first "to point" selection; then it'll create subsequent copies at the same interval. These techniques work as well for each of the five grips procedures.

- The rest of the options are self-explanatory.

Let's see what we can do with the *Stretch* grip procedure.

Do This: 9.3A	Editing with Grips – Stretch

I. Open the *Grips* file the C:\Steps\Lesson9 folder. (It's another star/circle drawing.)
II. Toggle the grid and snap on/off as needed.
III. Follow these steps.

9.3A: EDITING WITH GRIPS - STRETCH

1. Select the upper part of the star with a crossing window as shown.
   Command:

2. Notice that grips appear on each of the individual objects and that each highlights.
   Command:

3. Select 🖱 the topmost grip.
   Command:
   Notice that it turns red and that the command line presents the initial grips options.

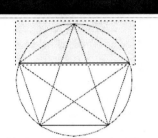

4. Pick a point three grid marks due north of the top of the star. (Turn on the snap to help you.)
   ** STRETCH **
   Specify stretch point or [Base point/Copy/Undo/eXit]:
   Your drawing looks like the figure at right. (Zoom or pan as necessary for a better view.) Notice which objects stretch.
   You've discovered the grips method of resizing a circle.

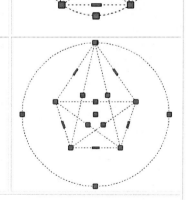

5. *Undo* the last modification ↩.
   Command: *u*

6. Repeat Step 1.

9.3A: EDITING WITH GRIPS - STRETCH

7. Let's remove the circle from the selection set. Hold down the SHIFT key and select it 🖱 (select the circle, *not a grip*). Notice that its grips disappear, and that it's no longer highlighted.	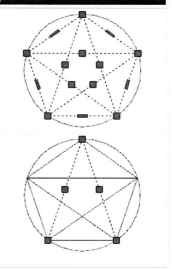
8. Hit the ESC key [Esc] to clear the grips.	
9. This time, select just the two lines that form the upper point of the star.	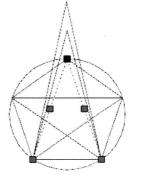
10. Select the topmost grip, then type *c* at the prompt (or select **Copy** [Copy] from the menu) to activate the **Copy** option. **** STRETCH **** **Specify stretch point or [Base point/Copy/Undo/ eXit]:** *c*	
11. Notice how the prompt changes. Going north, skip a grid mark and then pick the next. Do this twice. **** STRETCH (multiple) **** **Specify stretch point or [Base point/Copy/Undo/ eXit]:**	
12. Hit ENTER to complete the procedure. (Alternately, you can pick **Exit** from the cursor menu to complete the grips procedure). **** STRETCH (multiple) **** **Specify stretch point or [Base point/Copy/Undo/ eXit]:** *[ENTER]*	
13. Undo the last modification ↶. **Command:** *u*	

Figure 9.002

Thaw the **TXT** layer for this next exercise. The text *My Star* appears (Figure 9.001).

We can use the command line to access the various methods used by grips. Simply use the SPACEBAR to scroll through the options. But there's an easier way.

Any time the grip prompts are on the command line – that is, any time you've selected a grip – you can right click 🖱 in the drawing area to access the cursor menu that provides the same options as the command line. The menu is shown in Figure 9.002.

Figure 9.001

- The topmost option (**Enter**) is the equivalent of its keyboard counterpart.
- **Recent Input** lists the last several responses to the current command. You can select one if you wish, or enter something else.

- The next section includes the five grips procedures (the same ones you'll see when using the SPACEBAR to toggle through the grips procedures on the command line).
- The third section shows the options available for the current grip procedure.
- **Properties** in the fourth section will cause the Properties palette to appear.
- **Exit**, of course, returns you to the command prompt.

We'll use this menu, the dynamic input menu, and the SPACEBAR approach to toggle grips procedures in our next exercise.

Let's examine the *Rotate* and *Mirror* Grips procedures.

Do This: 9.3B	More Grips Editing

I. Be sure you're still in the *Grips* file the C:\Steps\Lesson9 folder. If not, please open it now.
II. Follow these steps.

9.3B: MORE GRIPS EDITING
1. Select all of the objects. Use a window or crossing window. **Command:**
2. Select 🖱 the grip at the tip of the star's upper right arm. **Command:** AutoCAD presents the grip options.
3. Hit the SPACEBAR twice to access the grips *Rotate* tool. **** STRETCH **** **Specify stretch point or [Base point/Copy/Undo/eXit]:** *[SPACEBAR]* **** MOVE **** **Specify move point or [Base point/Copy/Undo/eXit]:** *[SPACEBAR]*
4. Type *c* to access the **Copy** option. **** ROTATE **** **Specify rotation angle or [Base point/Copy/Undo/Reference/eXit]:** *c*
5. Tell AutoCAD you wish to create a rotated copy at an angle of *144°*. **** ROTATE (multiple) **** **Specify rotation angle or [Base point/Copy/Undo/Reference/eXit]:** *144*
6. Exit the procedure, and then clear the grips [Esc]. **** ROTATE (multiple) **** **Specify rotation angle or [Base point/ Copy/Undo/Reference/eXit]:** *[ENTER]* Your drawing looks like the figure at right. 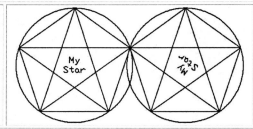
7. Undo the last modification ↶. **Command:** *u*

9.3B: MORE GRIPS EDITING

8. Let's try to rotate the objects using the menu. Repeat Steps 1 and 2.

9. Right click ▯ in the drawing area and select the *Rotate* tool.
   ** STRETCH **
   Specify stretch point or [Base point/Copy/Undo/eXit]: _rotate

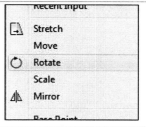

10. Right click again and select the **Copy** option.
    ** ROTATE **
    Specify rotation angle or [Base point/Copy/Undo/Reference/eXit]: _copy

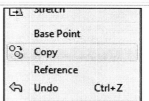

11. Tell AutoCAD you want a rotated copy at *180°*.
    ** ROTATE (multiple) **
    Specify new angle or [Base point/Copy/Undo/Reference/eXit]: *180*

12. Complete the command; clear the grips.
    ** ROTATE (multiple) **
    Specify new angle or [Base point/Copy/Undo/Reference/eXit]: *[ENTER]*
    Your drawing looks like the figure at right.

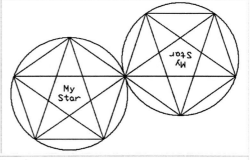

13. Undo the last modification ↶.
    Command: *u*

14. Now let's look at the Grips *Mirror* tool. Repeat Step 1.

15. Select the uppermost grip (at the upper tip of the star).
    Command:

16. Right click to display the Grips menu; pick the *Mirror* tool ⚠ Mirror.
    ** STRETCH **
    Specify stretch point or [Base point/Copy/Undo/eXit]: _mirror

17. Holding down the CTRL key on your keyboard (to force a copy), pick a point due west of the hot grip as shown. (Be sure to use Ortho.)
    ** MIRROR (multiple) **
    Specify second point or [Base point/Copy/Undo/eXit]:

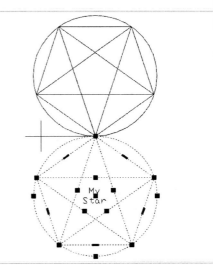

9.3B: MORE GRIPS EDITING

18. Pick the **Exit** option [eXit], then clear the grips.

   ** MIRROR (multiple) **
   Specify second point or [Base point/Copy/Undo/eXit]: _exit

Your drawing looks like the figure at right.

(Note: These objects were mirrored with the **Mirrtext** sysvar set to **1**. If yours is set to **0**, the text won't mirror.)

19. Undo the last modification.

   Command: *u*

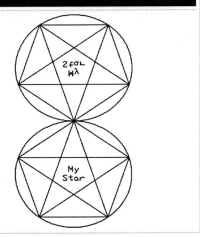

This last section will first walk you through the simple grips *Move* tool. Then we'll use the grips *Scale* tool to make sized copies of objects.

Let's proceed.

Do This: 9.3C	Move and Scale with Grips

I. Be sure you're still in the *Grips* file the C:\Steps\Lesson12 folder. If not, please open it now.
II. Follow these steps.

9.3C: MOVE AND SCALE WITH GRIPS

1. Select all of the objects. Use a window or crossing window.

2. Select one of the grips. (With the grips *Move* tool, all the objects with active grips will move.) I'm using the grip at the topmost point of the star. (We're going to redefine the **Base point** anyway.)

3. Right click to display the Grips menu. Select *Move* [Move].

   ** STRETCH **
   Specify stretch point or [Base point/Copy/Undo/eXit]: _move

4. Select the **Base point** option [Base point].

   ** MOVE **
   Specify move point or [Base point/Copy/Undo/eXit]: _base

5. Pick the center of the circle as the new base point.

   Specify base point: _cen of

6. Tell AutoCAD to move the objects to the absolute coordinate **6,4.5**. (Remember to toggle off dynamic input to use an absolute coordinate.)

   ** MOVE **
   Specify move point or [Base point/Copy/Undo/eXit]: *6,4.5*

Notice that all of the objects with grips showing move even though only one grip has been selected.

7. Save the drawing but don't exit.

   Command: *qsave*

217

9.3C: MOVE AND SCALE WITH GRIPS

AutoCAD provides an easier way to move one or even a few objects with grips. Once you've selected the object (any object with center or midpoint grips), select the center or midpoint grip. AutoCAD will move the object by default.

8. Now let's scale the objects. Select only the circle and the polygon.

9. Select the center grip.

10. Call the Grips menu and pick *Scale* 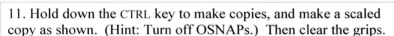.

   ** STRETCH **

   Specify stretch point or [Base point/Copy/Undo/eXit]: _scale

11. Hold down the CTRL key to make copies, and make a scaled copy as shown. (Hint: Turn off OSNAPs.) Then clear the grips.

   ** SCALE (multiple) **

   Specify scale factor or [Base point/Copy/Undo/ Reference/eXit]:

12. Save the drawing 💾 and exit.

   Command: *qsave*

So what do you think of grips?

After learning the basic modifying commands in earlier lessons, grips may be difficult at first. But practice, practice, practice! Grips will increase speed and productivity (excuse that nasty corporate word) ... and your earning potential!

9.4 Putting It All Together

We've covered several new modification tools in the last two lessons. Let's try a practical exercise (or two) and use several of these tools to create a graduated cylinder and a speedometer.

Do This: 9.4A A Practice Exercise

I. Open the *grad_cyl* drawing from the C:\Steps\Lesson\Steps\Lesson09 folder.
II. Save the drawing as *Pass1*.
III. Follow these steps.

9.4A: PRACTICE EXERCISE

1. Adjust your view to the one shown.

   Command: *z*

9.4A: PRACTICE EXERCISE

2. Use the *Array* command to make four copies (five lines in all) of the cyan line below the one shown. The spacing should be -0.1. **Command:** *ar*	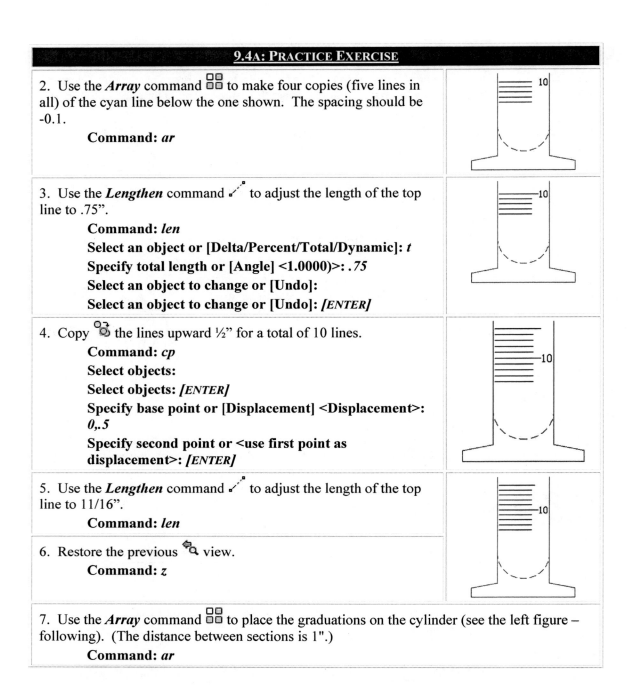
3. Use the *Lengthen* command to adjust the length of the top line to .75". **Command:** *len* Select an object or [Delta/Percent/Total/Dynamic]: *t* Specify total length or [Angle] <1.0000>: *.75* Select an object to change or [Undo]: Select an object to change or [Undo]: *[ENTER]*	
4. Copy the lines upward ½" for a total of 10 lines. **Command:** *cp* Select objects: Select objects: *[ENTER]* Specify base point or [Displacement] <Displacement>: *0,.5* Specify second point or <use first point as displacement>: *[ENTER]*	
5. Use the *Lengthen* command to adjust the length of the top line to 11/16". **Command:** *len*	
6. Restore the previous view. **Command:** *z*	
7. Use the *Array* command to place the graduations on the cylinder (see the left figure – following). (The distance between sections is 1".) **Command:** *ar*	

9.4A: PRACTICE EXERCISE

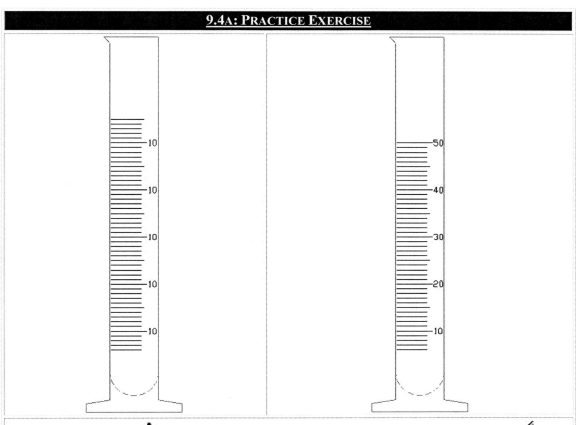

8. Use the text editor ✏️ to change the numbers to read 10, 20, 30, 40, and 50. Erase 🖊 the lines above 50.

    Command: *ed*

    Command: *e*

Your drawing looks like the right figure above.

Your graduated cylinder is finished. Good job! Now let's draw a speedometer.

9. Now adjust your view to get closer to the circle on the right.

    Command: *z*

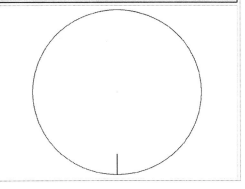

9.4A: PRACTICE EXERCISE

10. Use the *Offset* command to copy the circle inward twice at .17" increments.

   Command: *o*

   Current settings: Erase source=No Layer=Source OFFSETGAPTYPE=0

   Specify offset distance or [Through/Erase/Layer] <Through>: *.17*

   Select object to offset or [Exit/Undo] <Exit>: *[select the circle]*

   Specify point on side to offset or [Exit/Multiple/Undo] <Exit>: *m*

   Specify point on side to offset or [Exit/Undo] <next object>:

   Select object to offset or [Exit/Undo] <Exit>: *[ENTER]*

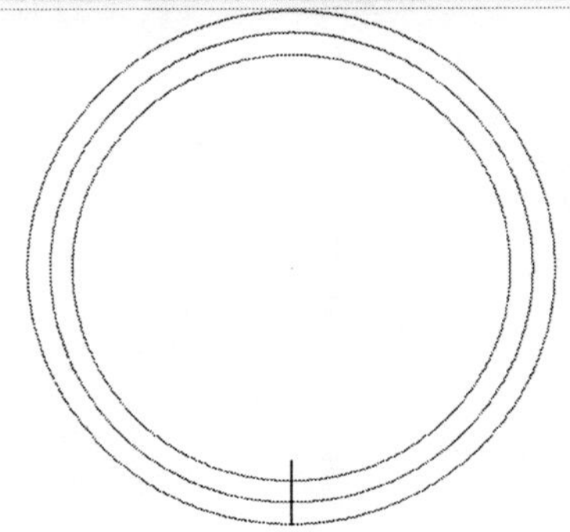

11. Use the *Array* command to make 10 copies (11 lines altogether) of the line along a 36° arc.

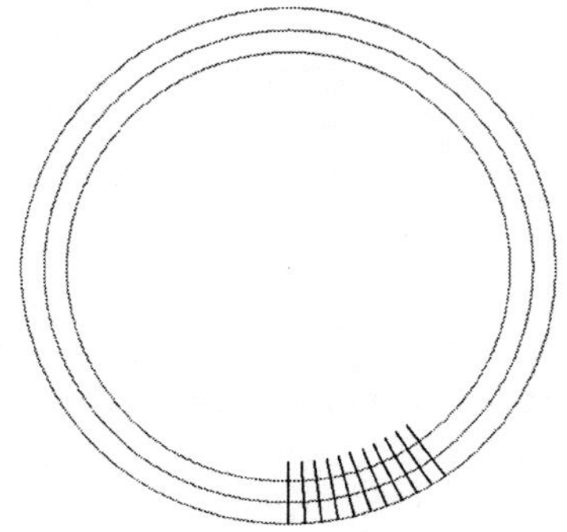

12. Using the inner circles as cutting edges, *trim* the lines, as shown. (Erase the right-most line.)

   Command: *tr*

13. *Erase* the two inner circles.

   Command: *e*

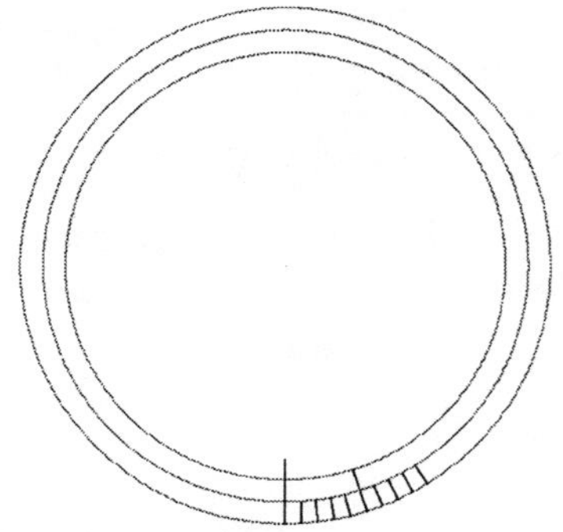

14. Array the lines as shown. (Make 9 copies – 10 sets – and fill a 360° angle.)

   Command: *ar*

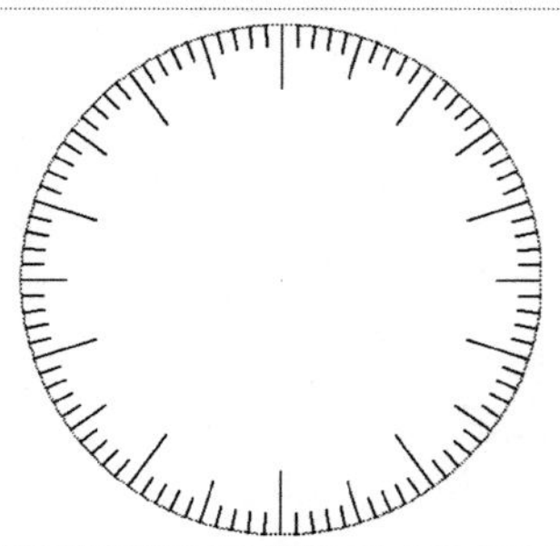

9.4A: PRACTICE EXERCISE

15. Erase the bottom set of lines as shown.
    Command: *e*

16. Add text and thaw the **Arrow** layer (use the appropriate layers and a text height of **0.2**) to complete the speedometer.

17. Save the drawing.
    Command: *qsave*

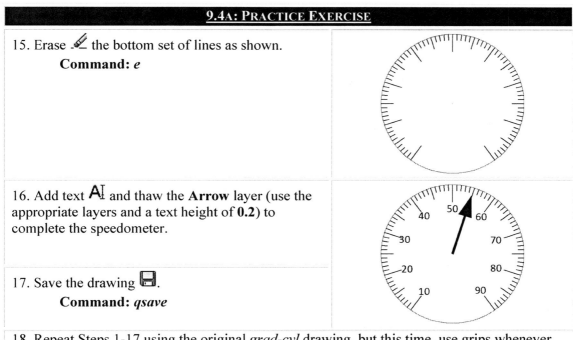

18. Repeat Steps 1-17 using the original *grad-cyl* drawing, but this time, use grips whenever possible.

9.5 Extra Steps

When All Else Fails – Ask for Help!
- My Wife (When I couldn't find the right exit)

I'm convinced that we learn more through exploration, experimentation, and error than by any other means.
- Anonymous

We're really making progress through our course! But what if you need help? Well, in this regard, you certainly have innumerable choices! (I mean besides emailing me!)

One of the smartest things you (or any student) can do is to know what resources are available to help you when you get into a bind. Make a list of the resources available to you – begin with this text. Add your instructor (if you're in a classroom environment), the smart guy sitting next to you, the InfoCenter, and any other classroom references (handouts, books, tests, etc.). Most importantly – add *your willingness to explore and experiment with the software* (never underestimate the power of trial and error as a learning tool).

Okay, now here are a couple tools that can help.

- The AutoCAD Welcome Screen, which appears when you start AutoCAD[*], contains links to short training videos. These can be quite useful for quick tips or to refresh your memory of selected tools.
- Now take a look at the Help pull-down menu. The first thing you may notice is the number of selections available to you. Look at some of these:
  - **Help** provides access to a standard Windows Help screen in Internet Explorer. I suggest starting with the subject index – simply enter the subject for which you're searching and pick the **Display** button (arrow). If that doesn't provide a solution, try the word search (**Search**) box in the upper right corner.

---

[*] If the Welcome Screen doesn't appear, you can call it with the *WelcomeScreen* command.

- o The **New Features Workshop** helps if you're already familiar with AutoCAD but new to this release. (You'll find new features for the last three releases!)
- o As so many software companies are doing now, AutoCAD also provides access to Online (**Additional**) Resources. These are useful mostly for upgrades and expansion of your software. Online support, while less expensive than telephone support, is generally quite slow.

9.6 What Have We Learned?

Items covered in this lesson include:

- *AutoCAD's Help tools*
- *AutoCAD Modify Commands:*
    - o ***Lengthen***
    - o ***Stretch***
    - o ***Rotate***
    - o ***Scale***
    - o ***Revcloud***
- *AutoCAD's Grip Tools*
    - o *Move*
    - o *Stretch*
    - o *Scale*
    - o *Rotate*
    - o *Mirror*
- *AutoCAD System Variables:*
    - o ***Angdir***
    - o ***Angbase***
    - o ***Delobj***

Have you noticed that the easiest of AutoCAD's tools are invariably modification commands?

With the conclusion of this lesson, you'll have learned all but one of the more common drawing and editing techniques AutoCAD has to offer the 2-dimensional drafter. We'll look at polylines next. Then we'll move on to a discussion of some fairly useful drawing tricks and toys.

9.7 Exercises

1. Create a drawing template with the following parameters:
    1.1. Lower left limits: 0,0
    1.2. Upper right limits: 17,11
    1.3. Grid: ½
    1.4. Snap: ¼
    1.5. Layers: At right
    1.6. Save the template as *BwLay.dwt* in the C:\Steps\Lesson09 folder.

LAYER	COLOR	LINETYPE
Obj	red	continuous
Cl	cyan	center2
Hidden	212	hidden2
Text	12	continuous
Dim	blue	continuous

2. Start a new drawing using the *BwLay.dwt* template. (You created this in Exercise #1.)
    2.1. Create the drawing at right. You're allowed to draw one box and the outline of the wall. You may use the following commands to help: *Copy, Rotate, Scale* and *Stretch*. The size of each window is 1.75x the size of the smaller window next to it. Feel free to add layers if you think it necessary.

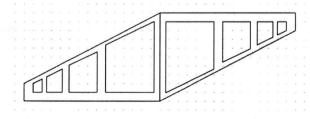

    2.2. Recreate the drawing using only grip modifying tools.

9.8 For Web-Based Review Questions and Additional Exercises, visit: http://foragerpub.com/AcadFiles/2011/2011.htm

Following this lesson, you will:
- ✓ Know how to
- ✓ Be familiar with these AutoCAD constraints
  - **Geometric Constraints**
  - **AutoConstraint**
  - **Inferred Constraints**
- ✓ Know how to use alternate selection techniques found in the **Selections** Tab of the Options dialog box, including:
  - Noun/Verb Selection
  - Shift to Add
  - Press and Drag
  - Implied Windowing
  - Isolate/Hide Objects
- ✓ Know how to use Select Similar and Selection Cycling

Geometric Constraints & Alternate Selection Methods

> My WordWeb[*] dictionary defines a constraint as "a device that retards something's motion". As you'll see, AutoCAD makes a full library of constraints available to assist you.

All geometry in a drawing exists in some relationship with all the other geometry in a drawing. For example, in simple terms, the corners of a rectangle join at the endpoints of the lines. In other words, the endpoints occupy the same location; they are coincident. *Other objects may be parallel or perpendicular to each other. The list goes on ...*

Imagine an environment where you can tell the objects to remain parallel or concentric or coincident with each other regardless of other changes made ... an environment where the objects do what you intend for them to do. Welcome a new AutoCAD phrase: design intent. *With the introduction of parametric constraints, you can restrict (constrain) objects to do what you intend them to do!*

AutoCAD provides two methods of constraining objects: Geometric *and* Dimensional. *We'll become familiar with geometric constraints in this lesson; we'll discover dimensional constraints in Lesson 17, p.394.*

Lesson 10 will also deal with some nifty techniques that, although not often used by anyone other than well-trained CAD operators, can enhance your drawing speed even beyond the techniques you've already learned!

10.1 Constraining Relationships: Constraint Symbols, Insertions, and Behaviors

Geometric constraints hold the geometry in a model together and make it behave properly. That is, they keep sides parallel or perpendicular to one another, they keep holes centered in arcs, and much more.

> You'll notice a familiarity about constraints. In fact, they look and function similarly to OSNAPs. But don't underestimate the differences between the two! Use OSNAPs to assist you when creating geometry; use constraints to anchor or restrain the geometry once you've created it.

To add constraints, use one of the methods shown in the WTFI table to begin the *GeomConstraint* command. (Select the specific constraint on the menu, toolbar, palette or ribbon to bypass having to select it on the command line.)

Where to Find It:	
Command Line:	*GeomConstraint*
Hotkey(s):	*gcon*
Ribbon (Tab/Panel):	Parametric – Geometric – **[select constraint]**
Menu:	Parametric – Geometric Constraints – **[select constraint]**
Toolbar:	Geometric Constraints – **[select constraint]**
Tool Palette:	Constraints – **[select constraint]**

The primary command looks like this:

Command: *geomconstraint*
Enter constraint type [Horizontal/Vertical/Perpendicular/ PArallel/Tangent/SMooth/Coincident/ CONcentric/COLinear/Symmetric/ Equal/Fix] <Coincident>: *[Select the type of constraint you wish to apply or hit ENTER to accept the default; see the following table for information about the specific type of constraint.]*
Select first point or [Object/Autoconstrain] <Object>: *[Follow the instructions to apply the constraint.]*
Select second point or [Object] <Object>:

---

[*] http://wordweb.info/

> That's the primary geometric constraint command, but each option has its own command as well. Just add a "gc" before the option name. Thus **GCConcentric** begins the **GeomConstraint** command's **CONcentric** option, and so forth. Unfortunately, you either have to use a button or type out the entire command; the "gc" commands have no hot keys.

Let's look at each type of constraint and how it behaves in your drawing or model.

CONSTRAINT	BUTTON	CURSOR ICON	BEHAVIOR
Coincident	↓	↓	A coincident constraint holds together (constrains) two points, one point to a curve, or an object to a point. (When you use a coincident constraint on two center points, it's the same as a concentric constraint.) *When selecting points, AutoCAD changes the second selected point or object to make it coincident with the first. It will infer the first as being fixed.*
Concentric	◎	◎	A concentric constraint holds together the center points of arcs, circles, or ellipses. AutoCAD makes the second selected object concentric to the first. It also infers that the first selected object is fixed.
Collinear	✓	✓	A collinear constraint holds together lines and ellipse axes to the same line. AutoCAD makes the second selected object collinear with the first.
Equal	=	=	An equal constraint forces radii or line lengths to be the same (resizing them as necessary to achieve equality). AutoCAD changes the radius or length of the second selected object to make it equal to the first selected object.
Fixed	🔒	🔒	A fixed constraint holds a selected point or curve to a fixed location relative to the WCS (0,0,0).
Horizontal or Vertical	─ / │	─ / │	Horizontal and vertical constraints force lines, ellipse axes or point pairs to be horizontal or vertical to the X (horizontal) or Y (vertical) axis. If you select a line segment, AutoCAD infers that the nearest point is fixed and adjusts the segment to become horizontal or vertical. If you select two points rather than a segment, AutoCAD infers that the first point is fixed.
Parallel	//	//	A parallel constraint forces two lines or ellipse axes to be parallel to each other. AutoCAD orients the second selected object parallel to the first. It infers that that the first object is fixed.
Perpendicular	⊾	⊾	A perpendicular constraint forces lines and ellipse axes to lie at right angles to each other. AutoCAD orients the second selected object perpendicular to the first. It infers that that the first object is fixed.

Constraint	Button	Cursor Icon	Behavior
Symmetrical	[¦]	[¦]	A symmetrical constraint orients geometry to a symmetrical position only (not size) about a selected line. AutoCAD makes the second selected object symmetric to the first selected object.
Smooth	⤳	⤳	A smooth constraint forces a spline to continue the curve begun by a line, arc or another spline. (We'll discuss splines in Lesson 13, p.295.) The splines must share a coincident endpoint.
Tangent	⌀	⌀	A tangent constraint adjusts a line, arc, or circle to make it tangent to another arc or circle. Geometry doesn't have to touch to be tangent. AutoCAD orients the second selected object tangent to the first. It infers that the first object is fixed.

To see existing constraints in a drawing, select **Show all** on the **Geometric** panel of the ribbon's **Parametric** tab or enter the *ConstraintBar* command and select the **Show** option. AutoCAD will display an icon – called the constraint bar – for each constraint. To hide the constraint bars after you've displayed them, select **Hide all** on the panel or repeat the *ConstraintBar* command and select the **Hide** option.

To remove a constraint, hover your mouse over its constraint bar and use the DEL key on your keyboard.

Let's set some constraints to see how they work.

Do This: 10.1A	Using Geometric Constraints

I. Open the *Constraint_Practice* drawing in the C:\Steps\Lesson09 folder. It looks like the figure at right. 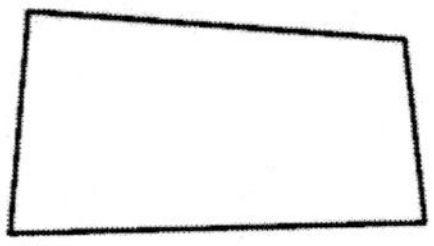
II. Follow these steps. (In this exercise, we'll use the buttons rather than the command line to keep it simple. But you can use the ribbon, toolbar, command line or tool palette if you prefer.)

10.1A: USING GEOMETRIC CONSTRAINTS

1. Move the right vertical line to the left about half an inch.
   Command: *m*
Notice that the lines separate.

2. Undo the move.
   Command: *u*

3. Open the **Parametric** tab on the ribbon. Our tools reside on the **Geometric** panel (right).

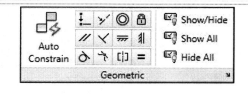

10.1A: USING GEOMETRIC CONSTRAINTS

4. Begin by creating coincident constraints at each of the four corners. Use one of the methods identified on the WTFI table on p.225. (I'll use the ribbon's **Coincident** button.)

   Command: _GcCoincident
   Select first point or [Object/Autoconstrain] <Object>: *[Select a line toward an endpoint.]*
   Select second point or [Object] <Object>: *[Select an adjoining line near the same endpoint.]*

5. Repeat Step 1. Notice that now, AutoCAD has constrained the endpoints to each other and forces the adjoining endpoints to accompany the move.

6. Undo the move.

   Command: u

7. Now let's create a perpendicular constraint at the upper left corner of the rectangle. (Again, I'll use the buttons throughout this exercise. The command prompts may be a bit different should you use the keyboard method.)

   Command: _GcPerpendicular
   Select first object: *[Select the top line ...]*
   Select second object: *[... and then the left.]*

   Notice (right) that a constraint bar appears at the corner indicating the presence of a Perpendicular constraint.

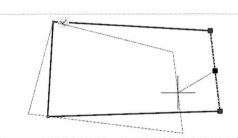

8. Using grips, move the right side of the rectangle left, right, up and down.

   Notice that the corners remain intact and that the perpendicular angles remain perpendicular.

9. Cancel the move with the ESC key.

10. Create parallel constraints // between the top/bottom and left/right lines.

    Command: _GcParallel
    Select first object: *[Select the top/left line.]*
    Select second object: *[Select the bottom/right line.]*

    How're we doing? (Check the figure at right.)

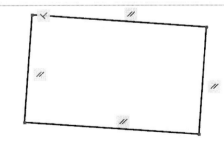

228

10.1A: USING GEOMETRIC CONSTRAINTS

11. Finally, make the top line horizontal.
    Command: *_GcHorizontal*
    Select an object or [2Points] <2Points>: *[Select the top line.]*

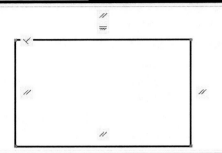

12. Thaw the **Circle** layer. Notice the circle that appears to the left of the rectangle.

13. Create a coincident constraint between the lower left corner of the rectangle and the center of the circle. (Pick the corner of the rectangle as the first object.)
    Command: *_GcCoincident*
    Select first point or [Object/Autoconstrain] <Object>: *[Select the left endpoint of the bottom line ...]*
    Select second point or [Object] <Object>: *_cen of [... then the center (use OSNAPs) of the circle.]*

14. Now move ✥ the left vertical line to the right about ½".
    Command: *m*
Notice that the circle moves with it.

15. Undo ↶ the move.
    Command: *u*

16. Copy the circle 6" to the right.
    Command: *co*

17. Now create a symmetric constraint [|] between the two circles, using the right vertical line as the symmetry line.
    Command: *_GcSymmetric*
    Select first object or [2Points] <2Points>: *[Select the original circle.]*
    Select second object: *[Select the new circle.]*
    Select symmetry line: *[Select the right vertical line.]*

18. Using grips, Stretch the right line to the left.
    Command: *s*
Notice that the right circle adjusts to remain symmetrical.

19. Undo ↶ the stretch.
    Command: *u*

20. Now let's delete a constraint. First hover the cursor over the symmetrical icon to the right of the object. Notice (following figure) that AutoCAD highlights the two circles and the symmetry line controlled by this constraint.

10.1A: USING GEOMETRIC CONSTRAINTS

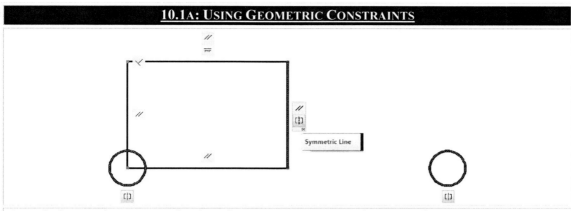

21. Hit the DEL key on your keyboard. AutoCAD removes the constraint.
22. Exit the drawing without saving your changes.

Constraints are just *soooo* cool!

Of course, you can only constrain certain points – depending upon the object type. Refer to the following chart.

OBJECT TYPE	AVAILABLE CONSTRAINT POINTS
Arc	center, endpoints, and midpoint
Blocks and Text	insertion
Circles and Ellipses	center
Line	endpoints and midpoint
PLines	endpoints and midpoint (center of arc segments)
Point	node
Splines	endpoints

An interesting point to make about AutoCAD constraints may serve to frustrate Inventor users while making AutoCAD novices sigh with relief; AutoCAD constraints work only on 2D objects! But I suppose that gives us something to anticipate in a future release!

10.2 Auto Constraints and Display Settings

Although manually creating constraints allows you more control over what gets constrained and what doesn't, you can have AutoCAD automatically constrain selected geometry. You'll use the Constraint Settings dialog box (Figure 10.001, p.231). See the WTFI table at right for ways to open the dialog box.

Where to Find It:	
Command Line:	*ConstraintSettings*
Hotkey(s):	*CSettings*
Ribbon (Tab/Panel):	Parametric – Geometric or Dimensional – Settings arrow
Menu:	Parametric – **Constraint Settings**

You have three tabs with which to work.

- The **Geometric** tab (Figure 10.001, p.231) lists the available constraints. When you pick the **Show All** button on the **Geometric** panel, AutoCAD will display the constraints with checks in their boxes (in the **Constraint bar settings** frame). Personally, I prefer to have AutoCAD display all constraints; if I want to **Show All** … then AutoCAD should do just that!

- o The **Only display constraint bars for objects in the current plane** check box controls the display in other working planes. We'll discuss working planes in the 3D text.
- o The **Constraint bar transparency** slider and number box control the visibility (transparency) of the constrain bars (the displayed icons).
- o A check next to **Show constraint bars after applying constraints to selected objects** means that AutoCAD will display the icon as soon as you've created the constraint. Doing so allows you to see that your constraint creation was successful, but not displaying the icon saves display clutter.
- We'll look at dimensional constraints (and the **Dimensional** tab) in Lesson 17, p.394.
- The **AutoConstrain** tab (Figure 10.002) controls which constraints AutoCAD will apply (and in what order it will apply them) in response to the *AutoConstrain* command.

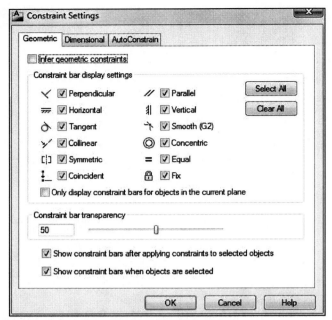

Figure 10.001

Where to Find It:	
Command Line:	*AutoConstrain*
Ribbon (Tab/Panel):	Parametric – Geometric – **Auto Constrain**
Menu:	Parametric – **Auto Constrain**

The command looks like this:
   Command: *autoconstrain*
   Select objects or [Settings]: *[Select the objects you wish to constrain.]*
   Select objects or [Settings]: *[ENTER to confirm the selection.]*
   8 constraint(s) applied to 4 object(s) *[AutoCAD reports on what it has done.]*

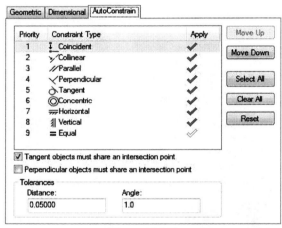

Figure 10.002

The list in the **AutoConstrain** tab lets you know which constraints AutoCAD will apply (if applicable) by turning the **Apply** check green. Deselect the check to let AutoCAD know not to apply this type of constraint. You can also move the constraints up and down in the list (**Move Up** and **Move Down** buttons) to change the order in which AutoCAD applies the constraints.

You'll also notice a couple more check boxes and a **Tolerances** frame.
- o Place a check next to **Tangent objects must share an intersection point** to ensure that two curves will have a common point before AutoCAD will apply the tangent constraint.

- A check next to **Perpendicular objects must share an intersection point** means that, before AutoCAD will apply a perpendicular constraint, the lines must either share a coincident endpoint or intersect each other.
- Finally, the **Tolerances** frame sets the proximity (**Distance** and **Angle**) within which objects must reside before AutoCAD will automatically apply a constraint.

Do This: 10.2A	Using AutoConstrain

I. Start a new drawing from scratch.
II. Open the Constraint Settings dialog box.
III. Follow these steps.

10.2A: USING AUTOCONSTRAIN

1. On the **Geometric** tab, move the **Constraint bar transparency** slider all the way to the right (so you can better see the bars).

2. **OK** the change and close the dialog box.

3. Create a rectangle. (The size doesn't matter.)
   Command: *rec*

4. Using the endpoint (corner) grips, move the various lines and watch the effect on the entire triangle.

5. Undo the changes.

6. Now begin the *AutoConstrain* command.
   Command: *autoconstrain*

7. Select the rectangle.
   Select objects or [Settings]:
   Select objects or [Settings]: *[ENTER]*
   4 constraint(s) applied to 1 object(s)
   Notice (right) the constrain bars.

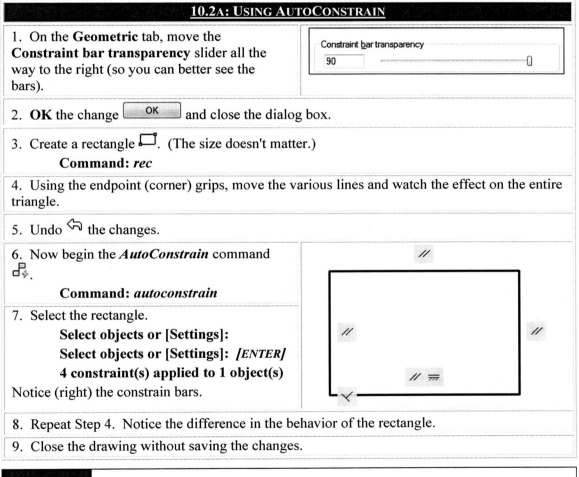

8. Repeat Step 4. Notice the difference in the behavior of the rectangle.
9. Close the drawing without saving the changes.

10.3	Inferred Constraints

What happens when you mate Running OSNAPs with *AutoConstrain*? You get *inferred constraints* (and big smiles on the faces of little AutoCAD users everywhere!).

Toggle inferred constraints on/off using the **Infer Constraints** button on the status bar (next to the other toggles). But like Running OSNAPs, remember to toggle them *off* when you're not using them or they'll cause some grief.

> AutoCAD will not place inferred constraints unless the objects meet the criteria for constraints. For example, it won't place coincident constraints on two lines which don't meet at their endpoints.

Take a look at the following table to see what type of constraint AutoCAD will create for which OSNAP.

OSNAP	Constraint	OSNAP	Constraint
Endpoint Midpoint Center Node Insertion	These OSNAPs create coincident constraints between the object you're creating or editing and the specific OSNAP target.	Intersection Apparent Intersection Extension Quadrant	*These OSNAPs don't create inferred constraints.*
Nearest	The Nearest OSNAP creates a coincident constraint between the object you're creating or editing and the nearest point on the OSNAP target	Parallel	This places a parallel constraint between the object you're creating or editing and the parallel OSNAP target.
Perpendicular	This OSNAP will result in a perpendicular constraint unless the object you're creating or editing ends on the OSNAP target. In that case, you'll get both the perpendicular constraint and a coincident constraint.	Tangent	The tangent OSNAP creates tangent constraints between the objects you're creating or editing.

In addition to creating inferred constraints in response to your OSNAP choices, AutoCAD will also place inferred constraints as follows:

- sequentially placed lines and polylines (Lesson 12, p.272) will have coincident constraints at the joining endpoints,
- the *Rectangle* command results in parallel constraints for opposing lines and coincident constraints at the conjoining endpoints,
- the *Fillet* command results in tangent and coincident constraints when using the extend or trim options.
- the *Chamfer* command result in coincident constraints at the conjoining endpoints when using the extend or trim options.

> AutoCAD will also create horizontal and vertical inferred constraints based on your Ortho and Polar settings. It will not, however, create these inferred constraints: fix, smooth, symmetric, concentric, equal, or collinear.

Let's see AutoCAD infer some constraints.

Do This: 10.3A	**Inferred Constraints**

I. Open the *Infer Constraint Practice* drawing in the C:\Steps\Lesson10 folder.
II. Set your Running OSNAPs to Endpoint, Midpoint, and Center. Clear all others.
III. Set your polar tracking to 45° and turn it on.
IV. Follow these steps.

10.3A: INFERRED CONSTRAINTS

1. Turn on the **Infer Constraints** toggle on the status bar.

10.3A: INFERRED CONSTRAINTS

2. Begin a line at point 1,2.
   Command: *l*

3. Using polar tracking and the direct distance entry method (p.52), draw lines (as indicated): 6" to the right, 3" at 45°, 2" to the left, 2" up, 2" to the left, 1.5" down, 4" to the left, and 3" at 225°.
   Notice that AutoCAD automatically creates (inferred) constraints.

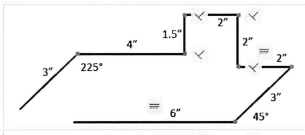

4. Add a line from the intersection of the 4" and 1.5" lines parallel to one of the 3" lines and crossing the bottom line as indicated.
   Notice the constraint.

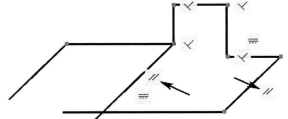

5. Fillet the top three lines; use a ¾" fillet.
   This time you have tangent constraints.

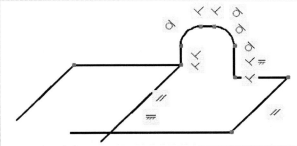

6. Finally, use the *Chamfer* command (use SHIFT-select) to corner the lower left part of the object.
   You should see a coincident constraint appear at the new intersection.

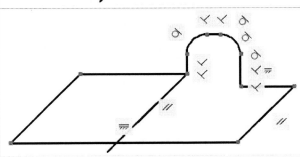

7. Exit the drawing without saving it.

10.4 Object Selection Tools

10.4.1 Selection Modes

It'll be easier to understand selection settings if we temporarily turn grips off. Follow this procedure:
   Command: *grips*
   Enter new value for GRIPS <2>: *0*
We'll reactivate them when we finish with this section of the text.

There are several different selection settings designed to enhance your use of AutoCAD, each with a different system variable that you can set at the command line. But AutoCAD also provides a tab on the Options dialog box that might be easier to use than half-a-dozen system variables. Figure 10.003 shows the dialog box with the **Selection** tab on top. Open it with the ***Options*** command or *op* hotkeys. (*Caution: Avoid changes to the other tabs at this time*.)

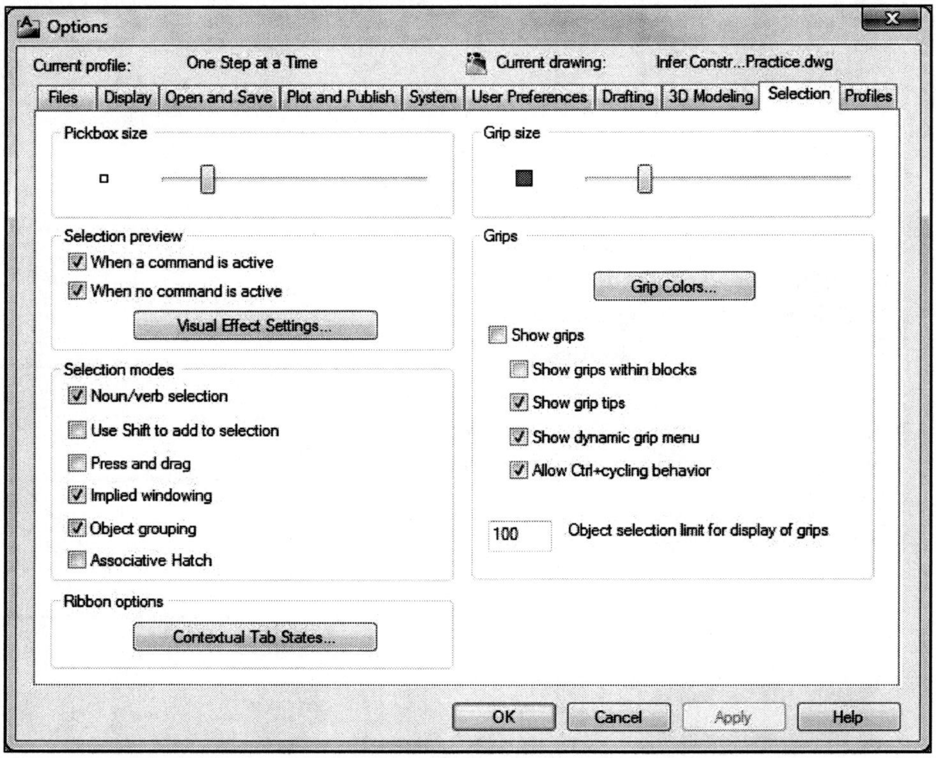

Figure 10.003

Look at the frames on the left first.

- **Pickbox size** contains a slider bar. Use this to adjust the size of the cursor box presented when AutoCAD uses a **Select objects** prompt.

  The system variable for **Pickbox Size** is **Pickbox**. I prefer a setting of **4** or **5** (an allowance for my aging eyes – many experienced users prefer **3**), but adjust it until you're happy.

- The **Selection Preview** frame (below the **Pickbox size** frame) has two check boxes:
  - Check **When a command is active** (the default) when you want AutoCAD to show you what you've selected in response to a **Select objects** prompt. (AutoCAD will highlight the selection.)
  - Check **When no command is active** (the default) when you want AutoCAD to show you what you've selected even if you made the selection at the command prompt. (You'll need this checked during the exercises in this lesson.)

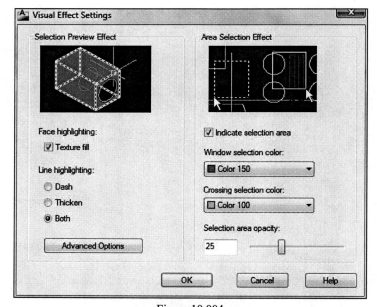

Figure 10.004

235

- o The **Visual Effect Settings** button in this frame presents the Visual Effect Settings dialog box (Figure 10.004, p.235).
    - Use the check box and radio buttons in the **Selection Preview Effect** frame to determine how AutoCAD will highlight selected objects. The **Advanced Options** button will present an Advanced Preview Options dialog box (Figure 10.005) where you can filter exactly what selections AutoCAD will preview.
    - Tools in the **Area Selection Effect** frame (Visual Effect Settings dialog box) allow you to customize the selection windows AutoCAD uses.
- We'll concentrate most of this section on the tools found in the **Selection Modes** frame (back on the **Selection** tab of the Options dialog box – Figure 10.003, p.235). Let's begin our study of these tools now.

Figure 10.005

The "normal" approach to object modification in AutoCAD is to issue a command (like *Erase*), and then select the object(s) of that command. This approach can be called *Verb/Noun* as the command is invariably a verb and the object is a noun. Placing a check in the box next to **Noun/Verb Selection** (as the default does) means that AutoCAD will also allow you to select the object of the command first, and then enter the command. The sequence (using the *Erase* command) looks like this:

    Command: *[Select an object(s).]*

    Command: *e*

The system variable for **Noun/Verb Selection** is **Pickfirst**. Its default setting is **1**.

Do This: 10.4.1A	Noun/Verb Selection

I. Open the *ObjSel* file the C:\Steps\Lesson10 folder. The drawing looks like the figure at right.
II. Follow these steps.

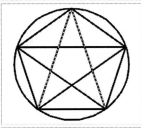

10.4.1A: NOUN/VERB SELECTION
1. Open the Options dialog box by typing *options* or *op*. Pick the **Selection** tab to put it on top (Figure 10.003, p.235).   **Command:** *op*
2. Verify that there is a check in the **Noun/verb selection** check box
3. Pick the **OK** button to return to the drawing.
4. Select the two lines that form the topmost point of the star.   **Command:**

10.4.1A: NOUN/VERB SELECTION

5. Enter the *Erase* command ✎. Notice that the lines are erased.
   Command: *e*

6. Save the drawing 💾 but don't exit.
   Command: *qsave*

In many Windows programs, selecting more than a single file or object requires that you hold down the SHIFT key. AutoCAD makes an allowance for users accustomed to this approach in the next option – **Use Shift to add to selection**. If this box contains a check, selecting a second object will remove all previous objects from the selection set unless you hold down the SHIFT key while selecting.

The system variable for **Use Shift to add to selection** is **Pickadd**. Its default setting is **1**.

Do This: 10.4.1B	Using Shift to Add

I. Be sure you're still in the *ObjSel* file the C:\Steps\Lesson10 folder. If not, please open it now.
II. Follow these steps.

10.4.1B: USING SHIFT TO ADD

1. Open the Options dialog box.
   Command: *op*
   Be sure the **Selection** tab is on top.

2. Place a check in the **Use Shift to add to selection** check box ☑ Use Shift to add to selection.

3. Pick the **OK** button [OK] to return to the drawing.

4. Enter the *Erase* command ✎.
   Command: *e*

5. Select the horizontal line and then the two angled lines. Notice that, as you select an object, the previously selected object is removed from the selection set (it's no longer highlighted).
   Select objects:

6. Now select the horizontal line. Then, while holding down the SHIFT key, select the angled lines. Notice that they all highlight.
   Select objects:

7. Complete the command. The view now looks like the figure at right.
   Select objects: *[ENTER]*

8. Save the drawing 💾 but don't exit.
   Command: *qsave*

Again, in many Windows programs, placing a window around objects requires you to hold down the mouse button between corner selections. AutoCAD doesn't require this, but it allows for those who

have become accustomed to it. A check in the **Press and Drag** box will change window creation to the following procedure:
1. Pick the first corner of the window.
2. Hold down the left mouse button while positioning the crosshairs at the opposite corner.
3. Release the mouse button.

The system variable for **Press and Drag** is **Pickdrag**. Its default setting is **0**.

Do This: 10.4.1C	Using Press and Drag

I. Be sure you're still in the *ObjSel* file the C:\Steps\Lesson10 folder. If not, please open it now.
II. Follow these steps.

10.4.1C: USING PRESS AND DRAG

1. Open the Options dialog box and place the **Selections** tab on top.
   Command: *op*

2. Remove the check from the **Use Shift to add to selection** check box, and add a check in the **Press and drag** check box.

3. Complete the procedure [OK].

4. Enter the *Erase* command.
   Command: *e*

5. Try to begin an implied window as you normally would – by selecting a point around the grid mark at coordinate **6,5**. Notice that you lose the window as soon as you release the mouse button.
   Select objects:

6. Try it again. But this time hold down the mouse button as you move the lower right corner of the window to the grid mark around coordinate **11,0**. Now you get the window.
   Select objects:

7. Release the mouse button to complete the selection window.

8. Complete the command.
   Select objects: *[ENTER]*
   The view is now empty.

9. Zoom to the limits of the drawing.
   Command: *z*

10. Repeat Steps 1 through 3, but this time, remove the check next to **Press and drag**.

11. Save the drawing but don't exit.
    Command: *qsave*

Some things to know about the other selection modes (Figure 10.003, p.235) include:
- We've been using implied windowing since Lesson 2. It enables us to pick an open area of the drawing to automatically begin a selection window. It's **On** by default. Removing the check will turn it off.
  The system variable for implied windowing is **Pickauto**. Its default setting is **1**.

- A check in the **Object Grouping** option box means that AutoCAD will recognize grouped objects as a single object. We'll learn more about grouping in Lesson 20, p.457.
  The system variable for **Object Grouping** is **Pickstyle**. Its default setting is **1**. (CTRL+A will also toggle object grouping On and Off.)
- We'll look at the **Associative Hatch** option in Lesson 20, p.448.

10.5 Selection Cycling

Another selection tool appears as a tab in the Drafting Settings dialog box (Figure 10.006) rather than the Options dialog box.

AutoCAD provides selection cycling for those busy drawings where it may be difficult to select a specific object because of the amount of clutter around it. Using the selection cycling tool, you can select the desired object from a dialog box (Figure 10.007) with a setting of 1, or with a setting of 2, you can scroll through the available objects using the tab key.

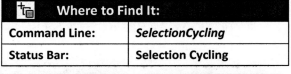

Where to Find It:	
Command Line:	*SelectionCycling*
Status Bar:	Selection Cycling

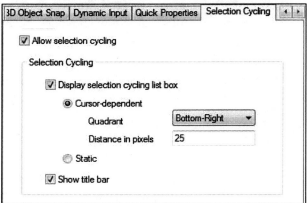

Figure 10.006

You have a few simple options.

- AutoCAD won't make selection cycling available if you don't place a check in the **Allow selection cycling** box. When you do, AutoCAD will display an icon next to your selection cursor whenever it finds more than one object available for selection.

- A check next to **Display selection cycling list box** means that AutoCAD will display a list box (Figure 10.007) next to your cursor whenever it finds more than one object available for selection. You can control where the icon appears with the **Cursor-depenent** and **Static** radio buttons and options.

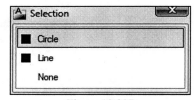

Figure 10.007

- A check next to **Show title bar** simply means that the Selection list box (Figure 10.007) will have a title bar. Probably a good idea to leave the title bar showing as it makes it easier to move the list box if it gets in your way.

> The **SelectionCycling** system variable has three possible settings:
>   **0** – means that AutoCAD will not allow selection cycling.
>   **1** – means that AutoCAD will allow selection cycling, but will not display the Selection list box (Figure 10.007).
>   **2** – means that AutoCAD will allow selection cycling and will display the Selection list box.

Let's check it out.

Do This: 10.5	Using Selection Cycling

I. Be sure you're still in the *ObjSel* file the C:\Steps\Lesson10 folder. If not, please open it now.
II. Thaw the **Obj3** layer. You'll see a series of lines and a circle where the star once lived.
III. Be sure the SelectionCycling sysvar is set to 2.
IV. Follow these steps.

10.5: USING SELECTION CYCLING

1. Open the Drafting Settings dialog box and place the **Selection Cycling** tab on top.
   Command: *ds*
   AutoCAD displays the Selection Cycling settings (Figure 10.006).

2. Place a check next to **Allow selection cycling** ☑ Allow selection cycling. Accept the rest of the defaults and close [OK] the Drafting Settings dialog box.

3. Enter the *Erase* command.
   Command: *e*

4. Hover your selection box over one of the circle/line intersections. Notice the icon.
   Select objects:

5. Try to pick the line. AutoCAD presents the Selection list box (Figure 10.007, p.239).

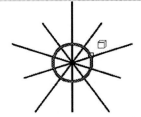

6. Select **Line** (Figure 10.007) and complete the command.
   Select objects: 1 found
   Select objects: *[Enter]*

7. Save and close the drawing.
   Command: *qsave*

Don't forget to turn grips back on before you continue!
   Command: *grips*
   Enter new value for GRIPS <0>: *2*

10.6 *SelectSimilar* and the Object Isolation Tools

Where to Find It:	
Command Line:	*SelectSimilar*
Cursor Menu:	Select Similar

Where to Find It:	
Command Line:	*IsolateObjects*
Cursor Menu:	Isolate – **Isolate Objects**

Where to Find It:	
Command Line:	*HideObjects*
Cursor Menu:	Isolate – **Hide Objects**

Where to Find It:	
Command Line:	*UnIsolateObjects*
Cursor Menu:	Isolate – **End Object Isolation**

You'll find *SelectSimilar* and the three object isolation commands on the cursor menu. (**Select Similar** appears after you've selected an object.) Their names are self-descriptive, but let's see them in action. (Once you get to know these simple tools, you may find them difficult to live without!)

If you enter *SelectSimilar* at the command promt, you'll notice that you have a **Settings** option.

      Command: *selectsimilar*

      Select objects or [SEttings]:

Selecting the **SEttings** option calls the Select Similar Settings dialog box (right). Here you can tell AutoCAD to select similar objects based on a variety of properties (or combination of properties). How convenient!

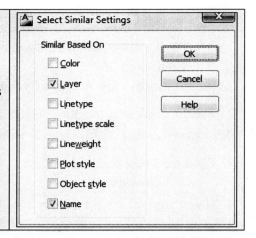

Do This: 10.6A	Using *SelectSimilar* and the Object Isolation Tools

  I. Open the *ObjSel2* file the C:\Steps\Lesson10 folder. It's similar to the *ObjSel* file, but the lines have different properties.

 II. Follow these steps.

10.6A: USING *SELECTSIMILAR* AND THE OBJECT ISOLATION TOOLS

1. Begin the *SelectSimilar* command and select the **SEttings** option.

      Command: *selectsimilar*

      Select objects or [SEttings]:

AutoCAD opens the Select Similar Settings dialog box.

2. Place a check next to **Linetype** and accept the rest of the defaults.

3. Close the dialog box [OK].

4. Without entering a command, select one of the solid red lines crossing the circle.

5. Pick **Select Similar** from the cursor menu.

Notice that AutoCAD selects the other solid red line. **Select Similar**, then, selects all the objects that are of the same type and have the same properties as the object(s) already selected.

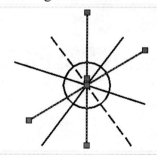

6. Clear the selection [Esc].

7. Select both a solid red line and a solid blue line.

8. Repeat Step 2.

Notice that Select Similar works on multiple selected objects.

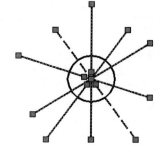

9. Clear the selection [Esc].

10. Select the circle.

10.6A: USING *SELECTSIMILAR* AND THE OBJECT ISOLATION TOOLS

11. Pick **Isolate Objects** from the cursor menu (**Isolate – Isolate Objects**).
Notice that AutoCAD hides (temporarily removes) all non-selected objects from the screen making it easier to work on the selected object(s) without distraction.

12. Pick **End Object Isolation** from the cursor menu (**Isolate – End Object Isolation**).
AutoCAD returns the hidden objects to view.

13. Select the circle.

14. Pick **Hide Objects** from the cursor menu (**Isolate – Hide Objects**).
Notice that AutoCAD hides the selected object(s) from the screen, this time, making it easier to work on the non-selected object(s) without distraction.

15. Repeat Step 9 to restore the circle to view.

Cool stuff, huh?!

10.7 Extra Steps

Reopen the samples file in the C:\Steps\Lesson06 folder. Experiment with the *SelectSimilar* and *AddSelected* commands while selecting the different objects.

- Note which properties force a selection and which do not when you use the *SelectSimilar* command.
- Note which commands the *AddSelected* command initiates when you select the various objects.

10.8 What Have We Learned?

Items covered in this lesson include:

- *Design Intent*
- *Geometric Constraints*
    - *Coincident*
    - *Concentric*
    - *Collinear*
    - *Equal*
    - *Fixed*
    - *Horizontal*
    - *Vertical*
    - *Parallel*
    - *Perpendicular*
    - *Symmetrical*
    - *Smooth*
    - *Tangent*
- *Constraint Show All*
- *Constraint Hide All*
- *Inferred Constraints*
- *Commands:*
    - **JGeomConstraint**
    - **GC[commands]**
    - **ConstaintBar**
    - **ConstraintSettings**
    - **AutoConstrain**
    - **SelectionCycling**
- *AutoCAD Selection Tools:*
    - *Noun/Verb Selection*
    - *Shift to Add*
    - *Press and Drag*
    - *Implied Windowing*
    - *Object Isolation*
    - *Selection Cycling*
    - *Select Similar*
    - *HideObjects*
    - *IsolateObjects & UnIsolateObjects*
    - *End*

- *AutoCAD System Variables:*
  - **Pickbox**
  - **Pickfirst**
  - **Pickadd**
  - **Pickdrag**

Who'd've thought such a short lesson could contain so much information?! But you may be surprised just how useful these tools can be!

In our next lesson, we'll look at *Some More Cool Tools* and finish off the "simple stuff" before moving on to the *advanced* lessons.

Practice an exercise or two, and then move on.

10.9 Exercises

1. Beginning with the *Handle* drawing found in the C:\Steps\Lesson10 folder, create the object shown at right.
   1.1. Use only the **Line** command to create new objects and the trim command to create the arcs from existing circles.
   1.2. Create the constraints shown.
   1.3. Save the drawing as *MyHandle* in the C:\Steps\Lesson10 folder.

2. See what you can do with this object.
   2.1. I started a new drawing from scratch.
   2.2. Dimensions are guides only.
   2.3. Fillets and chamfers are ½".

1. Start a new drawing from scratch.
   1.1. Repeat the setup you did in Exercise 1.
   1.2. Save the drawing as *MySpring* in the C:\Steps\Lesson12 folder.
   1.3. Create the drawing shown.

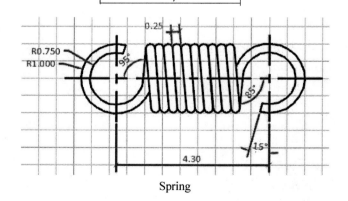

Spring

10.10 For Web-Based Review Questions and Additional Exercises, visit: http://foragerpub.com/AcadFiles/2011/2011.htm

Lesson 11

Following this lesson, you will:

✓ Know how to **Divide** and **Measure**

✓ Know how to work with these commands:

- **Point**
- **Solid**
- **Donut**
- **Wipeout**

✓ Be familiar with these Advanced Selection Methods:

- WindowPoly
- CrossingPoly
- Last
- SelectSimilar
- All
- Previous
- Add and Remove

✓ Know how to use Quick Select Filters

✓ Know how to use AutoCAD's Calculator

✓ Be familiar with AutoCAD's Measurement tools:

- MeasureGeom
- ID

Some More Cool Tools

You've come a very long way since learning the Cartesian Coordinate System so many lessons ago. You've learned the basic 2-dimensional tools for drawing and modifying most objects. You've learned to draw with a precision you probably never dreamt possible on a drawing board. Did you know that this drawing precision enables manufacturers to create products directly from your drawing? It's a system called CAM (Computer Aided Manufacturing). CAD-CAM is one possible direction your career might take if you pursue AutoCAD into the 3-dimensional world.

This lesson allows you to relax a bit and take a look at some of AutoCAD's tricks and toys meant to enhance the productivity of its users. Some of the toys you may never use; some you might rarely use. But all are full of possibilities.

Let's begin.

11.1 So Where's the *Point*?

We define a *point*, you will recall, as the place where an X-plane intersects a Y-plane. A *node* is an object that occupies a single point. It serves primarily as an identifier or locator in the drawing. CAD operators frequently use the terms *point* and *node* interchangeably when referring to a single point. It isn't that important to remember the difference, but this might help:

> I snap to a n**o**de (**o**bject) that occupies a point (**i**dea).

Nodes are a favorite tool of third-party software. I've even included a couple of Lisp routines that make use of nodes. But you can place nodes anywhere you think they might be useful with the *Point* command.

> Third-party software refers to any of a myriad of products designed to work within the AutoCAD environment to make life easier for you. Products are available for most industries.

The *Point* command sequence is

Command: *point*
Current point modes: PDMODE=3 PDSIZE=0'-0" *[AutoCAD tells you how the nodes are currently set to appear – more on this after the exercise.]*
Specify a point: *[Pick the location.]*

You really can't get any easier than this one. But remember to always use OSNAPs for precise placing of your nodes.

▪	Where to Find It:
Command Line:	*Point*
Hotkey(s):	*po*
Ribbon (Tab/Panel):	Home – Draw (subpanel) – Multiple Points (flyout) **Multiple Points**
Menu:	Draw – Point – [**Multiple** or **Single**] **Point**
Toolbar:	Draw – **Point**
Tool Palette:	Draw – **Point**

Do This: 11.1A	Making a Point

I. Open the *mea-div* drawing in the C:\Steps\Lesson11 folder. The drawing looks like the figure at right.
II. Set the **MARKERS** layer current; toggle inferred constraints off.
III. Follow these steps.

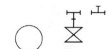

11.1A: MAKING A POINT

1. Zoom 🔍 in around the circle.
   Command: *z*

245

11.1A: MAKING A POINT

2. Place a node ° ...
   Command: *po*

3. ... in the center ⊙ of the circle.
   Current point modes: PDMODE=3 PDSIZE=0'-0"
   Specify a point: _cen of
   Your drawing looks like the figure at right.

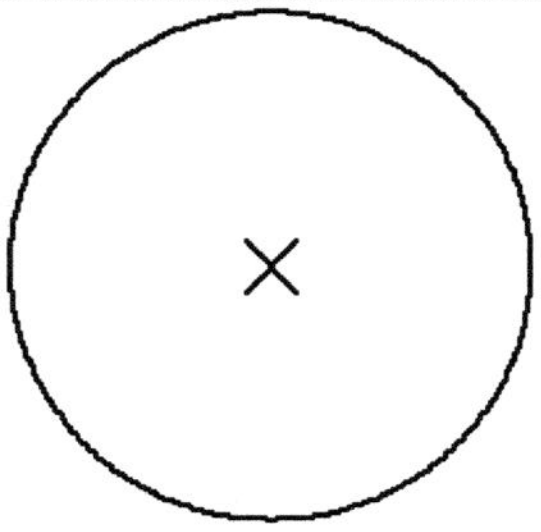

4. Save your drawing 💾, but don't exit.
   Command: *qsave*

📝	Where to Find It:
Command Line:	*DDPType*
Ribbon (Tab/Panel):	Home – Utilities (subpanel) – **Point Style**
Menu:	Format – **Point Style**

Figure 11.001

We've used an "X" to mark our node. I set this symbol in the drawing when I created it. But it isn't the only (or even the default) symbol available to show a node. To see the various symbols available, or to change the symbol, use the ***DDPType*** command. AutoCAD provides a dialog box (Figure 11.001) to help your selection.

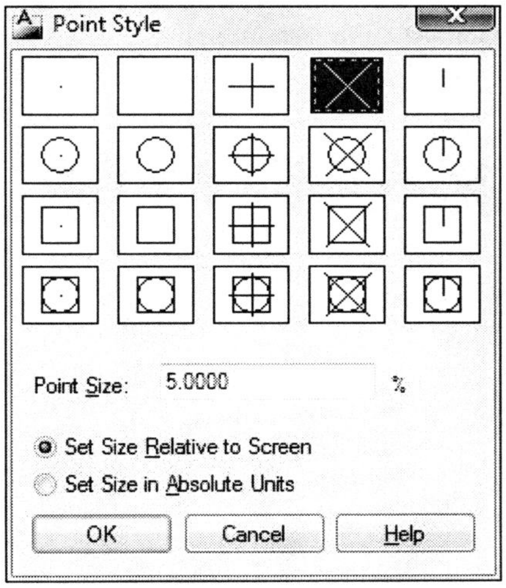

To change the node symbol you're using, simply pick the symbol you want to use.

Note that the second symbol on the top row is blank. This is an important "symbol" as it clears all the nodes in the drawing without removing them. *Always set the node symbol to blank before plotting a drawing!*

You can set the point (or node) size **Relative to Screen** or in **Absolute Units**. I recommend the former. AutoCAD will resize the nodes to keep them **Relative** when you alter views (zoom in or out). (If they don't automatically resize, regenerate the drawing.) **Absolute** nodes can be easily lost if you zoom out too far. I also recommend leaving the **Point Size** at its default. This is large enough to be seen but won't cause your nodes to dominate the screen.

Experiment with the different symbols to see which you prefer. Remember that you must ***regen*** to see each new setting.

AutoCAD stores the point type you select in a system variable called **PDMode**. If you know the number code of the symbol you want to use, you can set it like this:
   Command: *pdmode*
   Enter new value for PDMODE <0>: *3*

11.2 Equal or Measured Distances – The *Divide* and *Measure* Commands

Both these commands serve to place markers (nodes) at certain locations on an object. ***Divide*** places equally spaced nodes along the object. ***Measure*** places a node at user-set distances along the object. (Note: Neither command actually breaks the object. Rather, both place nodes along the object.)

The command sequences are very simple. Here's the ***Divide*** command:
   Command: *divide*

> **Select object to divide:** *[Select the object to divide.]*
>
> **Enter the number of segments or [Block]:** *[Tell AutoCAD how many segments you want.]*

The **Block** option allows you to place a predefined block rather than a node at the end of each segment. We'll study blocks in Lessons 21, p.468 and 22, p.500.

The *Measure* command sequence is

> **Command:** *measure*
>
> **Select object to measure:** *[Select the object to measure.]*
>
> **Specify length of segment or [Block]:** *[Tell AutoCAD how far apart to place the nodes.]*

Let's see these commands in action.

Where to Find It:	
Command Line:	*Divide*
Hotkey(s):	*div*
Ribbon (Tab/Panel):	Home – Draw (subpanel) – [Multiple Points flyout] **Divide**
Menu:	Draw – Point – **Divide**

Where to Find It:	
Command Line:	*Measure*
Hotkey(s):	*me*
Ribbon (Tab/Panel):	Home – Draw (subpanel) – [Multiple Points flyout] **Measure**
Menu:	Draw – Point – **Measure**

Do This: 11.2A	Dividing and Measuring

I. Be sure you're still in the *mea-div* drawing in the C:\Steps\Lesson11 folder.
II. Follow these steps.

11.2A: DIVIDING AND MEASURING

1. Enter the *Divide* command.
   Command: *div*

2. Select the circle.
   Select object to divide:

3. Tell AutoCAD you want five equal divisions marked off on the circle.
   Enter the number of segments or [Block]: *5*
   Your drawing looks like the figure at right.

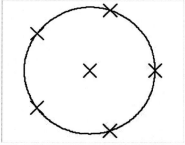

4. Set the Running OSNAP to **Node**. Clear all other settings.
   Command: *os*

5. Set the current layer to **OBJ2**.

247

11.2A: DIVIDING AND MEASURING

6. Draw lines connecting the nodes so that your drawing looks like the figure at right. **Command:** *l*	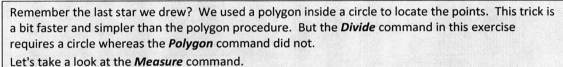
7. Erase the circle. **Command:** *e*	
8. Save your drawing, but don't exit. **Command:** *qsave*	

Remember the last star we drew? We used a polygon inside a circle to locate the points. This trick is a bit faster and simpler than the polygon procedure. But the **Divide** command in this exercise requires a circle whereas the **Polygon** command did not.
Let's take a look at the **Measure** command.

9. Zoom out to your previous screen.
   Command: *z*

10. Set the **MARKERS** layer current.

11. Enter the *Measure* command.
    Command: *me*

12. Select the horizontal line. (Select the left end of the line.)
    Select object to measure:

13. Tell AutoCAD you want to place the nodes 6" apart as shown below.
    Specify length of segment or [Block]: 6
AutoCAD placed the first mark 6" from the left, then placed marks every 6" thereafter. Notice there's a bit of leftover space at the other end. This space will always be equal to or shorter than the specified distance.

14. Move the *Tee* to the middle node as shown in the following figure.
    Command: *m*

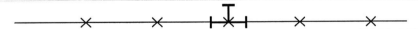

15. Now copy the *Tee and Valve* to each of the remaining nodes.
    Command: *co*

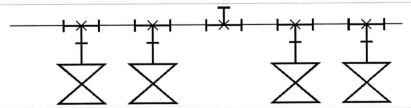

16. Using the tools you know, complete the header drawing, as shown in the following figure. (Note: The *pipe* uses a .3mm lineweight and goes on the **PIPE** layer.)

11.2A: DIVIDING AND MEASURING

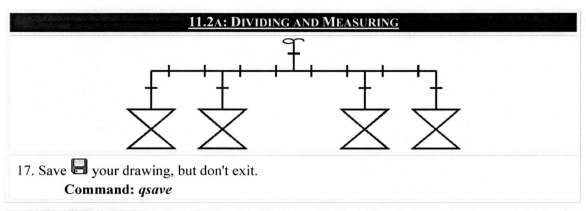

17. Save 💾 your drawing, but don't exit.

   Command: *qsave*

11.3	From Outlines to Solids – The *Solid*, *Donut*, and *Wipeout* Commands

Tired of drawing stick figures (outlines)?

Two easy tools for showing a solid surface in the 2-dimensional world are the ***Solid*** command and the ***Donut*** command. With the ***Solid*** command, you can fill triangular areas. With the ***Donut*** command, you can fill round areas.

Where to Find It:	
Command Line:	*Solid*
Hotkey(s):	*so*

Let's start with the ***Solid*** command. The sequence follows:

> **Command:** *solid*
> **Specify first point:** *[Pick the first corner of the area to fill.]*
> **Specify second point:** *[Pick the second corner of the area to fill.]*
> **Specify third point:** *[Pick the third corner of the area to fill – if there are four corners or more, select the corner nearest the first corner.]*
> **Specify fourth point or <exit>:** *[AutoCAD repeats the third and fourth corner until you end the command by hitting ENTER.]*
> **Specify third point:** *[ENTER]*

Let's try it.

Do This: 11.3A	Creating Solids

   I. Be sure you're still in the *mea-div* drawing in the C:\Steps\Lesson11 folder.
   II. Set the **OBJ1** layer current.
   III. Set the running OSNAP to **Node** and **Intersection**. Clear all other settings.
   IV. Follow these steps.

11.3A: CREATING SOLIDS

1. Enter the ***Solid*** command.

   Command: *so*

249

11.3A: CREATING SOLIDS

2. Select points 1, 2, and 3 as shown.
   Specify first point: *[Select point 1.]*
   Specify second point: *[Select point 2.]*
   Specify third point: *[Select point 3.]*

3. Then hit ENTER twice to exit the command.
   Specify fourth point or <exit>: *[ENTER]*
   Specify third point: *[ENTER]*

4. Add solids at each of the points of the star.
   Command: *so*
   Your drawing now looks like the figure at right.

5. Save your drawing 💾 but don't exit.
   Command: *qsave*

You see that drawing solids isn't difficult. Try drawing a rectangle next to the star, and then placing a solid inside. Pick your corners first in clockwise direction. Notice the hourglass shape of the solid (Figure 11.002).

Figure 11.002

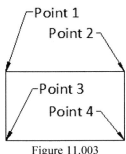

Figure 11.003

Figure 11.004

Now undo that and create your solid, picking the points in the order shown in Figure 11.003. Notice the difference? You should always reverse the direction taken from Points 3 to 4 from the direction taken from Point 1 to Point 2 (that is, work in triangles) in order to create a full solid, as shown in Figure 11.004.

You can draw donuts as easily as solids. But the *Donut* command does ask a couple questions.

> **Command:** *donut*
> **Specify inside diameter of donut <0'-0 1/2">:** *[Set the diameter of the donut hold.]*
> **Specify outside diameter of donut <0'-1">:** *[Set the outer diameter of the donut.]*
> **Specify center of donut or <exit>:** *[Place the donut.]*
> **Specify center of donut or <exit>:** *[ENTER to complete the command]*

◎ Where to Find It:	
Command Line:	*Donut*
Hotkey(s):	*do*
Ribbon (Tab/Panel):	Home – Draw (subpanel) – **Donut**
Menu:	Draw – **Donut**

The prompts are self-explanatory, so let's look at donuts in action.

Do This: 11.3B	Using Donuts

I. Be sure you're still in the *mea-div* drawing in the C:\Steps\Lesson11 folder.
II. Follow these steps.

11.3B: USING DONUTS

1. Enter the *Donut* command 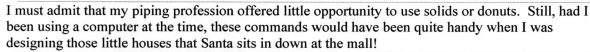.
   Command: *do*

2. Set the **inside diameter** to *0* ...
   Specify inside diameter of donut <0'-0 1/2">: *0*

3. ... and the **outside diameter** to *2.25*.
   Specify outside diameter of donut <0'-1">: *2.25*

4. Place the **center of donut** at the **node** in the center of the star.
   Specify center of donut or <exit>:

5. Complete the command.
   Specify center of donut or <exit>: *[ENTER]*
   Your star looks like the figure at right.

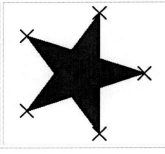

6. Let's place some other donuts. Repeat the command.
   Command: *[ENTER]*

7. Make the inside diameter *.25* and the outside diameter *.5*.
   Specify inside diameter of donut <0'-0">: *.25*
   Specify outside diameter of donut <0'-2 1/4">: *.5*

8. Place donuts at each of the points of the star.
   Specify center of donut or <exit>:
   Specify center of donut or <exit>: *[ENTER]*
   Your star looks like the figure at right.

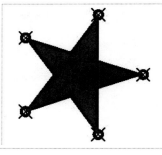

9. Save your drawing.
   Command: *qsave*

I must admit that my piping profession offered little opportunity to use solids or donuts. Still, had I been using a computer at the time, these commands would have been quite handy when I was designing those little houses that Santa sits in down at the mall!

Solids and donuts can cause a bit of a slowdown in regeneration time. They can also suck the ink right out of your plotter!

 But the programmers were clever; they provided a way to save time and ink. Until you're ready for that final plot, set your **Fillmode** system variable to *0*, then regenerate the drawing to see the results (right).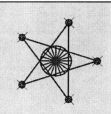

AutoCAD hides the filled area of solids and replaces the filled area of donuts with *wireframing* (more on wireframing in the 3D text). Set the **Fillmode** back to *1* for the final plot. (Note: The **Fillmode** system variable controls the fill on polylines as well.)

AutoCAD also includes a tool that works in a similar fashion to the *Solid* command – *Wipeout*. Use *Wipeout* to hide areas of a drawing or to provide a blank spot for notes on your printed drawing.

Where to Find It:	
Command Line:	Wipeout
Ribbon (Tab/Panel):	Home – Draw (subpanel) – Wipeout
Menu:	Draw – Wipeout

Here's how it works.

    Command: *wipeout*

    Specify first point or [Frames/Polyline] <Polyline>: *[Pick the first point much as you did with the Solid command.]*

    Specify next point: *[Continue defining the boundary of the area you wish to clear.]*

    Specify next point or [Undo]:

    Specify next point or [Close/Undo]: *[Once you define the area, hit ENTER to complete the procedure.]*

The only options you see appear in the first prompt line.

- **Frames** provides a toggle that will remove or add the defining outline of the wiped area from the display area.
- **Polyline** (the default) just let's AutoCAD know that you'll be defining the wiped area with a closed polyline.

Let's wipe out an area of our drawing.

Do This: 11.3C	Wiping an Area

    I. Be sure you're still in the *mea-div* drawing in the C:\Steps\Lesson11 folder.

    II. Set the **OBJ2** layer current and freeze **OBJ1**.

    III. Follow these steps.

11.3C: WIPING AN AREA
1. Enter the *Wipeout* command.   **Command:** *wipeout*

2. Trace the outline of the star. Be sure to use the **Close** option at the last point.
    Specify first point or [Frames/Polyline] <Polyline>:
    Specify next point:
    Specify next point or [Undo]:
    Specify next point or [Close/Undo]:
    Specify next point or [Close/Undo]: *c*
Your star looks like the figure at right. Notice that AutoCAD has "wiped out" the inner lines. 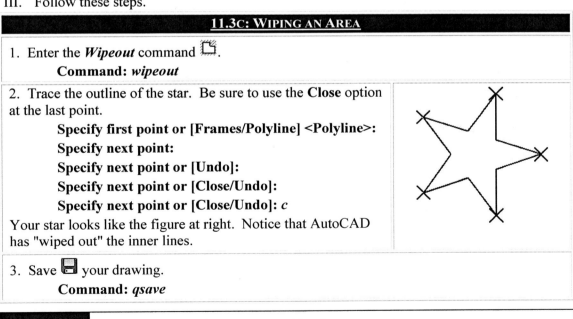

3. Save your drawing.   **Command:** *qsave*

11.4 AutoCAD's Measurement Tools

AutoCAD provides several commands or command options – *List*, *Dist*, *Radius*, *Angle*, *Area*, *Volume*, and *ID* – whose simplicity has led to their being almost completely forgotten. You already saw *List* in Lesson 6, p.138, and *ID* is a very simple command which we'll cover shortly. The others appear as either independent commands or options of the *MeasureGeom* command.

The *MeasureGeom* command sequence is

    Command: *measuregeom* [or *mea*]

    Enter an option [Distance/Radius/Angle/ARea/Volume] <Distance>: *[Select the option.]*

We'll look at each of the options in the following sections.

11.4.1	How Long or How Far – The *Dist* Tool

The first *MeasureGeom* tool inherits it sequence from the older *Dist* command. That command still resides within AutoCAD, but its hotkeys now call the *MeasureGeom's* **Distance** option. Both will help you to determine how long is a line or how far is it from one point to another.

The prompts look the same regardless of whether you begin with the *Dist* command or select the **Distance** option from the *MeasureGeom* command.

Where to Find It:	
Command Line:	*MeasureGeom* [Distance option]
Hotkey(s):	*di* [Distance command]
	mea [MeasureGeom command]
Ribbon (Tab/Panel):	Home – Utilities [Measure flyout] – **Distance**
Menu:	Tools – Inquiry – **Distance**
Toolbar:	Inquiry [Distance flyout] – **Distance**

    Specify first point:

    Specify second point or [Multiple points]:

The only real difference in behavior lies in the display of the distance. You'll see this in the exercise.

Do This: 11.4.1A	Measuring Distance

    I. Open *samples11* in the C:\Steps\Lesson11 folder.

    II. Follow these steps.

11.4.1A: MEASURING DISTANCE

1. Enter the *Dist* command. (The buttons call the *MeasureGeom* **Distance** option, so use the command line for this procedure.)

    Command: *di*

2. Using OSNAPs, pick the western endpoint of the cyan line. Then pick the center of the lower blue circle (right).

    Specify first point: _endp of

    Specify second point or [Multiple points]:

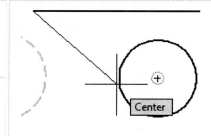

3. AutoCAD returns the following information on the command line. If you can't see all of the information, toggle to the text screen with the F2 key. (Close the text screen when you've finished.)

    Distance = 3.3232, Angle in XY Plane = 332, Angle from XY Plane = 0
    Delta X = 2.9230, Delta Y = -1.5811, Delta Z = 0.0000

Here, AutoCAD shows the true distance, the 2-dimensional angle *in* the XY plane, the 3-dimensional angle *from* the XY plane, the distance along the X-plane (shown as **Delta X**), the distance along the Y-plane (shown as **Delta Y**), and the distance along the 3-dimensional Z-plane (shown as **Delta Z**).

11.4.1A: MEASURING DISTANCE

4. This time start with the *MeasureGeom* command (use the button so you don't have to select an option).

   Command: _MEASUREGEOM

   Enter an option [Distance/Radius/Angle/ARea/Volume] <Distance>: _distance

5. Pick the same points you picked in Step 2.

   Specify first point: _endp of

   Specify second point or [Multiple points]: _cen of

   This time, you get the same command line information plus a terrific on-screen display of the data (right).

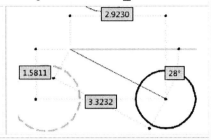

6. Complete the command.

   Enter an option [Distance/Radius/Angle/ARea/Volume/eXit] <Distance>: *x*

Don't quit yet – the best is yet to come!

11.4.2 The Radius Tool

MeasureGeom can't provide a simpler tool than its **Radius** option. With a single pick, it provides exactly what its name implies – the radius of an arc or circle. It prompts for an arc or circle and returns the radius and diameter on the command line.

Where to Find It:	
Command Line:	*MeasureGeom [Radius option]*
Hotkey(s):	*mea*
Ribbon (Tab/Panel):	Home – Utilities [Measure flyout] – **Radius**
Menu:	Tools – Inquiry – **Radius**
Toolbar:	Inquiry [Distance flyout] – **Radius**

   Select arc or circle: *[Select an arc or circle.]*

   Radius = 1.0661

   Diameter = 2.1322

It also displays the radius on your screen.

Do This: 11.4.2A	Measuring the Radius or Diameter

   I. Be sure you're still in *samples11* in the C:\Steps\Lesson11 folder. If not, please open it now.
   II. Follow these steps.

11.4.2A: MEASURING THE RADIUS OR DIAMETER

1. Start with the *MeasureGeom* command and select the **Radius** option, or simply pick the **Radius** button.

   Command: _MEASUREGEOM

   Enter an option [Distance/Radius/Angle/ARea/Volume] <Distance>: _radius

2. Select the dashed arc just below the left side of the polygon.

   Select arc or circle:

   AutoCAD displays the radius on the screen (right) and the radius and diameter on the command line.

   Radius = 1.0661

   Diameter = 2.1322

11.4.2A: MEASURING THE RADIUS OR DIAMETER

3. Select one of the circles in the drawing, too. AutoCAD displays the same things for the circles that it displayed for the arcs.

4. Complete the command.

   Enter an option [Distance/Radius/Angle/ARea/Volume/eXit] <Radius>: *x*

11.4.3 The Angle Tool

Determining an angle requires the simple selection of an arc, circle, line or a vertex and a couple points. As with the other **MeasureGeom** options, AutoCAD displays the results on both the screen and the command line.

It prompts

Select arc, circle, line, or <Specify vertex>: *[Tell AutoCAD what sort of angle you wish to measure.]*

The prompt which follows will depend upon your selection.

Where to Find It:	
Command Line:	*MeasureGeom [Angle option]*
Hotkey(s):	*mea*
Ribbon (Tab/Panel):	Home – Utilities [Measure flyout] – **Angle**
Menu:	Tools – Inquiry – **Angle**
Toolbar:	Inquiry [Distance flyout] – **Angle**

Do This: 11.4.3A Measuring an Angle

I. Be sure you're still in *samples11* in the C:\Steps\Lesson11 folder. If not, please open it now.
II. Follow these steps.

11.4.3A: MEASURING AN ANGLE

1. Start with the *MeasureGeom* command and select the **Angle** option, or simply pick the **Angle** button.

   Command: _MEASUREGEOM
   Enter an option [Distance/Radius/Angle/ARea/Volume] <Distance>: _angle

2. At the prompt, select one of the lower angled lines on the polygon …

   Select arc, circle, line, or <Specify vertex>:

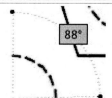

3. … then select the bottom line.

   Select second line:

   AutoCAD returns the angle on the command line and displays it on-screen (right).

   Angle = 108°

4. Continue in the command – reselect the **Angle** option.

   Enter an option [Distance/Radius/Angle/ARea/Volume] <Angle>: *[ENTER]*

5. Select the arc.

   Select arc, circle, line, or <Specify vertex>:
   Note the display.

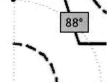

6. Repeat Step 4.

255

11.4.3A: MEASURING AN ANGLE

7. This time, hit ENTER or right click to access the **Specify vertex** option. **Select arc, circle, line, or <Specify vertex>:** *[ENTER]*	
8. Select the lower left endpoint of the rectangle as your vertex ... **Specify angle vertex: _endp of**	
9. ... the lower right endpoint as your first angle endpoint ... **Specify first angle endpoint: _endp of**	
10. ... and finally, the upper right as the second angle endpoint. As you move to select this one, notice that AutoCAD dynamically displays the angle (right). **Specify second angle endpoint: _endp of**	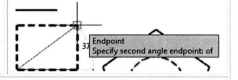
11. Complete the command. **Enter an option [Distance/Radius/Angle/ARea/Volume/eXit] <Angle>:** *x*	

You've seen several ways to measure angles, radii and distance; let's move on to something a bit more complicated. Let's look at area and volume.

11.4.4 Calculating the Area

The *List* command provides the area of closed rectangles, polygons, and circles as you saw in Lesson 6. But the boundaries in which we need to determine area aren't always closed objects. Sometimes, we need an area bounded by simple lines or even multiple objects. For this reason, AutoCAD provides the **Area** option of the *MeasureGeom* command – and it's predecessor, the *Area* command. As with the *Distance* command/option, the only difference lies in the display of the results – the *option* shades the area it's reporting on; the *command* doesn't.

Where to Find It:	
Command Line:	*MeasureGeom [Area option]*
Hotkey(s):	*aa [Area command]*
	mea [MeasureGeom command]
Ribbon (Tab/Panel):	Home – Utilities [Measure flyout] – **Area**
Menu:	Tools – Inquiry – **Area**
Toolbar:	Inquiry [Distance flyout] – **Area**

The *Area* command sequence is

    Command: *area*

    Specify first corner point or [Object/Add area/Subtract area] <Object>: *[Select the first corner of the area's boundary.]*

    Specify next point or [Arc/Length/Undo/Total] <Total>: *[Continue selecting corners – this prompt repeats until you hit ENTER (the Total option appears after you select the first two points).]*

    Area = 3.5494, Trimmed area = 0.0000, Perimeter = 7.6074

The options are fairly clear.

- The **Specify first corner point** option is the default and simply instructs you to select the first boundary point of the area to be calculated. AutoCAD follows with **Specify next point** prompts until the boundary is defined and you complete the command. AutoCAD then shows values for the **Area** within and **Perimeter** around the boundary. (Note: You must identify at least three points to define an area. AutoCAD will highlight the selected area to help guide you.)

- The **Object** option allows you to select an object – a circle, polygon, and so forth – and defines the boundary from the edges of the object.

- **Add area** and **Subtract area** are ways to keep a running total of several areas or to get the area of a bounded site minus a smaller site – such as the area of a plot of land minus the house sitting on it.

Remember, you're working with area so the total number will be in units squared.

Do This: 11.4.4A	Calculating Area

I. Be sure you're still in *samples11* in the C:\Steps\Lesson11 folder. If not, please open it now.
II. Follow these steps.

11.4.4A: CALCULATING AREA
1. Start with the *MeasureGeom* command and select the **ARea** option, or simply pick the **Area** button . **Command: _MEASUREGEOM** **Enter an option [Distance/Radius/Angle/ARea/Volume] <Distance>: _area**
2. Tell AutoCAD you want to use the **Object** option. **Specify first corner point or [Object/Add area/Subtract area/eXit] <Object>: *[ENTER]***
3. Select the circle in the center of the polygon (right). **Select objects:** AutoCAD returns this information on the command line while shading the selected area. **Area = 3.1416, Circumference = 6.2832** 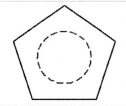
4. Repeat the **Area** option. **Enter an option [Distance/Radius/Angle/ARea/Volume/eXit] <ARea>: *[ENTER]***
5. Select the five points on the polygon (use OSNAPs). Notice (right) that AutoCAD highlights the selected area as you go. **Specify first corner point or [Object/Add area/Subtract area/eXit] <Object>:** **Specify next point or [Arc/Length/Undo]:** **Specify next point or [Arc/Length/Undo]:** **Specify next point or [Arc/Length/Undo/Total] <Total>:** **Specify next point or [Arc/Length/Undo/Total] <Total>:** 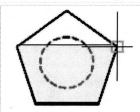
6. Hit ENTER to get the **Total** of your area. **Specify next point or [Arc/Length/Undo/Total] <Total>: *[ENTER]*** AutoCAD returns this information on the command line. **Area = 9.5106, Perimeter = 11.7557**
7. Repeat the **Area** option. **Enter an option [Distance/Radius/Angle/ARea/Volume/eXit] <ARea>: *[ENTER]***
8. Tell AutoCAD you want to use the **Add area** option. **Specify first corner point or [Object/Add area/Subtract area/eXit] <Object>:** *a*

11.4.4A: CALCULATING AREA

9. AutoCAD prompts again for points or objects. Pick the five points of the polygon as you did in Step 5.

   Specify first corner point or [Object/Subtract area/eXit]:
   (ADD mode)Specify next point or [Arc/Length/Undo]:

10. After selecting the fifth point, hit ENTER to complete the polygon.

    (ADD mode)Specify next point or [Arc/Length/Undo/Total] <Total>: *[ENTER]*

    AutoCAD tells you what the **Area** and **Perimeter** are so far, and the **Total area** defined during this command. It then prompts again as it did previously.

    Area = 9.5106, Perimeter = 11.7557
    Total area = 9.5106

11. Select the **Object** option.

    Specify first corner point or [Object/Subtract area/eXit]: *o*

12. Select the upper blue circle.

    (ADD mode) Select objects:

    AutoCAD tells you the **Area** and **Circumference** of the circle, and then adds the area of the circle to the **Total Area**.

    Area = 2.9741, Circumference = 6.1134
    Total area = 12.4846

    Then it prompts again.

13. Hit ENTER to leave the **Select objects** prompt.

    (ADD mode) Select objects: *[ENTER]*

14. Now tell AutoCAD you want to **Subtract area**.

    Specify first corner point or [Object/Subtract area/eXit]: *s*

15. We want to subtract an **Object**.

    Specify first corner point or [Object/Add area/eXit]: *o*

16. Select the circle inside the polygon.

    (SUBTRACT mode) Select objects:

    AutoCAD tells you the **Area** and **Circumference** of the circle, and then subtracts the area from the **Total Area** (the polygon and blue circle, less the area of the dashed circle).

    Area = 3.1416, Circumference = 6.2832
    Total area = 9.3430

17. Complete the command.

11.4.5 The Volume Tool

The volume tool works very much like the area tool, but of course, it includes a height (depth) calculation. In 2D space, you probably won't have much use for the volume tool, although you can use it and enter a height value to estimate a volume. Its real value, however, will appear when measuring a 3D object. In fact, you'll get your first glimpse of the 3D world in our volume exercise!

I'll show you how to estimate volume in 2D space, then we'll change to a 3D view I set up for this exercise and you'll get the volume of a box (cube). Check it out.

Where to Find It:	
Command Line:	*MeasureGeom [Volume option]*
Hotkey(s):	*mea [MeasureGeom command]*
Ribbon (Tab/Panel):	Home – Utilities [Measure flyout] – **Volume**
Menu:	Tools – Inquiry – **Volume**
Toolbar:	Inquiry [Distance flyout] – **Volume**

Do This: 11.4.5A   Calculating Volume

I. Be sure you're still in *samples11* in the C:\Steps\Lesson11 folder. If not, please open it now.
II. Follow these steps.

11.4.5A: CALCULATING VOLUME

1. Start with the *MeasureGeom* command and select the **Volume** option, or simply pick the **Volume** button.

   Command: _MEASUREGEOM

   Enter an option [Distance/Radius/Angle/ARea/Volume] <Distance>: _volume

2. Notice the similarity of the **Volume** prompt to the **Area** prompt. Select the **Object** option.

   Specify first corner point or [Object/Add volume/Subtract volume/eXit] <Object>: *[ENTER]*

3. Select the circle in the center of the polygon.

   Select objects:

4. Notice that AutoCAD now prompts for a height. Enter a value of two.

   Specify height: 2

5. AutoCAD returns the volume.

   Volume = 6.2832

   Exit the command.

   Enter an option [Distance/Radius/Angle/ARea/Volume/eXit] <Volume>: *x*

6. Now let's check out the volume tool on a 3D object. Restore the **Cube** view (right).

   Command: *v*

7. Repeat Steps 1 and 2.

   Command: *mea*

8. Select the cube.

   Select objects:

11.4.5A: CALCULATING VOLUME

This time AutoCAD returns the volume without asking for a height as height is part of the cube's definition.

Volume = 24.0000

9. Complete the command and close the drawing without saving it.

Enter an option [Distance/Radius/Angle/ARea/Volume/eXit] <Volume>: *x*
Command: *close*

We'll look at one final command to complete this section.

11.4.6 Identifying Any Point with *ID*

Although technically not a measuring device, the last of these simple tools enables you to identify any point in a drawing. This can prove particularly beneficial to the drafter who works in true coordinates (see the insert in Section 1.3, p.17 for an explanation of true coordinates) or someone working with the Ordinate system. We'll discuss the Ordinate system in Lesson 16, p.364.

Where to Find It:	
Command Line:	ID
Ribbon (Tab/Panel):	Home – Utilities (subpanel) – **ID Point**
Menu:	Tools – Inquiry – **ID Point**
Toolbar:	Inquiry – **Locate Point**

The command sequence for *ID* is

Command: *id*
Specify point: *[Select a point in the drawing.]*
X = 0.0000   Y = 0.0000   Z = 0.0000

As you can see, AutoCAD responds with the X,Y,Z coordinate location of the selected point.

Do This: 11.4.6A — Identifying Coordinates in a Drawing

I. Reopen *Samples11* in the C:\Steps\Lesson11 folder.
II. Toggle the **DYN** on and restore the **2D** view if necessary.
III. Follow these steps.

11.4.6A: IDENTIFYING COORDINATES

1. Enter the *ID* command.

   Command: *id*

2. Select the center point of the upper blue circle.

   Specify point: _cen of

   AutoCAD returns the coordinates locating the center of the circle.

   X = 3.5472   Y = 6.3395   Z = 0.0000

   Notice that, with **DYN** toggled on, this information also appears next to your cursor.

3. Exit the drawing. Don't save your changes.

   Command: *quit*

11.5 More Object Selection Methods

As our drawings get busier, selecting multiple objects for erasure or modification becomes more difficult. Fortunately, AutoCAD has provided several additional methods for creating a selection set.

These are **WindowPoly (wp)**, **CrossingPoly (cp)**, **Last**, **Fence**, **All**, **Previous**, **Add**, **Remove**, and **SelectSimilar**. You'll find it easy to master each.

- **WindowPoly** and **CrossingPoly** behave like **Window** and **Crossing Window**, except that you line out each side of the window. Neither is restricted to the four sides of their non-poly counterparts.
- **Last** refers to the last object drawn.
- **Fence** gives you a single line crossing "window".
- **Previous** refers to the last active selection set.
- Be careful with the **All** tool. **All** places *all* the thawed objects in the drawing in the selection set. Erasing **All**, then, may empty your drawing if you're not careful.
- **Remove** allows you to remove objects from the selection set. It's often easier to remove one or two objects from a selected group than it is to individually select multiple objects around the one or two you want to keep.
- **Add** enables you to put objects into a selection set. Use this when you've made your selection, removed an object, and then decided to add something else.
- **SelectSimilar** selects objects that are similar to something you've already selected. This handy tool means selecting one rather than several!

Let's see these in action.

Do This: 11.5A	Selection Practice

I. Open the *sel-prac* drawing in the C:\Steps\Lesson11 folder. It looks like the figure at right.

Note that the lines exist on different layers with different colors and linetypes. The objects between the lines are nodes.

II. Turn OSNAPs off.
III. Follow these steps.

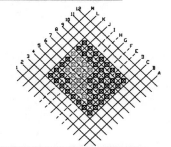

11.5A: SELECTION PRACTICE

1. Enter the *Erase* command.
   Command: *e*

2. Tell AutoCAD you wish to select objects using a **WindowPoly** by entering *wp* at the prompt.
   Select objects: *wp*

3. AutoCAD tells you to draw a polygon around the objects you wish to select. Draw the polygon around the brown nodes. (Be sure your OSNAPs are off.) AutoCAD will continue to prompt until you hit ENTER to tell it that you've completed the set.
   First polygon point:
   Specify endpoint of line or [Undo]:

11.5A: SELECTION PRACTICE

4. AutoCAD tells you how many objects it has found and repeats the **Select objects** prompt. Complete the command.
   14 found
   Select objects: *[ENTER]*
   Your drawing looks like the figure at right.

5. Undo the changes.
   Command: *u*

6. Draw a rectangle in the drawing, as shown.
   Command: *rec*

7. Now enter the **Trim** command.
   Command: *tr*

8. Select the rectangle as your cutting edge, but do it by typing *l* for *Last*.
   Current settings: Projection=UCS, Edge= Extend
   Select cutting edges ...
   Select objects or <select all>: *l*
   AutoCAD selects the last object you drew. Complete the selection.
   Select objects: *[ENTER]*

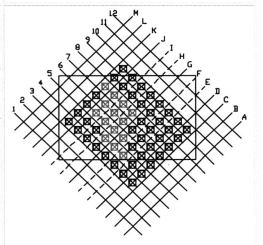

9. Tell AutoCAD you want to use a **Fence** to select the objects to trim. Then draw the fence line around the outside edge of the rectangle. Be sure to cross all the lines.
   Select object to trim or shift-select to extend or [Fence/Crossing/Project/Edge/eRase/Undo]: *f*
   Specify first fence point:
   Specify next fence point or [Undo]:
   Hit ENTER when you've finished.
   Specify next fence point or [Undo]: *[ENTER]*

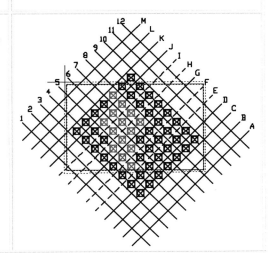

11.5A: SELECTION PRACTICE

10. AutoCAD trims the lines, then prompts for more selection. Hit ENTER to confirm completion of the command.

    Select object to trim or shift-select to extend or [Fence/Crossing/Project/Edge/eRase/Undo]: *[ENTER]*

    Does your drawing look something like the figure at right?

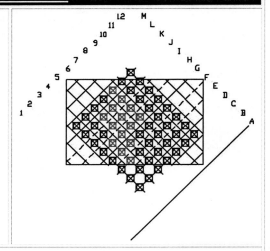

11. Erase the rectangle now, but tell AutoCAD you want to erase the **Previous** selection set.

    Command: *e*
    Select objects: *p*
    Select objects: *[ENTER]*

    AutoCAD removes the rectangle that you selected in your last modification.

12. Undo all the changes.

    Command: *u*

13. This time, let's use the **CrossingPoly** method. Enter the *Erase* command and tell AutoCAD to use a **CrossingPoly** as indicated.

    Command: *e*
    Select objects: *cp*

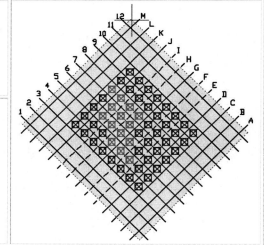

14. Draw a polygon that crosses all the lines but doesn't touch the numbers.

    First polygon point:
    Specify endpoint of line or [Undo]: *[This line repeats.]*
    Specify endpoint of line or [Undo]: *[ENTER]*
    81 found

15. Now tell AutoCAD to remove some objects from the selection set.

    Select objects: *r*

11.5A: SELECTION PRACTICE

16. Use a **Fence** to select the objects to remove, and place it across lines **E** through **I** as shown.
    Remove objects: *f*
    Specify first fence point:
    Specify next fence point or [Undo]:
    Specify next fence point or [Undo]: *[ENTER]*
    5 found, 5 removed, 76 total

17. Now tell AutoCAD you want to **Add** one of the lines back into the selection set, then select line **G**.
    Remove objects: *a*
    Select objects:

18. Complete the *Erase* command.
    Select objects: *[ENTER]*
    Your drawing looks like the figure at right.

19. Undo all the changes ⤺.
    Command: *u*

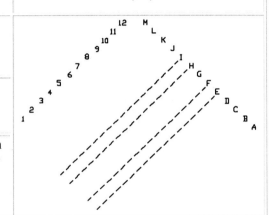

20. This last method is the easiest, but I must again caution you. It erases *everything*! Enter the *Erase* command ✏.
    Command: *e*

21. At the prompt, type *all* and hit ENTER at the next prompt. See that AutoCAD has removed all the objects in the drawing.
    Select objects: *all*
    Select objects: *[ENTER]*

22. Undo all the changes ⤺.
    Command: *u*

23. Quit the drawing without saving your changes.
    Command: *quit*

Well, now you've used all of AutoCAD's selection techniques. What do you think? They sure beat the simple windows we've used up to now, don't they? Certainly, you'll almost always use those simple windows and single-object selection boxes in your work. But occasionally, you'll be ever so glad to have learned these advanced selection tricks.

11.6 Object Selection Filters – Quick Filters

Selection filters have been available to CAD operators for quite some time; but the introduction of the Quick Select dialog box makes them friendly, convenient, and easy to use!

Use the *QSelect* command to access the dialog box (Figure 11.005, p.265). The dialog box may appear a bit frightening at first; but once mastered, you'll find it irreplaceable.

Where to Find It:	
Command Line:	*QSelect*
Ribbon (Tab/Panel):	Home – Utilities – **Quick Select**
Menu:	Tools – **Quick Select**
Cursor Menu:	**Quick Select**

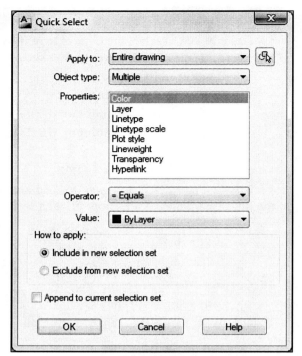

Let's take a look.

- At the top of the box, you find the **Apply to** control box. Here, AutoCAD allows you to apply the data listed in the rest of the dialog box to the **Entire drawing** or to the **Current selection** (once a selection has been made).
- Create a selection set to which you can apply the data listed in the dialog box by picking on the **Select objects** button to the right of the **Apply to** control box. AutoCAD will return you to the graphics area where you can select the objects with which you want to work.
- The **Object type** control box acts as your first filter. Picking the down arrow will produce a list of the objects currently in use in the drawing (or the selection set). You can select the type of object with which to work, or **Multiple** if you wish to apply the filters to more than one type of object.

Figure 11.005

- The **Properties** box allows you to filter the selection by specific properties.
- The **Operator** and **Value** control boxes work together. The **Operator** box allows you to set the filter: *equal to*, *not equal to*, *less than* or *greater than* the value set in the **Value** box. The properties shown in the **Value** box depend on the property selected in the **Properties** box.
- The **How to apply** frame allows you to use the filters above to include or exclude objects from the selection set.

It really isn't as difficult as it sounds. Let's use filters to repeat what we did in the first part of our last exercise.

Do This: 11.6A	**Selection Practice with Filters**

I. Reopen the *sel-prac* drawing in the C:\Steps\Lesson11 folder.
II. Follow these steps.

1. Open the Quick Select dialog box. **Command:** *qselect*	
2. Tell AutoCAD you want to select some nodes – pick the down arrow in the **Object type** control box and pick **Point**.	
3. We'll filter the nodes by layer. Select **Layer** in the **Properties** box.	

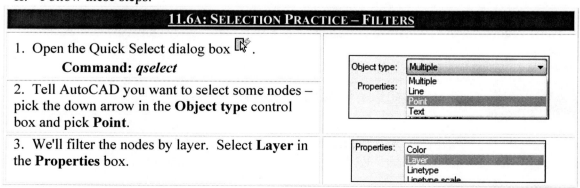

11.6A: SELECTION PRACTICE – FILTERS

4. Leave the **Operator** box set to = **Equals**, but change the **Value** to the **OBJ5** layer.

5. Pick the **OK** button to conclude the filter process. Notice that AutoCAD highlights the same nodes we selected in Steps 2, 3, and 4 of our Exercise 11.4A, p.261.

6. Enter the *Erase* command. (Alternately, you can use the DEL key on your keyboard.)
   Command: *e*
   AutoCAD erases the objects (right).

7. Quit the drawing. Don't save your changes.
   Command: *quit*

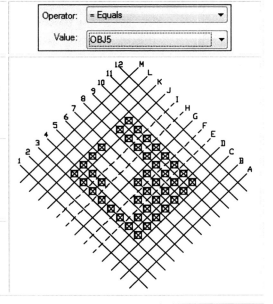

11.7 AutoCAD's Calculator

I won't provide basic calculator instruction, as you can use the QuickCalc in the same way you'd use your desktop calculator. I will, however, concentrate this section of our text on how best to use QuickCalc functions in an AutoCAD environment.

To learn more about how a calculator works, review the Help menu for QuickCalc. (Use the **Help** button on the calculator.)

Where to Find It:	
Command Line:	*QuickCalc*
Hotkey(s):	*qc*
Ribbon (Tab/Panel):	Home – Utilities – **Quick Calculator**
Menu:	Tools – Palette – **QuickCalc**
Cursor Menu:	QuickCalc
Toolbar:	Standard – **QuickCalc**

The folks at Autodesk really don't like taking their eyes off the screen! But their aversion has led to some marvelous tools – like the *QuickCalc* (Figure 11.006), which we'll examine next.

Designed to take the place of your desktop calculator, the QuickCalc sports both a standard and a scientific interface. Between these, you'll find all the functionality of your typical store-bought calculator.

Figure 11.006

In addition to standard calculator functions (including Trig functions in the **Scientific** panel), you can use QuickCalc to: convert units, modify object properties (in conjunction with the Properties palette), to calculate *and input* values at the various command prompts.

Let's take a look at the QuickCalc palette (Figure 11.006). We'll begin at the top.

- Notice the toolbar across the top. Read through the table below for the use of each button.

BUTTON	NAME	EXPLANATION
	Clear	Clears the **Input** area (below the **History** area).
	Clear History	Clears the **History** area (below the toolbar).
	Paste value to command line	Pastes the results of a calculation on the command line. This works well when you've calculated a value in response to a command line prompt.
	Get Coordinates	Gets coordinates via a mouse-pick on the screen.
	Distance Between Two Points	Gets the distance (X,Y,Z) between two user-identified points. Once you've picked this button, AutoCAD will hide the QuickCalc and allow you to pick the points on the screen.
	Angle of Line Defined by Two Points	Gets the angle of a line between two user-identified points. As with the **Distance** button, AutoCAD will hide the QuickCalc and allow you to pick the points on the screen.
	Intersection of Two Lines Defined by Four Points	Determines the intersection between two lines by prompting you for four points (to identify both lines).
	Help	Calls AutoCAD's help window.

- Below the toolbar lies the **History** area. Here you'll find a running account of completed calculations.

  The **History** area's cursor menu (Figure 11.007) allows you to customize the **color**s used in the area (options in the top frame), **Copy** listings to the Windows clipboard, **Append Expressions** or **Values to the Input Area** (useful for repeating calculations), **Clear** [the] **History** area, and **Paste** [a selection to the] **Command line**.

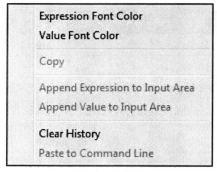

Figure 11.007

- You'll do most of your work in the **Input** area (the textbox below the **History** area). Input data (numbers and evaluation symbols) using QuickCalc's number pad, your keyboard, or the number pad on your keyboard (be sure **Num Lock** is on). Once you've entered data, tell AutoCAD to perform the calculation by picking the EQUALS key on QuickCalc's number pad, or by hitting the ENTER key on your keyboard.

- Below the **Input** area, AutoCAD displays one of three mode settings:
  o **Basic Calculator Mode** means that you can perform any calculations just as you would on your handheld or desktop calculator.
  o AutoCAD's calculator has two other approaches.
    - Accessing the calculator from the Properties palette means that AutoCAD can place solutions to your calculations in the appropriate boxes on the Properties palette. AutoCAD displays **Property Calculation** during this procedure.

- Access the calculator transparently (*'QuickCalc* or *'qc*) while in a command, means that AutoCAD can place solutions to your calculations as responses to the command prompt. AutoCAD displays **Active Command** during this procedure.
- Next to the mode display, you'll find an upward or downward pointing arrow button. Use this to display or hide the flyout menus and number keys. If you're comfortable with keystrokes on a scientific calculator, hiding the flyout menus will save some screen real estate. Otherwise, or if you need to perform units conversion or examine the variables list, leave the flyout menus displayed.

Let's try our calculator.

Do This: 11.7A	Using AutoCAD's Calculator

I. Open the *Calc-prac* drawing in the C:\Steps\Lesson11 folder. The drawing looks like the figure at right.
II. Set running OSNAPs to endpoint and center. Clear all others.
III. Follow these steps.

11.7A: USING AUTOCAD'S CALCULATOR

1. We'll begin by adding a circle that is 1¼ x the size of our existing circle. Enter the *Circle* command.

   Command: *c*

2. Select the center of the existing circle as our center point.

   Specify center point for circle or [3P/2P/Ttr (tan tan radius)]: _cen

3. We don't know the radius of our existing circle, so we'll let AutoCAD calculate our new radius for us. Call the calculator.

   Specify radius of circle or [Diameter]: *'qc*

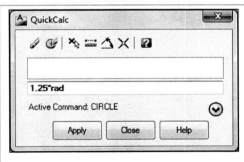

4. Tell AutoCAD to make the radius 1¼ x the radius of the existing circle as shown. When you hit ENTER, AutoCAD will ask you to select the circle on which to base its calculations. Do that now.

5. Select the existing circle.

   >>>> Select circle, arc or polyline segment for RAD function:

   AutoCAD returns to the calculator with the solution to your equation.

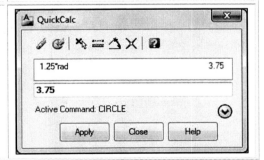

11.7A: USING AUTOCAD'S CALCULATOR

6. Pick the **Apply** button [Apply] to return the calculated value to the command prompt. AutoCAD returns with the radius. Complete the command.
   Resuming CIRCLE command.
   Specify radius of circle or [Diameter]:
   3.75 [ENTER]

7. Now let's use the Properties palette to recalculate an area. Double click on the inner circle to open the Properties palette with the circle's data displayed. (Alternately, you can use the **Quick Properties** panel .)

8. Pick in the value column next to **Area** [Area 28.2743]. (Scroll as needed to find **Area**.) Notice the **QuickCalc** button that appears.

9. We'll recalculate the area of the circle based on the area of the rectangle. Pick the **QuickCalc** button next to the **Area** display. AutoCAD opens the calculator.

10. To determine the area of the rectangle, we'll need to calculate the length x width. Start by picking the **Distance Between Two Points** button on the toolbar.

11. AutoCAD returns to the screen and asks for a first and second point. Pick the end points along the bottom of the rectangle.
   Enter a point:
   Enter a point:

12. AutoCAD returns to the calculator with the length of the line. Place an asterisk (*) after the length (for multiplication).   [6*]

13. Repeat Steps 10 and 11, but pick the end points of one of the sides of the rectangle.
   Enter a point:
   Enter a point:

14. AutoCAD displays the complete equation. Hit ENTER on your keyboard to complete the calculation.   [6*3.5]

15. The calculator now displays the solution to your calculation (the area of the rectangle). **Apply** it to the circle [Apply]. Notice the change in the circle.

16. Quit the drawing without saving.
   Command: *quit*

Of course, you can also use QuickCalc as a normal calculator!

> QuickCalc will take some getting used to. Some keys may be missing and you'll have to know to use keyboard entries for them. Other helpful entries are more obvious (like OSNAPs). I suggest spending some time with AutoCAD's Help menu browsing the QuickCalc entries if you're uncomfortable with calculator use.

11.8 Extra Steps

I've provided another Lisp routine – this one in the C:\Steps\Lesson11 folder. Simply called *Points.lsp*, this file (when loaded) creates two new commands that you may find useful.

The first command is *PT*. *PT* prompts you as follows:

> **Command:** *pt*
> **PICK A BASE POINT:** *[Pick a starting point.]*
> **HOW FAR:** *[How far away would you like to place an identifying point.]*
> **X, Y, OR P?** *[Would you like to place the point along the X or Y axis, or polar to the base point?]*
> **ANGLE?** *[If you selected P for polar, at what angle would you like to place the identifying point?]*

PT will create a **Markers** layer and set the node style (**PDMode**) to the X we used in our first exercises in this lesson. It then places a node at the requested location. This handy tool will help locate a point relative to another point without having to draw guidelines.

The other command is *MW*. This really cool tool will place a point midway between two user-identified points. It prompts:

> **Command:** *mw*
> **SELECT THE FIRST POINT:**
> **SELECT THE SECOND POINT:**

Is that simple enough? Use your OSNAPs to identify the points. This command is perfect for locating the center of rectangles or polyline ellipses. (Note: *MW* will also create the **Markers** layer and reset the **PDMode** to the X.)

Start a new drawing and play with these to get comfortable with them. See how much of what this routine does you can accomplish using the Action Recorder. (I warned you that it had limitations!)

11.9 What Have We Learned?

Items covered in this lesson include:

- *Commands:*
  - ***Divide***
  - ***Measure***
  - ***Point***
  - ***Solid***
  - ***Donut***
  - ***Wipeout***
  - ***Fillmode***
  - ***QuickCalc***
  - ***QSelect***
  - ***DDPType***
  - ***PDMode***
  - ***ID***
  - ***MeasureGeom***
- *Object Selection Methods*
  - ***WindowPoly***
  - ***CrossingPoly***
  - ***Fence***
  - ***Last***
  - ***All***
  - ***Previous***
  - ***Add and Remove***
- *Selection Filters*

We picked up some neat tricks in Lesson 11. But in your efforts to master so much material, you might easily forget some of the material in this lesson. Except for the calculator, you probably won't use most of these tools with much frequency; so, it'll take longer to fully master them. But once mastered, you'll probably wish you had done so sooner!

Work through the exercises, and then take some time to relax. The next three lessons are fairly easy to master, so this is a good time to explore and experiment with what you've learned.

11.10 Exercises

3. Open *MyRemote* from the C:\Steps\Lesson08 folder (or you can use the *Remote* drawing in the C:\Steps\Lesson11 folder).

    3.1. Using the **Solid** and **Donut** commands you learned in this lesson, fill in the buttons as shown.

    3.2. Save the drawing as *MyOtherRemote* in the C:\Steps\Lesson11 folder.

4. Setup a new drawing with the following parameters:

    4.1. Limits: [use default limits]

    4.2. Grid: .25

    4.3. Snap: as needed

    4.4. Units: architectural

    4.5. Text Height: .125

    4.6. The layers shown (below left).

    4.7. Create the drawing (below right).

    4.8. Save the drawing as *MyControlPanel* in the C:\Steps\Lesson11 folder.

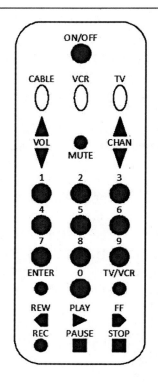

Layer Name	State	Col.	Line Type
0	On	black	Cont.
Button-A	On	blue	Cont.
Button-B	On	green	Cont.
Frame	On	red	Cont.
Infrared	On	212	Cont.
Lights	On	32	Cont.
Switch	On	84	Cont.
Text	On	12	Cont.
Vent	On	black	Cont.

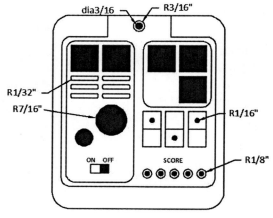

11.11 For Web-Based Review Questions and Additional Exercises, visit: http://foragerpub.com/AcadFiles/2011/2011.htm

Lesson 12

Following this lesson, you will:

- ✓ Know how to draw Polylines: The **PLine** command
- ✓ Know how to edit Polylines: The **PEdit** command
- ✓ Know how to convert Lines to Polylines
- ✓ Know how to convert Polylines to Lines: The **Explode** command
- ✓ Know how to use AutoCAD's Action Recorder

Polylines and the Action Recorder

Back in Lessons 6 and 7, we discussed ways of differentiating among the various parts of a drawing using colors, linetypes, and layers. We also used lineweights to add width, but we had problems with them because lineweight isn't a true WYSIWYG property. In that lesson, I promised to show you how to create lines with true WYSIWYG width. Once you've conquered width – through polylines – you can combine wide lines with linetypes and layers for more complex drawings.

*You can draw lines with width using the **PLine** command (pronounced "P-Line"). We call these lines polylines, not for their ability to show width, but for their ability to be drawn as multi-segmented lines. That is, polylines can contain many lines and still be treated as a single unit (much as you've already seen with polyline rectangles and polygons).*

*Polylines used to confuse the AutoCAD user because of the massive amounts of information they contained. They also dramatically increased the size of a drawing with stored information that was often never needed. For this reason, AutoCAD created the light weight polyline (lwpolyline). Capable of containing most of the information available to the polyline, the lwpolyline ("L-W-Polyline") is much more condensed, presents its information via the **List** command (Lesson 6, p.138) in a much more logical and understandable manner, and takes up much less drawing memory. The **PLine** command will actually draw an lwpolyline and convert automatically to a polyline if the need arises to function as the older polylines did.*

12.1 Using Polylines for Wide Lines and Multi-Segmented Lines

You've already made use of polylines in three commands – **Rectangle**, **Polygon** and **RevCloud**; so you have some idea of how polylines behave. Each group of lines acts as a single object when modified or manipulated. With **Rectangle**, you even saw that you could control the line width. Now we'll see how to create objects with a little more freedom than these commands allowed – by using the **PLine** command.

Where to Find It:	
Command Line:	PLine
Hotkey(s):	pl
Ribbon (Tab/Panel):	Home – Draw – **Polyline**
Menu:	Draw – **Polyline**
Toolbar:	Draw – **Polyline**
Tool Palette:	Draw – **Polyline**

One of AutoCAD's more difficult commands, **PLine** contains several options. We'll look at each, but the basic sequence is

    Command: *pline*
    Specify start point: *[Select a starting point.]*
    Current line-width is 0.0000
    Specify next point or [Arc/Halfwidth/Length/Undo/Width]: *[Select the next point.]*
    Specify next point or [Arc/Close/Halfwidth/Length/Undo/Width]: *[You can continue to create line segments or hit ENTER to complete the command.]*

The options appear more intimidating than they actually are.

- **Close** works just like it does with the **Line** command. That is, it closes the polyline. But remember, unlike lines, the entire polyline will be treated as a single unit.
- Both **Halfwidth** and **Width** allow you to tell AutoCAD how wide to make the next polyline segment. The ending width or half-width becomes the width for the remaining line segments.
    - **Halfwidth** allows you to define the distance from the center of the polyline to its two edges. It prompts:
        Specify starting half-width <0.0000>:
        Specify ending half-width <0.0000>:
    - **Width** let's you tell AutoCAD how wide you want the next segment of the polyline to be at its beginning and at its end. This'll come in handy when creating arrowheads, as we'll see in our exercises.

- **Length** allows you to enter the length of a line segment. AutoCAD defines the line direction by the direction of the previous line segment. You'll be happier if you ignore this option in favor of one of AutoCAD's many other methods of defining line length (direct distance, Cartesian coordinates, polar tracking, etc.). The **Length** option provides this prompt:
    Specify length of line:
- **Undo** will do (undo?) just that. Here at the *PLine* prompt, it undoes the last segment drawn.
- Notice that AutoCAD repeats the list of options, always with the **Specify next point** option as the default. This makes it simple to continue drawing – selecting points – until finished while making all the other options available for *each* line segment. In other words, you can change the width, close the polyline, or switch to an arc at any time during the creation process!
- **Arc** allows you to create polyline arcs – that is, sequential arcs or arcs with width. The **Arc** option provides a second tier of prompts that behave remarkably like the options of the *Arc* command:
    Specify endpoint of arc or [Angle/CEnter/CLose/Direction/Halfwidth/ Line/Radius/Second pt/Undo/Width]:
    o The **Angle** option allows you to specify an included angle to define the arc. To more easily understand this, think of *pieces* of a circle. A circle is 360°; 45° would be 1/8 of the circle, so specifying an angle of 45° will tell AutoCAD to draw 1/8 of a circle. AutoCAD will prompt as it did in the *Arc* command.
    o The **CEnter** option allows you to specify the center point of the arc followed by an **Angle**, the **Length** of the arc, or the **Endpoint** of the arc. Again, pline arc prompts resemble the *Arc* command.
    o The **CLose** option in this tier will close the polyline using an arc. You can use this method to create circles with line-width.
    o AutoCAD normally draws an arc in a counterclockwise direction. Once you draw an arc using the *PLine* command, AutoCAD reverses direction but stays in the **Arc** mode until told to switch back to **Line** or to **CLose** (or exit) the *PLine* command. If you wish to draw a series of arcs in the same direction – like tiles on a roof – the **Direction** option allows you to specify the direction of the next arc.
    o **Halfwidth** and **Width** provide a way to change the width of the polyline from within the **Arc** option. AutoCAD prompts the same way it did for these options in the first tier of *PLine* options.
    o The **Line** option returns you to the first tier of *PLine* options. From there, you can continue the polyline or exit the command.
    o One of the easier options is the **Radius** option. With this, AutoCAD allows you to specify the radius of the arc followed by the **Angle** or **endpoint**.
    o You can draw a three-point arc, just as you did with the *Arc* command, by using the **Second pt** option. AutoCAD prompts for the **second point** on the arc, then the **endpoint**.
    o **Undo** performs just as it did in the upper tier.
    o The default option is simply to **Specify endpoint of arc**. When that's done, AutoCAD repeats the options until you either select the **Line** option or exit the command by hitting ENTER.

It's taken a couple pages to explain the various parameters of the *PLine* command. It's a powerful tool, but don't let the scope of the *PLine* command deter you from its use. Most students get the hang of it fairly quickly. Let's look at some of the options in an exercise.

> The various options of the *PLine* command are also available in the cursor and dynamic input menus once the *PLine* command has been entered.

Do This: 12.1A	Drawing with Polylines

I. Start a new drawing from scratch.
II. Set the grid to *0.5* and the grid snap to *0.25*. Zoom all and turn on polar tracking.
III. Follow these steps.

12.1A: DRAWING WITH POLYLINES

1. Enter the *PLine* command ↵.
   Command: *pl*

2. Select a starting point at *6,5*.
   Specify start point: *6,5*

3. AutoCAD tells you what the **Current line-width** setting is, and then prompts you for the next step. Type *w* or pick **Width** on the menu to change the line width.
   Current line-width is 0.0000
   Specify next point or [Arc/Halfwidth/Length/Undo/Width]: *w*

4. Set the **starting width** to 1/16.
   Specify starting width <0.0000>: *1/16*

5. AutoCAD has set the **ending width** to match the starting width, but asks you if you want to change it. Accept the setting.
   Specify ending width <0.0625>: *[ENTER]*

6. Draw the first line segment *1.5* units upward. Use coordinate entry or polar tracking.
   Specify next point or [Arc/Halfwidth/Length/Undo/Width]: *@1.5<90*

7. Tell AutoCAD to draw an **Arc** next [Arc].
   Specify next point or [Arc/Close/Halfwidth/Length/Undo/Width]: *a*

8. Accept the **endpoint of arc** option and place the endpoint *1* unit to the left.
   Specify endpoint of arc or [Angle/CEnter/CLose/Direction/Halfwidth/Line/Radius/Second pt/Undo/Width]: *@1<180*

9. Select the **Line** option [Line] to return to the first tier of options. (Note: Remember, AutoCAD isn't case-sensitive – that is, it doesn't matter if you type a capital or small letter.)
   Specify endpoint of arc or [Angle/CEnter/CLose/Direction/Halfwidth/Line/Radius/Second pt/Undo/Width]: *l*

10. Continue the polyline *1.5* units downward as indicated.

11. Continue *1.5* units to the west.
    Specify next point or [Arc/Close/Halfwidth/Length/Undo/Width]: *@1.5<180*

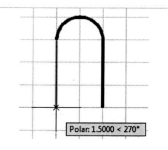

12. Use the **Arc** option again [Arc].
    Specify next point or [Arc/Close/Halfwidth/Length/Undo/Width]: *a*

12.1A: DRAWING WITH POLYLINES

13. Draw the arc with a ¼ unit **Radius** downward.
    Specify endpoint of arc or [Angle/CEnter/CLose/Direction/Halfwidth/Line/Radius/Second pt/Undo/Width]: *r*
    Specify radius of arc: *.25*
    Specify endpoint of arc or [Angle]: *[Select a point one grid mark due south of the current point.]*

14. Let's use the **Direction** option to repeat the arc.
    Specify endpoint of arc or [Angle/CEnter/CLose/Direction/Halfwidth/Line/Radius/Second pt/Undo/Width]: *d*

15. Pick a point due west of the current point, as indicated.
    Specify the tangent direction for the start point of arc:

16. Pick a point ½" to the south.
    Specify endpoint of the arc: *@.5<270*

17. Return to the **Line** option.
    Specify endpoint of arc or [Angle/CEnter/CLose/Direction/Halfwidth/Line/Radius/Second pt/Undo/Width]: *l*

18. Draw the line *1.5* units to the right using polar tracking.
    Specify next point or [Arc/Close/Halfwidth/Length/Undo/Width]:

19. Using the preceding details, see if you can complete the drawing. (Remember to type *c* to complete the polyline.) When you've finished, it'll look like the figure at right.

20. And now let's try something completely different. Zoom in around the center of the object you just drew. You'll be working in this area.
    Command: *z*

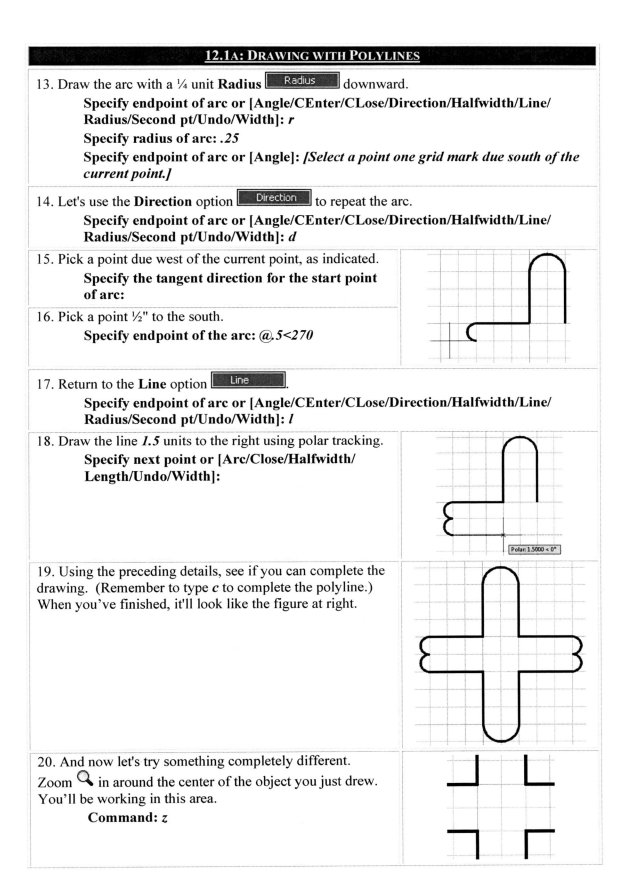

12.1A: DRAWING WITH POLYLINES

21. Repeat the *PLine* command.
    Command: *pl*

22. Pick the starting point indicated.
    Specify start point: *4.75,4.5*

23. Tell AutoCAD you want to change the **Width** [Width].
    Current line-width is 0.0625
    Specify next point or [Arc/Halfwidth/Length/Undo/Width]: *w*

24. Set different **starting** and **ending widths** as indicated.
    Specify starting width <0.0625>: *0*
    Specify ending width <0.0000>: *.25*

25. Draw the line segment. Notice the arrowhead effect of your width settings.
    Specify next point or [Arc/Halfwidth/Length/Undo/Width]: *@.5<0*

26. Continue the polyline. Now AutoCAD uses just the width setting of the endpoint of the last line segment drawn.
    Specify next point or [Arc/Halfwidth/Length/Undo/Width]: *@.5<0*

27. Change the **Width** again [Width]. Accept the **0.25** default of the starting width, but change the ending width back to *0*.
    Specify next point or [Arc/Close/Halfwidth/Length/Undo/Width]: *w*
    Specify starting width <0.2500>: *[ENTER]*
    Specify ending width <0.2500>: *0*

28. Complete the polyline.
    Specify next point or [Arc/Close/Halfwidth/Length/Undo/Width]: *@.5<0*
    Specify next point or [Arc/Close/Halfwidth/Length/Undo/Width]: *[ENTER]*
    Your drawing looks like the figure at right.

29. Complete the drawing by placing a final polyline as shown.
    Command: *pl*

30. Save the drawing as *MyPline* to the C:\Steps\Lesson12 folder and exit.
    Command: *qsave*

Polylines can be a bit frightening at first because of the depth of the command. But don't let too many options prevent you from using one of AutoCAD's best tools.

We'll need to spend some time with the polyline editing tools. Let's get right to it!

12.2 Editing Polylines – The *PEdit* Command

As with so many of AutoCAD's tools, we're faced with an easy way of doing things and the more traditional, complete (and difficult) way. With AutoCAD 2011, the makers-and-shakers have greatly enhanced the availability and ability of grips for editing polylines. Still, you can't do it all with these coveted tools. So we'll start with the hard stuff (don't we always?) – the *PEdit* command; then we'll look at how much easier grips will make your polyline editing.

If you thought drawing polylines was fun, you're in for a real treat now! Remember that second tier of options? The *PEdit* command has one, as well, *and several smaller third tiers!* Sort of makes you want to go back to the board, doesn't it? But to encourage you, let me just say that you'll probably never need the second (or third) tier – or at least only rarely in the 2-dimensional world.

> In the 3-dimensional world, it'll often be easier to edit a polyline. When you get there, it'll be useful if you already know how to add or move a vertex (the "corners" within a polyline). We'll see how to do that here, but you probably won't need it for a while.

Two-dimensional polylines are such that it's often easier to erase and redraw than it is to edit. Still, you should be familiar with the *PEdit* command for those benefits that it does provide. These include the ability to change the width of the polyline and to join several polylines into a single object.

The *PEdit* command sequence looks like this:

Command: *pedit*
Select polyline or [Multiple]: *[Select the polyline(s) to edit.]*
Enter an option [Open/Join/Width/Edit vertex/Fit/Spline/Decurve/Ltype gen/Reverse/Undo]: *[Tell AutoCAD how you want to edit the polyline.]*

Where to Find It:	
Command Line:	*PEdit*
Hotkey(s):	*pe*
Ribbon (Tab/Panel):	Home – Modify (subpanel) – **Edit Polyline**
Menu:	Modify – Object – **Polyline**
Toolbar:	Modify II – **Polyline**
Tool Palette:	Modify – **Edit Polyline**

This sequence includes the first tier of options. Let's stop here to examine these.

- Use the **Multiple** option to edit more than one polyline at a time. (Note: The **Edit vertex** option isn't available during multiple polyline editing sessions.)
- The **Open** option appears if a closed polyline is selected at the **Select polyline** prompt. Conversely, a **Close** option appears if an open polyline is selected. The **Open** option removes the last line segment – the one that closed the polyline. The **Close** option adds a polyline segment between two open endpoints.
- The **Join** option enables you to join one polyline to another to form one large polyline.
- The **Width** option allows you to modify the polyline's overall width.
- **Fit** and **Spline** will soften corners into curves. This was once the tool of choice for drawing contour lines for topographical maps. However, AutoCAD now provides a *Spline* command that was specifically designed for drawing contour lines (more on this in Lesson 13). The difference between **Fit** and **Spline** is that, although **Fit** will create curves that go through each point on the polyline, **Spline**'s curves go through only the first and last point. The rest of the points "pull" the curve but don't insist that the curve touch each point.

> The system variable **Splinetype** controls the amount of curve caused by the **Spline** option. A setting of **5** will cause a more pronounced curve (called a Quadratic B-spline) that's actually tangent to the original polyline. The default setting of **6** causes a softer, less pronounced curve (a Cubic B-spline). There are only the two settings available.

- **Decurve** removes all curves on the polyline whether put there as **Arcs**, **Fits**, or **Splines**.

- **Ltype gen** regulates the placement of dashes and spaces in linetypes.

  Let me make this as simple as possible. Through some fairly complicated mathematics, AutoCAD normally balances dashes and spaces in a line so that the amount of solid line at both ends is the same. If the length of the line doesn't allow enough room for the dashes and spaces defined by the linetype, no spaces are shown. When turned **on**, **Ltype gen** calculates the placement of dashes and spaces for the overall polyline. When turned **off** (the default), the placement is calculated for each individual line segment within the polyline. This will become clearer in our exercise.
- Use the **Reverse** option, like the *Reverse* command, to flip the direction of linetype dashes and spaces.
- By now you're familiar with the **Undo** option. It undoes the last modification made within the *PEdit* command.

We'll look in some detail at the **Edit vertex** option later in this lesson. But let's try an exercise on what we've learned so far.

> You'll find several ways to start the *PEdit* command in the WTFI table. The easy way, however, is to double click on the polyline you want to edit. AutoCAD will automatically start the *PEdit* command.
>
> The various options of the *PEdit* command are also available on the cursor and dynamic input menus once the *PEdit* command has been entered.

Do This: 12.2A	Working with Simple Polyline Editing Tools

I. Open the *MyPline* drawing you created earlier in this lesson. If this drawing isn't available, open the *pline* drawing in the C:\Steps\Lesson12 folder.

II. Follow these steps.

12.2A: SIMPLE POLYLINE EDITING TOOLS

1. Enter the *PEdit* command.

   Command: *pe*

2. Select the outer polyline.

   Select polyline or [Multiple]:

3. Let's start with the **Open** option [Open]. Type *o* or pick **Open** on the menu. Your drawing will look like the figure at right.

   Enter an option [Open/Join/Width/Edit vertex/ Fit/Spline/Decurve/Ltype gen/Reverse/Undo]: *o*

4. Notice how the prompt has changed from **Open** to **Close**. Use the **Close** option.

   Enter an option [Close/Join/Width/Edit vertex/ Fit/Spline/Decurve/Ltype gen/Reverse/Undo]: *c*

   Notice that the figure is closed and looks like it did when you started.

12.2A: SIMPLE POLYLINE EDITING TOOLS

5. Select the **Width** option [Width]. **Enter an option [Open/Join/Width/Edit vertex/Fit/ Spline/Decurve/Ltype gen/Reverse/Undo]:** *w*	
6. AutoCAD asks for a new width. Enter 1/32. **Specify new width for all segments:** *1/32* Notice (right) the difference in the figure.	
7. Note where the vertices (corners) are on the figure (use the grid if it helps), then select the **Fit** option [Fit]. **Enter an option [Open/Join/Width/Edit vertex/ Fit/Spline/Decurve/Ltype gen/Reverse/Undo]:** *f* Notice (right) that the polyline still goes through the vertices.	
8. Undo [Undo] the last modification. **Enter an option [Open/Join/Width/Edit vertex/Fit/ Spline/Decurve/Ltype gen/Reverse/Undo]:** *u*	
9. Now try the **Spline** option [Spline]. **Enter an option [Open/Join/Width/Edit vertex/Fit/ Spline/Decurve/Ltype gen/Reverse/Undo]:** *s* Notice (right) the differences between **Fit** and **Spline**.	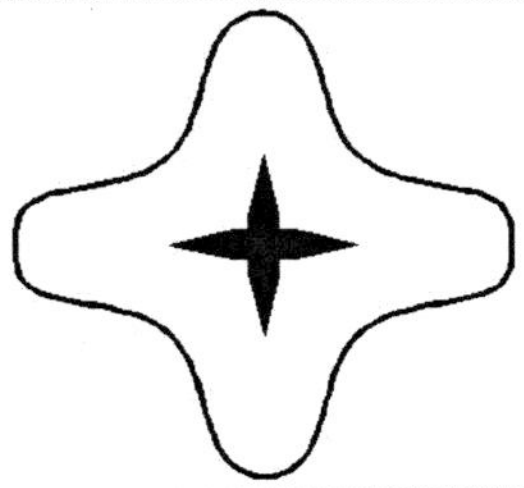
12. Decurve the polyline [Decurve]. **Enter an option [Open/Join/Width/Edit vertex/Fit/ Spline/Decurve/Ltype gen/Reverse/Undo]:** *d* Notice that AutoCAD removes all the polyline's curves.	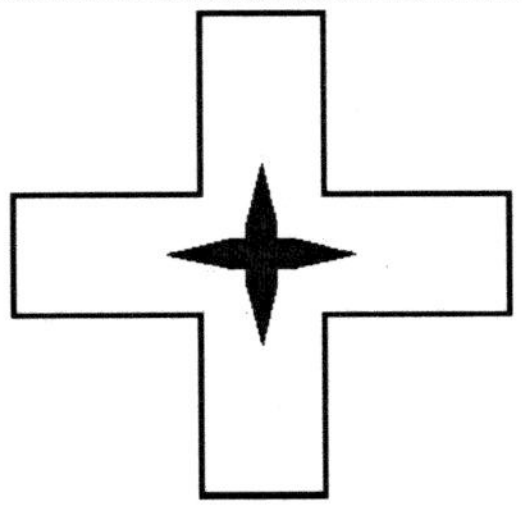
11. Undo the last two modifications. The drawing will look as it did in Step 6. **Enter an option [Open/Join/Width/Edit vertex/Fit/Spline/Decurve/Ltype gen/ Undo]:** *u*	
12. Exit the command. **Enter an option [Open/Join/Width/Edit vertex/Fit/Spline/Decurve/Ltype gen/ Undo]:** *[ENTER]*	

12.2A: SIMPLE POLYLINE EDITING TOOLS

13. Change the linetype of the polyline to **Center2**.

    Notice (right) that there are no dashes and spaces showing in the smaller arcs.

14. Repeat the *PEdit* command, but this time, do it by double clicking on the outer polyline.

15. Now turn the **Ltype gen** option 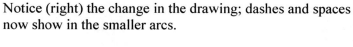 ...

    Enter an option [Open/Join/Width/Edit vertex/Fit/Spline/Decurve/Ltype gen/Undo]: *l*

16. ... **On** 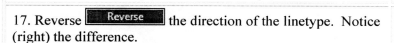.

    Enter polyline linetype generation option [ON/OFF] <Off>: *on*

    Notice (right) the change in the drawing; dashes and spaces now show in the smaller arcs.

17. Reverse 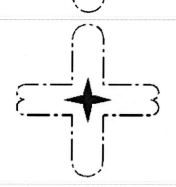 the direction of the linetype. Notice (right) the difference.

18. Exit the command.

    Enter an option [Open/Join/Width/Edit vertex/Fit/Spline/Decurve/Ltype gen/Reverse/Undo]: *[ENTER]*

19. Close the drawing without saving the changes.

    Command: *close*

As I mentioned earlier, the **Join** option allows you to join two (or more) polylines together to make a single polyline. You control how AutoCAD does this.

Command: *pedit*

Select polyline or [Multiple]: *m [Use the Multiple option to select more than one polyline.]*

Select objects: *[Select the polylines you wish to join together.]*

Select objects: *[Hit ENTER to complete the selection.]*

Enter an option [Close/Open/Join/Width/Fit/Spline/Decurve/Ltype gen/Reverse/Undo]: *j [Select the Join option.]*

Join Type = Extend *[AutoCAD tells you what type of join it'll perform.]*

Enter fuzz distance or [Jointype] <0.0000>: *[Tell AutoCAD how far apart the polylines may be.]*

> **1 segments added to polyline** *[AutoCAD tells you how many segments it has added to the polyline.]*
> **Enter an option [Close/Open/Join/Width/Fit/Spline/Decurve/Ltype gen/Reverse/Undo]:** *[Hit ENTER to complete the command.]*

You must make two decisions when joining polylines – the **Join Type** and the **fuzz distance**.

- **Fuzz distance** refers to how much distance separates the polylines. AutoCAD will only join polylines whose endpoints fall within the distance you specify – that is, the distance you specify is the outer limit of the polylines' proximity to each other. You can actually use this option to your advantage – joining only those polylines whose endpoints fall within a certain distance or even touch each other (a **0.0000** fuzz distance).
- The **Jointype** option will present the following prompt:
   Enter join type [Extend/Add/Both] <Extend>:
  Select the type of joining you would like between your polylines.
   o **Extend**, the default, will extend or trim the polylines as necessary to form the joint.
   o **Add** will add polyline segments between the endpoints of the selected polylines.
   o When using the **Both** option, AutoCAD will extend or trim the polylines whenever possible to form the joint; where this isn't possible, as in the case of parallel polylines, it'll add a polyline segment.

Let's try an exercise to see the **Join** option in action.

Do This: 12.2B	Joining Polylines

   I. Open the *Join* drawing in the C:\Steps\Lesson12 folder. It looks like the figure at right.
   II. Follow these steps.

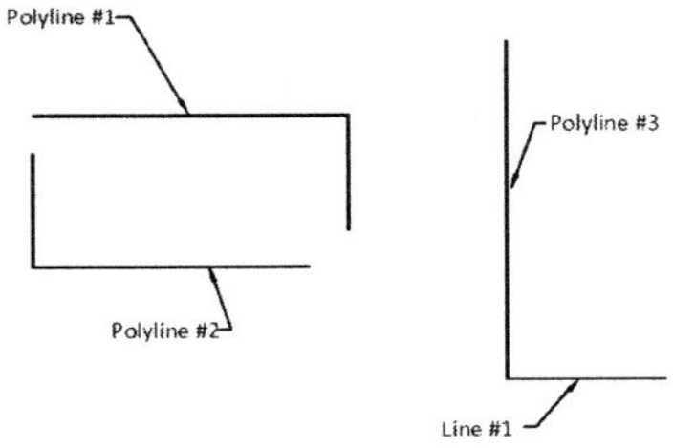

12.2B: JOINING POLYLINES

1. Enter the *PEdit* command.
   Command: *pe*

2. Use the **Multiple** option [Multiple], and select **Polyline #1** and **Polyline #2**.
   Select polyline or [Multiple]: *m*
   Select objects: *[ENTER]*

3. Use the **Join** option [Join].
   Enter an option [Close/Open/Join/Width/Fit/Spline/Decurve/Ltype gen/ Reverse/Undo]: *j*

4. The grid in this drawing is 0.5. The proximity of one set of endpoints is less than 0.6; the proximity of the other set is greater than 0.6. Set the **fuzz distance** to *0.6*.
   Join Type = Extend
   Enter fuzz distance or [Jointype] <0.0000>: *.6*

12.2B: JOINING POLYLINES

5. AutoCAD tells you how many segments it added to the polyline. Complete the command.

> **2 segments added to polyline**
>
> **Enter an option [Close/Open/Join/Width/Fit/Spline/ Decurve/Ltype gen/Reverse/Undo]:** *[ENTER]*

You now have a single polyline consisting of four segments (right).

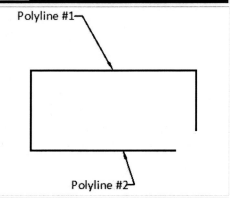

6. Try the same thing using Polyline #3 and the line. AutoCAD will prompt you to:

> **Convert Lines, Arcs and Splines to polylines [Yes/No]? <Y>** *[ENTER]*

Accept the offer. (This is how you convert from a line to a polyline.)

The objects don't look any different; but if you try to erase one, you'll notice that both are selected. Both segments are now part of a single polyline!

7. Close the drawing without saving it.

In addition to the **Join** option of the *PEdit* command, AutoCAD has a **Join** command in its repertoire. Use this to join objects such as:

- Line segments – Line segments must be collinear (that is, they should lie along the same infinite line whether real or imaginary), but they can have gaps.
- Polylines – Polylines must be touching at endpoints and be on the same XY-plane. Polyline line segments or arcs can be joined.

⊢⊣	Where to Find It:
Command Line:	Join
Hotkey(s):	j
Ribbon (Tab/Panel):	Home – Modify (subpanel) – **Join**
Menu:	Modify – **Join**
Toolbar:	Modify – **Join**
Tool Palette:	Modify – **Join**

- Arcs or Elliptical Arcs – Arcs which are part of the same (real or imaginary) circle can be joined or closed into a circle. Arc segments can have gaps.
- Splines – Splines must be in the same plane and have touching endpoints. (More on Splines in Lesson 13, p.295.)
- Helices – Helices must touch end-to-end. (More on helices in the 3D text.)

We've just about covered the first tier of options in the *PEdit* command. The only remaining option is the key to the next tier – **Edit vertex**. As a beginning CAD operator in a 2-dimensional world, you'll rarely find the need to edit a vertex (a corner or endpoint). But the option comes in quite handy for very complex polylines and 3-dimensional polylines. Choosing **Edit vertex** will result in the following list of second-tier options:

> **Enter a vertex editing option**
>
> **[Next/Previous/Break/Insert/Move/Regen/Straighten/Tangent/ Width/eXit] <N>:**

Figure 12.001

You'll notice that, when you enter this level of the *PEdit* command, a small "X" appears on the polyline you're editing (Figure 12.001). The X is a locator to let you know which vertex is being edited.

- The **Next** (default) and **Previous** options move the locator forward and backward to each vertex around the polyline.
- Use the **Break** option just as you used the *Break* command – to remove part of a polyline.
- **Insert** enables you to define a new vertex on the polyline and **Straighten** allows you to remove an existing vertex.
- **Move**, of course, enables you to move a vertex, thus reshaping the polyline.
- In the event that too much editing causes the polyline to display oddly on your screen, you can **Regen** just the polyline while still within the editing session. This saves you from having to leave the command, regen the drawing, and then return to the *PEdit* command.
- The **Tangent** option allows you to assign a tangent direction for AutoCAD to use when it fits the polyline (with the **Fit** option on the upper tier).
- The **Width** option on the first tier of choices allowed you to change the width for the entire polyline. The **Width** option on the second tier allows you to change the width for a specific segment of the polyline.

Let's look at some of these options.

Do This: 12.2C	More Complex Polyline Editing Tools

I. Open the *pline* drawing in the C:\Steps\Lesson12 folder. Refer to the figure at right for this exercise. (I've shown the numbers of the vertices on all of the graphics in this exercise to make it easier for you to follow. If you wish to display them in your drawing, thaw the **TEXT** layer.)

II. Follow these steps. (Again, feel free to use cursor or dynamic input menus to display the *PEdit* options.)

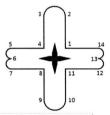

12.2C: MORE COMPLEX POLYLINE EDITING TOOLS

1. Enter the *PEdit* command by double clicking on the outer polyline.

2. Use the **Edit vertex** option [Edit vertex] to access the next tier of options.
   Enter an option [Open/Join/Width/Edit vertex/Fit/Spline/Decurve/Ltype gen/Reverse/Undo]: *e*

3. Reposition the locator to point 2.* Notice the default option is **N**, for the **Next** option, so just hit ENTER to reposition the locator.
   Enter a vertex editing option [Next/Previous/Break/Insert/Move/Regen/Straighten/Tangent/Width/eXit] <N>: *[ENTER]*

4. Select the **Break** option [Break].
   Enter a vertex editing option [Next/Previous/Break/Insert/Move/Regen/Straighten/Tangent/Width/eXit] <N>: *b*

5. Notice that AutoCAD drops to a third tier of options.
   - **Next** and **Previous** work as they do in the second tier.
   - **Go** executes the option that dropped you to this level (in this case, the **Break** option).
   - **eXit** to leave this level without executing the level two option.

   Hit ENTER to accept the **Next** default. The locator will move to point 3.
   Enter an option [Next/Previous/Go/eXit] <N>: *[ENTER]*

---

* If the locator doesn't initially appear at Point 1, use the *Pedit* command to reverse the direction.

12.2C: MORE COMPLEX POLYLINE EDITING TOOLS

6. Use the **Go** option [Go] to execute the **Break**.
    Enter an option [Next/Previous/Go/eXit] <N>: *g*
The segment between points 2 (where you began the **Break** option) and 3 (where you are now) is removed as shown.

7. Use the **eXit** option [eXit] to return to the primary tier, and then hit ENTER to leave the command.
    Enter a vertex editing option [Next/Previous/Break/Insert/Move/Regen/ Straighten/Tangent/Width/eXit] <N>: *x*
    Enter an option [Close/Join/Width/Edit vertex/ Fit/Spline/Decurve/Ltype gen/Reverse/Undo]: *[ENTER]*

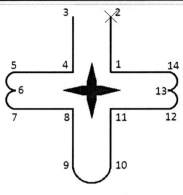

8. Repeat Steps 1 and 2.

9. Now let's **Insert** a vertex [Insert].
    Enter a vertex editing option [Next/Previous/Break/Insert/Move/Regen/ Straighten/Tangent/Width/eXit] <N>: *i*
Notice that AutoCAD provides a rubber band (a line from the currently selected vertex) to help guide you.

10. Place the new vertex midway between points 3 and 4 (right).
    Specify location for new vertex:
Notice the locator appears at this new point.

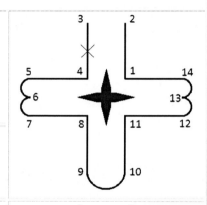

11. Now move [Move] the new vertex a single unit to the west (see figure).
    Enter a vertex editing option [Next/Previous/Break/Insert/Move/Regen/ Straighten/Tangent/Width/eXit] <N>: *m*
    Specify new location for marked vertex: *@1<180*

12. Now we'll remove the new vertex. Return the locator to point 3 (use the **Previous** option [Previous]).
    Enter a vertex editing option [Next/Previous/Break/Insert/Move/Regen/ Straighten/Tangent/Width/eXit] <N>: *p*

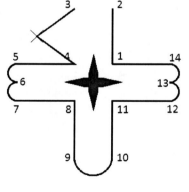

13. Use the **Straighten** option [Straighten].
    Enter a vertex editing option [Next/Previous/Break/Insert/Move/Regen/ Straighten/Tangent/Width/eXit] <P>: *s*

12.2C: MORE COMPLEX POLYLINE EDITING TOOLS

14. Hit ENTER until the locator moves to point 4. Then use **Go** [Go] to execute the **Straighten** option.

    Enter an option [Next/Previous/Go/eXit] <N>: *[ENTER]*
    Enter an option [Next/Previous/Go/eXit] <N>: *g*

The new vertex disappears and the drawing appears as it did in Step 6 (with the locator at point 3).

15. Now use the **Width** option [Width] to change the width of a single line segment.

    Enter a vertex editing option [Next/Previous/Break/Insert/Move/Regen/Straighten/Tangent/Width/eXit] <P>: *w*

16. Set the starting width to *0* and the ending width to *1/8*.

    Specify starting width for next segment <0.0313>: *0*
    Specify ending width for next segment <0.0000>: *.125*

Your drawing looks like the figure at right.

17. Exit this tier and undo the last modifications.

    Enter a vertex editing option [Next/Previous/Break/Insert/Move/Regen/Straighten/Tangent/Width/eXit] <P>: *x*
    Enter an option [Open/Join/Width/Edit vertex/Fit/Spline/Decurve/Ltype gen/Reverse/Undo]: *u*

18. Exit the *PEdit* command. We'll continue with this drawing shortly.

One other consideration of the *PEdit* command is its effect on non-polyline lines and arcs. You've seen what happens when you select a line or arc at the **Select polyline** prompt. AutoCAD will ask if you want to turn it into a polyline. This is the conversion method for lines or arcs to polylines.

To convert polylines to lines or arcs, simply use the *Explode* command. (Alternately, you can use the "X" hotkey or the **Explode** button on the ribbon's **Modify** panel). AutoCAD prompts:

    Select objects:

The prompt will repeat until you confirm the selection set. Be aware, however, that only a polyline can show WYSIWYG width. An exploded polyline becomes a line or arc and loses its width.

You may have noticed that polyline grips don't look exactly like line grips (Figure 12.2). AutoCAD makes some polyline-editing-specific tools available on its polyline grips.

- In addition to the standard grip tools, a *primary polyline grip* offers options to **Stretch**, **Add**, or **Remove** a vertex.

- Secondary grips (on the line segment) provide options to **Stretch** (moving the entire line segment), **Add Vertex** (as we did in Steps 9 and 10 of Exercise 12.2C), and **Convert to Arc** (or **Convert to Line** if the segment is an arc). This last option doesn't appear anywhere else and can be quite useful.

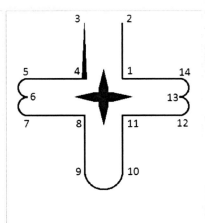

Figure 12.002

Give it a try.

Do This: 12.2D	And Now, the Easy Way: Working with Polyline Editing Grips

I. Continue in the same drawing you used in Exercise 12.2C.
II. Follow these steps.

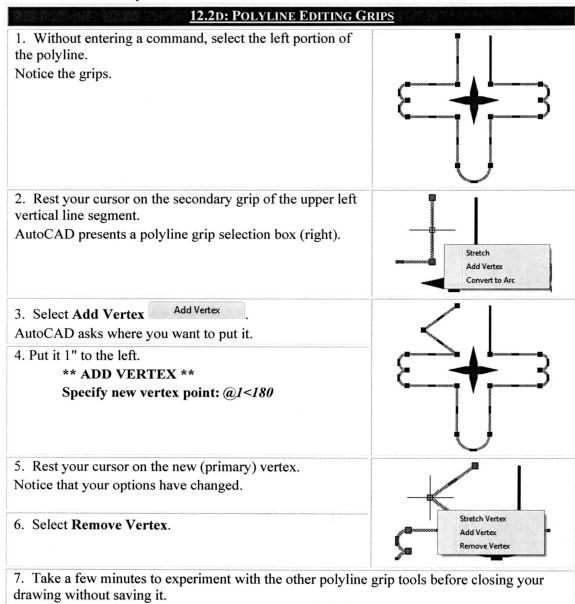

12.2D: POLYLINE EDITING GRIPS

1. Without entering a command, select the left portion of the polyline.
   Notice the grips.

2. Rest your cursor on the secondary grip of the upper left vertical line segment.
   AutoCAD presents a polyline grip selection box (right).

3. Select **Add Vertex** Add Vertex .
   AutoCAD asks where you want to put it.

4. Put it 1" to the left.
       ** ADD VERTEX **
       Specify new vertex point: *@1<180*

5. Rest your cursor on the new (primary) vertex.
   Notice that your options have changed.

6. Select **Remove Vertex**.

7. Take a few minutes to experiment with the other polyline grip tools before closing your drawing without saving it.

12.3 AutoCAD Macros – The Action Recorder

Say hello to AutoCAD's Action Recorder (Figure 12.003). (You'll find its options on the ribbon: Manage – **Action Recorder** panel, or on the menu: Tools – Action Recorder flyout.) Using this tool, you can record a sequence of commands or command inputs so you won't have to enter (and re-enter) them every time you wish to create a door or a nozzle or change the size of a bolt or hand wheel. You

Figure 12.003

287

can even have the recorder pause for user input ... just in case you don't want your commands to be exactly the same every time!

Look at the buttons.

BUTTON	FUNCTION
○	Before your first command, pick the **Record** button to start creating your macro. Alternately, you can enter the *ActRecord* command on the command line, or use the menu sequence: *Tools – Action Recorder – Record*.
☐	Stops the recording process and, if selected in the Action Recorder Preferences box (below) presents a dialog box asking for the macro's name. Alternately, you can enter the *ActStop* command on the command line, or use the menu sequence: *Tools – Action Recorder – Stop*.
 Arrow ▼	Select an existing macro to run from the **Available Action Macro** selection box.
▷	Use the **Play** button to play a selected macro. Alternately, you can use the menu sequence: *Tools – Action Recorder – Play – [select macro]*
📝	Use the **Preference** button to set up how your macros will behave. This button calls the Action Recorder Preferences dialog box. You can: • **Expand** [the Action Tree (Figure 12.004, p.289)] **on playback** [or] **recording** • Have AutoCAD prompt you for a **macro name** once you've completed the creation process 
📋	**Manage Action Macros** opens the Action Macro Manager (right). Here you can **Copy, Modify, Rename** or **Delete** a selected macro. The **Options** button opens the Options dialog box with the **Files** tab on top. Here you can redefine where AutoCAD stores the macros. (*Caution: Don't experiment with other settings in the dialog box.*) 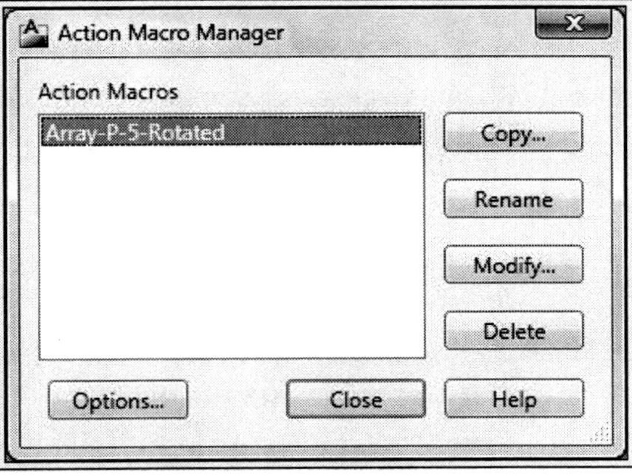
🔘	**Insert Base Point** allows you to create base points within your macro. You'll also find a cursor menu option to do the same thing. We'll place a base point in our exercise so you can see how it works.
💬	Use the **Insert Message** button to have AutoCAD deliver a message anywhere during the execution of your macro. Messages appear in (looming and often irritating) message boxes, so limit the use of these.
⏸	**Pause** the macro at any time so the user can provide input. Again, you'll also find a cursor menu option to do the same thing.

As you create a macro, AutoCAD will show you the sequence in the Action Tree (Figure 12.004). Here you can select a step for modification as you'll see in our next exercise. If you place a check next to **Expand on playback** in the Action Recorder Preferences dialog box, AutoCAD will also display the Action Tree during playback. This can help an unfamiliar user see where he's going with the macro, but isn't always necessary.

Let's try the Action Recorder and see what it can do. We'll create the Arrow macro shown in Figure 12.004.

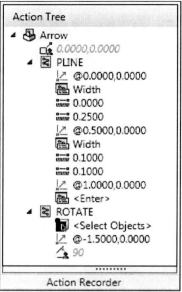

Figure 12.004

Do This: 12.3A	Creating Macros

I. Start a new drawing from scratch.
II. Turn the Orthomode on, and pan so that 0,0 is toward the center of your screen.
III. Follow these steps.

12.3A: CREATING MACROS

1. Open the **Manage** tab on the ribbon.

2. On the **Action Recorder** panel, pick the **Record** button. AutoCAD will record all the commands and responses you give from this point until you pick the **Stop** button.

3. Enter the *PLine* command.
   Command: *pl*

4. Begin by picking the start point (start at 0,0).
   Specify start point: *0,0*

5. Provide this input to the *PLine* prompts:
   Current line-width is 0.0000
   Specify next point or [Arc/Halfwidth/Length/Undo/Width]: *w*
   Specify starting width <0.0000>: *0*
   Specify ending width <0.0000>: *.25*
   Specify next point or [Arc/Halfwidth/Length/Undo/Width]: *@.5<0*
   Specify next point or [Arc/Close/Halfwidth/Length/Undo/Width]: *w*
   Specify starting width <0.2500>: *.1*
   Specify ending width <0.1000>: *.1*
   Specify next point or [Arc/Close/Halfwidth/Length/Undo/Width]: *@1<0*
   Specify next point or [Arc/Close/Halfwidth/Length/Undo/Width]: *[ENTER]*

12.3A: CREATING MACROS

6. Now let's allow the user to adjust the rotation of the arrow. Enter the **Rotate** command ⟳ and select the last object drawn.
   Command: *ro*
   Select objects: *l*
   Select objects: *[ENTER]*

7. Use the tip of the arrow as the rotation point (use the endpoint OSNAP).
   Specify base point: _endpt

8. And specify a rotation angle of 90°.
   Specify rotation angle or [Copy/Reference] <0>: *90*

9. Now pick the **Stop** button ☐ on the Action Recorder. AutoCAD prompts you to enter a name for your macro in the Action Macro dialog box.

10. Examine the Action Macro dialog box.
    - Enter the name of your macro in the **Action Macro Command Name** text box as indicated.
      AutoCAD repeats the name in the **File Name** box.
    - AutoCAD tells you where it will store the macro in the **Folder Path** box – this information will prove useful should you wish to share the macro.
    - You can enter a **Description** of the macro if you wish; it isn't required.
    - You may wish the **Restore pre–playback view** of the Action Tree if you thing it will be of benefit to users; but again, it isn't necessary.
    - Make sure you **Check for inconsistencies** to make sure everything works properly before you execute the macro.

11. Pick the **OK** button to complete the procedure.

12. Erase ✐ the arrow you drew during the macro creation.

13. Make sure *MyArrow* is the macro listed in the **Available Action Macro** box [MyArrow ▼], and pick the **Play** button ▷.
    When completed, AutoCAD presents a message box telling you the macro has finished. (Put a check in the **Do not show** box if this bothers you.)

12.3A: CREATING MACROS

14. Pick the **Close** button to complete the procedure.

Notice that AutoCAD draws the arrow exactly as you drew it before – in the same place! This doesn't give you much flexibility; let's fix that.

15. Let's allow our user to pick his own initial base point (the tip of the arrow). Right click on **PLINE** in the Action Recorder and pick **Insert Base Point** on the cursor menu.

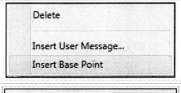

16. AutoCAD prompts you to select a base point. Make 0,0 the base point.

    Command: _ACTBASEPOINT
    Specify a base point: *0,0*

AutoCAD places a base point above the **PLINE** command.

17. Run the macro again (pick the **Play** button ▷). This time, AutoCAD asks for a base point and places the arrow at the selected site. But the user still can't reorient the arrow. Let's fix that now.

18. Right click on the angle setting (90) under the ***ROTATE*** command, and select Select **Pause for User Input** on the cursor menu.

Notice the change in the icon at the beginning of the line 90. This indicates that AutoCAD will pause for user input.

19. Run ▷ the macro again.

Notice that AutoCAD now pauses and asks where you wish to place the arrow, draws it, and then asks for an rotation angle! (So cool!)

20. Close the drawing without saving.

Your macro will be available in any drawings you open ... until you delete it in the Action Tree.

> For all its usefulness, the Action Recorder should not be considered a full programming tool; it does have its limitations as you may discover in time. If you like what you can do with it, however, I strongly encourage you to explore AutoLISP or even Visual Studio!

Admit it now, isn't the Action Recorder really cool?!

12.4 Extra Steps

I've included another Lisp routine – *pe-w.lsp* – in the C:\Steps\Lesson12 folder. This routine includes two commands – *W* and *PLW* – both designed to help you change a polyline's width without having to enter the frightening world of the ***PEdit*** command. *W* will prompt you for the desired width of a polyline, and then prompt you to select lines, arcs, or polylines to change. It saves a bit of time over the ***PEdit*** method. *PLW* allows you to select a polyline that has the desired width, and then select a polyline to change.

Take a few minutes to explore this routine in the *samples12.dwg*.

12.5 What Have We Learned?

Items covered in this lesson include:
- *Creating and Editing Polyline Commands:*
  - *PLine*
  - *PEdit*
  - *Explode*
  - *Splinetype*
- *Join*
- *Explode*
- *Action Recorder*
  - *ActRecord*
  - *ActStop*

This has been a difficult lesson. But polylines are exceptionally powerful tools, and I suppose the nature of the beast requires the multi-leveled structure of both the **PLine** and **PEdit** commands. Take some time to review the lesson. Repeat the exercises as needed to make yourself comfortable. I can guarantee that polylines will be an important part of your CAD future, so get comfortable with them now.

12.6 Exercises

1. Start a new drawing from scratch. Set the grid to ¼" and create the image at right. Save the drawing as *MyArrows* in the C:\Steps\Lesson12 folder.

2. Set up a new drawing with the following parameters:
   2.1. Lower left limits: 0,0
   2.2. Upper right limits: 16,16
   2.3. Grid: 1
   2.4. Snap: as needed
   2.5. Textsize: 5/8
   2.6. Text style: Arial font, Bold and Italicized
   2.7. Layers: as needed
   2.8. Create the backgammon board shown. Save the drawing as *MyBoard* in the C:\Steps\Lesson12 folder.

3. Create a new macro that will array a user selected object(s). Use these parameters:
   3.1. use a polar array with a user selected center point
   3.2. rotate the objects
   3.3. create four copies (five altogether)
   3.4. create a 360° array

   This macro presents a couple challenges. It'll be your first solo macro, but you'll discover that you can't create it using the Array dialog box(!). Here's a hint to get you started: remember that a dash before the command will present command line prompts rather than a dialog box.

   I placed the completed macro in the C:\Steps\Lesson12 folder – can you figure out how to access it?[*]

| 12.7 | **For Web-Based Review Questions and Additional Exercises, visit: http://foragerpub.com/AcadFiles/2011/2011.htm** |

Good morning, Dave.

---

[*] HINT: Use the **Options** button in the Action Macro Manager to see where the files on your computer are kept.

Lesson 13

Following this lesson, you will:

- ✓ Know how to draw contour lines with the **Spline** command
- ✓ Be familiar with the methods for editing a spline
- ✓ Know how to create guidelines (or construction lines) using:
  - ○ **XLine**
  - ○ **Ray**

Guidelines and Splines

*Lesson 13 will familiarize you with AutoCAD's version of guidelines – a tool missing through the earlier releases and welcomed by many when it first appeared. We'll also cover the use of the **Spline** command to draw contour lines like those found on topographical maps.*

13.1 Contour Lines with the *Spline* Command

In Lesson 12, we studied polylines and the *PEdit* command. I'm sure you found them as complicated as most people do the first time.

We learned that we can make a *spline* (a contour line) from a polyline using the *PEdit* command. Here we'll see how we can draw a spline more easily and in a way that shows a contour line as you draw it. Additionally, you'll see that a spline drawn using the *Spline* command takes up much less drawing memory and thus reduces the size of your drawing.

Where to Find It:	
Command Line:	*Spline*
Hotkey(s):	*spl*
Ribbon (Tab/Panel):	Home – Draw (subpanel) – **Spline**
Menu:	Draw – **Spline**
Toolbar:	Draw – **Spline**
Tool Palette:	Draw – **Spline**

The command sequence for the *Spline* command is

    Command: *spline*

    Current settings: Method=Fit

    Knots=Chord *[AutoCAD tells you about your current settings.]*

    Specify first point or [Method/Knots/Object]: *[Pick 🖱 the starting point or select an option.]*

    Enter next point or [start Tangency/toLerance]: *[Pick the second point.]*

    Enter next point or [end Tangency/toLerance/Undo/Close]: *[Continue selecting points until you've finished.]*

    Enter next point or [end Tangency/toLerance/Undo/Close]: *[Hit ENTER to complete the spline.]*

By default, you simply pick points along the spline until complete.

But let's look at some of the other options.

- The first options appear with the **Specify first point** prompt.
  - Use the **Method** option to switch between **Fit** and **CV** as shown.

    Enter spline creation method [Fit/CV] <Fit>:
    - The **Fit** method creates the spline through user-selected points.
    - The **CV** (control vertices) controls the spline's definition using invisible control points (useful tools when used with 3D NURBS surfaces – we'll see these in *3D AutoCAD 2011: One Step at a Time*).
  - **Knots** affect the shape of a spline as it passes through a fit point, and so, work only with the **Fit** method. Their chief function seems to be controlling how AutoCAD identifies the fit points.
  - Use **Object** to convert a splined polyline into an actual spline. This reduces the drawing size considerably as we'll see. (Note: To work properly, the polyline must have been converted to a spline using the **Spline** option of the *PEdit* command.)
- The **toLerance** option tells AutoCAD to draw the spline within a certain distance of the points selected. By default, it's set to zero (meaning, "draw the spline through the user-selected point"). Personally, I prefer leaving it at the default as it makes my drawing considerably more accurate. However, you can reset the tolerance as desired.
- The last two options – **start** and **end Ttangency** – enable you to control the curve of the spline at the beginning and ending points.

- Finally, the **Close** option works just as it does with the *Line* or *Pline* command. It simply closes the spline.

Let's take a look at the *Spline* command.

> *Spline*'s options are also available on the cursor and dynamic input menus once you've entered the command.

Do This: 13.1A	Working with Splines

I. Open the *Splines* drawing in the C:\Steps\Lesson13 folder. The drawing looks like the figure at right.
II. Set the Running OSNAPs to **Node**; clear all others.
III. Follow these steps.

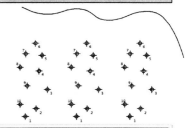

13.1A: WORKING WITH SPLINES

1. Perform a *List* on the existing spline.

   Command: *li*

   See that it's a polyline (following figure). Note, however, that this is actually a polyline and not an lwpolyline. When AutoCAD converted it to a spline, it automatically changed it to a polyline to hold the required information to define the object.

   ```
                   POLYLINE  Layer: "LAYER3"
                             Space: Model space
                 Handle = fd
          Open spline
   starting width    0.0000
     ending width    0.0000
             area   34.0237
           length   22.0284

                   VERTEX    Layer: "LAYER3"
                             Space: Model space
                 Handle = 145
           at point, X=   2.0414  Y=  14.3868  Z=   0.0000
   starting width    0.0000
     ending width    0.0000
                  (Spline control point)

   Press ENTER to continue:
   ```

2. You'll have to hit ENTER several times to view all the information attached to the polyline. Alternately, you can hit the ESC key on your keyboard to cancel the command.

3. Return to the graphics screen.

4. Enter the *Spline* command.

   Command: *spl*

5. Tell AutoCAD you want to convert an **Object** to a spline.

   Specify first point or [Method/Knots/Object]: *o*

13.1A: WORKING WITH SPLINES

6. Select the polyline spline...
    Select spline-fit polyline:
...and complete the command.
    Select spline-fit polyline: *[ENTER]*

7. Repeat Step 1. Note the differences (following figure). All the information concerning the spline fits onto a single screen. There is a corresponding reduction in the size of the drawing as well.

```
              SPLINE      Layer: "LAYER3"
                          Space: Model space
                 Handle = 228
                         Length: 22.0422
                          Order: 3
                     Properties: Planar, Non-Rational, Non-Periodic
              Parametric Range: Start    0.0000
                                End      5.0000
     Number of control points: 7
                 Control Points: X =  2.0414  , Y = 14.3868  , Z = 0.0000
                                 X =  4.3990  , Y = 13.0084  , Z = 0.0000
                                 X =  8.7692  , Y = 14.6740  , Z = 0.0000
                                 X = 11.4143  , Y = 12.2618  , Z = 0.0000
                                 X = 16.5321  , Y = 14.5017  , Z = 0.0000
                                 X = 19.2922  , Y = 11.8598  , Z = 0.0000
                                 X = 20.2123  , Y =  8.5287  , Z = 0.0000
```

8. Now let's draw a spline. Repeat the **Spline** command ∿.
    Command: *spl*

9. Select nodes 1 to 10 in the left group. Use OSNAPs °.
    Specify first point or [Method/Knots/Object]: *[Select node 1.]*
    Enter next point or [start Tangency/toLerance]: *[Select node 2.]*
    Enter next point or [end Tangency/toLerance/Undo/Close]: *[This prompt repeats – continue selecting through node 10.]*

10. After selecting node 10, **Close** the spline.
    Enter next point or [end Tangency/toLerance/Undo/Close]: *c*

11. AutoCAD prompts for a start/end tangent. Hold down the SHIFT key (Ortho override) and pick a point to the left.
    Specify tangent:
Your spline looks like the figure at right.

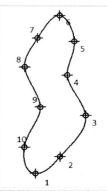

12. Repeat the **Spline** command ∿.
    Command: *[ENTER]*

13.1A: WORKING WITH SPLINES

13. Select nodes 1 and 2 in the middle group.
    Specify first point or [Method/Knots/Object]:
    Enter next point or [start Tangency/toLerance]:

14. Let's change the **toLerance** [toLerance] ...
    Enter next point or [end Tangency/toLerance/Undo/Close]: *l*

15. ... to *1*.
    Specify fit tolerance <0.0000>: *1*

16. Continue as in Steps 9 through 11.
    Your drawing looks like the figure at right. The spline has been drawn within one unit (a **tolerance** of 1) of the nodes selected.

17. Let's draw an open spline. Repeat the command ⌒.
    Command: *[ENTER]*

18. Select nodes 1 through 10 using the third group of nodes. Stop at node 10 (don't close the spline).
    Enter next point or [start Tangency/toLerance]:

19. Hit ENTER at the last prompt.
    Enter next point or [end Tangency/toLerance/Undo/Close]: *[ENTER]*
    Notice that AutoCAD doesn't prompt for tangents. We'll use some modification tools to set the start and end tangent.

20. Save the drawing 💾, but don't exit.
    Command: *qsave*

> In this exercise, I've provided the location of the spline points. On the job, surveyors will provide them. Learning to do the survey as well as the drafting might mean more income for the drafter/designer.

We've seen that drawing splines isn't difficult. Indeed, you've learned enough in this short exercise to draw some fairly complex topographical maps if the elevation points are provided. But what about editing the splines?

13.2 Changing Splines – The *Splinedit* Command

Like polylines, it's often easier to erase and redraw simple splines than it is to edit them. However, you'll find that some complex drawings require some knowledge of spline editing (particularly in 3-dimensional drawing). Also like polylines, editing can be quite a multi-tiered chore.

Luckily, you can often use cursor menus to bypass the primary editing command. The first of these, which contains several editing options, you'll find using the **Spline** flyout of the cursor menu (Figure 13.001) once you've selected a spline. Note that some options will be available when you **Display Fit Points** while others will be available when you **Display Control Vertices**. I'll refer to this menu as we proceed.

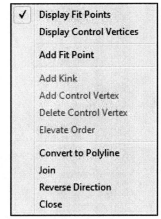

Figure 13.001

> You can also toggle between displaying **Fit Points** and **Control Vertices** using the selection grip's menu. The selection grip ▼ appears when you select a spline.

Here's the *Splinedit* command sequence:

> Command: *Splinedit*
> Select spline:
> Enter an option [Open/Fit data/Edit vertex/convert to Polyline/Reverse/Undo/eXit] <eXit>:

	Where to Find It:
Command Line:	*SplinEdit*
Hotkey(s):	*spe*
Ribbon (Tab/Panel):	Home – Modify (subpanel) – Edit Spline
Menu:	Modify – Object – **Spline**
Toolbar:	Modify II – **Edit Spline**
Tool Palette:	Modify – **Edit Spline**

Let's look at the options.

- **Open**, of course, opens a closed spline by removing the last segment. If the selected spline is already open, this prompt will read **Close**.

  You'll also find an **Open/Close** option on the cursor menu's **Spline** flyout (Figure 13.001).

- The **Fit Data** option enables you to edit the *fit points* of the spline. (Think of fit points as essentially the same thing as vertices on a polyline.) This option drops you into a second tier of options:

  > Enter a fit data option
  > [Add/Open/Delete/Move/Purge/Tangents/toLerance/eXit] <eXit>:

  o On this tier, you can **Add**, **Delete**, or **Move** fit points just as you did vertices in the *PEdit* command. You can also **Open** a closed spline or **Close** an open spline, again just as you did with polyline editing.

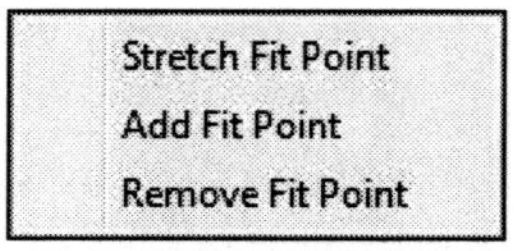

  You can accomplish the same results – and bypass the *SplinEdit* command altogether, by selecting the spline, resting your cursor on a fit point, and selecting an option from the menu (Figure 13.002). (**Add = Add Fit Point** on the menu; **Delete = Remove Fit Point**; **Move = Stretch Fit Point**.)

  Figure 13.002

  You can also begin the **Add** [a] **Fit Point** sequence using the cursor menu in Figure 13.001.

  o You can also change **Tangents** and **toLerances** on this level. An alternative to the *SplinEdit* command's **Tangents** option – **Tangent Direction** – lies on the cursor menu when you rest the cursor on a beginning or ending fit point.

  o The **Purge** option removes fit point information from the drawing's database. This makes editing very difficult. So if you must use it, wait until you've finished the drawing. Its only real benefit is that it'll reduce drawing size.

- Back on the first tier of *Splinedit* options, we find the **Edit Vertex** tool. This calls another second tier of options:

    Enter a vertex editing option [Add/Delete/Elevate order/add Kink/Move/Weight/eXit] <eXit>:

    You'll find several of these options on the cursor menu's **Spline** flyout (Figure 13.001, p.299) and on the menu which appear when you rest your cursor on a vertex (Figure 13.003).

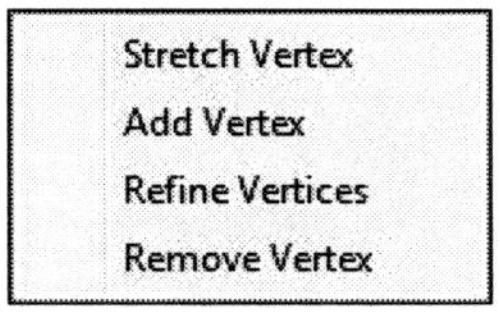

    | Stretch Vertex |
    | Add Vertex |
    | Refine Vertices |
    | Remove Vertex |

    Figure 13.003

    o Use **Add** and **Delete** [or **Remove**] to add or remove vertices. Move vertices allow greater flexibility in modifying your spline but also increase the complexity (and size) of your drawing. You'll also find these options on both menus (Figures 13.001, p.299, and 13.003).

    o **add Kink** prompts you to **specify a point on the spline**, and then removes any curvature at that point, giving you a nice sharp corner. This option also appears on the cursor menu's **Spline** flyout (Figure 13.001).

    o **Move**, of course, allows you to move an individual vertex. This option appears as **Stretch Vertex** on the menu in Figure 13.003.

    o **Elevate Order** also causes more vertices to appear. But this option adds vertices uniformly along the spline. You determine the number – up to 26. This option also appears on the cursor menu's **Spline** flyout (Figure 13.001).

    o A vertex's **Weight** is similar to tolerance. The **Weight** of vertex controls how much influence, or pull, that point has against the spline. Increasing the **Weight** of a point may cause the spline to pull away from adjacent vertices.

- The **Reverse** option switches the start point and endpoint of the spline. AutoCAD includes this option for programmers; you can ignore it. (It also appears on the cursor menu's **Spline** flyout.)

- **convert to Polyline** does just that with the selected spline, but it's a bit more complicated than converting a polyline to a spline. This option also appears on the cursor menu's **Spline** flyout (Figure 13.001).

    When you convert a spline to a polyline, you have two options controlled by the **PlineConvertMode** system variable. A **PlineConvertMode** setting of **0** means that AutoCAD will convert the spline to a series of polyline line segments; a setting of **1** means that AutoCAD will convert the spline to a series of polyline arcs.

Is this starting to feel like the *PEdit* command? Let's try an exercise. But let's try something different, let's see how much we can accomplish without actually entering the *Splinedit* command. That is, let's see how much we can do with the menus. [Okay, I'll put this exercise on the web – using the Splinedit command – if you just want to try it that way, too. Go to: www.uneedcad.com/Files/ModifyingSplinesExercise.pdf.]

| Do This: 13.2A | Modifying Splines |

I. If you're not still in the *Splines* drawing, please open it now. It's in the C:\Steps\Lesson13 folder.

II. Freeze **LAYER1** and the **TEXT** layer. The numbers and nodes disappear from the drawing.

III. Follow these steps.

13.2A: MODIFYING SPLINES

1. Without entering a command, select the spline on the left. Fit points appear as grips.
   Command:

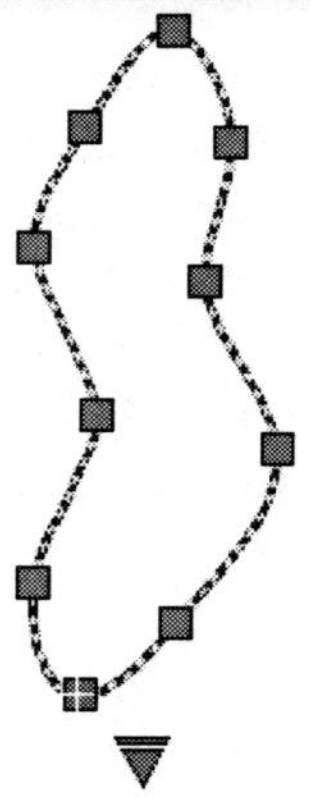

2. Remove the check next to the **Close** option ✓ Close on the cursor menu's **Spline** flyout (Figure 13.001) to open the spline.

3. Reselect the same spline.

4. Pick the selection grip ▽ and tell AutoCAD to **Show Control Vertices** ✓ Show Control Vertices .
   Vertices appear as dots connected by dashed lines (right).

5. Rest your cursor over the first dot on the bottom. AutoCAD will display a menu similar to Figure 13.003, (p.300).

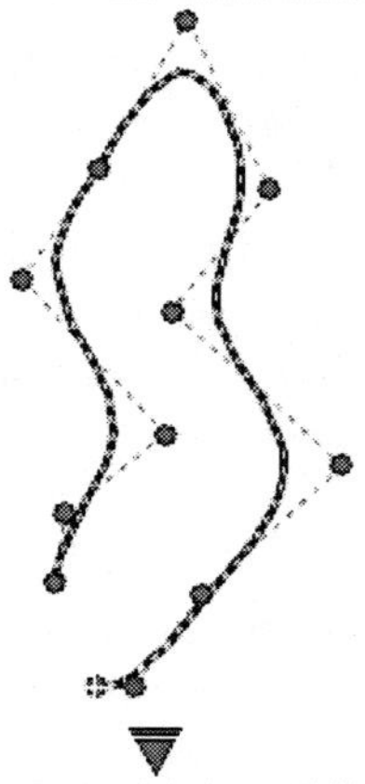

6. Select **Stretch Vertex** from the menu Stretch Vertex and the grip procedure to stretch it 2 units to the left.
   ** STRETCH **
   Specify stretch point or [Base point/Copy/Undo/eXit]: *@2<180*

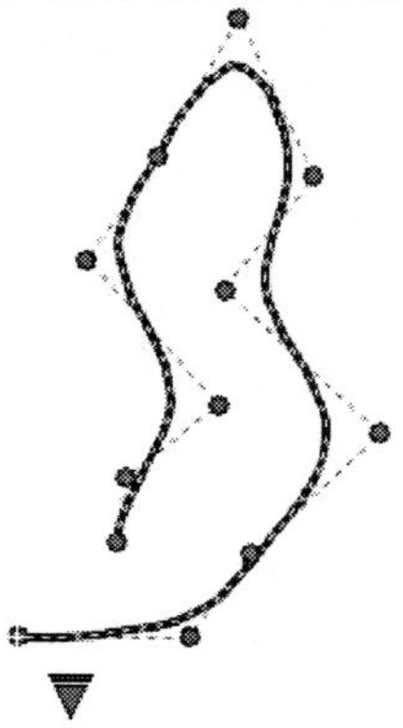

301

13.2A: MODIFYING SPLINES

7. Let's add a vertex (control point). Pick the **Add Control Vertex** [Add Control Vertex] option from the the cursor menu's **Spline** flyout (Figure 13.001, p.299). Notice that AutoCAD begins the *Splinedit* command and prompts for a point on the spline.

> **Command: _SPLINEDIT**
> **Enter an option [Close/Join/Fit data/Edit vertex/convert to Polyline/Reverse/Undo/eXit] <eXit>: _E**
> **Enter a vertex editing option [Add/Delete/Elevate order/add Kink/Move/Weight/eXit] <eXit>: _A**
> **Specify a point on the spline <exit>:**

8. Place the new vertex as shown, and then complete the command.

> **Specify a point on the spline <exit>:** *[Place the new vertex as shown]*
> **Specify a point on the spline <exit>:** *[Enter]*
> **Enter a vertex editing option [Add/Delete/Elevate order/add Kink/Move/Weight/eXit] <eXit>:** *[Enter]*
> **Enter an option [Close/Join/Fit data/Edit vertex/convert to Polyline/Reverse/Undo/eXit] <eXit>:** *[Enter]*

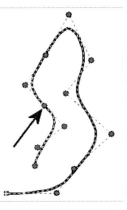

9. Reselect the same spline.

10. Use the cursor menu's **Spline** flyout to to **Elevate** the **order** [Elevate Order] of the spline's control vertices from 4 to 5. Notice that AutoCAD again takes you through the *Splinedit* command without forcing you to respond to a lot of prompts.

> **Command: _SPLINEDIT**
> **Enter an option [Close/Join/Fit data/Edit vertex/convert to Polyline/Reverse/Undo/eXit] <eXit>: _E**
> **Enter a vertex editing option [Add/Delete/Elevate order/add Kink/Move/Weight/eXit] <eXit>: _E**
> **Enter new order <4>: 5**

11. Complete the command.

> **Enter a vertex editing option [Add/Delete/Elevate order/add Kink/Move/Weight/eXit] <eXit>:** *[Enter]*
> **Enter an option [Close/Join/Fit data/Edit vertex/convert to Polyline/Reverse/Undo/eXit] <eXit>:** *[Enter]*

12. Reselect the spline. It looks like the figure at right. Notice all the new control vertices with which you can work.

13. Clear the grips and select the third spline.

13.2A: MODIFYING SPLINES

14. Rest your cursor on the beginning grip (the one with the cross in it). Notice that this menu includes an option to change the **Tangent Direction** [Tangent Direction]. Pick that one now.

15. AutoCAD prompts you to specify a new direction. Hold down the SHIFT key and pick a point to the right.

   ** TANGENT DIRECTION **
   Specify tangent direction: *[Pick a point to the right.]*

16. Clear the grips. The spline looks like the figure at right. Notice the new tangent at the end point.

17. Reselect the same spline.

18. This time we'll delete a fit point. Rest your cursor over the topmost grip and select **Remove Fit Point** [Remove Fit Point] from the menu. The spline now looks like the figure at right.

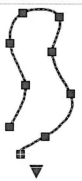

19. Let's add a kink in the spline. Reselect the same spline and tell AutoCAD to **Show Control Vertices** [✓ Show Control Vertices].

20. Now select **Add Kink** [Add Kink] from the cursor menu's **Spline** flyout.

21. AutoCAD begins the *SplinEdit* command and asks you where to place the kink. Place it about midway between the top two vertices as shown.

   Command: _SPLINEDIT
   Enter an option [Close/Join/Fit data/Edit vertex/convert to Polyline/Reverse/Undo/eXit] <eXit>: _E
   Enter a vertex editing option [Add/Delete/Elevate order/add Kink/Move/Weight/eXit] <eXit>: _K
   Specify a point on the spline <exit>:

303

13.2A: MODIFYING SPLINES

22. Complete the command.
    Specify a point on the spline <exit>: *[Enter]*
    Enter a vertex editing option [Add/Delete/Elevate order/add Kink/Move/Weight/eXit] <eXit>: *[Enter]*
    Enter an option [Close/Join/Fit data/Edit vertex/convert to Polyline/Reverse/Undo/eXit] <eXit>: *[Enter]*	

23. Reselect the same spline.	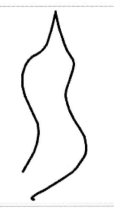
24. Rest your cursor over the new vertex and select **Stretch Vertex** Stretch Vertex from the menu.	

25. Use your grips procedures to stretch the new vertex two units straight up.
    ** STRETCH **
    Specify stretch point or [Base point/Copy/Undo/eXit]:
Specify stretch point or [Base point/Copy/Undo/eXit]: *@2<90*	

26. Clear the grips. The spline now looks like the figure at right.	

If you went through the *SplinEdit* procedure on the web (www.uneedcad.com/Files/ModifyingSplinesExercise.pdf), you'll see that we couldn't do a couple things using the menus that we could do with the command – we couldn't adjust the tolerances or the weight of a vertex. Still, considering the ease of use we find with the menus, it's a price worth paying! (Besides, you can always use the command procedure if you need to do those things.)

Before we finish, let's take a look at converting a spline to a polyline.

Do This: 13.2B	Converting Splines to Polylines

   I. If you're not still in the *Splines* drawing, please open it now. It's in the C:\Steps\Lesson13 folder.

  II. Set the **PlineConvertMode** system variable to **1** so that the conversion will create polyline *arc* segments.

 III. Follow these steps.

13.2B: CONVERTING SPLINES TO POLYLINES

1. Enter the *Splinedit* command.   **Command:** *spe*
2. Select the spline that you converted from a polyline (the long one across the top).   **Select spline:**
3. Tell AutoCAD you wish to **convert to [a] Polyline** convert to Polyline.   **Enter an option [Close/Join/Fit data/Edit vertex/convert to Polyline/Reverse/Undo/eXit] <eXit>:** *p*

13.2B: CONVERTING SPLINES TO POLYLINES

4. Now AutoCAD wants to know how precise you wish the conversion to be. You can enter any number from 1-99 (larger numbers are more precise but produce more segments). In order to see the differences between the types of conversions, we'll use a small number. Enter two.

   Specify a precision <10>: *2*

5. Now select the polyline without entering a command. Examine the results (following figure). Remember that AutoCAD places a grip at the endpoints and midpoint of polyline arcs.

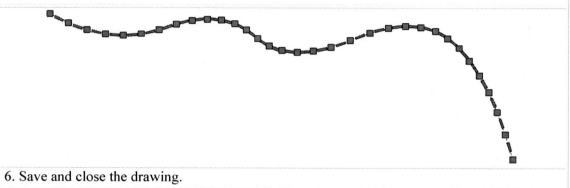

6. Save and close the drawing.

Try the previous exercise again with the **PlineConvertMode** system variable set to 0. Do you notice any difference? (You've coverted the spline to a pline made up of line segments this time.)

13.3 Guidelines

Beginning with this section, we'll create a multi-lesson project. We'll continue to have exercises at the end of each lesson; but in a series of *Do This* sample exercises, we'll create the floor plan of a house. We'll lay out the drawing in Lesson 13, place the walls in Lesson14, p.312, and then add some text in Lesson 15, p.328. Finally, in Lesson 17, p.376, we'll dimension the drawing.

On the drawing board, we use the *4H* lead to mark our guidelines – lines used to layout an area lightly before putting the heavier leads to paper. Precision is important, but the length of the line generally covers an area much larger than is eventually needed. This helps in locating other items on the drawing. Later, we might darken part of the guideline with an *F* or *HB* lead, or even erase it altogether.

The technical term used for these guidelines is *construction line*.

Create a construction line in AutoCAD with the *XLine* or *Ray* command. (In fact, the terms *xline* and *construction line* are used to mean the same thing.) The only real difference between an xline and a ray is that an xline is infinite in both directions, whereas a ray is infinite in only one direction. Often, the *Ray* command, with no options, is overlooked in favor of the more versatile *XLine*.

Of course, we can convert the xline or ray into a drawing object just as we could darken our construction line with a heavier lead. Simply trim away what you don't need and the xline/ray becomes an ordinary line!

	Where to Find It:
Command Line:	*XLine*
Hotkey(s):	*xl*
Ribbon (Tab/Panel):	Home – Draw (subpanel) – **Construction Line**
Menu:	Draw – **Construction Line**
Toolbar:	Draw – **Construction Line**
Tool Palette:	Draw – **Construction Line**

Let's look at each. The command sequence for *XLine* is

> **Command:** *xline*
> **Specify a point or [Hor/Ver/Ang/Bisect/Offset]:** *[Pick a point on the proposed line.]*
> **Specify through point:** *[Pick a second point on the proposed line.]*
> **Specify through point:** *[AutoCAD will continue to place xlines through the initially selected point and any second point you select – hit ENTER or 🖱 to complete the command.]*

- By default, AutoCAD requires you to select two points on the construction line. It then draws an infinite line (a line that continues infinitely in both directions) through those two points. You may, however, select the **Hor**, **Ver**, or **Ang** options to force AutoCAD to draw horizontal, vertical, or angular construction lines. Then you need only pick a single point on the construction line. **Hor** and **Ver** need no further input except the location. **Ang** will prompt for the desired angle:

    > **Enter angle of xline (0) or [Reference]:**

    You can enter the desired angle via the keyboard, or use the **Reference** option to get an angle from an existing object.

- The **Bisect** option allows you to find the bisector of an existing angle. It prompts:

    > **Specify angle vertex point:** *[Select the corner of the angle – use OSNAPs.]*
    > **Specify angle start point:** *[Select any point on one of the lines forming the angle – again, use OSNAPs.]*
    > **Specify angle end point:** *[Select any point on the other line.]*
    > **Specify angle end point:** *[ENTER to complete the command.]*

    Once you've used the **Bisect** option, you'll never want to go back to the old compass method!

- The last option – **Offset** – is peculiar. It replaces the *XLine* command prompt with the *Offset* command prompt. It then behaves like the *Offset* command. I recommend ignoring this option in favor of the actual *Offset* command.

The *Ray* command sequence is much easier:

> **Command:** *ray*
> **Specify start point:** *[Pick the start point.]*
> **Specify through point:** *[Pick a point through which the ray will pass.]*
> **Specify through point:** *[Continue or hit ENTER or 🖱 to complete the command.]*

	Where to Find It:
Command Line:	*Ray*
Ribbon (Tab/Panel):	Home – Draw (subpanel) – **Ray**
Menu:	Draw – **Ray**

> An interesting point about construction lines: although they're infinite in length, neither affects a *zoom extents* (zooming to include all the objects in a drawing). Construction lines will print; but otherwise, AutoCAD treats them as background images.

Let's begin our floor plan.

> You'll also find the options for these two commands on the cursor and dynamic menus once the command has been entered.

Do This: 13.3A	**Creating Construction Lines**

I. Open the *flr-pln 13* drawing in the C:\Steps\Lesson13 folder. This drawing has been set up to create a floor plan on a ¼"=1'-0" scale, on a C-size sheet of paper.

II. Follow these steps.

13.3A: CREATING CONSTRUCTIONS LINES

1. Be sure the **CONST** layer is current [CONST].

2. Enter the *XLine* command.
   Command: *xl*

3. Tell AutoCAD you want to draw vertical xlines [Ver].
   Specify a point or [Hor/Ver/Ang/Bisect/Offset]: *v*

4. Place them as indicated.
   Specify through point: *1',0*
   Specify through point: *87',0*

5. Complete the command.
   Specify through point: *[ENTER]*

6. Repeat the command.
   Command: *[ENTER]*

7. This time, let's draw horizontal xlines [Hor] ...
   Specify a point or [Hor/Ver/Ang/Bisect/Offset]: *h*

8. ... through these points.
   Specify through point: *0,1'*
   Specify through point: *0,67'*

9. Complete the command.
   Specify through point: *[ENTER]*

Your drawing looks like the figure at right. (I'll turn the grid off to make the images clearer.)

10. Use the *Offset* command to offset the bottom line upward as indicated. (Use the direct distance option. Place the crosshairs above the line and enter the following numbers followed by an ENTER: 18, 18, 21, 30.)
    Command: *o*
    Current settings: Erase source=No Layer=Source OFFSETGAPTYPE=0
    Specify offset distance or [Through/Erase/Layer] <Through>: *[ENTER]*
    Select object to offset or [Exit/Undo] <Exit>: *[Select the line.]*
    Specify through point or [Exit/Multiple/Undo] <Exit>: *m*
    Specify through point or [Exit/Undo] <next object>: *18 [As the prompt repeats, enter the numbers indicated above.]*
    Specify through point or [Exit/Undo] <next object>: *[ENTER]*
    Select object to offset or [Exit/Undo] <Exit>: *[ENTER]*

13.3A: CREATING CONSTRUCTIONS LINES

11. Now offset the right vertical line as indicated. (Follow the procedure in Step 10. Place your crosshairs to the left and enter the numbers: 48, 84, 84.)
    Command: *o*

The lower right corner of your drawing now looks like the left figure below.

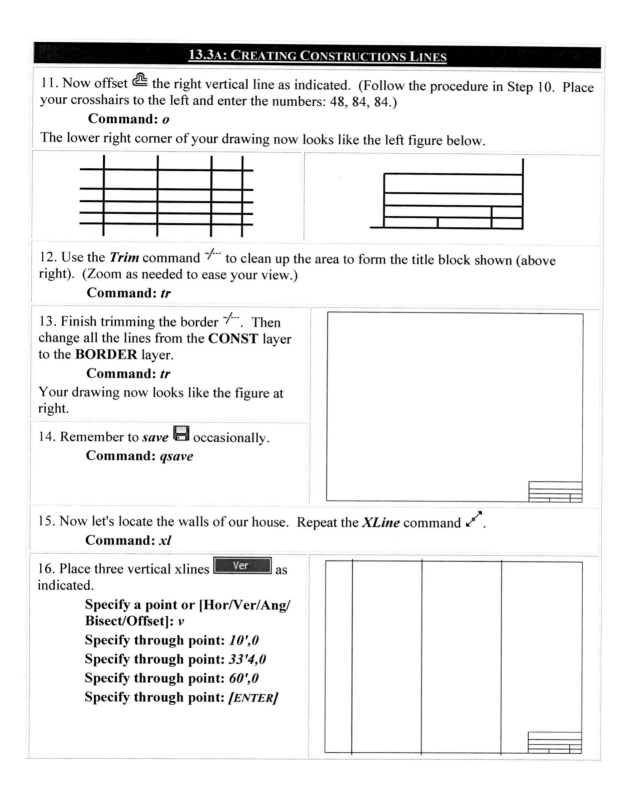

12. Use the ***Trim*** command to clean up the area to form the title block shown (above right). (Zoom as needed to ease your view.)
    Command: *tr*

13. Finish trimming the border. Then change all the lines from the **CONST** layer to the **BORDER** layer.
    Command: *tr*

Your drawing now looks like the figure at right.

14. Remember to ***save*** occasionally.
    Command: *qsave*

15. Now let's locate the walls of our house. Repeat the ***XLine*** command.
    Command: *xl*

16. Place three vertical xlines as indicated.

    Specify a point or [Hor/Ver/Ang/Bisect/Offset]: *v*
    Specify through point: *10',0*
    Specify through point: *33'4,0*
    Specify through point: *60',0*
    Specify through point: *[ENTER]*

13.3A: CREATING CONSTRUCTIONS LINES

17. Place three horizontal xlines **Hor** as indicated.

    Command: *[ENTER]*
    Specify a point or [Hor/Ver/Ang/Bisect/Offset]: *h*
    Specify through point: *0,10'*
    Specify through point: *0,33'*
    Specify through point: *0,60'*
    Specify through point: *[ENTER]*

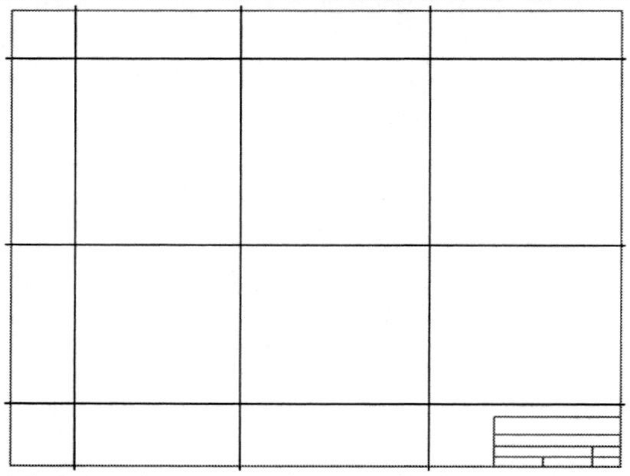

18. Trim away some of the excess lines so that your drawing looks like this.

    Command: *tr*

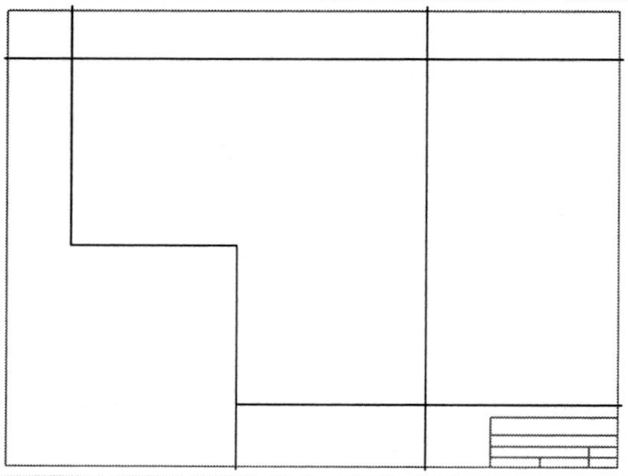

19. Now let's locate some inner walls. Make the **CONST2** layer current.

20. Create vertical xlines **Ver** at these coordinates: *26',0*; *40'4,0*; *43'10,0*; and *47'10,0*; create horizontal xlines at these coordinates: *0,19'10*; *0,23'4*; *0,28'10*; *0,39'10*; and *0,50'*.
    Now your drawing looks like the figure at right.

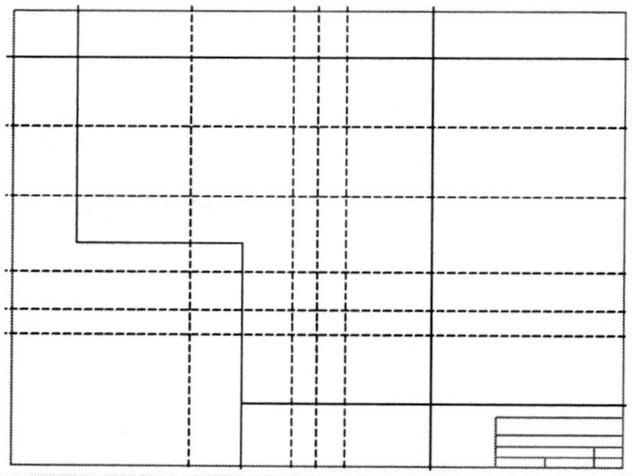

13.3A: CREATING CONSTRUCTIONS LINES

21. Let's add a bay window. Draw a ray due west from the point indicated.
    Command: *ray*
    Specify start point: *33'4,31'*
    Specify through point: *[Pick a point to the left.]*
    Specify through point: *[ENTER]*

22. Offset 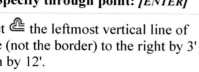 the leftmost vertical line of the house (not the border) to the right by 3' and again by 12'.
    Command: *o*
    The house now looks like the figure at right.

23. Use the **Angle** option to place an xline for the left wall of the bay window. Be sure to use OSNAPs to hit the intersection indicated.
    Command: *xl*
    Specify a point or [Hor/Ver/Ang/Bisect/Offset]: *a*
    Enter angle of xline (0) or [Reference]: *135*
    Specify through point: _int of
    Specify through point: *[enter]*

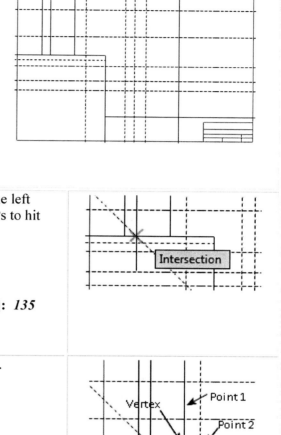

24. Use the **Bisect** option to place the other wall.
    Command: *[ENTER]*
    Specify a point or [Hor/Ver/Ang/Bisect/Offset]: *b*
    Specify angle vertex point: *[Select the intersection of the Vertex shown.]* _int of
    Specify angle start point: *[Select a point nearest to Point 1.]*_nea to
    Specify angle end point: *[Select a point nearest to Point 2.]*_nea to
    Specify angle end point: *[ENTER]*

25. Use the *Erase* and *Trim* commands to complete the drawing. Change the layer of the bay window to **CONST**. It'll look like this when you've finished.

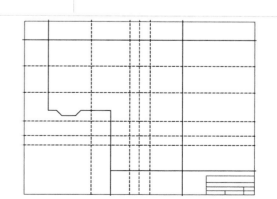

26. Save the drawing as *MyFlr-Pln14* in the C:\Steps\Lesson14 folder.
    Command: *saveas*

What do you think? It doesn't look much like a floor plan, does it?

Consider how a pencil layout might appear at this stage of production. All we've done is place some guidelines (construction lines) to help us locate our walls. We'll save the drawing for now and use it in our next lesson when we discuss the *MLine* (multiline) command.

13.4 Extra Steps

In the first half of this lesson, we discussed how to draw contour lines (splines). Contour lines generally represent changes in elevation on topographical or site maps. Go to your local library, check the encyclopedia or ask your employer for some samples of topographical or marine drawings to help familiarize yourself with these fascinating tools.

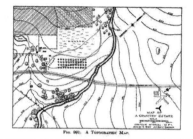

13.5 What Have We Learned?

Items covered in this lesson include:

- *System variables:*
  - *PlineConvertMode*
- *Commands:*
  - *Spline*
  - *Splinedit*
  - *Xline*
  - *Ray*

We've covered some interesting tools in this lesson. Splines will prove themselves surprisingly versatile in the fields where they are used. Construction lines, while not as critical to computer drafting as they are to board work, are nonetheless quite handy in most disciplines. By the time we've finished the next few lessons, this will become apparent.

In our next lesson, we'll add walls to the *MyFlr-Pln14* drawing. After that (in Lessons 15, p.328, and 17, p.376), we'll add some notes and dimensions. By the time we complete this section of the book, you'll have accomplished quite a drawing!

13.6 Exercises

1. Open *topography* from the C:\Steps\Lesson13 folder.

   We shot several elevations during a survey of an area golf course. The elevations for Hole # 3 are shown.

   Using the *Spline* command, draw contour lines connecting like elevations. This will afford you a topographical view of the area.

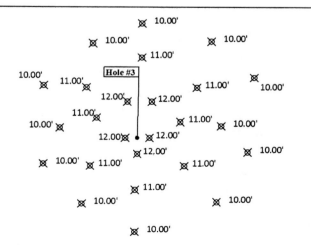

13.7 For Web-Based Review Questions and Additional Exercises, visit: http://foragerpub.com/AcadFiles/2011/2011.htm

Following this lesson, you will:

- ✓ Know how to draw several lines at once with the **MLine** command

- ✓ Be able to create multiline styles with the **MLStyle** command

- ✓ Be able to edit multilines with the **MLEdit** command

Advanced Lines - Multilines

Back in grade school, when I'd been naughty (and got caught), I was punished by being assigned to write "penance" sentences – "I will be good in school," one hundred times. Oh! The degradations of childhood!

But in my childish attempts to cut corners (there were always ways to cut corners), I would tape four pencils together. Then I only had to "be good in school" twenty-five times!

*Apparently, someone at Autodesk learned a similar childhood lesson. The result was the **MLine** command. The **MLine** command does just what taping four pencils together did – it enables you to create more than one line at a time.*

*AutoCAD's multiline procedures actually involve three commands: **MLine**, **MLEdit**, and **MLStyle**. The first actually draws the lines; the second enables you to edit, or change, the lines; the third enables you to define the lines.*

In Lesson 14, we'll look at each.

14.1 Many at Once – AutoCAD's Multilines and the *MLine* Command

The *MLine* command is an easy-to-use tool designed to enhance the efficiency of multiline drawing.

The command sequence is

	Where to Find It:
Command Line:	*MLine*
Hotkey(s):	ml
Menu:	Draw – **Multiline**

    Command: *mline*
    Current settings: Justification = Top, Scale = 1.00, Style = STANDARD
    Specify start point or [Justification/Scale/STyle]: *[Pick the start point.]*
    Specify next point: *[Pick the next point.]*
    Specify next point or [Undo]: *[Either continue picking points or hit ENTER to complete the command.]*

As you can see, the actual command sequence isn't very different from drawing any other line. AutoCAD prompts **Specify start point** and then repeats **Specify next point** until you've completed the line. There are, however, some simple options you must consider.

- AutoCAD makes three choices available when you select the **Justification** option:

    Enter justification type [Top/Zero/Bottom] <top>:

These involve where AutoCAD places the lines in relation to the user-identified point. (See Figures 14.001, 14.002, and 14.003 to see how these options work.)

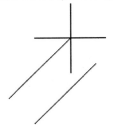

Figure 14.001: Top Justification   Figure 14.002: Zero Justification   Figure 14.003: Bottom Justification

This may seem a bit strange, but it really does have its uses. For example, most pipe is drawn using centerlines. For a piper, then, the **Zero** justification would be easiest to use. But for an architect who dimensions to outer walls, the **Top** or **Bottom** justification may be just the ticket.

- The **Scale** option enables you to insert a multiline at any width. For example, a piper creates a multiline style that has three lines – two outer walls and a centerline. The outer walls are one unit apart according to the style's definition. The piper needs a 3" diameter pipe, so he inserts the multiline at a **Scale** of 3.5 (the outer diameter, or OD, of the pipe). Similarly, the architect may insert a basic two-line wall at a **Scale** of 4 or 5.5 to cover different building materials for walls.
- The **STyle** option enables you to tell AutoCAD which multiline style is needed for this particular multiline. The default is **Standard**, which is a basic two-line multiline with a separation of 1 unit between the lines. We'll look at creating multiline styles in Section 14.2, p.315.

Let's draw a simple multiline using the **Standard** style.

> You'll find the *MLine* options on the cursor and dynamic input menus once the command has been entered.

Do This: 14.1A	Creating Multilines

I. Start a new drawing from scratch.
II. Follow these steps. (I use Absolute coordinates in this exercise; you might want to toggle dynamic input off and on as we go.)

14.1A: CREATING MULTILINES

1. Set the current color to **Blue** [Blue].

2. Enter the *MLine* command.
   Command: *ml*

3. Accept the defaults and draw a simple multiline as indicated.
   Current settings: Justification = Top, Scale = 1.00, Style = STANDARD
   Specify start point or [Justification/Scale/STyle]: *1,1*
   Specify next point: *1,6*
   Specify next point or [Undo]: *5,6*
   Specify next point or [Close/Undo]: *5,1*
   Specify next point or [Close/Undo]: *[ENTER]*

   Notice that the color of the multiline doesn't reflect the current color setting. This is because multiline colors are set according to the Multiline Style definition (Section 14.2, p.315).

   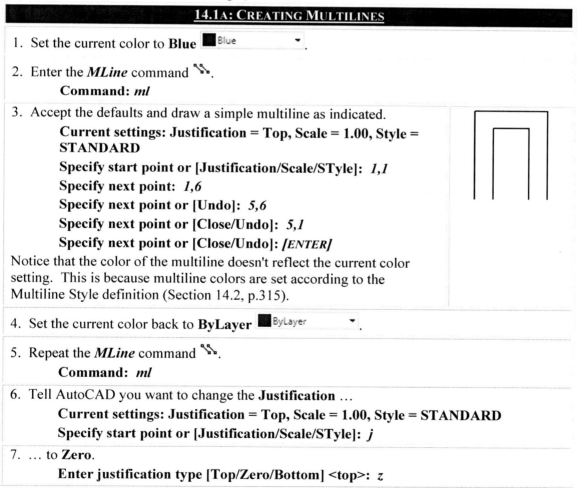

4. Set the current color back to **ByLayer** [ByLayer].

5. Repeat the *MLine* command.
   Command: *ml*

6. Tell AutoCAD you want to change the **Justification** …
   Current settings: Justification = Top, Scale = 1.00, Style = STANDARD
   Specify start point or [Justification/Scale/STyle]: *j*

7. … to **Zero**.
   Enter justification type [Top/Zero/Bottom] <top>: *z*

14.1A: CREATING MULTILINES

8. Repeat Step 3 using the same coordinates. You're now locating the multiline from the center.

   Current settings: Justification = Zero, Scale = 1.00, Style = STANDARD
   Specify start point or [Justification/Scale/STyle]:

   Your drawing (with both multilines) looks like the figure at right.

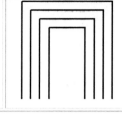

9. Repeat the *MLine* command.

   Command: *[ENTER]*

10. This time we'll change the **Scale** ...

    Current settings: Justification = Zero, Scale = 1.00, Style = STANDARD
    Specify start point or [Justification/Scale/STyle]: *s*

11. ... to *2*.

    Enter mline scale <1.00>: *2*

12. Draw the multiline as indicated.

    Current settings: Justification = Zero, Scale = 2.00, Style = STANDARD
    Specify start point or [Justification/Scale/STyle]: *7,1*
    Specify next point: *7,5*
    Specify next point or [Undo]: *12,5*
    Specify next point or [Close/Undo]: *12,1*
    Specify next point or [Close/Undo]: *[ENTER]*

    Your drawing looks like the figure at right.

13. Exit the drawing without saving.

    Command: *close*

So you see how easy it is to draw multilines. But we've only used AutoCAD's **Standard** style. Let's try something different. Let's define our own!

14.2 Options: The *MLStyle* Command

We define multiline styles using the *MLStyle* command. It can be a fairly complex task, but fortunately, AutoCAD provides dialog boxes to help. Upon receiving the *MLStyle* command, AutoCAD presents the Multiline Styles dialog box (Figure 14.004, p.316).

Where to Find It:	
Command Line:	*MLStyle*
Menu:	Format – **Multiline Style**

- First, you see a **Styles** list box. As with other list boxes, this one shows the styles available to you. You can select a style from this box by double clicking or by selecting and then using the **Set Current** button.

- Use the **Rename** and **Delete** buttons to rename or delete a selected multiline style. (Note that you can't delete a style that's in use.)

- **Load** calls the Load Multiline Styles dialog box (Figure 14.005). From here, you can load other *mln* files (multiline definition files).
- **Save** opens a standard Save File dialog box. Use this to save your drawing's multiline style definitions to a file. Then you can load the definitions into another file using the **Load** button.
- The **New** button begins the Multiline creation process by calling the Create New Multiline Style dialog box (Figure 14.006). Here you'll name your new style. You can also tell AutoCAD to begin with the setup associated with an existing style by selecting it from the **Start With** drop down menu.

Pick the **Continue** button to move on to the New Multiline Style dialog box (Figure 14.007, p.317). You'll set up your style here. (This information will become considerably clearer with an exercise.)

Figure 14.004

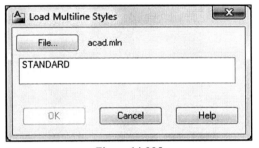

Figure 14.005

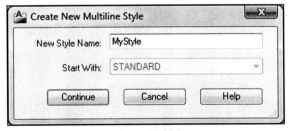

Figure 14.006

  - You don't have to add a **Description**, but if you anticipate using many multilines in your drawing, it might help identify which is which.
  - The **Caps** frame allows you to add or remove caps at the ends of your multiline. As you can see, you have options for a **Line, Outer arc**, and **Inner arcs**. (Note the difference; Outer arc – provides an arc joining the outermost lines at their endpoints; Inner arcs – provide arcs joining the endpoints of two *or more* inner lines.) You also control the **Angle** between the endpoints of the lines making up your multiline.
  - Use the **Fill** color option to fill the multilines (creating something like a polyline or a solid).
  - A check in the **Display joints** box tells AutoCAD to draw a line connecting the related vertices of your multiline.
  - You'll control the actual parts of your multiline in the **Elements** frame.
    - First you'll see another list box. This one shows the definitions for the individual lines making up your multiline. Notice that you can control the spacing between the lines, the color and the linetype for each. You can also modify each by selecting it and using the **Offset, Color,** and **Linetype** tools.
    - **Add** or **Delete** lines using the buttons below the list box.
    - Confirm your creation (**OK** button) or **Cancel** it to return to the Multiline Style dialog box.

- The last button on the Multiline Style dialog box (Figure 14.004, p.316) – **Modify** – calls the Modify Multiline Style dialog box. This is the same box as the New Multiline Style dialog box (Figure 14.007), but you'll use it to alter the definition of a selected multiline style.

Believe it or not, this approach to creating/modifying multiline styles beats the heck out of the old way! (We won't go into that!) Let's try an exercise.

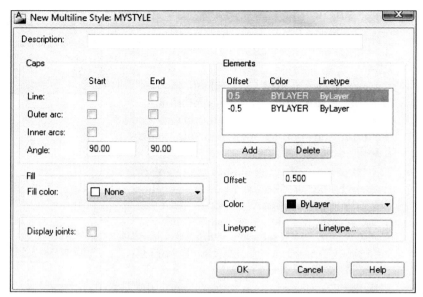

Figure 14.007

Do This: 14.2A	Defining Multilines

I. Start a new drawing from scratch. We'll create a basic style to be used in drawing pipe.

II. Follow these steps.

14.2A: DEFINING MULTILINES

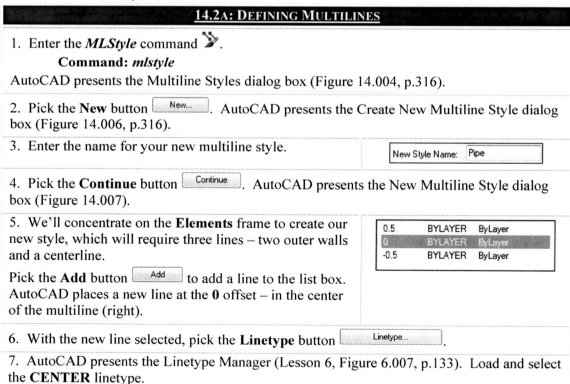

1. Enter the *MLStyle* command.
   Command: *mlstyle*
AutoCAD presents the Multiline Styles dialog box (Figure 14.004, p.316).

2. Pick the **New** button. AutoCAD presents the Create New Multiline Style dialog box (Figure 14.006, p.316).

3. Enter the name for your new multiline style.

4. Pick the **Continue** button. AutoCAD presents the New Multiline Style dialog box (Figure 14.007).

5. We'll concentrate on the **Elements** frame to create our new style, which will require three lines – two outer walls and a centerline.

Pick the **Add** button to add a line to the list box. AutoCAD places a new line at the **0** offset – in the center of the multiline (right).

6. With the new line selected, pick the **Linetype** button.

7. AutoCAD presents the Linetype Manager (Lesson 6, Figure 6.007, p.133). Load and select the **CENTER** linetype.

14.2A: DEFINING MULTILINES

8. With the center line still selected, set the **Color** to **Blue**. The **Elements** list box now displays the definition (right).

0.5	BYLAYER	ByLayer
0	blue	CENTER
-0.5	BYLAYER	ByLayer

9. Pick the **OK** button [OK] to return to the Multiline Style dialog box. Notice that your new definition appears in the list box.

10. Now let's save the definitions in this drawing to a file so that we can use them in other drawings. Pick the **Save** button [Save...].

11. AutoCAD presents a standard Save File dialog box. Save the file as *MyMLines* in the C:\Steps\Lesson14 folder.

12. Now pick the **OK** button [OK] to close the dialog box.

13. Close AutoCAD completely (don't save the drawing) and reopen it. (This will clear any definitions that might linger.) We'll use the default drawing.

14. Now reopen the Multiline Style dialog box.
    Command: *mlstyle*

Notice that the **PIPE** style you create isn't listed. You haven't loaded it into the current drawing. We'll fix that now.

15. Pick the **Load** button [Load...]. AutoCAD presents the Load Multiline Styles dialog box (Figure 14.005, p.316).

16. Pick the **File** button [File...]. AutoCAD presents a standard Open File dialog box.

17. Navigate to the C:\Steps\Lesson14 folder and open the *MyMLines* file you create in this exercise. (Note: if that file is unavailable, select the *MLines* file instead.)

18. AutoCAD now shows the **PIPE** definition in the Load Multline Styles dialog box. (If you selected the *MLines* file, it also shows a **DEMO** style – right). Select the **PIPE** style and pick the **OK** button [OK].

19. Almost done! AutoCAD returns to the Multiline Style dialog box. Select **PIPE** from the list box and pick the **Set Current** button.

20. Pick the **OK** button [OK] to complete the procedure.

You've done it! You've create a multiline style, saved it to a file, loaded the file into a new drawing, and set the style current in the new drawing. Any multiline you draw now will have the **PIPE** style! (Let's do that just to see. We'll create a couple of crossing, 1"ø pipes.)

21. Create a layer called *MyPipe*. Set the color to red and make it current.
    Command: *la*

14.2A: DEFINING MULTILINES
22. Enter the *MLine* command. **Command:** *ml*
23. Notice that **PIPE** is the current style. Set the scale for a 1"ø pipe as indicated. **Current settings: Justification = Top, Scale = 1.00, Style = PIPE** **Specify start point or [Justification/Scale/STyle]:** *s* **Enter mline scale <1.00>:** *1.375*
24. Draw two multilines – the first from point **6,1** to point **6,7**; the second from point **3,5** to point **11,5**. **Specify start point or [Justification/Scale/STyle]:** **Specify start point or [Justification/Scale/STyle]:** *[ENTER]* Your drawing looks like the figure at right. 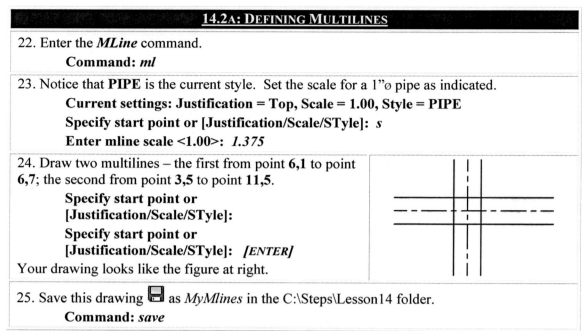
25. Save this drawing as *MyMlines* in the C:\Steps\Lesson14 folder. **Command:** *save*

Notice the colors and linetypes. Lines that had their colors setup as **ByLayer** reflect the layer color. Lines set up with a designated color reflect that color. The same is true for linetypes.

Now let's take a look at the other side of the New Multiline Styles dialog box. Let's set up some properties.

Do This: 14.2B	More on Multiline Styles

I. Open the *MLine Properties* file in the C:\Steps\Lesson14 folder.
II. Open the MLStyle dialog box. Notice that a style called **DEMO** is current.
III. Follow these steps.

14.2B: MORE ON MULTILINE STYLES
1. Pick the **Modify** button [Modify...]. AutoCAD presents the Modify Multiline Style dialog box (the same box as New Multiline Style seen in Figure 14.007, p.317).
2. Put a check in the **Display joints** check box. [Display joints: ✓] Pick **OK** to return to the Multiline Styles dialog box. Notice (right) the line in the middle of the demo box. This indicates that AutoCAD will put a line connecting associated vertices in the multiline joints. 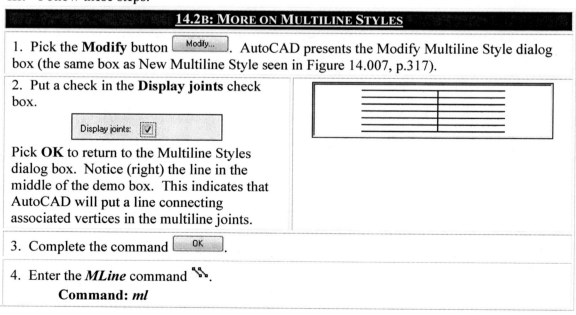
3. Complete the command [OK].
4. Enter the *MLine* command. **Command:** *ml*

14.2B: MORE ON MULTILINE STYLES

5. Accept the defaults and draw a multiline as indicated.
   Current settings: Justification = Top, Scale = 1.00, Style = DEMO
   Specify start point or [Justification/Scale/STyle]: *1,1*
   Specify next point: *1,6*
   Specify next point or [Undo]: *5,6*
   Specify next point or [Close/Undo]: *5,1*
   Specify next point or [Close/Undo]: *[ENTER]*

   Notice the joints.

6. Repeat the *MLStyle* command.
   Command: *mlstyle*

7. Create a **New** style based on the **Demo** style – call it **Demo1**. **Continue** to the New Multiline Style dialog box.

8. Make these adjustments:
   - Remove the check from the **Display joints** check box.
   - Add a check to the **Line** check box under the **Start** column.
   - Add a check to the **Outer** and **Inner arcs** check boxes under the **End** column.
   - Change the **Angle** under the **Start** column to *45°*.

9. Pick the **OK** button to return to the Multiline Styles dialog box (the image in the demo box looks like this).

10. Make the **DEMO1** style current 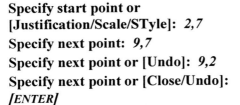 and complete the procedure.

11. Erase the last multiline you drew.

12. Draw a new multiline as indicated.
    Command: *ml*
    Current settings: Justification = Top, Scale = 1.00, Style = DEMO1
    Specify start point or [Justification/Scale/STyle]: *2,7*
    Specify next point: *9,7*
    Specify next point or [Undo]: *9,2*
    Specify next point or [Close/Undo]: *[ENTER]*

320

14.2B: MORE ON MULTILINE STYLES

Notice the angle of the start point is 45°. You can also see the line at the start point and the inner and outer arcs at the endpoint as you set in Step 8.	
13. Repeat Steps 6 and 7 (be sure to base your new style on **DEMO**, and call your new Style **Demo2**).	
14. Remove the check next to **Display joints** and set the **Fill color** to **Magenta**.	
15. Complete the command [OK]. (Be sure to set the **DEMO2** style current [Set Current].)	
16. Draw a new multiline. The multiline is filled.	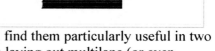
17. Exit the drawing without saving. **Command:** *quit*	

As you've seen, you can do a lot of things with multilines. You'll find them particularly useful in two fields – Piping and Architecture. But they are also quite useful in laying out multilane (or even single-lane) streets!

> And now the downside: Multilines have their limitations. The worst of these is the inability to set different layers for each of the individual lines. Because of this, centerlines can't be manipulated separately from outer lines. Another limitation is the need for a specific tool for some of the multiline modifications you'll want to make. This tool is called *MLEdit* and is the subject of the next section of our lesson.

14.3 Editing Multilines: The *MLEdit* Command

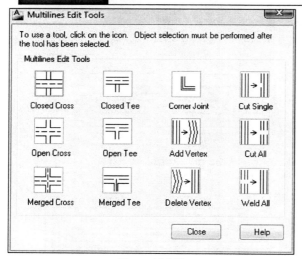

Figure 14.008

Where to Find It:	
Command Line:	*MLEdit*
Menu:	Modify – Object – **Multiline**

As I mentioned, editing multilines requires a special tool. Oh, some of AutoCAD's more basic modifying tools – such as *Copy* and *Move* – work on multilines, but the tools you might find most useful – *Trim* and *Extend* – work only on the multiline itself (collectively), not on the individual lines that compose it. When you try, AutoCAD automatically changes to a command line approach to the tools you'll see in this section.

This leaves you with two options – either *explode* the multiline (into individual objects as you would a polyline) or use the *MLEdit* command.

Exploding the multiline isn't a bad idea. In fact, once it's been drawn, there really isn't much reason to maintain its multiline definition except that, as a multiline, you can manipulate it as a single object. But let's look at the *MLEdit* command. The command calls the Multilines Edit Tools dialog box (Figure 14.008). The possibilities include three types of crosses and tees, corner creation, adding or removing a multiline vertex, breaking all or part of a multiline, and welding a broken multiline.

Let's see these in action.

Do This: 14.3A	Editing Multilines

I. Open the *MyMlines* file you created in the C:\Steps\Lesson14 folder. (If this file is unavailable, open *Mlines* in the same folder.)

II. For this exercise, we'll again forgo the usual format in the interest of brevity (and sanity). Follow these steps for each of the options:

   a. Enter the *MLEdit* command at the command prompt.
   b. Select the editing tool you wish to use (shown in the left column of our exercise table).
   c. At the **Select first mline** prompt, pick the lower part of the vertical multiline.
   d. At the **Select second mline** prompt, pick the right part of the horizontal multiline. (Hit ENTER after this to complete the command.) Your drawing will look like the image in the right column of our exercise table.
   e. *Undo* ↶ the last modification.
   f. Repeat Steps **a** through **e** for each of the tools in the left two columns of the Multilines Edit Tools dialog box, as well as for the Corner Joint tool.

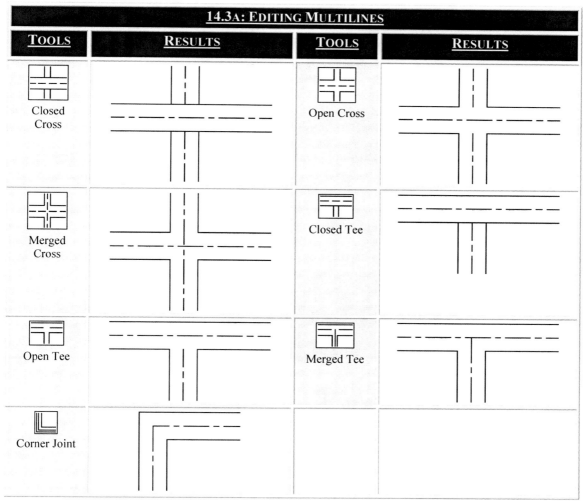

Are you getting a feel for the MLEdit tools? Let's take a more conventional look at the remaining tools.

Do This: 14.3B	More on Editing Multilines

I. Continue in the *MyMlines* file (or *Mlines*).
II. Follow these steps.

14.3B: MORE ON EDITING MULTILINES

1. Repeat the *MLEdit* command.
   Command: *mledit*

2. Select **Add Vertex** ‖→⟫.

3. Pick the midpoint of the horizontal line.
   Select mline: _mid of
   Complete the command.
   Select mline or [Undo]: *[ENTER]*

4. Stretch the middle of the horizontal line upward one unit.
   Command: *s*
   Your drawing looks like the figure at right. (Note: The **Stretch** procedure was to help you see the new vertex.)

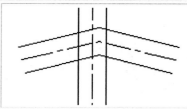

5. Repeat the *MLEdit* command.
   Command: *mledit*

6. Select **Delete Vertex** ⟫→‖.

7. Select the new vertex on the horizontal multiline.
   Select mline:
   Complete the command.
   Select mline or [Undo]: *[ENTER]*

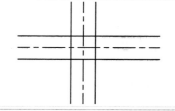

8. Repeat the *MLEdit* command.
   Command: *mledit*

9. Select **Cut Single** ‖→‖.

10. Select the left lower intersection of the two lines …
    Select mline: _int of

11. … then the right lower intersection.
    Select second point: _int of
    Select mline or [Undo]: *[ENTER]*
    Your drawing looks like the figure at right.

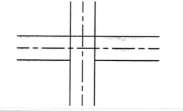

12. Undo the last command.
    Command: *u*

14.3B: MORE ON EDITING MULTILINES

13. Repeat the *MLEdit* command.
    Command: *mledit*

14. Select **Cut All**.

15. Select the left lower intersection of the two lines ...
    Select mline: _int of

16. ... then the right lower intersection.
    Select second mline: _int of
    Select mline or [Undo]: *[ENTER]*
    Your drawing looks like the figure at right.

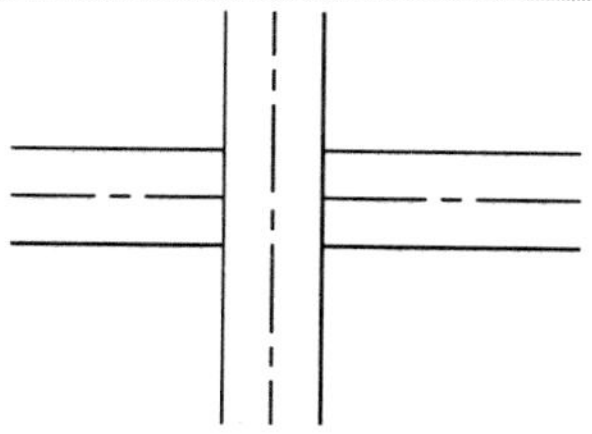

17. Repeat the *MLEdit* command.
    Command: *mledit*

18. Select **Weld All**.

19. Select the two endpoints of the multiline you just broke.
    Select mline:
    Select second point:
    Select mline or [Undo]: *[ENTER]*

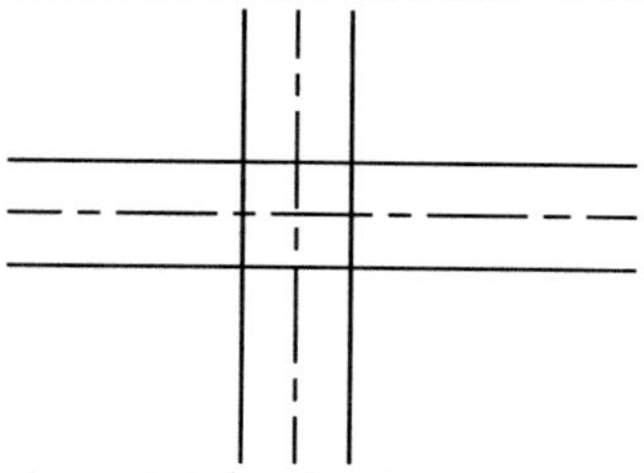

20. Quit the drawing without saving the changes.
    Command: *quit*

14.4 The Project

Now that you've had some practice in drawing and editing multilines, what do you think? Perhaps it's difficult to see the full benefits of these marvelous tools without some actual experience. Let's go back to our floor plan and add some walls!

Do This: 14.4A Add Some Walls

I. Open the *MyFlr-Pln 14* file you created in the C:\Steps\Lesson14 folder. (If this file is unavailable, open *flr_pln14* in the same folder.) The drawing looks like the figure at right.

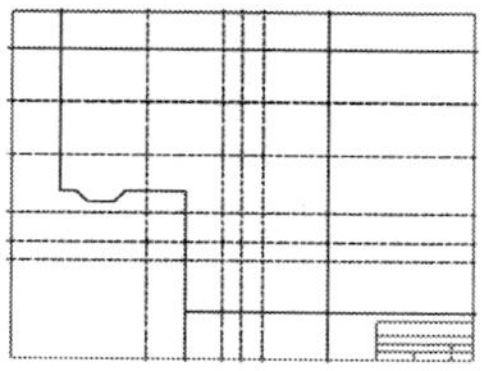

II. Be sure the **Walls** layer is current.

III. Follow these steps. (Do *not* explode the multilines!)

14.4A: ADD SOME WALLS

1. Enter the *MLine* command.
    Command: *ml*

2. Draw a 5½" wide multiline to show the outer walls. Use the construction lines to guide you. (Hint: Set the multiline scale [Scale] to *5.5* and use **Top** justification [Justification].)
    Specify start point or [Justify/Scale/STyle]:
    Your drawing looks like the figure at right.

3. Draw 4" wide multilines to show the inner walls. Use the construction lines to guide you. (Hint: Set the multiline scale [Scale] to *4* and use **Zero** justification [Justification].)
    Command: *ml*
    Your drawing looks like the figure at right.

4. Cut the walls for doors and windows, as shown in the following figure.

There are many possible ways to locate the openings. I suggest offsetting the construction lines according to the dimensions given, then using the *Trim* command to remove the doors and windows.

Floor plans just floor me!

14.4A: ADD SOME WALLS

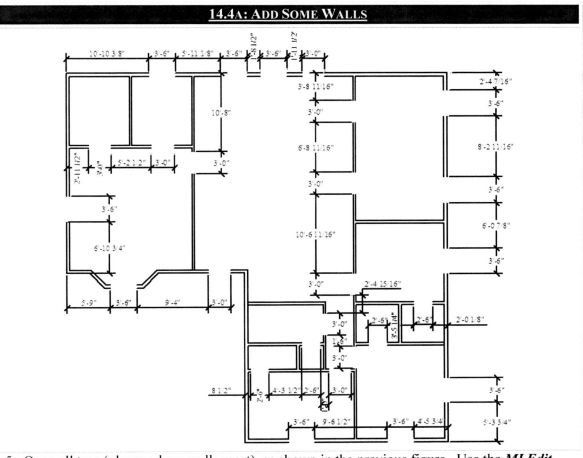

5. Open all tees (places where walls meet), as shown in the previous figure. Use the **MLEdit** command's **Open Tee** tool.

   Command: *mledit*

6. Erase the guidelines.

   Command: *e*

7. Save the drawing as *MyFlrPln15* in the C:\Steps\Lesson15 folder, and exit.

   Command: *saveas*

Well, that was fun! Does your drawing look more like a floor plan now? Are you beginning to feel like a CAD operator?

14.5 Extra Steps

- Go to a new subdivision in your area and pick up some sample floor plans from the sales office. Practice drawing them on the computer. You don't have to be exact on the dimensions (you'll only get rough dimensions on the plans anyway.)
- I know it takes a while to define styles for multilines, but now is a good time to establish a policy of never doing anything twice that can be done once and placed in a template (remember templates?). Anything that requires definition should be defined and put into a template to save time later. But does that mean you'll have thousands of templates from which to choose when creating a new drawing? Heavens, no! You should have a handful of templates – each with a lot

of information. (Remember, you can also use the Design Center to milk information from one drawing or template into another!)

Create some new templates that you might use on the job. Include units, grid and snap settings, limits, borders, and multiline definitions. But don't relax – there'll be more to add to your templates later!

14.6 What Have We Learned?

Items covered in this lesson include:

- *Commands*
    - *MLine*
    - *MLStyle*
    - *MLEdit*

This has been a fun lesson – and not nearly as difficult as others have been (or will be!).

I understand that you may be a bit frustrated that the last exercise didn't go quite as fast as you might like. My instructions weren't as detailed as some earlier exercises. I'm beginning to count on you to make a necessary transition.

> Many instructors rely heavily on the step-by-step approach. Others refuse to use it at all. They argue that, in the end, all the student learns is how to follow steps. So they rely on student self-reliance to accomplish the necessary tasks.
>
> We're using a combination of the step-by-step approach and self-reliance. I lead you at first, relying more and more on what you learn as we go.

After practicing the additional floor plans in the preceding *Extra Steps* section, you probably noticed that your speed increased as your comfort with multilines increased. Eventually, multilines will prove themselves to be a timesaving tool in your arsenal.

In our next lesson, we'll add some notes to our floor plan.

14.7 Exercises

1. Open the *FlrPln-HVAC* file in the C:\Steps\Lesson14 folder. Add the ductwork shown below to the floor plan. Don't add the detail – it's there as a guide only.

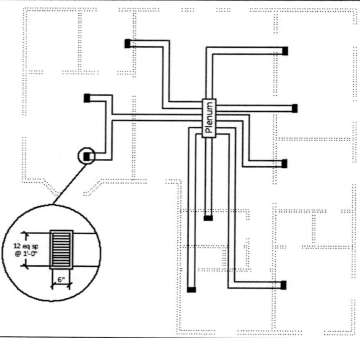

14.8 For Web-Based Review Questions and Additional Exercises, visit: http://foragerpub.com/AcadFiles/2011/2011.htm

Lesson 15

Following this lesson, you will:

- ✓ Know how to create and edit paragraph (multiline) text: The **MText** command

- ✓ Be familiar with AutoCAD's Spell Checker

- ✓ Know how to Find and Replace text in a drawing

- ✓ Know how to create Multiline Text Columns

Advanced Text - MText

In Lesson 4, we learned about text and text style. This information will remain quite useful – indeed, **Text** *will probably remain the primary method for creating call-outs on your drawings.*

But using **Text** *to create notes – that long list of construction information down the side of many drawings – can prove difficult. Editing a long list of notes can prove downright aggravating, especially if you must add or remove lines!*

When I was on the boards, we often sweet-talked the project secretary into typing our notes. We then photocopied them onto sticky transparencies that we attached to our drawings. This worked well

until revisions required additional notes or changing existing ones. Then we had to peel the sticky transparency from the drawing without damaging the paper.

With the **MText** *command, we have the benefits of the project secretary's typing abilities without the hassles of sticky transparencies. The secretary can type the notes using a computer's word processor (MS Word, WordPerfect, etc.), and then give us the file to import into our drawings.*

Of course, we can do the typing ourselves if the keyboard doesn't intimidate use. We can even share the secretary's dictionary to keep our spelling accurate (or at least consistent with the rest of the project). But why not take advantage of what's available – in many cases, this is a secretary's superior typing ability.

In this lesson, we'll see how to create, import, and edit notes in an AutoCAD drawing using the Multiline Text Editor.

15.1 AutoCAD's Word Processor: The Multiline Text Editor

AutoCAD took great pains to make their word processor behave like other word processors designed for the Windows environment (well, the latest processors anyway). Obviously, they couldn't incorporate the complete workings of an MS Word or WordPerfect; but the AutoCAD word processor works very much like Word despite its occasional shortcomings. So if you're familiar with MS Word, this lesson will move quickly and easily.

> A word processor is the computer equivalent of a typewriter. It's what you use to type letters, notes, résumés, and even AutoCAD books. There are several on the market, but the most common are Microsoft's Word and Corel's WordPerfect.

To access the Multiline Text Editor, follow this command sequence:

Command: *mtext*

Current text style: "Standard" Text height: 0.2000   Annotative: Yes

Specify first corner: *[You'll place a border around the area where you want the text to go; pick a corner of that border here.]*

Specify opposite corner or [Height/Justify/Line spacing/Rotation/ Style/Width/Columns]: *[Pick the other corner of the border.]*

A	Where to Find It:
Command Line:	*MText*
Hotkey(s):	*t* or *mt*
Ribbon (Tab/Panel):	Home – Annotation – **Multiline Text**
	Annotate – Text – **Multiline Text**
Menu:	Draw – Text – **Multiline Text**
Toolbar:	Draw – **Multiline Text**
	Text – **Multiline Text**
Tool Palette:	Draw – **MText**

AutoCAD first lets you know the style and text height you're using. Then it prompts for a border around the area in which to place the text. The **opposite corner** of the border is a bit tricky. There are a few options from which to choose – some refer to the bordered area, others refer to the text inside the border, and still others refer to both. Let's look at these options.

329

- The **Height** option allows you to set the text height. It prompts
   Specify height <0.2000>:
- The **Justify** option allows you to justify text within the border. The options are the same as the *Text* justification options, except that **Fit** and **Align** are missing. (See the text justification options in Figure 4.013, p.94.)

  The **Justify** option also controls the flow of the multiline text. For example, using the default **TL** (top left) option, AutoCAD anchors the text at the topmost and leftmost corner of the user-defined border. If the text entered is too large for the border, AutoCAD will automatically expand the border downward. Using the **BC** (bottom center) option causes the anchor point to be placed at the bottom and center of the border. Expansion of the border then occurs upward. Other justifications affect expansion in a similar manner.
- The **Line spacing** option provides an opportunity to control the spacing between lines of text. It prompts:
   Enter line spacing type [At least/Exactly] <At least>:
   Enter line spacing factor or distance <1x>:

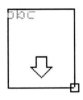

Figure 15.001

Figure 15.002

- The **Rotation** option allows you to control the rotation of the text just as it does within the *Text* command. It also controls the direction of expansion of the text border. For example, a downward-expanding border will actually expand toward the bottom of the text – or toward the right of the drawing when using a 90° text rotation (Figure 15.001 and 15.002).
- The next option is simple enough. **Style** allows you to specify the text style AutoCAD will use in the text box. Refer to Lesson 4, p.93, for more on text style.
- We'll look at **Columns** in some detail on p.337 and in Section 15.4, p.346.
- The **Width** option allows you to define the width of the text border more precisely. This option is more in keeping with the precision priority of CAD use.

You can reset most of these options from within the editor, as well.

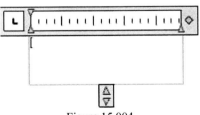

Figure 15.004

Figure 15.003

On completion of the text border, two things appear: the Multiline Text Ribbon (Figure 15.003) and Text Entry Window (Figure 15.004).

Let's start with the ribbon panels, which are packed with possibilities!

- The **Style** panel provides a selection box for picking the style of text you wish to use. The drop down menu provides convenient thumbnails of the available styles to assist you. Refer to Lesson 4, p.101, for more on setting up text styles.

You can also control the text height and annotation settings here.

- You'll probably use the **Formatting** panel more often than the others; it contains of the more common tools (so many, in fact, that it requires a subpanel to carry them all!).

- Use the **Font** control box to select from available fonts independent of the current style.
- Use the **Color** control box to assign a color independent of the current style.
- Use the remaining buttons on the panel – Bold **B**, Italic *I*, Underline U, Overline Ō, make Uppercase Aa, or make Lowercase aA – to assign formatting to your text.
- The **Background Mask** button calls a dialog box that allows you to set a background color for your text.
- Use the down arrow on the **Formatting** panel's title bar to access the subpanel (Figure 15.005), which includes a few more tools.

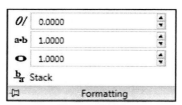

Figure 15.005

  - Use the **Oblique Angle** number box to set the slant of your lettering. Enter a value up to 85 – a positive value slants the text to the right, a negative value slants the text to the left. Again, all the tools on the Text Editor ribbon work independently of the current text style.
  - Use the **Tracking** number box to increase or decrease the spacing *between* individual letters independently of the text style you're using. Positive numbers increase the spacing; negative numbers decrease it. One is normal spacing.
  - Use the **Width factor** number box to increase or decrease the width *of* individual letters independently of the text style you're using. Again, positive numbers increase the width; negative numbers decrease it. One is normal width.
  - Use the **Stack** button to stack selected fractions.
- The **Paragraph** panel also has many useful tools that'll keep you coming back regularly. Here you'll set the overall layout of your text.
  - The **Justification** button calls a menu (Figure 15.006) which allows you to reset the justification of your text.
  - The **Paragraph settings** button calls the Paragraph dialog box (Figure 15.007) where you'll find standard word-processor definition tools for paragraphs.

Figure 15.006

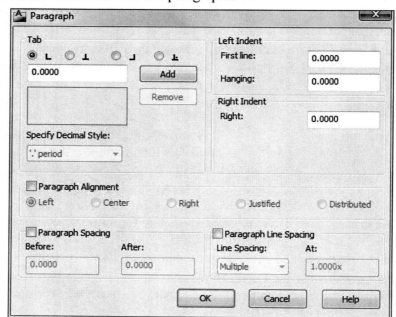

Figure 15.007

- The **Line Spacing** button calls another menu (Figure 15.008), which gives you a choice of standard predefined spacings or a link (**More…**) back to the Paragraph dialog box where you'll find more spacing tools.
- The **Bullets and Numbering** button calls yet another menu (Figure 15.009)! (You just thought the MText ribbon looked simple!) This menu bears closer scrutiny.
  - **Off** removes numbering from selected text.
  - **Numbered** enables you to use numbers rather than letters.
  - **Lettered** enables you to use letters rather than numbers. Use the flyout menu to toggle between upper and lower case letters.
  - **Bulleted** replaces numbers and/or letters with bullets (much as you're seeing on this page. Change the type of bullets with the TAB key on your keyboard.
  - **Start** restarts the numbers or letters of your list.
  - **Continue** does just that – it continues the previous sequence of numbers/letters in your list.
  - When **Allow Auto-list** has a check next to it, AutoCAD will attempt to recognize when you've manually begun a list and then continue the list for you. It watches for entries such as "a." or "1."
  - **Use Tab Delimiter Only** – limits the automatic creation of lists to use of the TAB key rather than the SPACEBAR.
  - **Allow Bullets and Lists** – automatically applies list formatting to anything that looks like a list. AutoCAD uses these criteria to identify lists: lines beginning with a number or symbol, use of a period after a number or letter, TAB spacing.
- The remaining buttons set the justification for the current line/paragraph – **Left**, **Center**, **Right**, **Full Justification**, and **Distribute** (evenly).
- Finally, the subpanel contains a single option – **Combine Paragraphs** – which makes a single paragraph out of multiple selected paragraphs.

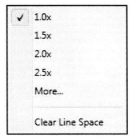
Figure 15.008

Figure 15.009

I know there're more panels to discuss, but let's put some text into our floor plan before we continue. First, however, if you're not familiar with word processors, please study the chart in Appendix D. This chart shows some keystrokes for maneuvering through and manipulating the text in the Multiline Text Editor. Familiarity with these keystrokes will save a considerable amount of time in Multiline Text Editing.

Do This: 15.1A	Using Multiline Text

I. Open the *FlrPln15* drawing the C:\Steps\Lesson15 folder (Vista or 7 users open the – *v* file). This is the same floor plan we've been working, but I've added a text style to expedite things.
II. Set the **Text** layer current.
III. Be sure the **Annotation Scale** for the drawing is set to ¼"=1'-0". (Use the **Annotation Scale** button on the status bar.)
IV. Follow these steps.

15.1A: USING MULTILINE TEXT

1. Enter the *MText* command **A**.
   Command: *t*

2. AutoCAD displays the **Current text style** and **Text height**, and then prompts for the **first corner** of the text border. Enter these coordinates.
   Current text style: "Calibri" Text height: 9" Annotative: Yes
   Specify first corner: *62',62'*

3. Tell AutoCAD you want to specify the **Width** [Width] of the text border …
   Specify opposite corner or [Height/Justify/Line spacing/Rotation/Style/Width/Columns]: *w*

4. … then set the **Width** to *23'*.
   Specify width: *23'*

AutoCAD presents the Multiline Text Editor with either the **Calibri** or **TIMES** style current.

5. Be sure AutoCAD is creating annotative text – the **Annotative Text** toggle on the **Style** panel should be lit or highlighted.

6. Set the font size (text height) to ¼" (**Style** panel).

7. Pick the **Center Justify** button (**Paragraph** panel), and then the **Bold** button **B** (**Formatting** panel). Notice that they appear lit when active.

8. Type *"The "*.

9. Pick the **Italic** button *I*.

10. Type *"Tara II "*.

11. Deselect the **Italic** button.

12. Type *"Floor Plan"*, and then hit ENTER to start a new line.

13. Pick the **Close Text Editor** button in the **Close** panel. The upper right corner of your drawing looks like the figure at right.

14. Remember to save occasionally.
    Command: *qsave*

    The *Tara II* Floor Plan

15. To return to the Multiline Text Editor for our notes, simply edit the text (double click on the text to edit it).

16. Hit the END key on your keyboard to go to the end of the current line; deselect the **Bold** button **B**.

17. Change the text height to 1/8".

18. Hit ENTER twice and pick the **Left Justify** button to locate your cursor on the second line below the text as indicated here.

    The *Tara II* Floor Plan

15.1A: USING MULTILINE TEXT

19. Now let's import some notes created by our summer intern. Pick **Import Text** from the cursor menu. AutoCAD displays a Select File dialog box. Open the *Notes.txt* file found in the C:\Steps\Lesson15 folder.

20. Complete the command ✖.
Your text looks like the figure at right.

The *Tara II* Floor Plan

NOTES:
SQUARE FEET:
BEDROOMS:
BATHS:
HVAC: Entire central heating and cooling system or heat pump ready to operate
PLUMBING: Builder to hook-up to MUD utilities
ELECTRICAL: Builder to instal range, range hood, bathroom exhaust fan, clothes dryer hook-up, and cable television and telephone jacks
INSULATION: To be fiberglass; R14 min. in all walls and R24 min in ceiling.
WALLS: Interior walls to be gipsum with builder to tape and float; Buyer to select paint from contractor options
DOORS & WINDOWS: Interior door to be hollow; exterior doors to be solid wood; windows to be sun-blocked; builder to provide storm doors
CABINETS: Buyer to select from builder options
FLOORING: Buyer to select fom builder options

21. Save the drawing 💾, but don't exit.
   Command: *qsave*

We've seen how easy it is to create Multiline Text both by keyboard entry and by importation. We created our text using the **Times/Calibri** style, but we could just as easily have used AutoCAD's default Standard style and changed fonts in the Multiline Text Editor itself.

We've also seen that we'll use the Multiline Text Editor both to create and to edit Multiline Text. Next, we'll look at some of our options for changing, or editing, our text.

Do This: 15.1B	Editing Multiline Text

I. Be sure your are in the *FlrPln15* drawing the C:\Steps\Lesson15 folder. If not, please open it now.
II. Review the material in Appendix D if you're unsure of the keystrokes required to move your cursor around the text. When unsure, use the arrow keys on the keyboard!
III. Follow these steps.

15.1B: EDITING MULTILINE TEXT

1. Return to the Multiline Text Editor. (Double click on the existing text.)

2. Select the text: **NOTES:**. (Be sure to get the colon, too.)

You can do this by holding down the SHIFT key and hitting the RIGHT ARROW key  six times. (Alternately, you can hold down the SHIFT key and hit the END key to select the whole line, or you can just hold down the left mouse button and drag over the text.)

3. Pick the **Bold B** and **Underline U** buttons.

4. Deselect the text (right).

5. Select the next three lines of text (including: **SQUARE FEET, BEDROOMS**, and **BATHS**).

An easy way to do this is to hold down the SHIFT key while hitting the DOWN ARROW ⬇ three times.

15.1B: EDITING MULTILINE TEXT

6. Make the text bold **B**. (You can pick the **Bold** button or use the CTRL+B keyboard method.)	
7. Place your cursor a space to the right of **SQUARE FEET:**, and deselect the **Bold** button.	
8. Type *1950*.	
9. Repeat Steps 7 and 8, adding the text *Four* next to **BEDROOMS**, and *Two* next to **BATHS**.	**The *Tara II* Floor Plan** NOTES: SQUARE FEET: 1950 BEDROOMS: Four BATHS: Two
10. Complete the command ✘. The text looks like the figure at right.	
11. Remember to save often 💾. **Command:** *qsave*	
12. Reopen the multiline text editor.	**The *Tara II* Floor Plan** NOTES: SQUARE FEET: 1950 BEDROOMS: Four BATHS: Two HVAC: Entire central heating and cooling system or heat pump ready to operate PLUMBING: Builder to hook-up to MUD utilities ELECTRICAL: Builder to instal range, range hood, bathroom exhaust fan, clothes dryer hook-up, and cable television and telephone jacks INSULATION: To be fiberglass; R14 min. in all walls and R24 min in ceiling. WALLS: Interior walls to be gipsum with builder to tape and float; Buyer to select paint from contractor options DOORS & WINDOWS: Interior door to be hollow; exterior doors to be solid wood; windows to be sun-blocked; builder to provide storm doors CABINETS: Buyer to select from builder options FLOORING: Buyer to select fom builder options
13. Make the rest of the capitalized words and colons bold **B**: **HVAC**, **ELECTRICAL**, **INSULATION**, **WALLS**, **DOORS & WINDOWS**, **CABINETS**, and **FLOORING**.	
14. Complete the command ✘. The multiline text looks like the figure at right.	
15. Now let's number our notes. Reenter the text editor.	NOTES: 1. SQUARE FEET: 1950 2. BEDROOMS: Four 3. BATHS: Two 4. HVAC: Entire central heating and cooling system or heat pump ready to operate 5. PLUMBING: Builder to hook-up to MUD utilities 6. ELECTRICAL: Builder to instal range, range hood, bathroom exhaust fan, clothes dryer hook-up, and cable television and telephone jacks 7. INSULATION: To be fiberglass; R14 min. in all walls and R24 min in ceiling. 8. WALLS: Interior walls to be gipsum with builder to tape and float; Buyer to select paint from contractor options 9. DOORS & WINDOWS: Interior door to be hollow; exterior doors to be solid wood; windows to be sun-blocked; builder to provide storm doors 10. CABINETS: Buyer to select from builder options 11. FLOORING: Buyer to select fom builder options
16. Select all the text below **NOTES**.	
17. Pick the **Bullets and Numbering** button ≔ and select **Numbered** from the menu. AutoCAD numbers the list, placing the tab at its default location.	

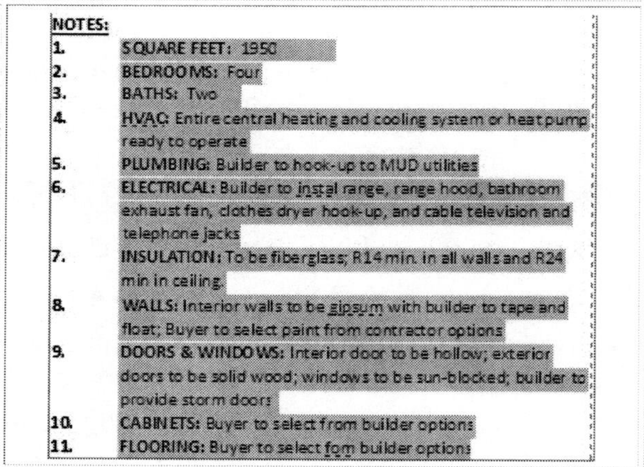

15.1B: EDITING MULTILINE TEXT

18. Let's fix the tab. (Refer to the figure at right.) Place your cursor over the Paragraph Indent marker. Move the marker to the 1'-0" location as shown.

19. Repeat Step 20 with the **Tab** marker.

20. Complete the procedure ✘. Your notes now look like the figure at right.

21. Save the drawing 💾 but don't exit.
    Command: *qsave*

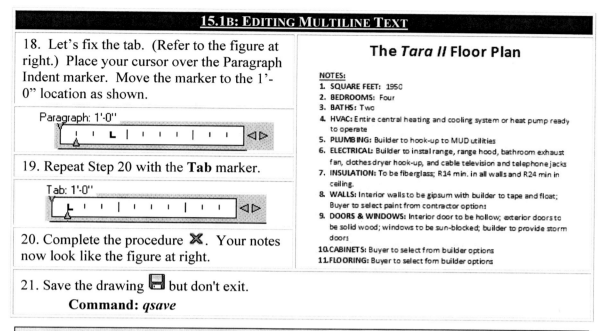

The *Tara II* Floor Plan

NOTES:
1. SQUARE FEET: 1950
2. BEDROOMS: Four
3. BATHS: Two
4. HVAC: Entire central heating and cooling system or heat pump ready to operate
5. PLUMBING: Builder to hook-up to MUD utilities
6. ELECTRICAL: Builder to install range, range hood, bathroom exhaust fan, clothes dryer hook-up, and cable television and telephone jacks
7. INSULATION: To be fiberglass; R14 min. in all walls and R24 min in ceiling.
8. WALLS: Interior walls to be gipsum with builder to tape and float; Buyer to select paint from contractor options
9. DOORS & WINDOWS: Interior door to be hollow; exterior doors to be solid wood; windows to be sun-blocked; builder to provide storm doors
10. CABINETS: Buyer to select from builder options
11. FLOORING: Buyer to select fom builder options

Note that you can also select large pieces of text by placing the cursor on one end of the text to select, holding down the left mouse button 🖱, and dragging to the other end.

Note also that Windows standard hotkeys also work in the Multiline Text Editor. For example, CTRL+B will make selected text bold/unbold, CTRL+I will italicize selected text, and so forth.

Now that you've seen the ease with which you can edit multiline text, what do you think? Let's see what else the Multiline Text Editor can do. We'll use the cursor menu in the next exercise.

Do This: 15.1C	Editing Multiline Text with the Cursor Menu

I. Be sure your are in the *FlrPln15* drawing the C:\Steps\Lesson15 folder. If not, please open it now.
II. Follow these steps.

15.1C: EDITING WITH THE CURSOR MENU

1. Return to the Multiline Text Editor. (Select the existing text.)

2. We want to move a section of text from one location to another. Let's take Note #4 and place it just above Note #9. First select Note #4.

3. Right click 🖱 anywhere in the text area of the Multiline Text Editor dialog box. AutoCAD will present a cursor menu. Pick **Cut** [Cut   Ctrl+X]. (Alternately, you can use the CTRL+X method.)

The selected text disappears. (Notice that AutoCAD automatically adjusts the numbering.)

4. Move your cursor just to the left of the word **CABINETS** (now Note #9).

5. Right click anywhere in the text area.

Pick **Paste** on the menu [Paste   Ctrl+V] (or CTRL+V).

The **HVAC** text appears. (If the numbering sequence appears off, put the cursor next to the first line that is out of sequence and select **Continue** [Continue] from the **Bullets and Numbering** flyout.)

> ### 15.1C: EDITING WITH THE CURSOR MENU
>
> 6. Close the Mtext Editor and save your drawing 💾 but don't exit.
>   **Command:** *qsave*

Okay. Let's look at the rest of the panels.

- The **Insert** panel has only three options.
    - The **Symbol** option @ calls a large menu (Figure 15.010) providing the most commonly used symbols in drafting. Not enough? Use the **Other...** option to call Windows' Character Map (which provides all the symbols available for every font on your computer!).
    - The next button allows you to **Insert [a] Field**. We'll discuss fields in Lesson 19, p.437.
    - The **Columns** button calls a menu (Figure 15.011) which allows you to define the columns you wish to use.

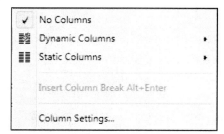

Figure 15.011

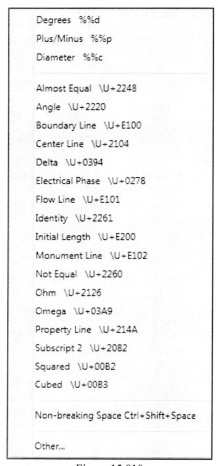

Figure 15.010

- **No Columns**, of course, means that AutoCAD won't divide your text into columns.
- **Dynamic Columns** – either **Manual height** or **Auto height** – allow you to resize the columns to accommodate your text. You can use the **Resize Height** tool in the text entry window to do this manually, or let AutoCAD do it with the **Auto height** option. I prefer to do it manually as I have a little more control over what I'm getting this way.
- **Static Columns** allows you to predetermine the number of columns AutoCAD will use when laying out your text. Select from a simple menu.
- **Insert Column Break** also gives you more control over your columns by providing a means for forcing a column break in your text.
- **Column Settings** calls a dialog box (Figure 15.012). This provides another way to determine what type of columns you'd like to use as well as some more specific tools for

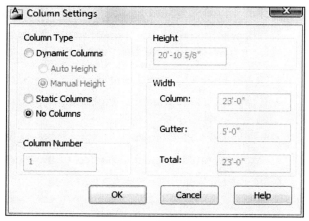

Figure 15.012

setting the number of columns (**Column Number**), **Height** of your columns (for **Auto Height Dynamic Columns** or **Static Columns**), and the **Width** of your **Column** and **Gutter**. (*Gutter* refers to a part of the page usually left empty for binding or spacing. Use it here to control the spacing between your columns.)

- The **Options** panel contains some simple, familiar tools – **Undo**, **Redo** and **Ruler** (to show the ruler across the top of the text entry window). This panel also includes a **More** button that calls the menu in Figure 15.013. Here you'll find tools to:
  o change your **Character Set** (ideal for oriental, Arabic, or other characters),
  o **Remove Formatting** from selected text.
  o change your **Editor Settings** for WYSIWYG, toolbar visibility, background, and text highlight color,

Figure 15.013

- You'll also find a **Spell Check** panel with an automatic **Spell Check** toggle to have AutoCAD check your spelling as you go, and an **Edit Dictionaries** button that's your ticket for changing your project dictionary. You'll see more on the dictionary in Section 15.5, p.348.
- The title bar arrow calls a dialog box which offers some control options for your spell checker. We'll discuss the dialog box in a moment.

A file created in most word processors can be saved in simple Text (*.txt*) format. This format strips away any formatting that may have been used in creating the file. You can, however, save the file in Rich Text Format (*.rtf*), which preserves the basic formatting. This was the format of choice in earlier releases.

Currently, however, AutoCAD allows us to copy text from many word processors (including MS Word) and insert it into AutoCAD *while maintaining the word processor's formatting (including paragraph alignment, bullets and lists, bold and italics, etc.)!*

Whenever possible, then, you should copy and paste text directly from the word processor; it will make the process easier! Unfortunately, it won't always be possible. So in our exercises, we'll use the more common *.txt* format when importing files so that we can study AutoCAD's methods for formatting text.

- Our final panel – **Tools** – offers some helpful tools.
  o **Find and Replace** calls the Find and Replace dialog box (Figure 15.014). Most of the tools in this box are fairly self-explanatory (but as with most things that are self-explanatory, they become much clearer with a bit of explanation).
    - **Find what** enables you to find a specific bit of text. The text for which you search must be located *after the flashing cursor in the Text Entry Window*.

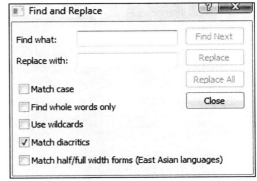

Figure 15.014

    - **Replace with** enables you to replace the text in the **Find what** box with a new string of text.
    - **Find whole words only** and **Match case** are filters that help streamline your search.
    - A check next to **Use wildcards** allows you to use standard windows wildcards (such as an asterisk – *) in your searches.
    - **Match diacritics** means that AutoCAD will include distinguishing marks in its search. (Diacritics are marks – such as accents – indicating special pronunciation. You frequently find them in Spanish, French, German, and other languages.)

- **Match half/full width forms** provides assistance for Asian language users.
- Once you've entered the text to locate in the **Find what** box and the appropriate text in the **Replace with** box (if desired), use the **Find Next** button to begin your search. If AutoCAD finds text to match your search, it'll highlight it.
- Use the **Replace** button to replace the located text with the text in the **Replace with** box.
- Use the **Replace All** button to replace all occurrences of the **Find what** text with the **Replace with** text.
- Use either the **Close** button or the exit **X** on the title bar to complete the procedure.

o The **Tools** panel subpanel offers options to **Import Text** (as you did in Exercise 15.1A) and an **AutoCAPS** tool that works like the CAPS LOCK on your keyboard – that is, it converts new or selected text to all caps.

Next, we'll look at the Find and Replace capabilities of the Multiline Text Editor. You'll really like this! The **Find** tool helps you find specific text. In a large group of notes, this is a real timesaver. The **Replace** tool allows you to replace that text with something new. Then we'll insert some symbols to wrap things up for this section.

Do This: 15.1D	Search and Replace Text

I. Be sure your are in the *FlrPln15* drawing the C:\Steps\Lesson15 folder. If not, please open it now.

II. Follow these steps.

15.1D: SEARCH AND REPLACE TEXT

1. Return to the Multiline Text Editor. (Select the existing text.)

2. Pick **Find and Replace** from the **Tools** panel. AutoCAD presents the Find and Replace dialog box (Figure 15.014, p.338).

3. Type *builder* into the box next to the words **Find what**. Then type *contractor* in the box next to the words **Replace with** as shown.

4. If we were searching for a word but didn't plan to replace that word with another, we would simply pick the **Find Next** button. But since we want to replace all instances of the word *builder* with the word *contractor*, we'll pick the **Replace All** button [Replace All]. (We won't need the filters at the bottom of the dialog box for this search.)

15.1D: SEARCH AND REPLACE TEXT

5. AutoCAD presents an information box telling you that it has completed its search. Pick the **OK** button [OK] to close that box, the **Close** button [Close] to close the Find and Replace dialog box, and then close the Multiline Text Editor ✖. The text now looks like the figure at right. 6. Save your drawing 💾 but don't exit. **Command:** *qsave*	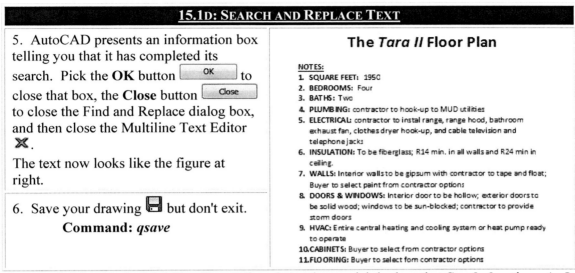

Before we finish with the Multiline Text Editor, let's take a quick look at that **Symbol** option. As I mentioned earlier, this comes in handy when you want to insert something a bit out of the ordinary.

Do This: 15.1E	Symbol Insertion

I. Be sure you're in the *FlrPln15* drawing the C:\Steps\Lesson15 folder. If not, please open it now.
II. Follow these steps.

15.1E: SYMBOL INSERTION

1. Return to the Multiline Text Editor. (Select the existing text.)
2. Move your cursor to the blank area just below **The *Tara II* Floor Plan**.
3. Reset the text height here to ¼".
4. Pick the **Symbol** button @ on the **Insert** panel.
5. Select the **Other** option [Other...].   AutoCAD presents the Windows Character Map.
6. Pick the down arrow in the **Font** control box. Scroll until you find the **Wingdings** font; select it.
7. Now select the happy face ☺ (it's located in different places depending upon the operating system you're using).
8. Pick the **Select** button [Select] 17 times, and then the **Copy** button [Copy]. Windows copies the string to the clipboard.
9. Close the Character Map ✖.

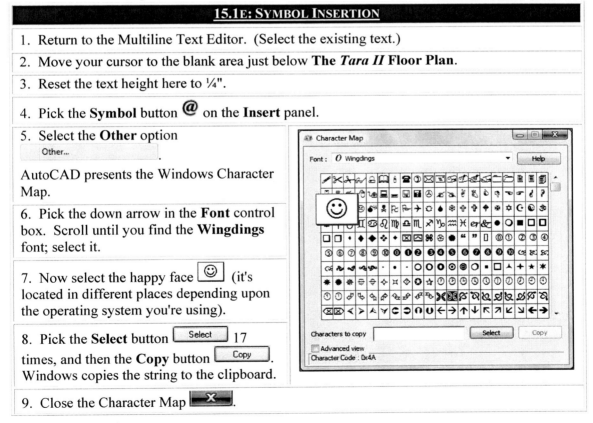

15.1E: SYMBOL INSERTION

10. With the cursor located on the blank line below the title, paste the clipboard image. (You can use the cursor menu we've discussed, or use the Windows CTRL+V method.) Make any adjustments needed to center the line and get rid of any extra lines the paste may have created.

11. Complete the command. The upper part of your text now looks like the figure at right. 12. Save your drawing 💾 but don't exit. **Command:** *qsave*	**The *Tara II* Floor Plan** ☺☺☺☺☺☺☺☺☺☺☺☺☺☺ NOTES: 1. SQUARE FEET: 1950 2. BEDROOMS: Four 3. BATHS: Two

15.2 Okay I Typed It, but I Don't Know If It's Right! – AutoCAD's *Spell* Command

The next aspect of text we must consider is AutoCAD's spell checkers. (You'll find two of them – an *on-the-fly* checker and a dialog box.)

ABC ✓	**Where to Find It:**
Command Line:	*Spell*
Hotkey(s):	*sp*
Ribbon (Tab/Panel):	Annotate – Text – **Check Spelling**
Menu:	Tools – **Spelling**
Toolbar:	Text – **Spell Check**

These remarkably simple tools can provide an inestimable service to those of us who never made it to the national spelling bee.

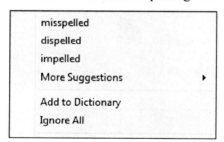

Figure 15.016

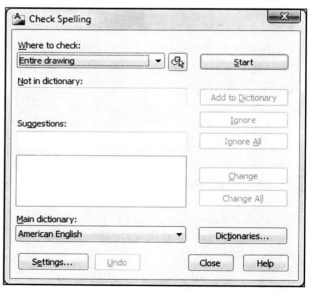

Figure 15.017

You may have noticed that AutoCAD underscores (with a dashed line) misspelled text. This is the on-the-fly checker. If you're familiar with MS Word, you probably know this to be (mostly) a useful tool. When the underscore appears, you can right click on the erroneous word for suggestions via a menu (Figure 15.016). In the menu, AutoCAD gives you the opportunity to select from a few suggestions (or a few **More Suggestions**), **Add** [the word] **to** [the] **Dictionary**, or **Ignore All** occurrences of the word.

AutoCAD also offers a more sophisticated dialog box approach to make it easy for you to perform a final spell check of the drawing before you issue it. It works in much the same way that spell checkers work in the major word processors.

The command opens the Check Spelling dialog box (Figure 15.017).

341

This box offers the same options as most spell checkers, but it also gives you the opportunity to change dictionaries. This can prove quite valuable to the project. It means that the project lead can assign a single person to control the project dictionary (usually that wonderful typist – the project secretary). That person creates the custom dictionary using a word processor. The secretary must then save the dictionary with a *.cus* extension.

AutoCAD needs this extension for recognition but your word processor should have no problem reading the file as well. (See Section 15.5, p.348, for a procedure to set up a custom dictionary.)

Let's take a look at the Spell Checker.

- Begin the spell checker by picking the **Start** button. AutoCAD will check what is indicated in the **Where to check** box (the **Entire drawing**, the **Current space/layout** or a **Selected object**). Use the Select objects button next to the **Where to check** control to select specific text to check.
- If AutoCAD finds a word that isn't in its dictionary, it presents the word in the **Not in dictionary** text box. It makes a few suggestions as to what word it thinks you may be trying to spell in the **Suggestions** list box with the word it thinks that you're most likely trying to spell in the **Suggestions** text box.
- You can pick a button to **Ignore** this word or **Ignore All** (ignore this spelling throughout this checking session), **Change** the word or **Change All** (change it every time it occurs in the selected text), or **Add** [it to the] **Dictionary**.
- You can also type a word into the **Suggestions** text box to replace the word AutoCAD has found.
- If you want to use the project dictionary, the **Dictionaries** button provides that opportunity.
- Use the **Settings** button to open the Check Spelling Settings dialog box (Figure 15.018). Here you can tell AutoCAD to include **Dimension text**, **Block attributes** (Lesson 22, p.501), or **External references** (Lesson 27, p.630). You can also tell AutoCAD what type of words to ignore with the **Options** check boxes.
- Finally, use the **Undo** button to undo an erroneous change within the spell checker.

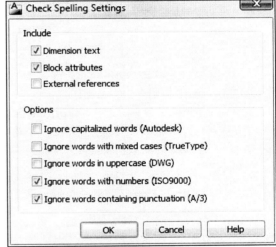

Figure 15.018

Let's take a look at AutoCAD's Spell Checker in action.

Do This: 15.2A	Checking Your Spelling

I. Be sure you're in the *MyFlrPln15* (or *FlrPln15*) drawing the C:\Steps\Lesson15 folder. If not, please open it now.
II. Follow these steps.

15.2A: CHECKING YOUR SPELLING

1. Enter the *Spell* command.
    Command: *sp*

AutoCAD presents the Check Spelling dialog box (Figure 15.017, p.341).

2. The spell checker works on all types of text in the drawing. Accept the default **Entire drawing** in the **Where to check** control box, and pick the **Start** button 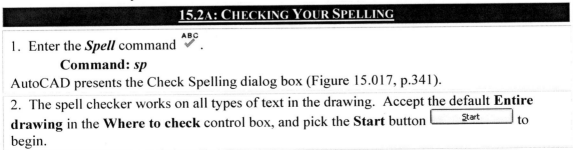 to begin.

15.2A: CHECKING YOUR SPELLING

3. (If AutoCAD identifies your happy faces as a misspelled word, pick the **Ignore** button.) AutoCAD finds **instal** and shows it in the **Not in dictionary** box with a suggested correction in the **Suggestions** box. (Notice that is adjusts your screen to show you where it found the error!) It also lists other possibilities in the list box. Notice that AutoCAD also highlights the found word in the drawing.

This was probably a typo so let's just pick the **Change** button [Change].

4. Change [Change] **gipsum** to **gypsum**.

5. **Ignore All** [Ignore All] occurrences of the abbreviation **HVAC**.

6. It looks like we omitted the **r** in the word **from**. Scroll down the list box to find the correct spelling. Pick on it and notice that it appears in the text box above.

7. Pick the **Change** button [Change] to fix the word.

8. Now AutoCAD presents a message box telling you that is has completed the spell check. Pick the **OK** button and then the **Close** button [Close] to complete the procedure.

9. Save your drawing 💾 but don't exit.
   Command: *qsave*

Now complete the text as seen in Figure 15.019, as follows:
- Place all text on the **Text** layer.
- Text heights are: ¼", 3/16", and 1/8".
- Put your own information into the title block (school name, your name, date, etc.).
- When you've finished, save the drawing as *MyFlrPln15* in the C:\Steps\Lesson15 folder.

We've come a long way, baby!

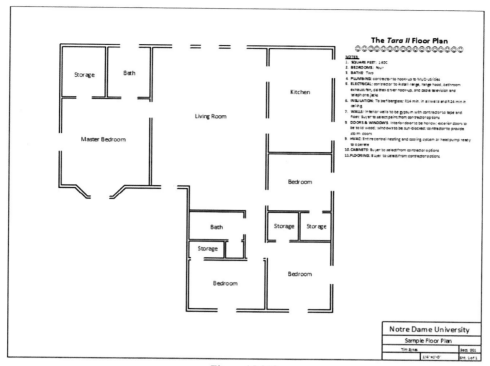

Figure 15.019

15.3 Find and Replace – without the Multiline Text Editor

Here's a piece of good news for those who spend a lot of time using text (rather than multiline text) for callouts – as with the spell checker, the Multiline Text Editor isn't required to perform a *Find and Replace*. The *Find* command will work regardless of the type of text it must search. It presents the Find and Replace dialog box (Figure 15.020, p.345).

Let's take a look.

Where to Find It:	
Command Line:	*Find*
Ribbon (Tab/Panel):	Annotate – Text – **Find Text [control box]**
Menu:	Edit – **Find**
Toolbar:	Text – **Find**

- Place the text you wish to locate in the **Find what** control box. Text that has been previously entered can be selected using the down arrow.
- Place the new text (the text you want to use as a replacement for found text) in the **Replace with** control box. Again, text that has been previously entered can be selected using the down arrow.
- Use the **Find where** control box or the **Select objects** button (next to the control box) to define your search area. (That is, where do you want to search?)
- The **More Options** button presents the Search Options and Text Types frames (Figure 15.020, p.345). Here you can filter the selection set to include certain types of text. A great benefit is the inclusion of the **Block Attribute Value** option that allows you to find and replace attribute values (more on attributes in Lesson 22, p.500).
- You can also tell AutoCAD to **Match case** (match capital letters in the text) or **Find whole words only** in the search.
- A check next to **List results** will tell AutoCAD to provide a list box showing the results of your searches.

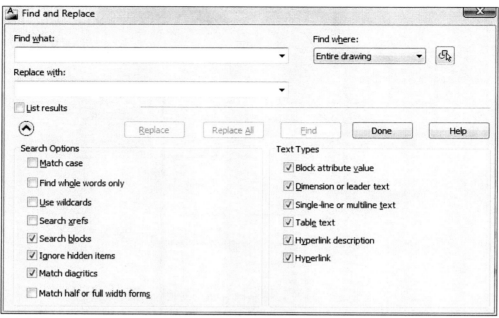

Figure 15.020

Let's try the Find and Replace tool.

Do This: 15.3A	Find and Replace Text without the Multiline Text Editor

I. Be sure your are in the *FlrPln15* drawing the C:\Steps\Lesson15 folder. If not, please open it now. If you haven't finished this drawing, open *FlrPln15.3.dwg* in the same folder.
II. Zoom all.
III. Follow these steps.

15.3A: FIND AND REPLACE TEXT WITHOUT THE EDITOR

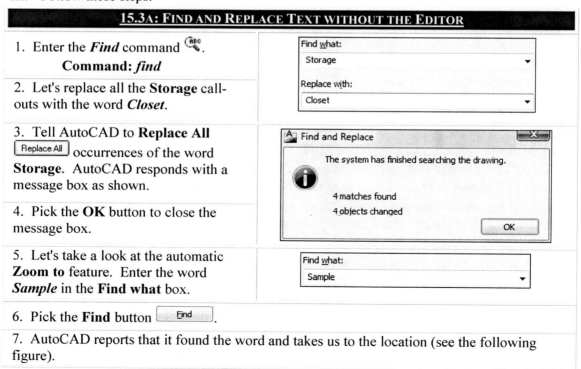

1. Enter the *Find* command.
   Command: *find*

2. Let's replace all the **Storage** call-outs with the word *Closet*.

3. Tell AutoCAD to **Replace All** occurrences of the word **Storage**. AutoCAD responds with a message box as shown.

4. Pick the **OK** button to close the message box.

5. Let's take a look at the automatic **Zoom to** feature. Enter the word *Sample* in the **Find what** box.

6. Pick the **Find** button.

7. AutoCAD reports that it found the word and takes us to the location (see the following figure).

15.3A: FIND AND REPLACE TEXT WITHOUT THE EDITOR

8. Pick the **Done** button to complete the procedure.

9. Zoom all to see the results of your work and then save the drawing .
   Command: *qsave*

15.4 Columns

Columns make placing and editing text a dream! In the olde days (oh, there he goes again – this guy must be a million-years-old!) ... *in the olde days* we had to manually move text from one column to the next when our editing increased or decreased the column length beyond acceptable limits. No More! Now your text will automatically adjust itself! In fact, you can create dynamic columns that can automatically or manually adjust themselves.

Access columns with the **Columns** button on the **Insert** panel of the ribbon's **MText** tab (note: this is *not* the ribbon's **Insert** *tab*). You'll see a small menu first (Figure 15.021). Here you can:

- Select **No Columns.**
- Select **Dynamic Column** options. Dynamic columns are text driven; columns are added or removed as the columns are adjusted. Options include **Auto Height** or **Manual Height.**
- Set from two to six **Static Columns.**
- Insert [a] **Column Break** which will force succeeding text into the next column.
- Call the Column Settings dialog box (Figure 15.012, p.337) with the **Column Settings** option.

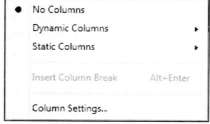

Figure 15.021

Let's import some text that requires columns to fit on the drawing. We'll need a lot of text so I decided to use the prelude from my last SF.[*] (I know, it's a shameless marketing ploy, but it does provide a few pages with which to work.)

Do This: 15.4A	Using Multiline Text Columns

1. Open the *ColumnsDemo* drawing in the C:\Steps\Lesson15 folder.

---

[*] *Sara: The Companion of God,* by Timothy Sean Sykes; Forager Publications; Copyright © 2007.

II. Follow these steps.

15.4A: MULTILINE TEXT COLUMNS

1. Enter the *MText* command A. (Accept the default annotation scale.)
   Command: *t*

2. Locate the text as indicated.
   MTEXT Current text style: "Calibri" Text height: 0.1875 Annotative: Yes
   Specify first corner: *1,15.75*
   Specify opposite corner or [Height/Justify/Line spacing/Rotation/Style/Width/Columns]: *w*
   Specify width: *4.75*

3. Set the **Text height** to 1/8" `.125` and the **Color** to **ByLayer** `ByLayer`.

4. Pick the **Columns** button. AutoCAD presents the column menu (Figure 15.021).

5. Select the **Column Settings** option. AutoCAD presents the Column Settings dialog box (see the Step 6 figure).

6. We'll let AutoCAD determine how many columns to use dynamically. Use the **Auto Height** option; assign a **Height** of 14", a **Column Width** of 4¾" and a **Gutter Width** of ½". AutoCAD will do the rest!

7. Pick the **OK** button to continue. Notice that AutoCAD begins with a single column.

8. Now let's import the text. Select **Import Text...** from the cursor menu...

9. ...and select the *Prelude-Sara.rtf* file in the C:\Steps\Lesson15 folder. (Be sure to search for RTF files.) If the **Text height** changes, select all the text – CTRL+A – and reset it to 1/8". Notice that AutoCAD provides additional columns.

10. Let's break up the column at the natural break (the row of stars). Place your cursor on the first line following the stars and pick **Insert Column Break** on the **Columns** menu (or you can hold down the ALT key while hitting the ENTER key on your keyboard).
Notice that AutoCAD forces the succeeding text into the next column.

11. Now pick on the Resize Height arrows in the lower left corner of the Text Entry Window and drag them upward until the text fills the last column.

12. Close the Multiline Text Editor and save the drawing which now looks like the following figure. (Enjoy the preview of *Sara!*)

15.4A: MULTILINE TEXT COLUMNS

Take some time to play with the other column options. I guarantee you'll find them useful!

15.5 Extra Steps

You can impress the boss by being the only one on the contract capable of setting up the project to use a shared dictionary. Here's how to tell AutoCAD to use a custom dictionary.

1. On the **Spell Check** panel of the **Text Editor** tab, pick the **Edit Dictionaries** button. AutoCAD presents a Dictionaries dialog box.
2. Pick the down arrow in the control box below the words **Current custom dictionary**.
3. Select **Manage custom dictionaries** from the menu. AutoCAD presents a Manage Custom Dictionaries dialog box.
4. Pick the **Add** button. AutoCAD presents a Windows standard Open Files dialog box (this one is called Add custom dictionary file).
5. Open the *Custom.cus* file in the C:\Steps\Lesson15 folder.
6. AutoCAD returns to the Manage Custom Dictionaries dialog box. Pick the **Close** button.
7. AutoCAD returns to the Dictionaries dialog box. Select the dictionary you'd like to use in the Current custom dictionary control box (select *custom.cus*). and close the dialog box.
8. AutoCAD returns to the Check Spelling dialog box. You can now continue with the spell check or exit the dialog box.

This little trick is guaranteed to impress the boss and save quite a bit of project time. Good luck!

15.6 What Have We Learned?

Items covered in this lesson include:

- Multiline text creation and editing
- AutoCAD's spell checker
- Commands:
  - *MText*
- How to find and replace text
- How to use Multiline Text Columns

  - *Spell*
  - *Find*

If you were already familiar with word processors, this has probably been a fairly easy – if not downright boring – lesson. The only new thing for you would have been the *Extra Steps* part of the lesson where we learned to share dictionaries.

But if you weren't familiar with word processors, this has probably been one of the easier lessons. The decision to use **MText** or **Text** won't always be an easy one. **Text** is less complicated (once the styles have been created); **MText** has more capabilities. I would suggest that any time it's a toss up, go with the easier **Text** (assuming, of course, that the style you need has already been created). AutoCAD, however, tends to favor the **MText** command. So it's really a question of preference. Do what is easier for you.

Take a break before you start the next lesson. In Lesson 16, we'll look at dimensioning!

15.7	Exercises

1. Open the *directions2* file in the C:\Steps\Lesson15 folder. Create the notes shown at right.
    1.1. Find the arrows at top in the Wingdings font. Make them 0.15" in height.
    1.2. Find the skull and crossbones images at the bottom in the Wingdings font. Make them 0.25" in height.
    1.3. All other text uses the Calibri or Times font at a height of 0.1".
    1.4. The width of the text box is 2.25".
    1.5. Save the drawing as *MyDirections2.dwg* in the C:\Steps\Lesson15 folder.

    ↑↓↖↗↙↘↑↓↖↗↙↘
    <u>Getting There</u>
    Slide eerily south on *Crane*, quietly sneaking up on *Ichobod*. Quickly now, turn right before the headless horseman sees you! Whew! You made it!
    You probably won't see the *Ghostly Drive* till it's too late, so turn and ooze down the *Vampire Lane* (watch for low flying bats).
    Ever so cautiously now, take a left on *Spooky St*. Don't go too far - people have gone down the *Graveyard Way* and never returned!
    We're at the end of *Spooky St.*. Park anywhere. Com'n in (just ignore the hooting of the owls)!

2. Open the *MyHolder2* file you created in Lesson11. It's in the C:\Steps\Lesson11 folder. (Alternately, you can use the *holder15* in the C:\Steps\Lesson15 folder.) Create the notes shown at right.
    2.1. Use the Calibri or Times font.
    2.2. Use a text height of 1/8".
    2.3. The width of the text box is 3.25".
    2.4. Save the drawing as *MyHolder15* in the C:\Steps\Lesson15 folder.

    <u>Notes:</u>
    1. All materials to be hot dipped galvanized steel.
    2. Tube to be 3/4"dia sch. 40 pipe.
    3. Slots to be clean and filed.
    4. Paint in accordance with HI spec. #HDGPS 10302871.
    5. All dimensions are +/- .002".
    6. Product to be stamped below with product code & model number: HSV5-Mod2A17.
    7. Product to be stamped below with serial number.
    8. Serial numbers to be sequential beginning with the number: SN:001001.

15.8	**For Web-Based Review Questions and Additional Exercises, visit: http://foragerpub.com/AcadFiles/2011/2011.htm**

Lesson 16

Following this lesson, you will:

- ✓ Know how to create dimensions in AutoCAD using a host of dimensioning tools
- ✓ Know the difference between associative and normal dimensions
- ✓ Know how to edit dimensions using AutoCAD's **DimEdit** and **DimTEdit** commands
- ✓ Know how to dimension an isometric

Basic Dimensioning

Let's see how you measure up!

Creating dimensions on a drawing board requires a certain amount of expertise, a bit of patience, a scale, a calculator, and a good eye. If the dimension doesn't fit between dimension lines just right, the drafter can always scrunch it in a bit. If the drafter is experienced, he or she can often do this without the results being flagrantly noticeable. The board drafter can also cheat on a dimension fairly easily. Indeed, the company for which I worked in the early eighties maintained a 3" plus or minus tolerance for dimensioning.

AutoCAD dimensioning also requires a certain amount of expertise. But in contrast to board work, the CAD operator needs a lot of patience, no scale or calculator (AutoCAD will perform the calculations), and AutoCAD's "eye" for precision. If a dimension doesn't fit between dimension lines just right, some (often) complex maneuvers are required to reposition it. You can cheat on a dimension fairly easily, but it involves overriding AutoCAD's precision and is rarely a good idea.

AutoCAD's dimensioning tools come in three types: Creation, Editing, and Customization.

In the dimension Creation category, AutoCAD provides several different dimensioning commands. Use these to actually draw the dimensions.

The Editing category includes two dimension-editing commands. Use these to change, reposition, or reorient the dimension.

The Customization category is one that'll allow you to create dimension styles. Creating dimension styles involves several tabs of a dialog box for each style you create. You'll use these tabs to set at least 60 dimension variables (dimvars).

But as complex as dimensioning is, it's not as complicated as it seems! We'll cover the Creation and Editing categories in this lesson. We'll save Customization for Lesson 17, p.376.

Let's get right to it!

16.1   First, Some Terminology

If you're an experienced drafter, you may already be familiar with dimension terminology. But I've provided the drawing in Figure 16.001 for the novice (and as a review for those who've been drafting for so long that the textbook terminology has been long forgotten).

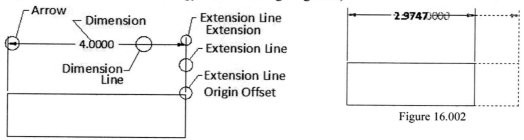

Figure 16.002

Figure 16.001

The dimension shown in this drawing, and those used throughout this lesson, use AutoCAD defaults for everything from location of the **Dimension** to the size of the **Extension Line Origin Offset**. (I have, however, substituted the default font with one that's a bit easier on the eyes.) You can see that AutoCAD uses **Arrows** by default rather than the slashes more often seen in architectural drafting. AutoCAD also defaults to decimal units (regardless of the units the drawing has been set up to use), a 0.18" text height, a 0.18" **Extension Line Extension** and **Arrow** length, and an **Extension Line Origin Offset** of 0.0625".

A new term with which you should become familiar is *Associative*. Associative dimensions are tied to the objects being dimensioned. As you stretch or move those objects, the dimension will change or move with the object. (Notice the dimension as the box is stretched in Figure 16.002.) Additionally, when associative dimensioning is active, AutoCAD uses true dimensions and doesn't prompt for verification.

The **DimAssoc** system variable controls the behavior of both associative and *normal* dimensions. There are three settings.

- Dimensions created with a **DimAssoc** value of **2** (the default) will be fully (truly) associative. That is, AutoCAD will invisibly connect the dimension to the dimensioned object. When you move or modify the object, the dimension will adjust accordingly. Treat these dimensions as single objects (much like polylines).

- Technically, dimensions created with a **DimAssoc** value of **1** will not be associative, although they'll behave as though they were. A setting of **1** means that AutoCAD actually associates the dimensions with definition points (*defpoints*) placed automatically at the extension line origin (see Figure 16.001, p.351). For AutoCAD to adjust the dimension when you modify the object, you must include the defpoints in the modification. You should also consider these dimensions to be single objects.

- AutoCAD will insert dimensions created with a **DimAssoc** value of **0** in an "exploded" format. That is, dimensions won't be associated with the object being dimensioned. We call these dimensions non-associative or *normal* dimensions. They are *not* single objects. Treat them as individual lines, text, and arrowheads.

> To convert from associative to normal, *explode* the dimension (just as you exploded a polyline into lines). You'll then find it necessary to change the layer of the separate objects to the dimension layer and the color of the object to **ByLayer**.

When associative dimensions were used in all releases prior to 2002, AutoCAD employed the defpoints method. To make legacy dimensions – dimensions created in earlier releases of AutoCAD – truly associative, use the *DimReAssociate* command. It looks like this:

> **Command:** *dimreassociate*
> **Select dimensions to reassociate ...**
> **Select objects:** *[Select the dimension object(s).]*
> **Select objects:** *[Enter to complete the selection set.]*
> **Specify first extension line origin or [Select object] <next>:** *[AutoCAD will identify it for you and you can accept it, or you can pick the first extension line origin.]*
> **Specify second extension line origin <next>:** *[Again, AutoCAD will identify it for you and you can accept it, or you can pick the second extension line origin.]*

AutoCAD provides two additional commands to assist you with associative dimensions:

- *DimDisAssociate* allows you to change the type of association from true (a **DimAssoc** setting of **2**) to the defpoints method (a **DimAssoc** setting of **1**). Use the *DimReAssociate* command to convert the other way.

- Use *DimRegen* to regenerate just the associative dimensions in a drawing (without regenerating other objects).

But perhaps the most important thing to remember about associative dimensioning is that, like polylines and splines, associative dimensions behave as a single object. That is, a single pick will select the dimension for erasure or modification. The problem with this lies in the way the single dimension object responds to modification commands like *Break* and *Trim*. Simply put, you need a special tool – *DimBreak* – to break or trim extension lines to allow room for text. We'll see this in action in Section 16.4.3, p.371. Associative dimensions require other special modification tools, too. We'll discuss these in Section 16.4, p.368.

Unlike associative dimensioning, AutoCAD creates normal dimensions as separate objects (lines, arrow, text, etc.). Therefore, you can break or trim the extension lines as needed. Normal dimensioning requires that you manually update the dimension when you edit the objects dimensioned. In other words, normal dimensions won't automatically update as the object is stretched. Additionally, although normal dimensioning reads the distance between the extension lines just as associative dimensioning does, AutoCAD will automatically prompt you with the true dimension and give you the opportunity to override it with another number.

See the following chart for a comparison of associative, defpoints, and normal dimensions.

MODIFICATION	ASSOCIATIVE OR DEFPOINT DIMENSION	NORMAL DIMENSION
Dimension will automatically update when stretched	Yes	No
Extension lines can be trimmed or broken to allow room for text in a crowded drawing	Yes	Yes
Automatically prompts to allow the true dimension to be overridden	No	Yes

The CAD coordinator for the job (the guru) and the project manager will usually decide the question of which type of dimension to use. The CAD operator should check with one of them before entering the wrong type of dimension since, although it can be tedious to convert from associative to normal, you can't convert the other way at all!

16.2   Dimension Creation: Dimension Commands

AutoCAD's dimension commands can be difficult to remember, but the ribbon has pictures to show what each does. I strongly suggest getting comfortable with that approach – or the older toolbar approach – rather than trying to memorize dimension commands.

16.2.1   Linear Dimensioning

The workhorse of AutoCAD dimensioning is the *DimLinear* command. Use it to draw all your horizontal and vertical dimensions. The command sequence is

> **Command:** *dimlinear*
> **Specify first extension line origin or <select object>:** *[Select the first dimension point.]*
> **Specify second extension line origin:** *[Select the second dimension line origin.]*
> **Specify dimension line location or [Mtext/Text/Angle/Horizontal/Vertical/Rotated]:** *[Locate the dimension line.]*
> **Dimension text = X.XXXX** *[AutoCAD reports the dimension value and places the dimension.]*

Where to Find It:	
Command Line:	*DimLinear*
Hotkey(s):	*dli*
Ribbon (Tab/Panel):	Home – Annotation – **Linear** (flyout)
	Annotate – Dimensions – **Dimension** (flyout)
Menu:	Dimension - **Linear**
Toolbar:	Dimension - **Linear**

- The first option AutoCAD presents is easily missed. AutoCAD prompts for the **First extension line origin**. Most users stop there, select the **first** and **second extension line origins** and proceed with the dimension. But the second part of that first prompt reads **or <select object>**. This option allows you to simply select an object to dimension. If you press ENTER and select an object, AutoCAD determines where the object's endpoints lie and places the extension lines accordingly.

The next set of options occurs after the extension line origins are located.

- AutoCAD will automatically determine whether the dimension should be **Horizontal** or **Vertical** by the location of the crosshairs. However, you can override AutoCAD's determination by typing **H** or **V**.
- The **Text** and **MText** options allow you to override AutoCAD's automatic determination of the actual dimension.

  The dimension won't automatically update if you override here. I don't recommend doing this as it defeats the purpose of associative dimensioning.
- The **Angle** option allows you to rotate the dimension text for a better fit (Figure 16.003). This handy tool is far too often overlooked.
- The **Rotated** option allows you to measure a dimension at an angle to the object other than 90°. (See Figure 16.003 for a comparison of the **Angle** and **Rotated** options.)

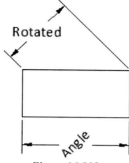

Figure 16.003

Let's look at the *Dimlinear* command.

> I'll use a dot-style grid for clarity in Lessons 16 and 17. Refer to Lesson 3 (Section 3.2, p.58) for details on using a dot-style grid.

Do This: 16.2.1A Linear Dimensioning

I. Open the *linear* file the C:\Steps\Lesson16 folder. The drawing looks like the figure at right.
II. Follow these steps.

16.2.1A: LINEAR DIMENSIONING

1. Enter the *Dimlinear* command .

   Command: *dli*

2. Pick the center of the circle at Point 1 ...

   Specify first extension line origin or <select object>:

3. ... then the endpoint at Point 2.

   Specify second extension line origin: _endp of

4. Locate the dimension line four grid points below the bottom line of the object.

   Specify dimension line location or [Mtext/Text/Angle/Horizontal/Vertical/Rotated]:
   Dimension text = 8.0000

   (Note: In these dimension examples, I've adjusted the text and arrowhead sizes for better viewing.)

5. Repeat the *Dimlinear* command.

   Command: *[ENTER]*

16.2.1A: LINEAR DIMENSIONING

6. Let's use the **select object** option.
   Specify first extension line origin or <select object>: *[ENTER]*

7. Select the angled line on the right end of the object.
   Select object to dimension:

8. Pick a point to the right.
   Specify dimension line location or [Mtext/Text/Angle/Horizontal/Vertical/Rotated]:
   Dimension text = 3.0000

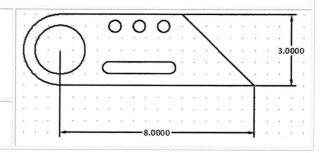

9. Save the drawing 💾 but don't exit.
   Command: *qsave*

You can use the Properties palette or Quick Properties panel to change one or both extension lines to centerlines if you'd prefer them (at the circle, perhaps?).

16.2.2 Dimensioning Angles

Our next dimension command is ***DimAngular***. You'll use this to dimension angles as well as angular dimensions on circles and arcs. The command sequence for dimensioning an arc or circle is

Where to Find It:	
Command Line:	*DimAngular*
Hotkey(s):	*dan*
Ribbon (Tab/Panel):	Home – Annotation – (Linear flyout) **Angular**
	Annotate – Dimensions – (Dimension flyout) **Angular**
Menu:	Dimension – **Angular**
Toolbar:	Dimension – **Angular**

    Command: *dimangular*
    Select arc, circle, line, or <specify vertex>: *[Show AutoCAD what you want to dimension or hit ENTER to select the vertex first.]*
    Specify second line: *[This prompt appears only if a circle is being dimensioned.]*
    Specify dimension arc line location or [Mtext/Text/Angle/Quadrant]: *[Locate the dimension line; Quadrant locks the extension lines into a specific area regardless of where you place the dimension.]*
    Dimension text = *[AutoCAD reports and places the dimension.]*

If you select a **line** at the first prompt, AutoCAD prompts:
    Select second line:

Select a line that forms an angle with the first selected line, and then locate the dimension as prompted.

If you opt to select the vertex first, AutoCAD prompts:
    Specify angle vertex: *[Select the vertex.]*
    Specify first angle endpoint: *[Select a point on the first line.]*
    Specify second angle endpoint: *[Select a point on the second line.]*

The **MText**, **Text**, and **Angle** options are the same as in the ***Dimlinear*** command.
Let's look at the ***Dimangular*** command.

Do This: 16.2.2A	Dimensioning Angles

I. Be sure you're in the *linear* file the C:\Steps\Lesson16 folder. If not, please open it now.
II. Follow these steps.

16.2.2A: DIMENSIONING ANGLES

1. Enter the *Dimangular* command ⌁.
   Command: *dan*

2. Select the arc on the left end of the object.
   Select arc, circle, line, or <specify vertex>:

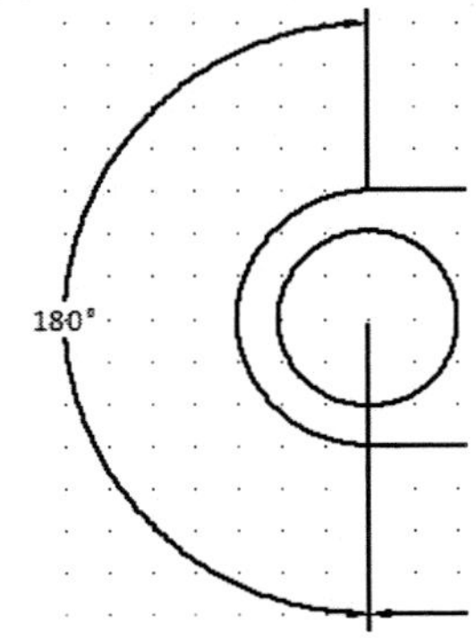

3. Place the dimension at the end of the bottom linear dimension as shown.
   Specify dimension arc line location or [MText/Text/Angle/Quadrant]:

4. Now let's dimension the angle at the other end of the object. Repeat the command ⌁.
   Command: *[ENTER]*

5. Select the bottom line on the object ...
   Select arc, circle, line, or <specify vertex>:

6. ... then the angled line.
   Select second line:

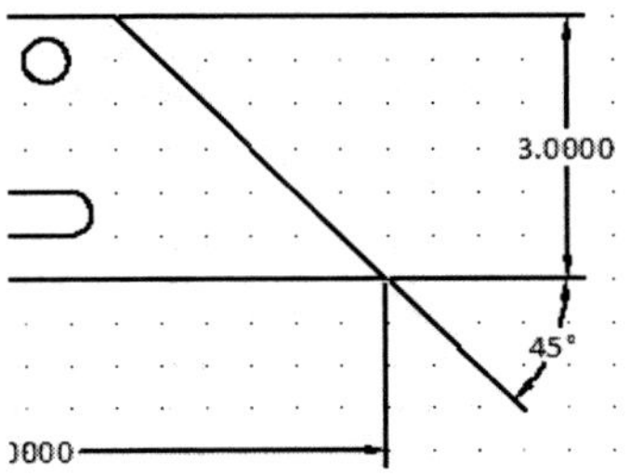

7. (Refer to the drawing next to Steps 5 and 6.) Move the cursor around and notice how you can reposition the dimension. Let's lock the dimension in the position shown. Select the Quadrant option [Quadrant].
   Specify dimension arc line location or [Mtext/Text/Angle/Quadrant]: *q*

8. Place the dimension in the position shown and pick with the left mouse button.
   Specify quadrant:
   Notice that now the extension lines remain in this quadrant regardless of where you move the cursor.

9. Now place the dimension.
   Specify dimension arc line location or [Mtext/Text/Angle/Quadrant]:
   Dimension text = 45

10. Save the drawing 💾 but don't exit.
    Command: *qsave*

16.2.3 Dimensioning Radii and Diameters

Very little difference separates dimensioning radii and dimensioning diameters. In fact, the command sequences are identical:

> **Command:** *dimradius* [or *dimdiameter*]
> **Select arc or circle:** *[Show AutoCAD what you want to dimension.]*
> **Dimension text =** *[AutoCAD reports the dimension.]*
> **Specify dimension line location or [Mtext/Text/Angle]:** *[Locate the dimension.]*

AutoCAD will automatically place either a diameter symbol (Ø) or an **R** in front of the dimension to indicate what has been dimensioned.

You have one other option for placing radial dimensions – the *jogged* (or *foreshortened radius*) dimension. Here is the sequence:

> **Command:** *dimjogged*
> **Select arc or circle:** *[Select the arc or circle to dimension.]*
> **Specify center location override:** *[Locate the dimension in terms of distance from the arc.]*
> **Dimension text = 15.5322** *[AutoCAD displays the dimension text.]*
> **Specify dimension line location or [Mtext/Text/Angle]:** *[Locate the dimension along the arc.]*
> **Specify jog location:** *[Locate the jog.]*

Let's take a look.

Where to Find It:	
Command Line:	*DimRadius*
Hotkey(s):	*dra*
Ribbon (Tab/Panel):	Home – Annotation – (Linear flyout) **Radius**
	Annotate – Dimensions – (Dimension flyout) **Radius**
Menu:	Dimensions – **Radius**
Toolbar:	Dimensions – **Radius**

Where to Find It:	
Command Line:	*DimDiameter*
Hotkey(s):	*ddi*
Ribbon (Tab/Panel):	Home – Annotation – (Linear flyout) **Diameter**
	Annotate – Dimensions – (Dimension flyout) **Diameter**
Menu:	Dimension – **Diameter**
Toolbar:	Dimension – **Diameter**

Where to Find It:	
Command Line:	*DimJogged*
Hotkey(s):	*jog*
Ribbon (Tab/Panel):	Home – Annotation – **Jogged**
	Annotate – Dimensions – (Dimension flyout) **Jogged**
Menu:	Dimension – **Jogged**
Toolbar:	Dimension – **Jogged**

Do This: 16.2.3A — Dimensioning Diameters and Radii

I. Be sure you're in the *linear* file the C:\Steps\Lesson16 folder. If not, please open it now.
II. Follow these steps.

16.2.3A: DIMENSIONING DIAMETERS AND RADII

1. Erase the 180° arc dimension.
   Command: *e*

16.2.3A: DIMENSIONING DIAMETERS AND RADII

2. Enter the ***Dimradius*** command.
   Command: *dra*

3. Select the arc at the left end of the object.
   Select arc or circle:

4. Locate the dimension as shown.
   Specify dimension line location or [Mtext/Text/Angle]:
   Notice that AutoCAD places the dimension and a *Center Mark* locating the center of the arc being dimensioned. You can place a center mark in a circle or arc (without creating a dimension) with the ***DimCenter*** command (Ribbon – Annotate – Dimensions (subpanel) – **Center Mark**).

5. Enter the ***Dimdiameter*** command.
   Command: *ddi*

6. Select the large circle to the left inside the object.
   Select arc or circle:

7. Place the dimension as shown.
   Dimension text = 2.0616
   Specify dimension line location or [Mtext/Text/Angle]:

8. Now let's try a jogged dimension. Thaw the **OBJ2** layer. Notice the arc above the object.

9. Enter the ***Dimjogged*** command.
   Command: *jog*

10. Pick the new arc.
    Select arc or circle:

11. Locate your dimension as shown.
    Specify center location override:
    Dimension text = 15.5322
    Specify dimension line location or [Mtext/Text/Angle]:
    Specify jog location:

12. Undo the last command and freeze the **OBJ2** layer.
    Command: *u*

13. Save the drawing but don't exit.
    Command: *qsave*

16.2.4	Dimension Arc Lengths

You've seen two approaches to dimensioning arcs (Sections 16.2.2, p.355, and 16.2.3, p.357) – one dimensions the angle of the arc, the other dimensions the radius or diameter. These won't help, however, if you need to dimension the actual length of an arc. Luckily, AutoCAD provides a dimensioning tool specifically for this task. It's called, appropriately, ***DimArc***.

The command sequence is

Command: *dimarc*
Select arc or polyline arc segment: *[Select the arc to dimension.]*
Specify arc length dimension location, or [Mtext/Text/Angle/Partial/Leader]: *[Locate the dimension.]*
Dimension text = X.XXXX

Where to Find It:	
Command Line:	*DimArc*
Hotkey(s):	*dar*
Ribbon (Tab/Panel):	Home – Annotation – (Linear flyout) **Arc Length**
	Annotate – Dimensions – (Dimension flyout) **Arc Length**
Menu:	Dimension – **Arc Length**
Toolbar:	Dimension – **Arc Length**

By now, you're familiar with the first three options of the **location** prompt. **Mtext/Text/Angle** options provide the same function throughout the dimension commands. Let's look at the other two.

- **Partial** enables you to dimension between two points on an arc.
- The **Leader** option tells AutoCAD to provide a leader from the dimension text to the arc being dimensioned.

Okay, let's try the *Dimarc* command.

Do This: 16.2.4A	Arc Length Dimensions

I. Be sure you're in the *linear* file the C:\Steps\Lesson16 folder. If not, please open it now.
II. Erase the radius and diameter dimensions.
III. Follow these steps.

16.2.4A: ARC LENGTH DIMENSIONS

1. Enter the *Dimarc* command.
   Command: *dar*
2. Select the arc on the left end of the object.
   Select arc or polyline arc segment:
3. Place the dimension as shown.
   Specify arc length dimension location, or [Mtext/Text/Angle/Partial/Leader]:
   Notice that AutoCAD automatically inserts the arc symbol.
4. Erase the dimension you created in Steps 1 through 3.
   Command: *e*
5. Thaw the **markers** layer. Notice the nodes that appear on the arc. Let's dimension between these.
6. Repeat Steps 1 and 2.
   Command: *dar*
7. Tell AutoCAD to use the **Partial** option [Partial].
   Specify arc length dimension location, or [Mtext/Text/Angle/Partial/Leader]: *p*
8. Dimension between the top and bottom nodes.
   Specify first point for arc length dimension: _nod of
   Specify second point for arc length dimension: _nod of

16.2.4A: ARC LENGTH DIMENSIONS

9. Finally, locate the dimension as shown.

   Specify arc length dimension location, or [Mtext/Text/Angle/Partial]:

10. Erase the last dimension you created and freeze the **Markers** layer.

    Command: *e*

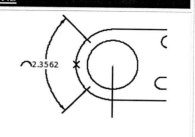

16.2.5	Dimension Strings

It's often preferable to string dimensions. This makes it easier for the contractor to find and read them, and it enhances the overall appearance of the drawing.

To string dimensions, we can repeat the *Dimlinear* command over and over, or we can begin our string with the *DimLinear* command and then follow it with the *DimContinue* command. The *DimContinue* command will place dimensions along the same line begun by the *DimLinear* command. With each selection, the second extension line from the previous dimension is used as the first extension line for the continued string. Some of the many benefits of this approach include:

	Where to Find It:
Command Line:	DimContinue
Hotkey(s):	*dco*
Ribbon (Tab/Panel):	Annotate – Dimensions – Continue
Menu:	Dimension – **Continue**
Toolbar:	Dimension – **Continue**

- Not overwriting the extension line.
- You don't have to locate the dimension line with each new selection. AutoCAD simply continues along the previous string.
- The command automatically repeats until you stop it.

The command sequence is

   Command: *dimcontinue*

   Specify a second extension line origin or [Undo/Select] <Select>: *[Select the origin of the second extension line; AutoCAD assumes the first extension line origin to be the second extension line of the last dimension entered.]*

   Dimension text = *[AutoCAD reports the dimension.]*

   Specify a second extension line origin or [Undo/Select] <Select>: *[AutoCAD repeats the command until you hit ENTER to exit.]*

   Select continued dimension: *[When you hit ENTER at the last prompt, AutoCAD will provide the opportunity to select a different dimension from which to continue; hitting ENTER at this prompt will exit the command.]*

Do This: 16.2.5A	String Dimensions

I. Be sure you're in the *linear* file the C:\Steps\Lesson16 folder. If not, please open it now.
II. Follow these steps.

16.2.5A: STRING DIMENSIONS

1. Create a linear dimension ⊢⊣ between the center of the larger circle and the center of the leftmost arc on the slot. Be sure to select the circle first.
   Command: *dli*

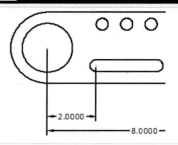

2. Enter the ***Dimcontinue*** command ⊢⊢⊣.
   Command: *dco*

3. Select the center of the arc on the other end of the slot.
   Specify a second extension line origin or [Undo/Select] <Select>:

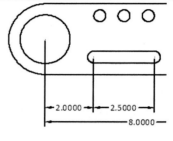

4. Hit ENTER twice to complete the command.
   Specify a second extension line origin or [Undo/Select] <Select>: *[ENTER]*
   Select continued dimension: *[ENTER]*

5. Now let's use what we know to dimension the upper circles. Put a linear dimension ⊢⊣ between the two left smaller circles (pick the middle one first).
   Command: *dli*
   Specify first extension line origin or <select object>: _cen of *[Middle small circle.]*
   Specify second extension line origin: _cen of *[Left small circle.]*

6. **Angle** the dimension at **60°**. Then place the dimensions as shown.
   Specify dimension line location or [Mtext/Text/Angle/Horizontal/Vertical/Rotated]: *a*
   Specify angle of dimension text: *60*
   Specify dimension line location or [Mtext/Text/Angle/Horizontal/Vertical/Rotated]:

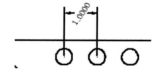

7. Continue the dimension ⊢⊢⊣ to the center of the large circle.
   Command: *dco*
   Specify a second extension line origin or [Undo/Select] <Select>: _cen of

8. We want to continue the dimension in the other direction now, so hit ENTER to go to the **Select continued dimension** prompt.
   Specify a second extension line origin or [Undo/Select] <Select>: *[ENTER]*

9. Select the right extension line (above the smaller middle circle).
   Select continued dimension:

10. Select the center of the smaller right circle.
    Specify a second extension line origin or [Undo/Select] <Select>: _cen of
    Dimension text = 1.0000

16.2.5A: STRING DIMENSIONS

11. Exit the command.

    Specify a second extension line origin or [Undo/Select] <Select>: *[ENTER]*

    Select continued dimension: *[ENTER]*

    The dimensions look like the figure at right.

12. Save the drawing.

    Command: *qsave*

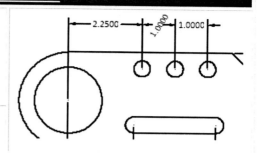

16.2.6 Aligning Dimensions

Nonlinear objects often require nonlinear dimensioning. This usually means that the dimension must be aligned with the object for clarity.

The *DimAligned* command works almost like the *DimLinear* command. The difference is that the aligned dimension parallels the first and second extension line origins.

The command sequence is

> **Command:** *dimaligned*
>
> **Specify first extension line origin or <select object>:** *[Select the first dimension point.]*
>
> **Specify second extension line origin:** *[Select the second dimension point.]*
>
> **Specify dimension line location or [Mtext/Text/Angle]:** *[Locate the dimension line.]*
>
> **Dimension text =** *[AutoCAD reports and places the dimension.]*

The options are identical to those used in the *Dimlinear* command.

	Where to Find It:
Command Line:	*DimAligned*
Hotkey(s):	*dal*
Ribbon (Tab/Panel):	Home – Annotation – (Linear flyout) **Aligned**
	Annotate – Dimensions – (Dimension flyout) **Aligned**
Menu:	Dimension – **Aligned**
Toolbar:	Dimension – **Aligned**

Do This: 16.2.6A	Aligned Dimensions

I. Be sure you're in the *linear* file the C:\Steps\Lesson16 folder. If not, please open it now.
II. Follow these steps.

16.2.6A: ALIGNED DIMENSIONS

1. Enter the *Dimaligned* command.

   Command: *dal*

2. Hit ENTER to select the object to dimension.

   Specify first extension line origin or <select object>: *[ENTER]*

3. Select the angled line on the right end of the object.

   Select object to dimension:

4. Locate the dimension line as shown.

   Specify dimension line location or [Mtext/Text/Angle]:

   Dimension text = 4.2426

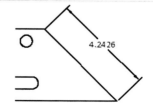

16.2.6A: ALIGNED DIMENSIONS
5. Save the drawing 💾 but don't exit.   Command: *qsave*

16.2.7	Baseline Dimensions & Spacing

Like the *Dimcontinue* command, *DimBaseline* works from an original linear dimension. But where continued dimensions were based on the *second* extension line origin of the last or selected linear dimension, the baseline dimension starts from the same *first* extension line origin of the last or selected linear dimension.

The command sequence parallels that of the *Dimcontinue* command as well.

Command: *dimbaseline*

Specify a second extension line origin or [Undo/Select] <Select>: *[Select the origin of the second extension line – AutoCAD assumes the first extension line origin to be the first extension line of the last dimension entered.]*

Dimension text = *[AutoCAD reports and inserts the dimension.]*

Specify a second extension line origin or [Undo/Select] <Select>: *[AutoCAD repeats the prompt until you hit ENTER to exit.]*

Select base dimension: *[When you hit ENTER at the last prompt, AutoCAD will provide the opportunity to select a different dimension from which to base the baseline dimensions; hitting ENTER at this prompt will exit the command.]*

	Where to Find It:
Command Line:	*DimBaseline*
Hotkey(s):	*dba*
Ribbon (Tab/Panel):	Annotate – Dimensions – (Continue flyout) **Baseline**
Menu:	Dimension – **Baseline**
Toolbar:	Dimension – **Baseline**

You may find the spacing for your baseline dimensions too narrow or too wide. You can fix that with the *DimSpace* command. It looks like this:

Command: *dimspace*

Select base dimension: *[Select the dimension from which you'll space.]*

Select dimensions to space: *[Select the dimension you wish to move.]*

Select dimensions to space: *[Confirm the selection – multiple dimensions should be linear to each other.]*

Enter value or [Auto] <Auto>: *[Enter your own spacing or let AutoCAD do it for you.]*

	Where to Find It:
Command Line:	*DimSpace*
Ribbon (Tab/Panel):	Annotate – Dimensions – **Adjust Space**
Menu:	Dimension – **Dimension Space**
Toolbar:	Dimension – **Dimension Space**

The only option – **Auto** – lets AutoCAD determine the new spacing. When you select **Auto**, AutoCAD will make the spacing twice the height of the dimension text.

Let's try these tools.

Do This: 16.2.7A	Baseline Dimensions and Spacing

    I. Be sure you're in the *linear* file the C:\Steps\Lesson16 folder. If not, please open it now.
   II. Erase all the dimensions below the object.
  III. Follow these steps.

16.2.7A: BASELINE DIMENSIONS AND SPACING

1. Draw a linear dimension as shown. Be sure to select the extension line origin on the right first.
   Command: *dli*

2. Enter the ***Dimbaseline*** command.
   Command: *dba*

3. Pick the center of the end of the slot. AutoCAD reports and places the dimension.
   Specify a second extension line origin or [Undo/Select] <Select>: _cen of
   Dimension text = **6.0000**

4. Pick the center of the large circle.
   Specify a second extension line origin or [Undo/Select] <Select>: _endp of
   Dimension text = **8.0000**

5. Hit *ENTER* twice to exit the command.
   Specify a second extension line origin or [Undo/Select] <Select>: *[enter]*
   Select base dimension: *[ENTER]*
   Your drawing looks like the figure at right.

6. Let's change the spacing of the last two dimensions. Enter the ***Dimspace*** command.
   Command: *dimspace*

7. Select the **6.000** dimension as our **base**.
   Select base dimension:

8. Select the **8.0000** as the dimension to move.
   Select dimensions to space:
   Select dimensions to space: *[ENTER]*

9. Space the two dimensions ¾" apart.
   Enter value or [Auto] <Auto>: .75
   The dimensions look like the figure at right.

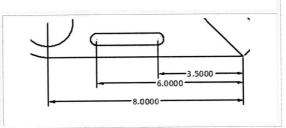

10. Save the drawing and exit.
    Command: *qsave*

| 16.2.8 | Ordinate Dimensions |

Ordinate dimensioning is a valuable tool in some types of manufacturing. ***DimOrdinate*** places a dimension that is actually a distance from the 0,0 coordinate of the drawing. To be useful then, 0,0 must be located somewhere on the object itself – usually the lower left corner.

The command sequence for the ***DimOrdinate*** command is

Command: *dimordinate*
Specify feature location: *[Select the feature to be dimensioned.]*
Specify leader endpoint or [Xdatum/Ydatum/Mtext/Text/Angle]: *[Specify the X or Y distance from 0,0 or identify the datum by mouse movement.]*
Dimension text = *[AutoCAD reports and places the dimension.]*

Where to Find It:	
Command Line:	DimOrdinate
Hotkey(s):	dor
Ribbon (Tab/Panel):	Home – Annotation – (Linear flyout) **Ordinate**
	Annotate – Dimensions – (Dimension flyout) **Ordinate**
Menu:	Dimension – **Ordinate**
Toolbar:	Dimension – **Ordinate**

- As indicated, you can specify an **Xdatum** (X-distance from 0,0) or a **Ydatum** (Y-distance from 0,0) by selecting that option.
- If you prefer to place **Text** or **MText** at the ordinate location, those options are also available.
- The **Angle** option behaves just as it does on other dimension tools – that is, you may define the angle of the dimension text.

Let's try an exercise.

Do This: 16.2.8A	Ordinate Dimensioning

I. Open the *ordinate* file the C:\Steps\Lesson16 folder. The drawing looks like the figure at right. (We've located the 0,0 coordinate at the lower left corner of the object.)

II. Follow these steps.

16.2.8A: ORDINATE DIMENSIONING

1. Enter the *Dimordinate* command.
    Command: *dor*

2. Select the left side of the lower indentation.
    Specify feature location:_endp of

3. Pick a point two grid marks due south of the point selected in Step 2.
    Specify leader endpoint or [Xdatum/Ydatum/Mtext/Text/Angle]:
    Dimension text = 4.5000

4. This time, specify the datum you'll use. Repeat the *Dimordinate* command.
    Command: *[ENTER]*

5. Select the lower right corner of the object.
    Specify feature location:_endp of

6. Tell AutoCAD that you want an **Xdatum** [Xdatum].
    Specify leader endpoint or [Xdatum/Ydatum/Mtext/Text/Angle]: *x*

7. Select a point two grid marks south and two grid marks to the left (see the figure in Step 11). (Toggle Ortho off to make this easier.)
    Specify leader endpoint or [Xdatum/Ydatum/Mtext/Text/Angle]:

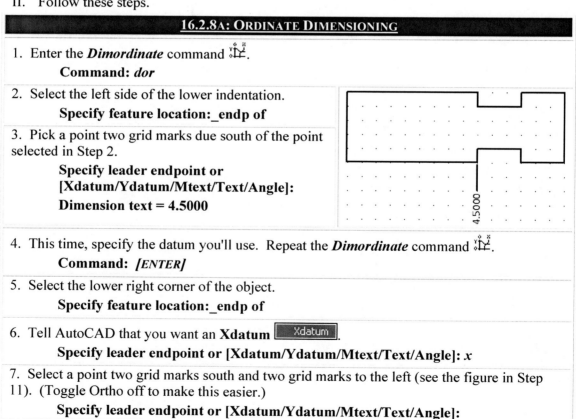

16.2.8A: ORDINATE DIMENSIONING

8. Repeat the command.
   Command: *[ENTER]*

9. Select the same point selected in Step 8.
   Specify feature location: _endp of

10. Tell AutoCAD that you want a **Ydatum**.
    Specify leader endpoint or [Xdatum/ Ydatum/Mtext/Text/ Angle]: *y*

11. Select a point two grid marks to the left and two grid marks upward.
    Specify leader endpoint or [Xdatum/Ydatum/Mtext/Text/Angle]:
    Dimension text = 0.0000

12. Complete the drawing as shown in the following figure.

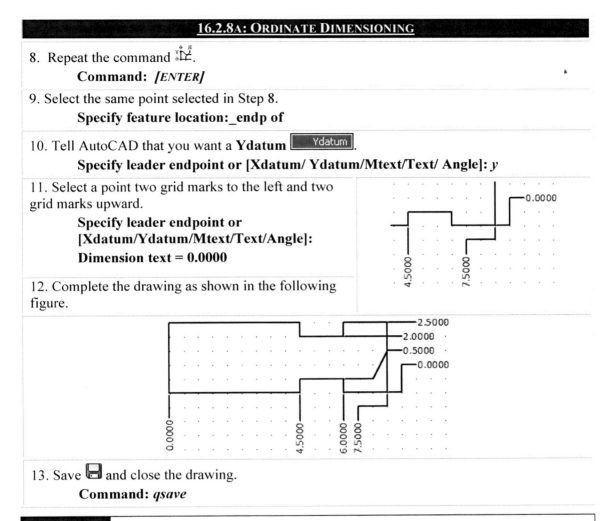

13. Save and close the drawing.
    Command: *qsave*

16.3 And Now the Easy Way: Quick Dimensioning (QDim)

AutoCAD includes a really cool dimensioning tool – *Quick Dimensioning* (**QDim**). With **QDim**, you can create a string of dimensions using two picks to select the objects (with a window), one click to confirm the selection and one pick to locate the dimension. That's only four mouse clicks!

The command sequence looks like this:

Where to Find It:	
Command Line:	QDim
Ribbon (Tab/Panel):	Annotate – Dimensions – **Quick Dimension**
Menu:	Dimension – **Quick Dimension**
Toolbar:	Dimension – **Quick Dimension**

Command: *qdim*
Select geometry to dimension: *[Use a window or crossing window to select the objects to dimension.]*
Select geometry to dimension: *[Hit ENTER to complete the selection process.]*
Specify dimension line position, or [Continuous/Staggered/Baseline/Ordinate/Radius/ Diameter/datumPoint/Edit/seTtings] <Continuous>: *[Tell AutoCAD where to place the dimension.]*

Most of the options are fairly obvious – **Continuous** for a continuous string, **Baseline** for a baseline string, **Ordinate** for ordinate dimensions, and so forth. But let's look at the last three.

- **datumPoint** allows you to define a selected point as the 0,0 coordinate of the dimension string (very handy when ordinate dimensioning).
- **Edit** prompts:
    Indicate dimension point to remove, or [Add/eXit] <eXit>:
    Using the **Edit** option then, you can add or remove extension line origins to the selection set.
- **seTtings** prompts:
    Associative dimension priority [Endpoint/Intersection] <Endpoint>:
    QDim automatically dimensions to endpoints of selected objects. Using this option, you can tell AutoCAD to dimension to intersections rather than endpoints.

Let's give this nifty tool a try.

Do This: 16.3A	Quick Dimensions

I. Be sure you're in the *linear* file the C:\Steps\Lesson16 folder. If not, please open it now.
II. Erase any existing dimensions.
III. Follow these steps.

16.3A: QUICK DIMENSIONS

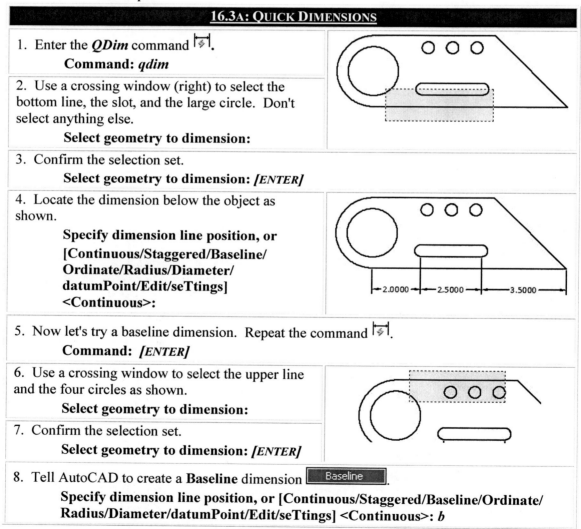

1. Enter the *QDim* command.
    Command: *qdim*

2. Use a crossing window (right) to select the bottom line, the slot, and the large circle. Don't select anything else.
    Select geometry to dimension:

3. Confirm the selection set.
    Select geometry to dimension: *[ENTER]*

4. Locate the dimension below the object as shown.
    Specify dimension line position, or [Continuous/Staggered/Baseline/Ordinate/Radius/Diameter/datumPoint/Edit/seTtings] <Continuous>:

5. Now let's try a baseline dimension. Repeat the command.
    Command: *[ENTER]*

6. Use a crossing window to select the upper line and the four circles as shown.
    Select geometry to dimension:

7. Confirm the selection set.
    Select geometry to dimension: *[ENTER]*

8. Tell AutoCAD to create a **Baseline** dimension.
    Specify dimension line position, or [Continuous/Staggered/Baseline/Ordinate/Radius/Diameter/datumPoint/Edit/seTtings] <Continuous>: *b*

16.3A: QUICK DIMENSIONS

9. Locate the dimensions as shown.

    Specify dimension line position, or [Continuous/Staggered/Baseline/Ordinate/Radius/Diameter/datumPoint/Edit/seTtings] <Baseline>:

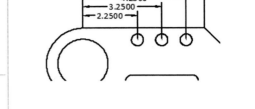

10. Exit the drawing without saving.

    Command: *close*

Doesn't that make you wonder why we ever did it any other way?

Still, by now, you may have noticed that dimensions don't always appear just the way we think they should. And sometimes (God forbid), the dimension has to change to accommodate some new design information. Our next section will show you how to modify dimensions.

16.4 Dimension Editing: The *DimEdit* and *DimTEdit* Commands

I'll show you how to use these editing tools, but frankly, I rarely use them. You'll find your best approach to editing most dimension properties in the Properties palette. Just double click on the dimension to edit, or use Quick Properties 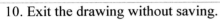.

AutoCAD includes two commands for editing associative dimensions and a third for breaking extension lines. The first two commands show a bit of overlap in their functions (redundancy again), but you'll generally change or rotate the text or change the angle of the extension lines with the *DimEdit* command. With the *DimTEdit* command, you can change the position of the text. With the third editing tool – *DimBreak* – you can automatically (or manually) break the extension lines around callouts!

Although AutoCAD provides tools for editing associative dimensions, normal dimensions don't require any special tools. Use basic modification commands on dimensions as you would on any other objects. Edit the text using the text editing command (*DDEdit*).

16.4.1 Position the Dimension: The *DimTedit* Command

The *DimTEdit* command sequence looks like this:

> **Command:** *dimtedit*
> **Select dimension:** *[Pick the dimension to reposition.]*
> **Specify new location for dimension text or [Left/Right/Center/Home/Angle]:** *[Pick the new position or select an option.]*

Where to Find It:	
Command Line:	DimTEdit
Hotkey(s):	dimted
Ribbon (Tab/Panel):	Annotate – Dimensions (subpanel) – [option]
Menu:	Dimension – Align Text – [option]
Toolbar:	Dimension – **Text Edit**

You can change the position of the dimension text dynamically by dragging it with the mouse, or by using AutoCAD's default **Left**, **Right**, or **Center** position. (Note: The **Left** or **Right** option redefines the **Home** position of the text next to the left or right arrow.) The **Angle** option allows you to rotate the dimension text, and the **Home** option returns the dimension text to its original position and rotation.

Notice (in the WTFI table) that the ribbon and menu approaches to *DimTEdit* end with *[option]*. This release of AutoCAD includes a series of options (and buttons) that significantly simplify the command. Consider these buttons/options as shortcuts to the dimension text options of the command: **Left Justify**, **Right Justify**, **Center Justify**, and **Text Angle**.

Do This: 16.4.1A	Repositioning Dimension Text

I. Open the *base2* file the C:\Steps\Lesson16 folder.

II. Follow these steps.

16.4.1A: REPOSITIONING DIMENSION TEXT

1. Begin with the ***DimTEdit's* Left** option. (I recommend using the button on the **Dimensions** subpanel or the **Left** option on one of the menus. Refer to the WTFI table for locations.)

    Command: *dimted*

2. Select one of the dimensions in the drawing.

    Select dimension:

    Specify new location for dimension text or [Left/Right/Center/Home/Angle]: _l

3. Repeat Steps 1 and 2 with the **Right** and **Center** options.

4. Now begin the **Angle** option.

5. Select the dimension below the right end of the object as indicated...

    Select dimension:

6. ...and give it an angle of 27°.

    Specify new location for dimension text or [Left/Right/Center/Home/Angle]: _a
    Specify angle for dimension text: *27*

7. Save the drawing but don't exit.

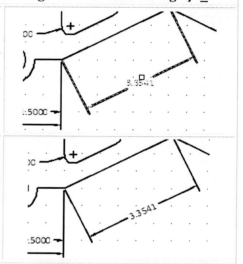

16.4.2	Changing Value of the Dimension Text: The *DimEdit* Command

The ***DimEdit*** command sequence is

Command: *dimedit*

Enter type of dimension editing [Home/New/Rotate/Oblique] <Home>:
[Hit ENTER to accept the default or tell AutoCAD what you want to do.]

Select objects: *[Select the object(s) to change.]*

Where to Find It:	
Command Line:	*DimEdit*
Hotkey(s):	*ded*
Toolbar:	Dimension – **Dimension Edit**

Notice that the ***DimEdit*** command uses a default (<**Home**>) for the initial prompt. Notice also that the option must be selected *before* the dimension objects (as opposed to the way ***DimTEdit*** required selecting the *object* first).

- Most of the options are simple. **Home** returns a rotated dimension to its original state. **Rotate** does what the **Angle** option did in the ***DimTEdit*** command.
- The **New** option presents the Multiline Text Editor. Just enter the desired text. When you pick the **OK** button, AutoCAD will prompt to **Select objects**. It'll then replace the dimension text

with what you've typed in the Multiline Text Editor. But be aware that once you've changed the dimension text in this manner, it'll no longer automatically update when the object/dimension is modified.

| You may find it easier to create a text override on the Quick Properties or Properties palette. |

- You'll find the **Oblique** option quite valuable when dimensioning isometrics. (In fact, it's the only option of the ***DimEdit*** command with it's own button !) To dimension an isometric, you'll use the ***DimAligned*** command, and then reposition the dimension in the correct isometric plane using this tool. (You'll see more on this in Section 16.5, p.372.)

| Do This: 16.4.2A | **Changing Dimension Text** |

I. Be sure you're still in the *base2* file in the C:\Steps\Lesson16 folder.
II. Follow these steps.

16.4.2A: CHANGING DIMENSION TEXT

1. Enter the ***Dimedit*** command.
   Command: *ded*

2. Tell AutoCAD to edit a dimension (select the **New** option).
   Enter type of dimension editing [Home/New/Rotate/Oblique] <Home>: *n*

3. AutoCAD displays the Multiline Text Editor with a number box. Type 4.125 in the highlighted area.

4. Close the text editor.

5. Select the **4.0000** dimension indicated. Then complete the command.
   Select objects:
   Select objects: *[ENTER]*

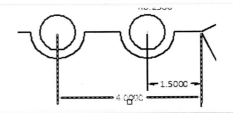

6. AutoCAD replaces the dimension text with the new text. Now try stretching the object one unit to the left.
   Command: *s*
   Select the objects as shown here.

7. Notice (following figure) that AutoCAD automatically updated the **5.5000** dimension but the **4.125** dimension didn't change. Remember, dimensions changed using this method are no longer associative!

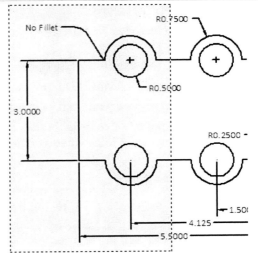

16.4.2A: CHANGING DIMENSION TEXT

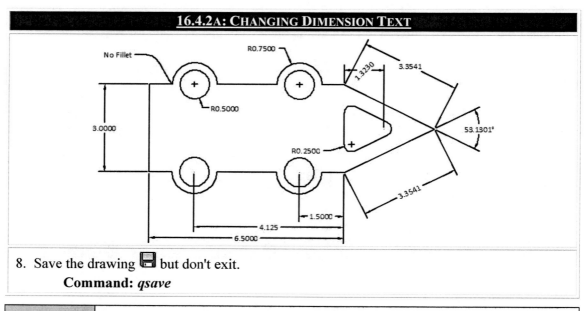

8. Save the drawing 💾 but don't exit.

   Command: *qsave*

16.4.3 Breaking an Extension Line - *DimBreak*

Now here's a command that was a long time in coming ... but well worth the wait! *DimBreak* automatically breaks an associative dimension's extension lines around a callout! Don't have the callout yet? No problem, you can use *DimBreak* to manually break an extension line, too.

The *DimBreak* command sequence is

    Command: *dimbreak*

    Select dimension to add/remove break or [Multiple]: *[Select the dimension whose extension lines you wish to break.]*

    Select object to break dimension or [Auto/Manual/Remove] <Auto>: *[Hit ENTER to let AutoCAD automatically break the extension line around text that overlaps it.]*

Where to Find It:	
Command Line:	*DimBreak*
Ribbon (Tab/Panel):	Annotate – Dimensions – Break
Menu:	Dimension – **Dimension Break**
Toolbar:	Dimension – **Dimension Break**

This command has some fairly simple options.

- **Auto**, of course, tells AutoCAD to automatically break the extension lines.
- Use **Manual** when you want to do it yourself.
- **Remove** takes out any extension breaks in selected dimensions. This comes in handy for restoring the extension lines when you determine that you don't actually want them broken after all.

Come on, let's give it a try.

Do This: 16.4.3A	Breaking an Extension Line

I. Be sure you're still in the *base2* file in the C:\Steps\Lesson16 folder.
II. Follow these steps.

16.4.3A: BREAKING AN EXTENSION LINE

1. Create a radius dimension ⊙ for the lower left hole. Be sure to position it as shown here.	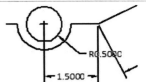
2. Enter the *Dimbreak* command. **Command:** *dimbreak*	
3. Tell AutoCAD you want to select more than one [Multiple] extension line … **Select dimension to add/remove break or [Multiple]:** *m*	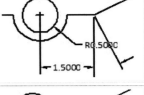
4. … and then put a crossing window around the extension lines as indicated. **Select dimensions:** **Select dimensions:** *[ENTER]*	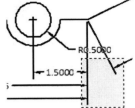
5. Leave the rest to AutoCAD! **Select object to break dimensions or [Auto/Remove] <Auto>:** *[ENTER]* Notice that AutoCAD automatically finds the interference and breaks the extension lines around it.	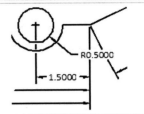
6. Use grips to move the radial dimension. Notice that AutoCAD automatically adjusts the break! (Okay, that deserves a "way, cool!")	
7. Save the drawing and exit. **Command:** *qsave*	

16.5 Isometric Dimensioning

It might relieve you to know that there's no actual isometric dimension command. (Then again, a *DimIsometric* command might make this section a bit easier!)

Dimension your isometrics using the *DimAligned* command. Then edit the aligned dimension using the **Oblique** option of the *DimEdit* command to adjust the aligned angle to the appropriate isometric plane.

Sound complicated? It isn't as complicated as it is tedious. Let's see how it works.

Do This: 16.5A	Dimensioning Isometrics

I. Open the *isodim* file in the C:\Steps\Lesson16 folder. The drawing looks like the figure at right.
II. Follow these steps.

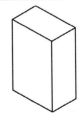

16.5A: DIMENSIONING ISOMETRICS

1. Use the *Dimaligned* command to place the dimensions shown.
   Command: *dal*

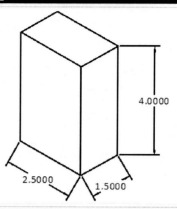

2. Begin the **Oblique** option of the *Dimedit* command.
   Command: *ded*

3. Select the **4.0000** vertical dimension.
   Select objects:
   Select objects: *[ENTER]*

4. Tell AutoCAD to angle the extension lines at **30°**.
   Enter obliquing angle (press ENTER for none): *30*
   Your drawing looks like the figure at right.

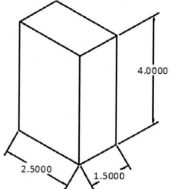

5. Now set the obliquing angle of the lower two dimensions to 270°.
   Command: *[ENTER]*
   Enter type of dimension editing [Home/New/Rotate/Oblique] <Home>: *o*
   Select objects:
   Select objects: *[ENTER]*
   Enter obliquing angle (press ENTER for none): *270*
   Your drawing looks like the figure at right.

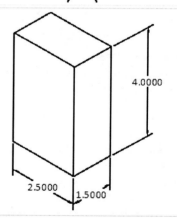

6. Save the drawing and exit.
   Command: *qsave*

To change the text to an isometric format, create appropriate text styles (Lesson 4) and then use the Properties palette (Lesson 6) to change the style of the dimension text.

16.6 Extra Steps

AutoCAD has one last dimensioning tool – **DimInspect** – used to "create" an inspection dimension.

An *inspection* dimension generally indicates a critical dimension – one that the manufacturer will refer to for control of the product's overall dimensions. Consider it a quality control dimension.

Where to Find It:	
Command Line:	*DimInspect*
Ribbon (Tab/Panel):	Annotate – Dimensions – **Inspect**
Menu:	Dimension – **Inspection**
Toolbar:	Dimension – **Inspection**

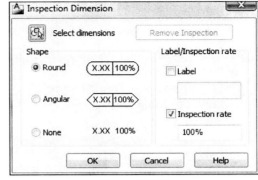

Figure 16.004

Okay, ***DimInspect*** won't actually create a dimension. It will, however, change the appearance of an existing dimension.

The ***DimInspect*** command calls the Inspection Dimension dialog box (Figure 16.004). Here, using the tools in the **Shape** and **Label/Inspection rate** frames, you can determine how you'd like your inspection dimension to look. Then use the **Select Dimensions** button to select the dimensions you'd like to appear as inspection dimensions.

Give it a try on one of the drawing we've used in this lesson.

16.7 What Have We Learned?

Items covered in this lesson include:

- *AutoCAD Dimension Terminology*
- *AutoCAD Dimension Commands:*
    - ***Dimlinear***
    - ***Dimangular***
    - ***Dimradius***
    - ***DimCenter***
    - ***Dimjogged***
    - ***DimJogLine***
    - ***Dimarc***
    - ***Dimdiameter***
    - ***Dimcontinue***
    - ***Dimaligned***
    - ***Dimbaseline***
    - ***Dimordinate***
    - ***QDim***
    - ***Dimedit***
    - ***DimTEdit***
    - ***Dimbreak***
    - ***DimRegen***
    - ***DimAssociate***
    - ***DimAssoc***
    - ***DimReAssociate***
    - ***DimDisAssociate***
    - ***DimInspect***
    - ***DimSpace***

You're now familiar with the various methods of creating and editing AutoCAD's dimensions. For the most part, you can place dimensions as you would on any board. But if you spend much time with dimensioning, you'll discover some shortfalls in your understanding. This is because you're only halfway through your study of AutoCAD's dimensioning tools.

In Lesson 17, we'll discuss customizing your dimensions. This sounds difficult, but it's where you'll learn:

- to use architectural units instead of decimals,
- how to change arrow styles and text sizes,
- how to have AutoCAD insert tolerances automatically,
- and much, much more!

16.8 Exercises

1. Dimension the *Brake* file in the C:\Steps\Lesson16 folder.
   Remember:
   - the *DimAssoc* command toggles associative and normal dimensioning
   - the *Explode* command converts an associative dimension to a normal dimension.
   - dimensions don't always fall where you want them, so remember your editing commands.

   When completed, your drawing will look like the figure at right.

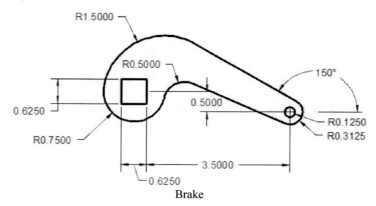

Brake

2. Dimension the *Pulley-slider* file in the C:\Steps\Lesson16 folder. The drawing will look like the figure at right when completed.

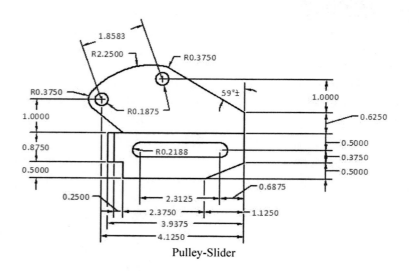

Pulley-Slider

16.9 For Web-Based Review Questions and Additional Exercises, visit: http://foragerpub.com/AcadFiles/2011/2011.htm

Lesson 17

Following this lesson, you will:

✓ Know how to create and use Dimension Styles

✓ Know how to override Dimension Styles

✓ Know how to create and use dimensional constraints

Customizing Dimensions and Using Dimensional Constraints

Every design industry has its own preferences about such things as units, dimension arrows, text sizes, tolerances, and so forth. If all industries had to accept decimal dimensions (with four decimal places), AutoCAD would probably lose its business fairly quickly – especially from architectural and petrochemical designers who rely on the ever-present feet and inches of architectural drafting. Indeed, even metric users might wash their hands of taking millimeters to four decimal places.

In this lesson, we'll look at customizing the way AutoCAD creates dimensions to fit different industrial standards.

17.1 Creating Dimension Styles

We'll use dialog boxes for our customization. AutoCAD will store each of our settings in a DIMension VARiable (**Dimvar**). You can adjust each dimvar manually at the command prompt (as we did in earlier releases). But take my word for it, only a bored programmer would want to customize dimensions this way!

Each drawing can incorporate several dimension styles, but there's rarely need for more than one. AutoCAD's clever use of the *Family* method of setup enables you to create overall dimension settings (*Parent* settings) for the drawing with different settings for each of its six children: **Linear, Radial, Diameter, Ordinal, Angular,** and **Leaders and Tolerances**. This is a remarkable accomplishment when compared with other CAD systems or even earlier releases of AutoCAD. The downside of this, of course, is that you must navigate the Dimension Style Manager to create the proper settings for the parent and each child. But remember, set it up once and save it to a template, then you don't have to do it again!

Where to Find It:	
Command Line:	*DimStyle*
Hotkey(s):	*ddim*
Ribbon (Tab/Panel):	Home – Annotation (subpanel) – **Dimension Style**
	Annotate – Dimensions – **Dimension Style**
Menu:	Dimension – **Dimension Style**
Toolbar:	Dimension – **Dimension Style**

Let's look at the first step. Start a new drawing from scratch and follow along on the screen.

The Dimension Style Manager appears in Figure 17.001. Let's examine it.

- In the **Styles** list box, AutoCAD will display a list of all the styles currently defined in the drawing. The left-justified names are the Parent styles. The children of that parent will be listed

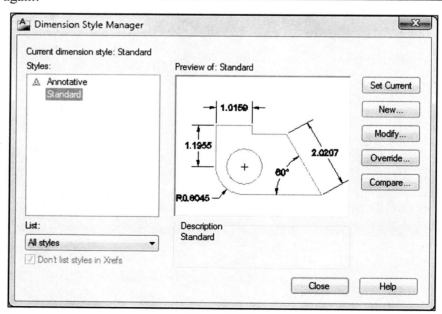

Figure 17.001

377

below it and slightly indented.
- The **List** control box controls what you'll see in the **Styles** list box. Here you can tell AutoCAD to list **All styles** or just the **Styles in use** in the current drawing.
- AutoCAD shows the current dimension setup in the **Preview** display box. Following the words "Preview of:" AutoCAD will list the dimension style being displayed. A written description of the dimstyle may appear in the **Description** frame below the display box.
- The five buttons down the right side of the Dimension Style Manager allow you to **Set Current** the dimstyle highlighted in the **Styles** list box, create a **New** or **Modify** an existing dimstyle, **Override** the settings in the current dimstyle, or **Compare** two dimstyles to find the differences. Let's look at these.
  o You must set a dimension style current before you can use it (just as you do with layers). You can do that by picking the **Set Current** button here. But it's easier to use the **Dimension Style** control box on the ribbon's **Dimensions** panel (as we'll see in our exercises).
  o The **New** and **Modify** buttons both present a tabbed dialog box used to set up the dimstyle (more on this in a moment).
  o Overriding a dimension style is rarely a good idea. But selecting the **Override...** button will present the same tabbed dialog box as the previous buttons. The difference is that **Override** won't use the changes to redefine the dimstyle, although you can use the changes in your dimensioning. (More on **Override** in Section 17.4, p.393).
  o The **Compare** button presents a dialog box that allows you to compare the differences between two dimension styles.

Now that you have some idea of what the Dimension Style Manager looks like, we need to get into the specifics of exactly how to define or create a new dimension style.

How do you think you'd start a new style? Of course, you'd pick the **New** button. When you do, AutoCAD presents the Create New Dimension Style dialog box (Figure 17.002). Let's look at our options.

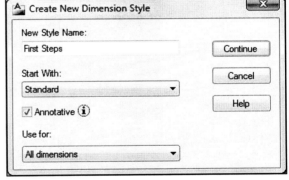

Figure 17.002

- Name your dimstyle in the **New Style Name** text box (here we've called our new style *First Steps*).
- In the **Start With** control box, you can select to use the settings from any existing dimstyle as the basis for your new style.
- Put a check next to **Annotative** to create annotative dimensions. Annotative dimensions behave just like annotative text – that is, they'll automatically size themselves according to the **Annotation Scale** of the drawing.
- The **Use for** control box is the key to the parent/child relationship of AutoCAD's dimension styles. By default, the settings you make for your new style will be used for **All dimensions** you create. However, you can create separate settings for **Linear**, **Angular**, **Radius**, **Diameter**, **Ordinate**, or **Leader and Tolerance** dimensions (the children).

Let's start a new dimension style.

Do This: 17.1A	Create a New Dimstyle

I. Open the *drill-gizmo-17* file the C:\Steps\Lesson17 folder (Vista users, open the *- v* file). The drawing looks like the figure at right.
II. Follow these steps.

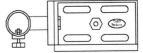

17.1A: CREATE A NEW DIMSTYLE
1. Open the Dimension Styles Manager.   **Command:** *ddim*
2. Start a **New** style.
3. Refer to Figure 17.002 (p.378) – create a style called ***First Steps*** based on the **Standard** style and make it Annotative. We'll set up this first style for **All dimensions**.
4. Pick the **Continue** button. AutoCAD presents the New Dimension Style: First Steps dialog box. (Read on before continuing.)

Once you've completed the selections in the Create New Dimension Style dialog box, you'll **Continue** to the tabbed New Dimension Style: First Steps dialog box (where *First Steps* is the name of your new style), as shown in Figure 17.003. Here you'll actually define the settings. We'll take a look at each tab over the next several pages.

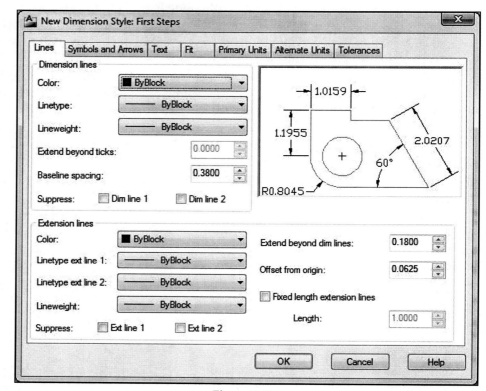

Figure 17.003

- The first tab – **Lines** – contains two frames. You'll define the appearance of your dimension's lines in these.
  o Preview your setup in the upper right **Preview** area.
  o Use the first frame to set up **Dimension lines**.
    ▪ The **Color, Linetype,** and **Lineweight** control boxes allow you to set the color, linetype, and lineweight of the dimension. The default for each is **ByBlock**.
    ▪ The **Extension beyond ticks** control box remains gray unless the arrowheads have been set to tick marks (we'll look at arrowheads in a moment). If tick marks have been selected, you can tell AutoCAD how far to extend the dimension beyond the extension line by typing a distance here (refer to Figure 17.004).

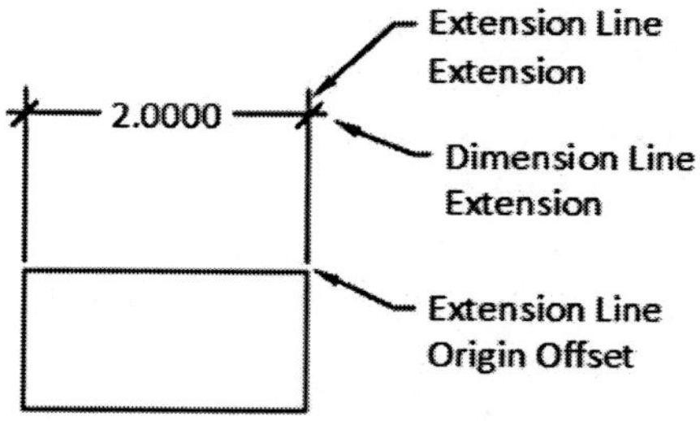

Figure 17.004

379

- **Baseline spacing** refers to the distance between baseline dimensions. It's probably best to leave this setting at its default unless your industry uses a different standard.
- The check boxes at the bottom of the frame allow you to **Suppress** (not draw) the dimension line on the 1st or 2nd side of the dimension. This can be useful in a crowded area.
  o Set up the extension lines in the **Extension lines** frame.
    - **Color**, **Linetype**, and **Lineweight** work the same as they did for dimension lines. Note that you can set a different linetype for each of the extension lines.
    - The **Suppress** options work on extension lines as they did on dimension lines. Use these to avoid placing one extension line atop another when stringing dimensions. I don't recommend this; AutoCAD has been programmed to avoid plotting one line atop another to prevent digging inky holes in paper (a problem in earlier releases). Besides, if you're placing more than a couple lines in one place, you might want to rethink your drafting strategy.
    - The **Extend beyond dim lines** setting determine the distance the extension lines continue beyond the dimension line (the **Extension Line Extension** as seen in Figure 17.004).
    - The **Offset from origin** setting determines the distance away from the origin the extension line will begin (see the **Extension Line Origin Offset** in Figure 17.004).
    - Place a check next to **Fixed length extension lines** to set the length of your extension lines to a fixed length. Use the **Length** number box to define the length.

Let's continue our exercise.

Do This: 17.1B	Setting up Dimension & Extension Lines

I. Continue the previous exercise.
II. Follow these steps.

17.1B: SETTING UP DIMENSION AND EXTENSION LINES

1. On the **Lines** tab:
   a. Accept the default settings in the **Dimension lines** frame.
   b. Set the extension lines to **Extend 0.125" beyond the dim lines**.

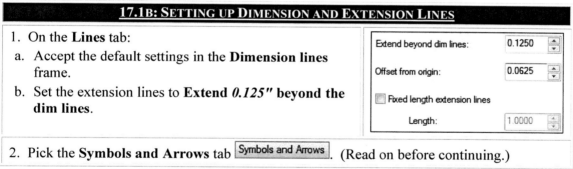

2. Pick the **Symbols and Arrows** tab [Symbols and Arrows]. (Read on before continuing.)

- The **Symbols and Arrows** tab (Figure 17.005, p.381) contains six frames for defining arrowheads, arcs, and radial symbols.
  o Use the **Preview** area to see what your dimension looks like as you create it.
  o You'll determine the size and style of the arrowheads in the **Arrowheads** frame.
    - AutoCAD provides sample images for each of the 1st, 2nd, and **Leader** arrowheads in the control boxes. AutoCAD also provides several selections as well as options to use a user-defined block or no arrowhead at all (**None**).

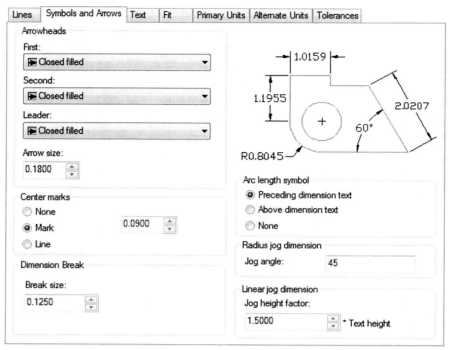

Figure 17.005

- Just below the control boxes is an **Arrow size** box in which you can adjust the size of the arrowhead.
  o The **Center marks** frame allows you to use a mark (a small "+"), centerlines, or no symbol when dimensioning the center of a circle or arc. It also allows you to change the size of the center marks or the extension of the centerlines outside the circle.
  o AutoCAD allows you to control the size of the break when you use the ***Dimbreak*** command with the **Break size** number box in the **Dimension Break** frame. I suggest you use a gap of about 1/8". Remember, annotative dimensions will adjust this setting according to the annotative scale.
  o Use the radio buttons in the **Arc length symbol** frame to define where AutoCAD will place an arc symbol when dimensioning an arc length.
  o Change the desired **angle** of a jogged dimension in the **Radius dimension jog** frame.
  o Finally, set the height of your dimension jog (the distance between the two jog vertices) in the **Linear jog dimension** frame. The default – 1.5 x the text height – should be good for most situations.

Let's set up our symbols and arrowheads.

Do This: 17.1C	Setting up Dimension Symbols and Arrowheads

I. Continue the previous exercise.
II. Follow these steps.

17.1C: SETTING UP DIMENSION SYMBOLS AND ARROWHEADS

1. On the **Symbols and Arrows** tab:
   a. Use **Dot small** for the 1st and 2nd arrowheads for the dimensions, but **Closed filled** arrowheads for the **Leader**.
   b. Make the arrowheads *0.125"*.
   c. Don't use **Center marks for Circles**.
   d. Accept the remaining defaults.

2. Pick the **Text** tab. (Read on before continuing.)

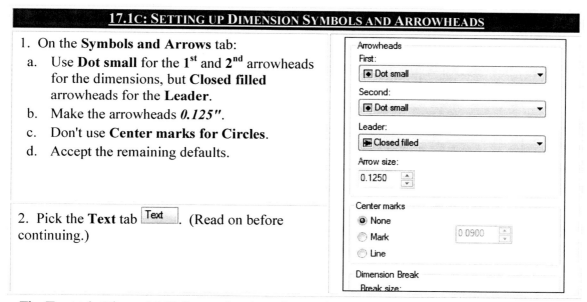

- The **Text** tab (Figure 17.006) contains three frames for defining how to use text in your dimensions.

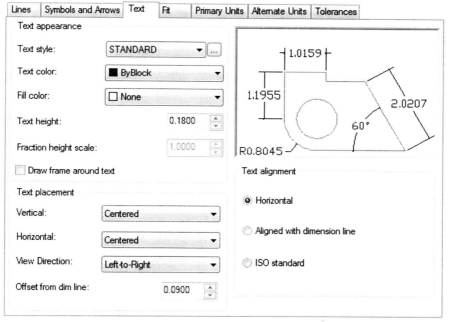

Figure 17.006

  o Use the **Preview** area to see what your dimension looks like as you create it.
  o The **Text appearance** frame contains settings for how the text will look.
    - Set the style of text for the dimensions using the **Text style** control box. If you haven't yet created the style, use the button to the right of the control box to access the Text Style dialog box (Lesson 4, p.101).
    - Set the **Text color** and **height** using the appropriate boxes.
    - When using fractional or architectural units, you can set the size of the fraction using the **Fraction height scale** number box.
    - Place a frame around the dimension text by checking the **Draw frame around text** box.

- The **Text placement** frame allows you to control where the text will be placed in relation to the dimension line and whether it will read **Left-to-Right** or **Right-to-Left**.
- In the **Text alignment** frame, determine whether to place dimension text **Horizontal, Aligned with** [the] **dimension line,** or in the **ISO standard** mode. (*ISO Standard* means that the text will be aligned with the dimension line if it falls between the extension lines or horizontally if it doesn't fit between the extension lines.)

Let's set up our text.

| Do This: 17.1D | Setting up Dimension Text |

I. Continue the previous exercise.
II. Follow these steps.

17.1D: SETTING UP DIMENSION TEXT

1. On the **Text** tab (refer to the following figure):
    a. Use the **TIMES** or **Calibri** text style (it's already been set up) and a text height of **.125"**.
    b. Place the text **Above** the line vertically but **Centered** horizontally.
    c. View the text from **Left-to-Right**.
    d. Use the **ISO standard** alignment mode.

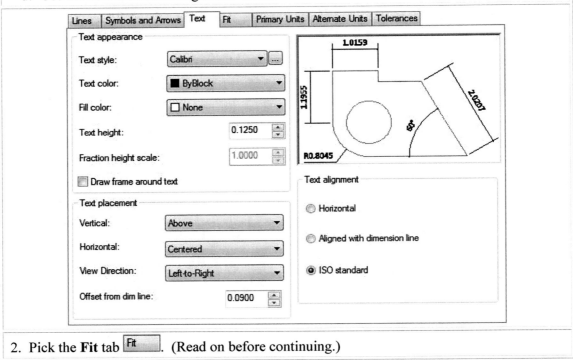

2. Pick the **Fit** tab. (Read on before continuing.)

- The **Fit** tab (Figure 17.007, p.384) has four frames to help fine-tune your dimension style.
  - The **Fit options** frame allows you to tell AutoCAD what to do when the dimension parts don't fit between the extension lines.
    - Move text and/or arrowheads from inside to outside the extension lines.
    - Don't place (**Suppress**) arrowheads on the dimension lines.
  - In the **Text placement** frame, tell AutoCAD where to place the text if it doesn't fit between the extension lines – either **beside** or **over** [the] **dimension line** (**with** or **without** [a] **leader**).

- o The **Scale for dimension features** frame contains some critical settings. Place a check next to **Annotative** for the best approach to handling dimension scale. As with text (and other tools) AutoCAD will size the text automatically using the **Annotation**

Figure 17.007

Scale for the drawing and the **Text height** you set on the **Text** tab.

(Ignore the **layout** option for now. We'll cover layouts in Lessons 24, p.560, and 25, p.584.)
- o The **Fine tuning** frame allows you to manually place text or to force dimension lines between extension lines (even when nothing else will fit).

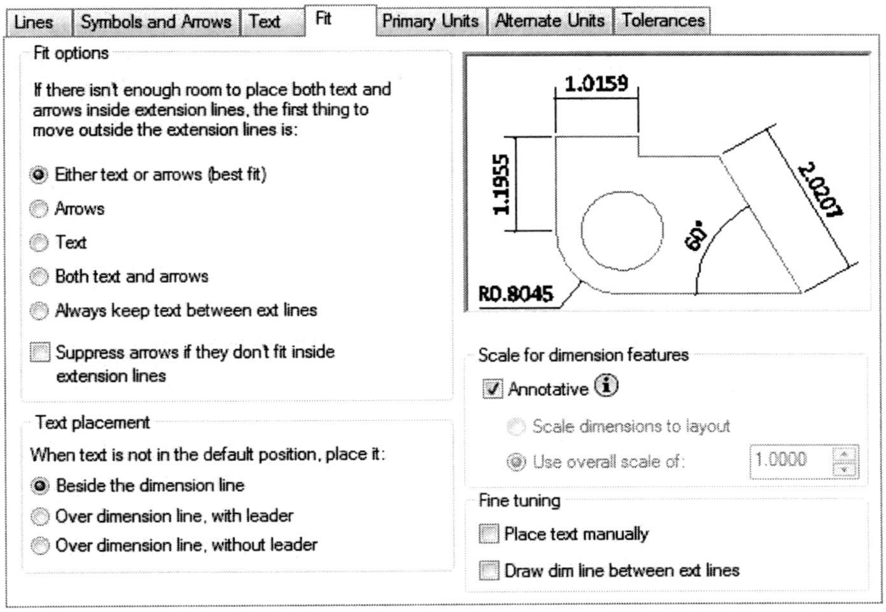

Let's continue.

Do This: 17.1E	Setting up the Fit Tab

I. Continue the previous exercise.
II. Follow these steps.

17.1E: SETTING UP THE FIT TAB

1. On the **Fit** tab:
    a. Allow AutoCAD to determine what to do if the text and arrows don't fit – leave the bullet in the first option of the **Fit options** frame.
    b. When text isn't in the default position, place it over the dimension with a leader (the **Text placement** frame).
    c. Put a check next to **Annotative** in the **Scale for dimension features** frame.
    d. Accept the rest of the defaults. 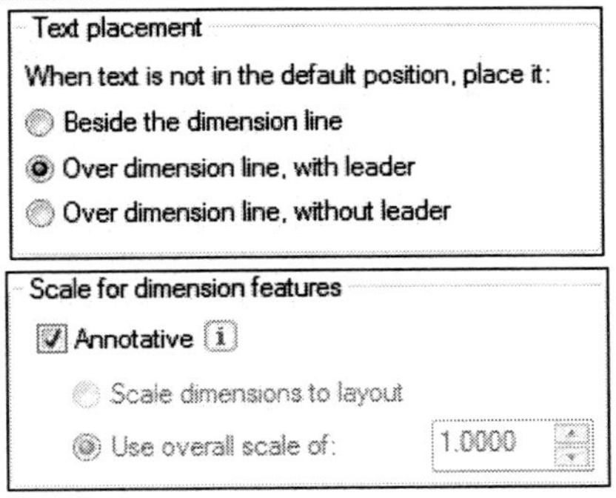
2. Pick the **Primary Units** tab . (Read on before continuing.)

- The **Primary Units** (Figure 17.008, p.385) and the **Alternate Units** tabs allow you to set the dimension units separately from the drawing units. The **Primary Units** tab has two large setup frames.
    - o The **Linear dimensions** frame has several options.
        - It begins with the actual **Unit format** control box. Select the type of units to use here.

384

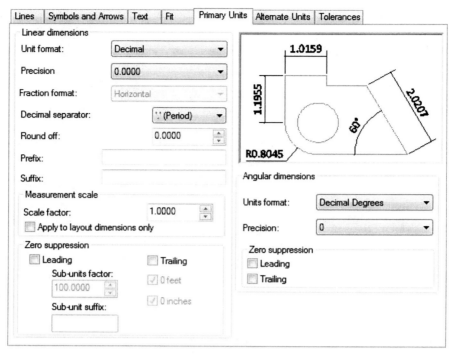

Figure 17.008

- Use the **Precision** control box to set the precision of your dimensions (how many decimal places or the fractional denominator to use).
- When using **Fractional** or **Architectural** units, you can make your fractions stack horizontally or vertically, or you can choose not to stack them at all. These options will be available in the **Fraction format** control box.
- Use the **Decimal separator** control box to separate decimals from whole numbers using a period (as done in the United States), a comma (as done in Europe), or with a space.
- Tell AutoCAD how to round your dimensions in the **Round off** box. AutoCAD will round the dimensions to the nearest unit in this box.
- The **Prefix** and **Suffix** boxes are provided to allow you to place leading or trailing text with your dimensions (like "mm" marks).
- You should probably leave the **Measurement scale** settings at their defaults. The **Scale factor** number box tells AutoCAD how to scale dimensions. Simply put, if the setting is 1.0000 (its default), dimensions reported are the same as the true distance between extension line origins. AutoCAD will take any other setting, multiply it by the true distance, and use the results as the dimension text. This is handy if you're using various details – drawn at different scales – in your Model Space drawing. On the other hand, you'll find a far better approach to different scales and detail work when we examine Paper Space in Lessons 24 and 25.
- **Zero suppression** enables you to dimension without unnecessary zeros. See the examples in Figure 17.009 to 17.012 (p.386).

Note that this frame also includes a couple sub-units entry boxes – for the sub-units factor and suffix. Use these to have AutoCAD automatically switch to a small unit when necessary. For example, when using decimal meters as your primary unit (and suppressing leading zeros), placing *100* in the sub-units factor box and *cm* in the suffix box tells AutoCAD that when a dimension drops below 1 meter, to automatically shift to hundredths of a meter and follow the dimension with a *cm* suffix. So what might have appeared as *.25* meters will now appear as *25cm*.

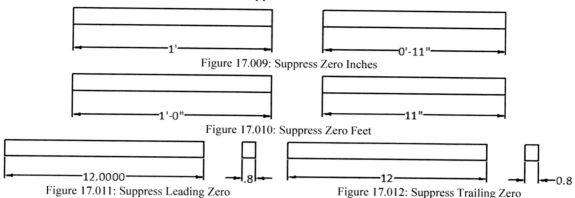

Figure 17.009: Suppress Zero Inches

Figure 17.010: Suppress Zero Feet

Figure 17.011: Suppress Leading Zero

Figure 17.012: Suppress Trailing Zero

- o The **Primary units** tab allows you to set up **Angular dimensions** as well. The **Alternate Units** tab doesn't allow the setup of angular dimensions, but it does provide a frame to allow you to determine the placement of alternate dimensions.

  Alternate units are useful when a company in the United States is working on a European or Asian project that must show metrics as well as feet and inches. A dimension may look something like the one in Figure 17.013.

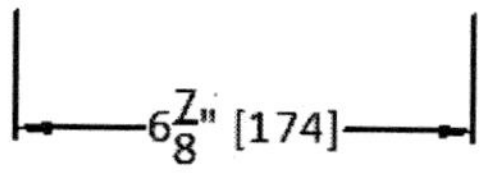

  Figure 17.013: Alternate Units

  Notice that AutoCAD places the alternate units in brackets after the primary units (by default). Also by default, AutoCAD sets the alternate to a scale of 25.4 – or *millimeters*.

Let's continue.

Do This: 17.1F	Setting up the Units

I. Continue the previous exercise.
II. Follow these steps.

17.1F: SETTING UP UNITS

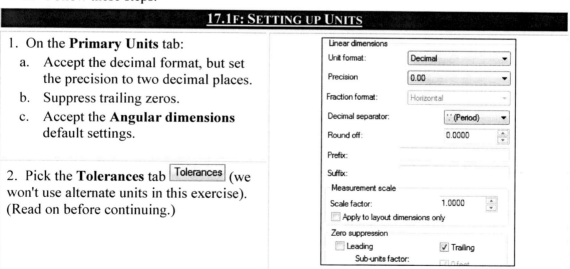

1. On the **Primary Units** tab:
   a. Accept the decimal format, but set the precision to two decimal places.
   b. Suppress trailing zeros.
   c. Accept the **Angular dimensions** default settings.

2. Pick the **Tolerances** tab (we won't use alternate units in this exercise). (Read on before continuing.)

- The last tab we must examine is the **Tolerance** tab (Figure 17.014, p.387). Here you set up any tolerances that your dimensioning may require. Again, we have two large frames.

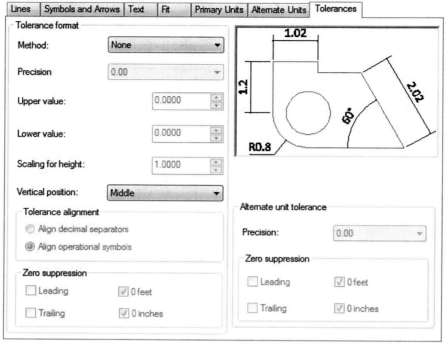

Figure 17.014

- o The first frame allows you to set up the **Tolerance format**.
  - See the examples in Figures 17.015 to 17.020 for **Method** and **Vertical position** values.

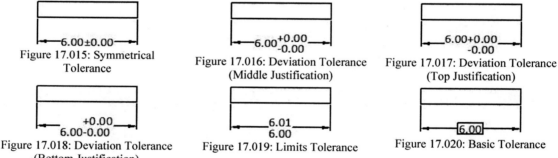

Figure 17.015: Symmetrical Tolerance

Figure 17.016: Deviation Tolerance (Middle Justification)

Figure 17.017: Deviation Tolerance (Top Justification)

Figure 17.018: Deviation Tolerance (Bottom Justification)

Figure 17.019: Limits Tolerance

Figure 17.020: Basic Tolerance

- **Precision** works just as it did on the **Units** tabs.
- **Upper** and **Lower value** identify the value of the tolerances.
- **Scaling for height** allows you to set a separate text height (in proportion to the height set on the **Text** tab) for the tolerances.
- **Tolerance alignment** allows you to line up tolerances according to either the **decimal separator** or the **operational symbols**.
- **Zero Suppression** works as it did on the **Units** tabs.
- o **Alternate unit tolerance** controls tolerances for alternate units.

Let's complete our setup.

Do This: 17.1G	Setting up Tolerances

I. Continue the previous exercise.
II. Follow these steps.

17.1G: SETTING UP TOLERANCES

1. On the **Tolerances** tab:
   a. Use the **Symmetrical** method ...
   b. ... with a precision of two decimal places ...
   c. ... and a value of **0.01**.
   d. Set the tolerance text height to ¾ the size of the dimension text ...
   e. ... and position the tolerances as shown.
   f. Suppress trailing zeros.

2. Pick the **OK** button to complete the setup.

3. Pick the **Close** button to exit the manager.

4. Save the drawing but don't exit.
   Command: *qsave*

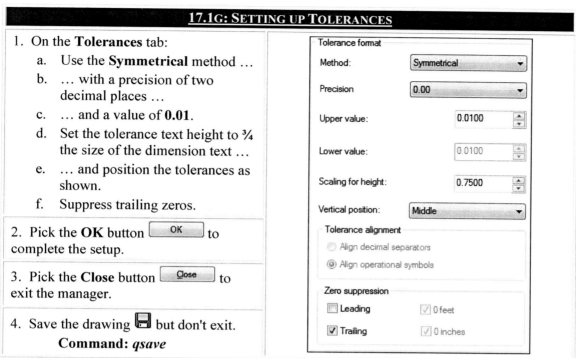

Remember: you can set each of the values you've studied in these last several pages separately for each member of the dimension *family*! This can prove quite handy when, for example, you want to place the dimension text *above* the dimension line, but you want to place radial text at the *end* of a leader. Alternately, you may want AutoCAD to automatically locate the dimension text for linear dimensions but allow you to manually locate it for angular dimensions.

Let's set up a child for radial dimensions.

Do This: 17.1H	Setting up a Child

I. Be sure you're still in the *drill-gizmo-17* file the C:\Steps\Lesson17 folder. If not, please open it now.
II. Follow these steps.

17.1H: SETTING UP A CHILD

1. Reopen the Dimension Style Manager.
   Command: *ddim*

2. Select the **First Steps** style in the list box.

3. Pick the **New** button to create a new child of the **First Steps** style.

4. Use the **Use for** control box to set the child for **Radius dimensions**. Notice that the **New Style Name** box is grayed and unavailable for child styles.
Pick the **Continue** button to continue.

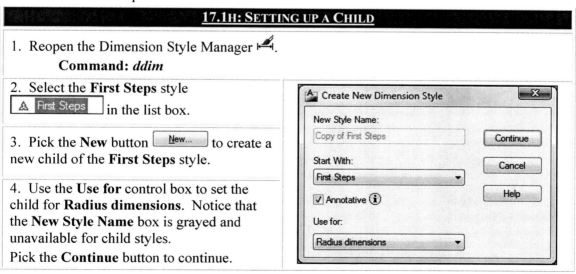

388

17.1H: SETTING UP A CHILD

5. On the **Symbols and Arrows** tab, change the arrowheads to **Closed filled**. Remember that this will only affect radial dimensions.

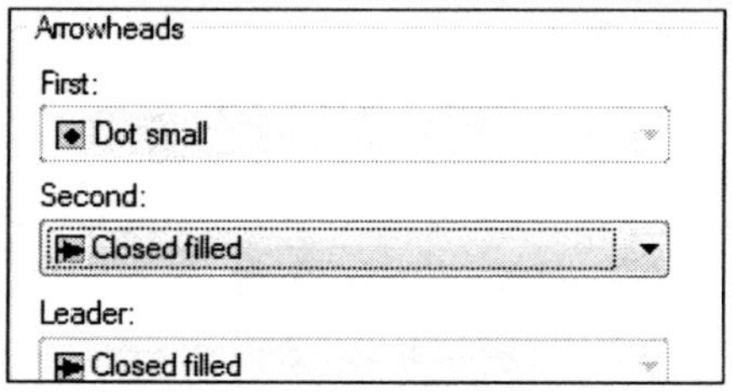

6. On the **Text** tab, set the vertical placement of the radial text as centered on the leader line …

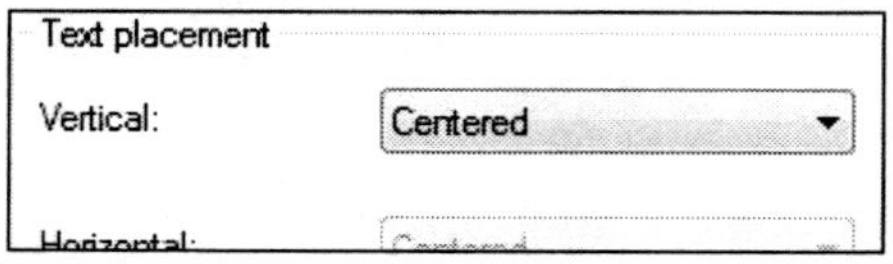

7. … and then set the text alignment to **Horizontal**. This will prevent radial dimensions appearing at oddball angles.

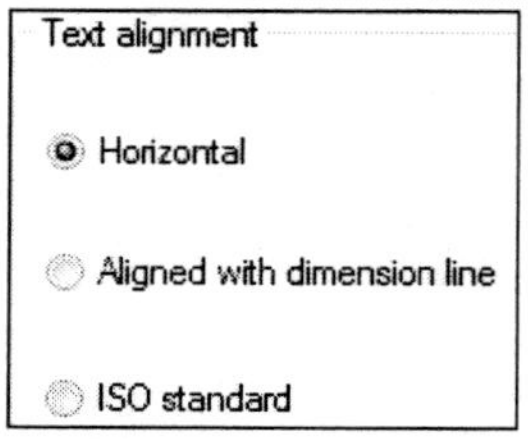

8. Complete the setup [OK].
AutoCAD returns to the Dimension Style Manager. Notice the **Radial** child in the **Styles** list box.

9. Pick the **Close** button [Close] to complete the command.

10. Save the drawing 💾 but don't exit.
   Command: *qsave*

Obviously, there's a tremendous amount of information to absorb if you want to master dimension styles. But unless you're determined to become the contract guru, it may not be necessary to memorize these dialog boxes. AutoCAD provides for everyone, even the casual user!

Your best approach to dimension styles may be to simply follow this text to set up the dimension style you prefer or the one you want to make standard for your contract. But be smart! Set up your dimension styles as part of your templates. Then you can forget them. (Another smart move will be to write down each setting as you go – computers crash and information gets lost!)

And remember – you can copy the dimstyle from one drawing to another using the AutoCAD Design Center (Lesson 7, p.169).

Let's see what our setup looks like.

Do This: 17.1I	Dimension the Drawing

I. Be sure you're still in the *drill-gizmo-17* file the C:\Steps\Lesson17 folder. If not, please open it now.

II. Dimension the drawing. (Watch your layers!) It'll look like Figure 17.021, p.390, when you've finished.

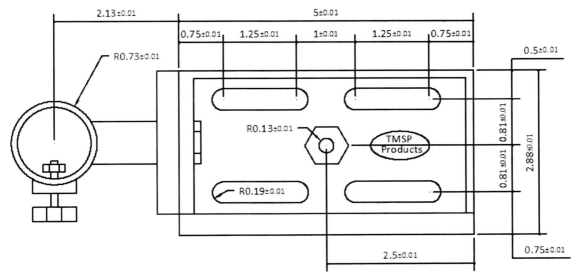

Figure 17.021

| 17.2 | **Miracles of Annotative Dimensioning** |

You've probably figured out that, while annotative text is really cool, changing the size of dimensions (arrows and text) by changing the annotation scale will cause a lot of problems with arrowheads/text not fitting properly between dimension lines. Well, the programmers anticipated the problem! (Praise the Programmers!)

You can actually put text in one location for one annotation scale and in another location for another scale! (No, really! Check it out!)

| **Do This:** 17.2A | **Annotative Dimensioning** |

I. Open the *drill-gizmo-17a* file from the C:\Steps\Lesson17 folder. It's a completed version of the last drawing you used. (Vista and 7 users, well, you know what to do by now.)

II. Follow these steps.

17.2A: ANNOTATIVE DIMENSIONING

1. Change the **Annotation Scale** to **1:2**. (Be sure the toggles next to Annotation scale show **On**.)

Wow! What a change that made in the drawing (following figure).

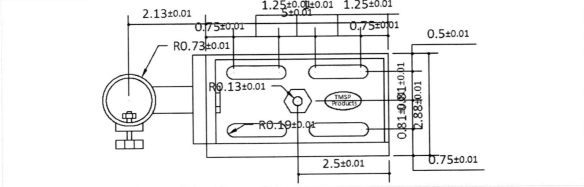

17.2A: ANNOTATIVE DIMENSIONING

2. Select the dimension text on the 2.13 dimension (upper left of drawing). Notice that AutoCAD highlights what you've selected with grips, but you can also see a ghost dimension, too. The ghost dimension is the other annotative scale available for this dimension! The grip you see will affect the current scale's dimension only!

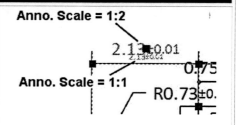

3. Use grips to move the dimensions as indicated in the following figure.

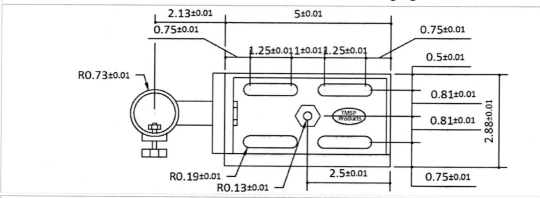

4. Okay, now the miracle. Change the **Annotation Scale** back to 1:1. (Toggle between 1:1 and 1:2 and watch the dimensions relocate automatically as they change size!)

5. Save the drawing and exit.
   Command: *qsave*

Now *that's* cool!

17.3 Try One

Now that you have some understanding of what it takes to create dimension styles and dimension a drawing, let's dimension our floor plan. As you're building experience with the software, I'm going to let you do as much as you can on your own. Refer back to the appropriate portion of the text as needed.

| Do This: 17.3A | Dimension the Floor Plan |

  I. Open your *MyFlrPln17* file from the C:\Steps\Lesson17 folder. If it isn't available, open *FlrPln17* in the same folder.
 II. Zoom all.
III. Create a dimension layer called **dim**. Assign a dark color to your new layer and make it current.
 IV. Follow these steps.

17.3A: DIMENSION THE FLOOR PLAN

1. Open the Dimension Style Manager.
   Command: *ddim*

17.3A: DIMENSION THE FLOOR PLAN

2. Create a new style [New...] called **Arch**. **Continue** [Continue] to the New Dimension Style: Arch dialog box.

3. On the **Lines** tab:
   a. accept the default settings in the **Dimension lines** frame,
   b. set **Extension lines** to **Extend beyond dim lines** by 1/8".

4. On the **Symbols and Arrows** tab:
   a. accept default settings for the **Arc length symbol** and the **Radius dimension jog**,
   b. use **Architectural ticks** for the **First** and **Second Arrowhead**,
   c. use **Closed filled** arrowheads for the **Leaders**,
   d. make the **Arrow size** 1/8",
   e. don't use **Center marks** on circles and arcs.

5. On the **Text** tab,
   a. use the **Times** or **Calibri Text style**,
   b. use a **Text height** of 1/8"
   c. place vertical text **Above** the dimension line and **Center** horizontal text,
   d. use the **ISO standard Text alignment**.

6. On the **Fit** tab,
   a. place the text **Over** the **dimension line, with leader**,
   b. use an **Annotative** dimension **scale**.

7. Set up the **Primary Units** tab as shown at right.

8. We won't need to do any setup on the **Alternate Units** or **Tolerances** tabs. Pick the **OK** button [OK] to continue …

9. … then the **Close** button [Close] to finish the setup.

10. Use the **Dimension Style** control box on the Dimensions panel to set the **Arch** dimstyle current [Arch].

11. Save the drawing 💾 but don't exit.
    Command: *qsave*

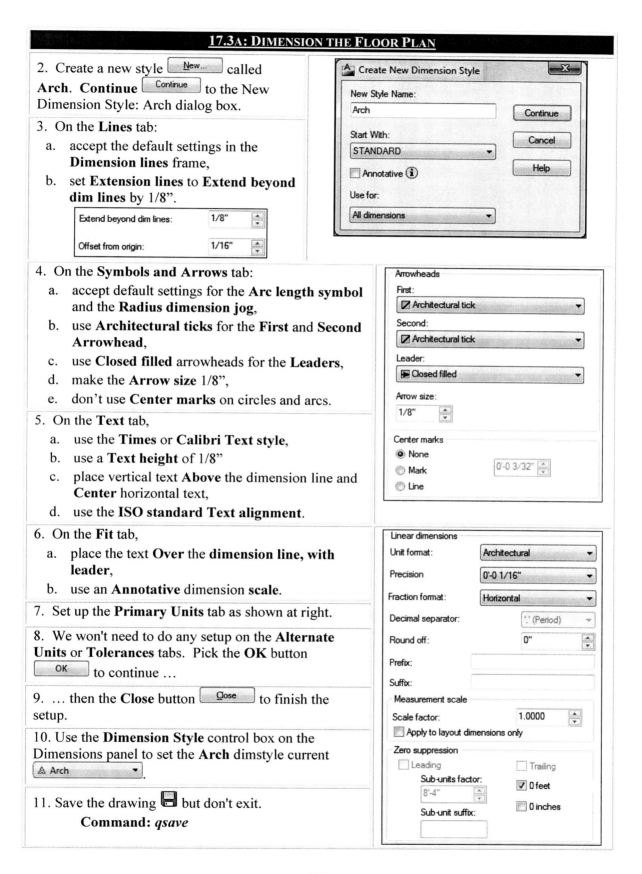

17.3A: DIMENSION THE FLOOR PLAN

12. Dimension the outer walls as shown in the figure following Step 13.

13. Now dimension the rest of the drawing. (Refer to the following figure.) Notice that inner walls and door and window openings are dimensioned to their centers. Use the **Mid Between 2 Points** OSNAP to do this.

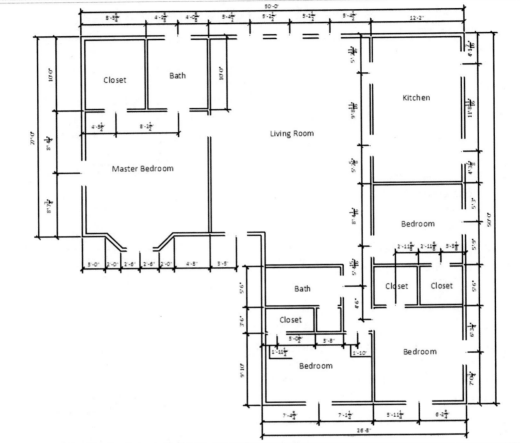

14. Save the drawing as *MyFlrPln18* to the C:\Steps\Lesson18 folder.

   Command: *saveas*

17.4 Overriding Dimensions

You can use the **Override...** button on the Dimension Style Manager to set up dimension variables that differ from the style's settings. But if you use dimension overrides, be aware that as soon as you set a different style as current, the override settings disappear.

I advise against doing this; after all, if you need different settings once, you might need them again. It's best to set up a new style.

A better approach might be to use the Properties palettes to adjust your dimensions' variables once you've placed them.

393

17.5　Dimensional Relationships (Constraints)

17.5.1　The Basics

So far, your dimensional efforts have involved creating dimensions for objects. With associative dimensions, these dimensions will automatically update when the geometry changes. Unfortunately, changes to the dimensions don't have the same effect on the geometry.

Welcome dimensional constraints!

With dimensional constraints, geometry will change when the dimension changes! Additionally, you can lock (constrain) two objects a specified distance apart (remember that term, *design intent*?). The possibilities are endless!

Where to Find It:	
Command Line:	*DimConstraint*
Hotkey(s):	*dcon*
Ribbon (Tab/Panel):	Parametric – Dimensional – (Linear flyout) **[option]**
Menu:	Parametric – Dimensional – **[option]**
Toolbar:	Dimensional Constraints – **[option]**
	Parametric – Aligned flyout – **[option]**

You have the same dimensional constraints that you found in basic dimensions: linear, (horizontal and vertical), aligned, angular, radial, and diameter. You can even use the *DimConstraint* command to convert an associative dimension to a dimensional constraint. You'll find these options within the *DimConstraint* command.

    Command: *dimconstraint*
    Current settings: Constraint form = Dynamic *[AutoCAD lets you know how it's currently set up to create dimensional constraints.]*
    Enter a dimensional constraint option
    [LInear/Horizontal/Vertical/Aligned/ANgular/Radius/Diameter/Form/Convert] <Aligned>: *[You can select an existing associative dimension for AutoCAD to convert or enter your choice of options.]*
    Specify first constraint point or [Object/Point & line/2Lines] <Object>: *[Using the default Aligned option, AutoCAD asks for the first constraint point just as it would ask for the first extension line origin in the DimLinear command; pick your points or hit ENTER to select an object for AutoCAD to automatically dimension.]*

You'll only find a few options with which you're unfamiliar.

- Use the **Form** option to create **Dynamic** or **Annotational** dimensional constraints. Either will control geometry, but they behave differently and you'll manage them in different ways.

  > You can also use the **CConstraintForm** system variable to control the **Form** of your dimensional constraints. A setting of zero means that AutoCAD will use **Dynamic** dimensional constraints; set it to 1 for AutoCAD to use **Annotational** dimensional constraints.

  o AutoCAD didn't intent **Dynamic** dimensional constraints as actual dimensional tools; or put more simply, don't use these if you want to plot the dimensional constraints as regular dimensions. They have a predefined style – much like constraint bars or grips – that you can't change. Using dynamic dimensional constraints means that you'll have to duplicate your dimensions for plotting.

  > Control the display of dynamic dimensional constraints with the **DynConstraintMode** system variable (set to 1 to display or to 0 to hide) or the **Show/Hide** button on the **Dimensional** panel of the **Parametric** ribbon tab.

- o Use **Annotational** dimensional constraints if you wish to avoid duplicating dimensions for plotting. These follow all the rules dimensions follow – including style.

> Although both **Dynamic** and **Annotational** dimensional constraints display a lock 🔒 next to the dimension, you can turn the lock off for the **Annotational** dimensional constraint in the Constraint Settings dialog box (Lesson 10, Section 10.2, p.230).

- Use the **Point & line** option to create a dimensional constraint between a point and a line. AutoCAD will constrain between the point and the nearest point on the selected line.
- Select the **2Lines** option to create a dimensional constraint between two lines. AutoCAD will make the lines parallel to each other and constrain the distance between them.

Let's take a look.

Do This: 17.5.1A	Creating Dimensional Constraints

I. Open the *DimConst* file in the C:\Steps\Lesson17 folder. It's a simple four line, dimensioned rectangle with a circle in the center.
II. Follow these steps.

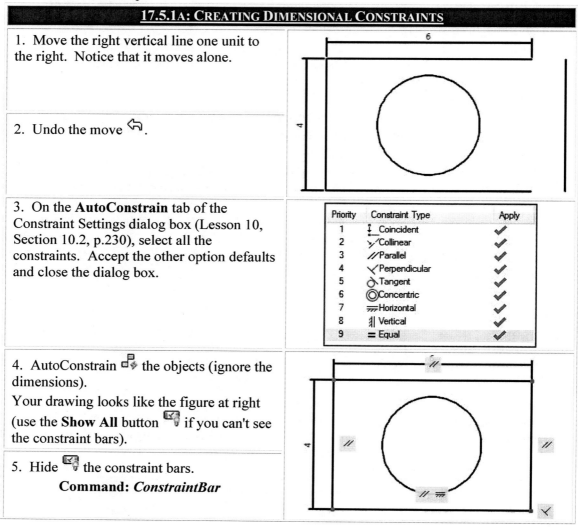

17.5.1A: CREATING DIMENSIONAL CONSTRAINTS

1. Move the right vertical line one unit to the right. Notice that it moves alone.

2. Undo the move.

3. On the **AutoConstrain** tab of the Constraint Settings dialog box (Lesson 10, Section 10.2, p.230), select all the constraints. Accept the other option defaults and close the dialog box.

4. AutoConstrain the objects (ignore the dimensions).
Your drawing looks like the figure at right (use the **Show All** button if you can't see the constraint bars).

5. Hide the constraint bars.
   Command: *ConstraintBar*

17.5.1A: CREATING DIMENSIONAL CONSTRAINTS

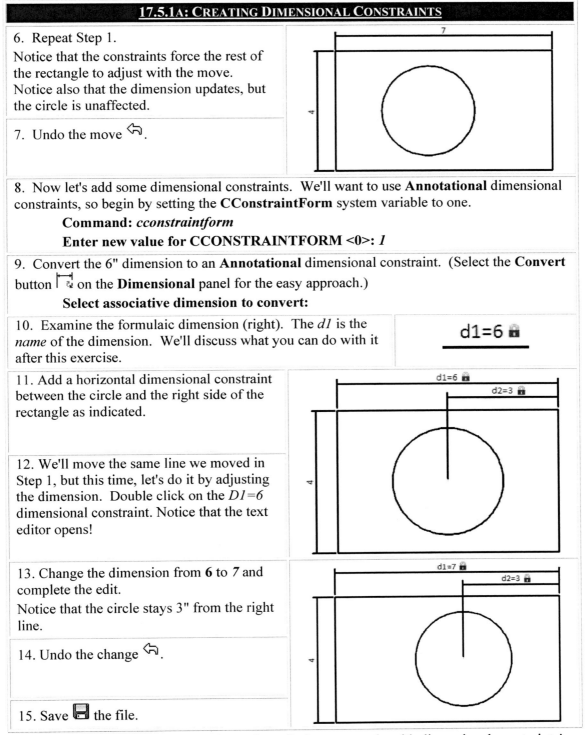

6. Repeat Step 1.
Notice that the constraints force the rest of the rectangle to adjust with the move. Notice also that the dimension updates, but the circle is unaffected.

7. Undo the move.

8. Now let's add some dimensional constraints. We'll want to use **Annotational** dimensional constraints, so begin by setting the **CConstraintForm** system variable to one.
   Command: *cconstraintform*
   Enter new value for CCONSTRAINTFORM <0>: *1*

9. Convert the 6" dimension to an **Annotational** dimensional constraint. (Select the **Convert** button on the **Dimensional** panel for the easy approach.)
   Select associative dimension to convert:

10. Examine the formulaic dimension (right). The *d1* is the *name* of the dimension. We'll discuss what you can do with it after this exercise.

11. Add a horizontal dimensional constraint between the circle and the right side of the rectangle as indicated.

12. We'll move the same line we moved in Step 1, but this time, let's do it by adjusting the dimension. Double click on the *D1=6* dimensional constraint. Notice that the text editor opens!

13. Change the dimension from **6** to **7** and complete the edit.
Notice that the circle stays 3" from the right line.

14. Undo the change.

15. Save the file.

But we've only just begun to scratch the surface of what we can do with dimensional constraints!

17.5.2 Managing Dimensional Constraints

You have two methods available for managing constraints – a simple dialog box (the **Dimensional** tab, Figure 17.023, of the Constraint Settings dialog box you met in Lesson 10, p.230) and a Parameters Manager (Figure 17.024, p.398), which appears in the form of a palette.

We'll look at each.

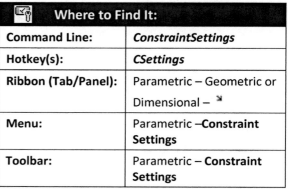

Where to Find It:	
Command Line:	*ConstraintSettings*
Hotkey(s):	*CSettings*
Ribbon (Tab/Panel):	Parametric – Geometric or Dimensional – ⌐
Menu:	Parametric –**Constraint Settings**
Toolbar:	Parametric – **Constraint Settings**

- The **Dimensional** tab of the Constraint Settings dialog box appears simple enough, but it provides some useful options.
    - When you hide the dynamic dimensional constraints, place a check next to **Show hidden dynamic constraints for selected objects** so that AutoCAD will display them temporarily for selected objects.
    - The **Dimensional constraint format** frame contains a tool that can save your sanity! Use the **Dimension name format** control box to tell AutoCAD exactly what to display:
        - **Name** tells AutoCAD to display only the name of the dimensional constraint – but no value,
        - **Value** tells AutoCAD to display only the dimensional value of the constraint,
        - **Name and Expression** displays the formula AutoCAD uses to determine the dimensional value, and its resulting value. You can enter an expression manually, basing a dimensional on other dimensions or values using algebraic formulae. We'll look at this in our next exercise.
        - If the lock next to the constraint in our last exercise bothered you, remove the check next to **Show lock icon for annotational constraints**. (The lock won't plot either way.)

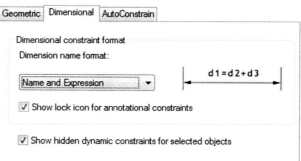

Figure 17.023

Do This: 17.5.2A Managing Dimensional Constraints with Constraint Settings

I. Be sure you're still in the *DimConst* file in the C:\Steps\Lesson17 folder. If not, please open it now.
II. Follow these steps.

17.5.2A: MANAGING DIMENSIONAL CONSTRAINTS WITH CONSTRAINT SETTINGS

1. First, let's see how we can use the name of the dimension to constrain one dimension to another using algebraic equations. Double click on the d2=3 dimension. AutoCAD opens the text editor.

17.5.2A: MANAGING DIMENSIONAL CONSTRAINTS WITH CONSTRAINT SETTINGS

2. Enter the expression d2=d1/2 and close the text editor. (This formula means to make the dimensional constraint equal to half the d1 dimension.) The display doesn't appear any different, does it?

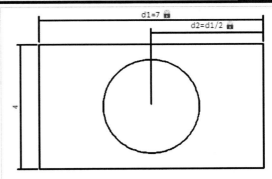

3. Okay, now change the 6" dimension to 7". Notice the difference now? The circle will remain horizontally centered in the rectangle regardless of how you change the dimension! (Cool?!)

4. Now we'll fix the appearance. Open the Constraint Settings dialog box ⁑ with the **Dimensional** tab on top (Figure 17.023, p.397).

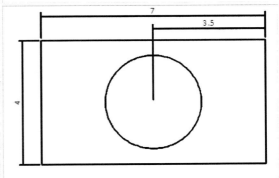

5. Change the **Dimension name format** to **Value** [Value ▼], remove the check next to **Show lock icon for annotational constraints**, and close the dialog box.

Notice the difference in the dimensions (they now appear just as they would if they were regular associative dimensions).

6. Save 💾 the file but don't exit yet.

- Open the Parameters Manager (Figure 17.024) using one of the methods indicated on the WFTI table. It'll be worth the effort!

  Notice that it appears (and behaves) just like any other palette – title bar, docking ability, **Properties** button, and so forth. But let's examine the things that make this palette unique.

  o Use the three buttons across the top to:
    - **create a new user parameter**
      *ƒₓ provides an opportunity for you to create your own parameters (an ideal method for basing your dimensions on a non-dimensional value),
    - **delete the selected parameters**
      ✗ allows you to do just that,
    - use the **New parameter group** button ▽ to create a new group of parameters, and the **Filters** column to sort the groups just as you did with layer filters in

ƒₓ	**Where to Find It:**
Command Line:	*Parameters*
Hotkey(s):	*par*
Ribbon (Tab/Panel):	Parametric – Manage – **Parameters Manager**
Menu:	Parametric –**Parameters Manager**
Toolbar:	Parametric –**Parameters Manager**

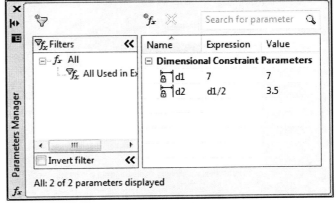

Figure 17.024

Lesson 7 (Section 7.2.1, p.153). (Use the **Double Arrow** buttons « to expand or contract the **Filters** column.)
- The manager's columns provide an easy method for renaming (in the **Name** column) or editing the formula (**Expression** column) of the selected parameter. The symbol shown in the **Name** column indicates the type of parameter you're editing and the figure in the **Value** column indicates the results of the formula in the **Expression** column.
- Of course, you can use the **Search for parameter** box to locate a specific parameter in a busy manager.

Don't you wish they were all that simple?

Do This: 17.5.2B	Using the Parameters Manager

I. Be sure you're still in the *DimConst* file in the C:\Steps\Lesson17 folder. If not, please open it now.
II. Follow these steps.

17.5.2B: USING THE PARAMETERS MANAGER

1. Convert the 4" dimension to a dimensional constraint. (The name of this dimension will be d3.)

2. Add a vertical dimensional constraint between the center of the circle and the top horizontal line.
Your drawing looks like the figure at right.

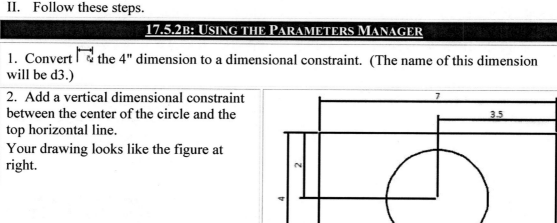

3. Examine the Parameters Manager (right).
- You have four linear-type dimensions, named (by default) **d1** through **d4**.
- Only **d2** currently uses a formula.

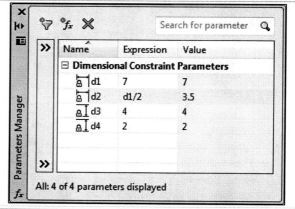

4. Pick **d1** in the **Name** column and rename it *Length*. Notice that the formula for d2 automatically updates to reflect the change.

5. Rename the remaining dimensional constraints as indicated (right).

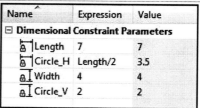

399

17.5.2B: USING THE PARAMETERS MANAGER

6. Now let's create some expressions that'll enable us to control the size of the rectangle through a single dimension – and keep the circle centered while we do it!
Pick in the cell where the **Expression** column and the **Width** row meet.

7. Enter the expression *Length*2/3* (two thirds of the **Length** dimension).
Notice the difference in the geometry.

8. Now center the circle vertically. In the cell where the **Expression** column meets the **Circle_V** row, enter the expression *Width/2*. The Parameters Manager looks like the following figure (left); the drawing looks like the following figure (right).

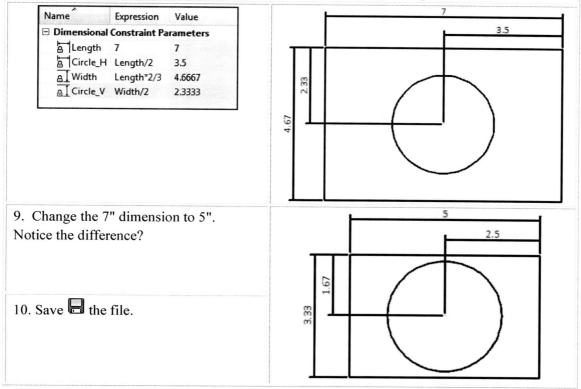

9. Change the 7" dimension to 5". Notice the difference?

10. Save the file.

Is that cool stuff or what?!
Suppose you wanted to make the circle resize as well; how would you do it?*

> Note that you can also modify all the same dimensional constraints parameters using tools in the Properties palette. But for ease of accessibility, use the Parameters Manager.

17.6 Extra Steps

Take a look at the command line approach to dimension styles. It looks like this:
    Command: *-dimstyle*
    Current dimension style: arch   Annotative: Yes *[AutoCAD tells you which style is current.]*
    Enter a dimension style option

---

* Create a diameter dimensional constraint and tie it to one of the dimensions with a formula.

[ANnotative/Save/Restore/STatus/Variables/Apply/?] <Restore>: *[Hit ENTER to make a different style current.]*
Enter a dimension style name, [?] or <select dimension>: *[Type the name of the dimension style you wish to make current.]*

- Use the **Save** option to create a new style based on the current dimvar settings. Use the **Restore** option (the default) to switch from one style to another.
- The **STatus** option will show you the current settings of the dimvars.
- Using the **Variables** option, you can select a dimension and read the values of the dimvars used to create that dimension.
- The **Apply** option will change user-selected dimensions so they use the current dimvars.

This should make you appreciate the Dimension Style Manager!

17.7 What Have We Learned?

Items covered in this lesson include:

- *Defining Dimension Styles*
- *Updating Dimensions*
- *System Variables*
  - *CConstraintForm*
- *Commands:*
  - *Ddim*
  - *Purge*
  - *DimConstraint*
- *Using Dimensional Constraints*
- *Using the Parameters Manager*

  - *DynConstraintMode*

  - *ConstraintSettings*
  - *Parameters*

It's important to take the time now to get comfortable with dimensioning. After all, drafting without dimensioning wouldn't be of much use to anyone. Dimensioning itself doesn't have to be difficult if you take the time to get familiar with the toolbar. But dimension variables, dimension styles, and dimensional constraints can easily overwhelm the AutoCAD novice.

If you don't feel comfortable with dimensioning, go back to Lesson 16, p.350, and do 16 and 17 again. Then say to yourself, "I've met the challenge and am wiser for it!"

17.8 Exercises

1. Open the *drillguide* file in the C:\Steps\Lesson17 folder. Create an appropriate dimension style to dimension the image as indicated in the Drill Guide figure, p.402.
    1.1. Use **small dot** arrowheads but no center marks.
    1.2. Use an overall **scale** of 1.
    1.3. Use decimal units accurate to three decimal places (suppress trailing zeros).
    1.4. Use standard 1/8" text.
    1.5. Allow a tolerance deviation of 1° on all angles, and a precision of zero decimal places.
    1.6. All dimensions should be above the dimension line except radii which should be centered on the leader.

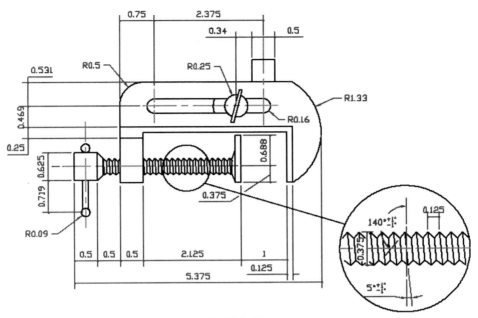

Drill Guide

2. Start with the *MyHandle* file you created in Lesson 7. If you don't have that one, use the *Handle* file instead. It's in the C:\Steps\Lesson17 folder. Add the dimensional constraints that appear in the following figure. Use the information in the Parameters Manager to constrain all the dimensions to the single length dimension.

Name	Expression	Value
⊟ **Dimensional Constraint Parameters**		
Length	10	10.0000
Lg_Arc	.25*Length	2.5000
Lg_Circle	1.25*Lg_Arc	3.1250
Sm_Arc	.25*Lg_Cir...	0.7813
Sm_Circle	.5*Lg_Circle	1.5625

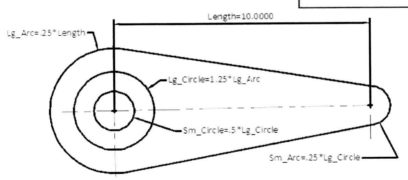

17.9 **For Web-Based Review Questions and Additional Exercises, visit: http://foragerpub.com/AcadFiles/2011/2011.htm**

So, how'd you do? Well, dimensions have lead us to leaders!
Com'n!

Lesson 18

Following this lesson, you will:

- ✓ *Know how to create leaders*
- ✓ *Know how to create and use Leader Styles*
- ✓ *Know how to purge unwanted styles from your drawing*

Leaders

What would drafting be without leaders – those tiny arrows that run from callouts or details to referenced objects? Why, without them, we might never know which holes to countersink or what part of the wall gets the chartreuse enamel!

*For such a tiny drafting tool, leaders sure present a challenge – as the number of commands involved indicates! These include **MLeader** (aka. MultiLeader – to create the leader), **MLeaderEdit** (to add or remove leader lines from callouts), **MLeaderAlign** (for, you guessed it, aligning leaders), and **MLeaderCollect** (for collecting several block callouts into one place using a single leader line). MLeaders also include a dialog box (accessed with the **MLeaderStyle** command) for defining/customizing leader styles!*

First, we'll concentrate on how a couple of the MLeader commands work; then we'll look at customizing dimensions and leaders in Section 18.2, p.413.

> As with so many commands, leaders present us with an Olde Way and a New Way. For details on the Olde Way – *QLeader* – refer to the *QLeaders.pdf* supplement found at *www.uneedcad.com/Files*.

So, before I "lead" you astray, let's look at leaders!

18.1 The *MLeader* Command

18.1.1 Placing Leaders

So let's take a look. First, the ***MLeader*** command:

Command: mleader

Specify leader arrowhead location or [leader Landing first/Content first/Options] <Options>: *[Begin by placing the leader head or select to place the tail first, the content (call out) first, or look at the other options.]*

Specify leader landing location: *[When using the leader head option, your next step is to pick the other end of your leader (the landing – a short, horizontal line that typically connects the leader to your callout).]*

[At this point, AutoCAD opens the MText editor.]

Where to Find It:	
Command Line:	*MLeader*
Hotkey(s):	*mld*
Ribbon (Tab/Panel):	Home – Annotation – **Multileader**
	Annotate – Leaders – **Multileader**
Menu:	Dimension – **Multileader**
Toolbar:	Multileader – **Multileader**
Tool Palette:	Leaders – [option]

The sequence looks deceptively simple – only three options. Accept the defaults for a quick and easy leader, but let's look at the other options. (The options are, as usual, also available on the dynamic and cursor menus.)

- The default – **leader arrowhead location** first – utilizes AutoCAD's traditional approach of identifying where you'd like the leader head to go (the referenced object) and then drawing the leader back to the callout.
- **leader Landing first** enables you to draw the leader starting at the callout and ending at the referenced object. This approach more closely resembles the way most people drew leaders on the drawing board.
- **Content first** also closely resembles the way most people drew leaders on the drawing board – but with **Content first**, you can create the callout and then create the leader! The **Content first** approach will even justify a multiline callout for you at the leader's landing.
- **Options** presents the following prompt:

405

Enter an option [Leader type/leader lAnding/Content type/Maxpoints/First angle/Second angle/eXit options] <eXit options>:

Most of these options can and probably should be set up using the MLeader Style dialog box (Section 18.2, p.413). That way, you can save the settings for use later. Still, let's see how they work here.

- **Leader type** enables you to determine what type of leader to use (line or spline).
- **Leader lAnding** gives you the chance to determine whether or not to use a landing and, if so, how long to make it.
- **Content type** lets you decide what type of content (if any) you wish to use. Possibilities include **Block**, **MText** (the default) or **None**. If you opt for **None**, AutoCAD won't open the MText editor after you insert the leader.
- **Maxpoints** specifies the maximum number of line-defining points you can use in creating your leader.
- **First angle** and **Second angle** restrict the angle of the leader lines.
- **eXit options** returns you to the **Specify leader** prompt.

I know it sounds like a lot of work for a leader, but the variety of leader types available requires a lot of options. The default settings will work fine for most industries. (I guess that's why AutoCAD made them defaults!)

Let's put some leaders into a drawing before we go on.

Do This: 18.1A	Creating Leaders

I. Open the *BoilerTank* file in the C:\Steps\Lesson18 folder. It looks like the figure at right.

This is the beginning of a pressure vessel's detail sheet. We'll add some callouts.

II. Be sure the **Text** layer is current and zoom in around the elevation of the vessel.

III. Follow these steps.

18.1A: CREATING LEADERS

1. Enter the *MLeader* command.

   Command: *mleader*

2. Begin at the coupling indicated here.

   Specify leader arrowhead location or [leader Landing first/Content first/Options] <Options>:

3. Put the landing up and to the left (refer to the figure in Step 4).

   Specify leader landing location...

4. AutoCAD presents the MText editor. Enter 1/8" text on two lines as indicated. (The text reads **1" 3000# CPLG for Level Gauge**.)

5. Close the text editor.

18.1A: CREATING LEADERS

6. Repeat Steps 1 – 5 for the remaining callouts. (Try to locate them as close as possible to their locations in the following figure.) Your drawing will look like the following figure when you've finished.

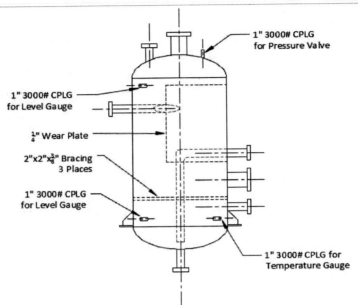

7. Adjust your view so you can see the plan.

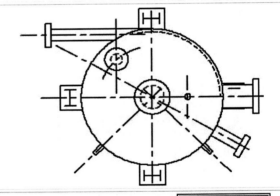

8. We'll create some bubble callouts here. Begin the *Mleader* command.
   Command: *mleader*

9. Use the **Options** option , and then the **Content type** option .
   Specify leader arrowhead location or [leader Landing first/Content first/Options] <Options>: *O*
   Enter an option [Leader type/leader lAnding/Content type/Maxpoints/First angle/Second angle/eXit options] <eXit options>: *C*

10. Select the **Block** option and enter the block name indicated (be sure to use the underscore before the name). (You'll see more on blocks in Lessons 21, p.468, and 22, p.500.)
    Select a content type [Block/Mtext/None] <Mtext>: *B*
    Enter block name: *_tagcircle*

11. Exit the options prompts.
    Enter an option [Leader type/leader lAnding/Content type/Maxpoints/First angle/Second angle/eXit options] <Content type>: *X*

18.1A: CREATING LEADERS

12. Place the first leader as indicated.

    Specify leader arrowhead location or [leader Landing first/Content first/ Options] <Options>:

    Specify leader landing location:

13. When prompted, identify the **tagnumber** as **O**.

    Enter attribute values

    Enter tag number <TAGNUMBER>: *O*

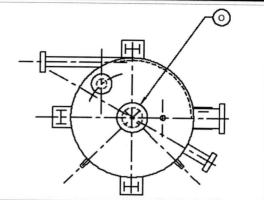

14. Place the remaining bubble leaders as indicated. (Some bubbles use the **None** leader type.)

15. Save the drawing 💾 but don't exit.

    Command: *qsave*

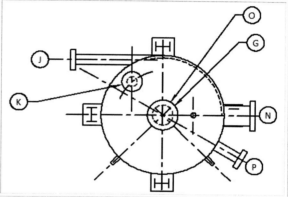

18.1.2 Aligning and Collecting Leaders

We'll use *MLeaderAlign* to align our callouts. *MLeaderAlign* works like this:

> **Command:** *mleaderalign*
>
> **Select multileaders:** *[Select the leaders you'd like to align – don't select the leader to which you'd like to align!]*
>
> **Select multileaders:** *[Confirm the selection set.]*
>
> **Current mode: Use current spacing** *[AutoCAD tells you how it'll align the leaders you select.]*
>
> **Select multileader to align to or [Options]:** *[Select the leader to which you'd like to align the others you've selected.]*
>
> **Specify direction:** *[In which direction would you like AutoCAD to align the leaders?]*

Where to Find It:	
Command Line:	*MLeaderAlign*
Hotkey(s):	*mla*
Ribbon (Tab/Panel):	Home - Annotation – (Multileader flyout) **Align**
	Annotate – Leaders – **Align**
Toolbar:	Multileader – **Align Multileaders**

This looks easy enough, but we haven't looked at those options yet (those infernal options – so many choices … so few callouts!). If you select the **Options** option, AutoCAD prompts:

> **Enter an option [Distribute/make leader segments Parallel/specify Spacing/Use current spacing] <Use current spacing>:**

- **Distribute** distributes the selected leaders between two user-defined points. This approach works well; but remember to use Ortho when selecting the points!

408

- **make leader segments Parallel** makes the last line segment created in each selected leader parallel to each other.
- **Specify Spacing** lets you determine the spacing between the selected leaders.
- **Use current spacing** lets you use the existing spacing and align the selected leaders with one you will select when you return to the **Select multileader to align to** prompt.

Finally, we'll use *MLeaderCollect* to gather our bubbles in a more orderly fashion. It works like this:

    Command: *mleadercollect*
    Select multileaders: *[Select the leaders you wish to collect – block leaders work best.]*
    Select multileaders: *[ENTER to confirm the selection set.]*
    Specify collected multileader location or [Vertical/Horizontal/Wrap] <Horizontal>: *[Place the collected leaders.]*

Where to Find It:	
Command Line:	*MLeaderCollect*
Hotkey(s):	*mlc*
Ribbon (Tab/Panel):	Home – Annotation – (Multileader flyout) **Collect**
	Annotate – Leaders – **Collect**
Toolbar:	Multileader – **Collect Multileaders**

The options include:
- stacking the blocks **Vertic**ally or **Horizontal**ly.
- **Wrap**ping the blocks within a specified width or number per row.

So, let's align and collect some of our leaders.

Do This: 18.1.2A	Aligning and Collecting Leaders

    I. Continue in the *BoilerTank* drawing, or open the *BoilerTank2* file in the C:\Steps\Lesson18 folder. It's the same drawing but is finished to this point.
   II. Zoom in around the vessel elevation.
  III. Follow these steps.

18.1.2A: ALIGNING AND COLLECTING LEADERS
1. Enter the *Mleaderalign* command.   **Command:** *mleaderalign*

2. Select the leaders indicated here.
     Select multileaders:
     Select multileaders: *[ENTER]*

3. Select the *1"-3000# CPLG* callout above those you just selected **to align to**.   **Select multileader to align to or [Options]:**
4. Using Ortho (hold down the SHIFT key for the Ortho override), pick a point straight down to align the leaders in that direction.   **Specify direction:**   Your leaders look something like the following figure.

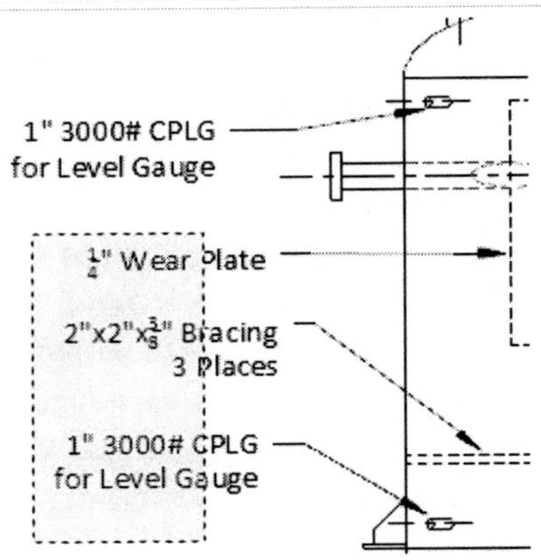

18.1.2A: ALIGNING AND COLLECTING LEADERS

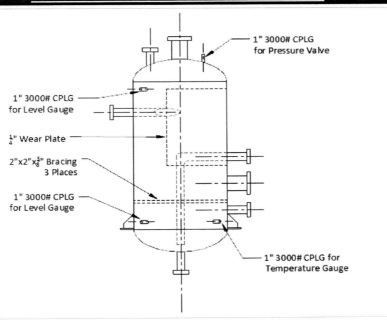

5. I know; that worked great. But let's try a different approach. Undo the changes ↶.
   Command: *u*

6. Repeat the *Mleaderalign* command ⟲.
   Command: *mleaderalign*

7. Select all the leaders left of the vessel.
   Select multileaders:
   Select multileaders: *[ENTER]*

8. Take the **Options** option [Options] and select **Distribute** [Distribute].
   Current mode: Use current spacing
   Select multileader to align to or [Options]: *o*
   Enter an option [Distribute/make leader segments Parallel/specify Spacing/Use current spacing] <Use current spacing>: *d*

9. Pick a point just above the "f" in the top callout…
   Specify first point or [Options]:

10. … and a second point about 2" directly below the first (use Polar tracking!).
    Specify second point:

Your drawing now looks something like the following figure.

18.1.2A: ALIGNING AND COLLECTING LEADERS

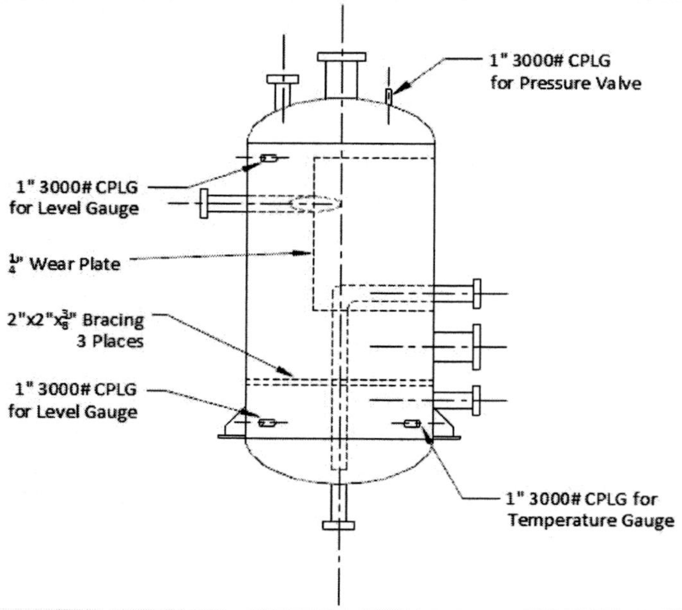

[Figure: Vessel elevation view with leader callouts:]
- 1" 3000# CPLG for Pressure Valve
- 1" 3000# CPLG for Level Gauge
- ¼" Wear Plate
- 2"x2"x⅜" Bracing 3 Places
- 1" 3000# CPLG for Level Gauge
- 1" 3000# CPLG for Temperature Gauge

11. Now let's collect a couple bubbles. Adjust your view so you can see the plan.
    Notice that bubbles **O** and **G** point to the same nozzle. Why not use a single leader?

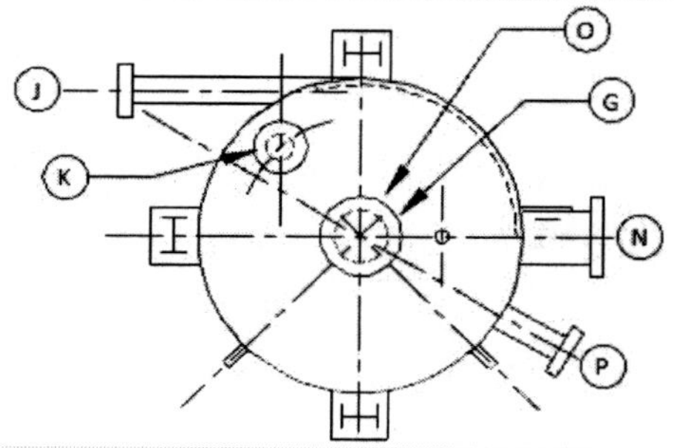

12. Enter the *MLeaderCollect* command 🔲.
    Command: *mleadercollect*

13. Select bubbles **O** and **G**.
    Select multileaders:
    Select multileaders: *[ENTER]*

14. Place the collected leader where bubble O had been.
    Specify collected multileader location or [Vertical/Horizontal/Wrap] <Horizontal>:
    Your drawing looks like the figure at right.

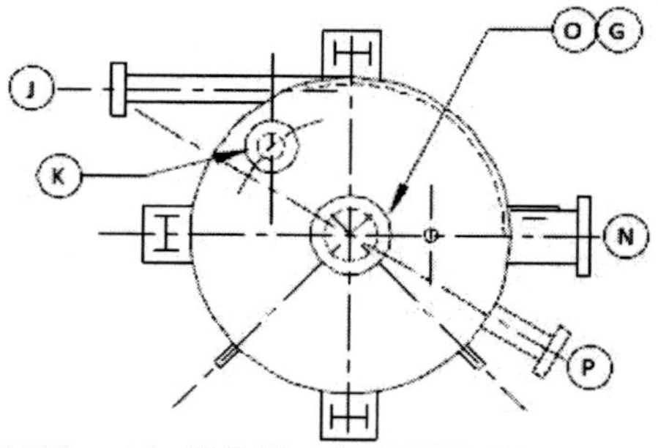

15. Save the drawing 💾.
    Command: *qsave*

18.1.3 Editing Leaders

Contrary to what you might think, you won't use the *MLeaderEdit* command to do most of your leader editing. Oh, it comes in quite handy for adding or removing additional leader heads to existing leaders (you can't beat it for that), but you'll do most other editing in the Properties or Quick Properties palettes.

But first, the *MLeaderEdit* command ...
The command sequence is quite simple:

> **Command:** *mleaderedit*
> **Select a multileader:** *[Select the multileader you wish to edit.]*
> **Specify leader arrowhead location or [Remove leaders]:** *[Add the new leader head or select the Remove leaders option.]*

We'll try this in our next exercise.

Editing anything else related to a multileader involves either double clicking on the leader to open the Properties palette or simply selecting it with Quick Properties toggled on ▣. Make your changes within the palette.

Where to Find It:	
Command Line:	*MLeaderEdit*
Hotkey(s):	*mle*
Ribbon (Tab/Panel):	Home – Annotation – (Multileader flyout) **[Add or Remove Leader]**
	Annotate – Leaders – **[Add or Remove Leader]**
Menu:	Modify – Object - Multileader – **[Add or Remove Leader]**
Toolbar:	Multileader – **[Add or Remove Leader]**
Tool Palette:	Leaders – [option]

You can also edit a specific line segment of the multileader; select what you wish to edit while holding down the CTRL key on your keyboard. AutoCAD will open the Properties palette with the selected line segment shown.

Do This: 18.1.3A — Editing Leaders

I. Continue in the *BoilerTank* or the *BoilerTank2* drawing.
II. Zoom in around the vessel plan.
III. Follow these steps.

18.1.3A: EDITING LEADERS

1. Add a 1" 3000# CPLG callout as indicated.

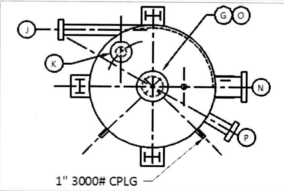

2. Begin the *MLeaderEdit* command (you can use the **Add Leader** button ⌐ for this exercise).

> **Command:** *mleaderedit*

3. Select the leader you just created.

> **Select a multileader:**

18.1.3A: EDITING LEADERS

4. And direct the new leader head toward the other couplings as indicated.

   Specify leader arrowhead location or [Remove leaders]:

   Specify leader arrowhead location or [Remove leaders]: *[ENTER]*

5. Save the drawing.

   Command: *qsave*

6. With Quick Properties toggled on, select the new callout. Examine the properties listed on the palette.

7. Change the **Leader type** to **Spline**. Notice the difference? (Wasn't that easy?!)

8. Undo the change.

9. Clear the selection and, holding down the CTRL key, select just one of the leader lines. The Quick Properties palette doesn't appear, but check the Properties palette (right). You can edit the individual lines using this approach.

10. Save the drawing and close it.

    Command: *qsave*

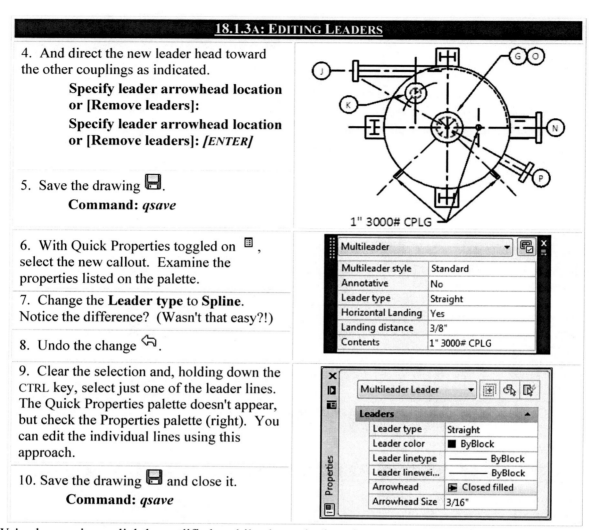

We've been using a slightly modified multileader style (it makes the graphics prettier). Wouldn't you like to know how we modified it?

Check out Section 18.2!

18.2 Customizing Leaders

You'll find the procedure for customizing leaders very similar to that for customizing dimensions – even the dialog boxes are similar (Figure 18.001, p.414) as you'll see when you enter the *MLeaderStyle* command.

I won't spend time explaining the buttons as they're identical to those found on the Dimension Style dialog box. You'll proceed to the Multileader Style dialog (Figure 18.002, p.415) by picking the **New** button, naming your new style in the Create New Multileader Style dialog box, just as you did with Dimension Styles, and using the **Continue** button.

Let's set up a new style as we explore the possibilities.

The right leader makes all the difference!

413

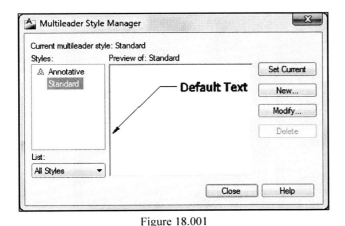

Figure 18.001

	Where to Find It:
Command Line:	*MLeaderStyle*
Hotkey(s):	*mls*
Ribbon (Tab/Panel):	Home – Annotation (subpanel) – **Multileader Style**
	Annotate – Leaders – **Multileader Style**
Menu:	Format – **Multileader Style**
Toolbar:	Multileader – **Multileader Style**

Do This: 18.2A	Create a New Leader Style

III. Open the *drill-gizmo-18* file in the C:\Steps\Lesson18 folder. This is the completed version of the file we used in Lesson 17.
IV. Set the **Annotation Scale** to 1:1.
V. Follow these steps.

18.2A: CREATE A NEW LEADER STYLE

1. Enter the *MLeaderStyle* command.
    Command: *mleaderstyle*
AutoCAD presents the Multileader Style Manager (Figure 18.001).

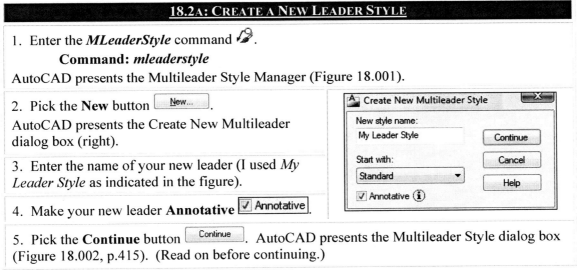

2. Pick the **New** button.
AutoCAD presents the Create New Multileader dialog box (right).

3. Enter the name of your new leader (I used *My Leader Style* as indicated in the figure).

4. Make your new leader **Annotative**.

5. Pick the **Continue** button. AutoCAD presents the Multileader Style dialog box (Figure 18.002, p.415). (Read on before continuing.)

In the Modify Multileader Style dialog box (Figure 18.002, p.415), you'll find three tabs (only three this time!) to help you define your leader style.

We'll start with the **Leader Format** tab (Figure 18.002).

- Options in the **General** frame allow you to set the **Type (Straight, Spline,** or **None), Color, Linetype** and **Lineweight** of your leader.
- **Arrowhead** options are the same as those found on the **Symbols and Arrows** tab of the Dimension Style dialog box (p.381).
- Use the **Break size** tool in the **Leader break** frame to set the size of the break when the *DimBreak* command is used on a leader.

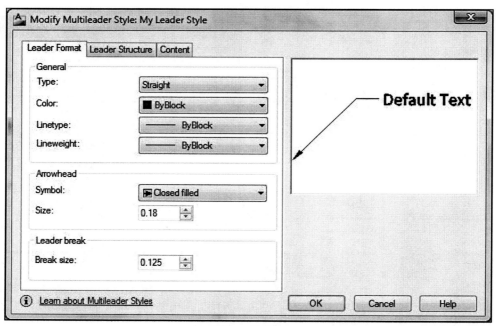

Figure 18.002

| Do This: 18.2B | Setting Up Your New Leader's Format |

I. Continue in the *drill-gizmo - 18* file.
II. Follow these steps.

18.2B: SETTING UP YOUR NEW LEADER'S FORMAT

1. Format the leader as follows:
   - Use a **Spline – Type**.
   - Tell AutoCAD to create leaders using the current layer's **Color**.
   - Accept the **Closed filled Arrowhead**, but make it **1/8"**.
   - Finally, use **.18** as the **Break size** for **Leader breaks**.

2. Pick on the **Leader Structure** tab 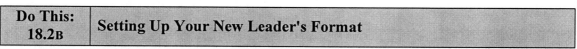. (Read on before continuing.)

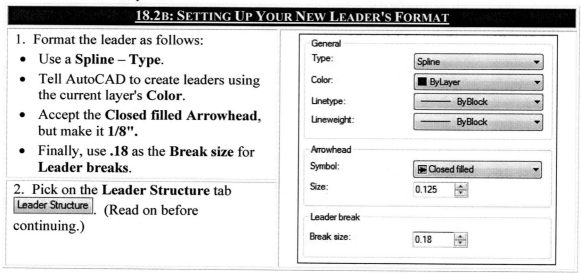

Use options on the **Leader Structure** tab (Figure 18.003, p.416) to set up the structure of your leader.
- Options in the **Constraints** frame provide a means to control how you draw the leader.
  - **Maximum leader points** is the number of picks you use to identify the points that define your leader. Set this to two or three – should you need more than that, I suggest you rethink your leader.
  - **First** and **Second segment angle**s restrict the angle of the leader lines.

- Use options in the **Landing settings** frame to control that tiny line that connects the leader to the callout. These options are only available for **Straight** leader **Types**.
- It's generally a good idea to stick with the **Annotative Scale** whenever creating an object that makes that option available to you. We'll look at layouts in Lessons 24, p.560, and 25, p.584, but even then, I'll recommend using the **Annotative Scale** whenever possible. If you don't use annotative scales, you can **Specify** [a] **scale** for your leaders. Use the Drawing Scale Factor for your drawing (see Appendix B).

Figure 18.003

| Do This: 18.2c | Setting Up Your New Leader's Structure |

I. Continue in the *drill-gizmo-18* file.
II. Follow these steps.

18.2C: SETTING UP YOUR NEW LEADER'S STRUCTURE

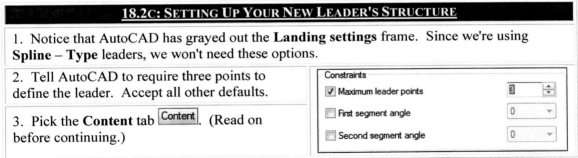

1. Notice that AutoCAD has grayed out the **Landing settings** frame. Since we're using **Spline – Type** leaders, we won't need these options.

2. Tell AutoCAD to require three points to define the leader. Accept all other defaults.

3. Pick the **Content** tab. (Read on before continuing.)

Use options on the **Content** tab (Figure 18.004) to set up the callout or blocks to use with your leader.
- Here, your **Multileader types** include **Mtext** or **Blocks**. We'll learn about blocks in Lessons 21, p.468, and 22, p.500.
- By now, items in the **Text options** frame are self-explanatory – except that you can enter a **Default text** if you like.
- **Leader connection** options, on the other hand, can really make the difference between professional-looking leaders and someone who just uses defaults. Here, you'll determine where to attach the leader (or landing) to your callout.
- When you select **Block** in the Multileader type control box, AutoCAD replaces the **Text options** and **Leader connection** frames with a **Block options** frame. Use this to select your **Source block**, the **Attachment** type (by the block's center point or its

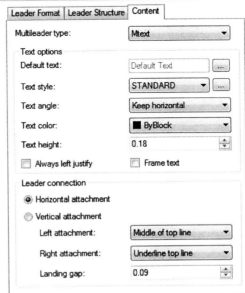

Figure 18.004

insertion point), and the block's **Color**.

Okay, one last set up to complete before adding some leaders to our drawing.

Do This: 18.2D	Setting Up Your New Leader's Content

I. Continue in the *drill-gizmo-18* file.
II. Fo these steps.

18.2D: SETTING UP YOUR NEW LEADER'S CONTENT

1. Set up the **Content** tab as follows:
   - We won't use **Default text**, so leave that box empty.
   - Use the **Times** or **Calibri Text style** and a **Text height** of 1/8". Accept the other defaults in the **Text options** frame.
   - Use a **Horizontal attachment**.

2. Pick the **OK** button to return to the Multileader Style Manager.

3. Pick the **Close** button to close the Multileader Style Manager.

4. Save the drawing but don't exit.
   Command: *qsave*

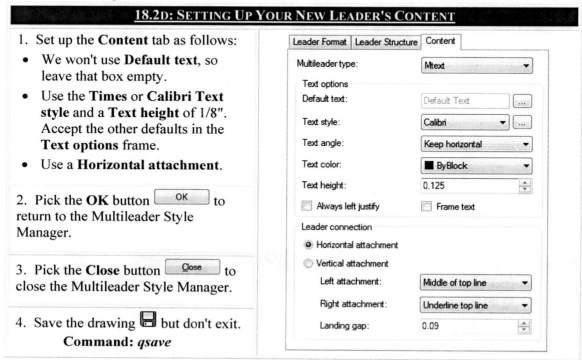

Now you've set up your own leader, add a few leaders to the drawing. It should look something like Figure 18.005 when you've finished.

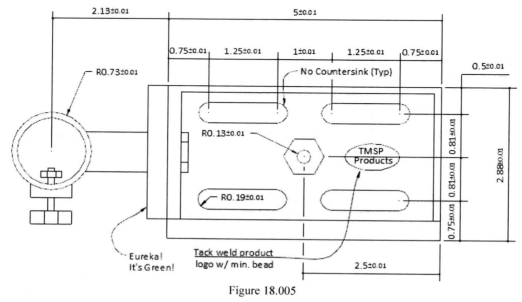

Figure 18.005

417

18.3 Extra Steps

Setting up dimensions and leaders is no easy chore. Fortunately, AutoCAD provides some tools for accidents, boo boos, and uh, ohs!

The easiest way to rid a drawing of unwanted (and unused) styles is to select the style in the list box of the style manager, and then hit the DEL key on your keyboard. AutoCAD will prompt with an Are You Sure message box. Pick **Yes** to remove the style.

Another alternative for removing styles – and unwanted linetypes, layers, and blocks (among other things) – is to use the Purge dialog box (Figure 18.006). Access the dialog box with the **Purge** command.

In the Purge dialog box, you can **View items you can purge** or **items you cannot purge**. AutoCAD will list the items in the list box making it easy for you to decide which to remove from the drawing. Remember, once removed the item can't be recovered. However, removing unused items will lower file size.

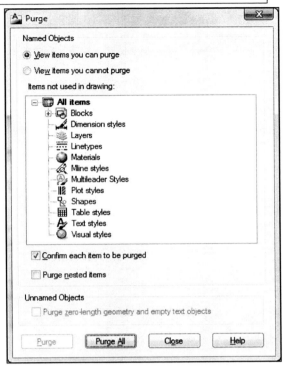

Figure 18.006

18.4 What Have We Learned?

Items covered in this lesson include:
- *Leaders*
- *How to purge a drawing*
- *Commands:*
    - **MLeader**
    - **MLeaderAlign**
    - **MLeaderCollect**
    - **MLeaderStyle**
    - **MLeaderEdit**
    - **Purge**

After two lessons on dimensions, you probably found leaders a bit redundant. But that's okay; without an occasional short and simple lesson, we'd both get burned out.

Try a quick exercise or two and then move on to tables. You may find them more challenging!

18.5 Exercises

1. Open the Soot Trap file in the C:\Steps\Lesson18 folder and add the text, dimensions, callouts, and leaders indicated in the following figure.

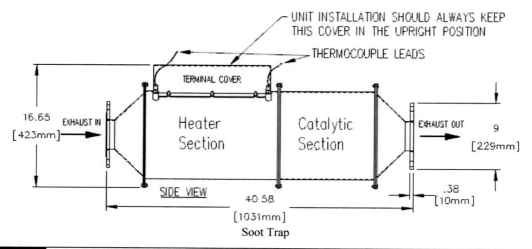
Soot Trap

18.6 For Web-Based Review Questions and Additional Exercises, visit: http://foragerpub.com/AcadFiles/2011/2011.htm

Well, let's table leaders for now and move on to, well, tables!

Lesson 19

Following this lesson, you will:

- ✓ Know how to use tables in your drawing
- ✓ Know how to use fields in your drawing
- ✓ Know how to use these commands:
  - **Table**
  - **TableStyle**
  - **DwgProps**
  - **Field**
  - **DataLink**

Tables and Fields

What we humans have worked for since the dawn of time has been a way to avoid work! (And oh, how hard we've worked to succeed at this most admirable goal!)

That's where things like tables and fields come in handy. Tables organize our world (nothing says lazy like pristine organizational skills, which of course, free up time to stare at the skyline or drift aimlessly in the pool); fields fill in those nicely organized blanks for us.

This section of our text will cover some razzle-dazzle tools. These are the tools that pushed CAD systems ahead of the drawing board as the preferred tool in design work. We'll begin with some text tools in this lesson; then, in subsequent lessons, we'll learn to create hatching, blocks of objects, and blocks of information.

So, stop yawning, and let's begin

19.1 Tables

It's amazing just how much work can be avoided by the innovative use of rows and columns! Of course, the collective name for rows and columns is *table*. Tables have saved time and effort in virtually every industry on earth (just ask any accountant).

AutoCAD makes tables available for legends, materials lists, revision blocks, spreadsheets, and just about any other use you might require. We'll look at creating a top-down table (a legend) and a bottom-up table (a revision block). More importantly, we'll look at the ease with which you can modify a list for your own uses.

> Just a quick note before we continue:
> AutoCAD doesn't yet incorporate Annotation Scale in its tables. I suppose there just wasn't time to get it done before the current release. If you haven't reviewed the Olde Way of entering text, please review the *TextSize-theOldeWay.pdf* supplement found at http://www.uneedcad.com/Files.

19.1.1 Creating a Table

To keep it simple, AutoCAD provides a dialog box (Figure 19.001, p.422) for creating tables.

Where to Find It:	
Command Line:	*Table*
Hotkey(s):	tb
Ribbon (Tab/Panel):	Home – Annotation – **Table**
	Annotate – Tables – **Table**
Menu:	Draw – **Table**
Toolbar:	Draw – **Table**
Tool Palette:	Draw – **Table**

- The first frame – **Table Style** – allows you to change the style you'll use to create your table. The default (**Standard**) is shown. If you'd like to change to another existing style, use the down arrow next to the **Table Style** control box. If you need to create a new style, you can use the **Table Style** button next to the control box. We'll look more at table style creation in Section 19.1.4, p.432.
- The **Insert options** frame contains three different approaches to inserting your table into a drawing:
    - A bullet next to **Start from empty table** means that AutoCAD will let you start "from scratch". You'll then need to fill in the rest of the dialog box to create a table to your desired specifications.

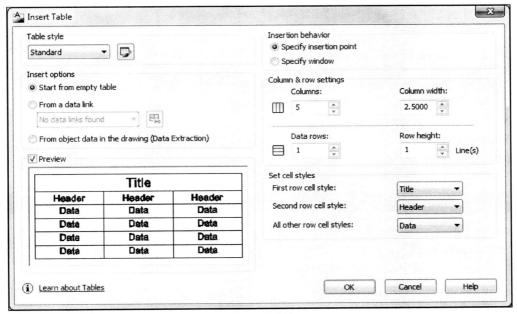

Figure 19.001

- o A bullet next to **From a data link** allows you to select a data link from the control box just below the option. If your data link isn't there, you can use the **Data Extraction Wizard** button to open the Select a Data Link dialog box (Figure 19.002). Use **the Create a new Excel Data Link** option to name your link and then open a File...Select dialog box where you can choose an existing Excel spreadsheet or CSV file (essentially, a comma-separated spreadsheet) to use as your table. (The good news is that you don't have to have Excel on your computer to use a spreadsheet in your drawing!) We'll see this in action in Exercise 19.1.1B, p.424.

Figure 19.002

- o Put a bullet next to **From object data in the drawing (Data Extraction)** to launch the Data Extraction Wizard. Using this wizard, AutoCAD will scan the drawing for block information to include in your table. We'll look at this in more detail in Lesson 22, Section 22.4, p.519.

- Place a check next to **Preview** ✓ Preview to display your current table settings in the **Preview** frame.
- The **Insert Behavior** frame provides two options – one to allow you to **Specify insertion point** of your new table and the other to **Specify window** to define the area for your table (much as you do with Mtext).
- The **Column and Row Settings** frame allows you to set up the number of columns and rows in your table. You'll also specify the width and height of your columns and rows here.
- Use the options in the **Set cell styles** frame to define the cell rows. You can define **Title**, **Header**, and **Data** rows and give each its own formatting using the Mtext editor.

Does that sound simple enough? Let's try a couple – we'll create a table from scratch, and then we'll bring in an Excel spreadsheet.

Do This: 19.1.1A	Creating a Table from Scratch

I. Open the *Floor Plan Data Sheet* drawing in the C:\Steps\Lesson19 folder. The drawing looks like the figure at right.
II. Set the **Text** layer current.
III. Follow these steps.

19.1.1A: CREATING A TABLE

1. Enter the *Table* command.
   Command: *tb*

AutoCAD presents the Insert Table dialog box (Figure 19.001, p.422).

2. Use the **Standard** table style (it's the only one available to you at this time).	

3. Use the **Start from empty table Insert option** for this table.	

4. Tell AutoCAD you'd like to use an insertion point to insert the table.	

5. Finally, give your table four **Columns** with a 1½" width, and two **Data Rows** with a single line height.	

6. Complete the setup [OK].	

7. AutoCAD asks where you'd like to insert the table. Use the coordinates indicated. **Specify insertion point:** *1,16* AutoCAD presents the **Text Editor** ribbon tab and highlights the title *cell* of your table for your input.	

8. Enter the name of the table (*Symbols Legend*). Press the TAB key on your keyboard to continue.

9. AutoCAD highlights the first cell in the column header row. Enter the word *Symbol* here.

19.1.1A: CREATING A TABLE

10. Using the TAB key to move to the next cells, complete the column header row as indicated. (Pick the **Close** button ✖ on the ribbon's **Text Editor** tab to complete the procedure.) Your table looks like the figure at right.	Symbols Legend table with columns: Symbol, Description, Symbol, Description
11. Save the drawing 💾 but don't exit. **Command:** *qsave*	

Now let's bring in that spreadsheet.

Do This: 19.1.1B	Inserting a Spreadsheet as a Table

I. Be sure you're still in the *Floor Plan Data Sheet* drawing in the C:\Steps\Lesson19 folder. If not, please open it now. (Familiarity with spreadsheets might be useful for this exercise; but you can manage without it.)

II. Follow these steps.

19.1.1B: INSERTING A SPREADSHEET AS A TABLE

1. Enter the *Table* command 🔲.
   Command: *tb*

AutoCAD presents the Insert Table dialog box (Figure 19.001, p.422).

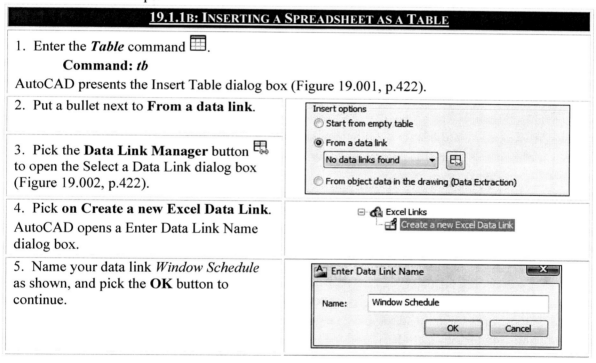

2. Put a bullet next to **From a data link**.	
3. Pick the **Data Link Manager** button 🔲 to open the Select a Data Link dialog box (Figure 19.002, p.422).	
4. Pick **on Create a new Excel Data Link**. AutoCAD opens a Enter Data Link Name dialog box.	
5. Name your data link *Window Schedule* as shown, and pick the **OK** button to continue.	

19.1.1B: INSERTING A SPREADSHEET AS A TABLE

6. Now AutoCAD opens the New Excel Data Link dialog box (shown). Pick the **Select** button next to **Browse for a file**. AutoCAD presents a standard Open ... File dialog box.

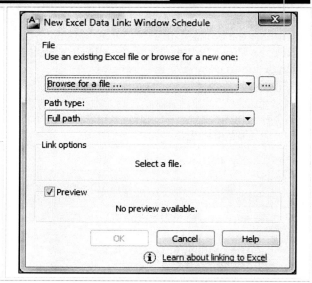

7. Open the *WindowSchedule.xls* file in the C:\Steps\Lesson19 folder.

AutoCAD returns to the New Excel Data Link dialog box – but notice the changes (see the figure after Step 8).

8. Take a moment to examine the dialog box. (Pick the **More options** button ⊙ next to the **Help** button to open the entire box.) You can link to (**Link options** frame):
   - a specific sheet by selecting it in the **Select Excel sheet to link to** box;
   - the **entire sheet**;
   - **a named range**;
   - a defined **range**.

AutoCAD even provides a **Preview** frame to help you decide which option to select.

The expansion area of the dialog box provides a couple more useful frames.

In the **Cell contents** frame, you can:
   - **Keep data formats and formulas** – this imports the spreadsheet as is; that is, everything comes in intact (including supported formulas)
   - **Keep data formats, [and] solve formulas in Excel** – this brings in the data format and formulae, then calculates the actual data using the imported formulae.
   - **Convert data formats to text, solve formulas in Excel** – this option brings in the data as text (formats not included)
   - **Allow writing to source file** – be careful of this check box! When checked, users can use the *DataUpLink* command to save changes back to the original file.

Cell formatting options include:
   - **Use Excel formatting** to bring any formatting associated with the spreadsheet into the AutoCAD table. Use the bullets below to either **Keep the table updated** with any formatting changes made in the spreadsheet or **do not update** the formatting.

19.1.1B: INSERTING A SPREADSHEET AS A TABLE

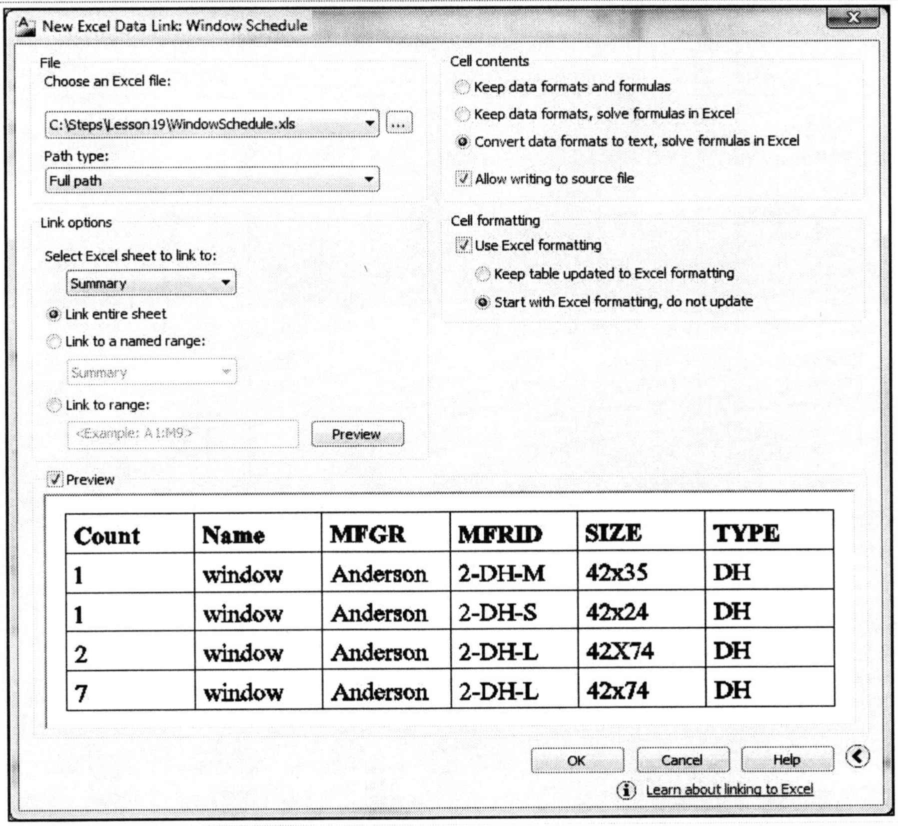

[Screenshot of New Excel Data Link: Window Schedule dialog box, showing file selection C:\Steps\Lesson19\WindowSchedule.xls, Full path, Link entire sheet option, and a preview table:]

Count	Name	MFGR	MFRID	SIZE	TYPE
1	window	Anderson	2-DH-M	42x35	DH
1	window	Anderson	2-DH-S	42x24	DH
2	window	Anderson	2-DH-L	42X74	DH
7	window	Anderson	2-DH-L	42x74	DH

9. Link the **entire sheet** ⦿ Link entire sheet.

10. Accept the other defaults and pick the **OK** button [OK] to close the New Excel Data Link dialog box and again [OK] to close the Select a Data Link dialog box.

11. Accept the other options in the Insert Table dialog box, and pick the **OK** button [OK] to continue.

12. Insert the table to these coordinates.
    Specify insertion point: 1,6

13. Save the drawing 💾 but don't exit.
    Command: *qsave*

We'll spend more time with this in Section 19.3, p.443 – after we've looked at fields (Section 19.2, p.437).

19.1.2 Editing a Table's Properties

The first thing you probably noticed about the Symbols Legend tables you inserted earlier was that the *Column Heads* didn't fit in the cells provided for them. In fact, both description columns are far too narrow for much of a description. Additionally, you might have noticed that the text in that table doesn't match the text we've used in the Floor Plan drawings through our last several lessons.

We'll use the Properties palette a bit, but AutoCAD makes use of cursor menus and a **Table** tab on the ribbon (Figure 19.003) for easier editing of a table's properties (as well as values).

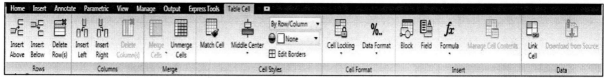

Figure 19.003

Let's look at the tools available on the ribbon. (Note: The **Table** tab appears when you pick once inside a table cell.)

	ROWS PANEL		
BUTTON	**DESCRIPTION**	**BUTTON**	**DESCRIPTION**
	Insert Above – insert a new row above the selected row/cell.		**Insert Below** – insert a new row below the selected row/cell.
	Delete Row – delete the selected row.		

	COLUMNS PANEL		
BUTTON	**DESCRIPTION**	**BUTTON**	**DESCRIPTION**
	Insert Left – insert a new column left of the selected column/cell.		**Insert Right** – insert a new column right of the selected column/cell.
	Delete Column – delete the selected column.		

	MERGE PANEL		
BUTTON	**DESCRIPTION**	**BUTTON**	**DESCRIPTION**
	Merge by Row – merges cells in a selected row into a single cell.		**Merge by Column** – merges cells in a selected column into a single cell.
	Merge All – merges all selected cells (rows & columns) into a single cell.		**Unmerge Cells** – separates merged cells into their constituent parts.

	CELL STYLES PANEL		
BUTTON	**DESCRIPTION**	**BUTTON**	**DESCRIPTION**
	Match Cell – match the properties of a selected cell to other cells.		**Edit Borders** – opens a dialog box allowing you to control: border insertion, **Linetype, Lineweight**, line **Color**, single/**double line**, **spacing** and **type**.
	Cell Style control box – options to control styles by definition – can open the Styles Manager here (more on that in Section 19.1.4, p.432)		**Cell Background Color** control box – options to select a cell's background color
	Align Bottom Left – cell text alignment		**Align Bottom Center** – cell text alignment
	Align Bottom Right – cell text alignment		**Align Middle Left** – cell text alignment
	Align Middle Center – cell text alignment		**Align Middle Right** – cell text alignment
	Align Top Left – cell text alignment		**Align Top Center** – cell text alignment
	Align Top Right – cell text alignment		

CELL FORMAT PANEL

BUTTON	DESCRIPTION	BUTTON	DESCRIPTION
🔒	**Cell Locking** – lock/unlock **content**, **format**, or **content and format**. Locked items cannot be changed.	%..	**Data format** – format the type of date for each cell. Formats include: **Angle, Currency, Date, Decimal Number, General, Percentage, Point, Text,** or **Whole Number**. There's even an option for customizing the format.

INSERT PANEL

BUTTON	DESCRIPTION	BUTTON	DESCRIPTION
📊	**Block** –calls the Insert a Block in a Table Cell dialog box. (You'll see this in Ex. 19.1.3A, p.430)	📋	**Field** – calls the Field dialog box to assist you in placing fields in your table.
fx	**Formula** – formulae include: **Sum, Average, Count, Cell** (for appending a cell's ID in a formula), and **Equation**.	📋	**Manage Cell Contents** – provides a dialog box that allows you to adjust the flow and text direction within the cell.

DATA PANEL

BUTTON	DESCRIPTION	BUTTON	DESCRIPTION
🔗	**Link Cell** – opens the Select a Data Link dialog box.	📥	**Download changes from source file** – downloads changes from linked data source file.

Let's see what we can do with our Symbols table before we start entering values.

Do This: 19.1.2A	**Editing a Table's Properties**

I. Be sure you're still in the *Floor Plan Data Sheet* drawing in the C:\Steps\Lesson19 folder. If not, please open it now.
II. Be sure the **Pickfirst** sysvar is set to **1**. Zoom in a bit on the Symbols table.
III. Follow these steps.

19.1.2A: EDITING A TABLE'S PROPERTIES

1. First, let's change the text style in the title and header cells. Pick once in the title cell. Notice that its grips and the **Table Cell** tab (Figure 19.003, p.427) on the ribbon display.

2. Now hold down the SHIFT key and select a cell in the header row. Notice that both rows display grips.

19.1.2A: EDITING A TABLE'S PROPERTIES

3. Now right click in one of the selected cells. Take a moment to look over the opportunities provided on the cursor menu (many options also appear on the ribbon's **Table** tab). We'll look at some of these, but you should make time later to experiment with the others.
Pick **Properties**.
AutoCAD opens the Properties palette.

4. Under the **Content** heading on the Properties palette, select **Text style**, and then select either **Calibri Bold** or **Times Bold** from the drop down menu. (The style has already been created for you.)

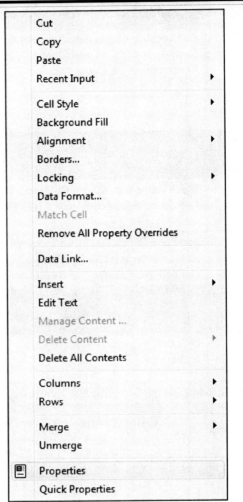

5. Close the Properties palette and clear the grips. Your table looks like the figure at right. (I've used **Calibri Bold** – if you used **Times Bold**, your font may be a bit different.)

6. We'll also need more rows than we originally set up. Pick inside any of the cells on the bottom row. Its grips will display.

7. Pick the **Insert below** button (**Row** panel). Repeat this until you have six empty rows. Clear the grips.

8. Now we'll make the **Description** columns wider. Select the entire table with a window. Notice that the entire table highlights and displays grips.

19.1.2A: EDITING A TABLE'S PROPERTIES

9. Pick on the line grip between the first **Symbol** and **Description** columns and move it to the left. Resize the other columns as well. (Move the grips left or right as far as possible without affecting the **Symbol** headers.)
Clear the grips. Your table looks like the figure at right.

Symbols Legend			
Symbol	Description	Symbol	Description

10. Let's change one last property. Repeat Steps 1 and 2; but this time, select the bottom row in Step 2 (the grips appear for the outer walls of the table).

11. Use the **Background Color** control box in the **Cell Styles** panel to change the background to **Cyan**.

12. Save the drawing 💾 but don't exit.
    Command: *qsave*

19.1.3 Table Column Tools and Adding Values to Table Cells

Now that you've set up your table, you'll need to add some values to the cells (otherwise, the table would be pretty ... but useless). Luckily, adding values is the easy part!
After we do that, let's take a quick look at some of the cool column tools you can use on your tables.

Do This: 19.1.3A	Inserting Table Values and Working with Table Columns

I. Be sure you're still in the *Floor Plan Data Sheet* drawing in the C:\Steps\Lesson19 folder. If not, please open it now.
   In this exercise, you'll get another preview of those wondrous timesavers called *blocks*. We'll look at blocks in much greater detail in Lessons 21, p.468, and 22, p.500, but enjoy the preview for now.

II. Follow these steps.

19.1.3A: INSERTING TABLE VALUES AND WORKING WITH TABLE COLUMNS

1. We'll start right off with a block insertion. Pick inside the first empty cell in the first **Symbol** column. Its grips appear.

2. Pick the **Block** button on the **Insert** panel. AutoCAD presents the Insert a Block in a Table Cell dialog box shown in Step 3.

19.1.3A: INSERTING TABLE VALUES AND WORKING WITH TABLE COLUMNS

3. Fill out the dialog box as follows:
   - Select the **230 volt outlet** block in the **Name** drop down list box.
   - Remove the check from the **AutoFit** check box.
   - Set a **Scale** of **1**.
   - Accept the **Rotation angle** default.
   - Use the **Middle Center cell alignment**.

4. Complete the procedure 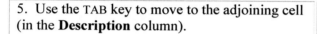.

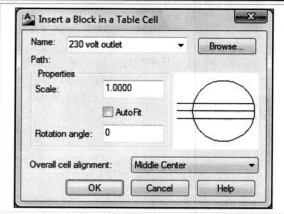

5. Use the TAB key to move to the adjoining cell (in the **Description** column).

6. Enter the text – **230 Volt Outlet**.

7. Repeat Steps 1 through 6 to complete the table (right). (Hint: To enter the text, simply pick in the appropriate cell and start typing. AutoCAD will automatically open the MText editor.)

8. Use the Properties palette to change the text style in each of the **Description** cells to **Calibri** or **Times**.

9. Use the **Middle Center** option ⊕ on the **Cell Styles** panel to properly justify the first Description cell.

10. With the same cell selected, pick the **Match Cell** button on the **Cell Styles** panel. AutoCAD returns to the graphics screen and asks you to select the destination cell. Pick the remaining Description cell.
    Select destination cell:
    Your table looks like the figure at right.

11. Save the drawing but don't exit.
    Command: *qsave*

19.1.3A: INSERTING TABLE VALUES AND WORKING WITH TABLE COLUMNS

12. Suppose your table ran long – would you have to move it? No. Let me show you why. Select the entire table. Notice the grips.

13. Pick the downward-pointing arrow in the bottom-middle of the table.

14. Drag the grip upward to just under row #6. Notice (following figure) that the bottom rows move to a new position beside the table!

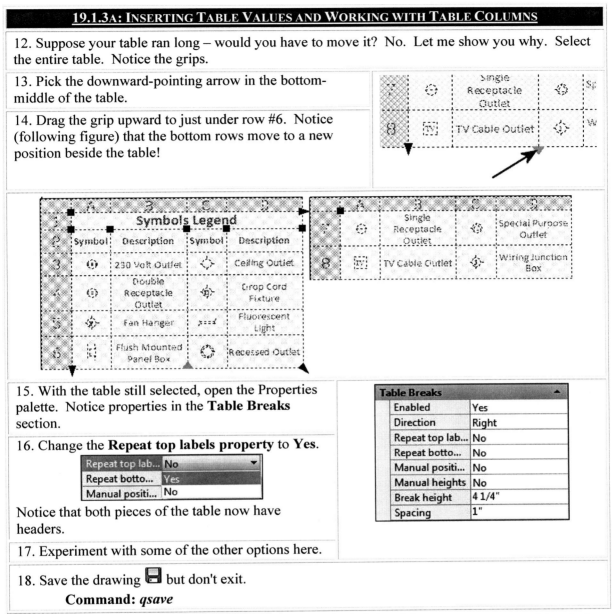

15. With the table still selected, open the Properties palette. Notice properties in the **Table Breaks** section.

16. Change the **Repeat top labels property** to **Yes**.

Notice that both pieces of the table now have headers.

17. Experiment with some of the other options here.

18. Save the drawing but don't exit.
    Command: *qsave*

Did you notice how AutoCAD automatically adjusted the height of the rows to accommodate what you inserted? Did you see the text automatically wrap within the cells? If you've ever had to create a table manually (with the *Line* command), you'll really appreciate these automated actions!

But there is more to tables than what you've seen. What about using a table for a revision block (one that grows from bottom to top)? Or how about setting up as defaults some of the things you had to do manually (text styles, justifications, etc.)?

Let's look next at table *styles*.

19.1.4 Customizing Tables

Since you've already mastered dimension and leader styles, you'll probably find table styles fairly easy to set up; they look as though the same programmers designed them.

432

The *TableStyle* command calls the Table Style dialog box (Figure 19.004). As you can see, the dialog box closely resembles the Dimension Style dialog box, p.377.

Where to Find It:	
Command Line:	*TableStyle*
Hotkey(s):	ts
Ribbon (Tab/Panel):	Home – Annotation (subpanel) – **Table Style**
	Annotate – Tables – ↘
Menu:	Format – **Table Style**
Toolbar:	Styles – **Table Style**

- The **Styles** list box indicates the names of styles available in this drawing. Luckily, however, AutoCAD doesn't inflict "children" styles here as it does with dimension styles, so your job is already easier!
- As with other tools, the **List** control box below the **Styles** box serves as a filter for what goes in to the **Styles** box (**All styles** or **Styles in use**).
- The **Preview** box, of course, shows a preview of the selected style.
- A total of six buttons reside on the dialog box. These are fairly self-explanatory:
    o **Set Current** sets the selected style as the one that AutoCAD will use when inserting a new table.

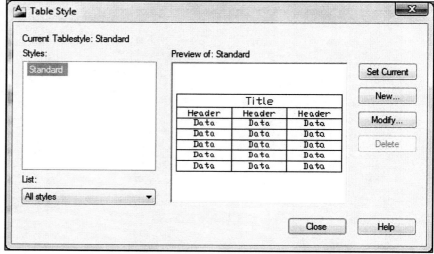

Figure 19.004

    o **Modify** allows you to alter settings for the selected style. You'll use the same dialog box as with the **New** style procedure.
    o **Delete**, of course, allows you to delete the selected style. Note that you can't remove a style that's in use.
    o **Close** closes the dialog box.
    o **Help** calls AutoCAD's help window with information on the Table Style dialog box.
    o Use the **New** button to create a new table style. It calls the Create New Style dialog box (Figure 19.005). Use this familiar tool to name your new style, and then to access the New Table Style dialog box (Figure 19.006) where you'll actually set up the style.

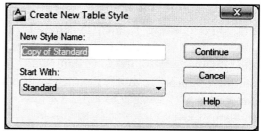

Figure 19.005

Let's spend some time with the New Table Style dialog box. We'll start with the simple frames on the left.

- Use the buttons in the **Starting table** frame to select an existing table. AutoCAD will then use the properties of that table as the starting point for creating your new table. This can be a real timesaver!

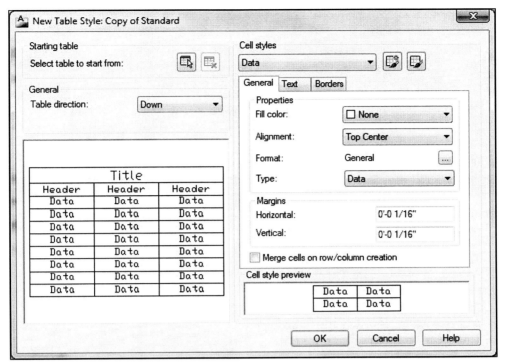

Figure 19.006

- Use the options in the **Table direction** control box (**General** frame) to have your table expand **Down** (data tables) or **Up** (revision blocks).
- The larger **Cell Styles** frame on the right has more options than you might initially guess. Use these to format your cells.
  - Select the type of cell you'll format – **Data**, **Header**, or **Title** – in the control box at the top. You can even use the buttons next to the control box to create and manage your own cell type. The options you select on the tabs below will apply to whatever cell type has been selected in the control box.

  You have three tabs full of options to define your selected cell type.

    - The **General** tab (Figure 19.006) includes formatting options for the cells, including **Fill color**, **Alignment**, **Format** and **Type** (**Label** – as with header or title cells, or **Data**). You'll also define your table's **Margins** here.
    - The **Text** tab (Figure 19.007) provides options for controlling the **Text style**, **Text height**, **Text color**, and **Text angle**.
    - The **Borders** tab (Figure 19.008) provides options for controlling the borders of your table. Here you can define the borders **Lineweight**, **Linetype**, and **Color**, whether or not the borders will have **Double lines** (and the double line **Spacing**). Use the buttons across the bottom to turn on/off specific lines within the table.

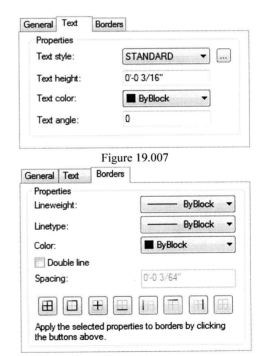

Figure 19.007

Figure 19.008

434

o Finally, the **Cell styles** frame contains a **Cell style preview** box at the bottom to help guide you through your setup.

Let's set up a table style to use in creating a revision block.

Do This: 19.1.4A	Defining a Table Style

I. Be sure you're still in the *Floor Plan Data Sheet* drawing in the C:\Steps\Lesson19 folder. If not, please open it now.
II. Zoom extents.
III. Follow these steps.

19.1.4A: DEFINING A TABLE STYLE

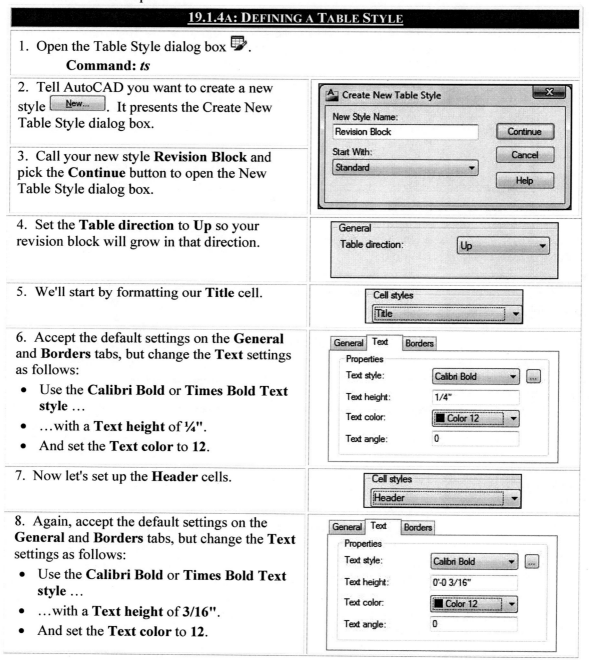

1. Open the Table Style dialog box.
   Command: *ts*

2. Tell AutoCAD you want to create a new style [New...]. It presents the Create New Table Style dialog box.

3. Call your new style **Revision Block** and pick the **Continue** button to open the New Table Style dialog box.

4. Set the **Table direction** to **Up** so your revision block will grow in that direction.

5. We'll start by formatting our **Title** cell.

6. Accept the default settings on the **General** and **Borders** tabs, but change the **Text** settings as follows:
   - Use the **Calibri Bold** or **Times Bold Text style** ...
   - ...with a **Text height** of ¼".
   - And set the **Text color** to **12**.

7. Now let's set up the **Header** cells.

8. Again, accept the default settings on the **General** and **Borders** tabs, but change the **Text** settings as follows:
   - Use the **Calibri Bold** or **Times Bold Text style** ...
   - ...with a **Text height** of **3/16"**.
   - And set the **Text color** to **12**.

435

19.1.4A: DEFINING A TABLE STYLE

9. Repeat Steps 7 and 8, but set up the **Data** cells to use the **Calibri** or **Times Text style** with a Text **height of 1/8"**. Set the **Text color** to **12**.

10. Pick the **OK** button [OK] to complete the setup.
Notice that your new table appears in the **Styles** list in the Table Styles dialog box.

11. Select **Revision Block** in the **Styles** list and pick the **Set Current** button [Set Current].

12. **Close** [Close] the Table Styles dialog box.

13. Save the drawing 💾 but don't exit.
    Command: *qsave*

Now let's take a look at our new style in action.

| Do This: 19.1.4B | Using Your New Table Style |

I. Be sure you're still in the *Floor Plan Data Sheet* drawing in the C:\Steps\Lesson19 folder. If not, please open it now.
II. Zoom in just to the left of the title block.
III. Set the **BORDER** layer current.
IV. Follow these steps.

19.1.4B: USING YOUR NEW TABLE STYLE

1. Begin a table 📋.
    Command: *tb*

2. Fill in the Insert Table dialog box as shown at right. Accept the other defaults, and pick **OK** to continue.

3. Insert the table next to the title block and fill in the **Title** and **Header** rows as shown.
    Specify insertion point:

4. Adjust the columns to match those shown at right. (Use the procedures you learned in Exercise 19.1.2A, p.428.)

5. Reposition ✥ the revision block next to the title block.
    Command: *m*

6. Fill in everything except the **Date** block with your own information.

7. Save the drawing 💾 but don't exit.
    Command: *qsave*

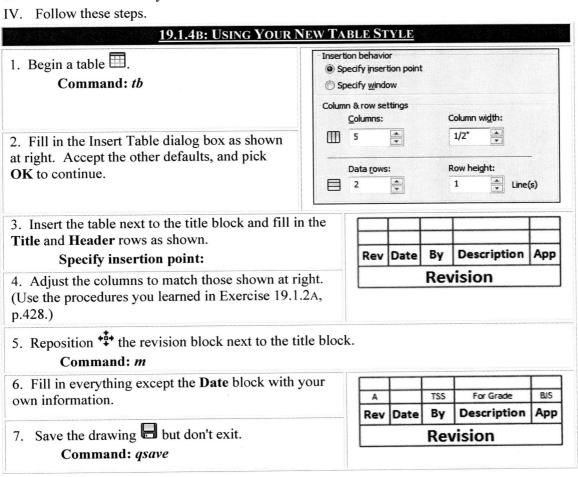

Okay, tables were pretty easy (all things considered). They may take a bit of work to set things up, but once that's done, the rest is child's play. Try to remember as you do the setup, that you won't see the real benefit to tables until you discover later that you need to change something or (Heaven forbid!) move the entire table!

Our next tool requires less setup, but it can be even more of a timesaver!

19.2 The Wonders of Automation – Fields

Fields (placeholders for text) require very little setup, but they relieve the operator of some of the more mundane chores of life – automatically filling in some of the common requirements of virtually all documents (author, date, etc.). In the case of the draftsperson, they can help with the revision block we just created as well as the title block, and even provide a modifiable plot stamp for your drawings. But the usefulness of fields in a drawing goes beyond the basic documentation benefits. You can also use fields to keep track of such things as length, area, and even elevation of an object!

First, we'll look at the source of some of the information AutoCAD will use to fill in the fields you'll create. Then we'll create some fields.

19.2.1 Drawing Properties

Where to Find It:	
Command Line:	*DwgProps*
Menu:	File – **Drawing Properties**
Application Menu:	Drawing Utilities – **Drawing Properties**

AutoCAD automatically sets up many of the things you'll want to use for field values – dates, object information, and so forth. But you'll have to set up some of the other things – drawing name, author name, and so forth. Luckily, AutoCAD limits your input to a single dialog box (with four fairly simple tabs). See the [Drawing Name] Properties dialog box in Figure 19.009. (Access the dialog box with the *DwgProps* command.)

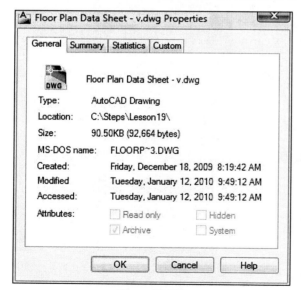

Figure 19.009

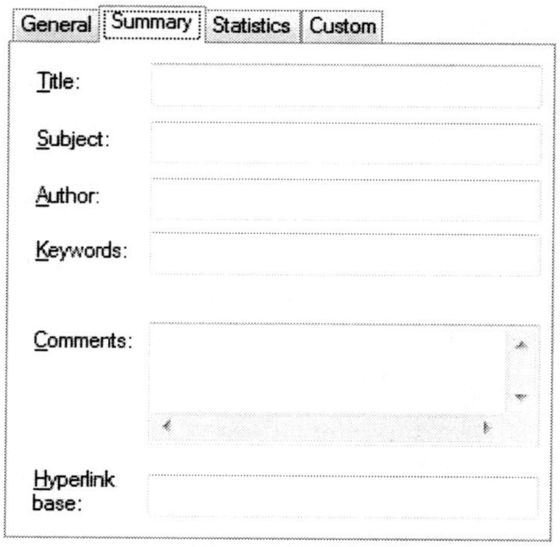

Figure 19.010

- The **General** tab provides information – file attributes, recent modification dates, and so forth. You can't change any of this information here, but some of it can prove useful in tracking things you may need to know (such as file size).
- The **Summary** tab (Figure 19.010) is your workhorse. You'll need to fill in this information. Remember, if you don't fill it in, it won't be available to you; so use your best judgment as to what to include.

Most of the blanks are self-explanatory. But this information might help:
- o Use the **Keywords** box to include words that might be useful when searching a group of drawings for this one.
- o Use the **Comments** box to add any useful comments you'd like to make.
- o The **Hyperlink base** can be a folder or website – anywhere that drawings including or related to this one might be located.
- Like the **General** tab, you won't be able to alter the information provided on the **Statistics** tab (Figure 19.011). This one, however, has some really useful information that an operator can use to keep track of his time on a specific drawing, or when the drawing was edited last. (There's no defense before an irate employer like computer-generated facts!)

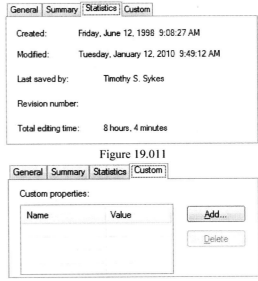

Figure 19.011

Figure 19.012

- The last tab – **Custom** (Figure 19.012) – provides an opportunity for including some of your own information.

You'll be able to see the benefits of this tool once you've set it up and we examine fields in more detail. Let's set up our drawing's properties.

> You can find an additional benefit of the properties dialog box shown here by right clicking on the file in Windows Explorer. Select **Properties** from the cursor menu, and Windows will show you the information you've provided without having to open the file!

Do This: 19.2.1A	Setting Up Drawing Properties

I. Be sure you're still in the *Floor Plan Data Sheet* drawing in the C:\Steps\Lesson19 folder. If not, please open it now.
II. Follow these steps.

19.2.1A: SETTING UP DRAWING PROPERTIES

1. Open the *Floor Plan Data Sheet* Properties dialog box. It opens with the **General** tab on top as shown in Figure 19.009.
    Command: *dwgprops*

2. Pick the **Summary** tab to move it to the top.

3. Enter the data shown.

4. Examine the **Statistics** and **Custom** tabs. These can be useful, but we won't include any custom information for this drawing.

5. Complete the procedure OK.

6. Save the drawing but don't exit.
    Command: *qsave*

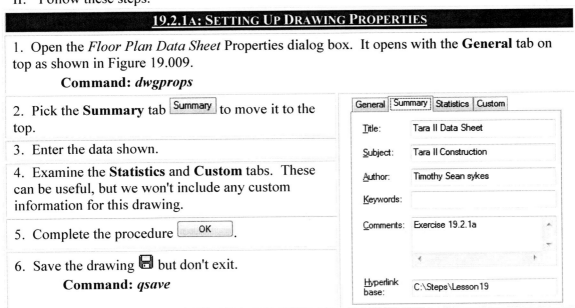

Nothing to it, right?

Now let's take a look at why we set up this dialog box. Let's look at fields.

19.2.2 Inserting Fields

You'll find two ways to insert a field – insertion via the ribbon's **MText** or **Table** tabs and simple insertion. (We'll use both in our next exercise.) Both methods involve the Field dialog box (Figure 19.013).

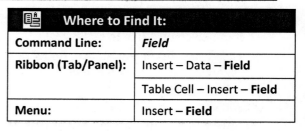

Where to Find It:	
Command Line:	*Field*
Ribbon (Tab/Panel):	Insert – Data – **Field**
	Table Cell – Insert – **Field**
Menu:	Insert – **Field**

This dynamic dialog box, despite its simple appearance, allows for a great number of possibilities.

- The **Field category** drop down box provides a list of the various categories available for your field. If you aren't sure which category to use, leave the **All** default, which will provide a list of all the available field names in the list box below.

- The **Field names** list box provides a list of fields available in the selected category.

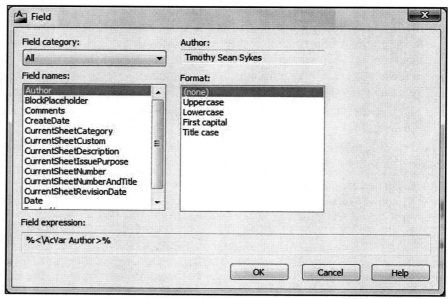

Figure 19.013

- The **Field value** box (shown here as **Author**) lets you know the current value of the selected field. This box changes with the selection of different fields in the **Field names** list box. It will show dashes if no value has been assigned to the selected field or pound symbols (#) if the field is invalid.

- The **Format** box allows you to select from a list of available formats for your field. Formats change with the selection of different fields.

- To the right of the **Field value** and **Format** boxes, you'll see a large empty space. AutoCAD will fill this space with hints or additional options as your selection requires.

- The **Field expression** box just above the buttons is informational only. It shows the coding required to create the selected field. Unless you're a programmer, you can ignore this area.

AutoCAD provides the usual buttons at the bottom of the dialog box, but I'd like to point out the **Help** button just to make sure you know it's there. This very intuitive dialog box may require some support as you get used to it.

Let's insert some fields.

Do This: 19.2.2A	Inserting Fields

I. Be sure you're still in the *Floor Plan Data Sheet* drawing in the C:\Steps\Lesson19 folder. If not, please open it now.
II. Set the **TEXT** layer and the **Calibri** or **TIMES** text style current.
III. Set the **TextSize** system variable to 1/8".
IV. If you're not already zoomed in around the title/revision blocks, please do so now.
V. Follow these steps.

19.2.2A: INSERTING FIELDS

1. We'll begin by putting a date in our revision block. Pick in the cell set aside for the date. AutoCAD displays the ribbon's **Table Cell** tab.

2. Pick the **Field** button on the **Insert** panel. AutoCAD presents the Field dialog box (Figure 19.013, p.439).

3. Set up the dialog box as shown in the following figure.
   - Select **Date & Time** in the **Field category** drop down box. Notice that the **Field names** list shows just those fields in the selected category.
   - Select **Date** from the **Field names** list.
   - Select the format shown.

   Notice the **Hints** shown to the right of the selection boxes. This box can help you make an educated decision about your field.

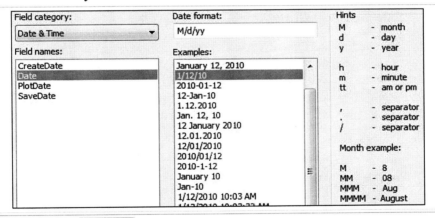

4. Pick the **OK** button to close the field dialog box.

5. Complete the entry (pick outside the table). AutoCAD removes the toolbar and puts the field in the table. Notice that the field has a colored background. AutoCAD does this to make it easier to see the fields in a drawing; the background won't print/plot.

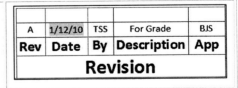

6. The title in the title block is incorrect. Erase it (the part that reads *Sample Floor Plan*).
   Command: *e*

7. Reset the **TextSize** system variable to 3/16".
   Command: *textsize*
   Enter new value for TEXTSIZE <0'-0 1/8">: *3/16*

19.2.2A: INSERTING FIELDS

8. This time we'll enter the *Field* command at the command prompt.
   Command: *field*

AutoCAD presents the Field dialog box (Figure 19.013, p.439).

9. In the **Document** category, select **Title**. Notice the title is the one you entered in the drawing properties dialog box (Exercise 19.2.1A, p.438). Use the **Title case** format and pick the **OK** button to continue.

10. Place the field in the title block as shown.

11. Save the drawing as *Tara II Data Sheet* in the C:\Steps\Lesson19 folder.
    Command: *saveas*

12. Close all open drawings.
    Command: *closeall*

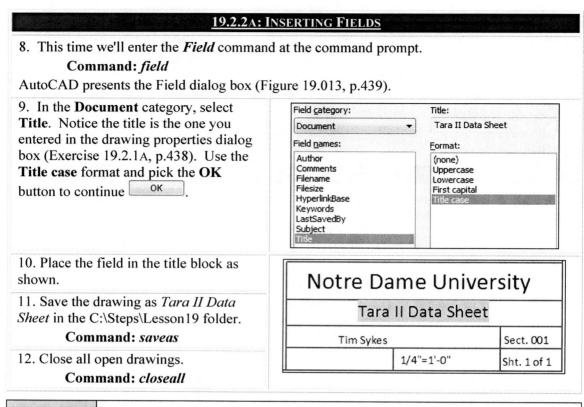

19.2.3 Calculating Fields

You've really just scratched the surface of what you can do with tables and fields. AutoCAD has incorporated some of the more useful elements of spreadsheets into their tables, too. If you've ever used MS Excel or any other spreadsheet program, you might see the potential!

We're going to use fields in an AutoCAD table to calculate the square footage of each of the rooms of our Tara II floor plan. Then we'll total the square footage … and even calculate the average square footage per room!

Let's get started right away.

Do This: 19.2.3A	Calculating with Fields

I. Open the *FlrPln19* drawing in the C:\Steps\Lesson19 folder. This is the same floor plan we've been using except that I've added a Square Footage table in the lower left corner.
II. Zoom in around the Square Footage table.
III. Follow these steps.

19.2.3A: CALCULATING WITH FIELDS

1. Pick in the D3 cell (last column in the first data row).

2. Pick the **Field** button on the **Insert** panel.

3. (Refer to the figure that follows.)
   a. Select **Objects** in the **Field Category** selection box,
   b. Select **Formula** in the **Field names** list box,
   c. Enter *B3*C3* in the **Formula** text box,

19.2.3A: CALCULATING WITH FIELDS

    d. Pick the **Evaluate** button (AutoCAD displays options in the **Format** list box),

    e. Select **Fractional** in the **Format** list box.

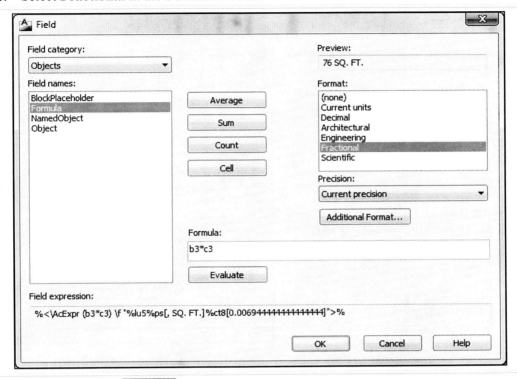

4. Pick the **OK** button to complete the procedure. (Adjust the column width as needed.) Notice that AutoCAD places the product of the values of cells B3 and C3 into cell D3. (Or put more simply, AutoCAD multiples 8 by 9.5, reduces the answer to **Fractional** format, and puts the answer in the location you selected in Step 1.)

5. Now here's a really cool trick. Pick the rotated square grip at the lower right corner of the D3 cell...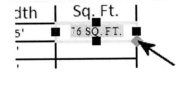

6. Now pick in cell D14. AutoCAD automatically uses the formula in cell D3 in the other selected cells! (Way cool!)

7. Now select cell D15. Here we'll total (sum) the square feet column.

8. Pick the **Field** button.

9. Pick the **Sum** button.

10. AutoCAD returns to the graphics screen and asks you to place a window through the **table cell range** you wish to total. Put the window through the cells you've been using (D3 through D14).

        Select first corner of table cell range:

        Select second corner of table cell range:

11. AutoCAD returns to the Field dialog box. Select the **Fractional** format you've been using, and then pick the **OK** button to complete this procedure.

442

19.2.3A: CALCULATING WITH FIELDS

12. Finally, we'll give our client an average figure for the rooms in our house. Move to cell D16.

13. Pick the **Field** button.

14. Again, we'll let AutoCAD do the work. Pick the **Average** button and select the same cells (D3 through D14).

15. AutoCAD again returns to the Field dialog box. Again, use the **Fractional** format, and pick the **OK** button to complete this procedure.

16. We've finished our calculations! Clear the grips.
Your table looks like the figure at right. (Adjust your column/row size if necessary.)

17. Save the drawing.
    Command: *qsave*

Square Footage			
Room	Length	Width	Sq. Ft.
Closet - Mstr	8'	9.5'	76 SQ. FT.
Closet #1	10'	3'	30 SQ. FT.
Closet #2	5.5'	5'	27 1/2 SQ. FT.
Closet #3	5.5'	5'	27 1/2 SQ. FT.
Bath - Mstr	8'	9.5'	76 SQ. FT.
Bath - Com.	10'	5'	50 SQ. FT.
BR - Mstr.	16'	16.5'	264 SQ. FT.
Br #1	11.5'	10.5'	120 3/4 SQ. FT.
Br #2	11.5'	12.5'	143 3/4 SQ. FT.
Br #3	13.5'	9'	121 1/2 SQ. FT.
Kitchen	11.5'	19.5'	224 1/4 SQ. FT.
LR	21'	30.5'	640 1/2 SQ. FT.
Total			1801 3/4 SQ. FT.
Rm. Avg.			150 1/8 SQ. FT.

Cool, huh? Want to see something even cooler? Change the value in one of the **Length** or **Width** cells. AutoCAD will automatically adjust not only the value in the **Sq. Ft.** column, but the **Total** and **Rm. Avg.** values as well!

19.3 Altogether Now: Tables, Fields, and MS Excel

In Section 19.1, we saw a fairly simply procedure for inserting an Excel spreadsheet into AutoCAD. But now that we know something about fields, we should take another look at working with Excel.

In our next exercise, we'll use a different (easier?) method for inserting the spreadsheet and at some of the really cool things you can do when you link the spreadsheet to your AutoCAD table. (You might want to let your material tracking/ordering people in on this one!)

> This exercise requires the use of MS Excel. If you don't have MS Excel on your computer (or know how to use it), just read through this one.

Do This: 19.3A — Tables, Fields, and MS Excel

I. Start a new drawing from scratch. Save it as *NewTable* in the C:\Steps\Lesson19 folder.

II. Open the *SquareFootage.xls* file in MS Excel. You'll find it in the C:\Steps\Lesson19 folder.

> Users of AutoCAD 2011 on a 64-bit Windows 7 system may have a problem with the older .xls format. In the unlikely even that you're using AutoCAD 2011, Windows 7, and a Microsoft Excel edition that predates the Office 2007 package, this exercise might give you some compatibility problems. In this case, just read through this one.
>
> I've included the *SquareFootage* file in an .xlsx format as well. Should you have problems with the .xls file, try this one before you give up!

III. Follow these steps.

19.3A: TABLES, FIELDS, AND MS EXCEL

1. Select the active cells in MS Excel as shown.

2. Copy the information to the Windows clipboard. (Use CTRL+C or pick **Copy** from the Edit pull down menu.)

3. Move to the AutoCAD window. (Pick AutoCAD on the taskbar or use the ALT+TAB approach.)

4. Pick **Paste Special** from the Edit pull down menu.
AutoCAD presents the Paste Special dialog box (following Step 5).

5. Put a bullet next to **Paste Link** and select **AutoCAD Entities** from the list box.

	A	B	C	D
1		Square Footage		
2	Room	Length	Width	Sq. Ft.
3	Closet - Mstr	10.00	9.50	95.00
4	Closet #1	10.00	3.00	30.00
5	Closet #2	5.50	5.00	27.50
6	Closet #3	5.50	6.00	33.00
7	Bath - Mstr	8.00	9.50	76.00
8	Bath - Com	10.00	5.00	50.00
9	BR - Mstr	16.00	16.50	264.00
10	BR #1	11.50	10.50	120.75
11	BR #2	11.50	12.50	143.75
12	BR #3	13.50	9.00	121.50
13	Kitchen	11.50	19.50	224.25
14	LR	21.00	30.50	640.50
15	Total			1826.25
16	Rm. Avg.			152.19

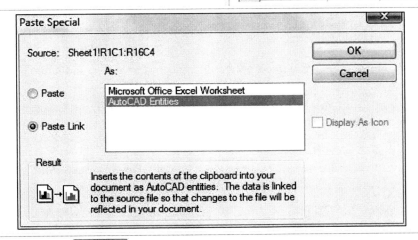

6. Pick the **OK** button to continue.

7. Put the table anywhere in the drawing and zoom in around it.
   Specify insertion point:
   Command: *z*

8. Return to Excel and close the program, then return to AutoCAD.

9. Enter the *DataLink* command or pick the **Data Link** button in the ribbon's Annotate – **Table** panel.
   Command: *datalink*
AutoCAD presents the Data Link Manager we saw in Section 19.1.1 (Figure 19.002, p.422).

19.3A: TABLES, FIELDS, AND MS EXCEL

10. Right click on the data link and select Edit from the menu.
AutoCAD presents the Modify Excel Link dialog box. (Be sure you can see the full dialog box ⊘.)

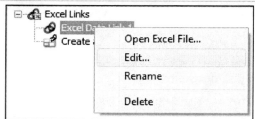

11. In the **Cell contents** frame
    - Place a check next to **Keep data formats and formulas** (so you can use the Excel formulas in your AutoCAD table)
    - Tell AutoCAD to **Allow** you to write to the **source file** so you can save the changes you make here back to the spreadsheet
    - **Keep** [the] **table updated to Excel formatting**

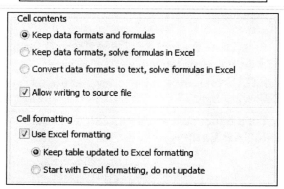

12. Pick the **OK** button to complete this part of our setup, and again to close the Data Link Manager.

13. You may see a **Data Link Has Changed** information bubble in the lower right corner of your screen. Pick the **Update Tables** link to keep your table up to date when this appears.

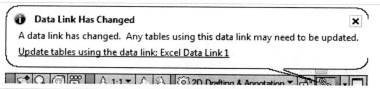

14. Notice that fields now appear in your table, but you still can't do much with them. Let's fix that.

Select the entire table with a window. Notice the two icons that follow your cursor 🔒🔗. The left one tells you that the table is locked; the right tells you that it is linked. We'll have to unlock it to be able to work on it.

15. Pick once in the **B3** cell. Hold down your SHIFT key and pick in the **D16** cell. AutoCAD selects these cells and all those between them.

16. Right click anywhere in the selected area and pick **Locking – Unlocked** from the menu. Notice that your cursor's Lock icon disappears.
Now we can make some changes.

17. Let's change a room size. Double click in the **B3** cell and change the number to **9**.

18. Pick the **MText** panel's **Close** button ✕ to complete the change. Notice that fields update the **D3**, **D15**, and **D16** cells.

19.3A: TABLES, FIELDS, AND MS EXCEL

19. Now the cool part. Reselect the entire table and pick **Write Data Links to External Source** from the right click cursor menu.
   1 data link(s) written out succesfully.
   AutoCAD tells you that it has updated the spreadsheet.

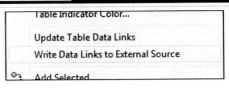

20. Reopen the *SquareFootage.xls* file in MS Excel. Notice that your AutoCAD changes now appear here. This is a Way Cool way to keep your materials people always up-to-date with your drawing! But wait! (Don't order yet!)

21. In MS Excel, change the width of Closet #2 to 6'.

22. Save the changes and return to AutoCAD. AutoCAD should inform you (with an information bubble) of changes to your table. If it does, pick the link it provides. If it doesn't, right click on the **Data Link** button on the status bar and pick **Update All Data Links** from the menu.
   AutoCAD updates its table with the information provided on the spreadsheet! (Use this procedure for information within the material people's control – such as pricing and availability of items.)

23. Exit the drawing and MS Excel without saving.

19.4 Extra Steps

1. Take a look at the **Search** tool on the right end of the menu bar (Figure 19.014).
2. Where it promts you to **Type a keyword or phrase**, enter *Table*.
3. AutoCAD opens a drop down list of possible assistance sources. (Nifty, huh?)
4. Take some time to explore some of these.

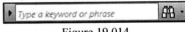
Figure 19.014

Remember to take advantage of AutoCAD's various training tools when you have a few extra minutes between lessons or projects at the job site; the effort will prove occasionally beneficial (but never harmful)!

There's help out there ... just ask!

Figure 19.015

19.5 What Have We Learned?

Items covered in this lesson include:
- *Field tools:*
  - **Field**
  - **DwgProps**
- *Table tools:*
  - **Table**
  - **TableStyle**
  - **DataLink**

AutoCAD makes dozens of combinations of fields and tables available to you. Hopefully, this lesson has whetted your appetite for knowledge and you'll spend some time exploring these new and innovative tools.

Try some of the exercises in Section 19.6 and the web, and then relax for a minute before tackling the lesson on hatching. It isn't difficult material, but it's always best to start a new lesson with a fresh mind.

19.6 Exercises

1. Add drawing properties and a revision block to the Piping Plan you've been working on. If you're not up to date on it, use *Piping Plan 19* in the C:\Steps\Lesson19 folder. (Refer to the figure below.)
    1.1. Try setting up a new field (Scale) on the **Custom** tab of the [drawing] Properties dialog box. Give it a value of 3/8"=1'-0". Insert it as a field in the title block as shown.
    1.2. When you set up the new table style for the revision block, remember to take the drawing scale factor into account when you assign text sizes. (Hint: I gave you the scale for this drawing in the previous step.)
    1.3. Save the drawing.

0	12/19/07	TSS	For Construction	BB
Rev	Date	By	Description	App
Revision				

| North Harris College |
| Houston, Texas |
| Sample Piping Plan |
| Tim Sykes | PLN-002 |
| July 4, 1776 | 3/8"=1'-0" | Sht. 1 of 1 |

19.7 For Web-Based Review Questions and Additional Exercises, visit: http://foragerpub.com/AcadFiles/2011/2011.htm

This stuff rocks!

Lesson 20

Following this lesson, you will:

✓ *Know how to add Hatching and Section Lines to your AutoCAD drawing through:*
- *The **Hatch** dialog box*
- *Tool Palettes*

✓ *Know how to edit hatch patterns using the **HatchEdit** command*

✓ *Know how to create and manipulate Groups*

Hatching and Grouping

Remember all those templates I mentioned in Lesson 5 – the ones I needed to draw ellipses? Remember how much simpler (and more accurate) it was to draw ellipses with AutoCAD?

Here's another drawing tool that puts those pieces of plastic to shame. No longer will the drafter need to spend hours drawing each line or symbol to show section lines, a brick façade, concrete, and so forth. With AutoCAD's hatch commands, you can fill a large area with lines or symbols in a matter or seconds!

When we've finished with hatching, we'll get a head-start on manipulating multiple objects as one; we'll look at grouping.

20.1 Hatching and Filling

Where to Find It:	
Command Line:	*Hatch*
Hotkey(s):	h
Ribbon (Tab/Panel):	Home – Draw – **Hatch**
Menu:	Draw – **Hatch**
Toolbar:	Draw – **Hatch**
Tool Palette:	Draw – **Hatch**

There are actually two ways to draw hatch patterns. (Section lines and fills are created as styles of hatch pattern. So, I'll refer to all as hatch patterns.) There's a ribbon approach and a simple drag-n-drop approach that utilizes the tool palettes. We'll begin with the **Hatch Creation** ribbon tab (Figure 20.001, p.450). Open it with the *Hatch* command. Let's start from the left and look at the basics.

- Use the tools in the **Boundaries** panel to identify the edges (boundaries) of your hatch pattern.
    o When you use the **Pick points** button, AutoCAD returns to the graphics screen and prompts:

    Pick internal point or [Select objects/seTtings]:

    Pick anywhere within the boundaries (where you want to place your hatching). Be sure the area is closed; AutoCAD will *not* hatch an open area unless the gap tolerance has been properly set (more on that in a few minutes)!

    > Unless you tell it to do something else, AutoCAD assumes the **Pick points** button procedure when you begin the *Hatch* command.

    Picking the **seTtings** option opens the Hatch and Gradient dialog box, is another way of presenting the robbon tools.

    o Use the **Select** button to go straight to the **Select objects** prompt with a **picK internal point** option.

    Select objects or [picK internal point/seTtings]:

    Use this tool to select objects that will define the boundary of your hatching.

    o Use the **Remove** button to remove selected objects from your boundary definition.

    o Use **Recreate** to redefine existing hatching. (This option becomes available when editing an existing hatch or gradient.)

    o You'll find some more options in the **Boundaries** subpanel.
        - AutoCAD makes the **Display Boundary Objects** tool available only when you're editing an existing hatch or gradient pattern. Use this to identify the boundaries of the selected pattern.
        - The **Retain Boundary Objects** control box offers you the chance to save the selected boundary as a **Polyline** or a **Region** (an object similar to a 2-dimensional solid).

- Use the **Specify Boundary Set** control box to select what AutoCAD will evaluate when defining a boundary with the pick point procedure.
- The **Pattern** panel displays available hatching or gradient patterns for your selection. Pick the down arrow ▼ below the scroll bars to see more.
- Next to the **Pattern** panel you'll find the real workhorse of the **Hatch Creation** tab – the **Properties** panel.
    - The **Hatch Type** control box allows you to select a **Pattern**, a **User defined** pattern, a **Solid** or a **Gradient**.
        - **Pattern**s and **Solid** accompany AutoCAD. You see these by default in the **Pattern** panel.
        - The **User defined** option allows you to set the **Angle** and **Spacing** of hatch lines using the other tools on the **Properties** panel. Use these tools to define your own hatch pattern. The **User defined** option also makes the **Double** option available (in the **Properties** sub panel). When selected, the **Double** option will cause AutoCAD to duplicate your hatch pattern within the same boundaries *but at 90° to the first pattern*.
        - The **Gradient** option allows you to create some interesting gradient fills similar to the solid option but with more clout.

    > You can access the **Gradient** tools directly with the **Gradient** command.

    - Use the **Hatch Color** control box to select the color of your hatching. Select **ByLayer** to create a pattern with a layer-defined color.
    - Like the **Hatch Color** control, you'll use the **Hatch Background Color** control box to control the background color (if any) of your pattern.
    - You can control the visibility of your pattern with the **Transparency** slider or value entry box.
    - Between the **Pattern** and **Transparency** controls, you'll find a small control box where you can tell AutoCAD how to create the transparency: **ByLayer**, **ByBlock**, or by **Value**.
    - You can also control the **Angle** of your pattern with a slider (a marked improvement from previous methods) or by value entry box.
    - Finally, control the scale of your pattern with the **Scale** spinner. Note that you can also enter a value manually.
    - The **Properties** panel includes a subpanel (Figure 20.002). Here you can:
        - control the layer on which your pattern will appear,
        - tell AutoCAD to scale the pattern **Relative to Paper** Space (more on paper space in Lessons 24, p.560, and 25, p.584),
        - tell AutoCAD to **Double** your user-defined pattern, and
        - set an **ISO Pen Width** for ISO patterns.

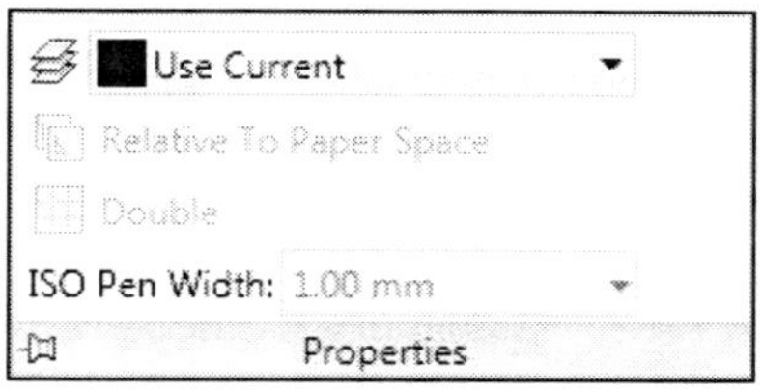

Figure 20.002

Figure 20.001

- Use options in the **Origin** panel and subpanel to locate the pattern within the selected boundaries. Normally, the hatch origin defaults to the 0,0 coordinates. Theoretically, bricks, grating, etc. will begin there even if you only see the portion of the hatch within the boundaries you've designated. This can result in partial bricks (or grating, etc.) along the bottom of the "wall" you're bricking. I doubt that a mason will cut bricks in half to start a new row just to make a building look like your hatching, so AutoCAD decided to allow you to start with a whole brick – by allowing you to redefine the hatch origin.
    - You can set a new origin for your hatching by picking the **Set origin** button. But this is the hard way (unless you want to put the origin in an odd place).
    - Use the buttons in the subpanel to locate your hatch origin in relation to the boundary (**Bottom left, Bottom right, Top left, Top right,** or **Center**). Alternately, you can opt to **Use the Current Origin**.
    - Finally, you can select **Store as default origin** if you wish to use these settings by default.
- Our next panel – **Options** – also contains a subpanel, several options, and a **Properties** button.
    - **Associative** hatching means that hatch patterns will update automatically when the boundaries change (much as associate dimensions do).

    > You can convert a hatch pattern (associative or non-associative) into its constituent lines using the *Explode* command. You cannot convert the other way, however, so do this with caution.

    - If you select **Annotative**, AutoCAD will adjust your scale according to the drawing's **Annotation Scale**.
    - Use the **Match Properties** button to duplicate any existing hatching in your drawing – even if you don't know how it was set up! Pick the button; AutoCAD offers choices to **Use [the] current origin** or to **Use [the] source hatch origin**. When you make a selection, AutoCAD will return to the graphics screen and prompt you to **Select hatch object**. Pick the one you wish to duplicate and voila! AutoCAD automatically adjusts the ribbon settings to duplicate it! (Way cool!)
    - The **Options** subpanel contains some very useful tools, too.
        - **Gap tolerance** can prevent temper tantrums! (I know, designers don't have tempers, right?) When trying to hatch an open grouping of objects, AutoCAD used to run into problems (okay, it wouldn't work). But set this to a number greater than 0 (use the slider or the value input box). If the ends/edges of the objects fall within that tolerance, AutoCAD will hatch it without grief.

        > When AutoCAD finds an open boundary, it'll try to show you where the problem occurs by placing red circles around the open end points. Then you can close the boundary and proceed with your hatching.

        - Normally, if you hatch within several boundaries at once, AutoCAD treats the different hatches as a single object. (That is, if you erase one, they all go.) Select **Create Separate Hatches** to avoid problems that this might present.
        - Islands are boundaries within boundaries. By default, AutoCAD recognizes islands and uses an island hopping approach to hatching. That is, hatch – skip – hatch. It does, however, allow other approaches using the **Islands** menu.
            - **Normal Island Detection** works just as I've described; that is, hatch – skip – hatch.
            - **Outer Island Detection** means that AutoCAD will hatch between the outer boundary and the first inner boundary, ignoring any boundaries within that one.
            - When AutoCAD **Ignores Island Detection** (or uses **No Island Detection**), it hatches everything within the outer selected boundary.

- Use options on the **Send Behind Boundary** menu to place the hatching in relation to the boundaries.
  o Finally, you'll notice an options arrow on the right end of the **Options** panel's title bar. This calls the Hatch and Gradient dialog box – aka. the *olde way* of doing everything we've discussed in these last three pages! Your best bet here is just to ignore the dialog box in favor of the ribbon tools.

Are you thoroughly confused? Let's try an exercise.

Do This: 20.1A	Hatching

I. Open the *demo-hatch* file the C:\Steps\Lesson20 folder. The drawing looks like the figure at right.
II. Follow these steps.

20.1A: HATCHING

1. Start the *Hatch* command.
   Command: h

2. On the **Pattern** panel (**Hatch Creation** tab), select the desired pattern (**ANSI33**). (Scroll down as necessary until you see it.)

3. In the **Properties** panel, set the **Scale** to *1.5* and the **Angle** to *0*.
   Accept the **Color** and **Transparency** defaults.

4. On the **Options** subpanel, select the **Outer Island detection** option.

5. Pick any point between the rectangle and the polygon. (Remember, AutoCAD assumes the **Pick points** procedure when you begin the *Hatch* command.
   Pick internal point or [Select objects/seTtings]:

6. AutoCAD tells you what it's doing, and then repeats the prompt.
   Selecting everything...
   Selecting everything visible...
   Analyzing the selected data...
   Analyzing internal islands...
   Pick internal point or [Select objects/seTtings]:
   Notice the preview.

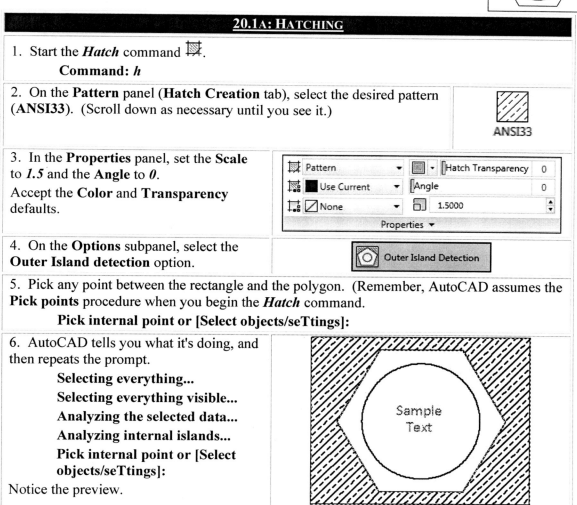

20.1A: HATCHING

7. Complete the command.
   Pick internal point or [Select objects/seTtings]: *[Enter]*
   AutoCAD returns to the graphics screen and the command prompt.
   Your drawing looks like the figure at right.

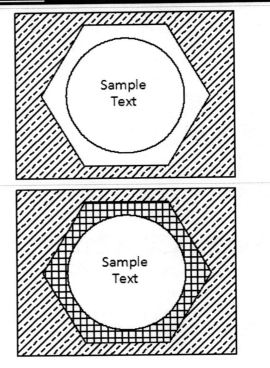

8. Repeat Steps 1 through 7, but hatch between the polygon and the circle with the ANSI37 pattern. Use a **Scale** of *2.0* and an **Angle** of *45°*.

9. Now we'll use the **Match Properties** button to hatch the circle using the ANSI33 pattern we used between the rectangle and the polygon. Repeat the *Hatch* command.
   Command: *[ENTER]*
   Notice that the settings default to the last settings used.

10. Pick the **Match Properties** button. AutoCAD prompts you to
    Select hatch object:

11. At the prompt, pick a line on the hatch pattern between the rectangle and the polygon. AutoCAD adjusts the settings according to what you've selected.

12. Now pick a point inside the circle.
    Pick internal point or [Select objects/seTtings]:

13. Change the **Angle** setting to *90°*.

Angle	90

14. Hit ENTER.
    Pick internal point or [Select objects/seTtings]:
    Your drawing looks like the figure at right.

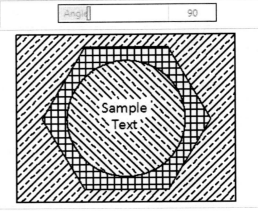

15. Now for a real treat, try stretching the rectangle or the polygon. Note the effects of *Associative* hatching!
    Command: *s*

20.1A: HATCHING

16. Undo the *Stretch* command and save the drawing.
    Command: *qsave*

17. Now let's look at some other options. Thaw the **OBJ2** layer (but leave **HATCH** current). Notice the new rectangles. (Adjust your view as necessary.)

18. Repeat the **Hatch** command.
    Command: *h*

19. Select the AR-B816 pattern from the **Pattern** panel.

20. Set the **Angle** and **Scale** as indicated.

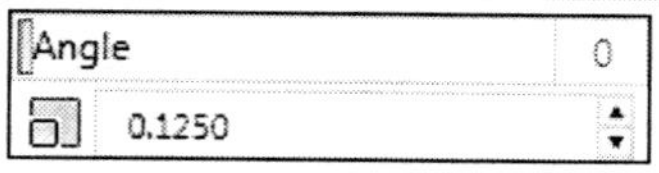

    Angle: 0
    0.1250

21. Tell AutoCAD to use **Normal Island detection** ...

    Normal Island Detection

22. ... and Use [the] **current origin**.

23. Now pick a point between the two outer rectangles.
    Pick internal point or [Select objects/seTtings]:

24. Finally, complete the procedure.
    Pick internal point or [Select objects/seTtings]: *[Enter]*
    Your rectangles look like the figure at right. Notice the **Normal** hatching (island hopping: hatch – skip – hatch). Notice also where the bricks start.

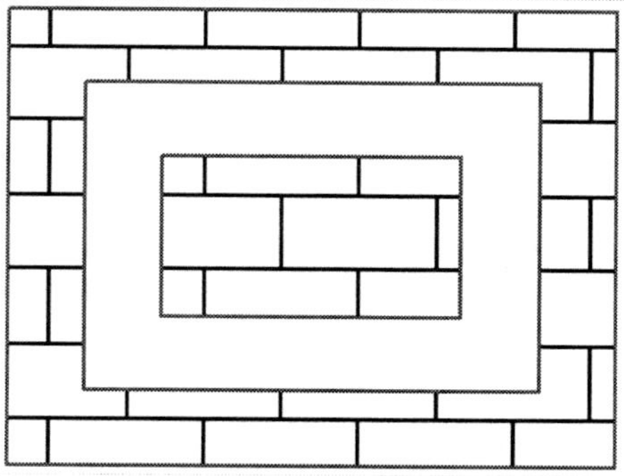

25. Erase the brick pattern.
    Command: *e*

26. Repeat Steps 18 through 25, but this time, select the **Bottom left** boundary. See the difference (right)?

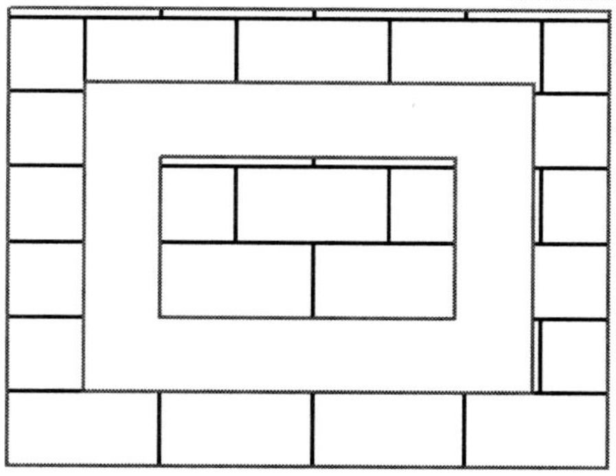

27. Save the drawing.
    Command: *qsave*

20.2	Editing Hatched Areas
20.2.1	Normal Hatch Editing

We have another way to apply hatching, but before we look at them, let's take a quick look at editing the pattern once you've applied it. (This may seem a bit out of order, but it'll help in later explanations.)

Remember when we edited text? We found that the easy way to start it was simply to double click on the text to be edited. AutoCAD automatically reopened the text editor (for dynamic text) or the MText editor (ribbon). Once there, editing the text was as easy – and followed the same rules and procedures – as creating the text.

Well editing hatches works the same way. Just double click on the pattern you wish to edit, and AutoCAD will reopen the **Hatch Creation** ribbon tab (note that it renames the tab *Hatch Editor*, but it's the same interface). Then you can proceed as though you were creating the hatching for the first time.

Let's try it.

> AutoCAD does have a *HatchEdit* command. Unfortunately, however, it calls the old HatchEdit dialog box. As I mentioned earlier, the smart designer will opt for the ribbon interface in lieu of the dialog box.

Do This: 20.2.1A	Normal Hatch Editing

I. Be sure you're still in the *demo-hatch* file in the C:\Steps\Lesson20 folder. If not, please open it now.
II. Turn **Quick Properties** off on the status bar.
III. Follow these steps.

20.2.1A: NORMAL HATCH EDITING

1. Double click on the hatching inside the circle. AutoCAD presents the Hatch Editor ribbon tab. (It may also present the Properties palette. If it does, just close it for now.)
Notice that theHatch Editor settings reflect those you used to create the pattern.

2. On the **Hatch Type** control box (**Properties** panel), select the **Gradient** tool
`Gradient`.
Notice how the selections in the **Pattern** panel change to gradients.

3. Select the GR_SPHER pattern .

4. Clear the grips.
The drawing now looks like the figure at right. (Nifty, huh?)

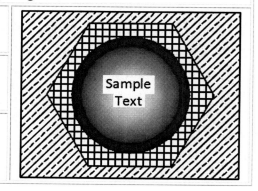

5. Save the drawing.
   Command: *qsave*

20.2.2 Non-Associative Hatch Editing

You can edit non-associative hatching, but it takes a bit more effort. You'll need to use grips. You'll see this best with an exercise.

Do This: 20.2.2A — Non-Associative Hatch Editing

I. Be sure you're still in the *demo-hatch* file in the C:\Steps\Lesson20 folder. If not, please open it now.
II. Erase the brick hatching in the rectangles on the right.
III. Hatch the rectangles on the right with an ANSI31 pattern at a scale of 2. Be sure to deselect the **Associative** option before completing the hatching.
IV. Follow these steps.

20.2.2A: NON-ASSOCIATIVE HATCH EDITING

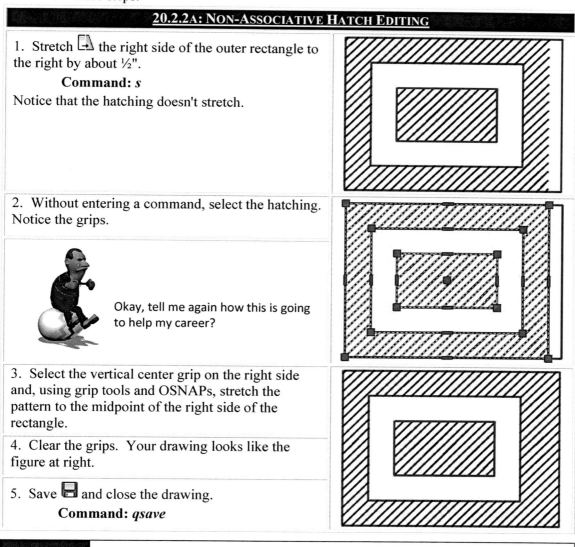

1. Stretch the right side of the outer rectangle to the right by about ½".
   Command: *s*
   Notice that the hatching doesn't stretch.

2. Without entering a command, select the hatching. Notice the grips.

   Okay, tell me again how this is going to help my career?

3. Select the vertical center grip on the right side and, using grip tools and OSNAPs, stretch the pattern to the midpoint of the right side of the rectangle.

4. Clear the grips. Your drawing looks like the figure at right.

5. Save and close the drawing.
   Command: *qsave*

20.3 Using Tool Palettes to Hatch

Using tool palettes is easy, but it relies on the Properties palette to modify hatch pattern definitions before you insert them, or the **Hatch Editor** ribbon tab to modify patterns after they've been inserted.

Let's see how this works.

Do This: 20.3A	Drag-and-Drop Hatching with Tool Palettes

I. Open the *d&d-hatch* file the C:\Steps\Lesson20 folder. It looks like the figure at right.
II. Open the Tool Palettes and place the **Hatches and Fills** tab on top.
III. Follow these steps.

20.3A: DRAG-AND-DROP HATCHING WITH TOOL PALETTES

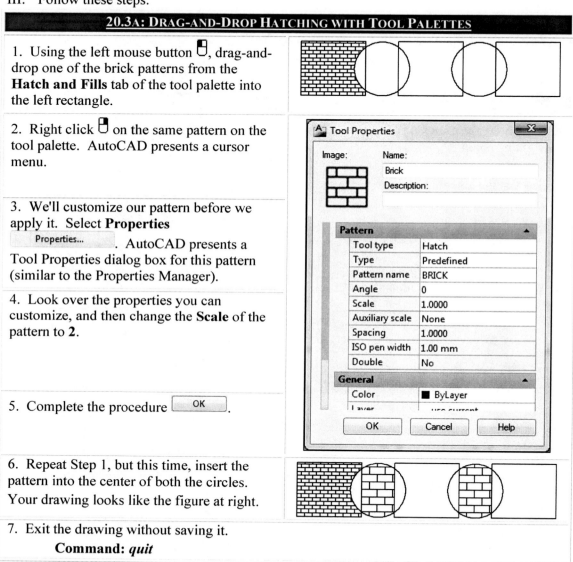

1. Using the left mouse button, drag-and-drop one of the brick patterns from the **Hatch and Fills** tab of the tool palette into the left rectangle.

2. Right click on the same pattern on the tool palette. AutoCAD presents a cursor menu.

3. We'll customize our pattern before we apply it. Select **Properties**. AutoCAD presents a Tool Properties dialog box for this pattern (similar to the Properties Manager).

4. Look over the properties you can customize, and then change the **Scale** of the pattern to **2**.

5. Complete the procedure OK.

6. Repeat Step 1, but this time, insert the pattern into the center of both the circles. Your drawing looks like the figure at right.

7. Exit the drawing without saving it.
    Command: *quit*

So which method do you prefer? Obviously, you prefer the easiest!

20.4 Paper Dolls: The *Group* Command

Have you ever done a furniture layout for an office or an equipment layout for a shop or plant? You may know then that the easiest way to accomplish the layout (in such a fashion that the equipment can be moved around as the plan develops) is to use the "paper dolls" approach. That is, you create a scaled drawing of the room or work area, make a blue line of it, and then move scaled representations

(paper dolls) of the equipment around the blue line until you're satisfied with the locations. Once you're satisfied, you tape the dolls to the blue line for reference and draw the objects in your layout.

It's easy work *if* you keep the windows closed and nobody slams a door. A slight breeze, however, sends people scrambling for the paper dolls and mumbling things that might make a sailor blush. Of course, this is only after the hours spent reminiscing about kindergarten while cutting out paper dolls.

Once again, AutoCAD has created tools that provide the benefits of construction paper and tape (combined with a few new benefits) while removing the hassles. Enter ***Groups***.

What is a group? Simply put, like hatching, a group is a collection of objects treated as a single unit. But groups have considerably more flexibility than hatches.

Let's take a look.

Do This: **20.4A**	**Working with Groups**

I. Open the *groups* file the C:\Steps\Lesson20 folder. The drawing looks like the figure at right.

II. Turn Selection Cycling off (Refer to Lesson 10, Section 10.5, p.239, or just set the **SelectionCycling** system variable to **0**.)

III. Follow these steps.

20.4A: WORKING WITH GROUPS

1. We'll switch the tree (in the upper left corner of the screen) with the easel pad (in the upper right corner). Enter the *Move* command 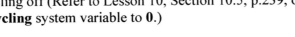.

    Command: *m*

2. Select anywhere in the tree as shown. Complete the selection.

    Select objects: 80 found, 1 group

    Select objects: *[ENTER]*

    (Notice how easy it is to select these 80 objects!)

3. Move the tree 21 feet to the right.

    Specify base point or [Displacement] <Displacement>: *21',0*

    Specify second point or <use first point as displacement>: *[ENTER]*

4. Repeat Steps 1 through 3 to move the easel pad 21 feet to the left. Your drawing looks like the following figure.

Notice that, in each instance, you selected the objects with a single pick and AutoCAD indicated the number of objects *and the number of groups* found.

20.4A: WORKING WITH GROUPS

5. Now rotate the easel pad 90° to face the correct direction. (Use grips or any other method you've learned.)
This part of your drawing now looks like the figure at right.

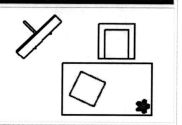

6. Save the drawing 💾 but don't exit.
   Command: *qsave*

You can see how easy it is to manipulate grouped objects. And before you ask, yes, you can manipulate individual objects within a group (but *not* within a block – well, at least not easily ... more on blocks in Lesson 21, p.468).

Let's take a look at how you can create groups yourself and at how you can work with individual parts of the group.

When you enter the ***Group*** command (or ***g***), AutoCAD provides the Object Grouping dialog box to assist you (Figure 20.003).

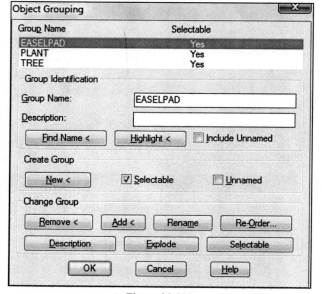

- The upper list box shows the names of existing groups (under **Group Name**) and whether or not the group is currently **Selectable (Yes** or **No)**. Selectable simply means that you can manipulate the group as a single object. You'll want the group to be selectable to do what we did in the last exercise; but you won't want it selectable if you need to modify something within the group (like pruning the branches of the tree we moved). We'll see how to change the **Selectable** setting shortly.

Figure 20.003

- The **Group Identification** frame holds several useful items.

  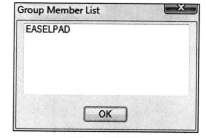

  Figure 20.004

  o Appearing in the appropriate text box, you'll find the name and description (if any) of the group highlighted in the list box. You'll begin to create a new group by typing a new **Group Name** and **Description** (if desired) into these boxes. (We'll create some new groups in our next exercise.)

  o To find the name of an existing group, begin by picking the **Find Name** button. AutoCAD returns to the graphics screen and prompts you to **Pick a member of a group**. It then displays the Group Member List dialog box (Figure 20.004) showing a list of all the groups to which that object belongs.

  o To find which objects belong to a certain group, begin by selecting the name of the group in the upper list box, and then pick the **Highlight** button. AutoCAD returns to the graphics screen, highlights all the objects belonging to that group, and presents a **Continue** button to return to the Object Grouping dialog box.

  o Sitting inconspicuously in the lower right corner of the frame, you'll find a check box next to the words **Include Unnamed**. When a group is copied, the copy is also a group. You haven't named the copy, but AutoCAD has assigned a temporary name to it. Placing a check in this box tells AutoCAD to display the temporary names as well as the user-assigned names. You can then rename the copies to something more appropriate.

- Use tools in the **Create Group** frame to create a new group. After typing a name for your new group in the **Group Name** text box (or placing a check in the **Unnamed** check box if you want AutoCAD to assign the name), pick the **New** button. AutoCAD will return to the graphics screen and prompt **Select objects for grouping**. When you've finished selecting the objects, hit ENTER to return to the Object Grouping dialog box. Be sure there's a check in the **Selectable** check box if you want to manipulate the selected objects as a group.

- Use the seven buttons in the **Change Group** frame to modify a group. First, select the group to be modified in the list box. Then pick one of the buttons in the **Change Group** frame.
    o Add or remove objects from a group with the **Add** or **Remove** buttons.
    o Use the **Rename** button to change the name of a group. For example, to change the temporary name assigned by AutoCAD to a more recognizable name, highlight the temporary name in the list box, type the new name in the **Group Name** text box, then pick the **Rename** button.
    o Type some text into the **Description** text box, and then pick the **Description** button to update the description of the highlighted group.
    o The **Explode** button will remove the definition of the group from the drawing's database. All objects in the group will then behave as individual objects (not as part of a group).
    o The **Selectable** button is a toggle for treating a group as a group or for suspending the group

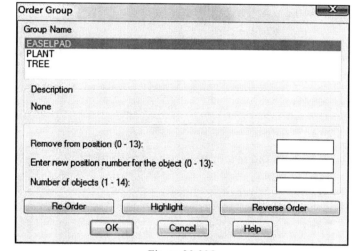

Figure 20.005

definition so you can modify one or more of the objects within the group. When **Selectable** is toggled off, the word **No** appears in the **Selectable** column of the list box, and all the objects within a group behave as individual objects. To manipulate the objects as part of the group, simply toggle **Selectable** back on.

    o The **Re-Order...** button [Re-Order...] presents the Order Group dialog box (Figure 20.005). Here you can change the order in which AutoCAD reads the objects in a group. Type the number of the object you wish to change in the **Remove from position** text box. Type the position to which you wish to move the object in the **Enter new position number for the object** text box. Type the number or range of objects to reorder in the **Number of objects** text box.

    Pick the **Reverse Order** button to simply reverse the order of the objects within the group.

> The **Order** option of the *Group* command proves useful when ordering the tool cutting sequence in a CAM system.

If you're unsure of the position of the object you wish to reorder, pick the **Highlight** button. AutoCAD will present the Object Grouping dialog box shown in Figure 20.006, and highlights the objects making up the group you selected.

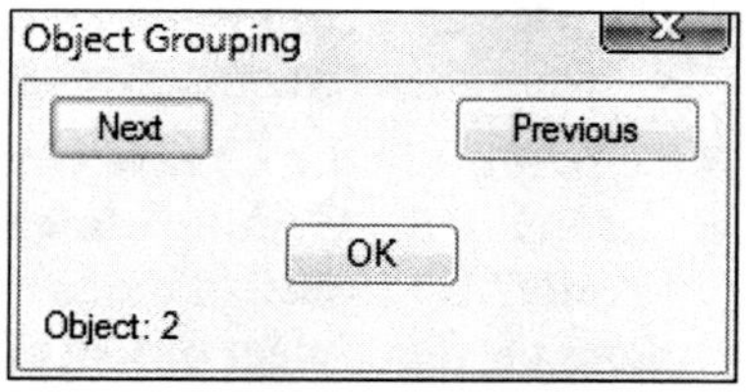

Let's play with the *Group* command!

Figure 20.006

Do This: 20.4B	**Creating and Manipulating Groups**

I. Be sure you're still in the *groups* file the C:\Steps\Lesson20 folder. If not, please open it now.
II. Follow these steps.

20.4B: CREATING AND MANIPULATING GROUPS

1. Enter the *Group* command. AutoCAD presents the Object Grouping dialog box (Figure 20.003, p.459).

   Command: g

2. We'll group the teacher's desk, chair, computer, and the plant on the desk. Enter the name and description shown.

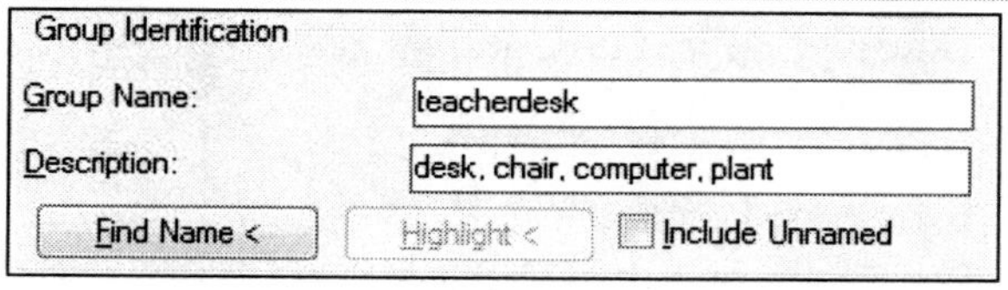

3. Pick the **New** button.

4. Select the teacher's desk and the other objects. Hit ENTER or right click to return to the dialog box.

   Select objects for grouping:
   Select objects: 86 found, 1 group
   Select objects: [ENTER]

5. AutoCAD now lists the **TEACHERDESK** in the list box as shown.

   Complete the command OK.

6. Move the **TEACHERDESK** group next to the tree as indicated.

   Command: m
   Select objects: 86 found, 1 group
   Select objects:
   Specify base point or [Displacement] <Displacement>: 13',0

   Your drawing looks like the figure at right.

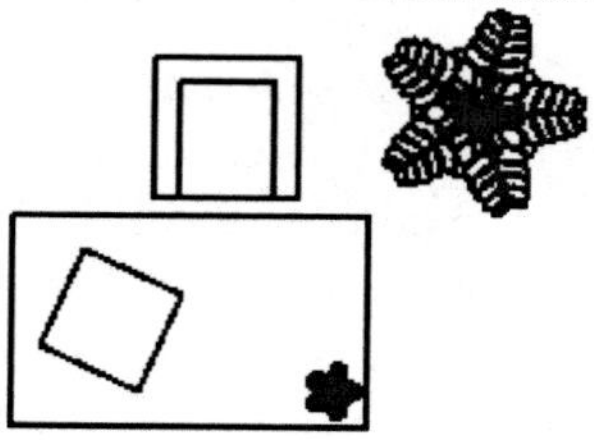

7. Reopen the Object Grouping dialog box.

   Command: g

8. Now we'll list all the groups to which an object belongs. Pick the **Find Name** button.

20.4B: CREATING AND MANIPULATING GROUPS

9. AutoCAD returns to the graphics screen. Pick the plant on the teacher's desk.
    Pick a member of a group.

10. AutoCAD presents the Group Member List dialog box showing that the object selected belongs to the **Plant** and **Teacherdesk** groups. You see now that groups can be *nested* – that is, one group can contain another group. Objects can also be shared by more than one group.
Pick the **OK** button [OK] to return to the Object Grouping dialog box.

11. Let's see which objects are in the **Plant** group. Pick **PLANT** in the list box. Notice that the name appears in the **Group Name** text box.

12. Pick the **Highlight** button [Highlight <]. AutoCAD returns to the graphics screen and highlights the plant on the teacher's desk.

13. Return to the Object Grouping dialog box [Continue].

14. Complete the command [OK].

15. Let's give the teacher two new plants. Try to copy the **Plant** group.
    Command: *co*
Notice that the entire desk highlights when you select the plant. We must first make the desk *unselectable*.

16. Hit the ESC key [Esc] to cancel the command.

17. Return to the Object Grouping dialog box.
    Command: *g*

18. Pick **TEACHERDESK** in the list box and then pick the **Selectable** button [Selectable] in the **Change Group** frame. Notice (right) that the word **No** appears in the **Selectable** column of the list box.

19. Complete the command [OK].

20. Now make two copies of the plant on the teacher's desk.
    Command: *co*
Your drawing now looks something like the figure at right.

21. Return to the Object Grouping dialog box.
    Command: *g*

22. Place a check in the **Include Unnamed** check box [✓] Include Unnamed in the Group Identification frame. Notice that two new groups appear in the list box – ***A1** and ***A2**.

20.4B: CREATING AND MANIPULATING GROUPS

23. Select ***A1**.	*A1 Yes *A2 Yes EASELPAD Yes
24. Enter the name *PlantB* into the Group Name text box.	Group Name: PlantB
25. Now pick the **Rename** button [Rename]. AutoCAD renames the group.	
26. Repeat Steps 23 through 25 to rename ***A2** to *PlantC*.	
27. Now let's add these new plants to the **Teacherdesk** group. Select the **Teacherdesk** group in the list box (you may have to scroll down a bit to find it).	
28. Pick the **Add** button [Add <] in the **Change Group** frame.	

29. AutoCAD returns to the graphics screen and prompts you to **Select objects to add to group**. Select the two new plants, and then return to the Object Grouping dialog box.
  Select objects to add to group...
  Select objects: 80 found, 1 group
  Select objects: 80 found, 1 group, 246 total
  Select objects: *[ENTER]*	

30. Pick the **Selectable** button [Selectable] to make the **TEACHERDESK** group selectable again. Notice that the word **Yes** appears in the **Selectable** column of the list box.	
31. Complete the procedure [OK].	
32. Save the drawing 💾 but don't exit. Command: *qsave*	

We've seen how groups can benefit our work by simplifying object selection and manipulation. We've also seen how to manage our groups through some of the buttons in the **Change Group** frame of the Object Grouping dialog box. Now let's remove a group definition from a drawing.

Do This: 20.4C	Removing a Group Definition

  I. Be sure you're still in the *groups* file the C:\Steps\Lesson20 folder. If not, please open it now.
  II. Follow these steps.

20.4C: REMOVING A GROUP DEFINITION

1. Enter the *Group* command. Command: *g*	
2. Select the **EASELPAD** group in the list box.	EASELPAD Yes PLANT Yes PLANTB Yes
3. Pick the **Explode** button [Explode] in the **Change Group** frame. Notice that **Easelpad** disappears from the list box.	
4. Complete the command [OK]. AutoCAD returns to the graphics screen.	

20.4C: REMOVING A GROUP DEFINITION

5. Try to move the easel pad now.

   Command: *m*

   Notice that you can't select the entire pad with a single pick. The pad is no longer a group.

6. Save the drawing and exit.

   Command: *qsave*

I've frequently used groups in my capacity as a piping designer to manipulate (particularly moving and copying) items such as control stations and pipe configurations. Architects and interior designers can also use groups as we've done in these exercises. Other disciplines can make use of groups when dealing with layouts, circuitry, boltholes, windows, and many other items.

The only limitations to groups are simple – their use is limited to a single drawing (you can't share groups between drawings), the group definition is permanently lost once the group is exploded or erased from a drawing, and you can't attach information to a group. To overcome these limitations, use blocks instead of groups. (How's that for a lead-in to our next lesson?!)

20.5 Extra Steps

We've now used the AutoCAD Design Center, the Tool palettes, and the Properties palette with some frequency. Have you been opening and closing them as we go?

There's an easier way. Try this,

1. Open each of the three. Notice the similarities in their title bars.

2. Pick the **Properties** button at the bottom of one of the title bars and select **Anchor Left** from the menu. (If it remains open, pick the dash at the left end of the double lines at the top of the palette.) Notice that the palette's title bar appears docked on the left side of the graphics area. Move your cursor over it to open it.

3. Repeat Step 2 for each of the other two palettes. Notice that each now appears in a single line along the left side of the graphics area. AutoCAD has made room for all three.

4. Now move your cursor over each. Notice that when AutoCAD opens the palette, it is full size – you haven't lost anything by stacking them!

You may get tired of my saying, "Way cool!" But com'n, this really is Way Cool!

> A little bonus to docking the palettes: if you right click on the docked palettes and select **Icons only** on the menu, AutoCAD will stack icons for the palettes rather than names. Try it; it looks so much neater!

| 20.6 | **What Have We Learned?** |

Classroom acquired skills are to the mind what paint is to pottery – nothing but surface fluff that fades and disappears with the passing of time. For the skills to become actual knowledge, they must be tempered into the mind's clay through the repeated pounding of experience.

PRACTICE PRACTICE PRACTICE

Anonymous

Items covered in this lesson include:

- *Working with Groups (the **Group** command)*
- *Hatching Tools:*
  - ○ ***Hatch***
  - ○ ***Hatch Editing***
  - ○ ***Drag and Drop Hatching***
  - ○ ***Gradient***

By now you should be growing more comfortable with AutoCAD procedures and routines. You've passed beyond basic drawing and can create more complex objects in a few keystrokes or mouse picks. The one thing you still need – and need plenty of – is practice!

Work through the exercises at the end of this lesson and in the additional exercises (Section 20.7). They may take a bit longer than earlier exercises, but that's because the drawings are becoming more complex. Additionally, I'm expecting more from you while providing less.

Our next two lessons will teach you to use the first tools that make AutoCAD something more than a very expensive drafting tool. We'll cover blocks in Lesson 21, p.468; we'll attach *information* to blocks in Lesson 22, p.500.

| 20.7 | **Exercises** |

1. Start a new drawing from scratch. Using the same information you used in the *Exterior Slab* drawing in Exercise 2, create the *Concrete Pier Footing* drawing.

   1.1. Use this information to hatch the concrete:

   Pattern: AR-CONC

   Scale: 0.5000

   Angle: 0

   1.2. Save the drawing as *MyFooting* in the C:\Steps\Lesson20 folder.

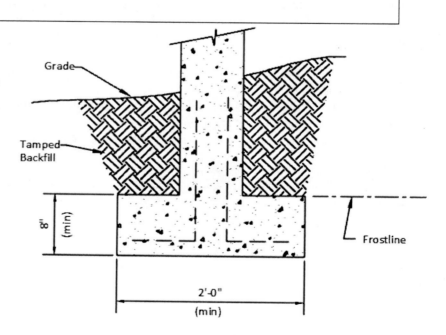

Concrete Pier Footing

2. Start a new drawing from scratch.
    2.1. Create the following setup:
        2.1.1. Use a 1½"=1'-0" scale on an A-size sheet of paper (8½x11)
        2.1.2. Grid: 1" (snap as needed)
        2.1.3. The layers in the table.
    2.2. Use this information for the Grade hatching:
        Pattern: EARTH
        Scale: 8.0000
        Angle: 45
    2.3. Use this information for the sand hatching:
        Pattern: AR-SAND
        Scale: 0.5000
        Angle: 0
    2.4. Use the Times New Roman or Calibri font.
        Large text should plot at ¼"; small text should plot at 1/8".
    2.5. Create the *Exterior Slab* drawing shown here. Save it as *MySlab* in the C:\Steps\Lesson20 folder.

LAYER NAME	COLOR	LINETYPE
0	black	Continuous
Border	blue	Continuous
Cl	212	Center2
Dim	12	Continuous
Const	red	Continuous
Hidden	42	Hidden
Obj1	blue	Continuous
Obj2	green	Continuous
Obj3	red	Continuous
Obj4	212	Continuous
Text	12	Continuous

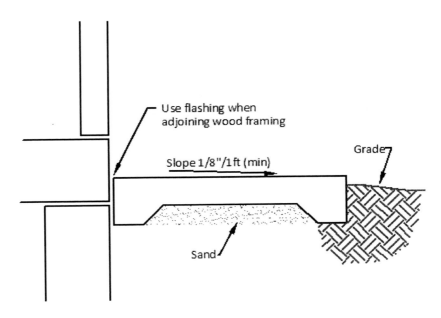

Exterior Slab

20.8 For Web-Based Review Questions and Additional Exercises, visit: http://foragerpub.com/AcadFiles/2011/2011.htm

It's Block Time!

Lesson 21

Following this lesson, you will:

✓ Know how to use blocks in a drawing
  - The **Block** command
  - The **WBlock** command
  - The **Insert** Command

✓ Know how to create libraries of blocked objects

✓ Know how to create Dynamic Blocks

✓ Know how to iDrop a block into a drawing from a website

Many as One – Blocks

One of the great benefits of using a computer to create design plans lies in the ability of the operator to create something once, and then duplicate or manipulate it as desired. We've used several modification commands, techniques, and procedures throughout the course of this text and have become fairly comfortable with the basic tools. But there are still two modification routines that can increase speed and make your work easier. These are the **Group** *and* **Block** *commands.*

We saw groups in Lesson 20, p.457, so we'll look at blocks in Lesson 21. Although they often behave in a similar manner, each has its niche in the drawing world and it'll be important to know when to use them. Then, in Lesson 22, p.500, we'll learn why blocks are one of the most valuable design tools ever created.

21.1   Groups with Backbone – The *Block* Commands

No single aspect of CAD has served as a stronger selling point than blocks. In fact, all quality CAD systems have blocks (or cells, or something that behaves like blocks). No other single command or routine can speed the drawing process as well, and none has a greater potential for cost cutting or streamlining design management.

What is a block? A block is a single object made up of several other objects. You can manipulate blocks, like groups, as single units. But you can share blocks *between* drawings. More importantly, *blocks can contain user-defined data*. We'll discuss creation and manipulation of blocks here. In Lesson 22, p.519, we'll discuss attaching data to blocks and exporting that data to other computer applications (such as spreadsheets or databases).

There are two important block commands – ***Block*** (which creates blocks within a drawing) and ***WBlock*** (which creates a block as a separate drawing file). Mercifully, both use dialog boxes. But to keep it simple, you'll use the ***Block*** command to create blocks and the ***WBlock*** command to copy your blocks into a *Folder library*.

What's a library?

Groups of blocks are often assembled in useful packages called libraries. A library is a group of blocks used as inserts in a drawing. Use of blocked objects saves the time you might otherwise need to create the objects.

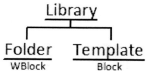

There are two types of libraries: *Template* – associated with the ***Block*** command, and *Folders* – associated with the ***WBlock*** command. Each has its pluses and minuses, but each project should use only one type of library.

The decision as to which type of library to use should be universal for the project. Let the guru or the design supervisor make this one. That doesn't mean that the operator can't create and use blocks on the fly. But take care not to duplicate (or overwrite) project standards.

> Most libraries consist of drawings associated with a single design discipline. For example, an architectural library may contain drawings of doors, windows, toilets, tubs, and so forth, whereas a piping library will contain drawings of valves, elbows, tees, etc. Library creation has become quite a business alongside CAD industries.

Another thing that must concern the CAD operator, both in the creation and insertion of blocks, is how blocks relate to layers. This can sound complicated, so let's use a table (after all, a table is worth a thousand words – give or take).

IF THE OBJECTS USED TO MAKE THE BLOCK ARE ON:	WHEN INSERTED, THE BLOCK HAS THE CHARACTERISTICS OF:	WHEN INSERTED, THE BLOCK EXISTS ON:
Layer 0	The current layer	The current layer
Any other layer	The layer on which it was created	The current layer

Let me try to explain this by using some examples (and fewer than a thousand words).

Generally speaking, you should always create objects for blocks on layer **0**. Then when inserting the block on another layer (let's say, layer **obj**), all the objects that went in to creating the block will appear on – and share the characteristics of – the current layer (layer **obj**). When this layer (**obj**) is frozen, turned off, or locked, the block will be affected accordingly.

Text might be an exception to this rule (and is rarely a good idea in a block – generally speaking). When part or all of the objects used to create a block exists on a layer other than 0 (say, **text**), those objects in the inserted block will appear with the characteristics of the layer on which they were created. (If that layer doesn't exist in the drawing, AutoCAD will create it.) This will be true regardless of the layer that's current (say, **obj**) when you insert the block. Now, when the **obj** layer is frozen, turned off, or locked, the block will be affected accordingly because the block exists on (was inserted on) layer **obj**. *Additionally*, when the **text** layer is frozen, turned off, or locked, the block will also be affected because that is the layer on which it was created.

Sound complicated? Use the chart!

21.1.1	Template Library Creation

We use the ***Block*** command to create template libraries or to create blocks on the fly. We'll use the ***WBlock*** command to convert template library blocks to folder libraries.

When completed, the newly defined block has become part of the current drawing's database and can be inserted into the current drawing at any time. Save this drawing as a template, and the block will be available to any drawing using this template.

Block presents the Block Definition dialog box (Figure 21.001) where you can enter the information you'll need to create a block.

Where to Find It:	
Command Line:	***Block* or *BMake***
Hotkey(s):	***b***
Ribbon (Tab/Panel):	Home – Block – **Create**
	Insert – Block – **Create**
Menu:	Draw – Block – **Make**
Toolbar:	Draw – **Make Block**
Tool Palette:	Draw – **Make Block**

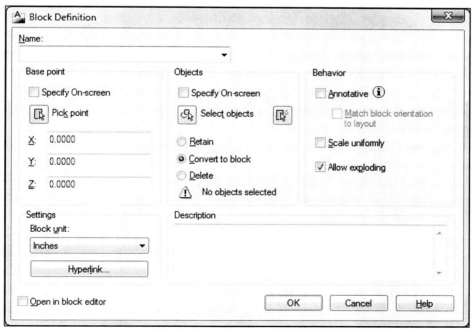

Figure 21.001

- Place the name of the block in the **Name** control box.

- Although you can enter an insertion point's coordinates in the **Base Point** frame (X, Y, and Z text boxes), it's easier to use the **Pick point** button. AutoCAD hides the dialog box and prompts:
   Specify insertion base point:
   You can then pick a base point on the screen.
- A check next to **Specify On-screen** will cause AutoCAD to prompt you to select objects for the block after you pick the **OK** button. Alternately, you can use the **Select Objects** button (in the **Objects** frame) to return to the graphics screen immediately and select objects to include in the block. You can also use Quick Select filters by picking the **Quick Select** button to the right of **Select objects**.
   Once you've selected the objects, you must tell AutoCAD what to do with them after it makes the block. Simply place a bullet next to the appropriate option:
   - **Retain** tells AutoCAD to keep the selected objects in their current state. AutoCAD won't automatically delete the objects after creating the block.
   - **Convert to block** tells AutoCAD to convert the objects from individual objects to the first insertion of the newly defined block.
   - **Delete** will delete the objects used to create the block.
- In the **Block unit** control box (**Settings** frame), you should define what units to use when inserting the block (preferably the units used by the drawing in which the block will be inserted). AutoCAD will use these units when it inserts a block via the ADC or tool palettes.
   The **Hyperlink** button at the bottom of the frame allows you to associate a hyperlink with the block using an Insert Hyperlink dialog box.
- Options in the **Behavior** frame restrict how the block will act when it's inserted.
   - An **Annotative** block will use the drawing's **Annotation Scale** to determine its insertion size.
   - If you intend to use a layout when plotting your drawing (more on Layouts in Lessons 24, p.560, and 25, p.584), it's a good idea to put a check next to **Match block orientation to layout** to keep the annotative block properly oriented.
   - Place a check next to **Scale uniformly** to restrict scaling of the block to uniform X, Y, and Z scaling.
   - Place a check next to **Allow exploding** if you wish to allow this option. Exploding a block loses all the block's definitions and leaves only the objects that went in to creating it.
- Enter a description of the block in the **Description** text box.
- The inconspicuous check box at the bottom of the dialog box – **Open in block editor** – provides access to a wonderful new world of block control! We'll discuss this in more detail in Section 21.2, p.476.

First, let's create some simple blocks. You can use simple blocks on the fly, but these instructions will also help to get your feet wet.

Do This: 21.1.1A	Creating Simple Blocks

   I. Open the *blocks* file the C:\Steps\Lesson21 folder. The drawing looks like the figure at right.
   II. Refer to the door seen in the center of the drawing.
   III. Follow these steps.

21.1.1A: CREATING SIMPLE BLOCKS

1. Enter the *Block* command.
   Command: *b*
   AutoCAD presents the Block Definition dialog box (Figure 21.001, p.470).

2. Enter the name (*door*) in the **Name** text box.

3. Pick the **Pick point** button in the **Base point** frame. AutoCAD returns to the graphics screen.

4. Pick the insertion point indicated.
   Specify insertion base point: _mid of
   AutoCAD returns to the dialog box.

5. Pick the **Select objects** button.
   AutoCAD returns to the graphics screen.

6. Select the arc, the vertical lines, and the angled line. Do *not* select the horizontal line. Confirm your selections.
   Select objects:
   Select objects: *[ENTER]*
   AutoCAD returns to the dialog box.

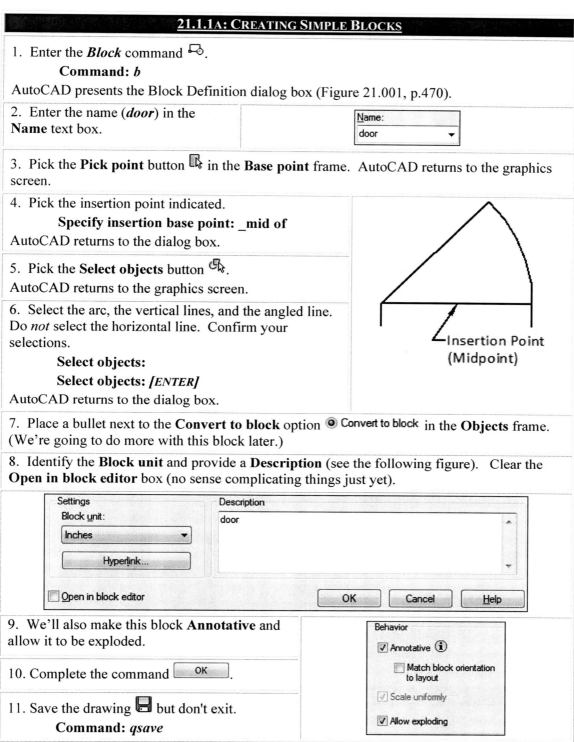

7. Place a bullet next to the **Convert to block** option in the **Objects** frame. (We're going to do more with this block later.)

8. Identify the **Block unit** and provide a **Description** (see the following figure). Clear the **Open in block editor** box (no sense complicating things just yet).

9. We'll also make this block **Annotative** and allow it to be exploded.

10. Complete the command OK.

11. Save the drawing but don't exit.
    Command: *qsave*

In this exercise, we've begun a Template library. We could continue to add dozens of drawings as needed (deleting the drawings as we go). Then, when we've finished, we would save this drawing with the .dwt extension and use it as a template for any number of future drawings. But let's use what we have to start a Folder library.

21.1.2 Folder Library Creation

Use the *WBlock* (**Write Block**) command to create folder libraries. The procedure can be very similar to creating blocks, but it also allows you to convert template library files to folder library files.

As with the *Block* command, AutoCAD also provides a dialog box interface to ease creation of WBlocks. Access it by entering the *WBlock* command (or the *W* hotkey). The Write Block dialog box appears (Figure 21.002).

Does it look familiar? It should. Two of the frames are almost identical to the Block Definition dialog box. Let's examine the other frames.

Figure 21.002

- In the **Source** frame, AutoCAD gives you three options:
  o The **Block** option opens the control box next to it. Here you can select from a list of blocks existing in the drawing to be saved as separate drawing files (WBlocks).
  o Of course, the **Entire drawing** option allows you to create a separate drawing file using all of the objects in the drawing.
  o The **Objects** option (the default) behaves like the *Block* command and allows you to select objects just as you did using the Block Definition dialog box.
- In the **Destination** frame, you'll:
  o name and locate the new block (**File name and path** text box).
  o identify the **Insert units** of the new block (this is the same thing as the **Block unit** found on the Block Definition dialog box).

In our next exercise, we'll create a WBlock to begin a folder library by converting the block created in our template library to a file we can use in our folder library.

Do This: 21.1.2A	Saving Blocks to a Folder Library

I. Be sure you're still in the *blocks* file the C:\Steps\Lesson21 folder. If not, please open it now.
II. Follow these steps.

21.1.2A: SAVING BLOCKS TO A FOLDER LIBRARY

1. We'll use the *WBlock* command to export our earlier block to our Folder Library. Enter the *WBlock* command.

   Command: *w*

21.1.2A: SAVING BLOCKS TO A FOLDER LIBRARY

2. (Refer to the figure at right.) Do this:

   a. Place the **Source** bullet in the **Block** option and then select **door** from the control box. (This is the block you created in our last exercise.) Notice that the **Base point** and **Objects** frames are no longer available. (This information was provided when you defined the original block.)

   b. Be sure the **Destination** information in your dialog box saves the new file as *door* in the C:\Steps\ Lesson21 folder.

   c. Pick the **OK** button [OK] to complete the command.

   How simple!

3. Save 💾 and close the drawing.
   Command: *qsave*

The block you just created will appear in the C:\Steps\Lesson21 folder as a new drawing that you can insert into other drawings as blocks!

21.1.3   Using Blocks in a Drawing – The *Insert* Command

Before we continue, we must know how to insert a block into a drawing once we've created it. We'll do this, appropriately enough, with the *Insert* command.

The *Insert* command calls the dialog box shown in Figure 21.003, p.475.

Where to Find It:	
Command Line:	*Insert*
Hotkey(s):	*i*
Ribbon (Tab/Panel):	Home – Block – **Insert**
	Insert – Block – **Insert**
Menu:	Insert – **Block**
Toolbar:	Draw – **Insert Block**
Tool Palette:	Draw – **Insert Block**

- The **Name** control box will list all of the blocks currently associated with the drawing (by definition, previous insertion, or template). To insert a WBlock (or drawing file) not currently associated with the drawing, use the **Browse** button to locate the file. AutoCAD will show the path next to the word **Path**.

- In the **Insertion point** frame, you can identify the **X/Y/Z** coordinates for the insertion or (preferably) leave a check in the **Specify On-screen** box. Then you can identify the insertion point on the screen with a mouse pick and OSNAP.

- The **Scale** frame also has a **Specify On-screen** box. But it may be preferable (unless you must see the scaled block to be sure of its size) to enter the **X/Y/Z** scale in the text boxes provided. A check in the **Uniform Scale** box will ensure the **X/Y/Z** scales remain proportional to the original block definition.

- The **Rotation** frame also has a **Specify On-screen** box. Like the **Scale** data, it's up to you to decide what's easiest to use.
- The **Block Unit** frame displays information about the block currently identified in the **Name** control box. This frame is informational only.
- The last item in the Insert dialog box – the **Explode** check box in the lower left corner – enables you to place all the objects of a block without the definition of the block. In other words, you can insert the block pre-exploded.

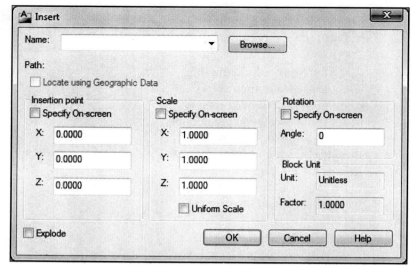

Figure 21.003

Let's insert a block into a drawing.

Do This: 21.1.3A	Inserting Blocks

I. Open the *cab-pln* file the C:\Steps\Lesson21 folder. The drawing looks like the figure at right.
II. Follow these steps.

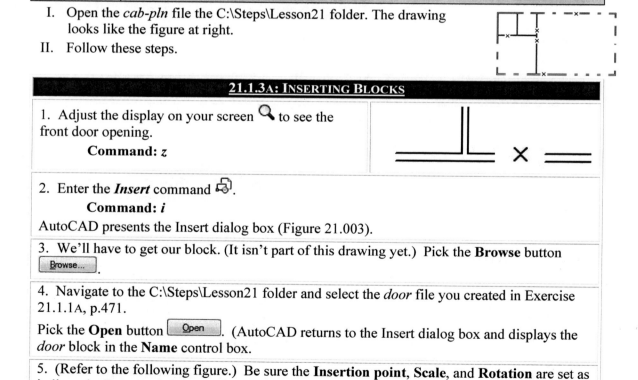

21.1.3A: INSERTING BLOCKS
1. Adjust the display on your screen to see the front door opening. **Command:** *z*
2. Enter the *Insert* command. **Command:** *i* AutoCAD presents the Insert dialog box (Figure 21.003).
3. We'll have to get our block. (It isn't part of this drawing yet.) Pick the **Browse** button.
4. Navigate to the C:\Steps\Lesson21 folder and select the *door* file you created in Exercise 21.1.1A, p.471. Pick the **Open** button. (AutoCAD returns to the Insert dialog box and displays the *door* block in the **Name** control box.)
5. (Refer to the following figure.) Be sure the **Insertion point**, **Scale**, and **Rotation** are set as indicated. (Be sure the **Explode** check box is empty.)

21.1.3A: INSERTING BLOCKS

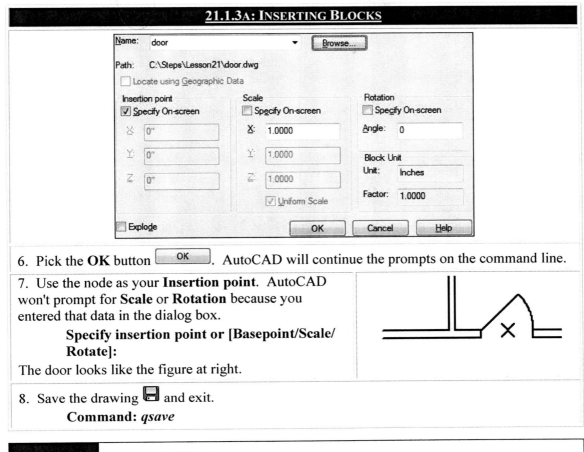

6. Pick the **OK** button. AutoCAD will continue the prompts on the command line.

7. Use the node as your **Insertion point**. AutoCAD won't prompt for **Scale** or **Rotation** because you entered that data in the dialog box.

   Specify insertion point or [Basepoint/Scale/Rotate]:

   The door looks like the figure at right.

8. Save the drawing and exit.

   Command: *qsave*

21.2 Dynamic Blocks

Wow! Look what I made it do!

Novice CAD Operator

I've been working with AutoCAD now for about a million years. They don't come up with new things anymore that really evoke a "way cool!" out of me (at least not very often), but dynamic blocks even made me capitalize the expression – *WAY COOL!* Dynamic blocks will take a bit of time to master, but the time will be well spent.

What is a dynamic block?

Well, a dynamic block has at least one *parameter* (a tool that generally looks like a dimension but acts as a control for selected object properties). The

Where to Find It:	
Command Line:	*BEdit*
Hotkey(s):	*be*
Ribbon (Tab/Panel):	Home – Block – **Edit**
	Insert – Block – **Block Editor**
Menu:	Tools – **Block Editor**
Toolbar:	Standard – **Block Editor**

CAD operator can manipulate the block objects associated with the parameter through grips which drive user-defined actions which, in turn, can change the geometry of the block.

By making them *dynamic*, AutoCAD has given us the power to manipulate the objects that compose a block. By manipulate, I mean that you can insert a single table & chairs combination (block), adjust its size, adjust its *style* (that's right, change the drafting symbol to any predefined style), flip parts of it, rotate parts of it, move parts around (like the chairs), and more! The same holds true for valve stations, doors and windows, electrical boxes, landscaping, and so forth! (*The dickens, you say!*)

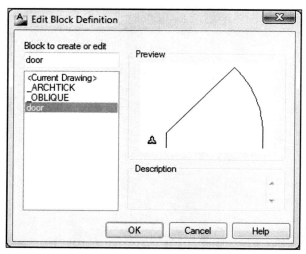

Figure 21.004

The size of your libraries just got radically smaller!

As you can imagine, creating these blocks requires a bit more effort than the simple blocks you've already done. We'll do the work in the *block editor*. Remember that check box at the bottom of the Block dialog box (Figure 21.001, p.470)? It takes you directly to the block editor. Alternately, you can get there by entering the *BEdit* command or just double clicking on the block you wish to edit.

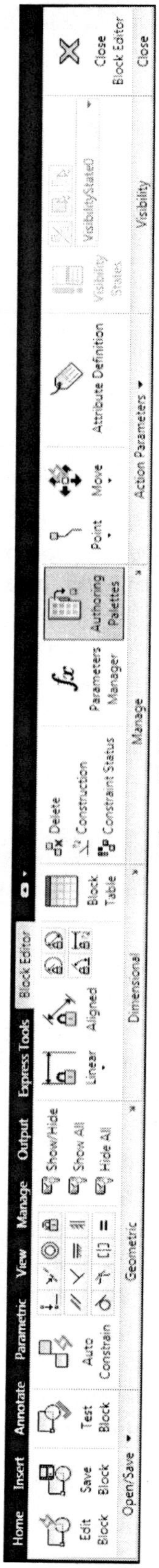

AutoCAD responds to the *BEdit* command with the Edit Block Definition dialog box (Figure 21.004). Pick the block you wish to edit from the list, and then pick the **OK** button to open the editor.

You'll immediately notice a marked change on your screen – the **Block Editor** ribbon tab appears (Figure 21.005) and the Block Authoring Palettes appear (Figure 21.006, p.479). Let's spend a few minutes with each of these treasure houses of tools.

We'll begin with the ribbon and use a table for clarity. (Use the table as a reference but don't dwell on it. You'll get more from the exercises than from trying to memorize tables!)

Figure 21.005

OPEN/SAVE PANEL	
ITEM	DESCRIPTION
	The **Edit Block** button returns you to the Edit Block Definition dialog box (Figure 21.004).
	Use the **Save Block Definition** button (*BSave* command) to save your block work before exiting the block editor.
	The **Text Block** button (*BTestBlock* command) opens the block in a new, temporary drawing window. Use this as a running check on your block. (Close the new window to return to the block editor.)
	Use the **Save Block As** button (*BSaveAs*) in the Open/Save subpanel to save your work as a different block name before exiting the editor.
GEOMETRIC AND DIMENSIONAL PANELS	
ITEM	DESCRIPTION
Although the command sequences are slightly different, the **Geometric** and **Dimensional** panels contain the same constraint tools we discussed in Lessons 10 (p.224) and 17 (p.394). Use these within the block editor to constrain block geometry. (Note: you'll find the same tools available on the **Constraints** tab of the Block Authoring palettes.)	
	The **Block Table** button (*BTable*) displays a table where you can define or edit the properties of your block. We'll look at this in more detail in Lesson 22, p.507.

MANAGE PANEL	
ITEM	DESCRIPTION
	The **Delete** button calls the ***DelConstraint*** command. Use this one to remove all the constraints associated with selected objects.
	Construction (***BConstruction***) converts existing geometry to construction geometry. This can be handier than it sounds; consider that: • construction geometry won't appear as part of the block insertion (or plot), and • you can constrain block geometry to the construction geometry.
	The **Constraint Status** button toggles the BConStatusMode system variable, which controls constraint display. A setting of **0** (default) means that AutoCAD will not display the constraints; a setting of **1** means that it will.
f_x	The **Parameters Manager** button calls the Parameters Manager we discussed in Lesson 17, p.397.
	The **Authoring Palettes** button toggles the visibility of the block authoring palettes (Figure 21.006, p.479). We'll discuss these in some detail shortly.)

ACTION PARAMETERS PANEL	
ITEM	DESCRIPTION

	The **Point** parameter flyout contains several parameter tools, which AutoCAD duplicates on the **Parameters** tab of the **Block Authoring** palettes. We'll discuss them individually, beginning at the end of this table (p.479). Collectively, each calls an option of the ***BParameter*** command. This is how you create parameters in your block. The prompt looks like this:
	    Enter parameter type [Alignment/Base/pOint/Linear/Polar/Xy/
	    Rotation/Flip/Visibility/looKup]: *[Select an option.]*
	You're probably better off using the specific tool on the Block Authoring Palettes (which we'll examine shortly) or the **Action Parameters** panel, but all the options appear in both locations.

	The **Move** action flyout contains several action tools, which AutoCAD duplicates on the **Action** tab of the **Block Authoring** palettes. We'll discuss these individually after the parameters exercise (p.484). Collectively, each calls an option of the ***BAction*** command. This is how you create actions for your parameters. The prompt looks like this:
	    Select parameter: *[You must associate actions with parameters.]*
	    Enter action type [Array/Move/Scale/Stretch/Polar Stretch]: *[Select an action to associate with the parameter.]*
	The prompts that follow will depend on the action you've selected.

	The **Define Attribute** button (***AttDef***) is the key to even richer possibilities within the world of blocks. We'll dedicate most of Lesson 22, p.500, to these!

CLOSE PANEL	
ITEM	DESCRIPTION
	Close Block Editor closes the block editor and returns you to your drawing environment. If you haven't saved your work, AutoCAD will display a dialog box asking if you wish to save or discard your changes.

We'll look at the **Visibility** tools a little later in this lesson (p.490). For now, let's look at Block Authoring Palettes (Figure 21.006, p. 479).

Let's start with the Parameters palette (shown). Remember, for a block to be dynamic, it must contain at least one parameter (control object). The type of action you intend will govern which type of parameter you'll need to add. (All items on this palette invoke the associated option of the *BParameter* command.)

We'll start at the top.

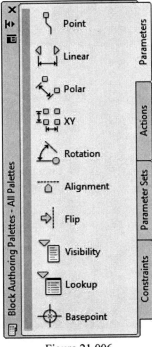

Figure 21.006

> Read through this material, but bear in mind that you'll have a better understanding of it after some exercises and your study of Actions beginning on p.484.

- Associate a **Point Parameter** with a **Move** or **Stretch** operation, although it also works quite well for defining additional insertion points for your block. You can move or stretch specific points of the block when you've added a point parameter.

    The **Point Parameter** will prompt:

    Specify parameter location or [Name/Label/Chain/Description/Palette]: *[Locate the point.]*

    Specify label location: *[Locate the label – this can go anywhere but for organizational purposes, should go fairly close to the point.]*

    Enter number of grips [0/1] <1>: *[Would you like AutoCAD to provide a grip for editing this parameter; without a grip, you can edit the parameter using the Properties palette or quick properties. The grips prompt may not appear when you initiate a command through the Authoring palette.]*

    The first prompt offers you several options (which repeat for some of the other parameters).

    o Assigning a **Name** to the parameter helps you locate it later in the Properties palette where you'll go to edit it.

    o The **Label** should reflect the **Name**. The label helps you keep track of where you've located the parameter.

    o You may find the **Chain** concept a bit difficult to understand at first. When you include the point parameter in the selection set of an action associated with a different parameter (call it parameter B), an edit of parameter B in the block will affect the point parameter. Now suppose you have a separate action associated with the point parameter. (That's two actions – the point parameter action and the parameter B action.) If the **Chain** option setting is **Yes**, any editing done with parameter B will also cause the action associated with the point parameter. If the **Chain** option setting is **No** (the default), the parameter B action will not cause the point parameter action to occur.

    o Any **Description** you enter will appear in the Properties palette when you select the parameter in the block editor.

    o The **Palette** option controls whether or not the parameter will appear in the Properties palette when the block is selected.

- Associate a **Linear Parameter** with a **Move**, **Stretch**, **Scale**, or **Array** action. You can move, stretch, scale, or array a group of objects when you've included a **Linear Parameter**. (Like a linear dimension, this is a workhorse. You'll use it as one way to resize tables, windows, doors, and other objects.) It prompts:

    Specify start point or [Name/Label/Chain/Description/Base/Palette/Value set]: *[Pick the first endpoint of the parameter or another option]*

    Specify endpoint: *[Pick the other endpoint of the parameter.]*

    Specify label location: *[Locate the label.]*

Enter number of grips [0/2] <2>: *[Would you like AutoCAD to provide a grip for editing this parameter; without a grip, you can edit the parameter using the Properties palette or quick properties.]*

You've seen most of the options of the first prompt, but let's look at the rest.

- o The **Base** option will ask you for a **Base location** (either a **start point** or a **midpoint**). The **start point** will remain fixed when you edit the end points in the block editor. The **midpoint** will also remain fixed and both endpoints will react simultaneously to any editing.
- o The **Value set** option provides a valuable key to block insertion. Its options include:
    - **None** means that a block edit of this parameter has no limits.
    - **List** means that, when a user edits the block, he can select from a list of available options (such as: window sizes, valve sizes, etc.).
    - **Increment** means that the user doesn't have to select from a list of different sizes, but resizing is restricted to incremental adjustments defined by the block editor. The editor can define the increment and a minimum and maximum adjustment.

- Associate a **Polar Parameter** with a **Move, Scale, Stretch, Polar Stretch,** or **Array** action. It's prompt is simple:

    Specify base point or [Name/Label/Chain/Description/Palette/Value set]: *[Pick base point (or anchor) of the parameter or another option.]*
    Specify endpoint: *[Pick the other endpoint of the parameter.]*
    Specify label location: *[Locate the label]*
    Enter number of grips [0/1/2] <2>: *[Would you like AutoCAD to provide a grip for editing this parameter; without grips, you can edit the parameter using the Properties palette or quick properties.]*

    You've already seen these options.

- Associate an **XY Parameter** with a **Move, Scale, Stretch,** or **Array** action. Use this parameter to move, scale, stretch, or array objects a set distance along the X- or Y-axis. (Set the distance with the **Value set** option.) It prompts:

    Specify base point or [Name/Label/Chain/Description/Palette/Value set]: *[Pick base point of the parameter or another option]*
    Specify endpoint: *[Pick a point that you'll use to define both the X- and Y-distance.]*
    Enter number of grips [0/1/2/4] <1>: *[Would you like AutoCAD to provide a grip for editing this parameter; without grips, you can edit the parameter using the Properties palette or quick properties.]*

- Associate a **Rotation Parameter** with a **Rotation** action only. Use it to rotate objects within a block (like the hands of a clock). It prompts:

    Specify base point or [Name/Label/Chain/Description/Palette/Value set]: *[Pick the base point (or anchor) of the parameter or another option.]*
    Specify radius of parameter: *[Specify the radius or pick anywhere on the arc of the rotation.]*
    Specify default rotation angle or [Base angle] <0>: *[Identify the base angle; zero indicates the current location of the object which will be rotated.]*
    Specify label location: *[Locate the label.]*
    Enter number of grips [0/1] <1>: *[Would you like AutoCAD to provide a grip for editing this parameter; without a grip, you can rotate the block using the Properties palette or quick properties.]*

- You won't associate an **Alignment Parameter** with an action. Its location, however, is important as it will align the identified points with an object upon insertion. It prompts:

    Specify base point of alignment or [Name]: *[Pick a point along the path which will align with the drawing object.]*
    Alignment type = Perpendicular *[The type can be Perpendicular or Tangent.]*
    Specify alignment direction or alignment type [Type] <Type>: *[Pick a second point along the path for a* **Perpendicular** *alignment or use the* **Type** *option to create a* **Tangent** *alignment.]*

- Associate a **Flip Parameter** with a **Flip** action only. Use this to mirror a block insertion. It prompts:

    Specify base point of reflection line or [Name/Label/Description/Palette]: *[Pick a base point of the parameter (the first point of your mirror line) or another option.]*
    Specify endpoint of reflection line: *[Pick a second point to identify the mirror line.]*
    Specify label location: *[Locate the label.]*
    Enter number of grips [0/1] <1>: *[Would you like AutoCAD to provide a grip for editing this parameter; without a grip, you can flip the block using the Properties palette or quick properties.]*

- Like the **Alignment Parameter**, you won't associate a **Visibility Parameter** with an action. It is, however, required to use Visibility States, which makes it required for a multiple styles block. (We'll discuss visibility states on p.490.) It prompts:

    Specify parameter location or [Name/Label/Description/Palette]:
    Enter number of grips [0/1] <1>: *[Would you like AutoCAD to provide a grip for editing this parameter; without a grip, you can change the parameter's visibility using the Properties palette or quick properties.]*

- Associate a **Lookup Parameter** with a **Lookup** action. This cool tool allows the user to select the size of an object from a lookup box. (We'll see this one in action in our exercise, too.) It prompts:

    Specify parameter location or [Name/Label/Description/Palette]: *[In this case, the location is fairly important as the grip for the lookup box will appear here.]*
    Enter number of grips [0/1] <1>: *[Would you like AutoCAD to provide a grip for editing this parameter; without a grip, you can look up the parameter's values using the Properties palette or quick properties.]*

- The **Base Point Parameter** is another parameter that isn't associated with an action. It locates a point within the block which other objects in the block will use as a base point. It prompts:

    Specify parameter location:

Believe it or not, we have miles to go before we finish dynamic blocks. But let's pause for some practice of what we've covered so far.

Do This: 21.2A	Creating Dynamic Block Parameters

I. Reopen the *blocks* file the C:\Steps\Lesson21 folder.
II. Erase the horizontal line associated with the door block you created in Ex. 21.1.1A, p.471.
III. Follow these steps.

21.2A: CREATING DYNAMIC BLOCK PARAMETERS

1. Let's start with the easy one (the window), enter the **Block** command.
AutoCAD presents the Block Definition dialog box (Figure 21.001, p.470.)
   Command: *b*

21.2A: CREATING DYNAMIC BLOCK PARAMETERS

2. Enter the information shown (use the upper left corner of the window as the insertion point). Be sure the **Open in block editor** check box is checked. Select the window objects.

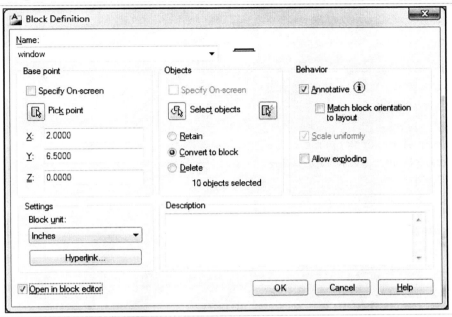

3. Pick the **OK** button to continue. Because you checked **Open in block editor**, AutoCAD opens the new *window* block in the block editor. (Resize the drawing for a better view.)

4. We want to be able to stretch the window, so pick the **Linear** tool on the Parameters palette or the **Action Parameters** panel (**Point** flyout).

5. AutoCAD begins the parameter prompt. Tell it you'd like to set up a **Value set**.

    Specify start point or [Name/Label/Chain/Description/Base/Palette/Value set]: *v*

6. We'll use the **Increment** option to limit the size of our window.
    Enter distance value set type [None/List/Increment] <None> : *i*

7. We'll allow windows from 24" to 42", sized in 3" increments.
    Enter distance increment: *3*
    Enter minimum distance: *24*
    Enter maximum distance: *42*

8. Now AutoCAD needs to know where to place the parameter. Place it as shown (start point on the left). Accept the default number of grips.
    Specify start point or [Name/Label/Chain/Description/Base/Palette/Value set]:
    Specify endpoint:
    Specify label location:
    Enter number of grips [0/2] <2>: *[ENTER]*

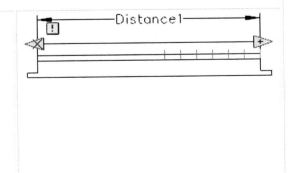

21.2A: CREATING DYNAMIC BLOCK PARAMETERS

Notice the increment identifiers locating the sizes you identified in Step 7.
The exclamation symbol indicates that you haven't identified an action yet. We'll look at that shortly.

9. Save the block and close the block editor.

Note: Nothing will actually happen with the block until we associate an action with the parameter. We'll do that in our next exercise, but first, let's create some different parameters.

10. Save the drawing but don't exit.
   Command: *qsave*

11. Let's edit our door block to make it resizable ... and flip-able. Double click on the door block you created earlier to open it in the Edit Block Definition dialog box.
Select the **door** block from the Edit Block Definition dialog box. AutoCAD opens it in the block editor.

12. We'll make the door flip-able. Select the **Flip** parameter tool (Parameters palette or **Action Parameters** panel – **Point** flyout).

13. Use the midpoint of the left vertical line as the **base point of reflection**.
   Specify base point of reflection line or [Name/Label/Description/Palette]:

14. Place the **endpoint of reflection line** (second point of the mirror line) to the right (use ORTHO).
   Specify endpoint of reflection line:
   Specify label location:
   Enter number of grips [0/1] <1>:
   [ENTER]
Locate the label as shown and accept the default number of grips.

15. Repeat Steps 12 through 14 using a **base point** midway between the midpoints of the two vertical lines. Make your **reflection line** vertical.

16. Now we'll make the block resizable and list the available sizes in a lookup box. Begin a **Linear** parameter.

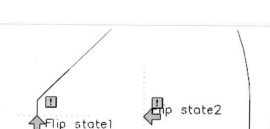

17. This time, we'll use a **List Value** set.
   Specify start point or [Name/Label/Chain/Description/Base/Palette/Value set]: *v*
   Enter distance value set type [None/List/Increment] <None> : *l*
List is very similar to **Increment**, but it makes a list available for our use in the lookup box.

21.2A: CREATING DYNAMIC BLOCK PARAMETERS

18. Enter the values shown.

    Enter list of distance values (separated by commas): *24,30,36*

19. Finally, locate the parameter as shown, and accept the default number of grips.

    Specify start point or [Name/Label/Chain/Description/Base/Palette/Value set]:
    Specify endpoint:
    Specify label location:
    Enter number of grips [0/2] <2>: *[ENTER]*

20. Now place the **Lookup** parameter

21. Place it just above the door as indicated.

22. Save the block and close the block editor.

23. Save the drawing but don't exit.

    Command: *qsave*

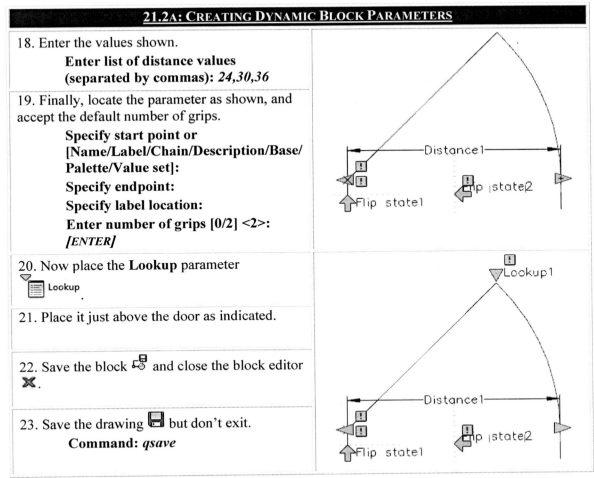

Most parameters won't do anything without an associated action, so we'll look at the Actions Palette next (Figure 21.007). (You'll find the same tools on the **Action Parameters** panel, **Move** flyout.) Here, we'll assign actions to the parameters we added using the Parameters palette and/or the **Action Parameters** panel. This is the actual block modification setup. (All items on this palette invoke the associated option of the *BAction* command.)

- You can associate the **Move** action with a **Point, Linear, Polar,** or **XY** parameter. Use it to allow a user to move a block object(s). It prompts:

    Select parameter: *[Select the parameter with which you'll associate the action.]*
    Specify parameter point to associate with action or enter [sTart point/Second point] <Start>: *[Pick one of the points of the parameter with which you'll associate the action – hit ENTER to use the second parameter point.]*
    Specify selection set for action
    Select objects: *[Select the objects that will be affected by the action.]*
    Select objects: *[ENTER]*

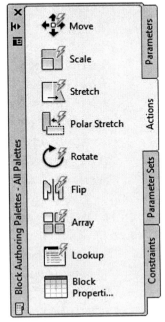

Figure 21.007

The second prompt allows you to select a **sTart point** or a **Second point**. You can ignore these options if you wish as the **sTart point** is the default and hitting ENTER will automatically select the **Second point** option.

- You can associate the **Scale** action with a **Linear, Polar,** or **XY** parameter. Use it to allow the CAD operator to scale selected objects within the block. **Scale Action** prompts:

    Select parameter: *[Select the parameter with which you'll associate the action.]*
    Specify selection set for action
    Select objects: *[Select the objects that will be affected by the action.]*
    Select objects: *[ENTER]*

- You can associate the **Stretch** action with a **Point, Linear, Polar,** or **XY** parameter. Use it to allow the CAD operator to stretch selected objects within the block. **Stretch Action** prompts are similar to both the **Scale** and **Move Action** prompts:

    Select parameter: *[Select the parameter with which you'll associate the action.]*
    Specify parameter point to associate with action or enter [sTart point/Second point] <Start >: *[Pick one of the points of the parameter with which you'll associate the action.]*
    Specify first corner of stretch frame or [CPolygon]: *[Create a typical crossing window.]*
    Specify opposite corner:
    Specify objects to stretch
    Select objects: *[Select the objects to stretch.]*
    Select objects: *[ENTER]*

- You can associate the **Polar Stretch** action only with a **Polar** parameter. Use it to allow the CAD operator to stretch and rotate selected objects within the block. **Polar Stretch Action** prompts are nearly identical to **Stretch Action** prompts:

    Select parameter: *[Select the Polar parameter with which you'll associate the action.]*
    Specify parameter point to associate with action or enter [sTart point/Second point] < Start >: *[Pick one of the points of the parameter with which you'll associate the action – hit ENTER to use the second parameter point.]*
    Specify first corner of stretch frame or [CPolygon]: *[Create a typical crossing window.]*
    Specify opposite corner:
    Specify objects to stretch
    Select objects: *[Select the objects to stretch.]*
    Select objects: *[ENTER]*
    Specify objects to rotate only
    Select objects: *[Select objects that will rotate only (not stretch).]*
    Select objects: *[ENTER]*
    Specify action location or [Multiplier/Offset]: *[Locate the action symbol.]*

- You can associate the **Rotate** action only with a **Rotation** parameter. Use it to allow the CAD operator to rotate selected objects within the block. Compared to other actions, **Rotate Action** prompts are simple (and familiar):

    Select parameter: *[Select the Rotation parameter with which you'll associate the action.]*
    Specify selection set for action
    Select objects: *[Select the objects that will rotate.]*
    Select objects: *[ENTER]*

- You can associate the **Flip** action only with a **Flip** parameter. Use it to allow the CAD operator to "flip" or mirror objects within the block. This works much like the **Mirror** command (with the original objects removed). These prompts are also simple:
    Select parameter: *[Select the Flip parameter with which you'll associate the action.]*
    Specify selection set for action
    Select objects: *[Select the objects that will flip.]*
    Select objects: *[ENTER]*
- You can associate the **Array** action with a **Linear, Polar,** or **XY** parameter. Use it to allow the CAD operator to array selected objects within the block. **Array Action** prompts:
    Select parameter: *[Select the parameter with which you'll associate the action.]*
    Specify selection set for action
    Select objects: *[Select the objects that will array.]*
    Select objects: *[ENTER]*
    Enter the distance between rows or specify unit cell (---): *[As with the* **Array** *command, you must enter a distance between rows and cells; AutoCAD will limit the block object's array to these distances.]*
    Enter the distance between columns (|||):
- You can associate the **Lookup** action only with a **Lookup** parameter. Use it to allow the operator to select specific property values from a predefined list. **Lookup Action** prompts are very simple:
    Select parameter: *[Select the Lookup parameter with which you'll associate the action.]*

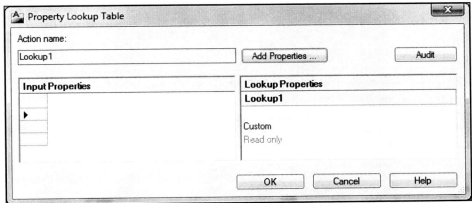

Figure 21.008

Once you've selected the Lookup parameter, AutoCAD presents the Property Lookup Table dialog box (Figure 21.008). Here you'll define the properties that will appear in the lookup box.
  o The name of the action associated with the table appears in the **Action name** box (upper left corner of the dialog box).
  o The **Input Properties** list box contains columns of user-selected parameters associated with the dynamic block. When a parameter appears here, the corresponding name (found in the **Lookup Properties** list box) will appear in the lookup box.
  o Use the **Add Properties** button to call the Add Properties dialog box. (Figure 21.009, p.487). Here you'll find a list of parameters already associated with the dynamic block. Select the property you'd like to add to the lookup box. (Use the **Add ... properties** radio buttons to set what can be added to each of the list boxes.)

We'll see lookup boxes in action in our next exercise.

After taking so many pages to describe the Parameter and Action palettes, you might expect more from the Parameter Sets palette (Figure 21.010). These clever tools combine a parameter with an action, thus saving you the hassle of using both the Parameter and Action palette tools. Notice the slider bar to the left; there are more tools on the palette than can be shown here. (You'll find a chart to help you keep track of which actions and parameters can be paired in Appendix G.)

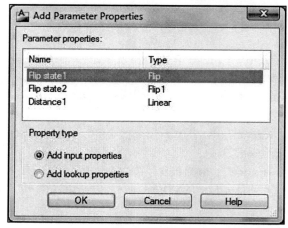

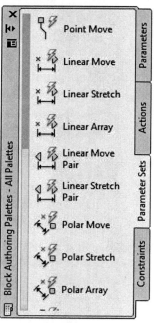

Figure 21.009

Well that was a lot of pages of fun! I know, most of you fully understand dynamic blocks now, but let's try a couple exercises for those suffering that glazed look in their eyes.

Take a deep breath, put on some soft mood music, clear your calendar, and let's go …

Figure 21.010

Do This: 21.2B	Creating Dynamic Block Actions

I. Be sure you're still in the *blocks* file the C:\Steps\Lesson21 folder.
II. Follow these steps.

21.2B: CREATING DYNAMIC BLOCK ACTIONS

1. We'll start by adding a **Stretch** action to our window's **Linear** parameter. Open the window block in the block editor.

2. Begin a **Stretch** action .

3. Select the **Linear (Distance1)** parameter.
   Select parameter:

4. We'll work with the second parameter point .
   Specify parameter point to associate with action or enter [sTart point/Second point] <Start>: *s*

5. Create the crossing window shown.
   Specify first corner of stretch frame or [CPolygon]:
   Specify opposite corner:

21.2B: CREATING DYNAMIC BLOCK ACTIONS

6. Then you'll actually select the objects to stretch. Recreate the crossing window in Step 5.
   Specify objects to stretch
   Select objects:
   AutoCAD places a stretch action bar as shown.

7. Save the block and pick the **Test Block** button .
   AutoCAD opens the block in a test window.

8. Pick the window bloc. Notice the grips. The arrow grip is a *block stretch* grip. Select it.

9. Notice that when you selected the block stretch grip, AutoCAD showed you the available increments.
   Move the arrow back and forth. Notice that you can only resize the block to the available increments! (Turn off your OSNAPs!)

10. Close the **Test Block Window** and the block editor . (Save the changes.)

11. Save the drawing . Don't exit.
    Command: *qsave*

That was easy enough, but we have several parameters with which to associate actions in our door block – including a lookup action! Let's do that now.

| Do This: 21.2C | Creating Dynamic Block Actions - Continued |

I. Be sure you're still in the *blocks* file the C:\Steps\Lesson21 folder.
II. Follow these steps.

21.2C: CREATING DYNAMIC BLOCK ACTIONS - CONTINUED

1. Open the door in the block editor.

2. Let's start by adding a **Scale** action .

3. Select the **Linear (Distance1)** parameter.
   Select parameter:

4. Create the crossing window shown.
   Specify selection set for action
   Select objects:
   AutoCAD places a stretch action label next to the **Linear** parameter. (See the Step 8 figure.)

5. Now begin a **Flip** action .

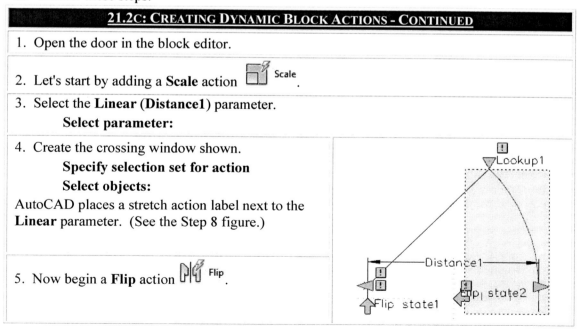

21.2C: CREATING DYNAMIC BLOCK ACTIONS - CONTINUED

6. Select the first **Flip state** you created.
   Select parameter:

7. Put a window around all the objects in the block.
   Specify selection set for action
   Select objects:
   AutoCAD places a **Flip action** bar next to the **Flip state1** parameter (right).

8. Repeat Steps 5 through 7 for the other **Flip state**. Your drawing looks like the figure at right.

9. Those were easy – let's try the **Lookup** action [Lookup]. This one will be more involved.

10. Begin the **Lookup** action command and select the **Lookup** parameter.
    Select parameter:
    AutoCAD opens the Property Lookup Table dialog box (Figure 21.008, p.486).

11. Pick the **Add Properties** button [Add Properties...] to open the Add Parameter Properties dialog box (Figure 21.009, p.487).

12. Select **Distance1 – Linear** from the parameters list and pick the **OK** button [OK] to continue. AutoCAD returns to the Property Lookup Table dialog box which now shows **Distance1** as an **Input Property**.

13. Pick the arrow in the row just below the **Distance1** listing. AutoCAD presents a control box which carries the list of **Distance1 values** we gave this parameter back in Exercise 21.2A, p.481.

14. Select the first value.

15. Repeat Steps 13 and 14 for the two rows below the **24.0000** row. The **Input Properties** column now looks like this. These are the values which AutoCAD will use when you select an option from your Lookup List.

16. Pick the **24.0000** row in the **Lookup Properties** column. AutoCAD makes it available for your input. Enter the words *24" Door*. You'll see this callout in the Lookup List.

17. Repeat Step 16 for the other two rows.

21.2c: CREATING DYNAMIC BLOCK ACTIONS - CONTINUED

18. Change **Read only** to **Allow reverse lookup**, as shown. Your dialog box looks like the figure at right.	Input Properties	Lookup Properties
	Distance1	**Lookup1**
	24.0000	24" Door
	30.0000	30" Door
	36.0000	36" Door
19. Pick the **OK** button [OK] to complete the procedure.	<Unmatched>	Custom
		Allow reverse lookup

20. Erase the two vertical lines below the door. (They resize with the door and aren't needed for the block.)

21. Save the block and open the Test Block Window.

22. Pick the **Door** block. Notice the grips.

23. Experiment with the two flip grips (the two bottom arrows) to see how they affect the block.

24. Experiment with the scale grip (the right pointing arrow). See that you can resize the block as you defined (to 24" and 30").

25. Now experiment with the list box grip (the downward pointing arrow above the block). Notice that it shows the available sizes for the door. Notice also that you can resize the door by selecting the size you want.

26. Experiment with selecting the other listings.

27. Clear the grips and close the Text Block Window and the block editor. Save the drawing but don't exit.
   Command: *qsave*

Are you beginning to see the benefits of dynamic blocks? Catch your breath and then move on to the next exercise where you'll get the chance to put what you've learned into action – and see another really cool tool!

The other end of the toolbar you've been using contains some additional tools located on the **Visibility** panel. Here they are:

VISIBILITY PANEL	
ITEM	**DESCRIPTION**
	Use the **Visibility States** button to access the Visibility States dialog box (Figure 21.011). You'll set up your visibility states there.
	Make Visible and **Make Invisible** enable you to "add" and "remove" objects from your blocks. To enable your user to toggle between different styles of sinks (or other objects) within a single block, you'll draw both sinks and make one visible and the other invisible. You'll then associate the visibility with a visibility state.

VISIBILITY PANEL	
ITEM	DESCRIPTION
⌗	**Visibility Mode** serves as a toggle for the **BVMode** system variable. **On** (a setting of **1**) means that hidden objects will appear visible but dimmed. **Off** (a setting of **0**) means that hidden objects will not appear at all. This is a working tool and won't affect the actual insertion of the block.
VisibilityState0 ▼	Use the **Block Visibilities** control box much as you use the Layer and Properties control boxes – in this case to make a selected visibility state current. As with other layers, colors, and linetypes, work performed while in a selected visibility state is associated with that visibility state.

Luckily, you won't find visibility states as complicated as layers. Take a look at the Visibility States dialog box (Figure 21.011). (You can also reach this dialog box by double clicking on the Visibility parameter.) Most of it is fairly clear.

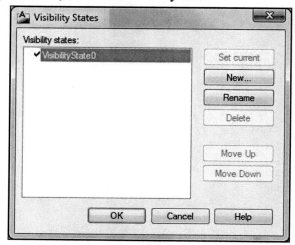

Figure 21.011

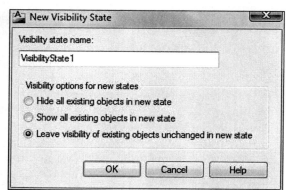

Figure 21.012

- The **Visibility states** list box lists all the states available in the block. Select one and use the **Set current** button to make it current.
- Most of the buttons down the right side and across the bottom of the box are self-explanatory.
- The **New** button calls the New Visibility State dialog box (Figure 21.012). Here you'll name your new state and decide what, if anything, should be hidden in it. I recommend leaving the bullet next to **Leave visibility of existing objects unchanged in new state**. You can then adjust what is visible using the **Make Visible** and **Make Invisible** buttons on the toolbar.

Let's see what we can do with that table block. We'll use the Parameter Sets palette this time – to save time. Parameter Sets create both a parameter and an accompanying action, but you'll have to double click on the action label to actually provide the action sequence.

Do This: 21.2D	**Creating Dynamic Blocks with Visibility States**

I. Be sure you're still in the *blocks* file the C:\Steps\Lesson21 folder. If not, please open it now.
II. Restore the **Table** view.
III. Follow these steps.

21.2D: CREATING DYNAMIC BLOCKS WITH VISIBILITY STATES

1. Open the *table* block in the block editor.

2. (Refer to the Step 7 figure.) Create a **Linear Stretch** parameter set **Linear Stretch** with listed increments (**Value Set – List** of 36, 48, 60, 72, and 84). Once you've created the parameter, right click on the Stretch bar and pick **Action Selection Set – Modify Selection Set** on the menu. Select the right end of the table to stretch.

3. Create a **Lookup Set** **Lookup Set** to go with the **Linear** parameter list you created in Step 2. (Right click on the look up icon and select **Display Lookup Table** from the menu to create the lookup table.) Be sure to **Allow reverse lookup**.

4. Now let's set up our block so we can move the chair around after it's inserted. Begin a **Point Move** parameter set **Point Move**.

5. Call the parameter, *Chair Location*.
   Specify parameter location or [Name/Label/Chain/Description/Palette]: *L*
   Enter position property label <Position1>: *Chair Location*

6. Place the parameter at the northern quadrant of the chair.
   Specify parameter location or [Name/Label/Chain/Description/Palette]: _qua of
   Specify label location:

7. Right click on the **Move** action icon and pick **Action Selection Set – New Selection Set** on the menu. Create a selection set consisting of the objects that make up the chair.
   Specify selection set for action
   Select objects:
   The block looks like the figure at right.

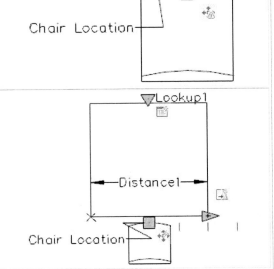

8. Add a **Flip** action for the chair **Flip Set**. Use these guidelines:
   a. Select a **base point of reflection** in the center of the table.
      Specify base point of reflection line or [Name/Label/Description/Palette]:
   b. Pick an endpoint of reflection on the horizontal.
      Specify endpoint of reflection line:
   c. Locate the label next to the reflection line.
      Specify label location:
   d. Use the cursor menu to create a new selection set and include the objects that compose the chair.
      Specify selection set for action
      Select objects:

21.2D: CREATING DYNAMIC BLOCKS WITH VISIBILITY STATES

9. Let's enable our operator to align the table upon insertion. Place an **Alignment Parameter** along the top of the table.

   Specify base point of alignment or [Name]:
   Alignment type = Perpendicular
   Specify alignment direction or alignment type [Type] <Type>:

Now we're going to provide for two really cool tricks. First, we'll tell AutoCAD to increase the number of chairs when the table is resized. Then, we'll set up our block so the operator can select a different style of chair.

10. Before we begin, draw a two point circle over the chair using the midpoint of the chair sides as the two points.

    Command: c

11. Now here's a handy trick. We need to add the circle to the selection set for the **Flip** action.

Right click on the **Flip action** bar and pick **Action Selection Set – Modify Selection Set** on the menu. AutoCAD assumes you wish to add objects to the selection set and prompts you to do so. Select the circle.

   Specify selection set for action object [New/Modify] <New>: _m
   Select object to add to action set or [Remove]:
   Select object to add to action set or [Remove]: *[ENTER]*

12. Repeat Step 11 to add the circle to the **Move** action next to the chair. (Isn't editing actions a breeze?!)

13. We'll use the **Array** action with an existing parameter to have AutoCAD increase the number of chairs as the table stretches. Pick **Array** on the Actions palette.

14. Select the **Linear** parameter (**Distance1**) you created in Step 2.
    Select parameter:

15. Select the objects that compose the chair and the circle you drew in Step 10.
    Specify selection set for action
    Select objects:
    Select objects: *[ENTER]*

16. Tell AutoCAD to place the chairs 24.1" apart.
    Enter the distance between columns (|||): *24.1*

AutoCAD places an array icon next to the linear parameter (right).

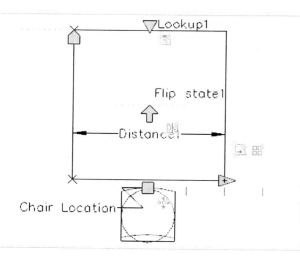

21.2D: CREATING DYNAMIC BLOCKS WITH VISIBILITY STATES

17. Finally, we'll set up our block so an operator can select a stool rather than a chair. Begin by adding a **Visibility** parameter from the Parameters palette. Place it next to the **Lookup** symbol.

    Specify parameter location or [Name/Label/ Description/Palette]:

18. Now we'll create some Visibility States. Double click on the **Visibility** symbol Visibility1 to display the Visibility States dialog box (Figure 21.011, p.491).

19. Select the only existing visibility state and rename it to *Chair*.

20. Pick the **New** button to display the New Visibility State dialog box.

21. Create a new **Visibility state** called *Stool*. Leave the **Visibility options for new states** bullet next to **Leave visibility of existing objects unchanged in new state**, and pick the **OK** button twice to complete the command.

22. Set the **Chair** visibility state current in the **Visibility State** control box (right end of the **Visibility** panel).

23. Pick the **Make Invisible** button (next to the **Visibility State** control box), and select the circle. Make it invisible for the **Current state**. (Pick the **Visibility Mode** button to see a ghost image of invisible objects.)

    Select objects to hide:
    Select objects:
    Select objects: *[ENTER]*
    Hide for current state or all visibility states [Current/All] <Current>: *[ENTER]*

24. Repeat Steps 22 and 23, making the chair objects invisible on the **Stool** visibility state.

25. Save the block and close the block editor.

26. Select the table. Your grips should look something like these.

27. Experiment with flipping the chair (use the arrow grip in the center of the table).

28. Experiment with the chair location (use the grip on the chair).

29. Select the different table sizes from the list box grip. Notice that AutoCAD adds more chairs as the table gets longer.

30. Use the other down arrow grip to select the stool. Notice how the type of chair changes.

21.2D: CREATING DYNAMIC BLOCKS WITH VISIBILITY STATES
31. Save the drawing 💾.   **Command:** *qsave*
32. Use the *WBlock* command to save the blocks to the C:\Steps\Lesson21 folder. (Refer to Exercise 21.1.2A, p.473 if you need help.)

The last tool we'll consider in our exploration of dynamic blocks (yea!) is a simple but handy one. Within the Block Manager, you'll use the Parameters Manager (Figure 21.013) to modify or delete action parameters or even to add some user-defined parameters of your own.

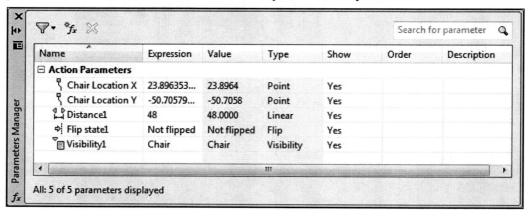

Figure 21.013

Notice that you have several columns with which to work – some you can edit and one (**Value**) is for reference. Editing a cell often requires nothing more than selecting it and typing ... or selecting it and picking from a list of alternate entries.

You'll find three buttons across the top of the manager:

- Use the **Create new user parameter** button *fx* to create your own reference parameter. Use this to display something in the Properties palette or to create a reference for some equation.
- The **Delete** parameter button ✕ does just that to the selected parameter.
- The **Parameter Display Filter** button ▽ allows you to filter which parameters appear in the manager. You can show **All Parameters** or **All Parameters Used in Equations**.

Check it out.

Do This: 21.2E	Editing Parameters with the Parameter Manager

   I. Be sure you're still in the *blocks* file the C:\Steps\Lesson21 folder. If not, please open it now.
   II. Follow these steps.

21.2E: EDITING PARAMETERS WITH THE PARAMETER MANAGER
1. Open the *table* block in the block editor.
2. Display the Parameters Manager (pick the button *fx* on the **Manage** panel). It looks like Figure 21.013, p.495.
3. Pick on **Distance1** in the **Name** column. AutoCAD makes the name available for editing.
4. Change the name to **Table Width**   Table Width .

It's really just that simple.

Now use the blocks you've created to complete the *cab-pln* file you began earlier. It will look like Figure 21.014 when you've finished.

Let's take a look now at another cool tool that makes blocks nearly indispensable.

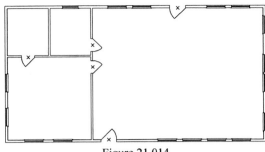
Figure 21.014

| 21.3 | **Getting Blocks from a Folder Library on a Web Site via AutoCAD's iDrop** |

AutoCAD has a marvelous toy called *iDrop*, and it bears closer examination before we proceed.

What is iDrop?

iDrop is the name given to a simple procedure for grabbing drawings (Write Blocks) from an iDrop enabled web site (folder) library and then dragging and dropping them into your drawing.

Why bother when I can create my own blocks?

Taking iDrop drawings from a manufacturer's site means that you have the exact item you want. It greatly reduces any possibility of error during the exchange of information between manufacturer and designer, including dimensional errors and engineering data provided through attributes. (We'll look at attributes in Lesson 22, p.500.) It also saves the designer/draftsperson drawing time as well as research time.

Before you get too excited, however, let me give the downside. Not all manufacturers have iDrop-enabled blocks for you to use (although the list is growing). Additionally, the programmers at Autodesk designed the plug-in for Internet Explorer 5 or better. So you may need to change or upgrade your web browser. Finally, all insertions use a command line interface.

Let's take a look at that command line interface.

When you insert a iDrop, AutoCAD will prompt:

Specify insertion point or [Basepoint/Scale/X/Y/Z/Rotate]: *[Using coordinates or picking a point with the mouse, tell AutoCAD where to put the block (you're precisely locating the block's insertion point).]*

We're presented with a few new options.

- AutoCAD asks for an **insertion point**. This sounds simple enough but the list of options that follows the prompt can be quite confusing. Let's take a look at these:
  o Selecting the **Basepoint** option will allow you to temporarily drop the block into the drawing where it's currently located. It then prompts:

  Specify base point: *[Select a new base point.]*

  AutoCAD will then allow you to insert the block using your new base point. (This procedure *does not* redefine the base point of the block.)
  o The **X/Y/Z** options allow you to scale along the selected axis.
  o **Scale** allows you to scale the block uniformly.

Once you're ready, proceed with the next exercise. (This exercise requires Internet access and Internet Explorer.)

| **Do This: 21.3A** | **iDrop and Block Insertions** |

I. Open the *cab-pln* file the C:\Steps\Lesson21 folder. We'll iDrop some bathroom fixtures from our uneedcad site.

II. Freeze the **MARKER** layer and create a new layer for fixtures. Make the new layer current.
III. Open Internet Explorer and go to this site: *http://www.uneedcad.com/Webfile*.
IV. Follow these steps.

21.3A: IDROP AND BLOCK INSERTIONS

1. If your browser is maximized, pick the **Restore** button in the upper right corner of the window to float it over the AutoCAD window.

2. Scroll down until you see the tub, sink, and WC.

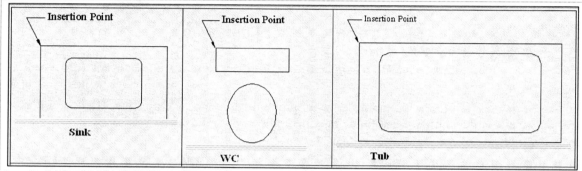

3. Notice the cursor when you pass over one of these items. It takes the shape of a tiny eyedropper. (You may have to pick on an object to activate the iDrop control.)

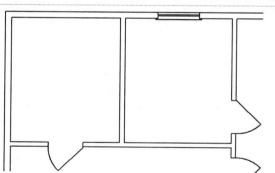

4. Pick anywhere in the AutoCAD window to make it active.

5. Zoom in around the bathroom. Your screen should look something like the figure at right.

6. Hold down the ALT key on your keyboard and press the TAB key until your browser highlights on the menu. Release the ALT key. Windows floats the Internet Explorer window over AutoCAD again. (Use this procedure for toggling between AutoCAD and Internet Explorer.)

7. Place the iDropper over the tub. Holding the left mouse button down, drag the object into the bathroom. Release the button.

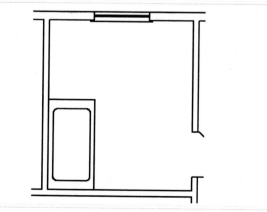

8. Place the tub as shown. (The block has an alignment action attached to it to make insertion easier.)

Specify insertion point or [Basepoint/Scale/X/Y/Z/Rotate]:

21.3A: IDROP AND BLOCK INSERTIONS

9. Repeat Steps 6, 7, and 8 to insert the sink and WC. Then change the sink to **oval** and the WC to **Smooth**. Your drawing looks like this.

10. Save your work 💾.
    Command: *qsave*
    Close Internet Explorer.

21.4 Extra Steps

Familiarize yourself with the following chart to help you know when to use groups and when to use blocks.

ABILITY	BLOCKS	GROUPS
Can be shared between drawings	Yes	No
You can work on objects within the block/group	Yes	Yes
Can be treated as a single object	Yes	Yes
Can carry information (data)	Yes	No
Retrievable after erasure (without the Undo command)	Yes	No
May contain (other) blocks	Yes	Yes
May contain (other) groups	No	Yes

21.5 What Have We Learned?

Items covered in this lesson include:

- *Folder & template libraries*
- *AutoCAD Commands:*
  - **Block**
  - **Explode**
  - **WBlock**
  - **Insert**
  - **BEdit**
  - **BParameter**
  - **BAction**
  - **BVMode**
  - **BTable**
  - **BSave**
  - **BTestBlock**
  - **DelConstraint**
- *Block creation, insertion, and editing*
- *Authoring Palettes*
- *Using iDrop*
- **BConstruction**

After 20 lessons of learning *How*, we've begun to answer the question *Why*. *Why* is CAD a better design tool than paper and pencil? *Why* should I spend thousands of dollars to do on a computer what I can do for a lot less on a board? *Why* must I educate myself (or in many cases, reeducate myself) to work with a computer?

Hopefully, the answers began to dawn on many of you in this last lesson.

We'd already seen throughout this book that CAD drawings are neater and easier to read than many pen or pencil drawings. But up to the beginning of Lesson 20, it all seemed like a lot of work – often more than a comparable board drawing might require.

Then in Lesson 21, we found ways to manipulate large numbers of objects at once. With blocks, once created (or purchased), *no further drawing is necessary!* Additionally, a project (often a company and occasionally an entire industry) can expect standard symbols!

But believe it or not, we've only begun to scratch the surface of what we can accomplish using blocks. Wait until you see what we do in Lesson 22!

21.6 Exercises

1. Start a new drawing from scratch.
    1.1. Set the limits for a full scale drawing on an 8½" x 11" sheet of paper.
    1.2. Grid: ¼"
    1.3. Create the layers indicated (all layers use a continuous linetype).
    1.4. Use the Times New Roman or Calibri font. Text sizes are ¼", 0.2", and 1/8."
    1.5. Create the blocks shown in the *Legend* (below right). (Remember to create all blocks on layer 0.)
    1.6. Create the drawing (be sure to insert blocks on the appropriate layer). Use a table for the legend. Save it as *MyBridge* in the C:\Steps\Lesson21 folder.

LAYER NAME	COLOR
Battery	56
Border	black
Coil	blue
Copperwire	green
Galvanometer	black
Pin	black
Resistor	222
Switch	red
Text	12

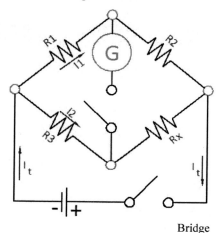

Bridge

2. Start a new drawing from scratch.
    2.1. Repeat the setup you used for Exercise 2 with the following exceptions:
        2.1.1. Add a layer called *Loop* using color 32.
        2.1.2. Grid: ½
        2.1.3. Use the standard text style with ¼" and 0.2" text heights.
        2.1.4. Use the battery block you created in Exercise 2 and create a block for the wire loop.
    2.2. Create the *ElectroMagnet Circuitry* drawing. Save it as *MyMagnet* in the C:\Steps\Lesson21 folder.

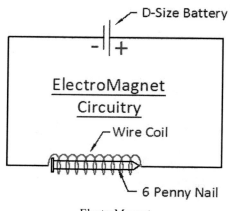

ElectroMagnet

21.7 For Web-Based Review Questions and Additional Exercises, visit: http://foragerpub.com/AcadFiles/2011/2011.htm

Lesson 22

Advanced Blocks

Following this lesson, you will:

- ✓ Know how to create Block Attributes with the **AttDef** command
- ✓ Know how to insert blocks with attributes
- ✓ Know how to use the Block Table
- ✓ Know how to control the display of attributes with the **AttDisp**, **AttReq**, and **AttDia** system variables
- ✓ Know how to edit attribute information with the **-AttEdit** command and the Properties palette
- ✓ Know how to redefine a block with attributes with the Block Attribute Manager
- ✓ Know how to extract attribute data to another program or for a bill of materials

One of the most important jobs a CAD operator may have involves the politics of convincing his or her supervisor (and often the supervisor's supervisor) of the importance of using AutoCAD as its creators intended. This will inevitably mean that the initial job setup will take more time than a non-CAD-oriented person might consider necessary. But the delay will be repaid "seven-fold" at the end of the project.

The operator might explain that AutoCAD isn't, as is commonly believed, simply a very expensive drafting tool. Rather, AutoCAD should be considered the backbone of the overall project. Indeed, a design properly created in AutoCAD serves not only as an outline for construction but also reduces material take-off and purchasing chores from weeks to minutes.

To cut large pieces of time from the end of the project, smaller pieces of time must be spent during the setup phase. Part of this time is required to create your libraries or to adjust purchased libraries to project standards. This adjustment should involve the addition of project-specific attributes to your blocks. You'll use these attributes to generate bills of materials and to share material information with material take-off (MTO) and purchasing personnel and programs.

This lesson will cover how to create and edit attributes and how to share attribute data.

Let's begin.

22.1	**Attributes**
22.1.1	**Creating Attributes**

What exactly is an attribute?

Think of an attribute as a vessel that carries information. This information can be AutoCAD generated or user defined. We've learned that all objects in a drawing contain information kept in the drawing's database. This information identifies the object using such things as type, style, color, linetype, layer, position, and so forth. An attribute allows you to attach this (and other) information to a block and to retrieve it later.

AutoCAD provides a simple, straightforward dialog box (Figure 22.001) for creating attributes. Access it from within Model Space or the block editor with the *AttDef* command. Understanding this dialog box will go a long way toward helping you understand how attributes work. Let's take a look.

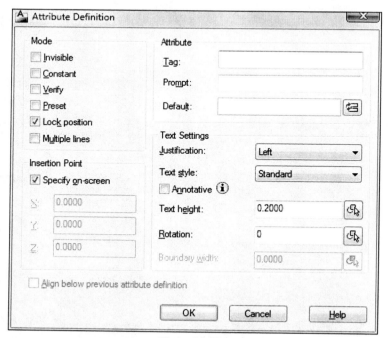

Figure 22.001

501

- As you can see in the upper left frame (**Mode**), AutoCAD provides several modes from which to choose. These control how the attribute functions. By default, most modes are toggled off (the check boxes are clear). Let's see what happens when you toggle each one on.
    - An **Invisible** attribute holds data that won't appear on the screen or drawing. This setting is ideal for most information. (Use visible data for information you wish to see on the drawing – such as valve sizes or title block data.)
    - A **Constant** attribute contains information that doesn't change. AutoCAD won't prompt for this information when you insert the block, and you can't edit the value of the constant attribute after inserting the block.
    - When you insert a block with an attribute that was created in **Verify** mode, AutoCAD will first prompt you to enter a value for the attribute and then prompt you again to *verify* that value.

Where to Find It:	
Command Line:	*AttDef*
Hotkey(s):	*att*
Ribbon (Tab/Panel):	Home – Block (subpanel) – **Define Attributes**
	Insert – Attributes – **Define Attributes**
	Block Editor – Action Parameters – **Attribute Definition**
Menu:	Draw – Block – **Define Attributes**
Palette:	Draw – **Define Attribute**

    - AutoCAD won't prompt for an attribute value (except those requiring verification) when you create it in the **Preset** mode. AutoCAD assumes default values for all the attributes when the block is inserted.
    - You won't be able to move an attribute created with a check in the **Lock position** box. You must, however, put a check here for an attribute which will become part of a dynamic block's action.
    - Placing a check next to **Multiple lines** tells AutoCAD to allow you to create multiline attribute values. Depending on the settings of the **AttDia** system variable, you can even use the Multiline Text Editor to enter your attribute values!

> Two system variables contribute to how a multiline text attribute behaves:
> **AttDia** controls whether or not AutoCAD will prompt you for attribute values with a dialog box (more on this in Section 22.2, p.509), and **ATTipe** controls the Multiline Text Editor's toolbar. A setting of **0** (the default) offers an abbreviated MText ribbon panel; a setting of **1** offers the full toolbar.

- The **Attribute** frame provides a convenient location for you to easily place important information.
    - Enter the name of the attribute in the **Tag** text box. Make this something simple but descriptive so that the operator can easily identify it. Don't use spaces or special characters in the name.
    - Place the prompt – what you want AutoCAD to say when asking you for a value to assign to this attribute – in the **Prompt** text box. Follow the advice of the old English professor: "Be brilliant; be brief." (In other words, keep your prompts short, concise, and to the point.)
    - You can place a default value for the attribute in the **Value** text box. AutoCAD doesn't require a default value, but it'll provide you with the ability to respond to the prompt with an ENTER keystroke or a right click of the mouse.
    - To the right of the **Value** box, you'll find an **Insert field** button. This handy tool can even increase the incredible usefulness of attributes! Add a field into the **Value** box and AutoCAD will automatically update the attribute value even after you've inserted the block!
- The **Insertion Point** frame works exactly as it did on the Insert dialog box. But remember, this insertion is for the attribute not the block.

- In the **Text Settings** frame, you'll use control boxes to select the **Justification** and **Style** of the attribute's text. You can also place the **Height** and **Rotation** of the text into the appropriate text box, or pick a button to select each from the screen. You can also set the **Boundary Width** here for multiline text when you've checked the **Multiple Lines** box in the **Mode** frame.

  When creating an **Annotative** block (one that resizes automatically with the **Annotation Scale** of the drawing), put a check next to **Annotative** to be sure the attribute also resizes.
- When creating several attributes, a check in the **Align below previous attribute** box will help keep the attributes organized.

> To edit or change an attribute before creating the block or within the block editor, use the standard text editor command (**DDEdit**) or double click on it. AutoCAD will present the Edit Attribute Definition dialog box shown here.
>
>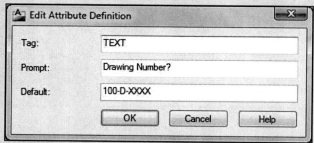
>
> Change the **Tag**, **Prompt**, or **Default** value from here, but you can't change the mode of the attribute. You can, however, change all the attribute's properties using the Properties palette.
>
> Remember this handy method when you have several similar attributes to create. Simply copy your attributes to the blocks you wish to create, edit them as needed, and then create the block.

Let's add some attributes to a few blocks.

> You can create attributes before creating a block – just include them in the objects selected to create the block. Alternately, you can create attributes from within the Block editor.

Do This: 22.1.1A	Creating Attributes

I. Open the *blocks-pipe* file the C:\Steps\Lesson22 folder. The drawing looks like the figure at right.

II. Follow these steps.

Remember: If you discover an error in an attribute after you've created it, edit the definition with the Properties palette.

22.1.1A: CREATING ATTRIBUTES

1. Open the left valve in the block editor.
   Command: *be*

2. Change the current layer to **TEXT**.

3. We're going to create five attributes to attach to this valve. Enter the *Attdef* command.
   Command: *att*

4. Fill in the dialog box as shown in the following figure.
   a. Clear all the **Mode** check boxes for this block.
   b. Call the attribute *Size*.

22.1.1A: CREATING ATTRIBUTES

  c. The prompt should read as shown.
  d. Give the attribute the default value indicated.
  e. Center-justify the text.
  f. Use a text height of 1/8".
  g. Enter the **Insertion Point** shown.

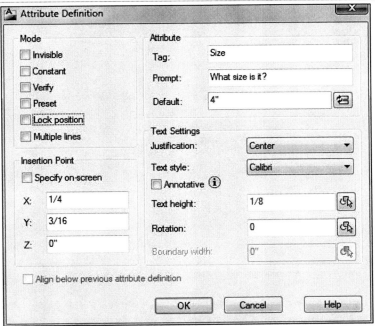

5. Pick the **OK** button to complete your first attribute definition. Your screen should look like the figure at right.	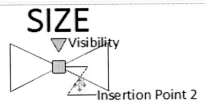
6. Save the block and drawing but don't exit. (Update the reference when prompted.) **Command:** *qsave*	
7. Repeat Steps 3 through 6 to create a **Rating** attribute. All the modes for the attribute should be toggled off, the tag should be *rating*, the prompt should read, *What is the rating?*, and the default attribute value should be *150#*. Center the tag below the valve at coordinates .25,-.25.	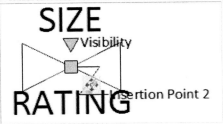

8. Now let's create a **Constant, Invisible** attribute. Repeat the *Attdef* command.
   Command: *[ENTER]*

9. Create the settings shown (figure "A" – following).
   a. Begin by placing checks next to **Invisible** and **Constant**. Notice that the other modes become unavailable.
   b. Call the attribute *Filename*. Notice that, with a constant attribute, you don't need a prompt.

22.1.1A: CREATING ATTRIBUTES

    c. Pick the **Insert Field** button next to the **Default** input box.

       Select **Filename** in the **Field Names** box and use the **Lowercase** format (Figure "B"). Don't **display** [the] **file extension**.

       Pick the **OK** button to close the Field dialog box.

    d. Align the attribute below the last one you created.

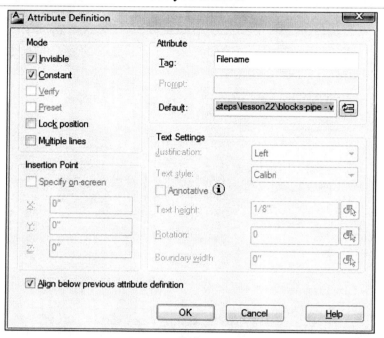

"A"

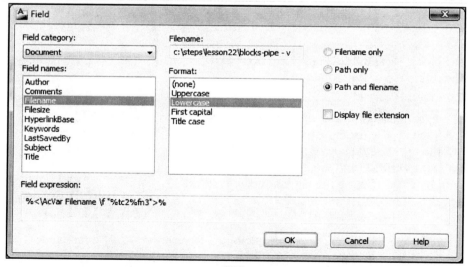

"B"

10. Complete the definition OK.

11. Save the block and drawing but don't exit.

       Command: *qsave*

22.1.1A: CREATING ATTRIBUTES

12. Now create an **Invisible** *Type* attribute as shown.
Put the tag beneath the **Filename** tag.

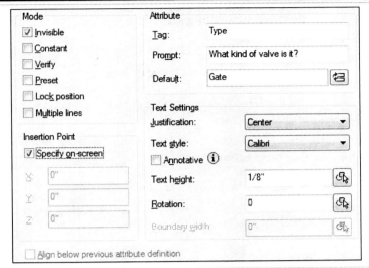

13. Create a final attribute – this one for the price of the valve.
   a. Call the attribute *Price*.
   b. Have it prompt for the cost as indicated.
   c. It should be invisible, but you should verify the value entered.
   d. Align it below the previous attribute.

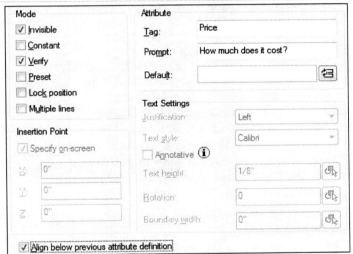

14. Save the block and drawing but don't exit.
   Command: *qsave*

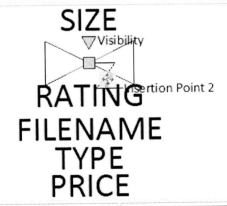

15. Notice that you've been working in the **Gate Valve** visibility state. (Look in the **Visibility State** control box.) Use the control box to change to the **Globe Valve** visibility state.
Notice that all the attributes have disappeared! Let's fix that.

506

22.1.1A: CREATING ATTRIBUTES

16. Toggle the **Visibility Mode** so that you can see the attributes. (They'll appear transparent.)
    Command: *bvmode*
    Enter new value for BVMODE <0>: *1*

17. Now, with the **Globe Valve** visibility state current, use the **Make Visible** button to make the attributes visible.

18. Repeat Step 17 for the **Check Valve** and **Control Valve** visibility states. Make the **Gate valve** visibility state active.

19. Save the block and close the block editor.

20. WBlock the valve block as *Valve* to the C:\Steps\Lesson22 folder.
    Command: *w*

21. Save the drawing.
    Command: *qsave*

22.1.2 Block Tables

In Lesson 21, p.476, we learned how to use lookup tables in our blocks to simplify the selection of what we wanted to see. Lookup tables work well, but you're limited as to how much information you can provide to help make the proper selection. AutoCAD provides a tool that can carry even more information! We'll use Block Tables to ensure that our user knows exactly what he's inserting.

Where to Find It:	
Command Line:	*BTable*
Ribbon (Tab/Panel):	Block Editor – Dimensional – **Block Table**
Menu:	Tools – **Block Editor**

Use the ***BTable*** command *within the Block Editor* to create the table.

Command: *btable*

Specify parameter location or [Palette]: *[Locate the selection grip for the table; the user will use this "down arrow" to open the table.]*

Enter number of grips [0/1] <1>: *[Accept the default to display the selection grip or enter 0 if you wish to use the Properties palette to open the table.]*

At this point, AutoCAD presents an empty Block Properties Table (Figure 22.002). Although this table is easy to use, you'll need some instruction before you try to use it.

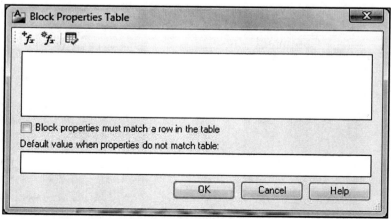

Figure 22.002

- The only new option in the prompts – **Palette** – will offer you the opportunity to **Display property in Properties palette**. The default answer is Yes, so unless you don't want to display the property in the palette, you can ignore this option.
- You'll find three buttons across the top of the dialog box. Use these to create your table.
  - Use the **Add Properties** button $^+\!f_x$ to include properties in the table.
  - Use the **New User Parameters** button $^*\!f_x$ to add your own reference parameters.
  - Finally, before you complete the table, use the **Audit** button to make sure everything works properly.
- Place a check next to **Block properties must match a row in the table** to restrict your user to only using values you provide.
- The **Default value when properties do not match table** will offer entry cells where you can provide default values.

Let's create a block table with our attributes.

Do This: 22.1.2A	Creating a Block Table

I. Be sure you're still in the *blocks-pipe* file the C:\Steps\Lesson22 folder. If not, please open it now.
II. Reopen the valve in the block editor.
III. Follow these steps.

22.1.2A: CREATING ATTRIBUTES

1. Begin the **BTable** command.
   Command: *btable*

2. Pick a point above the valve around the coordinates shown, and accept the default number of grips.
   Specify parameter location or [Palette]: *3/16,1/8*
   Enter number of grips [0/1] <1>:
AutoCAD opens the Block Properties Table (Figure 22.002, p.507).

3. Pick the **Add Properties** button $^+\!f_x$. AutoCAD opens the Add Parameter Properties dialog box (right) with the available parameters listed.

4. Select the **SIZE** attribute and pick the **OK** button. Notice that AutoCAD places the attribute in the table.

5. Repeat Steps 3 and 4 to add the **RATING**, and **TYPE** attributes and the **Visibility State** to the table.

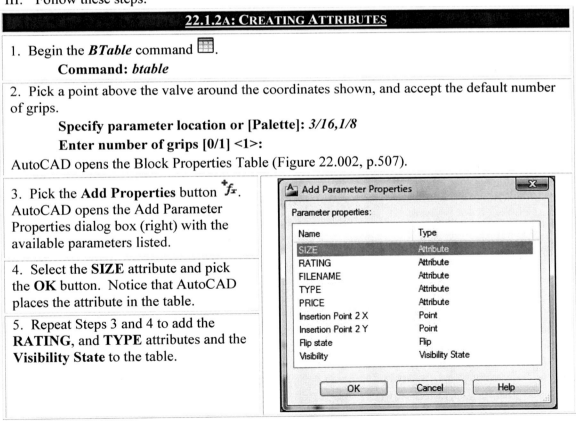

22.1.2A: CREATING ATTRIBUTES

6. Now fill in the table as indicated in the figure at right. (Adjust the row sizes and locations using drag-n-drop.)

7. **OK** the table and save the block and drawing.

8. Make the table available for the Globe, Check, and Control valves using the Visibility states tools as you did in Steps 15-18 of Exercise 22.1.1A (p.506).

9. Close the Block Editor, WBlock the valve to the C:\Steps\Lesson22 folder (replacing the existing valve), and close the drawing.

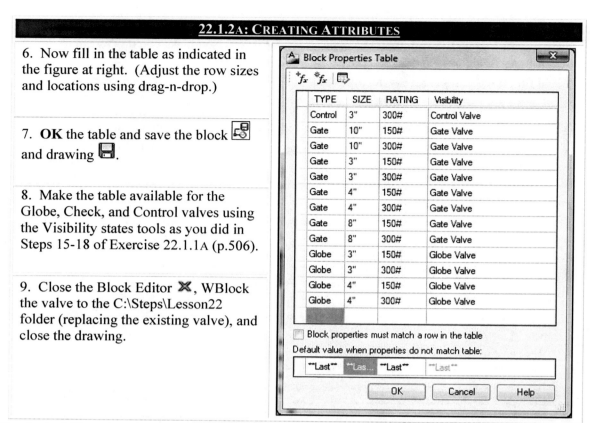

That looked like a lot of work for a few steps! But wait until you see the benefit in our next section!

22.2 Inserting Attributed Blocks

When you insert an attributed block into a drawing, the attribute prompts will follow the standard insertion prompts. This can occur in one of two ways – command line prompts or the Edit Attributes dialog box (Figure 22.003).

Let's look at the command line method first; then we'll look at the dialog box in Exercise 22.2B, p.511.

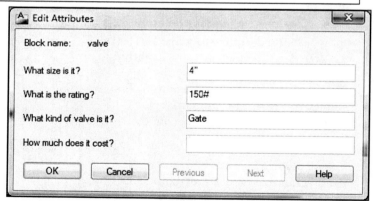

Figure 22.003

AutoCAD includes a system variable – **AttReq** – which controls whether or not you'll be prompted for attribute values. If the variable is set to **1** (the default), AutoCAD prompts on the command line or with a dialog box. If you set this system variable to **0**, however, AutoCAD won't prompt at all for attribute values and you must use the attribute editor or the Properties palette to add the values.

Whether AutoCAD prompts for attribute values at the command line or with a dialog box is controlled by the **AttDia** system variable. A setting of **0** (the default) means that you'll receive prompts on the command line, whereas a setting a **1** calls a dialog box (Figure 22.003).

Do This: 22.2A	**Inserting Attributed Blocks Using the Command Line**

I. Open the *pid-22* file the C:\Steps\Lesson22 folder. The drawing looks like the figure at right.
II. Be sure the *AttDia* system variable is set to **0**. (This means that attribute prompts will appear on the command line.)
III. Set **VA** as the current layer.
IV. Follow these steps.

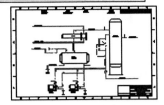

22.2A: INSERTING ATTRIBUTED BLOCKS – COMMAND LINE

1. Restore the **Cont_Station1** view. The view looks like the figure at right.

    Command: *v*

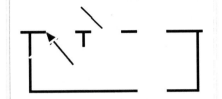

2. Insert the block you created in the last exercise. (Find it in the C:\Steps\Lesson22 folder.) Insert it at the leftmost endpoint (arrow in the Step #1 figure). Accept the default size and orientation.

    Command: *i*

3. Once you've located the block, AutoCAD asks you to **Enter attribute values**. Accept the default **rating**, **size**, and **type**, and enter a price of **55.00**. (Note that you assigned these defaults in our last exercise.)

The prompts may appear in a different order. They'll reflect the order in which you created them.

    Enter attribute values
    What size is it? <4">: *[ENTER]*
    What is the rating? <150#>: *[ENTER]*
    What kind of valve is it? <Gate>: *[ENTER]*
    How much does it cost?: *55.00*
    Verify attribute values
    How much does it cost? <55.00>: *[ENTER]*

4. Save the drawing but don't exit.

    Command: *qsave*

It looks like the figure at right.

Notice that the **Type**, **Filename**, and **Price** attributes aren't visible. Notice also that the valve assumed the characteristics defined by the **VA** layer, and the attribute text assumed the characteristics defined by this drawing's **Text** layer.

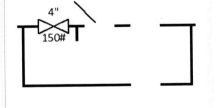

22.2A: INSERTING ATTRIBUTED BLOCKS – COMMAND LINE

5. Repeat Steps 2 through 4 to place the other block valve. The drawing looks like the figure at right. (Can you think of an easier way to have achieved the same results?*)

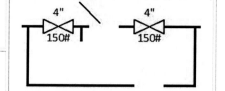

6. Save the drawing 💾 but don't exit.
   Command: *qsave*

We'll look at the dialog box approach next. Note that, although the dialog box presents a comfortable user interface, familiarity with the command line approach will prove quite handy when you use AutoCAD's iDrop tools.

Do This: 22.2B	Using a Dialog Box to Insert Attributed Blocks

   I. Be sure you're still in the *pid-22* file the C:\Steps\Lesson22 folder.
   II. Be sure the **Cont_Station1** view is still current.
   III. Set the *AttDia* system variable to **1**. (This tells AutoCAD to use a dialog box for its attribute insertions.)
   IV. Follow these steps.

22.2B: INSERTING ATTRIBUTED BLOCKS – DIALOG BOX

1. Insert the valve 🗗. Put it at the endpoint of the opening in the bypass line. Accept the size and orientation defaults.
   Command: *i*

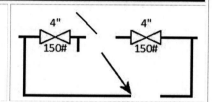

2. AutoCAD presents the Edit Attributes dialog box (following figure). The box shows the command line prompts with their defaults (yours may appear in a different order). Notice that no prompts are given for the **Filename** attribute. Remember that you created this one as a **Constant**; it can't be changed.

Enter the data shown, in the appropriate text boxes.

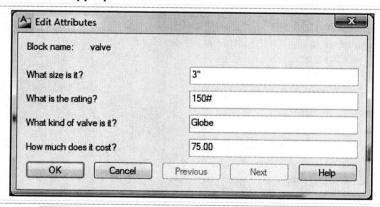

3. Pick the **OK** button [OK] to complete the command.

---

* You could simply have copied the existing attributed valve

22.2B: INSERTING ATTRIBUTED BLOCKS – DIALOG BOX

4. Pick the valve and select **Globe Valve** from the list box. (Pick the down grip on the right to display the list box.) 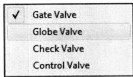 Your drawing looks like the figure at right.	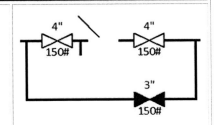
5. Save the drawing 💾 but don't exit. **Command:** *qsave*	
6. Insert the valve in the remaining opening and accept the default attribute values, but make the price 175.00 (we'll use the table to make the other adjustments).	
7. Select the valve and pick the down arrow you created for the table. AutoCAD presents the menu shown here.	
8. Select the **Control** valve from the menu. Notice how AutoCAD adjusts the valve. (Repeat this step and select **Properties Table…** to select a valve directly from the table.)	
9. Insert 🗎 the drain. (I created this already as *DrainVent*; it's in the C:\Steps\Lesson22 folder.) • Use grips to move the attributes as needed. (Hold down the CTRL key when selecting the attribute you wish to move.) • Use a **Drain with Plug** (list box). • The price of the drain is 35.00. The drawing looks like the figure at right.	
10. Save the drawing 💾 but don't exit. **Command:** *qsave*	
11. Use these blocks to complete the drawing. (The blocks have already been created and can be found in the C:\Steps\Lesson22 folder.) See the completed drawing sections in the following figures. Use the price list shown here to assign attribute values.	

SIZE/VALVE	GATE	GLOBE	CONTROL	CHECK
¾"	35.00			
2"	55.00			
4"	85.00	175.00	275.00	
6"	127.00			195.00
8"	385.00			
10"	585.00			

22.2B: INSERTING ATTRIBUTED BLOCKS – DIALOG BOX

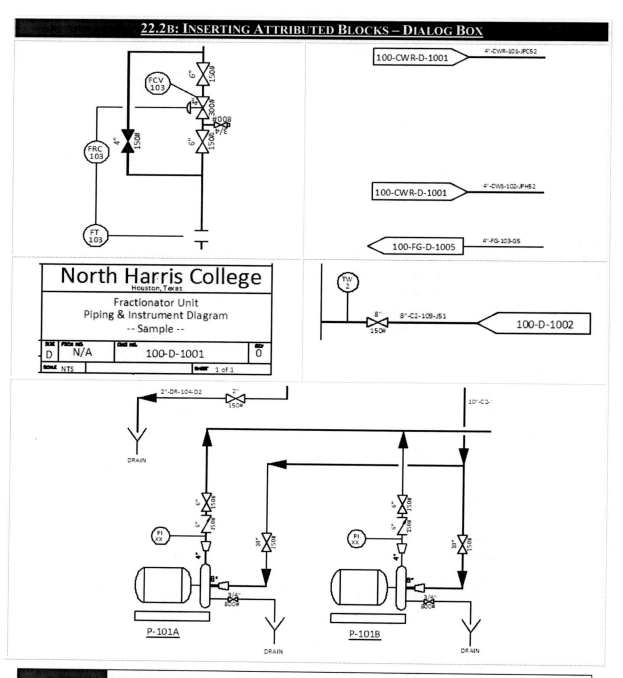

22.3 Editing Attributes

To keep life simple, I'm going to show you how to accomplish all the attribute value editing you'll need using a familiar tool – the Properties palette – and one really old command line tool – *-AttEdit* (we'll need that one for universal changes). Then we'll look at editing attribute definitions with the BAttMan!

22.3.1 Editing Attribute Values

Editing attribute values is as easy as calling the Properties palette.

Do This: 22.3.1A	Editing Attribute Values

When we inserted the gate valves, we assigned a price value of 55.00. According to the chart at the end of our last exercise, the price should be 85.00. Let's fix that.

I. Be sure you're still in the *pid-22* file the C:\Steps\Lesson22 folder. If you haven't finished this one, open *pid-22 Phase 2* instead.
II. Be sure the **Cont_Station1** view is current.
III. Open the Properties palette and move it to the side of the screen. Alternately, you can use Quick Properties.
IV. Follow these steps.

22.3.1A: EDITING ATTRIBUTE VALUES

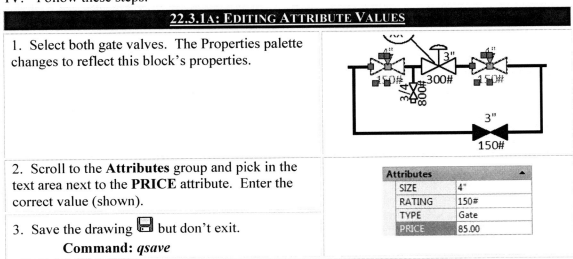

1. Select both gate valves. The Properties palette changes to reflect this block's properties.

2. Scroll to the **Attributes** group and pick in the text area next to the **PRICE** attribute. Enter the correct value (shown).

3. Save the drawing but don't exit.
    Command: *qsave*

Was that easy enough for you? Take a moment and look over the other properties listed in the palette. Remember, you can edit the white values; the gray values are informational.

But what if you have several attributes to edit at once?

It's possible, using an older command line AutoCAD tool, to edit the values and some of the properties of attributes in multiple insertions of a block. The command sequence is:

> **Command:** *-attedit* (or *–ate*)
> **Edit attributes one at a time? [Yes/No] <Y>:** *[ENTER]*
> **Enter block name specification <*>:** *[ENTER]*
> **Enter attribute tag specification <*>:** *[ENTER]*
> **Enter attribute value specification <*>:** *[ENTER]*
> **Select Attributes:** *[Select the attributes to change.]*
> **Select Attributes:** *[ENTER to complete the selection set.]*
> **X attributes selected.** *[AutoCAD reports how many attributes were selected.]*
> **Enter an option [Value/Position/Height/Angle/Style/Layer/Color/Next] <N>:** *[Tell AutoCAD what you want to do.]*

	Where to Find It:
Command Line:	–AttEdit
Hotkey(s):	–ate
Ribbon (Tab/Panel):	Home – Block – Edit Attributes (flyout) – **Multiple**
	Insert – Attributes – Edit Attributes (flyout) – **Multiple**
Menu:	Modify – Object – Attribute – **Global**

- The first option – **Edit attributes one at a time?** – allows you to change attributes globally (or several at once).
- The next three lines ask for some specifics about what you wish to edit. You can simply hit ENTER at each prompt if you'll edit the attributes individually. The default is to accept all selected attributes for possible editing. But you can use these prompts to act as filters in a global selection set. For example, you can edit just the gate valves in our *PID-22* drawing by responding *valve* to the **Enter block name specification** prompt.
- After you select the attributes to edit, AutoCAD presents a line of options to help specify what type of editing to perform.
  - The **Value** option allows you to change the value of the attribute. Use this if you need to globally change a value. For example, assume the price of 4" globe valves has just changed. You could edit your drawing globally to change the value of the **Price** attribute for all 4" globe valves. (It's easier to use the Properties palette to change individual values.)

> An easy way to see the values of all the attributes attached to a block is to set the **AttDisp** system variable to **On**. The command sequence looks like this:
>
> **Command:** *attdisp*
>
> **Enter attribute visibility setting [Normal/ ON/OFF] <Normal>:** *on*
>
> It can be useful to display all the attributes when searching for errors, but the display can quickly become quite crowded.
> - The **Normal** option tells AutoCAD to display only those attributes created with the **Invisible** mode off.
> - Turning **Off** the *AttDisp* tells AutoCAD to hide all the attributes regardless of the **Invisible** mode setting. This setting will speed regeneration, but be careful not to place objects where they'll overlap the attributes when they're displayed again.

  - The **Position** option comes in quite handy when the value of an attribute is physically too large to fit in the area allotted for it (it overlaps something else). The *Move* command will move the entire block, but the **Position** option allows you to move just a single attribute. (Then again, so do grips!)
  - The **Angle** option is also quite handy for making attributes read properly despite the insertion rotation of the block.
  - The other options – **Height**, **Style**, **Layer**, and **Color** – allow you to edit these properties of attributes. If it's necessary to change any of these, however, it may be better to redefine the block(s) so that it inserts properly in the first place.

Let's try multiple attribute editing.

Do This: 22.3.1B	Editing Attribute Values Globally

  I. Be sure you're still in the *pid-22* file or *pid-22 Phase2* file in the C:\Steps\Lesson22 folder.
  II. Be sure the **Cont_Station1** view is still current.
  III. Follow these steps.

22.3.1B: EDITING ATTRIBUTE VALUES GLOBALLY

1. Our project engineer has determined that this control station requires a higher rating for the block and bypass valves. Enter the *-AttEdit* command.
   Command: *-ate*

22.3.1B: EDITING ATTRIBUTE VALUES GLOBALLY

2. We want to edit all the 150# valves. Tell AutoCAD that you don't want to **Edit attributes one at a time**.

   Edit attributes one at a time? [Yes/No] <Y>: *n*

   AutoCAD responds that this will be a global edit.

   Performing global editing of attribute values.

3. We'll want only the valves on the screen to be affected by this editing.

   Edit only attributes visible on screen? [Yes/No] <Y>: *[ENTER]*

4. We'll be editing more than one type of block, so leave this default at the global setting.

   Enter block name specification <*>: *[ENTER]*

5. We want to edit only the **RATING** attributes. If we tell AutoCAD this, it won't change any other type of attribute.

   Enter attribute tag specification <*>: *rating*

6. We don't need any further filtering for our edit.

   Enter attribute value specification <*>: *[ENTER]*

7. Put a window around the entire control station; let AutoCAD sort things out for the **RATING** attributes.

   Select Attributes:

   Select Attributes: *[ENTER]*

8. AutoCAD wants to know what to change …

   Enter string to change: *150#*

9. … and what to make it.

   Enter new string: *300#*

   Your drawing looks like the figure at right.

10. Save the drawing 💾 but don't exit.

    Command: *qsave*

Let's take a look at how we can edit an attribute's *definition* after creating the block.

22.3.2 Editing Attribute Definitions

With the Block Attribute Manager (Figure 22.004, p.517), AutoCAD has made editing attribute properties after block creation almost easier than editing them before you make the block! Access the manager with the **BAttMan** command, but exercise some caution when editing block properties. Remember that *modifications may affect all insertions of the block (current and future).*

Let's take a look.

Where to Find It:	
Command Line:	**BAttMan**
Ribbon (Tab/Panel):	Home – Blocks (subpanel) – **Manage Attributes**
	Insert – Attributes – **Manage**
Menu:	Modify – Object – Attribute – **Block Attribute Manager**
Toolbar:	Modify II – **Block Attribute Manager**

- You'll probably notice the large list box first. Here AutoCAD lists the various attributes associated with the block shown in the **Block** control box above it. You can select another block from the control box, or you can use the

Select block button to select a block from the drawing. The button is handy when you don't know the name of a particular block that needs editing.

- For its size, this dialog box contains quite a few buttons.
  - Use the **Sync** button to update all insertions of the selected block with the currently defined attribute definitions. AutoCAD does this automatically when you edit but provides this manual procedure for those who prefer it.
  - AutoCAD lists the attributes in the list box in the order in which it prompts for values (when you insert the block). Use the **Move Up** and **Move Down** buttons to change the order of the prompts.

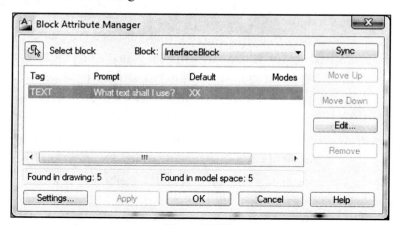

Figure 22.004

  - Use the **Remove** button to remove an attribute from a block definition. Be careful with this one. By default, AutoCAD will remove the attribute definition *and values* from all the insertions in the drawing; you may lose some information.
  - The **Apply** button (along the bottom of the box) applies your changes but leaves the dialog box open.
  - The **OK** button also applies your changes but closes the dialog box.
  - **Cancel** closes the dialog box without saving your changes.
  - **Help** calls the Help dialog box.
- The two buttons we omitted – **Edit** and **Settings** – each call additional dialog boxes.
  - The **Edit** button calls the Edit Attribute dialog box (Figure 22.005). Its three tabs provide access to the attribute definitions.
    - The **Attribute** tab (Figure 22.005) allows you to change the **Mode**, **Tag**, **Prompt**, and **Default** settings.
    - The **Text Options** tab (Figure 22.006) allows you to change the **Text Style**, **Justification**, **Height**, **Rotation**, **Width Factor**, **Oblique Angle**, **Annotative**, and Multiline text **boundary width**.

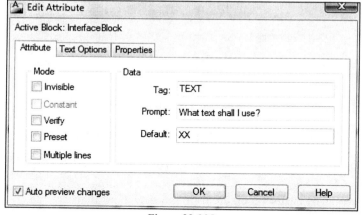

Figure 22.005

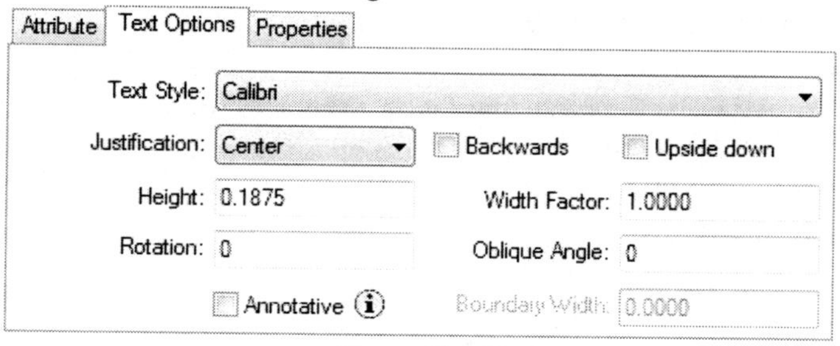

Figure 22.006

- The **Properties** tab (Figure 22.007) allows you to change the **Layer**, **Linetype**, **Color**, **Lineweight**, and **Plot style**.
  o The **Settings** button (back on the Block Attribute Manager) calls the Settings dialog box (Figure 22.008). Here you can tell AutoCAD exactly what (and what not) to show in the Block Attribute Manager.

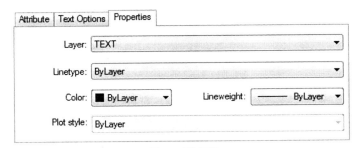

Figure 22.007

The bottom of the Edit Attributes dialog box contains a simple check box that might save you some time and grief. Put a check next to **Auto preview changes** and you can see the changes take place on your screen as you make them. This nifty tool enables you to catch errors before locking yourself into them.

- Use the check boxes in the **Display in list** frame to customize the list box of the Block Attribute Manager. A check next to an item tells AutoCAD to display that information in BAttMan's list box; a clear check box tells AutoCAD not to display it. When too many items are checked for AutoCAD to display them all, a scroll bar will appear in the bottom of the list box to allow you to see everything.
- Use the **Emphasize duplicate tags** check box to have AutoCAD let you know if a block has attributes with duplicate tags in it. AutoCAD will highlight the duplications where they occur. When that happens, you might want to change one of the tags for clarity.
- The purpose of **Apply changes to existing references** might seem a bit confusing, but understanding its use is quite important. A check next to this tool means that AutoCAD will update all current insertions of the block with the modifications you're making. Clear this box and AutoCAD will use the modifications for any new insertions but won't update existing blocks.

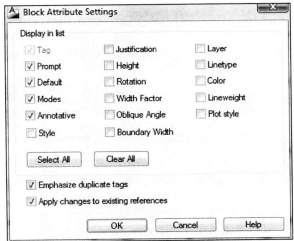

Figure 22.008

Let's see what BAttMan can do!

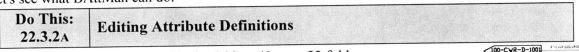

Do This: 22.3.2A	**Editing Attribute Definitions**

I. Open the *pid-22-done* file the C:\Steps\Lesson22 folder.
II. Restore the I_O_LEFT view. Your drawing looks like the figure at right.
III. Follow these steps.

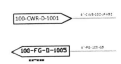

22.3.2A: EDITING ATTRIBUTE DEFINITIONS

1. Open the Block Attribute Manager.
   Command: *battman*

2. On the Block Attribute Manager, pick the **Select block** button.

3. AutoCAD returns you to the graphics screen and prompts you to select a block. Select the upper block with the text that reads *100-CWR-D-1001*.
   Select a block:

4. Select the **TEXT** attribute.

5. Pick the **Edit** button.

Tag	Prompt	Default	Modes
TEXT	Drawing Number?	100-D-XXXX	
SUPPLEME...	Supplementary infor...	xx-xxx	

6. AutoCAD presents the Edit Attribute dialog box (Figure 22.005, p.517). Go to the **Text Options** tab.

7. On the **Text Options** tab, change the **Text Style** to **TIMES** or **Calibri**.
   Text Style: Calibri

8. Go to the **Properties** tab.

9. Notice that the attribute was placed on the **txt** layer rather than the **TEXT** layer that is standard in this drawing. Change it to the proper layer. (Be sure the **Color** control box is set to **ByLayer**.)
   Layer: TEXT

   If there's a check next to **Auto Preview Changes**, you can already see your changes.

10. Pick the **OK** button to return to the Block Attribute Manager.

11. Now let's do something about that extra attribute. Select the **Supplementary** attribute in the list box and then pick the **Edit** button.

12. Go to the **Attribute** tab.

13. Rather than delete the extra attribute (there's always a chance we might need it later), we'll simply make it invisible. Place a check in the proper **Mode** box.

    Mode
    ☑ Invisible
    ☐ Constant
    ☐ Verify
    ☐ Preset
    ☐ Multiple lines

14. Pick the **OK** button to return to the Block Attribute Manager, and again to complete the procedure.

15. Save the drawing.
    Command: *qsave*

Oh, yeah; notice that the changes appeared in both the OFF-DRAWING-LEFT blocks!

22.4 The Coup de Grace: Using Attribute Information in Bills of Materials, Spreadsheets, or Database Programs

This nifty stuff is guaranteed to move you to the head of the class!

One of the most useful and timesaving devices available to AutoCAD users lies in AutoCAD's ability to save steps along the road to project completion. In this section of our lesson, we'll discover how to use the information we've attached to our blocks. We'll extract the data into a table, *and* we'll simultaneously export the table and import it into a Microsoft Excel spreadsheet (for the folks in the Materials department).

Where to Find It:	
Command Line:	*DataExtraction*
Hotkey(s):	*dx*
Ribbon (Tab/Panel):	Insert – Linking & Extraction – **Extract Data**
Menu:	Tools – **Data Extraction**
Toolbar:	Modify II – **Data Extraction**

After so much work creating blocks, dynamic blocks, and attributes, extracting the data might seem almost anticlimactic. You just have to follow instructions in a wizard! (Access the wizard using the *DataExtraction* command.)

Let's give it a shot.

Do This: 22.4A	Extracting Attribute Data

I. Reopen the *pid-22 Phase 2* file the C:\Steps\Lesson22 folder.
II. Set the **TEXT** layer current and zoom all.
III. Follow these steps.

22.4A: EXTRACTING ATTRIBUTE DATA

1. Enter the *DataExtraction* command.
    Command: *dx*

AutoCAD begins the Attribute Extraction wizard.

2. You'll find two options on the Begin page (following figure):
   - **Create a new data extraction**
   - **Edit an existing data extraction**

When creating a new data extraction, you can base it on a **previous extraction**.

We'll create a **new data extraction**, but as we haven't created one before, we won't base it on a previous extraction.

Pick the **Next** button to continue.

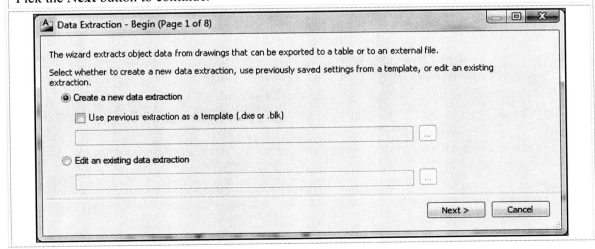

22.4A: EXTRACTING ATTRIBUTE DATA

3. AutoCAD presents a standard Select ... File dialog box to save the extraction. Save it as *Valve List* in the C:\Steps\Lesson22 folder.

4. On the Define Data Source page (following figure), AutoCAD needs to know from where your data will come. We haven't discussed Sheet Sets yet (Lesson 26, p.608), so put a bullet next to **Select objects in the current drawing**, and pick the **Select objects** button . At the **Select objects** prompt, enter all and accept the selection set.

    Select objects: *all*

    Select objects: *[ENTER]*

The **Settings** button calls an Additional Settings dialog box where you can filter for: **Nested blocks**, **Xrefs** (more on Xrefs in Lesson 27, p.630), or **Model Space**. We'll ignore these for now.

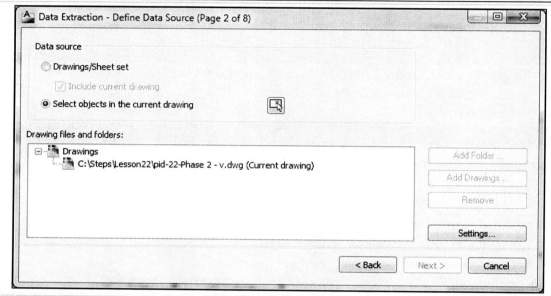

5. Pick the **Next** button to continue.

6. Now we come to the key page – Select Objects (figure following Step 7). The **objects to extract from** list is too long, so let's begin with the **Display options** frame:

- We're currently displaying **all object types** in the drawing. Clear that checkbox and AutoCAD presents a couple other options.
  - We can **Display blocks only** or
  - **Display non-blocks only**.

  We're working with blocks, so put a bullet in that circle as indicated.
- Next, as we're working with attributes, put a check next to **Display blocks with attributes only**.
- You can put a check next to **Display objects currently in-use only** if you wish, but our filters have given us the list we need.

7. Now we can **Select the objects to extract data from**. Remove the check next to the **ANSI_D** (title block), **instbubble**, and **InterfaceBlock** blocks and accept the others. (We only want to work with the valves.)

Pick the **Next** button to continue.

22.4A: EXTRACTING ATTRIBUTE DATA

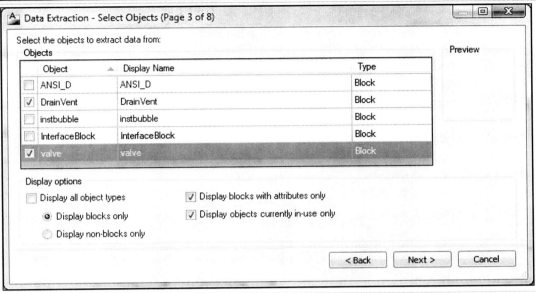

8. Now AutoCAD needs to know which properties of our selected objects we wish to extract. Notice the **Category** filter to the right (following figure) – clear everything except **Attribute**. Notice that this makes our **Properties** list more manageable. Remove the check next to **Filename** (this attribute is redundant with the block name, and we know it's a valve list anyway). Pick the **Next** button to continue.

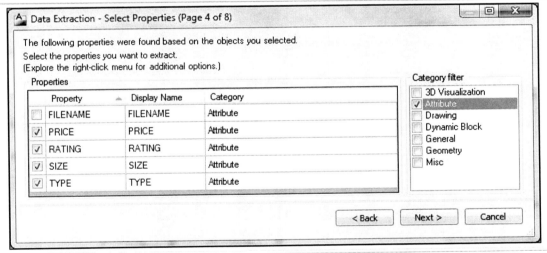

9. On the Refine Data page (following figure), you'll set up your data table.
   - Start with the check boxes below the grid:
     - If you don't **Combine identical rows**, AutoCAD will make a separate row for each object. This increases the size of your table.
     - If you do combine identical rows, be sure you leave a check next to **Show count column** so you'll know how many of each object the row references.
     - **Show name column** can be cleared for our table as the **Type** column covers the data we need better than the block's name would.
   - The **Link External Data** button calls the Link External Data dialog box. We discussed

22.4A: EXTRACTING ATTRIBUTE DATA

Data Linking in Lesson 19, p.443.

- Use the **Sort Columns Option** button to define the sort order of your table. Alternately, you can simply pick the title of a column to tell AutoCAD to sort by that column – a single pick will cause AutoCAD to sort in ascending order, picking again will cause AutoCAD to sort in descending order.
- The **Full Preview** button will tell AutoCAD to let you see the full table before you commit to its setup.
- Adjust the grid by dragging the column headers to the desired position.
- Adjust the column width by placing your cursor on the line separating columns and dragging the double-arrow.

Create the setup indicated. Adjust your table to look like the following image, and then pick the **Next** button to continue.

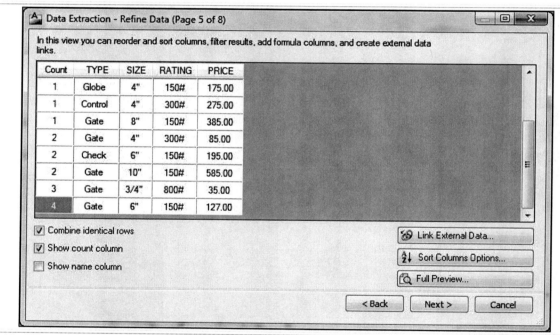

10. Where do you want to put the table? Tell AutoCAD to both **Insert data extraction table into drawing** and to **Output data to external file**. Output the file as an Excel spreadsheet to the C:\Steps\Lesson22 folder. (Your table will be linked to the spreadsheet.) AutoCAD will give it a default filename corresponding to the drawing's filename. Accept that default.

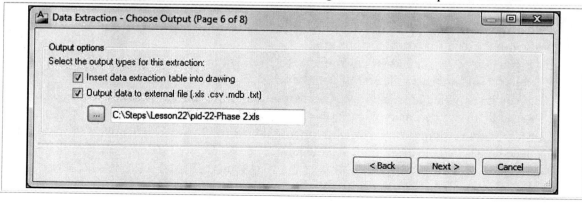

22.4A: EXTRACTING ATTRIBUTE DATA

11. Pick the **Next** button to continue.

12. Now you'll select the Table Style. In the **Table Style** frame of the Table Style page (following figure):
   - Select **Valve table** from the control box. (I set this one up earlier.)

In the **Formatting and structure** frame:
   - Enter a title for the table, we'll call ours *Valve List*
   - Put a check next to **Use property names as additional column headers** so that AutoCAD will create columns for your data using the proper headings.

(This really is easy, isn't it?) Pick the **Next** button to continue.

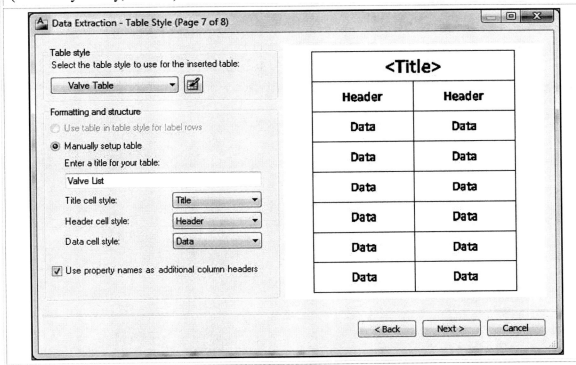

13. Finally, **Finish** the extraction.

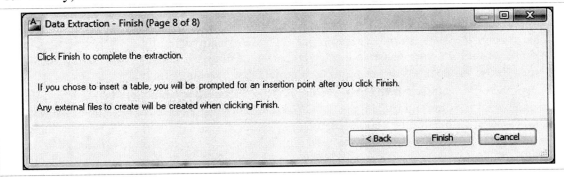

22.4A: EXTRACTING ATTRIBUTE DATA

14. Locate the table around coordinates 3,7.

    External file "C:\Steps\Lesson22\pid-22-Phase 2.xls" was successfully created.

    Specify insertion point: 3,7

    The table looks like the figure at right.

15. Save the drawing.

    Command: *qsave*

Valve List				
Count	TYPE	SIZE	RATING	PRICE
2	Check	6"	150#	195.00
1	Control	3"	300#	175.00
1	Control	4"	300#	275.00
1	Gate	3/4"	800#	35.00
1	Gate	2"	150#	55.00
2	Gate	4"	150#	55.00
2	Gate	10"	150#	585.00
3	Gate	3/4"	800#	35.00
4	Gate	6"	150#	127.00
1	Gate	8"	150#	385.00
1	Globe	3"	150#	75
1	Globe	4"	150#	175.00

22.5 Extra Steps

Check out the Excel spreadsheet you created in Exercise 22.4A (if you have a copy of MS Excel). You'll notice that both number columns (**Quantity** and **Price**) are formatted as text. You'll have to select the columns and use the Excel converter to turn the data into numbers so you (or your Materials rep.) can total the **Price** column.

22.6 What Have We Learned?

Items covered in this lesson include:

- Block Tables (**BTable**)
- Block Attributes
  - **AttDef**
  - **Insert**
  - **AttDia**
  - **Attdisp**
  - **AttReq**
  - **Attedit** and **BAttMan**
  - The Properties Palette
  - **DataExtraction**
  - **ATTipe**

In this lesson, we learned what makes AutoCAD worth the price you (or your company) paid for it. Here we saw how we can shave weeks from a project by spending a day or two in additional setup time. This savings translates into increased profit for the company by cutting production time (thus increasing the amount of work possible in a given period of time). It means more money for the trained operator because it's his or her training that makes the savings possible.

What we've done in this lesson is to cover the methods and techniques available that can accomplish these savings. But it's often up to you, the CAD operator and designer, to explain, demonstrate, and even sell these possibilities to the people and companies for whom you work. Remember that CAD is still relatively new to many people. They are relying on your expertise to show them what can be done!

22.7 Exercises

1. Using what you've learned, create the drawing at right.
   1.1. Place the drawing on an appropriate title block (it was created on a 34"x22" sheet of paper and is NTS).
   1.2. Using attributes, create a BOM.

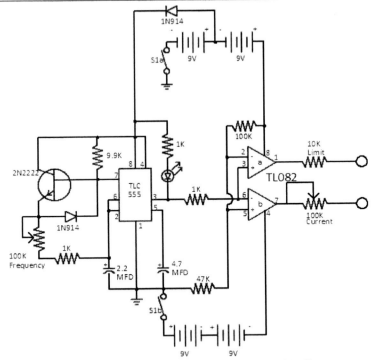

Design Courtesy of Thomas Miller or Richmond, Indiana

Bill of Materials					
Component	Type	Resistance	Volts	Manufacturer	Price
Switch	SPST			Switches R Us	1.55
Switch	SPST			Switches R Us	1.55
Resistor		100K		Radio Shack	3.45
Resistor		10K		Radio Shack	1.56
Resistor		100K		Radio Shack	3.45
Resistor		47K		Radio Shack	2.97
Resistor		1K		Radio Shack	.95
Resistor		1K		Radio Shack	.95
Resistor		1K		Radio Shack	.95
Resistor		100K		Radio Shack	3.45
Resistor		9.9K		Radio Shack	.95
Photo Diode				Hammond Electronics	25.00
NPN Transmitter				Livingston Electronic Systems	5.95
Multicell Battery			9	VoltFlo	3.97
Multicell Battery			9	VoltFlo	3.97
Multicell Battery			9	VoltFlo	3.97
Multicell Battery			9	VoltFlo	3.97
Capacitor				Capacitors Unlimited	7.95
Capacitor				Capacitors Unlimited	7.95
Amplifier				Radio Shack	23.75
Amplifier				Radio Shack	23.75

22.8 For Web-Based Review Questions and Additional Exercises, visit: http://foragerpub.com/AcadFiles/2011/2011.htm

Eventually, it still has to go on paper!

Lesson 23

Following this lesson, you will:

- ✓ Know how to plot a drawing
  - Setting up a drawing (page)
  - Plotting
- ✓ Know how to plot multiple drawings at one time
- ✓ Know how to create an eTransmittal
- ✓ Know how to publish a drawing to the web
- ✓ Know how to publish electronically as a DWF, DWFx, or PDF
- ✓ Be familiar with Autodesk Design Review

Sharing Your Work with Others

> *A General Note to Students and Instructors:*
> Many instructors prefer to cover plotting early in a course; others prefer to cover it at the end. For this reason, we've made this lesson completely modular. The instructor might choose to cover this material earlier in the course.
> However, I don't recommend covering this material later than this. Material in the next few chapters assumes familiarization with plotting.

Goods which are not shared are not goods.

Vernando de Rojas, **La Celestina**, *Act 1*

Sharing. This simple word might carry different meanings for different people. Indeed, it might mean something different for the same person depending on the circumstances.

Certainly, you must share the results of your labor – at least if you want to get paid for it – and for this purpose, AutoCAD provides the tried-and-true **Plot** *command as well as the eTransmit tools and the* **Publish** *command.*

We'll start with the **Plot** *command and explore the ways AutoCAD has provided for you to share the work and then share the fruit.*

Let's begin.

> The terms print and plot are used interchangeably in industry – and in this lesson.

23.1 The Old-Fashioned Way – Putting It on Paper (Plotting)

It seems that we'll never escape our paper world!

Lamentations for a Pine Tree

Anonymous

The ironic thing about CAD is that, after all the wonders of the computer world have created this marvelous drawing; in the end, you still need that paper for construction. After all, it might be a bit clumsy to fold a PC and stick it in your tool belt down at the job site (and not many companies are ready to provide those cool pocket computers to their field workers … yet!)

Fortunately, AutoCAD has made it fairly painless to make the transfer from computer to paper. We'll use a series of dialog boxes that will guide us through the process – *One Step at a Time*.

Let's get started!

23.1.1 First Things First – Setting Up Your Printer (or Plotter)

Before actually printing a document in any application, you must first tell the application which printing device you'll use. The reasons are technical and involve those driver gizmos about which you've probably read. All printers ship with their own drivers. You must help the application – in this case, AutoCAD – to figure out which one to use.

Fortunately, that isn't as complicated as it might seem, and it only needs to be done once for each plotting device.

Since most people won't have plotters available to them, I've moved the printer/plotter setup section of this chapter to a supplemental pdf file. (Download *PlotterSetup.pdf* from here: *http://www.uneedcad.com/Files/PlotterSetup.pdf*. You'll find the appropriate file in C:\Steps\Lesson23 rather than 21 as the supplement says; otherwise, this procedure hasn't changed beyond recognition in the last several releases.) If you don't have a plotter, you can ignore it. If you or your company will use a plotter, you might want to go through those instructions first.

> I must distinguish between printers and plotters here (as opposed to printing and plotting). *Printers* are comparatively inexpensive devices that usually sit on a desk or table and use letter or legal size paper. *Plotters* are much larger and are generally used to create drawings on C, D, or E size sheets of paper.
>
> As a rule, AutoCAD can print to a printer using the computer's settings; you won't need to do any additional set up. Plotters, on the other hand, will require some effort on your part.

23.1.2 Plot Styles

Believe it or not, there will be times when you must ask yourself, "Do I want to print it the way I drew it or some other way – different lineweights, linetypes, colors, etc.?" Personally, I've always been a firm advocate of *WYSIWYG* technology (see the insert), but there will be times when a change at plot time is necessary. (Some folks even prefer to set lineweights and linetypes at plot time!?) For example, my printer uses color. I like that, but I can't afford the cost of printer cartridges when I print working drawings in full color. So I prefer to print in black-and-white until the final product is ready. With plot styles, I can do that without affecting the drawing. (Of course, for those of you who would rather wait until you create a plot to see what it looks like, you can always set weights and types using plot styles.)

> **WYSIWYG** (pronounced "whiz-ee-wig") – literally, "What You See Is What You Get" – simply means that what appears on your screen is what will appear on your paper. This may seem obvious (and generally is for other types of documents). But some gurus prefer to assign (or change) such things as lineweights, linetypes, or colors at plot time rather than during the drawing setup.

As with plotters, you're fairly unlikely to need this information. However, I understand that once I say that .... Anyway, I've put details on setting up plot styles in a supplemental pdf file for those who'll need it. (*Download PlotStyleSetup.pdf* from here: *http://www.uneedcad.com/Files/PlotStyleSetup.pdf*.)

23.1.3 Setting Up the Page to Be Plotted

As with plotter setups and table setups, you don't absolutely have to know how to set up a page for printing/plotting. However, in this case, it's a good thing to know as it may save you a lot of time as using named page setups means that you won't have to repeat the setup every time you plot.

It isn't difficult; although it isn't as easy as the initial dialog box might indicate.

When you access the Page Setup Manager with the *PageSetup* command, AutoCAD presents Figure

Where to Find It:	
Command Line:	*PageSetup*
Ribbon (Tab/Panel):	Output – Plot – **Page Setup Manager**
Menu:	File – **Page Setup Manager**
Application Menu:	Print – **Page Setup**
Toolbar:	Layouts – **Page Setup Manager**

23.001, p.531. Here you see two frames – one in which you'll work and one for information. Additionally, there are two fairly standard buttons and a check box.

- Most of your work will occur in the **Page setups** frame. Here, AutoCAD begins by telling you the **Current page setup** (**None** in our figure) and presents a list of existing setups from which you can select. Selecting a setup from the list means that your work here is finished. But if that was all there was to it, you wouldn't need those buttons down the side of the frame.
    - Use the **Set Current** button to use the selected setup to do your printing.
    - The **New** and **Modify** buttons present dialog boxes. **New** gives you an opportunity to name the new setup before continuing (Figure 23.002, p.531). After you name the new setup, AutoCAD presents the Page Setup dialog box (Figure 23.003, p.532). (**Modify** takes you

directly to the Page Setup dialog box and makes the selected setup's information available for modification.)
- We'll look at the Page Setup dialog box in more detail in a moment.
- **Import** allows you to take the page setup definition from another drawing and make it a definition in your drawing. You'll use a common Select File dialog box for this and select a DWG, DWT, or DXF file.

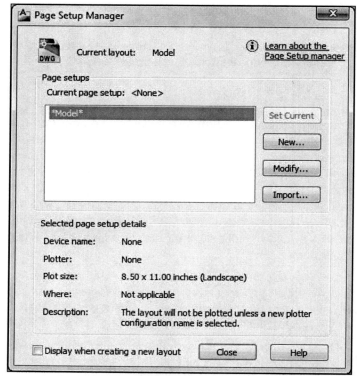

Figure 23.001

- The **Selected page setup details** frame provides information about the selected setup. Use this information to be sure you've selected the correct setup for your plot.
- Placing a check next to **Display when creating a new layout** won't mean anything to you at this point – even if you've gone through the chapters ahead of this one. It will, however, prove useful when you begin using layouts (Paper Space). We'll look at layouts in Lessons 24, p.560, and 25, p.584.

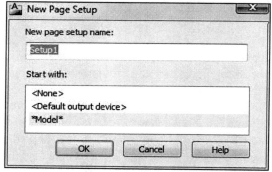

Figure 23.002

You can also remove a setup using the Page Setup Manager. Simply select the setup and use the DELETE key on your keyboard.

The important thing to remember about the manager you've just seen is that it provides only the opportunity to select an existing setup for use, or to create or modify a setup. It doesn't create the setup for you. You'll need the Page Setup dialog box for that (Figure 23.003, p.532) ... and it's a ten-frame doozy! Let's begin in the upper left corner and take a look.

- The first frame – **Page setup** – is harmless enough. It tells you the name of the setup you're creating or modifying.
- The next frame down – **Printer/plotter** – allows you to select the device you'll use for this setup. Select the device from the **Name** drop down box and set any properties you want on the device via the **Properties** button just to the right of the control box. This is fairly simply stuff (and standard for most computer applications) as long as the device you want to use is on the list. If it isn't, you'll need to set it up. Refer to the plotter setup supplement for this (Section 23.1.1, p. 529).

AutoCAD fills the rest of the frame with useful information about the selected device and the orientation of the setup.

531

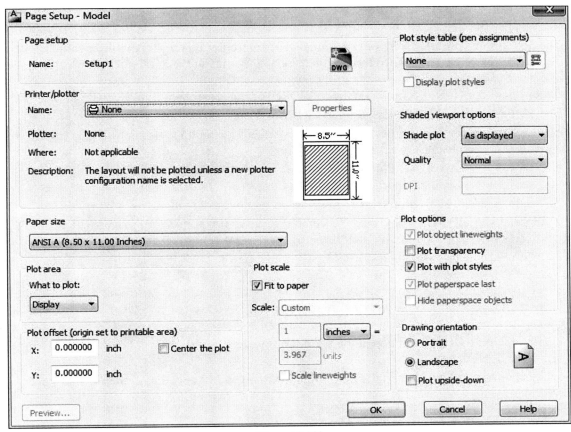

Figure 23.003

- Below the **Printer/plotter** frame, you'll find the **Paper size** frame, which you'll use for selecting the size of the paper on which you wish to print.
- The **Plot area** frame also contains only a single drop down box. Use it to select exactly what you want to plot on the drawing. AutoCAD allows you to plot just the currently displayed portion of the drawing, the drawing's extents, limits, or an area defined by a window. (If you've opted to complete this lesson out of order, these options will become clearer as you complete the first few lessons.)
- Use the tools within the handy **Plot offset (origin set to printable area)** frame to adjust where your drawing lands on the page. I've found that a check in the box next to **Center the plot** works well for my printing, but you can adjust the position manually using the **X** and **Y** boxes if you wish.
- The **Plot scale** frame contains some critical options.
    - A check next to **Fit to paper** eliminates any effort to plot to scale and forces the drawing onto the page. The results are reduced\enlarged by ratio to fit on the paper. Use this setting for creating small, working drawings.
    - If you don't place a check in the **Fit to paper** box, AutoCAD allows you to select the scale from the **Scale** drop down box, or to enter a customized scale in the boxes below the drop down box.
    - **Scal**[ing] **lineweights** is a good idea if you've used lineweights (Lesson 6, p.137).
- The **Plot style table (pen assignments)** frame (back at the top, on the right) allows you to select a style table. You can use plot style tables to convert a color drawing to black and white or to assign specific colors to specific pens on a multi-pen plotter. Use the **Edit** button next to the drop

down selection box to modify the selected table. For more information on plot style tables, see the supplement mentioned in Section 23.1.2 (p.530).
- The next frame – **Shaded viewport options** – contains controls for plotting three-dimensional drawings. We'll discuss shaded three-dimensional drawings in the *3D AutoCAD 2011: One Step at a Time* text. Don't mistake the **Quality** option as something that might be useful in a two-dimensional drawing; it controls quality of shaded or rendered three-dimensional objects.
- The **Plot options** frame provides four check boxes.
  - Use **Plot object lineweights** only if you've set the drawing up to use lineweights (Lesson 6, p.137). If you haven't, you can ignore this box.
  - You can have AutoCAD **Plot transparency** with a check the next box. Leave it blank and AutoCAD will ignore the transparency property when it plots.
  - Again, use **Plot with plotstyles** only if you've set up the drawing to use plot styles.
  - Normally, AutoCAD plots Paper Space objects first. If you prefer to plot Model Space objects first, place a check in the **Plot paperspace last** box. I haven't seen where this is necessary, but you might discover a need for it. (As I've mentioned, we'll discuss Paper Space in Lessons 24 and 25.)
- You'll tell AutoCAD how to position the drawing on the paper in the last frame (Whew!) – **Drawing orientation**. Your options are the standard **Portrait** and **Landscape** orientations, and a **Plot upside-down** option that you can use to print sepias.
- Finally, there's a **Preview** button at the bottom of the dialog box that will help you see what your settings will produce.

I know this has been a lot of material, but it might hearten you to know that it's less that has been required in previous releases. Let's set up a page and plot a drawing.

Do This: 23.1.3A	Setting Up a Page and Plotting

I. Open the *pid-23* file in the C:\Steps\Lesson23 folder.
II. Follow these steps.

23.1.3A: SETTING UP A PAGE AND PLOTTING

1. Open the Page Setup Manager.
   Command: *pagesetup*

2. Create a new page setup [New...].

3. AutoCAD presents the New Page Setup dialog box. Select the **Model** setup in the **Start with** list box, and then enter the name of your new setup as shown. Pick the **OK** button to continue.

4. AutoCAD presents the Page Setup dialog box (Figure 23.003, p.532). We'll go through the frames to set up our printer.
Begin in the **Printer/Plotter** frame and select the printer you wish to use from the **Name** drop down selection box. (I'll use my HP Photosmart.)

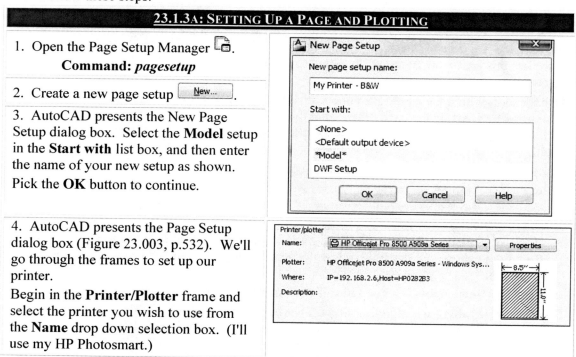

23.1.3A: SETTING UP A PAGE AND PLOTTING

5. Select the paper size you wish to use. I'll use **Letter** for this setup.

6. I want to fit the entire drawing on the page, so I'll plot the **Extents**. (More on **Extents** in Lesson 4, p.79.) I'll **center the plot** on the page as well.
Since I designed the drawing for a D-size sheet, I'll have to make it **Fit** on this smaller size. (This will suit my purposes for a working drawing.)

7. I could've use the **Properties** button in the Printer/plotter frame to access my printer's settings, but I'll use tables to tell AutoCAD to convert all the geometry in my drawing to black and white.

8. AutoCAD wants to know if I'd like to assign this table to all the layouts in my drawing. We're not using layouts, but I may decide to use some Paper Space technology in later lessons, so I'll tell it to go ahead by picking the **Yes** button.

9. Since I told AutoCAD to use a styles table in Step 7, I need to select that in the **Plot options** frame.

10. Finally, I'll let AutoCAD know how to position the drawing on the paper.

11. That's it for my page setup OK ! I'm ready to print.

12. Back at the Page Setup Manager, set the new style current Set Current (select it and pick the appropriate button).

13. Close the Page Setup Manager Close .

14. Now we can print the drawing. Enter the command at the command line. Alternately, you can pick the **Plot** button on the **Output** tab's **Plot** panel.
     Command: *plot*

23.1.3A: SETTING UP A PAGE AND PLOTTING

15. Notice that the Plot dialog box is just a simplified version of the Page Setup dialog box. This time, however, because we set our new page setup as current (Step 12), all of our settings have been made for us. Use the **Preview** button [Preview...] to be sure that this setup will work for our drawing.

16. Our drawing looks okay on the paper, so we'll pick the **Close Preview Window** button ⊗ to leave the preview.

17. Pick the **OK** button [OK] to start the print.

18. AutoCAD will let you know that it has completed your printing job. It'll also let you know of any problems! Pick the **X** to close the bubble.

19. Save your drawing 💾 but don't exit.
    Command: *qsave*

| 23.2 | **Sharing Your Drawing with the *Plot* Command – and *No Paper*!** |

"Gettin' paid a check's okay, I s'pose, 'cept it don't leave me nuttin' to jingle in my pocket."

from A Geezer's Lament
Anonymous

Ah, progress! Just as our poor geezer lost the joy of jingling pocket change, so we must eventually say goodbye to paper drawings – at least as much as we've said goodbye to cash!

You can plot to an Adobe PDF file – useful if you wish to share a file with some (limited) capabilities such as the ability to manipulate layers (Lesson 7, p.148). Adobe's reader is free and commonly used so you can be sure your client will be able to read the file.

But for real power, AutoCAD has provided a special type of file that you can use to share design information via the Internet – without the need to send paper. This new file type also loads considerably faster than a drawing file, contains the ability to manipulate layers and views (and more), and can be printed like any other web document or PDF.

AutoCAD calls the file a Drawing Web Format file (DWF, pronounced "DWIF"), gives it the ability to contain hyperlinks (just like any other web document), and makes its creation as simple as plotting a document. (You'll find several approaches to creating a PDF or a DWF – including the plot command!) The DWF file is a wonderful innovation that allows AutoCAD users – *and clients without access to the AutoCAD program* – to view AutoCAD drawings!

> Autodesk also includes a new DWFx driver with its plot devices. Use this one when you plan to distribute your DWFs to VISTA or 7 systems.

Let's see how it works.

| 23.2.1 | **Quickly Creating a Drawing Web Format (DWF) or Adobe PDF File with the *Export* Command** |

The menu's **Output** tab contains an **Export to DWF/PDF** panel (Figure 23.004) that can make DWF/DWFx/PDF creation fast and easy. Each file follows the same procedure, making the learning curve easy as well! Look at it closely.

- The flyout on the left contains three **Export** buttons:

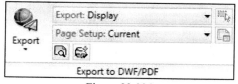

Figure 23.004

535

PDF, **DWF**, and **DWFx**. Each will begin the *Export* command for the appropriate file type. (You'll also find an **Export** button under the File pull down menu and the menu browser. This button will start a simplified version of the *Export* command.) Each of these buttons calls a Save As dialog box with some plot/print-specific tools. We'll create a file in our next exercise and take a look at these tools.

- Use the **Export** control box to control what display you will export (the current **Display**, the drawing's **Extents**, or a selection **Window**).
- Use the **Page Setup** control box to select either the **Current** setup or an **Override**. **Override** will call a simple setup dialog box where you can make the adjustments you wish.
- The **Preview** button will call a preview window to give you a sneak-peak of what your exported file will look like.
- Finally, you can preset your PDF/DWF options in the Export to DWF/PDF Options dialog box (Figure 23.005) called by the **Export to DWF/PDF Options** button. Take a moment to familiarize yourself with this dialog box before starting the exercise.

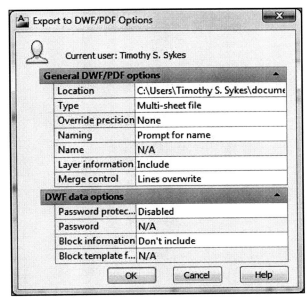

Figure 23.005

Do This: 23.2.1A	Exporting a PDF File

I. Be sure you're still in the *PID-23* file. If not, please open it now.
II. If you don't have the latest version of the Adobe Reader installed on your computer, please install it now. (It's a free download at www.adobe.com.)
III. Follow these steps.

23.2.1A: EXPORTING A PDF FILE

1. Let's begin with a basic setup. Open the Export to DWF/PDF Options dialog box (Figure 23.005).

2. Fill in the dialog box as indicated at right.
- Locate the exported file in the C:\Steps\Lesson23 folder,
- Use a **Single-sheet file**,
- **Include Layer information**,
- **merge** overlapping lines (the color of the lines on top will blend with lines on bottom)
- Don't use a **Password** or include **Block information**.

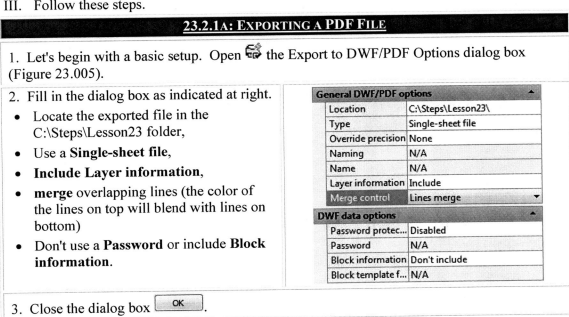

3. Close the dialog box [OK].

23.2.1A: EXPORTING A PDF FILE

4. Pick the **Export PDF** button on the **Export to DWF/PDF** panel. (Alternately, you can enter the *ExportPDF* command on the command line.)

AutoCAD opens the Save As PDF dialog box (following figure).

5. Notice that the **Current Settings** frame contains the information you provided in Step 2. Let's fill out the remaining options:

   - In the **Output Controls** frame, tell AutoCAD to **Open in viewer when done** to open the PDF in Adobe's Reader and to **Include plot stamp**. (You can adjust what will appear on the stamp through the dialog box called by the **Plot Stamp Settings** button next to the check box. See the insert on p.543 for more information on plot stamps.)
   - In the lower part of the **Output Controls** frame, tell AutoCAD to **Export** the drawing's **Extents** and to use the **Current Page Setup**.

6. Save [Save] the document.

AutoCAD creates the PDF and opens it in the Adobe Reader! (If it doesn't open the file in the reader, please open it now.) It looks like the following figure (Adobe Reader 9 shown).

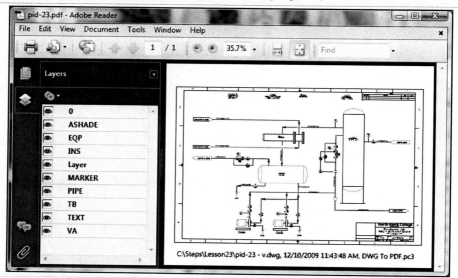

537

23.2.1A: EXPORTING A PDF FILE

7. Notice the layers list down the left side of the window. Pick the "eye" next to each and see the effects on the drawing. (Pretty cool, huh?) If you're doing this lesson out of order, you'll learn more about layers in Lesson 7.

Notice also the plot stamp across the bottom. This one uses the default stamp settings; did you examine the Plot Settings dialog box in Step 5?

8. Close the Adobe Reader and any open AutoCAD drawings.

That's about all you'll get from the PDF, but the DWF file can do so much more! Let's take a look at viewing a DWF next.

23.2.2 Viewing a Drawing Web Format (DWF) File

The AutoCAD installation includes a tool called the Autodesk Design Review. This tool, like its many predecessors, enables you to view an AutoCAD DWF file.

> We could easily fill a small textbook explaining all the intricacies of Autodesk Design Review, but we just don't have the space. We'll skim the basics and encourage you to continue to explore this nifty tool on your own.
>
> For a more detailed look at Autodesk Design Review, just hit the F1 key from within the Design Review window.

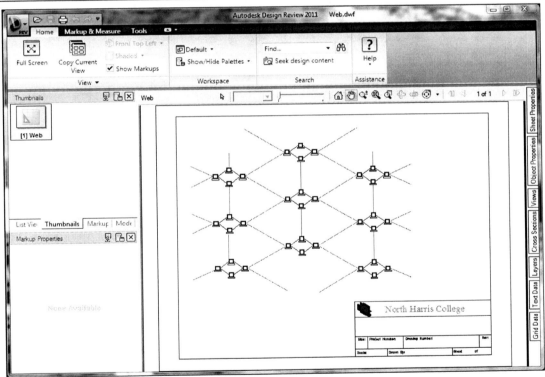

Figure 23.006

Design Review ships with the AutoCAD program, but for those clients who don't have access to AutoCAD, it's a free download at the Autodesk web site: http://www.autodesk.com/. With Design Review, you can view files from within your browser or through the viewer program itself – both approaches use the same tools (although the interface may appear a bit different).

Let's take a look. Open Design Review. (From your desktop, follow this path: Start – All Programs – Autodesk – Autodesk Design Review.) Design Review (Figure 23.006, p.538) contains the same basic structure as AutoCAD.

- At the top, you'll find the familiar menu browser and a Quick Access toolbar containing the save basic tools you find on the AutoCAD QA toolbar.
- Below the Quick Access toolbar, you'll find Design Review's ribbon, which works just like AutoCAD's. Here you'll find three tabs, each with several panels.
  - The **Home** tab contains these panels:
    - **View** tools enable you to view the drawing **Full Screen**, in typical 3D views (more on those in our 3D text), in various shading modes (also in our 3D text), and with or without **Markups** (viewer notes). You can also **Copy** the **Current View** to Windows' clipboard for pasting into another document.
    - **Workspace** provides a toggle between the Design Review workspaces and provides a tool for controlling which palettes you see on the left side of the window.
    - **Search** tools include a box for searching the current file for text or for searching the Autodesk Seek website for information.
    - **Assistance**, of course, provides access to Design Review's help dialog box as well as some other useful assistance tools.

Figure 23.007

  - The **Markup & Measure** tab (Figure 23.007) contains these panels:
    - **Clipboard** contains Windows standard clipboard tools.
    - **Formatting** contains tools you'll use to format your markup text, callouts, leaders and such.
    - **Callouts** includes several tools – many with leaders – to put your comments into boxes. These boxes (circles, etc.) include a form of grip (see Lesson 9, p.212) to help resize and relocate them once they've been created. You can also access the revision clouds here.
    - **Draw** contains tools to draw directional lines (pointers), ellipses or circles. This is *not* an AutoCAD precision tool but rather a markup tool. Don't expect 12-decimal place accuracy!
    - **Measure** provides several handy tools which work with true AutoCAD accuracy!
    - Finally, you'll find several typical **Stamps & Symbols** in the last panel.
  - The **Tools** tab (Figure 23.008) contains these panels:

Figure 23.008

    - **Canvass** contains tools for rotating the drawing (**Left** or **Right**) and for comparing different versions of the same sheets.
    - **3D Tools** includes tools for moving and rotating 3D objects and creating **Section Faces**.

- Finally, you can use the tools on the **Create Sheet** panel to create your own sheet **From [a] Snapshot** or **From [a] Grid**.

You'll find even more tools on the cursor menu (Figure 23.009)! Here you'll find several frames of tools, as well.

- The top frame contains the **Select** tool. Design Review uses this by default, but you can return to it here after using another tool.
- The next frame contains viewing tools:
  o The **Go** flyout contains tools for navigating through your views (**Next**, **Previous**, etc.).
  Find uses a palette to find items just as the Search panel did.
  View offers you the options to view the drawing in **Color**, **Grayscale**, or **Black & White**.
  o The next frame offers you a tool for copying the **Current View** to the Windows clipboard.
  o **Save As** and **Print** are self-explanatory.
    - (**Show/Hide Full Screen** – let's you use the entire screen to display your DWF. Hit ESC to return to the normal screen.
    - Use the options in the next frame to navigate through the drawing. The **Pan** and **Zoom** tools behave like their realtime counterparts in the AutoCAD program (see Chapter 4, p.79). **Fit in Window** behaves like a zoom extents (also Chapter 4).
  o Design Review contains the same steering wheels found in AutoCAD, and an additional **2D Navigation Wheel** that provides quick access to the **Zoom** and **Pan** tools.
  o **Compare Sheets** works just as the panel tool worked.
  o Show provides some 3D tools as well as tools for viewing a **Canvass Background**, **Hyperlinks** and **Markups**. You can reset the **2D Units & Scale** with the next option and have Design Review **Snap to** [your] **Geometry**. **Options**, of course, opens the Options dialog box.

Figure 23.009

Now, you have to admit, Adobe has a long way to go to match Design Review! Let's take just a moment to look at a DWF file with Design Review.

Do This: 23.2.2A	Viewing a DWF File

This will be a unique exercise – we won't use AutoCAD at all!
  I. Open Design Review. (From you desktop, follow this path: Start – All Programs – Autodesk – Autodesk Design Review.)
  II. Follow these steps.

23.2.2A: VIEWING A DWF FILE

1. Pick the **Open** button on the Quick Access toolbar. Design Review presents a standard Open ... File dialog box.

2. Open the file *C:\Steps\Lesson23\Web.dwf*. The viewer now looks like the Figure 23.006, p.538.

3. Pick **Zoom Rectangle** on the cursor menu.

23.2.2A: VIEWING A DWF FILE

4. Notice that your cursor changes to a magnifying glass with a rectangle behind it. Pick just northwest of the title block and drag (holding the mouse button down) to the south and east. Release the mouse button when the entire title block has been highlighted. Design Review presents the view shown.

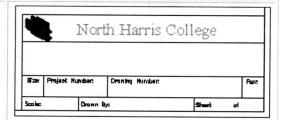

5. Notice that the magnifying glass cursor remains. You could continue to use **Zoom Rectangle** if you wished.

Pick the **Fit to Window** option to return to the original view.

6. Let's take a look at the **Layers** palette (shown). (Move your cursor over the Layers title along the *right* side of the window.)

Notice the light bulb symbol. When yellow, objects residing on that layer are visible. Pick on a light bulb to turn it gray (**Off**). Objects residing on that layer are now hidden. Toggle the other layers **On** and **Off** to see how the display is affected.

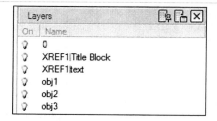

7. Take a moment and explore the rest of the panels and palettes. (Note: Use some of the other PDF and DWFx files in the C:\Steps\Lesson23 folder for some other opportunities. There's even a 3D drawing for you to play with!)

8. Okay. Now let's take a look at printing your DWF file. Pick the **Print** button on the Quick Access toolbar. Design Review presents a Print dialog box (see the following figure) very similar to the Page Setup dialog box we saw on p.532.

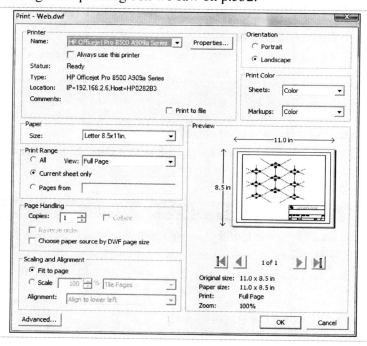

541

23.2.2A: VIEWING A DWF FILE

9. Select your printer and make whatever other adjustments you wish to make (I usually set my prints to **Print in Black and White** to save my printer's color cartridges.) Then pick the **OK** button [OK]. Design Review prints your drawing.

So, now you can see what a remarkable tool a DWF file can be – shared information without compromising the security of your drawing files!

> Again, I strongly encourage you to spend some time with the assistance tools Autodesk provides for this spiffy tool. From within the Design Review window, pick the **Help** button on the ribbon, and select any of the Content headings to get started.

> Note also that you can view a DWF or a DWFx file within Internet Explorer. IE makes many of the same tools available.

Let's create our own DWF file.

23.2.3 Multiple Plots and Creating a DWF File – with Hyperlinks!

In previous versions of AutoCAD, we created DWF files using some different settings in our plotting procedure. It was simply a question of selecting the DWF device on the **Plot Device** tab of the Plot dialog box.

Where to Find It:	
Command Line:	*Publish*
Ribbon (Tab/Panel):	Output – Plot – **Batch Plot**
Menu:	File – **Publish**
Toolbar:	Standard – **Publish**

We can still use this approach – simply select the **DWF6** or the newer **DWFx** plotting device. (Select the **DWG to PDF** printer to do the same for PDF files.) Alternately, we can opt for the *Publish* command and dialog box (Figure 23.010), which will plot one or several drawings at once!

> The DWFx plot driver produces files that you can distribute on the Windows Vista or 7 systems and the XPS Viewer.

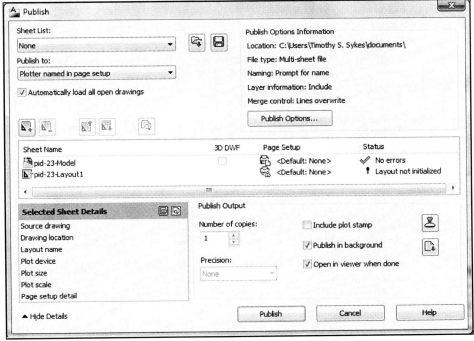

Figure 23.010

Let's take a look.
- The first thing you may notice – the **Sheet List** box – shows the various sheets associated with the current or open drawing file(s), their current page setups, and their status. As we haven't discussed Paper Space and **Layout** tabs yet, we'll concentrate on the **Model** sheet.
- [The Publisher contains a host of tools; we'll start in the top left corner and work our way down.] The **Sheet List** is just that – a list of the drawing sheets you wish to include in your batch plotting. We'll discuss adding/removing files from here in a moment. Once you have your list, however, you can save it using the Save button 💾. Reload it using the **Load Sheet List** button 📂.

  Place a check next to **Automatically load all open drawings** if you wish to include all the currently opened drawings in your batch plot.
- Just above the **Sheet List**, you'll find several buttons. Let's look at these.

BUTTON	DESCRIPTION
	Add sheets enables you to add drawing sheets to a multi-sheet DWF file.
	Remove allows you to remove sheets from the list.
	Move Sheet Up and **Move Sheet Down** allow you to change the order in which the sheets will appear in the DWF file.
	Preview enables you to preview the plot.

- The **Publish Options Information** frame tells you about the plotting you're about to perform. Pick the **Publish Options** button to open the Publish Options dialog box (similar in form and function to the Export to DWF/PDF dialog box shown in Figure 23.005, p.536).
- The two frames below the list box include:
  - A **Selected Sheet Details** frame with information about the selected sheet and the plot you're about to perform. Use the buttons to the right of the title bar to toggle between the **Selected Sheet Details** frame shown in Figure 23.010 (p.542) and a **Preview** of the plot.
  - The **Publish Output** frame contains some final details concerning the number of copies you wish to plot, plot stamps (see the following insert), background publishing (while you do something else), and whether or not to open the file(s) in a viewer when you've finished.

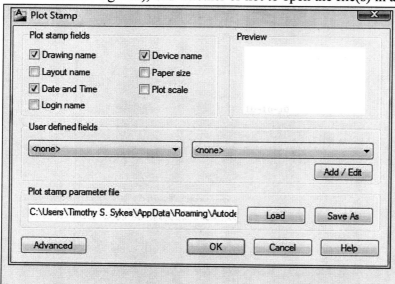

Stamping a plot provides useful information for tracking the particulars of when and what was plotted. Picking the **Plot Stamp Settings** button in the Publish dialog box produces the Plot Stamp dialog box (shown).

Let's look at this.
- Select what your stamp should include in the **Plot stamp fields** frame. Preview your selections in the **Preview** frame.
- Pick the **Add / Edit** button to add **User defined fields**. AutoCAD provides a simple dialog box to assist you.

(continued on next page)

543

> - Simplify your efforts by saving your setup and then loading it in the **Plot stamp parameter file frame**.
> - The **Advanced** button calls another useful dialog box where you can set the **Location**, **Orientation**, and **offset** of your stamp, as well as **Text properties**. You can even tell AutoCAD to keep a log of your plots. Note that the plot stamp is *not* annotative.

We'll use the Publish dialog box in our next exercise, but first, let's take a look at creating hyperlinks.

23.2.4 Hyperlinks

If you plan to use hyperlinks in your DWF/PDF file, you must assign them before plotting. Let's look at how to do that now.

Begin by entering the *Hyperlink* command and selecting the object(s) to which you want to assign the jump (the thing you want to pick to go to another web address). AutoCAD will present the Insert Hyperlink dialog box (Figure 23.011).

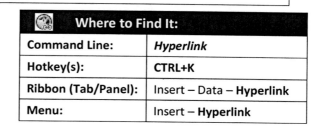

Where to Find It:	
Command Line:	*Hyperlink*
Hotkey(s):	CTRL+K
Ribbon (Tab/Panel):	Insert – Data – **Hyperlink**
Menu:	Insert – **Hyperlink**

The dialog box looks considerably more complicated than it actually is. Let's look at one piece at a time.

- At the top of the dialog box, you'll find a text box with the words **Text to display** next to it. What you enter here will display in a small tooltip whenever a viewer passes his cursor over the hyperlink.
- Down the left side of the box, you'll see three **Link to** picks. These control the methods you'll use to identify the target hyperlink (the place to which you wish to jump).

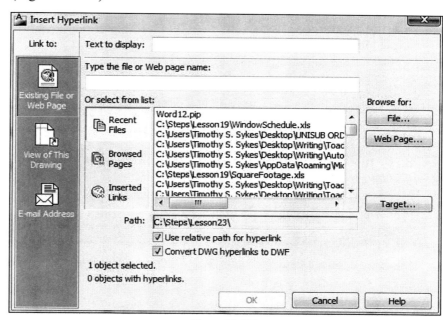

Figure 23.011

- The first **Link to** pick – **Existing File or Web Page** – provides the dialog shown in Figure 23.011. Use it to create a link to a file, layout, or web site that already exists. (The other links change the dialog box as you'll see shortly.) Let's look at how to do this.
  o Below the **Text to display** box, you see a **Type the file or Web page name** text box. This provides the key to your hyperlink and the easy way to identify your web site or target file. Enter the address of the target site in this box.
  o In the **Or select from list** box, AutoCAD has provided a selection list from which you can pick the data that will go into the **Type the file or Web page name** text box. Selecting a target here saves you from having to type in a target (and deal with typos). Three option buttons help organize your choices into these groups:
    Recent Files – files you've recently opened in AutoCAD.

Browsed Pages – Internet pages to which you've browsed.
Inserted Links – recently inserted hyperlinks.
Each of these option buttons changes the list to accommodate the choice.
- To the right of the list box are three buttons that also provide an easy method of filling the **Type the file or Web page name** text box.
  - The **File** button opens a standard Open dialog box where you can select the target file.
  - The **Web Page** button works just like the **File** button but opens AutoCAD's browser where you'll select the target web site.
  - The **Target** button allows you to select a layout or preset view within the drawing as a target.
- The second **Link to** pick – **View of This Drawing** – provides the same tree view the **Target** button provides when the **Existing file or Web page** button has been selected.
- With the last **Link to** pick – **Email Address** – you can place a hyperlink in the drawing that will allow the reader to send an email to a predefined address. AutoCAD provides the simple interface shown in Figure 23.012 for this task. (It even provides a list of recently used email addresses from which to choose.)

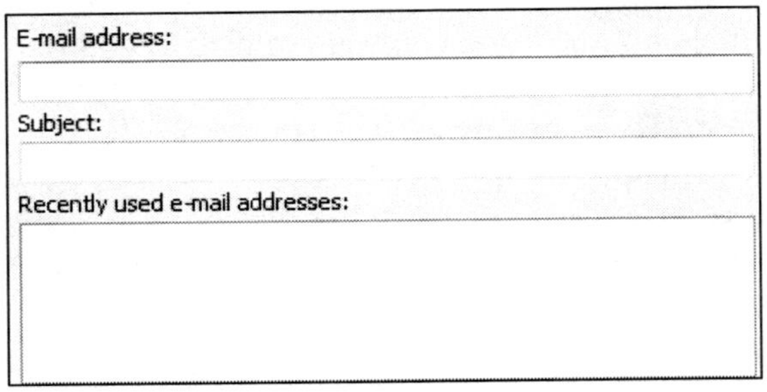

Figure 23.012

Let's create a DWF file with a hyperlink.

Do This: 23.2.4A	Creating a DWF File with Links

I. Be sure you're still in the *pid-23* file in the C:\Steps\Lesson23 folder. If not, please open it now.
II. Zoom in around the title block and save the drawing.
III. Follow these steps.

23.2.4A: CREATING A DWF FILE WITH LINKS

1. First, we must create our hyperlinks. Enter the *Hyperlink* command .
   Command: *hyperlink*

2. Select the **North Harris College** text in the title block. (Remember to hit ENTER to complete the selection set.)
   Select objects:

3. Link to the North Harris College web site by entering the web address in the **Type the file or Web page name** text box as shown. (Use the address shown: *http://northharris.lonestar.edu*.) Notice that the name appears in the **Text to Display** text box as well. You can change this to something more descriptive or simply leave it as is. (I'll call it *North Harris College Website*.)

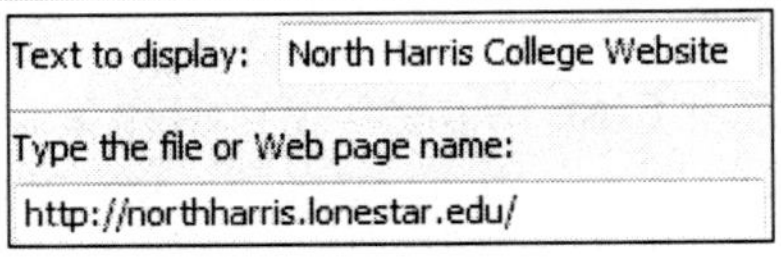

4. Pick the **OK** button to complete the command.

5. Now let's provide an email hyperlink. Repeat Step 1, but select the name in the title block. (Zoom as necessary.)

23.2.4A: CREATING A DWF FILE WITH LINKS

6. Pick **Email Address**  in the **Link to** column.

7. Enter an email address as shown. (Note: The address shown isn't real – enter one of your own.) AutoCAD will place the *mailto:* part before the address.
Place a subject in the appropriate text box.

8. Pick the **OK** button [OK] to complete the command.

9. Run your cursor over the North Harris College text in the title block. Notice (right) how it changes to indicate the presence of a hyperlink jump. The tooltip gives you the description of the link and tells you how to follow it.

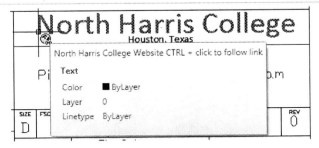

10. Now we'll publish the drawing to create the DWF file. (Hint: Save the drawing first.) Zoom all, and then enter the ***Publish*** command 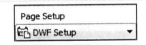.
    Command: *publish*
AutoCAD presents the Publish dialog box (Figure 23.010, p.542).

11. Use the **Remove sheets** button to remove everything in the List box except the *PID-23-Model* listing.

12. In the **Page setup** column of the list box, select **DWF Setup** from the drop down selection box.

13. Pick the **Publish Options** button 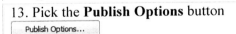.

14. On the Publish Options palette:
 • locate the output directory as indicated
 tell AutoCAD that you want to plot only a single sheet and that you want to **Include Layer information**

15. Return to the Publish dialog box [OK].

16. Pick the **Publish** button [Publish] to complete the command. (Don't save the current list of sheets if prompted.)

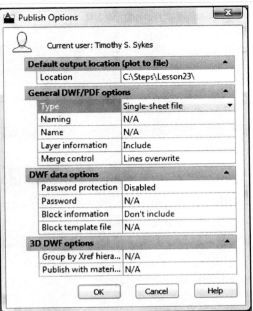

23.2.4A: CREATING A DWF FILE WITH LINKS

17. AutoCAD lets you know when it has finished the publishing job. Close the bubble.	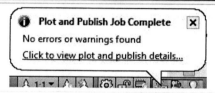
18. Open Design Review.	
19. Open 📂 the file you just created: *C:\Steps\Lesson23\pid-23-Model.dwf*.	
20. Run your cursor over the text in the title block. Notice that it changes to indicate a hyperlink. Hold down the CTRL key and pick on the text. Your browser opens the North Harris College web site!	
21. CTRL+🖱 the name in the title block. Your email software will launch with the address and subject filled in as you indicated back in Step 7!	
22. Close all the windows except AutoCAD.	
23. Save your drawing 💾 but don't exit. **Command:** *qsave*	

Wow!

23.2.5 AutoCAD can Create Full Web Pages, Too!

Your efforts so far in this section have been to create and manipulate files that a programmer can place at a web site for you. But AutoCAD has also provided a way to cut that middleman – the extremely well paid web-programmer – out of the loop and save yourself (or The Company) some money.

Where to Find It:	
Command Line:	*PublishToWeb*
Hotkey(s):	*ptw*
Menu:	File – Publish to Web

With the *Publish to Web* tool, AutoCAD enables you to create (or edit) your own web page! The tool even works in the form of a wizard so that you can do the job in no time at all!

Let's create a web page.

Do This: 23.2.5A	Creating Your Own Web Page

I. Be sure you're still in the *pid-23* file in the C:\Steps\Lesson23 folder. If not, please open it now.
II. Save the drawing. (Note: This tool will work only on a saved drawing.)
III. Follow these steps.

23.2.5A: CREATING YOUR OWN WEB PAGE

1. Begin by entering the ***PublishToWeb*** command 🖨.
   Command: *ptw*

AutoCAD begins the Publish to Web wizard with the dialog box shown following Step 2.

2. Notice (following figure) that AutoCAD allows you the option of creating a new web page or editing an existing page. The second option will come in handy when you need to make adjustments to an existing page. We'll **Create** [a] **New Web Page** for now. Be sure the bullet is next to that option and pick the **Next** button to continue.

23.2.5A: CREATING YOUR OWN WEB PAGE

![Publish to Web - Begin dialog box showing Create New Web Page selected]

3. AutoCAD presents the Create Web Page dialog box (following figure). Here you'll name your web page. AutoCAD will create a new subfolder with the name of the web page.

- To begin, enter the name of your web page as indicated (I'll call mine *PID23WebPage*). Don't place a file extension on the name; AutoCAD will do that for you. (Note: Don't use spaces in the name of the page. Although they won't bother AutoCAD, the Internet won't like them.)
- Use the **Browse** button ··· to locate the web page folder in the C:\Steps\Lesson23 folder.
- Add the description shown in the figure.
- Pick the **Next** button to continue.

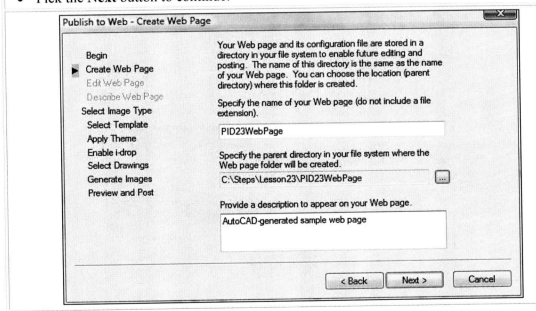

23.2.5A: CREATING YOUR OWN WEB PAGE

4. AutoCAD presents the Select Image Type dialog box (following figure). Here you'll decide what type of image to place on your web page. You've already seen the many uses of DWF files, but AutoCAD allows you to use a JPEG (.jpg) file or a PNG (.png) file as an alternative to DWF. These files are generally smaller and load faster. Their chief benefit is that they require no special programming or plug-in for proper viewing. However, they lack the quality of a DWF file.

Select the layout you wish to use for your web page; we'll use the **DWF** (or **DWFx**) image option.

Pick the **Next** button to continue.

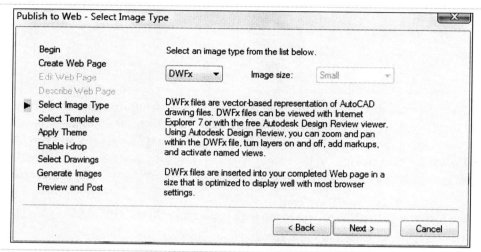

5. Now AutoCAD wants to know how to lay out your web page. Pick each of the four options to view samples of the layouts, and then select **List Plus Summary**.

Pick the **Next** button to continue.

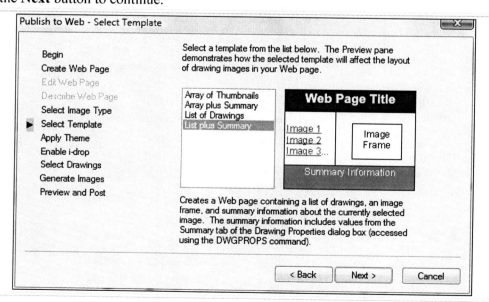

549

23.2.5A: CREATING YOUR OWN WEB PAGE

6. Select the **Theme** (color scheme and fonts) you'd like to use. I'll use **Rainy Day** as indicated.

Pick the **Next** button to continue.

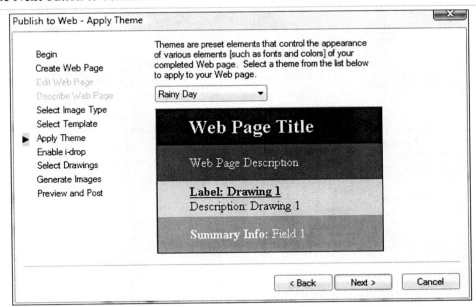

7. AutoCAD offers you an opportunity to make your drawings I-Drop available. We won't use iDrop for now; its value lies in the ability to drop blocks into the current drawing. (Be sure you read the explanation at the top of the dialog box.)

Pick the **Next** button to continue.

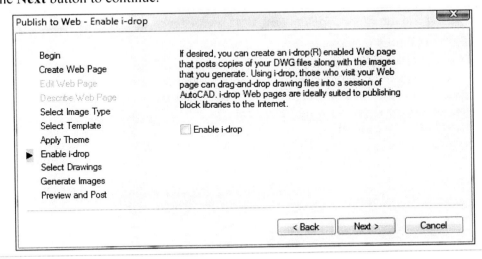

23.2.5A: CREATING YOUR OWN WEB PAGE

8. AutoCAD needs to know what to include in your web page.
   - In the **Label** text box, enter a label for the image selected.
   - The **Description** box allows you to annotate the image. Place a description there as indicated.

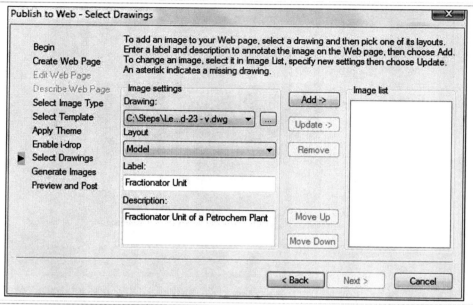

9. Once you've finished the **Image settings**, pick the **Add** button to add the image to the **Image list**.

You can add more images if you wish by picking the **Browse** button ··· next to the **Drawing** control box. I'll add the *Floor Plan* drawing from the C:\Steps\Lesson23 folder.

10. Pick the **Next** button to continue.

11. Now AutoCAD will generate the web page. First it would like to know how much regeneration to do (refer to the following figure). Remember, regeneration insures that what's on the screen is what's in the drawing's database (what you see is what's actually there). It's safest to regenerate the entire drawing before creating the web page. Put a bullet there and pick the **Next** button to continue.

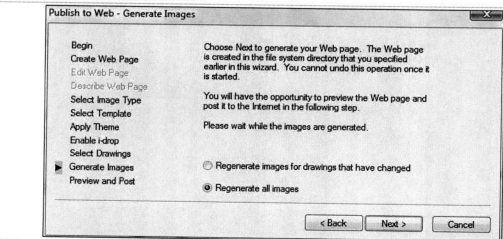

551

23.2.5A: CREATING YOUR OWN WEB PAGE

12. Now AutoCAD tells you it has finished creating the web page and offers you the chance to view it or post it to a web site (see the following figure).

Posting the page requires an ftp site (a place to put the files on the Internet or an Intranet) – usually involving permissions and passwords. If you don't post it now, you can always post it later by copying all the files created during this procedure (found in the folder you told AutoCAD to create earlier) to the web site.

For now, pick the **Preview** button to see your new web page.

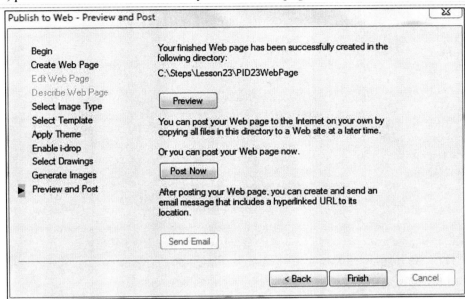

13. Take a few minutes to explore the web page. Notice that the hyperlink jumps you created in our last exercise still work here (you can create hyperlinks prior to creating a web page just as you did prior to creating the DWF file).

14. When you've finished with the web page, close your browser and pick the **Finish** button 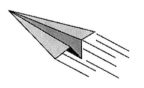 on the wizard.

15. Save your drawing 💾 but don't exit.
   Command: *qsave*

23.3 Sending the Package over the Internet with *eTransmit*

Where to Find It:	
Command Line:	ETransmit
Application Menu:	Send - eTransmit
Menu:	File – ETransmit

Times will undoubtedly occur when you need to share an actual drawing file with someone – that is, you'll need to send the drawing to someone for changes or even as a final product to a client. In the past, difficulties have occasionally arisen with incompatible or missing support files (such as fonts, linetypes, and so forth). At other times, you may need to share secure data with a client or co-worker and posting information to a web site simply won't do. With AutoCAD's **eTransmit** tool, these difficulties need no longer concern you.

ETransmit groups a drawing file with all the necessary support files (and any other files you wish to attach) into a single self-extracting executable or .zip file. You can protect this data with an encrypted password and transmit it to a friend or client via web site, email, CD, or even a good old-fashioned floppy disk!

To make things even easier, AutoCAD uses a dialog box (Figure 23.013) to help create the transmittal. Just answer a few questions!

Let's take a look.

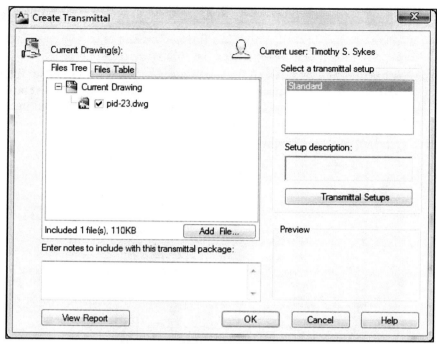

Figure 23.013

The Create Transmittal dialog box presented by the *eTransmit* command contains two tabs. (Note: There will be a third tab – Sheets – when sheet sets are open. More on sheet sets in Lesson 26, p.608.)

- The **File Tree** tab displays all the files included in the transmittal.
- The **Files Table** tab displays the same thing, but in table format.
- Use the **Add File** button at the bottom of the tab to add files to the transmittal using a standard Open File dialog box.
- Below the tabs, you'll find a text box. Use this to **Enter notes to include with this transmittal package**.
- Below the notes box, AutoCAD has placed a **View Report** button. This will present the report that'll accompany your eTransmittal. The report contains details about what AutoCAD has included in the transmittal. You can save this file or print it for your records.
- Atop the right side of the dialog box, you'll find the **Select a transmittal setup** frame. Here, you should select the format for your transmittal or pick the **Transmittal Setups** button (below the **Setup description** box) to create a new one.

To set up a transmittal, you'll follow a procedure very similar to that which you used to set up a page for plotting.

The first thing you'll see after you pick the **Transmittal Setups** button is the Transmittal Setups dialog box (Figure 23.014). Use the buttons on the right of this box to create a **New** transmittal, **Rename** an existing transmittal, **Modify** an existing transmittal, or **Delete** an existing transmittal.

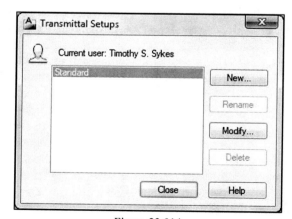

Figure 23.014

- The **New** button calls the New Transmittal Setup dialog box (Figure 23.015). Use this to name the new transmittal (use something that you can trace later). You can even base the settings for the new transmittal on those of an existing transmittal by selecting the existing one from the **Based on** list box.
- The **Continue** button on the New Transmittal Setup dialog box calls the Modify Transmittal Setup dialog box (Figure 23.016). (Note: The **Modify** button on the Transmittal Setups dialog box will take you directly to the Modify Transmittal Setup dialog box.) Let's examine this frightening behemoth!

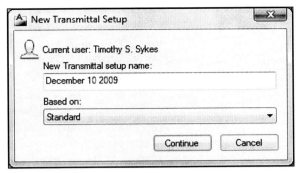

Figure 23.015

  o The **Transmittal type and location** frame (top left) provides many options:
    - Select the type of transmittal (how it will be compressed for electronic transmittal) from the **Transmittal package type** drop down list box. Your options include:
      Self-extracting executable (*.exe) files. This option compresses all files into a single file with an EXE extension. Executing this file (double clicking on it) will cause it to extract (decompress) the files necessary to open and edit your drawing.

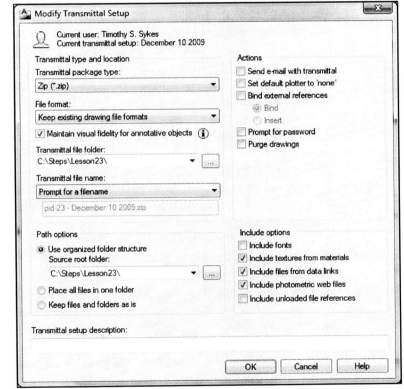

Figure 23.016

      Folder (set of files). This option copies all the necessary files for opening and editing your drawing into a single folder. You can copy this folder to a disk or web site for your client or co-worker to retrieve.
      Zip (*.zip) files. Like self-extracting executable files, zip files compress the appropriate files into a single file (this time with a ZIP extension). Most Windows operating systems can read the zip file, or you can pick up an application (*WinZip* or *Zip*) to extract the files.
    - Use the **File format** list box to change the drawing file type to be more compatible with previous releases of AutoCAD (if necessary).

554

- **Maintain visual fidelity for annotative objects** provides a necessary check box for maintaining the integrity of annotative objects when you open the drawing in an earlier release of AutoCAD.
- The **Transmittal file folder** is where AutoCAD will create the transmittal. Use the **Browse** button to select an alternate location.
- Your best option in the **Transmittal file name** box is the default (**Prompt for a file name**). The **Overwrite if necessary** option means that you'll run the risk of accidentally overwriting an existing transmittal; the **Increment file name if necessary** option means that you may end up with a lot of unnecessary files on your computer.
- The last box in the **Transmittal type and location** frame simply lists the name of the transmittal.
- o The second frame – **Path options** – requires that you make some organizational decisions.
    - **Use organized folder structure** forces AutoCAD to use the existing folder structure for the files being transmitted.
    - The **Source root folder** is the relative path for drawing-dependent files (like Xrefs or images – we'll see more on Xrefs and images in Lessons 27, p.630, and 28, p.656).
    - **Place all files in one folder** forces AutoCAD to include all transmittal files in a single folder.
    - **Keep files and folders as is** tells AutoCAD to retain the existing hierarchical structure of transmitted files. (Don't use this one if you intend to save the transmittal package to an Internet location.)
- o Use options in the **Actions** frame to help you organize actions associated with your transmittal.
    - **Send email with transmittal** means that AutoCAD will launch your email program so that you can send an email with the transmittal as an attachment.
    - It's a courtesy to **Set default plotter to none** before sending the transmittal. Your printer/plotter settings won't mean anything to your client (unless he happens to be using the same printer/plotter).
    - We'll learn about Xrefs and binding Xrefs in Lesson 27. For now, you can ignore the **Bind external references** option.
    - **Prompt for password** is yet another acknowledgement of the paranoia of most companies. Adding a password to your package insures that no one but your designated recipient can open it.
    - **Purge** your **drawing** to remove any excess material – dimension styles, text styles, etc. This ensures that your transmittal will be as small (economical) as possible.
- o Use items in the **Include options** frame to define what will go with the transmittal.
    - A check next to **Include fonts** tells AutoCAD to transmit the fonts used in the drawing with the package. This isn't necessary unless you've used some non-standard fonts in your drawing.
    - You'll learn about **materials** and **textures** in the 3D text. Ignore them for now.
    - You have the option to include **data link** files, **web files**, and **photometric web files** if you've used these in creating your drawing.
    - Again, we'll discuss **File references** in Lesson 27.
- o Use the **Transmittal setup description** box to give a brief description of this setup. AutoCAD will present this in the Create Transmittal dialog box when you select this setup.

JEEPERS; that was a lot of information!

Let's create an eTransmittal.

Do This: 23.3A	Creating an eTransmittal

I. Be sure you're still in the *pid-23* file in the C:\Steps\Lesson23 folder. If not, please open it now.

II. Follow these steps.

23.3A: CREATING AN ETRANSMITTAL

1. Begin by entering the *eTransmit* command.
   Command: *etransmit*
   AutoCAD presents the Create Transmittal dialog box (Figure 23.013, p.553).

2. We'll create a new transmittal setup; pick the appropriate button **Transmittal Setups...**.

3. AutoCAD presents the Transmittal Setups dialog box (Figure 23.014, p.553). Pick the **New** button **New...** to continue.

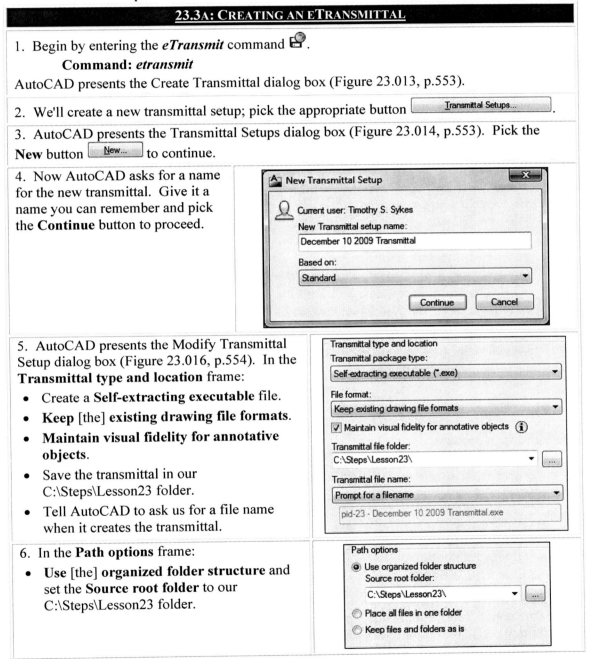

4. Now AutoCAD asks for a name for the new transmittal. Give it a name you can remember and pick the **Continue** button to proceed.

5. AutoCAD presents the Modify Transmittal Setup dialog box (Figure 23.016, p.554). In the **Transmittal type and location** frame:
   - Create a **Self-extracting executable** file.
   - **Keep** [the] **existing drawing file formats**.
   - **Maintain visual fidelity for annotative objects**.
   - Save the transmittal in our C:\Steps\Lesson23 folder.
   - Tell AutoCAD to ask us for a file name when it creates the transmittal.

6. In the **Path options** frame:
   - **Use** [the] **organized folder structure** and set the **Source root folder** to our C:\Steps\Lesson23 folder.

23.3A: CREATING AN ETRANSMITTAL

7. In the **Actions** frame:
   - Send our transmittal with an email.
   - Courteously set our **default plotter** to **none**.

   Actions
   - ☑ Send e-mail with transmittal
   - ☑ Set default plotter to 'none'
   - ☐ Bind external references
     - ◉ Bind
     - ○ Insert
   - ☐ Prompt for password
   - ☐ Purge drawings

8. In the **Include options** frame:
   - **Include** [our] **fonts** with the transmittal just in case our client needs them.
   - We won't need to include **materials** or **web files**, but let's include **data links**.

   Include options
   - ☑ Include fonts
   - ☐ Include textures from materials
   - ☑ Include files from data links
   - ☐ Include photometric web files
   - ☐ Include unloaded file references

7. Let's give our setup a description so we'll know which one it is in the future.

   Transmittal setup description:
   One Step at a Time Exercise

8. Now complete the setup... [OK]

9. ... and close [Close] the Transmittal setups dialog box.

10. In the Create Transmittal dialog box, notice that the new setup is available and that a description appears below the list box in the **Select a transmittal setup** frame.

    Select a transmittal setup
    December 10 2009 Transmittal
    Standard

    Setup description:
    One Step at a Time Exercise

    [Transmittal Setups...]

11. Select your new transmittal in the **Select a transmittal setup** frame, and pick the **OK** button [OK] to create your transmittal.

12. As instructed in Step 5, AutoCAD prompts for a name for the transmittal file. Accept the default (below) and pick the **Save** button to continue.

    File name: pid-23 - December 10 2009 Transmittal.exe   [Save]
    Files of type: Self-extracting executable file (*.exe)   [Cancel]

13. AutoCAD creates the transmittal and, again as instructed in Step 7, opens your email software with the transmittal and a text file attached. (The text file is the report we discussed at the beginning of this section.)

    All you have to do is fill in the recipient's address and send the email!

The recipient of the transmittal will receive a file that, once executed, will create a folder with all the files and data required for your client to open, read, and modify the drawing you've sent him! (If you want to check this out, the files are located in the C:\Steps\Lesson23 folder. Feel free to execute it; for this exercise, I'd extract the files to a separate directory just to see exactly what was included in it.)

| 23.4 | **Extra Steps** |

- If you haven't looked at the supplementary material for this lesson, it would be helpful for you to do so.
- Autodesk offers a unique approach to sharing project drawings and other information. It works something like a web host and operates for an addition fee. Visit their Buzzsaw site (http://usa.autodesk.com/) for more information.

| 23.5 | **What Have We Learned?** |

Items covered in this lesson include:

- *Setting up a printer/plotter*
- *Setting up a Plot Style*
- *Setting up a drawing page*
- *Creating an eTransmittal*
- *Creating a DWF or DWFx file*
- *AutoCAD plot commands:*
  - ***Pagesetup***
  - ***Plot & Print***
  - ***Browser***
  - ***Hyperlink***
  - ***PublishToWeb***
- *Creating a PDF file*
- *Adding hyperlinks to a drawing*
- *Creating a web page through AutoCAD*
- *Design Review*
- *Plot Stamps*
  - ***Publish***
  - ***eTransmit***
  - ***Export***
  - ***ExportPDF***

I strongly suggest printing at least two of the assignments for each of the lessons you'll complete throughout the remainder of this text. Then, when you've finished the course, review this lesson once again. After all, you really can't roll up the PC and stick it in your tool belt down at the job site. Your drawings must, eventually, be put on paper.

| 23.6 | **Exercises** |

*Before trying these exercises, familiarize yourself with your printer and / or plotter by reading the appropriate manuals.

1. If you've a plotter available, plot the *Floor Plan* drawing found in the C:\Steps\Lesson23 folder. Use these parameters:
   - Plot to a ¼"=1'-0" scale;
   - Plot to a C-size (23"x17") sheet of paper.
2. Using a printer, plot the *Floor Plan* drawing to fit on a 17" x 11" sheet of paper.
3. Using a printer, plot the *Floor Plan* drawing to fit on a 11" x 8.5" sheet of paper.
4. Create a hyperlink from the title of the *Floor Plan* drawing to your company's or school's web site.
5. Create a DWF file for the *Floor Plan* drawing.
6. Create a web page for the *Floor Plan* drawing.

| 23.7 | **For Web-Based Review Questions and Additional Exercises, visit: http://foragerpub.com/AcadFiles/2011/2011.htm** |

Paper Space! Let's hop right to it!

Lesson 24

Following this lesson, you will:

- ✓ Be familiar with Viewports
- ✓ Know the difference between Model Space and Paper Space (aka. Layouts)
- ✓ Know how to set up a drawing in the Paper Space environment

Space for a New Beginning

By now, you've printed (or plotted) a drawing in Model Space. You understand, then, the complexity of the mathematics involved ...

- *How large is the area to plot?*
- *How large is the paper?*
- *At what scale will I want to plot?*

... and a host of other questions that must be answered. And then you have to consider stuff like text and dimensions – plotted size and annotation scale. AGHH!

Does it have to be that difficult!?

The answer is a resounding, NO! ("Now I tell you!")

We begin this section of our text by simplifying these tasks with a remarkable tool called Paper Space. (Okay; that's an older term for "layouts". But you'll still hear it used quite often!)

When it comes to drawing display and arrangement, there exist two distinct groups of CAD operators – those who've used layouts and would never use anything else, and those who (generally for lack of training) haven't used layouts.

This lesson will familiarize you (painlessly) to the wonders of layouts and Viewports so that you may join the ranks of enlightened operators! To keep it simple, we'll remain in the two-dimensional world throughout this lesson. However, you'll continue to discover the benefits of these tools even as you explore the third dimension in our next text – 3D AutoCAD 2011: One Step at a Time.

Let's begin by answering the basic questions.

24.1 Understanding the Terminology

What is a viewport?

My *New American Dictionary* defines a port as a window in the side of a ship. Similarly, a viewport is a window into your drawing.

If you imagine viewing an object that's resting in the center of a box through holes in the sides, top, and bottom of the box, you'll get a fairly good idea of what viewports are. Essentially, viewports are openings into your drawing, each presenting a different view of the drawn object(s).

AutoCAD provides two types of viewport from which to choose – simple *tiled* viewports and more complex *floating* (or untiled) viewports. We'll look at each of these.

What is a layout (Paper Space)?

Paper Space is a plotting tool. (You'll like this one!) It provides a method for creating a finished drawing that uses more than one scale and/or view of a drawing. In other words, this is how you can create a drawing with separate details shown at larger scales for ease of viewing (and dimensioning).

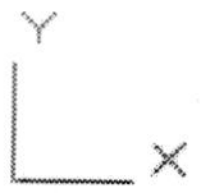

You'll create the drawing as you always have – in *Model Space* (the *space* where you create your drawing or three-dimensional *model*). But before you plot, you'll place the (layout) drawing and details in their own spaces on an imaginary sheet of *paper*.

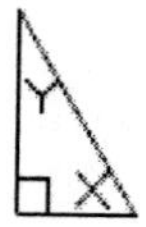

Figure 24.001

How do I know which space I'm using?

There are several ways to know.

First, look at the UCS icon. The standard XY icon (Figure 24.001) indicates Model Space; the triangular icon (Figure 24.002) indicates Paper Space.

Figure 24.002

If you have the UCS icon turned off, you can look at the **Model/Layout** toggles on the status bar or the tabs at the bottom/left end of the graphics. If **MODEL** appears lit, you're in Model Space. Conversely, if **Layout** appears lit, you're in Paper Space.

> If the Model/Layout toggles don't appear, right click on the status bar and put a check next to **Layout/Model** on the cursor menu.

How do I switch between Model Space and Paper Space?

You can pick the toggle on the status bar. Alternately, if you're in a layout, you can double click in the desired space or enter *MSpace* (*MS*) for Model Space or *PSpace* (*PS*) for Paper Space. But for a more visual approach, use the Quick View Layouts tool.

Quick View Layouts works just like Quick View Drawings (Lesson 1, Section 1.5). When toggled on (use the button on the status bar), AutoCAD presents thumbnails (Figure 24.003) of Model Space and the layouts that currently exist in the drawing. Pick on the thumbnail of what you wish to view and AutoCAD will bring that layout/model to the graphics area. (See Lesson 1, p.26, for more details and an exercise on how Quick View works.)

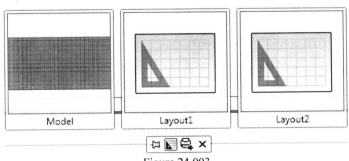

Figure 24.003

All of this will become clearer as we proceed.

24.2 Using Tiled Viewports

Before we jump into paperspace, we need to look at tiled viewports. Use tiled viewports in Model Space (what you might at this point consider "normal" drawing space) to enhance your ability to see several parts of the drawing at once. This ability will become particularly important when drawing in three dimensions. You can place a three-dimensional view of the object in one viewport while drawing on a single, two-dimensional plane (one side of the object) in another.

Create tiled viewports using the Viewports dialog box (Figure 24.004, p.563). Access the dialog box using the *VPorts* command.

	Where to Find It:
Command Line:	*Viewports*
Hotkey(s):	*vports*
Ribbon (Tab/Panel):	View – Viewports – **Named** [or **New**] **Viewports**
Menu:	View – Viewports – **Named** [or **New**] **Viewports**
Toolbar:	Viewports – **Vports**

Let's look at our options.

The names of the two tabs indicate their function; use **New Viewports** to create new viewports and **Named Viewports** to activate or set current a saved viewport configuration.

We'll begin with the **New Viewports** tab.

- Place a name or title in the **New name** text box at the top if you wish to save a current configuration. (You won't need to use a name if you don't want to save the configuration.) You'll be able to recall your configuration later using the **Change view to** control box (lower-right) or the **Named Viewports** tab.

- The **Standard viewports** list box provides a list of the more common viewport setups. Select each and see the configuration in the **Preview** frame.

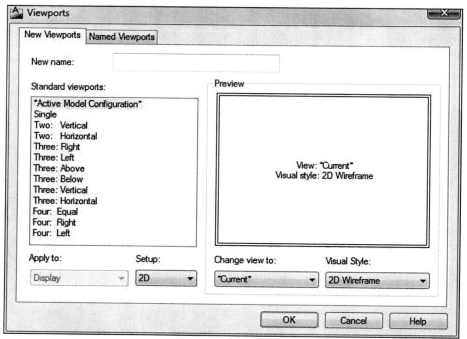

Figure 24.004

- Once you've selected a viewport configuration, you can choose to apply your selection to the **Display** or the **Current Viewport** using the **Apply to** control box. Using the **Display** option will replace the drawing's current configuration with the new selection. Using the **Current Viewport** option will place the new configuration inside the currently active viewport. This is how you customize your viewports.
- The **Setup** control box offers two choices: **2D** or **3D**.
    o The **2D** option will place the drawing's current view in each of the viewports.
    o The **3D** option will create the viewports using standard 3D views (top, front, side, isometric). Once selected, you can adjust the view using display commands (***Zoom***, ***Pan***, and ***View***).
- Leave the **Visual Styles** control box set to **2D Wireframe** for your 2D work. We'll discuss the other options in the 3D text.
- The **Named Viewports** tab (Figure 24.005) presents the **Named viewports** list box and a **Preview** frame.

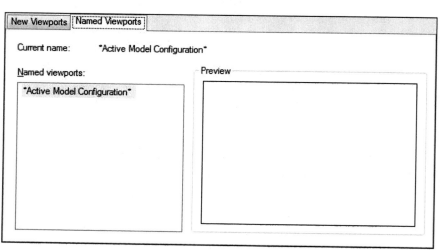

Figure 24.005

563

- The **Named viewports** list box offers the names of any viewport configurations that have been saved. You can set a viewport configuration current by selecting its name and picking the **OK** button.
- The **Preview** frame allows you to see the setup of the selected viewport configuration.

Let's experiment a bit.

Do This: 24.2A	Working with Tiled Viewports

I. Open the *flr-pln24a* file in the C:\Steps\Lesson24 folder. The drawing looks like the figure at right.
II. Turn off the view cube (*Cube*) and the navigation bar (*NavBar*) for clarity.
III. Follow these steps.

24.2A: WORKING WITH TILED VIEWPORTS

1. We'll begin by creating some viewports. Enter the *Viewports* command.

    Command: *vports*

2. Select the **Three: Right** configuration. Notice the **Preview** window (following figure).

3. Pick the **OK** button to continue. Your drawing looks like the following figure.

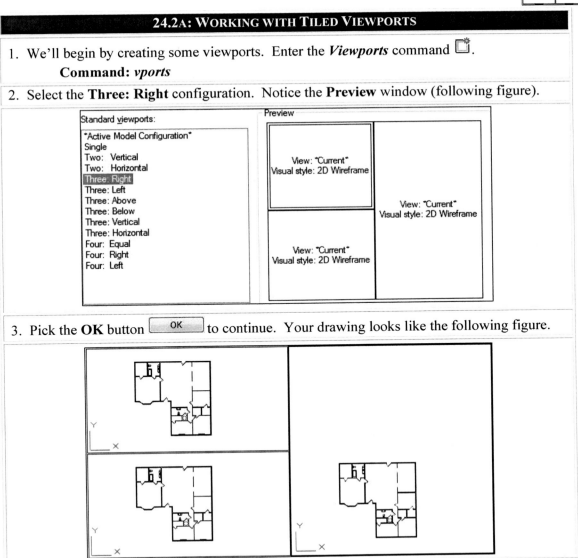

24.2A: WORKING WITH TILED VIEWPORTS

4. Place your cursor into each of the viewports. Notice that only the right (active) viewport presents crosshairs. Other (inactive) viewports present a cursor arrow. Activate a viewport by placing the cursor in the desired viewport and picking once with the left mouse button. Notice that you can activate only one viewport at a time.

Notice also that each viewport contains the UCS icon (Figure 24.001, p.561). You can manipulate each viewport separately as though it were your entire drawing screen.

5. Be sure the right viewport is active. (Place the cursor in the right viewport. If it doesn't show crosshairs, pick once with the left mouse button.)

6. *Zoom* extents.
   Command: *z*

Notice that the *Zoom* command affects only the active viewport. If necessary, *Pan* to center the floor plan in the viewport.

7. Pick anywhere in the upper left viewport to activate it.

8. Zoom in around the master bath.
   Command: *z*

The viewport will look like the figure at right.

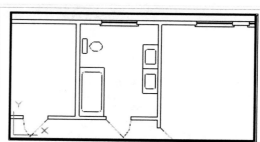

9. Now activate the lower left viewport.

10. Zoom in around the common bath.
    Command: *z*

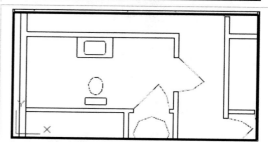

11. Let's place a tub in the common bath. Enter the *Copy* command.
    Command: *co*

12. Pick once in the upper left viewport to activate it; then select the bathtub. Select the lower left endpoint of the tub as your base point.
    Select objects:
    Specify base point or [Displacement/mOde] <Displacement>: _endp of

13. Pick once in the lower left viewport to activate it, and then place the tub in the lower left corner of the bathroom.
    Specify second point or <use first point as displacement>: _endp of

Your screen now looks like the following figure. Notice that the changes are reflected in the right viewport as well as the lower left.

24.2A: WORKING WITH TILED VIEWPORTS

14. Remember to save the drawing 💾 occasionally.
    Command: *qsave*

15. Now let's save our viewport configuration. Open the Viewports dialog box.
    Command: *vports*

16. Enter the name *Baths* in the **New name** text box.

    | New name: | Baths |

17. Complete the procedure [OK].

18. Now let's reset the screen to a single viewport. Reopen the Viewports dialog box.
    Command: *vports*

19. Select the **Single** configuration and then pick the **OK** button [OK] to continue. AutoCAD presents a single viewport displaying the view of the current (lower left) viewport.

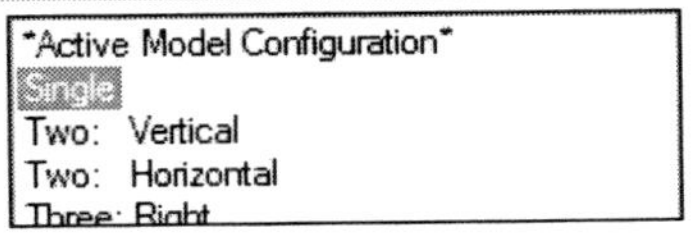

    ```
    *Active Model Configuration*
    Single
    Two:   Vertical
    Two:   Horizontal
    Three: Right
    ```

20. *Zoom* extents.
    Command: *z*

21. Now restore the **Baths** configuration. Open the Viewports dialog box. (This time use the **Named** button.)
    Command: *vports*

22. Select the **Baths** configuration in the **Named viewports** list box (following figure).

24.2A: WORKING WITH TILED VIEWPORTS

[Dialog box showing Named Viewports tab with Current name: Baths, Named viewports list containing "*Active Model Configuration*" and "Baths", and a Preview pane showing three rectangular viewports.]

23. Pick the **OK** button to complete the command [OK]. AutoCAD presents the **Baths** configuration (see the Step 13 figure, p.566).

24. Save the drawing 💾 but don't exit.
    Command: *qsave*

You can see the benefits of using viewports when manipulating several smaller parts of your drawing at one time. But when we consider the limitations of tiled viewports, we can see that they were designed as drawing aids, not plotting tools.

Some things to remember about tiled viewports include:

- Tiled viewports were designed for use in Model Space.
- Tiled viewports are not objects and won't plot. Your plot will show only the view in the *active* viewport.
- Only one viewport is active at a time. This means that you can draw in only one viewport at a time, but you can activate a new viewport transparently (while using another command).
- The active viewport will show crosshairs while inactive viewports will show a cursor. Place the cursor in the viewport you wish to activate and click once 🖱 to activate it.
- Layers behave universally in tiled viewports. You can't freeze or thaw a layer just for one viewport as you can with floating viewports. (You'll see how to use layers in floating viewports in Lesson 25, p.588.)
- **Annotation Scale** also behaves universally in tiled viewports. That is, all tiled viewports reflect the drawing's **Annotation Scale**. (This doesn't hold true for Floating Viewports as you'll soon see.)
- While you can control the number and position of tiled viewports, you can't easily control the size. You can't control the shape at all (all tiled viewports are rectangular).
- You can create viewports within viewports, but you should consider the size of the viewports and the size of your monitor before attempting this.
- Manipulation of viewports doesn't affect the actual drawing.

To see how you can use viewports to assist in plotting a drawing that uses multiple scales, we'll take a look at floating viewports in Section 24.4, p.570. But first, we must set up our layout.

24.3 Setting Up a Layout Environment (Paper Space)

Entering Paper Space is as easy as picking the **Layout** toggle on the status bar. When you open a **Layout** for the first time, AutoCAD will create a single floating viewport in the center of the page. Like Model Space, Paper Space must be set up. But it's quite a bit easier with Paper Space. Let's set up a Paper Space layout for our floor plan.

> You're setup may display Model/Layout tabs just above the command line rather than Model/Layout toggles on the status bar. Use these in the same way you'd use the toggles. (Right click on the tab and select **Hide Layout and Model tabs** to display the toggles rather than the tabs; right click on the toggle and select **Display Layout and Model tabs** to display tabs rather than toggles.)

> You'll occasionally see the term *Paper Space* written as *paperspace*. Both refer to the same thing, though *paperspace* is generally a programmer's term.

> Note: This procedure will utilize my printer. You may have to adjust the settings to match your printer or plotter setup.

Do This: 24.3A — **Creating a Paper Space Layout**

I. Be sure you're still in the *flr-pln24a* file in the C:\Steps\Lesson24 folder. If not, please open it now.
II. Set **VPORTS** as the current layer.
III. Reset the drawing to a single viewport and zoom extents.
IV. Follow these steps.

24.3A: CREATING A PAPER SPACE LAYOUT

1. Pick the **Layout1** toggle on the status bar or the **Layout** tab beneath the graphics area. (If the toggle isn't there, right click on the status bar and place a check next to **Layout/Model** on the menu.) AutoCAD opens the layout and creates a floating viewport (see the following figure).

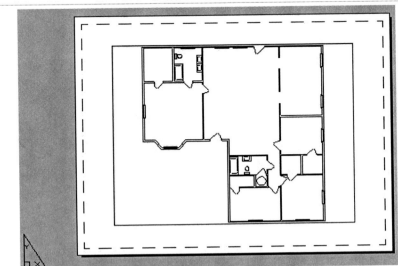

24.3A: CREATING A PAPER SPACE LAYOUT

2. Let me show you a trick to make organization of multiple layouts a bit easier. Open Quick View Layouts .

3. Right click on the Layout 1 thumbnail and select **Rename** from the menu.

4. Call the layout *Tara II Layout* (as indicated), and then hit ENTER to complete the procedure. Notice the new name on the thumbnail.

5. Save the drawing .
   Command: *qsave*

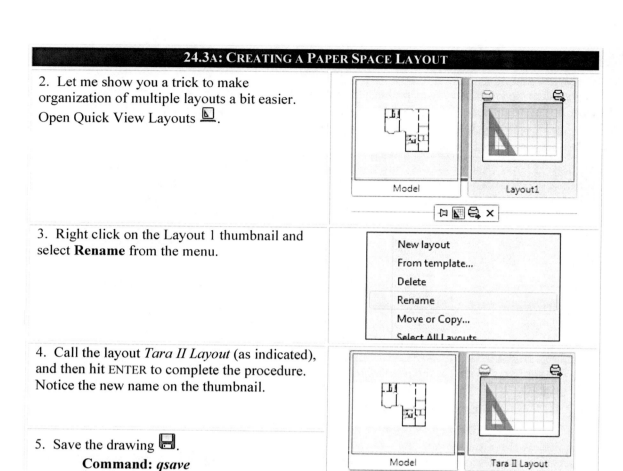

6. Try erasing different parts of the drawing. Notice that only the frame around the floor plan (the floating viewport) is selectable.

   Erase the floating viewport.
   Command: *e*

   Notice that the entire floor plan disappears. There is no longer a hole (a viewport) in the paper through which to see the model (your drawing).

7. Save the drawing but don't exit.
   Command: *qsave*

Name each new layout to better organize your work. Keep the names simple but descriptive.

You've seen how to start a Paper Space layout – just go to the **Layout** toggle and let AutoCAD do the rest. AutoCAD presents a sheet of paper with a floating viewport (a hole in the page) that's the only thing that exists on the page. (The dashed outline shows the margins of the working area on the page.)

You can place objects on the paper that won't affect the model (title block, text, and so forth). Alternately, you can return to Model Space to work on the model. You can even work on the model through the viewport! We'll look at this in more detail in later sections of this lesson; but first, let's examine floating viewports.

You can have more than one layout in your drawing – each set up differently for different printing needs. Use the *Layout* command to add, remove, or copy layout tabs. It presents the following options:

Command: *layout*

Enter layout option [Copy/Delete/New/Template/Rename/SAveas/Set/?] <set>:

Where to Find It:	
Command Line:	*Layout*
Hotkey(s):	*lo*
Menu:	Insert – Layout – **New Layout**
Toolbar:	Layouts – **New Layout**

Most of these options are self-explanatory. However, the **Saveas** option might not do what you expect. It saves the layout as a drawing template (.dwt).

The command line isn't difficult, but you'll probably find it easier simply to right click on a tab and select the desired option from the cursor menu.

24.4 Using Floating Viewports

Floating, or untiled, viewports, although more complex than tiled viewports, offer considerably more flexibility. But how are floating viewports different from tiled viewports? Let's take a look.

As we saw in Section 24.2 (p.562), AutoCAD designed tiled viewports as *drawing* tools for a Model Space environment. Tiled viewports act like holes in an imaginary box through which you view

different parts of your drawing. The location and size of the hole determine what you see, how much you *see*, and at what angle you see the drawing objects. Tiled viewports don't affect how the drawing will be plotted.

Floating viewports were designed as *plotting* tools in a Paper Space environment. Floating viewports also act like holes, but this the location and size of the holes determine what will be *plotted*, how much will be plotted, and at what angle the drawing objects will be plotted.

Some things to remember about floating viewports include:

- Floating viewports are objects and can be moved, stretched, and erased like any other object on a drawing. Consequently, you can place them on a separate layer that can be frozen at plot time.
- Unlike tiled viewports, floating viewports don't have to be rectangular.
- Floating viewports can overlap.
- You can control layers within each viewport independently of the rest of the drawing. This is a key benefit since you often won't want details shown in one viewport reflected in another.
- You can assign different scales or annotation scales to individual floating viewports. This is another key benefit to these handy tools!
- Use floating viewports to show different views of one or more objects on a single plot.
- Like tiled viewports, only one floating viewport is active at a time.
- Floating viewports live in Paper Space. Although they're not available in Model Space, you can work in Model Space *through* a floating viewport (much as a doctor works on a kidney or gall bladder through a hole in the skin).
- Sizing text on Paper Space is really quite easy. Since you're working on the actual plotted page, simply enter the text at the desired plotted size. No need for scale factors!

24.4.1 Creating Floating Viewports Using *MView*

Creating floating viewports using the *MView* command appears more daunting than it actually is. The command sequence looks like this:

> Command: *mview*
> Specify corner of viewport or [ON/OFF/Fit/Shadeplot/Lock/Object/Polygonal/Restore/LAyer/2/3/4] <Fit>: *[Begin a window.]*
> Specify opposite corner: *[Complete the window.]*

Where to Find It:	
Command Line:	*MView* [or *-vports*]
Hotkey(s):	*mv*
Menu:	View – Viewports – [X] Viewport(s)

Look at each option.

- The **ON/OFF** options allow you to "open" or "close" the viewport through which you see the drawing. If the viewport is **OFF**, it's closed and you can't see through it. Thus, that part of the drawing is hidden.

When you select one of these options, AutoCAD will prompt you to select the viewport to turn **ON** or **OFF**.

- The **Fit** option will create a single, maximum-size floating viewport within the display area.
- **Shadeplot** controls exactly what plots within a viewport. It prompts

> Shade plot? [As Displayed/Wireframe/Hidden/Visual styles/Rendered] <As Displayed>:

AutoCAD presents several options (mostly stuff you'll see in the 3D text). For our 2D purposes, leave the **As Displayed** default. If you change it accidentally, re-enter the **2D Wireframe** setting.

- The **Lock** option allows you to lock a viewport. It prompts

> Viewport View Locking [ON/OFF]: *[Enter ON or OFF.]*
> Select objects: *[Select the viewport to lock or unlock.]*

When a viewport is locked, you can't (intentionally or accidentally) change the scale factor within it. This is a handy tool if you intend to work on the Model Space within a viewport.

- The **Object** option allows you to select any existing closed polyline, ellipse, spline, region, or circle that exists in Paper Space and turn it into a floating viewport.
- The **Polygonal** option allows you to create a floating viewport with any number of sizes and in any shape. It prompts

> Specify start point: *[Pick a start point.]*
> Specify next point or [Arc/Length/Undo]: *[Create the shape just as you would a polyline.]*

- The **2/3/4** options tell AutoCAD to create two, three, or four viewports. AutoCAD will prompt you to locate the viewports.
- Each viewport can have its own layer settings. **LAyer** resets layers in a selected viewport to their global settings.
- **Restore** is a very useful tool. It allows you to translate tiled viewports into floating viewport objects. When selected, AutoCAD allows you to enter the name of a saved tiled configuration or to translate the current tiled configuration into floating viewports.
- Of course the default option – **Specify corner of viewport** – allows you to manually create viewports one at a time.

Let's insert a title block into our drawing and then set up some floating viewports.

Do This: 24.4.1A	**Creating Floating Viewports**

I. Be sure you're still in the *flr-pln24a* file in the C:\Steps\Lesson24 folder. If not, please open it now.
II. Set the **BORDER** layer current.
III. Follow these steps.

24.4.1A: CREATING FLOATING VIEWPORTS

1. Insert the *title block* drawing in the C:\Steps\Lesson24 folder. Put it at the 0,0 coordinates.
   Command: *i*

2. Set the **VPORTS** layer current.

3. Enter the *MView* command.
   Command: *mv*

4. Place the first floating viewport as indicated.
   Specify corner of viewport or [ON/OFF/Fit/Shadeplot/Lock/Object/Polygonal/ Restore/LAyer/2/3/4] <Fit>: *.5,7.5*
   Specify opposite corner: *4.75,4.25*

5. Place a second floating viewport as indicated.
   Command: *[ENTER]*
   Specify corner of viewport or [ON/OFF/Fit/Shadeplot/Lock/ Object/Polygonal/Restore/ LAyer/2/3/4] <Fit>: *5,6.5*
   Specify opposite corner: *10,2.5*

6. Now let's make a nonrectangular viewport – the easy way. Draw a circle as indicated.
   Command: *c*
   Specify center point for circle or [3P/2P/Ttr (tan tan radius)]: *2.25,2.25*
   Specify radius of circle or [Diameter]: *1.75*

7. Now we'll convert the circle to a floating viewport. Enter the *MView* command.
   Command: *mv*

8. Use the **Object** option.
   Specify corner of viewport or [ON/OFF/Fit/Shadeplot/Lock/Object/Polygonal/ Restore/LAyer/2/3/4] <Fit>: *o*

24.4.1A: CREATING FLOATING VIEWPORTS

9. Select the circle.

    Select object to clip viewport:

    Your drawing now looks like the figure at right. Notice that the round viewport shows only part of the drawing. It began with the full drawing and then clipped away the part that didn't fit into the circle.

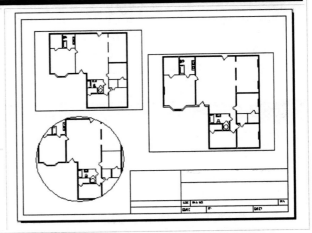

10. Save the drawing 💾 but don't exit.

    Command: *qsave*

24.4.2 The Viewports Panel

*The one common thread that permeates the thrills of life – your first kiss, your first child, your first paycheck ...– is that each results from the **discovery** of something new and wonderful.*

Never miss an opportunity to explore or you may miss an opportunity to discover.

Anonymous

We've made a good start with Paper Space and floating viewports. But have you really look at the viewports panel yet (Figure 24.006)? (It's on the ribbon's **View** tab.)

This handy item provides single-button selection of some of the *MView* (*-VPorts*) options as well as some other tools. Let's take a look.

Figure 24.006

- **Named** 📇 and **New** 🗔 display the Viewports dialog box with the corresponding tab on top. You can also use this dialog box to create *floating* viewports in Paper Space! Simply follow the same procedures you learned for tiled viewports.

- **Create Polygonal** 🗔 allows you to create a viewport with any number of sizes and in any shape by executing the **Polygon** option of the *MView* command.

- **Create from Object** 🗔 (below the **Create Polygonal** flyout) executes the **Object** option of the *MView* command as we did in our last exercise.

- The next button – **Clip** 🗔 – is a modifying tool used to redefine viewport boundaries. We'll learn more about it in Lesson 25, p.596.

- Finally, the **Join** button 🗔 combines selected viewports into one. (The **Join** option only works with tiled viewports.

- Use the **Viewport Configurations List** menu 🗔 to choose between available viewport configurations.

24.4.3 Adjusting the Views in Floating Viewports

Once you've created floating viewports, you'll need to adjust what you see through them. This means adjusting the scale and panning to achieve the appropriate view for your plot. Before we do either, we must tell AutoCAD to allow us to work in the Model Space we see through the floating viewport. Do this by simply entering *MSpace* or *MS* at the command prompt and then selecting the viewport you wish to use (see the following insert).

> To toggle between Paper Space and Model Space, simply type *PSpace* (or *ps*) or *MSpace* (or *ms*) at the command prompt. Alternately, you can pick on **Paper** |PAPER| or **Model** |MODEL| on the status bar or use the Quick View Layouts method. (To display this toggle, right click on the status bar and select **Paper/Model** on the menu.) *You can also double click in a Paper Space area to activate Paper Space, or in a viewport to activate Model Space (simultaneously making that viewport current).*

We'll begin by setting the scale for the viewport. AutoCAD provides three methods of doing this – the **XP** option of the *Zoom* command, the **Viewport Scale** control box on the Viewports toolbar, or the **VP Scale** control which appeared beside the **Annotation Scale** control on the status bar when you entered Paper Space.

- The Zoom Approach: While at the *Zoom* prompt, simply type *1/[SF]xp* (where *SF* is the scale factor for the scale you wish to use in this particular viewport – refer to Appendix B).
- The Control Box Approach: Pick the down arrow (**Viewport Scale** control box on the Viewports toolbar) and select the appropriate scale. Alternately, you can type a scale into the box.
- Use the same approach to set **VP Scale** on the status bar as you used to set the **Annotation Scale**.

Let's try it.

Do This: 24.4.3A	**Scaling Floating Viewports**

I. Be sure you're still in the *flr-pln24a* file in the C:\Steps\Lesson24 folder. If not, please open it now.

II. Follow these steps.

24.4.3A: SCALING FLOATING VIEWPORTS

1. Double click in the upper left viewport to open it in Model Space.
Notice that the boundary darkens to indicate that it's active.

2. *Pan* to center the master bath in the viewport.
   Command: *p*

3. We'll set the scale for this viewport using the *Zoom* approach. Enter the command.
   Command: *z*

4. We wish to set the scale of this viewport to ¼"=1'-0". The scale factor for this scale is 48 (see Appendix B). So enter the scale factor, as shown, using the *xp* suffix to indicate that you are scaling Paper Space.

   Specify corner of window, enter a scale factor (nX or nXP), or [All/Center/Dynamic/Extents/ Previous/Scale/Window/Object] <real time>: *1/48xp*

   AutoCAD scales the viewport. (You may have to pan slightly to center the view again.) This viewport now looks like the figure at right.

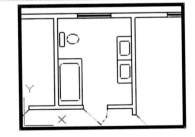

5. Now let's use the control box method. Pick anywhere in the right viewport to activate it.

24.4.3A: SCALING FLOATING VIEWPORTS

6. Pick the down arrow in the **Viewport Scale** control box on the Viewport toolbar, and then scroll down until you can see the 1/16"=1' selection. Pick that one.

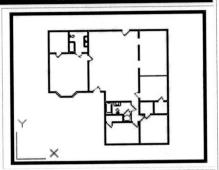

Easy? This viewport now looks like the figure at right.

7. Activate the round viewport and center the bay window in the view. Use the *Zoom – Center* option.
   Command: *z*

8. Use the **Viewport Scale** control on the status bar to set the scale in this viewport to ¼"=1'-0". Note that AutoCAD automatically sets the Annotation Scale when you use this method.

9. Return to Paper Space.
   Command: *ps*

10. On the **TEXT** layer, add the geometry shown in the following figure.

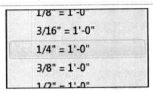

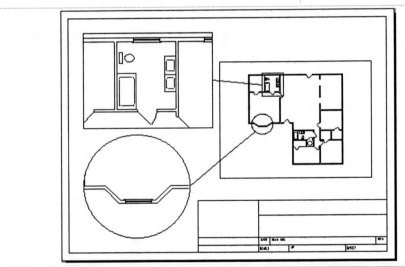

11. Save the drawing as *MyFlrPln25a* in the C:\Steps\Lesson25 folder.
    Command: *saveas*
    Close the drawing.

The drawing now shows the full plan at a 1/16"=1'-0" scale and the common bathroom and bay window at a ¼"=1'-0" scale.

We've completed the Paper Space setup of our drawing. Let's review what we've accomplished.

- We began with a floor plan drawn in Model Space. The limits of Model Space had been set to produce a ¼"=1'-0" drawing on a C-size sheet of paper.

- We opened the layout tab of our drawing and set it up to print our Model Space drawing on an A-size sheet (AutoCAD's default). We used an AutoCAD title block designed for this page size and added three floating viewports – two rectangular and one converted object (circle).
- We set different views of the same model in each of the floating viewports – each viewport having its own scale.

24.4.4	Adjusting the Position of the Floating Viewport

Adjusting the position of a floating viewport doesn't require any special tools – you can use the move or rotate commands without a problem. But you should be aware of a new system variable when rotating a viewport. **VPRotateAssoc** comes with two settings:

- 0 means that the view within the viewport will *not* rotate with the viewport, and
- 1 (the default) means that the view within the viewport will rotate with the viewport.

24.5	And Now the Easy Way – The *LayoutWizard* Command

We've explored several steps in creating and setting up a Paper Space layout. But as you know, I always save the best for last.

I briefly discussed the *Layout* command in Section 24.3, p.568. That command allowed you to create new layouts. But AutoCAD's Layout Wizard does the same thing and more!

Where to Find It:	
Command Line:	*LayoutWizard*
Menu:	Insert – Layout – **Create Layout Wizard**

Using the *LayoutWizard* command, you can create a new layout and set it up as well.

Let's take a look at the wizard.

Do This: 24.5A	Using the Layout Wizard

I. Open the *flr-pln24b* file in the C:\Steps\Lesson24 folder. This is the same file as *flr-pln22a*, but hasn't had the layout set up yet.
II. Set the **VPORTS** layer current.
III. Follow these steps.

24.5A: USING THE LAYOUT WIZARD

1. Enter the *LayoutWizard* command.
   Command: *layoutwizard*

Just do what the wizard tells you to do!

24.5A: USING THE LAYOUT WIZARD

2. AutoCAD presents the Create Layout – Begin dialog box (following figure). Enter the name *Tara II Layout* in the text box as shown.

Pick the **Next** button to continue.

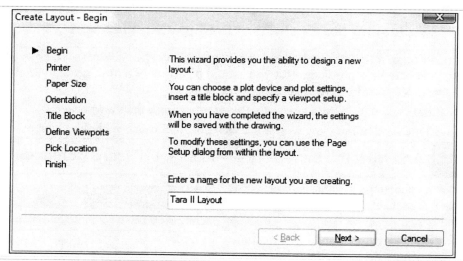

3. Select the printer or plotter you wish to use. (Your options may differ from mine.) Pick the **Next** button to continue.

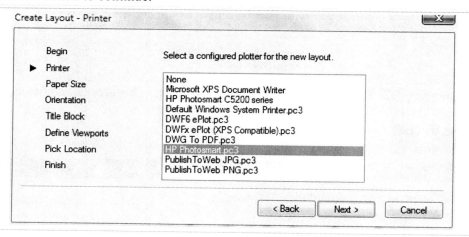

577

24.5A: USING THE LAYOUT WIZARD

4. Select a paper size for the layout (I'll use the **8½" x 11"** size). Remember to tell AutoCAD to set up the layout in **Inches** or **Millimeters**. Here we'll use **Inches** (following figure).
Pick the **Next** button to continue.

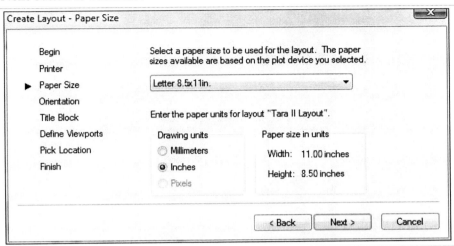

5. You'll probably want to leave the layout set to **Landscape** as we will here. Use **Portrait** when you wish to print to an upright sheet of paper (the longer dimension vertical rather than horizontal).
Pick the **Next** button to continue.

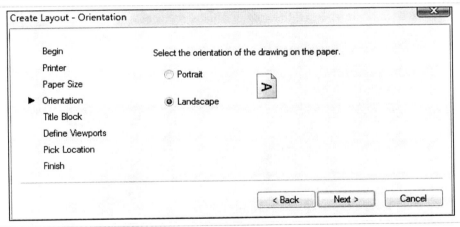

Whew! This can wear an old wizard out!

24.5A: USING THE LAYOUT WIZARD

6. Select the title block you wish to use or choose **None** if you want to insert one later. I'll insert one later.

Pick the **Next** button to continue.

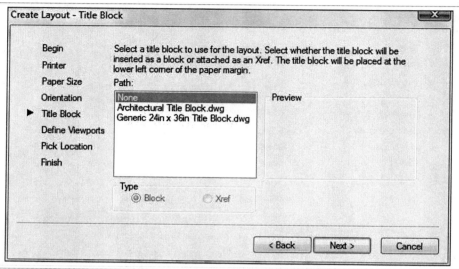

7. Now we can create a rough setup for our viewports.

In the **Viewport setup** frame (see the following figure), select the **Single** option to create a single viewport. You can use the other options to create the standard three-dimensional engineering views (four viewports – three with orthographic views and one three-dimensional view) or an **Array** of viewports. The **Array** option displays the **Rows/Columns** and **Spacing** boxes below the **Viewport setup** frame.

In the **Viewport scale** control box, set the scale for our single viewport as 1/16"=1'-0".

Pick the **Next** button to continue.

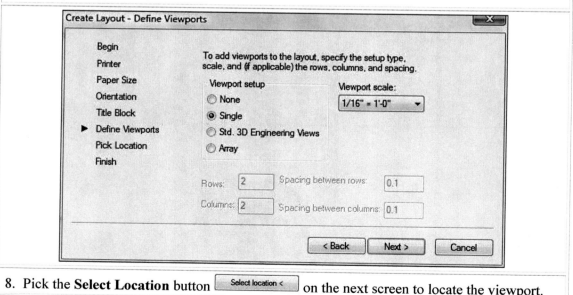

8. Pick the **Select Location** button on the next screen to locate the viewport.

24.5A: USING THE LAYOUT WIZARD

9. AutoCAD takes you to the graphics screen of your new layout, shows the title block you selected (if any), and prompts for the corners of the viewport. Enter the coordinates indicated.
   Specify first corner: *4.75,6.5*
   Specify opposite corner: *10,2.25*

10. AutoCAD returns to the wizard. Pick the **Finish** button [Finish] to complete the command.
    AutoCAD returns to the graphics screen with the new layout visible.

11. Save the drawing 💾.
    Command: *qsave*

Wasn't that easier than our previous exercises? You can use the *MView* command to add additional viewports as desired, but the bulk of the work has been accomplished in fewer steps and with less effort.

How will you set up your layouts?

We've accomplished quite a lot in such a short lesson. But we haven't finished learning the basics of Paper Space. We must consider how dimensions, layers, text, and plotting relate to Paper Space, as well as some methods of editing our layout. We'll look at these in Lesson 25, but for now, let's take time for some projects.

> We've used the *MView* command to set up our floating viewports. We've also seen an easier approach to creating layouts – using the *LayoutWizard* command. But wait! There is another approach.
>
> AutoCAD provides a method that incorporates all aspects of Paper Space setup into a single command – the *MVSetup* command. You may experiment with this command if you like, but be warned that consolidation doesn't always make a task easier! We'll explore *MVSetup* as an editing tool in Lesson 25.

24.6 Extra Steps

If you have access to a printer or plotter, try to plot the results of the last exercise. Don't worry if you have problems; we'll cover plotting in Paper Space in Lesson 25. This is just a preview.

Some important things to remember when plotting include:

- Be sure to plot the **Tara II Layout** tab.
- Plot to a 1=1 scale.
- Use an A-size sheet of paper (ANSI A is 11" x 8½").

Examine the scale. Designers use Paper Space to create drawings at industry standard scales while showing details at whatever scale is necessary for clarity.

24.7 What Have We Learned?

Items covered in this lesson include:

- *Tiled viewports (Model Space)*
- *Floating viewports (Paper Space)*
- *The Layout Wizard (**LayoutWizard**)*
- *Commands:*
    - *PSpace*
    - *MSpace*
    - *Viewports*
    - *MView*
    - *VPRotateAssoc*
    - *Layout*

We've seen the basic setup of Paper Space in two-dimensional drawings.

While the setup may seem somewhat confusing at first, the benefits of practice can't be overemphasized. In the two-dimensional world, Paper Space provides convenience when using multiple drawing scales and details. In the three-dimensional world, you'll find Paper Space an essential tool for printing the same object from different angles simultaneously!

Try some of the exercises at the end of this lesson to make yourself more comfortable with Paper Space. Then move on to the next lesson where you'll discover more about how to work in Paper Space.

24.8 Exercises

1. Open the *needle* file in the C:\Steps\Lesson24 folder. Create the drawing configuration for plotting found in the *needle* drawing. Some helpful information includes:

    1.1. The page size is A4 (metric – 210mm x 297mm).

    1.2. The title block is the *ISO-A4* file found in the C:\Steps\Lesson24 folder. (You may need to adjust its position on the sheet.)

    1.3. Watch your layers (place viewports on the **VPorts** layer).

    1.4. The radius of the large circle is 60mm; the radius of the smaller circle is 36mm.

    1.5. Remember that floating viewports can overlap.

    1.6. The scale of each viewport (from top to bottom) is 10:1, 8:1, 2:1.

    1.7. Fill in the title block as desired.

    1.8. Save the drawing as *MyNeedle* in the C:\Steps\Lesson25 folder.

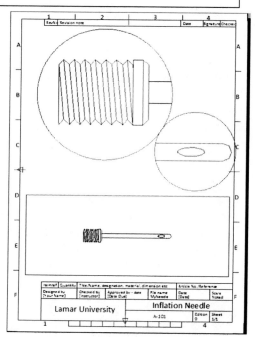

Inflation Needle

2. Open the *cable splitter* file in the C:\Steps\Lesson24 folder. Create the drawing configuration for plotting found the *cable splitter* drawing. Some helpful information includes:

   2.1. The page size is 11 x 8½".
   2.2. Watch your layers (place viewports on the **VPorts** layer).
   2.3. The radius of the circle is 1.75".
   2.4. The title block is the *ANSI A title block* file found in the C:\Steps\Lesson24 folder.
   2.5. The scale of each viewport is 4:1 (upper left), 4:1 (lower left), and 1:1 (right).
   2.6. Fill in the title block as desired.
   2.7. Save the drawing as *MySplitter* in the C:\Steps\Lesson25 folder.

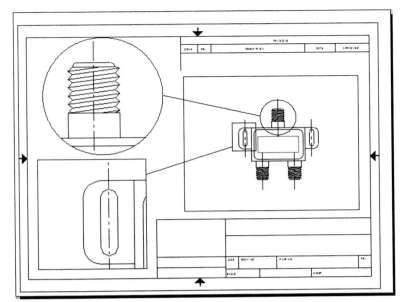

Cable Splitter

| 24.9 | **For Web-Based Review Questions and Additional Exercises, visit: http://foragerpub.com/AcadFiles/2011/2011.htm** |

Paper space can do more than this?!
Wow; check out the next lesson!

Lesson 25

Following this lesson, you will:

✓ *Know how to work with a layout*
- *Know how to use layers in Paper Space*
- *Know how to use text in Paper Space*
- *Know how to dimension in Paper Space*
- *Know how to plot a Paper Space drawing*

✓ *Know how to use **MVSetup** as a floating viewport editor*

✓ *Know how to use **Regenall** and **Redrawall** to refresh viewports*

The New Beginning Continues ...

You've seen how to show a drawing (a model) at different scales on the same page. But what'll happen to text and dimensions when the scale changes? What'll happen when you need to show information in a detail that you don't want to see on the plan (after all, that's why we use details)?

In creating Paper Space, AutoCAD took these problems into consideration and provided some clever solutions. In Lesson 25, we'll investigate these solutions as well as some tricks to control the display both inside and outside of your viewports.

Let's begin.

25.1 Dimensioning and Paper Space

AutoCAD has two ways to create dimensions when Paper Space is involved – the *Annotative Way* and the *Other Way*. The *Annotative Way* takes advantage of AutoCAD's annotative tools – automatically adjusting dimensioning, text, hatching, etc. to the annotation scale. The *Other Way* allows you to place your dimension in Paper Space.

Let's take a look at both.

25.1.1 Dimensioning and Paper Space – the Annotative Way

The Annotative Way of setting up dimensioning for a Paper Space drawing is almost identical to setting up dimensioning for Model Space. But you must be sure to check off that you want to use Annotative dimensions on the **Fit** tab of the Dimension Style Manager.

This will become clearer with an example.

Do This: 25.1.1A Adding Paper Space Dimensions – The Annotative Way

I. Open the *Flr-pln25a* file in the C:\Steps\Lesson25 folder. (Alternately, you can continue with the *MyFlrPln-25a* file you worked in Lesson 24.)
II. Set the **DIM** layer current.
III. Follow these steps.

25.1.1A: ADDING PAPER SPACE DIMENSIONS – THE ANNOTATIVE WAY

1. Double click in the upper left viewport to activate it in Model Space.

2. Call the Dimension Style Manager.
   Command: *ddim*

3. Select the **Arch** parent style…

4. …and pick the **Modify** button.

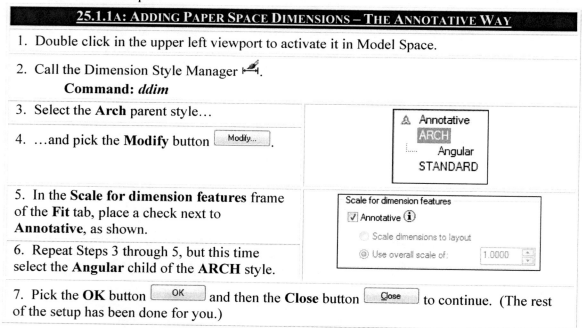

5. In the **Scale for dimension features** frame of the **Fit** tab, place a check next to **Annotative**, as shown.

6. Repeat Steps 3 through 5, but this time select the **Angular** child of the **ARCH** style.

7. Pick the **OK** button and then the **Close** button to continue. (The rest of the setup has been done for you.)

25.1.1A: ADDING PAPER SPACE DIMENSIONS – THE ANNOTATIVE WAY

8. Place the dimensions shown. (Use grips or the *Stretch* command to adjust the size and shape of the viewport as necessary.)

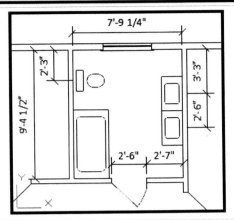

9. Now place the dimensions in the round viewport as shown. (Hint: You may have to use *DimTEdit* or grips to help you.)

10. Return to Paper Space by double clicking outside the viewports.

11. Save the drawing as *MyFlr-pln25a*, but don't exit.

   Command: *saveas*

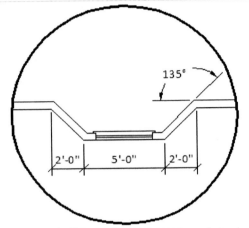

You'll find a useful tool on the status bar. Use the **Minimize (or Maximize) Viewport** button to toggle between the view in an active viewport and the same view on the **Model** tab. Use the arrows on the sides to toggle between viewport views while still on the **Model** tab.

As you can see, the only problem presented by dimensioning in Paper Space was one of room. But if dimensions didn't fit into the viewport, you could use standard modifying tools (grips or the *Stretch* command) to resize it. (Remember that the viewport exists in Paper Space. So any modifying on the viewport itself must be done there.)

Did you notice that the dimension appeared in the ¼" scaled viewports but not in the 1/16" scaled viewport? The dimensions you've created have an Annotative Scale of ¼"=1'-0" (they've automatically adopted the annotation scale of the viewport through which they were created), and because of the setting of the **AnnoAllVisible** system variable, won't appear in a viewport unless that viewport's scale matches their own scale! Check it out.

Do This: 25.1.1B	Annotative Object Visibility

   I. Be sure you're still in the *MyFlr-pln25a* file in the C:\Steps\Lesson25 folder.
   II. Follow these steps.

25.1.1B: ANNOTATIVE OBJECT VISIBILITY

1. Change the value of the **AnnoAllVisible** system variable to **one**. You can do this on the command line or you can pick the **Annotation Visibility** button which lies next to **Annotation Scale** on the status bar.

   Command: *annoallvisible*
   Enter new value for ANNOALLVISIBLE <0>: *1*

   Notice that you can now see the annotative dimensions in the 1/16" viewport.

2. Reset the **AnnoAllVisible** system variable to zero.

   Command: *annoallvisible*
   Enter new value for ANNOALLVISIBLE <1>: *0*

But what if you didn't want the dimensions to show up and both viewports had the same annotation scale? Well, it's a bit more complicated, but you can use layers to accomplish the same task. We'll take a look at how to do that in Section 25.2, p.588. First, let's look at that Other Way for dimensioning in a layout.

25.1.2 Dimensioning and Paper Space – the Other Way

The other way of Paper Space dimensioning really involves no new techniques at all – it doesn't even involve annotation scales! You simply place dimensions in Paper Space rather than Model Space (regardless of where the objects to be dimensioned reside). AutoCAD automatically interprets in which space the objects exist and adjusts the dimension accordingly. Using this approach, all dimensions exist in Paper Space – even if the object dimensioned exists in Model Space.

You should remember some general rules when using this method:

- The **Dimassoc** system variable must be set to **2**. Since this is the default for new drawings, you won't have a problem with them. However, if you're working with a drawing created in an earlier release, you'll have to change the variable setting.
- The new programming may have a problem recognizing multilines. If this occurs, you'll find it necessary to either explode the multilines or use the Annotative Way as detailed in Section 25.1.1, p.585.

Do This: 25.1.2A Adding Paper Space Dimensions – The Other Way

I. Open the *flr-pln25a-new* file in the C:\Steps\Lesson25 folder. Here, we've set up the viewport scale, but we haven't set up an annotation scale.
II. Be sure the **Dimassoc** system variable is set to **2** and the **DIM** layer is current.
III. Be sure you're in Paper Space.
IV. Follow these steps.

25.1.2A: ADDING PAPER SPACE DIMENSIONS – THE OTHER WAY

1. Place the dimensions indicated for the upper left viewport. Do *not* change to Model Space except to make adjustments in the display as needed. Use grips to resize the viewport as needed.

   Command: *dli*

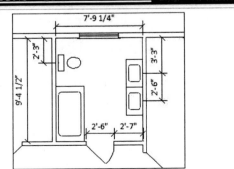

25.1.2A: ADDING PAPER SPACE DIMENSIONS – THE OTHER WAY

2. Now dimension the bay window as shown. Again, don't change to Model Space except to make adjustments in the display.

3. Now Zoom all and look at the drawing. Compare it to the drawing you did in the last exercise (reopen that drawing if necessary and view the two side by side).

4. Save the drawing and exit.
   Command: *qsave*

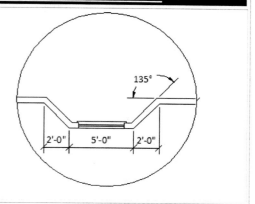

Did you notice any difference between the two drawings?

It won't be necessary to manipulate layers to hide dimensions in the second drawing, but you may have to do so in the first. You'll see how in our next section.

25.2 The Benefits of Layers in Paper Space

One of the real wonders of Paper Space has to be the ability to manipulate layers in a single viewport independently of other viewports. In other words, you can freeze (or thaw) layers (or change the Color, Linetype, Lineweight, or Plot Style) in one viewport while leaving them in their global state in another! The trick lies in using the new columns that appear in the Layer Properties Manager when you open a layout (Figure 25.001). These columns work in essentially the same manner as their non-viewport-specific tools. (The only completely new column – **New VP Freeze** – indicates that the selected layer will be frozen in all new viewports.)

Figure 25.001

Thawed/ Frozen Layer in Current Viewport

Note that, in the **Layer** control box of the **Layers** panel, AutoCAD provides a slightly different symbol (right) to differentiate between globally frozen/thawed layers and viewport-specific frozen thawed layers.

Let's take a look at how layers interact with viewports.

As with so many other procedures, this one has changed over the years. To see how we used to do it, please review the *VPLayer* command in the posted supplement: http://www.uneedcad.com/Files/VPLayer.pdf. (This section appeared in our 2007 text.)

Do This: 25.2A	Manipulating Layers in Floating Viewports

I. Go back to the *MyFlr-pln25a* file in the C:\Steps\Lesson25 folder. Reopen it if necessary.
II. Be sure you're in Paper Space.
III. Follow these steps.

25.2A: MANIPULATING LAYERS IN FLOATING VIEWPORTS

1. We'll start with a simple procedure. Open Model Space in the right viewport.

25.2A: MANIPULATING LAYERS IN FLOATING VIEWPORTS

2. Freeze the fixtures layer just in the current viewport. (Pick the appropriate icon in the **Layers** panel's control box as indicated.)

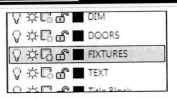

Notice (in the following figure) that the fixtures in the bathrooms disappear just in this viewport.

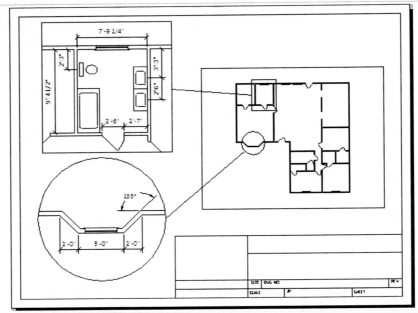

3. Let's crosshatch the master bath area (the area shown in the detail) in the right viewport. First, create a **hatch** layer. Make it a dark color, and make it current.

   Command: *la*

4. Be sure the right viewport is active in Model Space.

5. Now enter the *Hatch* command.

   Command: *h*

6. On the **Properties** panel of the **Hatch Creation** tab, set the **Scale** to ¼" and select **Relative to paper space** [Relative To Paper Space] on the subpanel. Pick on the **Annotative** button (**Options** panel) to turn it on. (Both settings will affect the final hatching.) Accept all other defaults.

7. Pick the **Pick points** button and select a point inside the master bath. Hit ENTER to complete the procedure.

   Pick internal point or [Select objects/setTings]:

Your drawing looks like the following figure. Notice that the hatch pattern appears in the 1/16" viewport only (because of the **Annotative** and the **AnnoAllVisible** settings) and properly scaled (because of the **Relative to paper space** setting).

25.2A: MANIPULATING LAYERS IN FLOATING VIEWPORTS

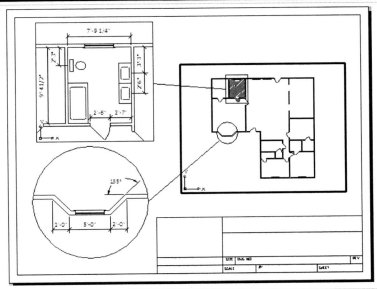

8. We don't need the doors and windows to stand out quite so much in the right viewport, so let's change the color to match the walls – but just in this viewport. Open the Layer Properties Manager.

   Command: *la*

9. Change the color of the **Windows** and **Doors** layers to blue, but use the tool in the **VP Color** column VP Color .

10. Return to Paper Space.

    Command: *ps*

11. Change the two viewports on the left to the **TEXT** layer.

    Command: *props*

12. Now freeze the **VPORTS** layer. Your drawing now looks like the following figure.

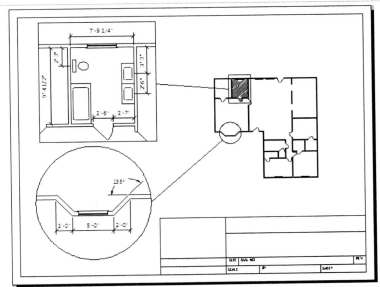

25.2A: MANIPULATING LAYERS IN FLOATING VIEWPORTS

13. Save the drawing and exit.
    Command: *qsave*

This is a much finer image for plotting, but there's still more to do. We must add some callouts to the drawing before we plot it.

25.3 Using Text in Paper Space

In a drawing that uses Paper Space, place all text in Paper Space unless it's part of the model. The reasoning behind this general rule is this: *Placing text in Paper Space is easier than placing text in Model Space* (and it doesn't affect the model).

Let's add a bit of text to our drawing.

| Do This: 25.3A | Adding Text in Paper Space |

I. Open the *flr-pln25b* file in the C:\Steps\Lesson25 folder.
II. Be sure you're in Paper Space.
III. Follow these steps.

25.3A: ADDING TEXT IN PAPER SPACE

1. On the **TEXT** layer, add the callouts shown in the following figure. Large text is 3/16" and smaller text is 1/8". (Use the **Times** or **Calibri** text style that has already been set up for you.)

    Command: *dt*

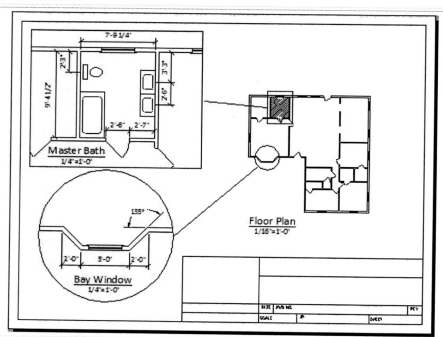

2. Now fill out the title block as shown in the following figure. Text sizes are ¼", 3/16", and 1/8". Feel free to use your favorite school's name.

25.3A: ADDING TEXT IN PAPER SPACE

	Texas Tech University			
	Tara II Building Design Sample Layout			
AutoCAD Text	SIZE A	DWG NO A-231		REV 0
One Step at a Time	SCALE Noted	BY [Your Name]	SHEET 1 of 1	

3. Save the drawing 💾 but don't exit.

   Command: *qsave*

Take a moment and toggle between the model and the layout using Quick View Layouts. Notice which objects exist as part of the model and which don't. Only those objects that you created through the viewports in Model Space exist in the **Model**. Everything else exists as part of the paper on which you'll plot your model.

Once the dimensions are drawn and the text placed – the *i*'s dotted and the *t*'s crossed – it's time to plot the drawing. Let's look at that next.

25.4   Plotting the Layout

It's funny that, after all the setting up, creating, and manipulating, it always comes back to getting it on paper. Perhaps that's why AutoCAD has put so much time and money into making printing as user-friendly as possible.

Printing a layout isn't very different from printing Model Space (although it's a bit easier). Let's print our floor plan layout.

Do This: 25.4A	Printing/Plotting the Layout

I. Be sure you're still in the *flr-pln25b* file in the C:\Steps\Lesson25 folder. If not, please open it now.

II. Follow these steps.

25.4A: PRINTING/PLOTTING THE LAYOUT

1. Enter the *Plot* command 🖨.

   Command: *plot*

AutoCAD presents the Plot dialog box.

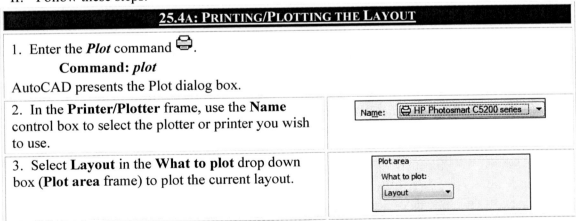

2. In the **Printer/Plotter** frame, use the **Name** control box to select the plotter or printer you wish to use.

3. Select **Layout** in the **What to plot** drop down box (**Plot area** frame) to plot the current layout.

25.4A: PRINTING/PLOTTING THE LAYOUT

4. Be sure your **Plot scale** is set to 1:1.

5. Use the **Preview** button [Preview...] to help set the Plot offset to properly center your drawing on the page.

6. When you're satisfied with the offset, pick the **OK** button [OK] to print the drawing.

7. Save the drawing 💾.

Command: *qsave*

Use a drafting scale to check the dimensions on your plot. How did you do?

As with any print job, it's inevitable that something must be modified after the plot. Despite the ability to preview a drawing before plotting, some things simply don't show up until you see it on paper.

You can make changes in Model Space through the viewports or on the model. You can make changes in Paper Space on the layout. However, changes to the viewports themselves may require some additional tools. We'll look at these next.

25.5 Tweaking the Layout

Generally speaking, modifying the layout is as easy as modifying any other part of a drawing. The basic modifying tools – ***Move***, ***Copy***, ***Stretch***, and so forth – will work in Paper Space as well as Model Space. However, adjusting the view through a viewport or adjusting the shape or scale of a viewport will require some new modifying procedures. These include two commands: ***MVSetup*** and ***VPClip***.

25.5.1 Modifying Viewports with the *MVSetup* Command

AutoCAD originally designed the ***MVSetup*** command to perform the same function as the layout wizard. Using the single ***MVSetup*** command, we can insert the title block and create/scale the viewports just as we did with the wizard. But the wizard is so much easier to use than ***MVSetup***'s command line approach that ***MVSetup*** might have been removed. But AutoCAD chose to leave it (it's so hard to lose a good tool) because of its use as a modifying tool.

We'll look at all parts of the ***MVSetup*** command, but we'll use the command as a modifying tool. The command sequence looks like this:

    Command: *mvsetup*

    Enable paper space? [No/Yes] <Y>:

When entered while in the model, the ***MVSetup*** command initially asks if you wish to enable Paper Space. What follows depends on your answer.

- If you respond *No*, you'll set up the drawing for Model Space. AutoCAD prompts as follows:

    Enter units type [Scientific/Decimal/Engineering/Architectural/Metric]:

Tell AutoCAD which type of units you wish to use. AutoCAD will respond with a list of scale factors available. Select one. Then AutoCAD will ask for the width and height of the sheet of paper to which you will plot. When it has all the information, it will place a polyline border around the limits of the drawing.

- If you respond *Yes* to the **Enable paper space** prompt (or if you enter the command while in a layout), AutoCAD flips to the first layout, creates a single viewport (if none currently exist), and then prompts as follows:

Enter an option [Align/Create/Scale viewports/Options/Title block/Undo]:
You have several options. These include:
- The **Align** option allows you to align a view in one viewport with the view in another.
- **Create** allows you to create (or delete) floating viewports. It prompts:
    Enter option [Delete objects/Create viewports/Undo] <Create>:
    - The **Delete objects** option, of course, allows you to remove a viewport.
    - The default option – **Create** – prompts:
        Available layout options: . . .
        0: None *[For no viewports.]*
        1: Single *[For a single viewport.]*
        2: Std. Engineering *[This option creates four viewports with standard three-dimensional views (plan, front, side, isometric).]*
        3: Array of Viewports *[This option allows you to create a rectangular array of viewports – you define the number of rows and columns to use and the spacing between them.]*
        Enter layout number to load or [Redisplay]: *[Make your selection or hit ENTER to return to MVSetup's first option line.]*
- Scaling the viewports can be done one at a time or uniformly. You'll find setting the scale one viewport at a time more easily accomplished using the **Viewport Scale** control box on the Viewports toolbar, or the ***Zoom*** command. If you select more than one viewport to scale, AutoCAD will prompt as follows:
    Set zoom scale factors for viewports. Interactively/<Uniform>:
    Set the ratio of paper space units to model space units...
    Enter the number of paper space units <1.0>: *[Usually hit ENTER.]*
    Enter the number of model space units <1.0>: *[Enter a scale factor.]*
- The **Options** prompt allows you to set the following:
    Enter an option [Layer/LImits/Units/Xref] <exit>:
    You can set the current layer and Paper Space units, or you can allow AutoCAD to set the limits according to the drawing you've created. (We'll look at Xrefs in Lesson 27, p.630.)
- The **Title Block** option allows you to insert a title block in Paper Space.

Let's change the scale in one of our viewports. We'll use the *MVSetup* command to change the actual scale. Then we'll use more conventional commands to adjust the viewport.

Do This: 25.5.1A	Using MVSetup as a Modifying Tool

I. Be sure you're still in the *flr-pln25b* file in the C:\Steps\Lesson25 folder. If not, please open it now.
II. Follow these steps.

25.5.1A: USING MVSETUP AS A MODIFYING TOOL

1. Enter the *MVSetup* command.
    Command: *mvsetup*

2. Choose the **Scale viewports** option [Scale viewports] and select the upper left viewport (the master bath). Hit ENTER to complete the selection.
    Enter an option [Align/Create/Scale viewports/Options/Title block/Undo]: *S*
    Select the viewports to scale...

25.5.1A: USING MVSETUP AS A MODIFYING TOOL

      Select objects: *[Select the upper left viewport.]*
      Select objects: *[ENTER]*

3. Now you'll set the ratio of Paper Space to Model Space. Accept **1** as the number of **paper space units**.

      Set the ratio of paper space units to model space units...
      Enter the number of paper space units <1.0>: *[ENTER]*

4. We'll change the scale from ¼"=1'-0" to 3/16"=1'-0". Enter the scale factor for the 3/16" scale (**64**). Notice how everything is rescaled.

      Enter the number of model space units <1.0>: *64*

5. Complete the command.

      Enter an option [Align/Create/Scale viewports/Options/Title block/Undo]:
      [ENTER]

6. We'll have to make some adjustments for the dimensions. Open Model Space in the viewport.

7. Select all the dimensions and open the Properties palette.

      Command: *props*

8. In the **Misc** section, select the button ··· next to **Annotative scale**.

AutoCAD opens an Annotation Object Scale dialog box which lists the Annotation Scales available to the selected objects.

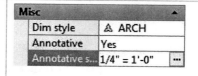

9. Pick the **Add** button 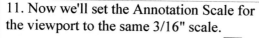. AutoCAD presents an Add Scales to Objects dialog box.

10. Select the 3/16"=1'-0" scale and pick the **OK** button twice to continue.

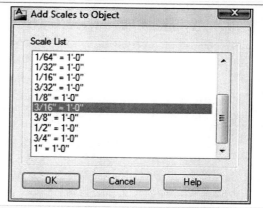

11. Now we'll set the Annotation Scale for the viewport to the same 3/16" scale.

Begin by deselecting the dimensions [Esc].

12. With the upper left viewport still open in Model Space, open the Properties palette and select the scale next to **Annotation Scale** under the **Misc** header. Set the scale to 3/16"=1'-0".

Notice that the viewport adjusts the dimensions.

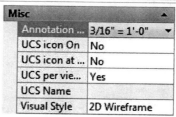

25.5.1A: USING MVSETUP AS A MODIFYING TOOL

13. Use ***DimTEdit*** or grips to adjust the locations of the dimensions as needed. **Command:** *dimtedit*	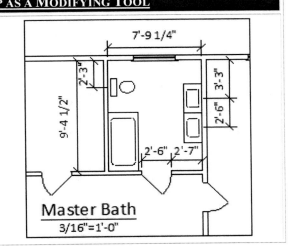
14. Using grips or other modifying tools, resize and reposition the viewport. Then edit the text as shown.	
15. Save the drawing 💾 but don't exit. **Command:** *qsave*	

> When working with several viewports, you may notice that the *Redraw* and *Regen* commands only affect the one that is currently active. To use these commands to refresh all of the viewport simultaneously, use the *Redrawall* and *Regenall* commands.

Okay, we've created our viewports and set up Paper Space; let's see what comes next.

25.5.2 Changing the Shape of a Viewport with the *VPClip* Command

Occasionally you'll discover that the shape you chose for your viewport doesn't satisfy the needs of that particular view. If you created the viewport using the **Object** option of the *MView* command to convert a polygon, closed polyline, or spline to a viewport, you can use the *PEdit* or *Splinedit* commands to reshape the viewport. If you used any other method to create the viewport, modifying the shape won't be so easy.

Fortunately, AutoCAD has provided the *VPClip* command. You can use this to reshape the view in a standard rectangular viewport. The command sequence looks like this:

🖻	Where to Find It:
Command Line:	*VPClip*
Ribbon (Tab/Panel):	View – Viewports – **Clip**
Menu:	Modify – Clip – **Viewport**
Cursor Menu:	[With viewport selected] – **Viewport Clip**
Toolbar:	Viewports – **Clip existing viewport**

    Command: *vpclip*
    Select viewport to clip: *[Pick the viewport you wish to reshape (clip).]*
    Select clipping object or [Polygonal/Delete] <Polygonal>: *[ENTER]*
    Specify start point:
    Specify next point or [Arc/Length/Undo]: *[This line operates like the* **PLine** *command and repeats until the polyline is closed.]*

Let's look at each of the prompts.
- The first line is simple enough. It asks which viewport you wish to clip. Select one.
- The next line gives you a couple options:
  - The default option – **Polygonal** – allows you to place the clipping border manually with a closed polyline (a polygon).
  - The **Delete** option allows you to remove a clipping border. It'll return the view to its original shape.
  - Enter **S** to use the **Select clipping object** option (or just select the object) if you've created a new boundary around the view using a polyline, spline, and so forth. It prompts:

Select object to clip viewport:

Simply select the object you wish to use as the new viewport.

Let's clip the round viewport in our drawing.

Do This: 25.5.2A	Reshaping a Viewport

I. Be sure you're still in the *flr-pln25b* file in the C:\Steps\Lesson25 folder. If not, please open it now.

II. Be sure you're in Paper Space and the **Text** layer is current.

III. Follow these steps.

25.5.2A: RESHAPING A VIEWPORT

1. Draw an ellipse around the bay window similar to the one shown.
   Command: *el*

2. Enter the *VPClip* command.
   Command: *vpclip*

3. Select the viewport (the circle).
   Select viewport to clip:

4. Select the ellipse.
   Select clipping object or [Polygonal/ Delete] <Polygonal>:
   AutoCAD replaces the circle with a new viewport based on the shape of the rectangle.

5. Adjust the location of the text.
   The viewport now looks like the figure at right.

6. Save and close the drawing.
   Command: *qsave*

25.6 Putting It All Together

We've learned quite a bit about Paper Space and viewports. Let's try a project from the beginning (well, almost the beginning – I'll provide the drawing).

Do This: 25.6A	From Setup to Plot – A Project

I. Open the *table saw* file in the C:\Steps\Lesson25 folder. The drawing looks like the figure at right. (I've hidden the grid for clarity.)

II. Set the **VPORTS** layer current.

III. Follow these steps.

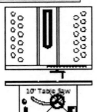

597

25.6A: FROM SETUP TO PLOT – A PROJECT

1. Let's start the easy way. Enter the *LayoutWizard* command.
   Command: *layoutwizard*

2. Call the layout *Master Layout*.
   Pick the **Next** button [Next >] to continue.

3. Select a plotter if one is available. Otherwise, select **None**.
   Pick the **Next** button [Next >] to continue.

4. Select an **ANSI D** size sheet and use **Inches** for your drawing units.
   Pick the **Next** button [Next >] to continue.

5. Tell AutoCAD to plot a **Landscape** drawing.
   Pick the **Next** button [Next >] to continue.

6. Select an appropriate title block for the sheet of paper you selected in Step 4 (**ANSI D**). Use a **Block** insertion.
   *NOTE: If ANSI D isn't available, pick **None** – I'll give you a chance to insert it after the wizard.*
   Pick the **Next** button [Next >] to continue.

7. Define the viewports by **Array**. Use **2** rows and **3** columns and accept the default spacing.
   Pick the **Next** button [Next >] to continue.

8. Pick the **Select Location** button [Select location <].

9. Specify the area for the viewports.
   Specify first corner: *1.5,19*
   Specify opposite corner: *31,3*

10. Pick the **Finish** button [Finish] to complete the command.

598

25.6A: FROM SETUP TO PLOT – A PROJECT

If you didn't get the chance to insert the title block in Step 6, use the *Insert* command and do so now. It's the **ANSI-D title block** found in the C:\Steps\Lesson25 folder. Insert it at the lower left corner of the sheet, on layer **0**, and accept the other defaults.

11. Remember to save occasionally.

    Command: *qsave*

Your drawing looks like the following figure.

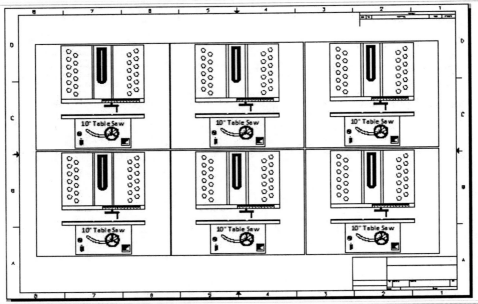

12. Let's turn the UCS icon off in all of the viewports (so they don't appear when we activate Model Space). Enter the *UCSIcon* command.

    Command: *ucsicon*

13. Use the **All** option ...

    Enter an option [ON/OFF/All/Noorigin/ORigin/Properties] <ON>: *a*

14. ... and turn the icon **OFF** .

    Enter an option [ON/OFF/Noorigin/ORigin/Properties] <ON>: *off*

15. Resize and relocate each viewport using grips and/or the *Stretch* and *Move* commands. The lower left and upper right coordinates for each viewport follow. (Hint: Grips and absolute coordinates make this job much easier.)

VIEWPORT (ROW/COL)	HANDWHEEL PLAN (1/1)	SAW PLAN (1/2)	SWITCH (1/3)	HANDWHEEL ELEV. (2/1)	SAW ELEV. (2/2)	SPEC. PLATE (2/3)
Lower left coordinate	1.5,15.1	7.25,8.25	2,1.75	1.5,10	7.25,1	25.75,4
Upper right coordinate	7.75,19	25.5,20	5.75,7.75	7.75,15	25.5,11	31,11.75

25.6A: FROM SETUP TO PLOT – A PROJECT

16. Set the viewport scale and the annotation scale in each viewport as follows: (use the status bar controls, Properties palette, or the *MVSetup* command): Handwheel (plan and elevation) – 1:1, Saw (plan and elevation) – 1:2, Switch – 2:1, Spec. Plate – NTS (zoom in to make it visible).

(*Readjust the views in each port as shown in the following figure.*)

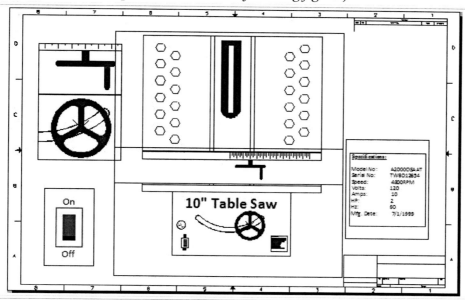

17. Freeze unwanted layers as follows. (Hint: It might be faster to freeze all layers and then thaw the few you wish visible.)

VIEWPORT:	HANDWHEEL – PLAN & ELEV.	SWITCH	SAW – PLAN & ELEV.	SPECS
Freeze All Layers Except:	Obj3 Obj3a	Obj2 Obj2a Obj3 Text	(Don't freeze any layers)	Text

18. Remember to save 💾 occasionally.
    Command: *qsave*

Your drawing looks like the following figure.

Still With Me?

25.6A: FROM SETUP TO PLOT – A PROJECT

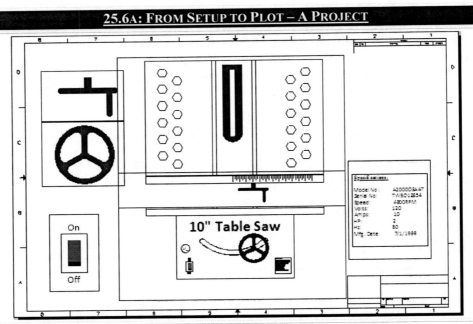

19. Now use the ***MVSetup*** command to align the views in the handwheel plan and elevation. Begin by entering the command.

    Command: *mvsetup*

20. Select the **Align** option [Align].

    Enter an option [Align/Create/Scale viewports/Options/Title block/Undo]: *a*

21. We'll align the objects vertically [Vertical alignment].

    Enter an option [Angled/Horizontal/Vertical alignment/Rotate view/Undo]: *v*

22. Select the leftmost quadrant of the handwheel in the plan view …

    Specify basepoint: _qua of

23. … and the corresponding quadrant in the elevation view.

    Specify point in viewport to be panned: _qua of

24. Complete the command.

    Enter an option [Angled/Horizontal/Vertical alignment/Rotate view/Undo]: *[ENTER]*

    Enter an option [Align/Create/Scale viewports/Options/Title block/Undo]: *[ENTER]*

25. Repeat Steps 19 through 24 to align the plan and elevation views of the saw.

26. We'll use the ***VPClip*** command to set the saw plan and elevation viewports so that all you see is the plan in the plan viewport and the elevation in the elevation viewport.

    Begin by drawing a rectangle around the elevation in the bottom viewport. (Be sure you're in Paper Space and that the **vports** layer is current.)

    Command: *rec*

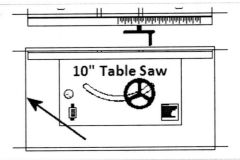

601

25.6A: FROM SETUP TO PLOT – A PROJECT

27. Enter the **VPClip** command.
    Command: *vpclip*

28. Select the viewport with the saw elevation in it.
    Select viewport to clip:

29. Select the polyline (the rectangle) you drew in Step 26.
    Select clipping object or [Polygonal] <Polygonal>:

30. Set the **text** layer current and freeze the **vports** layer.

31. Remember to save occasionally.
    Command: *qsave*

Your drawing looks like the following figure.

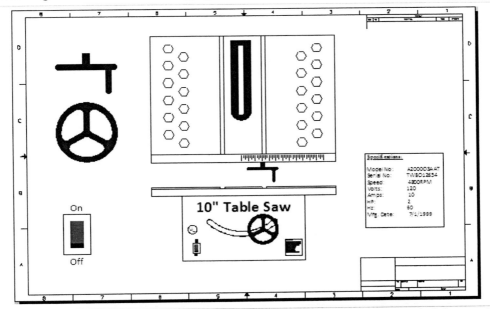

32. Add the text and detail markings as shown in the following figure. Text heights are ¼" and 3/16". The text style is **simple** (or **Calibri**). (Place the detail markings on the **details** layer.)

25.6A: FROM SETUP TO PLOT – A PROJECT

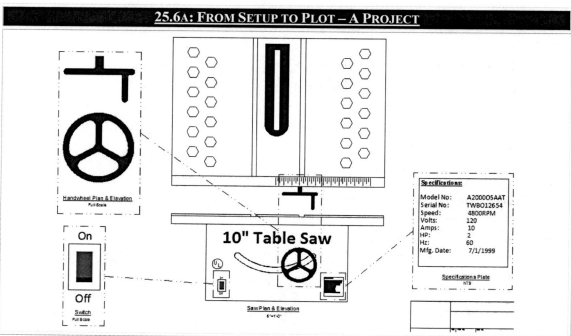

33. Fill in the title block as shown in the following figure. Text sizes are ¼", 3/16", and 1/8". Styles are **Timesbold** and **Times** (or **Calibribold** and **Calibri**).

34. Remember to save occasionally.
    Command: *qsave*

35. Now we must plot the drawing. (Note: If you haven't a plotter available, you can print to your own printer – just make sure you print to fit.) Begin by zooming all and then entering the *Plot* command.
    Command: *plot*

36. Your Plot dialog box should resemble the one in the following figure. (Note: I'm printing to fit on a printer, as I don't have a plotter available.) Your plotter may vary and the **Plot offset** may vary. Use the **Preview** button to be sure you have set up the plot correctly and then plot the drawing.

25.6A: FROM SETUP TO PLOT – A PROJECT

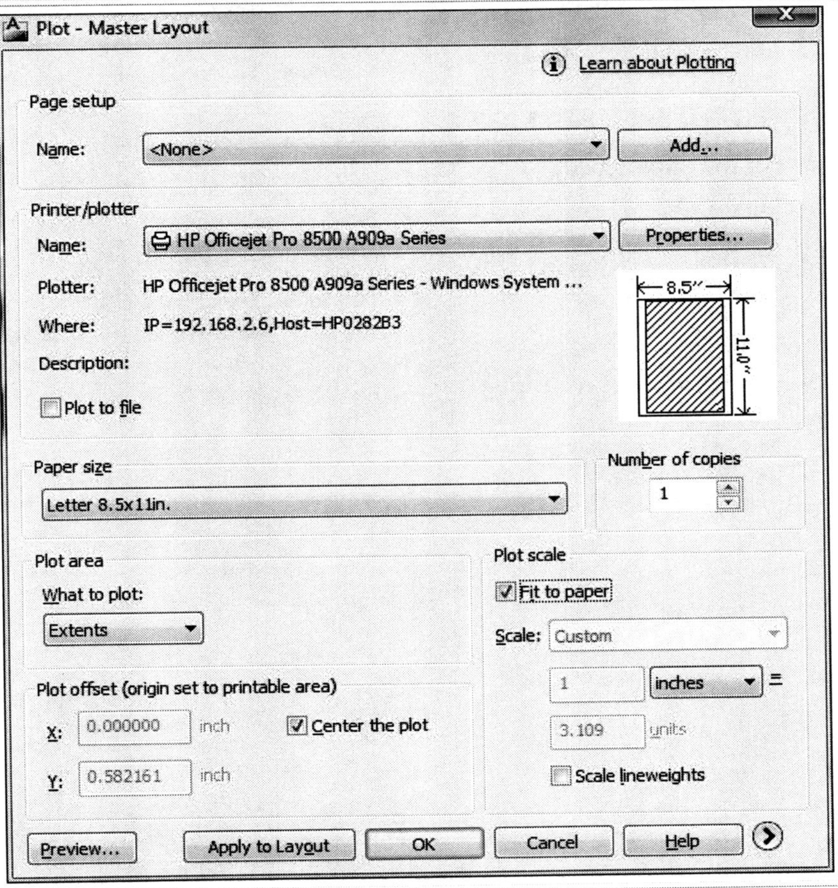

37. Save the drawing 💾 and exit.
    Command: *qsave*

How was that? Just 37 easy steps to completion!

I know what that sounds like, but most of these steps will become second nature in time. Where it probably took 45 minutes to complete the exercise, with experience it'll take about 15 minutes to complete a similar project. This text can help you acquire some of that experience. Try some of the projects at the end of this lesson!

25.7 Extra Steps

This is a cool trick that might save you some space on a travel computer or in your archives. It involves the ***ExportLayout*** command.

We'll use a quick exercise to demonstrate this one.

	Where to Find It:
Command Line:	*ExportLayout*
Menu:	File – **Export Layout to Model**

Do This: 25.7A	Exporting a Layout

I. Reopen the *table saw* file in the C:\Steps\Lesson25 folder.
II. Follow these steps.

25.7A: EXPORTING A LAYOUT

1. Start by running a *DwgProps* on the drawing. It should look something like this.

2. Note the size of the drawing and then close the dialog box.

3. Enter the *ExportLayout* command.

   Command: *exportlayout*

   AutoCAD presents a standard save file dialog box.

4. Accept the default name and pick the **Save** button.

5. Open the new drawing.

6. Repeat Step 1. Compare the size of the two drawings.

7. Close both drawings without saving them.

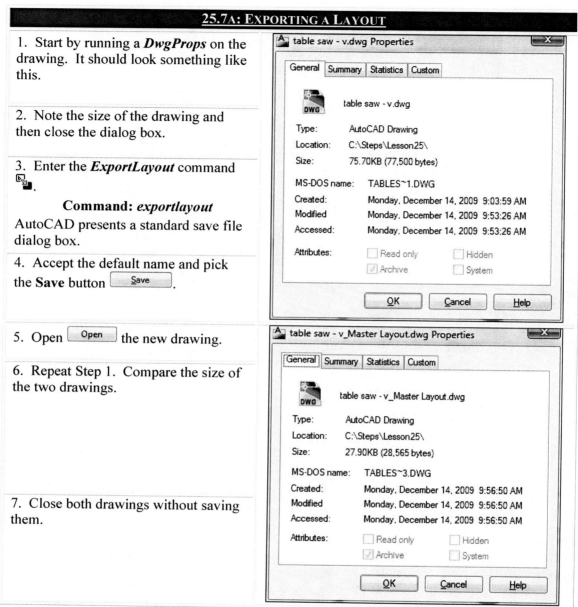

So what have you accomplished?
Well, a couple things:
- You've reduced the drawing size considerably (by nearly 2/3).
- You've moved all the Model and Paper Space objects to a Model Space drawing.

Use this tool with some caution – you'll notice that the viewports are no longer viewports; you can't change what you see in them (although you can modify the geometry).

25.8 What Have We Learned?

Items covered in this lesson include:
- *Using layers in Paper Space*
- *How to use text in Paper Space*
- *How to dimension in Paper Space*
- *How to plot a Paper Space drawing*

- *Commands*
  - *MVSetup*
  - *VPClip*
  - *Regenall*
  - *Redrawall*
  - *AnnoAllVisible*
  - *ExportLayout*

Remember the first paragraphs of Lesson 24? You may now step into that group of CAD operators who have used Paper Space and would never use anything else!

In Paper Space, you've taken your first steps into a whole new AutoCAD world! With experience, you'll soon wonder why we don't teach Paper Space from the beginning. But think back to what you knew when you began your first AutoCAD class. So much of what you learned in the last two lessons requires that you already have a foundation in the basics. And now, with Paper Space (and viewports) under your belt, you have a foundation for what comes next.

25.9 Exercises

1. Open the *MyNeedle* file in the C:\Steps\Lesson25 folder. (If that drawing isn't there, use the *needle 25* file in the same folder.) Create the drawing configuration for plotting found in the figure at right. Some helpful information includes:

   1.1. The text height is 5mm and 2.5mm; the font is Times New Roman or Calibri.
   1.2. The title block text is attributed.
   1.3. I used two dimension layers (create new layers as required).
   1.4. Dimension text is 3mm.
   1.5. Remember that Floating viewports can overlap.
   1.6. Save the drawing as *MyNeedle25* in the C:\Steps\Lesson25 folder.

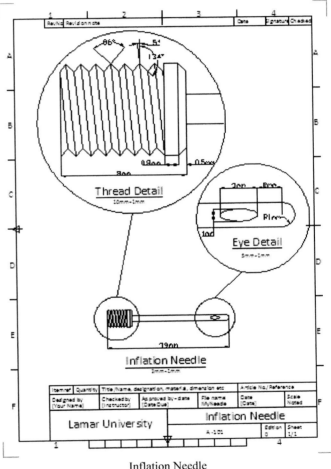

Inflation Needle

2. Open the *MySplitter* file in the C:\Steps\Lesson25 folder. (If this drawing isn't there, open the *cable splitter 25* file instead.) Create the drawing configuration for plotting found in the following figure. Some helpful information includes:

   2.1. The text height is 3/16" and 1/8"; the font is Times New Roman or Calibri.
   2.2. The title block text is ¼", 3/16", and 1/8"; the font is Times New Roman or Calibri.
   2.3. I used two dimension layers (create new layers as required).
   2.4. Dimension text is 1/8".
   2.5. Remember that floating viewports can overlap.
   2.6. Save the drawing as *MySplitter25* in the C:\Steps\Lesson25 folder.

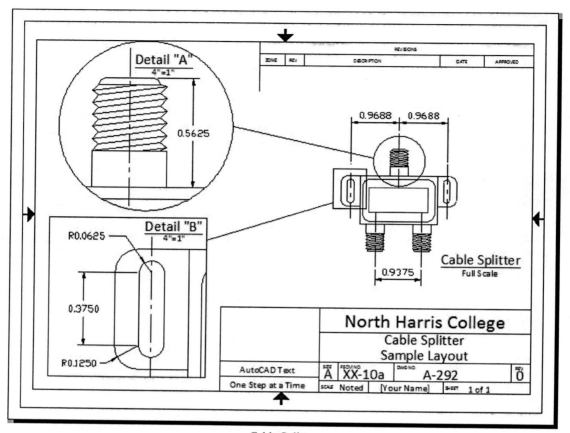

Cable Splitter

25.10 For Web-Based Review Questions and Additional Exercises, visit: http://foragerpub.com/AcadFiles/2011/2011.htm

I caught this bunch of nuts still foolin' around in Model Space!

Lesson 26

Following this lesson, you will:

✓ Know how to organize your Paper Space Drawings into Drawing (Sheet) Sets
  - **SheetSet**
  - **NewSheetSet**
  - **OpenSheetSet**

✓ Know how to transmit your sheet set to a client

✓ Know how to archive a project

✓ Know how to create multiple-sheet DWF files

More Sharing Tools – Drawing Sheet Sets

If the last couple lessons frightened you or you didn't see where Paper Space would do you any good, you can skip this chapter altogether. If, on the other hand, you saw the potential of this remarkable "plotting tool", then by all means, please continue. What you find in Lesson 26 will place the icing on the cake of your Paper Space aspirations! (Be warned, however; if you have an employer who lacks foresight – as many do – he or she will no doubt resist sheet sets. They are, after all, the bane of the sedentary boss ... they're an innovation – something completely different!)

26.1 Sheet Sets – A Primer

Knowledge isn't what you know ... it's what you know how to find out!

Hickory

> Let me warn you before we start; the Sheet Sets palette, although benign in appearance, contains an amazing amount of opportunity (as you can tell from the amount of explanation before we even begin our first exercise). But don't feel as though you have to learn it all at once; take your time and repeat a section if you become confused. Remember what Hickory said at the beginning of this section. In other words, keep this text handy and make reference to it when you need help!

AutoCAD doesn't make it easy to find a definition for sheet sets. In fact, the Preview Guide refers both to drawing sets and sheet sets, but further reading indicates that both names essentially refer to the same thing (but like *print* and *plot*, don't expect them to settle on one name any time soon!). That being said, "sets" actually include both drawings and "sheets" (or Paper Space layouts).

So, what exactly are drawing/sheet sets?

Often, you'll find yourself with a bundle of drawings that need to be numbered and catalogued properly so that you (and your client) can keep track of them. Usually, this means keeping a log of some sort with drawing numbers assigned by discipline, and then running plots that we store in separate drawers of a large cabinet ... or folders on someone's hard drive (or network).

Sheet sets (we'll settle on one name to keep it simple) are a systematic approach to organizing the package you're creating for your client – plans, details, electrical, plumbing, data sheets, etc. Further, the Sheet Set Manager greatly simplifies locating, opening, and editing any drawing or any layout in the set.

In this lesson, I'll walk you through the set up and use of sheet sets using the Sheet Set Manager. But bear in mind that this tool warrants greater exploration than we can do in these few pages. Once you've gotten your feet wet, feel free to head on into the deep end of the pool. (By now, you should be able to swim fairly well!)

26.2 Using the Sheet Set Manager to Organize Your Project

26.2.1 The Sheet List Tab

The Sheet Set Manager appears as a palette (Figure 26.001, p.610). This handy palette will prove its value in more complex projects.

We'll look at the **Sheet List** tab first.

- The **Sheet List** tab (Figure 26.001, p.610) presents a list of the sheets in the set under the **Sheets** heading. Right click on any selected sheet and pick open to open that drawing layout.

> When you let your cursor hover over a listing, AutoCAD displays an informational tip about that listing.

Above the **Sheets** list, you'll find a selection box and three buttons.

609

Where to Find It:	
Command Line:	*SheetSet*
Hotkey(s):	*ssm* or CTRL+4
Ribbon (Tab/Panel):	View – Palettes – **Sheet Set Manager**
Menu:	Tools – Palettes – **Sheet Set Manager**
Toolbar:	Standard – **Sheet Set**

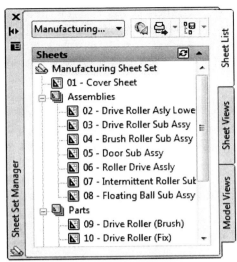

Figure 26.001

- o Use the selection box to **Open** an existing set, create a **New Sheet Set** (to launch the New Sheet Set wizard – see Exercise 26.2.1A, p.614), or view a list of **Recent**[ly] opened sets.
- o The first button – **Publish to DWFx** – works much like a print button except that it tells AutoCAD to create a DWFx file. This button works automatically and uses the defaults set up for the **Publish** command.
- o The second button – **Publish** – calls a menu (Figure 26.002) with several options.

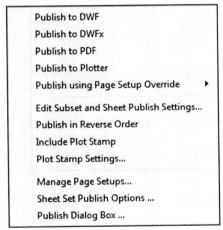

Figure 26.002

  - **Publish to DWF** (or **DWFx**, or **PDF**) operates the same as the button. **Publish to Plotter** does the same thing, but plots to the default plotter.
  - **Publish using Page Setup Override** publishes the sheet using the selected overrides. The overrides come from drawing templates with their own setups.
  - **Publish in Reverse Order** prints the drawings/sheets in reverse order of how they're listed in the Sheet Set Manager.
  - The toggle **Include Plot Stamp** tells AutoCAD whether or not to place a stamp on the plot you're about to make.
  - **Plot Stamp Settings** calls the Plot Stamp dialog box that will help you set up the stamp. (For more on the Plot Stamp dialog box, see Lesson 23, p.543.)
  - **Manage Page Setups** displays the Page Setup Manager. (For more on the Page Setup Manager, see Lesson 23, p.529.)
  - **Sheet Set Publish Options** calls the Publish Options dialog box but the settings are specific to the current sheet set. (For more on the Publish Options dialog box, see Lesson 23, Section 23.2.3, p.542.)
  - **Publish Dialog Box** calls the Publish dialog box (Lesson 23, p.542).
  - Finally, the **Edit Subset and Sheet Publish Settings** option calls the dialog box (Figure 26.003, p.611). Place a check next to the drawings and sheets you wish to include in the sheet set.
- o The last button on the **Sheet List** tab – **Sheet Selections** – enables you to either **Manage** or **Create** sheet selections. Use this tool to group several selected sheets for simultaneous

publishing or transmitting action. AutoCAD will ask you for a name for the sheet selection. (Use something descriptive as there is no opportunity to describe the set anywhere else.) Once you've identified the sheet selection, you can publish or transmit the grouping. Use the **Manage** option to rename or delete the sheet selection.

- The **Sheets** list box presents a list of the drawings and sheets (layouts) currently listed in the sheet set. AutoCAD presents the cursor menu shown in Figure 26.004 when you right click on one of the listings. (Note: The options will vary according to where your cursor is when you right click. I produced the menu shown by right clicking on the sheet set name – at the top of the palette.) Let's look at these options.

  - Use **Close Sheet Set** to close the current sheet set. You can't delete the DST file (sheet set data file – the file that contains the setup information for your sheet set) while the sheet set is open.
  - Create a **New Sheet** set with the next option. First, create your new drawing with a layout, and then this option will present the New Sheet dialog box (Figure 26.005), which will allow you to number and identify (**File name**) the new sheet in the current set.
  - Use the **New Subset** option to call the Subset Properties dialog box (Figure 26.006, p.612). It'll help organize your list with a subset of an existing set. The Subset Properties dialog box provides options for naming (**Subset name**), storing (**New Sheet Location**), and even providing a template to create the new subset (**Sheet Creation Template**).
  Prompt for Template tells AutoCAD to ignore the default template and ask you which template to use. **Create Folder Hierarchy** helps you better organize your efforts by placing subfolders in the **New Sheet Location**.
  - **Import Layout as Sheet** presents another dialog box (Figure 26.007). Use the **Browse for Drawings** button to select drawings to include in your sheet set. Once you've selected drawings, the list box will provide the layout names

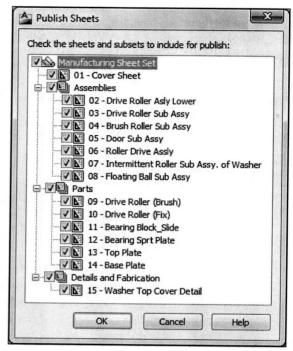

26.003

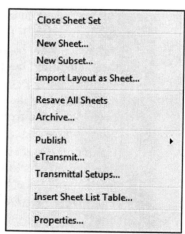

Figure 26.004

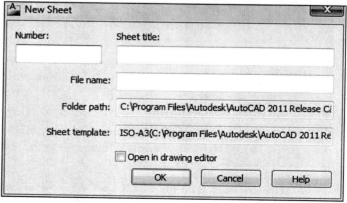

Figure 26.005

611

available for import. Put a check next to those rows that contain layouts you want to use.

- **Resave All Sheets** opens and saves all the sheets associated with the sheet set. Use this tool to save any new sheet set information to the sheets, but be sure all the sheets are closed first.
- The **Publish** option presents the same cursor menu seen in Figure 26.002, p.610.
- **ETransmit** presents the Create Transmittal dialog box we discussed in Lesson 23, p.552.
- **Transmittal Setups** presents the Transmittal Setup dialog box. We also discussed this one in Lesson 23, p.552.
- Use **Insert Sheet List Table** to insert a table listing the sheets in the set. You'll usually insert this table on the data sheet for a project. AutoCAD simplifies this procedure with a dialog box.
- The **Properties** option will display a Properties dialog box with information about the sheet set or the selected sheet.
- **Archive** (just below **Resave All Sheets**) presents the Archive a Sheet Set dialog box (Figure 26.008). (Reach the same dialog box with the *Archive* command.) You can use this to archive (store) all the drawings associated with a project at critical stages (Issued for Approval, Issued for Construction, Issued for Review, and so forth).

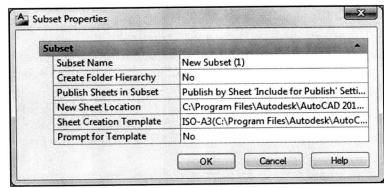

Figure 26.006

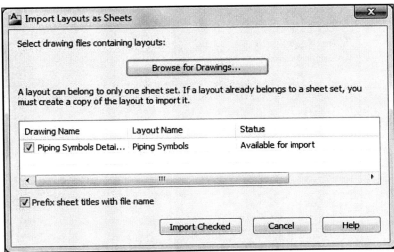

Figure 26.007

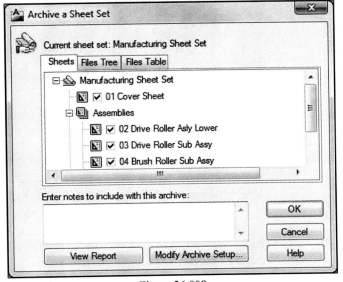

Figure 26.008

You'll notice three tabs in this dialog box. The first, **Sheets**, creates an archive package from your sheet set. The other two list the files in that package – the **Files Tree** tab does it in a tree view while the **Files Table** tab does it in a table view. You can remove any sheet or file by removing the check in the box to its left.

Use the **Enter notes to include with this archive** text box to add your own notes to the report that accompanies the archive. (The **View Report** button will display the report for you.)

The **Modify Archive Setup** button calls the Modify Archive Setup dialog box (Figure 26.009). This box is almost identical to the Modify Transmittal Setup dialog box we discussed in Lesson 23, p.554. Please refer to that lesson for explanations of the various options.

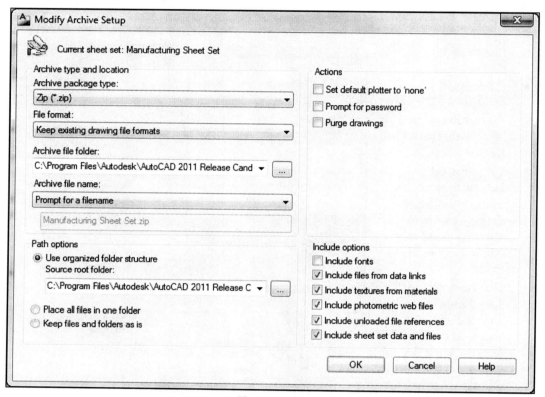

Figure 26.009

- o The cursor menu presented when you right click over a sheet in the sheet set looks like Figure 26.010. We haven't discussed two of these options.
    - The **Rename and Renumber** option calls a dialog box similar to the New Sheet dialog box (Figure 26.005, p.611). Here, however, you have a couple check boxes that enable you to **Rename layout to match Sheet title** (with or without the **Sheet number** prefix) or to **Rename drawing file to match Sheet title** (again, with or without the **Sheet number** prefix).
    - **Remove Sheet** removes the selected sheet from the set; it does not delete the drawing.

That's a tremendous amount of material for the first tab of a manager! But wait, there's more to come! (This might be a good time to shake

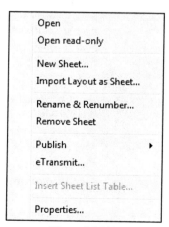

Figure 26.010

the cobwebs loose and go get some coffee before you continue.)

Let's try an exercise with what we've learned so far.

Do This: 26.2.1A	Creating Sheet Sets

I. Open the *Tara II Data Sheet* drawing in the C:\Steps\Lesson26\Sheet Sets folder. Notice that I've set up this drawing in Paper Space. We'll create a sheet set for this project.

II. Follow these steps.

26.2.1A: CREATING SHEET SETS

1. Open the Sheet Set Manager.

   Command: *ssm*

2. We'll use the Create Sheet Set wizard. You can start it by either entering the *NewSheetSet* command at the command line, or selecting **New Sheet Set** from the control box at the top of the palette.

   Command: *newsheetset*

3. AutoCAD begins the New Sheet wizard with an option to create your new sheet set using **Existing drawings** or **An example sheet set**.

   Using existing drawings allows you to select a folder(s) with files in it. AutoCAD will use the layouts from these drawings in the new sheet set.

   Using **An example sheet set** tells AutoCAD to use the structure from the sample in creating the structure for the new set. It won't add any drawings to your set.

   We'll use the **Existing drawings** option for this exercise.

   Pick the **Next** button to continue.

4. On the second page of our wizard, we can name the new set (call it *Tara II Project*). Add a description if you wish and tell AutoCAD to store the data file for the sheet set in the C:\Steps\Lesson26\Sheet Sets folder.

   Notice the **Sheet Set Properties** button. Use this to create custom properties for your project. (We'll ignore it for this exercise.)

   Pick the **Next** button to continue.

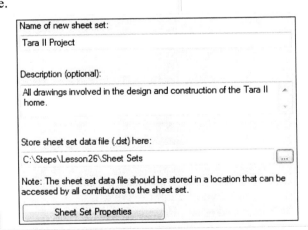

26.2.1A: CREATING SHEET SETS

5. Pick the **Browse** button on the next screen to select the folder where the drawings for your sheet set reside (they're in C:\Steps\Lesson26\Sheet Sets).
[NOTE: Remove the checks next to the
- *v* drawings if you're not using Windows Vista or 7; remove the checks next to the non - *v* drawings if you are using Vista or 7.]
Notice the **Import Options** button. This will call a dialog box where you can tell AutoCAD to **Prefix the sheet titles with the file name** (a good idea to help keep track of things) and/or to **Create subsets based on the folder structure**.
We'll ignore the import options for now.
Pick the **Next** button to continue.

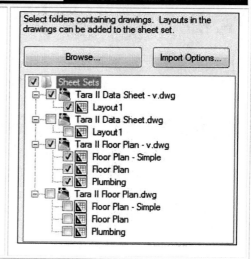

6. AutoCAD finishes with a preview of your sheet set properties (following figure) and asks you to confirm your setup. Pick the **Finish** button to complete the wizard.

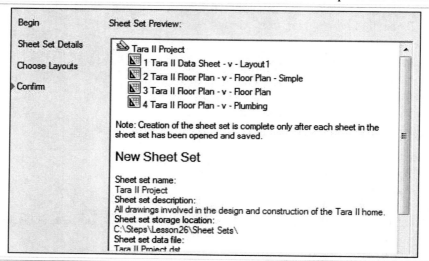

7. Notice (right) that AutoCAD has created your sheet set.

8. Now let's see what we can do with the cursor menus. Right click on **Tara II Data Sheet** and select **Rename and Renumber**.

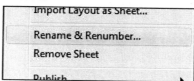

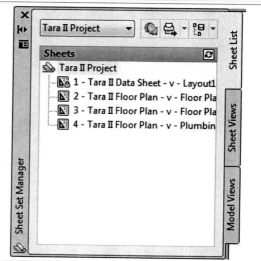

26.2.1A: CREATING SHEET SETS

9. Rename the sheet *Data Sheet* and:
   - be sure it's drawing #1,
   - put a check next to **Sheet title** under **Rename layout to match:**
   - Then pick the **Next** button to continue.

10. Repeat Step 9 to rename/renumber the rest of the sheets as follows:
    - sheet #2 *Simple Floor Plan*
    - sheet #3 *Floor Plan*
    - sheet #4 *Plumbing*

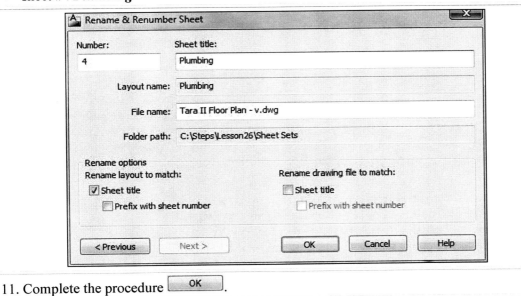

11. Complete the procedure [OK].

In this exercise, AutoCAD has created a sheet set using the two drawings available in the C:\Steps\Lesson26\Sheet Sets folder (*Tara II Data Sheet* and *Tara II Floor Plan*). Bear in mind that I created both with Paper Space layouts (these are versions of the files you've been using), views, and blocks. But this simple four-step wizard has done so much more than is readily visible.

Take a moment and explore the **Sheet List** in the light of what we've discussed so far. Then let's continue our discussion and see what else this complex tool has to offer!

26.2.2 The Sheet Views Tab

The **Sheet Views** tab of the Sheet Set Manager (Figure 26.011, p.617) presents a list of available views. In the **View by sheet** area, AutoCAD organizes the views under the name of the sheets in the set. Use the two buttons on the section's title bar to toggle between **View by category** and **View by sheet**. You can even use the button at the top of the palette (**New View Category**) to create your own categories. Once you have your own category, just drag and drop the views where you want them.

You can also use this tab to mine drawings for their blocks.

Let's look at the cursor menus for this tab.

- The first (Figure 26.012, p.617) displays when you right click on the title of the sheet set (when in **View by category** mode). Use this one to either create a **New View Category** (just as the button does) or to check the **Properties** of the sheet set.

616

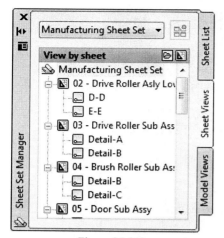

Figure 26.011

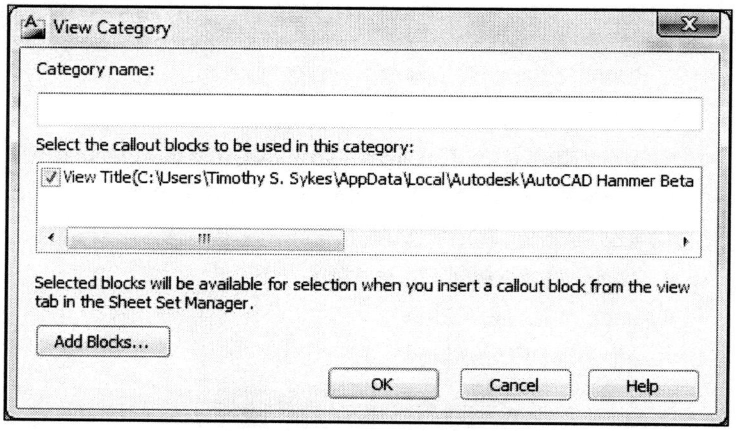

Figure 26.013

Figure 26.012

If you select the **New View Category** option (or pick the button at the top of the palette), AutoCAD will present the View Category dialog box (Figure 26.013). Here you can name your new category (**Category name** text box) or **Select the callout blocks to be used in this category**. The list box shows the blocks that are currently available in the sheet set. Use the **Add Blocks** button to add additional blocks or remove existing ones (you'll see this in our next exercise).

- The second cursor menu (Figure 26.014) appears when you right click on one of the folders. Use it to **Rename** a category, **Remove** [a] **Category**, or view the **Properties** of that category. You can also add blocks to the category via the Properties dialog box.

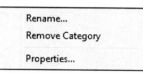

Figure 26.014

- The last (Figure 26.015) appears when you right click on one of the views.
  - The **Display** option is one of AutoCAD's finest. It opens the appropriate drawing with the selected view displayed. How's that for handy?! (Show some restraint with this one. You can wind up with several drawings open at once ... and wonder why your system has slowed so much!)

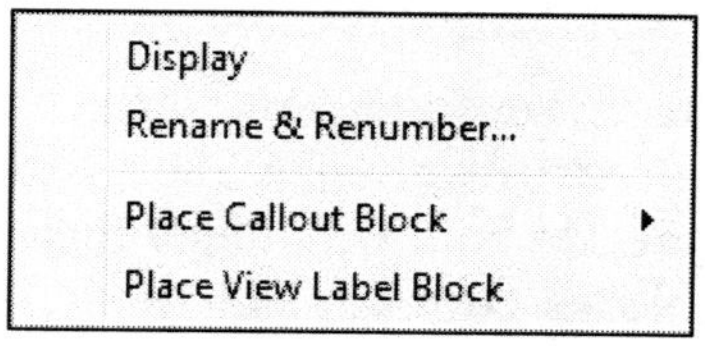

Figure 26.015

  - When you select **Rename and Renumber**, AutoCAD will present the Rename and Renumber dialog box seen in Figure

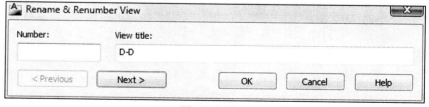

Figure 26.016

  26.016. Here you can rename and/or renumber the selected view.
  - **Place Callout Block** presents a list of blocks available from the sheet set drawing that contains the selected view. You can insert these blocks into the current drawing. If no blocks are available, AutoCAD displays a **Select Blocks** option. Use this to call the View Category dialog box (Figure 26.013). From there, pick the **Add Blocks** button to add blocks to the list.
  - **Place View Label Block** inserts a standard view label (we'll see this in our next exercise).

Let's try another exercise before we continue to the last tab.

Do This: 26.2.2A	**Working with the Sheet Views Tab**

I. Begin a new drawing from scratch.

II. Use architectural units and set the limits for a ¼" scale, C-size sheet of paper (upper left limits of 88',68').

III. Follow these steps.

26.2.2A: WORKING WITH THE SHEET VIEWS TAB

1. If the Tara II sheet set isn't open, use the selection control box at the top of the palette to open it. (Select **Open** from the list, navigate to the C:\Steps\Lesson26\Sheet Sets folder, and then select the *Tara II Project.dst* file.) Alternately, you can use the *OpenSheetSet* command.
   Command: *opensheetset*

2. Pick the **Sheet Views** tab and then the **View by category** button to adjust the list. Notice the categories. Let's make these more manageable.

3. Use the cursor menu to **Rename and Renumber** the three views in the Border Data category as follows:
   - rename **FP - Block Border Data** to **Floor Plan Title Block**; make it #**001**;
   - rename **FP – Simple Title and Revision Block** to **Simple Plan Title Block**; make it #**002**;
   - rename **Plumbing – Block Border Data** to **Plumbing Title Block**; make it #**003**.

4. Right click on the **Project** title and select **New View Category**.

5. Create a new category called **Details**. (Don't add any blocks at this time.)

6. Move both of the views listed under the **Plumbing Views** category to the **Details** category. (Drag the listing with the left mouse button and release it under the **Details** category title.)

7. Right click on the empty **Plumbing Views** category and pick **Remove Category** from the cursor menu.
   The **Sheet Views** tab looks like the figure at right.

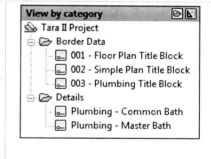

8. Use the cursor menu to open the View Category dialog box of the **Details** category (pick **Properties**).

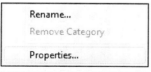

9. Pick the **Add Blocks** button.

10. AutoCAD presents a List of Blocks dialog box. The blocks we need aren't in this list, so pick the **Add** button.

11. AutoCAD presents the Select Block dialog box (following figure). Pick the **Browse** button ··· next to the **Enter the drawing file name** text box.

26.2.2A: WORKING WITH THE SHEET VIEWS TAB

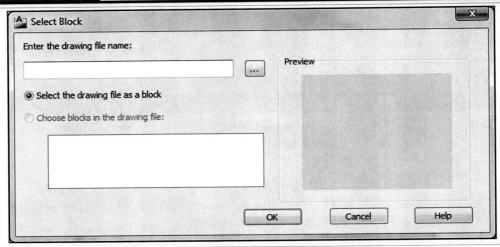

12. AutoCAD presents a standard Open File dialog box. Navigate to the C:\Steps\Lesson26\Sheet Sets folder and open [Open] the *Tara II Floor Plan* drawing.

13. AutoCAD returns to the Select Block dialog box. Notice (following figure) the difference. Now AutoCAD lists the source file for your blocks in the **Enter the drawing file name** text box.
Put a bullet next to **Choose blocks in the drawing** file. This makes the individual blocks in the listed drawing available for your selection.
Select the **Sink**, **Tub**, and **WC**, as shown.
Pick the **OK** button here and on the List of Blocks dialog box to return to the View Category dialog box.

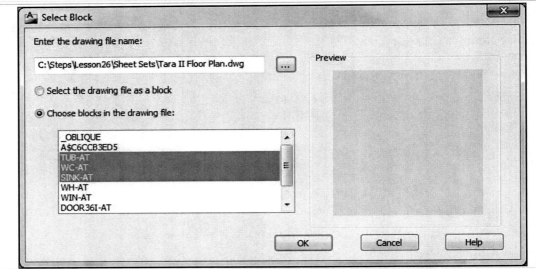

14. AutoCAD tells you that the blocks are now associated with your sheet set. Pick the **OK** button [OK] to continue.

15. Put checks in the boxes next to our new blocks (following figure) and pick the **OK** button to complete the procedure. Your blocks are now available for insertion into the current drawing (or any drawing while the Tara II Project is open in the Sheet Set palette).

26.2.2A: WORKING WITH THE SHEET VIEWS TAB

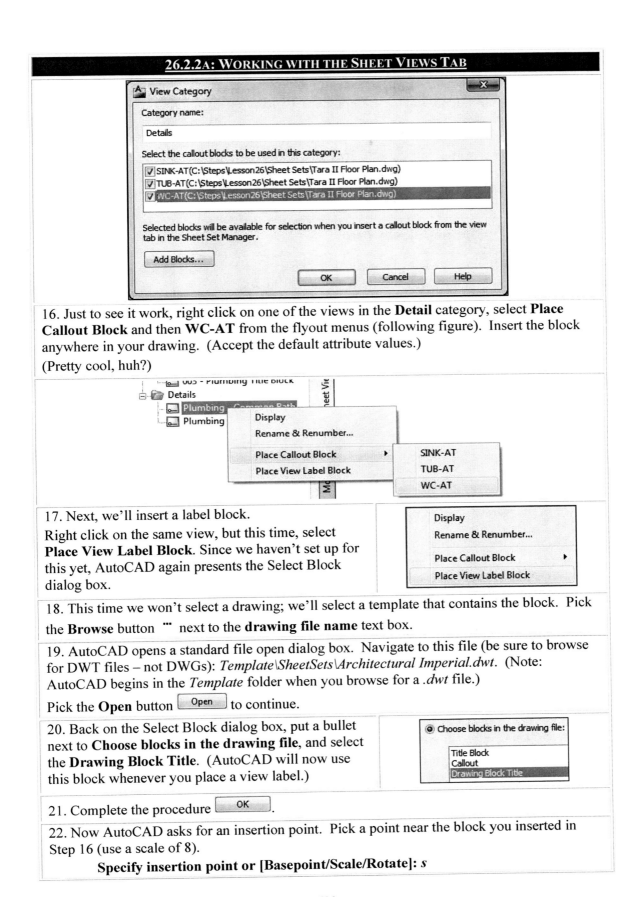

16. Just to see it work, right click on one of the views in the **Detail** category, select **Place Callout Block** and then **WC-AT** from the flyout menus (following figure). Insert the block anywhere in your drawing. (Accept the default attribute values.)
(Pretty cool, huh?)

17. Next, we'll insert a label block.
Right click on the same view, but this time, select **Place View Label Block**. Since we haven't set up for this yet, AutoCAD again presents the Select Block dialog box.

18. This time we won't select a drawing; we'll select a template that contains the block. Pick the **Browse** button ``...`` next to the **drawing file name** text box.

19. AutoCAD opens a standard file open dialog box. Navigate to this file (be sure to browse for DWT files – not DWGs): *Template\SheetSets\Architectural Imperial.dwt*. (Note: AutoCAD begins in the *Template* folder when you browse for a *.dwt* file.)
Pick the **Open** button to continue.

20. Back on the Select Block dialog box, put a bullet next to **Choose blocks in the drawing file**, and select the **Drawing Block Title**. (AutoCAD will now use this block whenever you place a view label.)

21. Complete the procedure.

22. Now AutoCAD asks for an insertion point. Pick a point near the block you inserted in Step 16 (use a scale of 8).
   Specify insertion point or [Basepoint/Scale/Rotate]: *s*

26.2.2A: WORKING WITH THE SHEET VIEWS TAB

Specify scale factor for XYZ axes <1>: *8*

Specify insertion point or [Basepoint/Scale/Rotate]:

The label looks like the following figure.

⌒ Plumbing – Common Bath
SCALE: ×××××××××××××

23. You can use the Properties palette to edit the attributes of this block.

Wasn't that nifty? Of course, you won't have to do the setup in Steps 18 through 21 every time.

| 26.2.3 | **The Model Views Tab** |

One more tab! And we can relax … this one's fairly easy.

The **Model Views** tab (Figure 26.017) contains a single option – **Add New Location**. Use this to add the location (folder) where you've located additional drawings that you wish to be associated with the current sheet set.

In our next exercise, we'll use this third tab to add an additional folder to our sheet set. Then we'll import a layout into our sheet set and place a sheet list onto our *Tara II Data Sheet*.

Let's begin.

Figure 26.017

| Do This: 26.2.3A | **Working with the Model Views Tab & Some Final Touches** |

I. Be sure you're still in the *Tara II Data Sheet* drawing in the C:\Steps\Lesson26\Sheet Sets folder. If not, please open it now.

II. Set the **TEXT** layer current, and pick on the **Model Views** tab to put it on top of the Sheet Set palette.

III. Follow these steps.

26.2.3A: WORKING WITH THE MODEL VIEWS TAB & SOME FINAL TOUCHES

1. First, let's associate the new folder. Double click on the **Add New Location** option in the **Locations** list box. Alternately, you can pick the **Add New Location** button 🗄 at the top of the palette.

2. AutoCAD presents a standard Open dialog box. Navigate to the C:\Steps\Lesson26\Resource Drawings folder and pick the **Open** button [Open].

Notice (right) that the palette now shows the new folder and the drawing file within it. This drawing is now associated with the sheet set and may be plotted, archived, or transmitted with it.

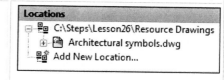

3. Next, we'll import a new sheet into our set. Return to the **Sheet List** tab.

26.2.3A: WORKING WITH THE MODEL VIEWS TAB & SOME FINAL TOUCHES

4. Right click on the **Tara II Project** title (at the top of the **Sheets** section) and pick the **Import Layout as Sheet** option [Import Layout as Sheet...].
AutoCAD presents the Import Layouts as Sheets dialog box (Figure 26.007, p.612).

5. Pick the **Browse for Drawings** button [Browse for Drawings...]. AutoCAD presents a standard Open File dialog box.

6. Navigate to the C:\Steps\Lesson26\Resource Drawings folder and select the *Architectural Symbols* file. AutoCAD returns to the Import Layouts as Sheets dialog box, which now shows the Architectural symbols drawing.

7. Complete the procedure [Import Checked]. Notice that the new sheet appears in the **Sheets** section of the palette.

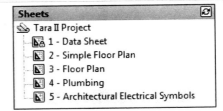

8. Use the **Rename and Renumber** option in the cursor menu to give this sheet the number *5* and the title *Architectural Electrical Symbols*.

9. Finally, let's place a drawing sheet list on our *Tara II Data Sheet*. Begin by picking the **Insert Sheet List Table** option [Insert Sheet List Table...] on the cursor menu presented when you right click on the **Tara II Project** title.
AutoCAD presents an Insert Sheet List Table dialog box (following figure).

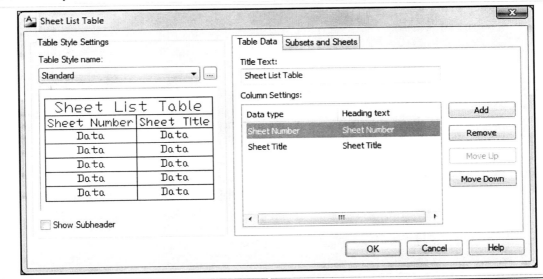

10. This dialog box is very similar to others you've used to insert tables. But this one has a **Subsets and Sheets** tab. Pick that one now (right).

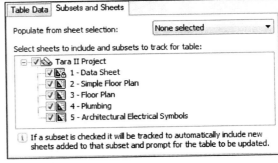

11. This tab allows you to select which sheets will go into your table. (Uncheck any you wish to exclude.)
We want all the sheets, so accept the defaults and pick the **OK** button [OK] to proceed.

622

26.2.3A: WORKING WITH THE MODEL VIEWS TAB & SOME FINAL TOUCHES

12. AutoCAD removes the dialog box and asks you to specify an insertion. Place the table below the text (on the right side of the sheet). Your drawing now looks like the following figure.

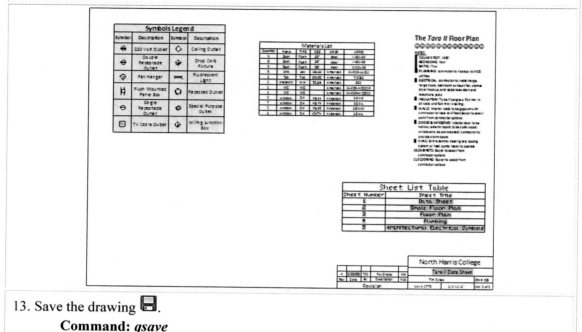

13. Save the drawing 💾.

    Command: *qsave*

The table is filled with fields that contain hyperlinks. These will link to the DWFs you'll create in Section 26.3.

26.3 Using Your Sheet Sets to Share Information

You've seen how to set up your sheet sets, but what's the purpose? Aside from making the transition between drawings and sharing blocks (which you do as easily with other tools), what good are sheet sets?

Sheet sets help you share *project* information (as opposed to information about a specific drawing or discipline). Use them to communicate project status to clients or employers and to archive your project at predetermined stages of development.

Let's do some exercises that'll walk you through plotting (as we don't have a plotter, we'll plot to a DWF), transmitting information to our client/employer, and archiving our project.

Do This: 26.3A	Sharing Information with Sheet Sets – Creating a Multi-Sheet DWF File

I. Be sure you're in the *Tara II Data Sheet* drawing in the C:\Steps\Lesson26\Sheet Sets folder. If not, please open it now.
II. Be sure the Sheet Set Manager and the **Tara II Project** are open.
III. Put the **Sheet List** tab on top of the manager.
IV. Follow these steps.

26.3A: SHARING INFORMATION WITH SHEET SETS – DWFs

1. We'll begin by creating a DWFx file. Right click on the project title and select the **Publish** option `Publish ▶` from the cursor menu. Then pick the **Publish to DWFx**

26.3A: SHARING INFORMATION WITH SHEET SETS – DWFS

`Publish to DWFx` from the **Publish** menu.
AutoCAD presents a standard Select File Dialog box.

2. Save the DWFx file as *Tara II Project* in the C:\Steps\Lesson26\DWFs folder. (Create the DWFs folder if it isn't there.)

3. Pick the **Select** button `Select` to complete the procedure. (AutoCAD may take a few minutes to complete the procedure, but it works in the background so you can continue to work on the drawing if you wish.)

4. AutoCAD let's you know when it's finished, and it gives you a chance to see the details of what it did. You can look at this if you wish, then close the bubble.

5. Let me show you a quick way to view your most recently published DWF. Right click on the **Plot/Publish Details** … icon on the right end of the status bar and select **View Plotted File** `View Plotted File...`. AutoCAD will launch Design Review with the DWF file (see the following figure).

Take a few minutes to look thing over. Scroll down to the **Sheet Properties** area at the left side of the window. This offers details about the currently selected sheet.

You can select different sheets to view in the **Thumbnails** frame at the left area of the window. (You can also pick the titles in the Sheet List Table you created in the last exercise!)

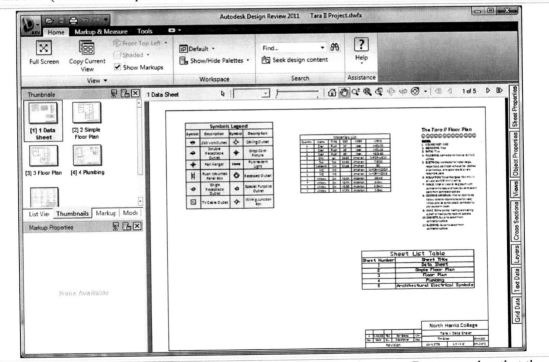

You see that creating a DWF file for a multi-sheet set is really quite easy. But remember that the **Publish to DWF(x)** option uses defaults set up in the Publish dialog box, so you'll need to go there to change any settings.

The client liked our project; so let's transmit it to him.

| Do This: 26.3B | Sharing Information with Sheet Sets – Using eTransmit for a Multi-Sheet Set |

I. Be sure you're in the *Tara II Data Sheet* drawing in the C:\Steps\Lesson26\Sheet Sets folder.
II. If not, please open it now.
III. Close Design Review.
IV. Be sure the Sheet Set Manager and the **Tara II Project** are open.
V. Follow these steps.

26.3B: SHARING INFORMATION WITH SHEET SETS - ETRANSMIT

1. First, we'll have to create a setup for our transmittal. Right click on the project title and select **Transmittal Setups** `Transmittal Setups...` from the cursor menu. AutoCAD presents the Transmittal Setups dialog box (Lesson 23, p.553).

2. Pick the **New** button `New...` to create a new setup.

3. AutoCAD asks for a name for the new setup. I'll call mine *Construction Issue Transmittal*.
Continue to the Modify Transmittal Setup dialog box. (You're familiar with this from Lesson 23, p.554.)

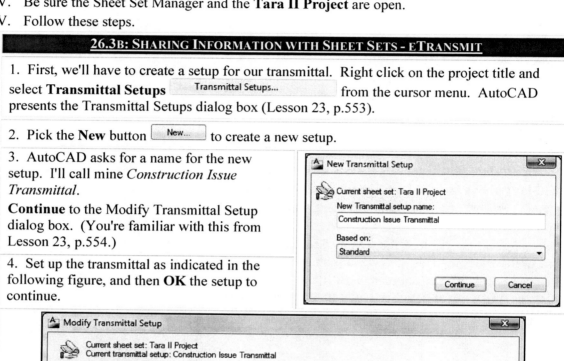

4. Set up the transmittal as indicated in the following figure, and then **OK** the setup to continue.

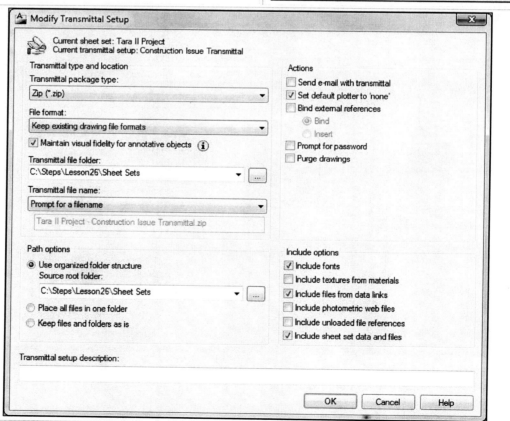

26.3B: SHARING INFORMATION WITH SHEET SETS - ETRANSMIT

5. **Close** the Transmittal Setups dialog box.

6. Save the drawing.
   Command: *qsave*

7. Now we can create the transmittal. (This is the easy part.) Select **eTransmit** from the project title's cursor menu.

8. AutoCAD prepares the transmittal, and then presents the Create Transmittal dialog box (see the following figure).

   You're familiar with this dialog box from your studies in Lesson 23, but notice that AutoCAD has included a **Sheets** tab. This tab lists the drawings that AutoCAD will include in the transmittal (you can remove the check from the check box to exclude a drawing).

   Notice that not all of the sheets have checks. These sheets are included as layouts in a checked drawing.

   Select the **Construction Issue Transmittal** setup (or whatever setup you created) and pick the **OK** button to continue.

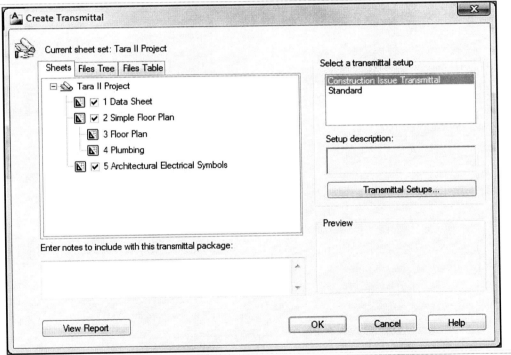

9. AutoCAD asks for a name for the transmittal file. Accept the default and pick the **Save** button to continue.

   AutoCAD creates the transmittal and closes the dialog box.

10. You should check your transmittal before sending it. Go to the C:\Steps\Lesson26\Sheet Sets folder to make sure your transmittal file was created. (You can open it if you have a program capable of opening a zip file, and check the files that were included.)

Okay, we've plotted our project and sent it to the client. Now we need to archive it – put it away so we can start working on something else.

Here we go.

| Do This: 26.3C | Sharing Information with Sheet Sets – Archiving Your Project |

I. Be sure you're in the *Tara II Data Sheet* drawing in the C:\Steps\Lesson26\Sheet Sets folder. If not, please open it now.
II. Be sure the Sheet Set Manager and the **Tara II Project** are open.
III. Follow these steps.

26.3C: SHARING INFORMATION WITH SHEET SETS - ARCHIVING

1. Select **Archive** [Archive...] from the project title's cursor menu. AutoCAD presents the Archive a Sheet Set dialog box (Figure 26.008, p.612).

2. We'll need to set up the archive so pick the **Modify Archive Setup** button [Modify Archive Setup...].

3. AutoCAD presents the Modify Archive Setup dialog box. Set it up as indicated and pick the **OK** button to continue.

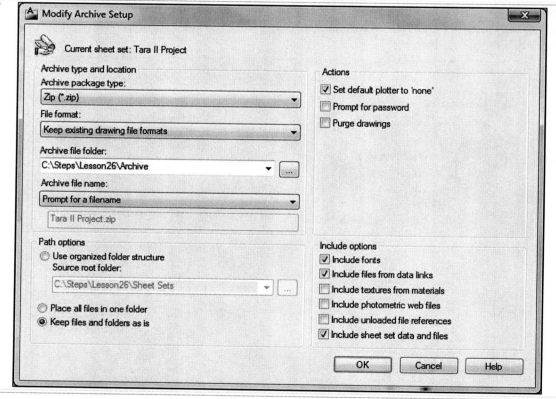

4. AutoCAD returns to the Archive a Sheet Set dialog box. Pick the **OK** button [OK] to complete the procedure.

5. AutoCAD presents a standard Save File dialog box. Accept the file name and pick the **Save** button [Save].

It takes a moment or two for AutoCAD to complete the procedure, but when it's finished, close the drawing. (Don't make any changes or save anything after archiving or you may have to repeat the procedure!)

26.4 Extra Steps

Figure 26.018

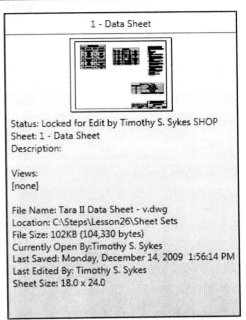

Figure 26.019

We didn't look at one other cursor menu; you'll find this one (Figure 26.018) when you right click on the border of the Sheet Set Manager.

Here you can toggle on/off the **Preview** and **Display** panes or **Close** the manager. That's simple enough, but let's take a look at the **Tooltip Style** options.

First, let your cursor rest on one of the sheets in your Sheet Set Manager. Notice the tooltip. Let it rest a bit longer and notice that the tip changes to something like Figure 26.019. The **Tooltip Style** options control what you see here. You're currently viewing the **Previews & Details** settings – the default. What other settings are there?

- **Name** – for viewing just the name of the sheet.
- **Preview** – for seeing a preview thumbnail of the actual sheet. You can select from **Small**, **Medium** and **Large** thumbnails.
- **Details** – for seeing just the name and details of the sheet without the thumbnail.

These settings will appear again in the Xrefs manager you'll see in our next lesson!

Find the best available sheet set references at: http://heidihewett.blogs.com/my_weblog/files/Sheets_Happen.pdf. In fact, I strongly recommend any of the myriad references you'll find at Heidi Hewett's website. Heidi is one of the shiny stars in Autodesk's brass. She writes well, clearly, and knows her material. If you ever get a chance to hear her speak, it'll be worth the effort, too! (Heidi frequently speaks at AUGI – AutoCAD User Group International - meetings.)

26.5 What Have We Learned?

Items covered in this lesson include:

- *Sheet Sets*
  - *creation*
  - *transmittals*
  - *archiving*
  - *publishing*
- *Commands:*
  - ***NewSheetSet***
  - ***OpenSheetSet***
  - ***SheetSet***

This has been a tough lesson. For a palette with such a simple appearance, AutoCAD has created a bit of a monster … but like Godzilla, it's a monster that can be quite helpful. (Okay, so Godzilla helped mostly in urban renewal efforts for a model Tokyo … but you get the idea.)

I do encourage you to take some time to get comfortable with sheet sets. Like most complex tools, those who demonstrate mastery are the ones who make the money.

26.6 Exercises

Open the *Sample Unit Data Sheet* in the C:\Steps\Lesson26\Exercises folder. Follow these instructions.

1. Create a Sample Unit Project sheet set in the C:\Steps\Lesson26\Exercises folder.
2. Use the drawings found in the C:\Steps\Lesson26\Exercise folder.
3. Rename the sheet names to *Data*, *Plan*, and *Details*.
4. Import the drawings/sheets found in the C:\Steps\Lesson26\Exercise Resource Drawings folder.
5. Rename *Piping Symbols Data Sheet* to *Piping Symbols* and the *Welding Symbols Data Sheet* to *Welding Symbols*.
6. Insert a Sheet List Table.
7. Archive the sheet set.

Go back to the Tara II project and publish to a PDF. Compare the PDF with the DWFx you created in Ex 26.3A (p.623).

26.7 For Web-Based Review Questions and Additional Exercises, visit: http://foragerpub.com/AcadFiles/2011/2011.htm

Don't stop now! Check this out!

Lesson 27

Following this lesson, you will:

- ✓ *Know how to reference a drawing or PDF/DWF/DGN from another drawing*
  - *Know how to use the External Reference Palette & Reference Panel*
  - *Know how to attach, detach, overlay, and underlay a reference*
  - *Know how to clip a reference to see just what you want to see*
  - *Understand reference drawings' dependent symbols*
  - *Know how to load, unload, and reload a reference*
  - *Know how to edit a referenced drawing from within the primary drawing*
- ✓ *Know how to open a reference from within the primary drawing*
- ✓ *Know how to permanently bind a referenced drawing to the primary drawing*

Externally Referenced DWGs, PDFs, DWFs, & DGNs

Think way back to our discussion of layers in Lesson 7. I explained layers by referring to the presentation method Encyclopedia Britannica *used to detail the human body (plastic overlays with the different systems of the body, p.149). This is also the idea behind external references – Xrefs.*

There is, however, quite a difference between layers and external references. We use layers, as you know, to control the display of, and differentiate between, objects in a drawing (much as we used linetypes and widths on the drawing board). On the other hand, we use Xrefs to save drawing time and computer memory by sharing information (much as Britannica shared the outline of the human body among the various system overlays). We use attached PDFs, DWFs and DGNs much as we use referenced drawings, but they're easier (if not as accessible) and considerably more secure.

> When working with Xrefs, file location is a critical consideration. For the exercises in this lesson to work properly with the files provided, the files must be located in one of two places:
> 1. The best location is the C:\Steps\Lesson27 folder. This site has been thoroughly tested to be sure the files work properly.
> 2. The files may be placed on a network or other location *provided the path is defined in the* **Project File Search Path** (**Files** tab of the Options dialog box). If you're not comfortable with the Options dialog box, however, and don't have a CAD guru who can help, *please don't attempt to make changes in the Options dialog box.* Use the first choice instead.

27.1 Working with Externally Referenced DWGs, PDFs, DWFs, and DGNs

We've seen PDFs and DWFs before, but what are Xrefs?

Xrefs – externally referenced files – are typically drawings (but can be PDFs, DWFs, or DGNs) called into a drawing as a referenced background or object. They are, in fact, quite similar to blocks with some notable exceptions:

- Unlike blocks, Xrefs occupy very little space within a drawing. That is, you'll find the drawing's increase in size negligible with the inclusion of an Xref.

- Whereas a block may contain attributed information, an Xref can't. However, the referenced drawing file may contain blocks with attributes, and the values of these attributes may be extracted from the primary drawing (the one doing the referencing). This doesn't hold true, however, for PDF/DWF/DGNs.

- The primary drawing automatically reloads the referenced file whenever the primary drawing is accessed. Additionally, when the reference is updated, AutoCAD will let you know with an information bubble in the lower right corner of the window. So, the reference is as current as the last time the primary drawing was opened.

> Microstation – another CAD system – produces DGN files in much the same way AutoCAD produces DWGs. AutoCAD allows the underlaying (more on this as we go) of 2D, V7 and V8 DGN files for reference.

What are the benefits of using Xrefs?

- An important benefit of using Xrefs is that another operator may edit the referenced drawing without tying up your primary drawing. The changes he makes become available to you as soon as you reload the reference. (This may prove more difficult with a PDF, DWF, or DGN unless you have software to edit them.)

- Like blocks, any number of primary drawings may use a single referenced file. But unlike blocks, updating the references is automatic.

    Consider a refinery unit. Each unit requires civil, structural, piping, and electrical plans. Generally, the structural designer will copy the civil drawing for a "foundation" for his work. Likewise, the piper and electrical designer will use the structural plan. Changes to the structural

plan may go unnoticed for days or weeks by others (unless some extraordinary communication occurs). However, if the pipers and electrical designers Xref the structural plan,
- they won't have to draw the background themselves, and
- they'll be automatically notified of any changes made to the referenced drawing, and
- any changes made by the structural designer will automatically show on the others' drawings with the next loading or reloading.
- On the other hand, your client may want you to use an existing drawing as a background but may be uncomfortable with actually giving you the drawing. The client can, instead, give you a PDF/DWF "print" of the drawing (thus keeping his pristine original) and allow you to use the reference as an unchangeable background for your work.

Let's take a look at how to use Xrefs.

27.1.1 Attaching and Detaching External References to Your Drawing

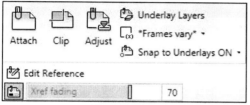

Figure 27.001

	Where to Find It:
Command Line:	*ExternalReferences* or *Xref*
Hotkey(s):	xr
Ribbon (Tab/Panel):	Insert – Reference – [Settings arrow ⌐]
Menu:	Insert – **External References**
Toolbar:	Reference – **External References**

You'll discover two major methods for working with referenced files: the ribbon (Figure 27.001) and the External References palette (Figure 27.002) which you'll open with the *Xref* command. You

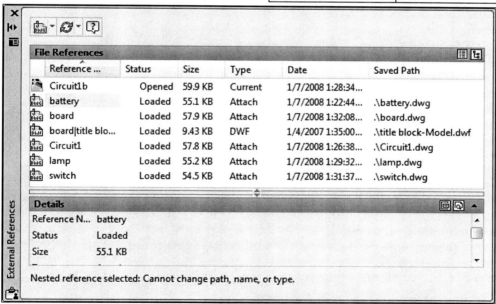

Figure 27.002

can also use the command line for most commands (refer to the WTFI tables as we go). Let's take a look.
- At the top of the External Reference palette (Figure 27.002), you'll find a couple flyout buttons. You can reach the commands these buttons call a bit faster using their ribbon counterparts.

- o The **Attach** button begins the *Attach* command[*].

  The *XAttach* command begins with a standard Select File dialog box. (Accessing the command via the palette button causes AutoCAD to filter the Select File dialog box for the selected file type. Using the button

	Where to Find It:
Command Line:	*Attach*
Ribbon (Tab/Panel):	Insert – Reference – **Attach**
Menu:	Insert – [file type]

  means that you may have to change the type of file the dialog box displays.) There you'll follow the path to the file you wish to reference. Once you've selected a file and picked the **Open** button, AutoCAD presents the Attach External Reference dialog box (Figure 27.003).

  - The **Name** control box indicates the file you're attaching. You may pick the **Browse** button to select another file.
  - In the **Reference Type** frame, tell AutoCAD if you want your reference to be an **Attachment** or **Overlay** (more on this in Section 27.1.4, p.644).
  - The **Path type** provides three options to make location of the reference a bit easier.

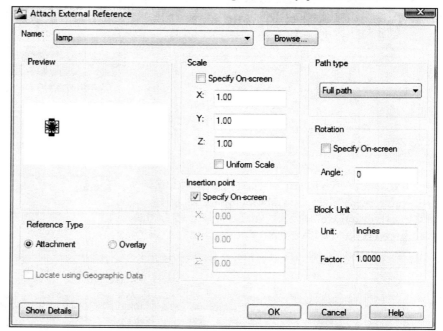

Figure 27.003

    ❖ Using **Full path** (the default) means that AutoCAD will use the full path (drive & location) when searching for the reference.
    ❖ The **Relative path** option tells AutoCAD that the referenced file may be located in a subfolder of the folder in which the primary drawing is located. Relative paths won't work if the referenced drawing is on another drive.
    ❖ **No path** means that the referenced drawing must be located in the same folder as the primary drawing or in a path identified in the **Project File Search Path** in the Options dialog box.
  - The **Insertion point**, **Scale**, **Rotation**, and **Block Unit** frames are identical to their counterparts on the Insert dialog box.

---

[*] A similar, older command – *XAttach* – is identical to the *Attach* command, but attaches only .dwgs. The *Attach* command follows the same procedures to attach .dwgs, .dwf, .dgns, and .pdfs.

- Place a check next to **Locate using Geographic Data** if both your primary and referenced files have geographic data available. AutoCAD will use the geographic date as a referent to insert the file.

We'll continue now with the External References palette and discuss the remaining ribbon tools as we proceed through the lesson.

- The **Attach DWF** button (reached via the **Attach** flyout on the palette) begins the *DWFAttach* command; the **Attach DGN** button begins the *DGNAttach* command, and the **Attach PDF** button begins the *PDFAttach* command.

  These commands also begin with a standard Select File dialog box where you'll follow the path to the file you wish to reference. Once you've selected a file, AutoCAD presents the Attach [file type] Underlay dialog box (Figure 27.004). The options presented here correspond to their counterparts on the External Reference dialog box already discussed (Figure 27.003, p.633). Notice, however, that you cannot overlay or attach the DWF/PDF/DGN; these can only be "underlain" – that is, they're for background reference only.

You'll find another button under the **Attach** flyouts – **Attach Image** – which we'll discuss in Lesson 28, p.658.

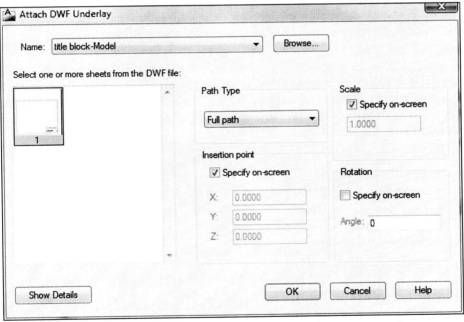

Figure 27.004

- The flyout buttons next to the palette's **Attach** flyouts contains two buttons – **Refresh** and **Reload All Xrefs**. These do exactly what their names imply but provide a convenient way to perform these tasks without resorting to the keyboard.

- The final button along the top of the palette calls AutoCAD's help window.

- AutoCAD allows you to view the items in the **File References** frame (of the External Reference palette – Figure 27.002, p.632) using either a **List View** (the default -) or a **Tree View** . Toggle between types of views using these buttons in the upper right corner of the frame. (We'll use the default list view in this lesson).

- In the **File References** frame, AutoCAD provides several columns of information:

- The **Reference Name** is the file name of the referenced file.
- The drawing's **Status** will fall into one of several categories:
  - **Loaded** simply means that the reference is currently attached.
  - An **Unloaded** reference is removed from the primary drawing. This isn't a permanent condition. The reference remains intact and may be reloaded as desired.
  - An **Unreferenced** file is attached to the primary drawing but has been erased. You can reattach it.
  - A file that no longer resides in the folder from which it was originally referenced (or in the search path defined in the Options dialog box) is marked in the **Status** column as **Not Found**.
  - If AutoCAD can't read the reference, it marks it as **Unresolved**.
  - Xrefs can be nested. An **Orphaned** reference was nested into a reference that is unreferenced, unloaded, or not found.
- The **Size** column indicates the size of the referenced file.
- You can either **Attach** or **Overlay** (see the **Type** column) a referenced drawing (but not a PDF, DWF, or DGN). An attached drawing will go with the primary drawing if another drawing references it; an overlaid drawing won't (more on this in Section 27.1.4, p.644).
- The **Date** column indicates when the reference was last modified.
- The **Saved Path** column is a bit misleading. The path refers to the location of the drawing when it was originally referenced but *not necessarily where the reference is found*. Let me explain.

When AutoCAD begins a drawing, it searches for a referenced drawing in the **Saved Path** location. If it doesn't find it there, it searches the **Project File Search Path** defined in the Options dialog box. Then it searches the folder in which the current drawing resides. It'll use the first file it finds with the appropriate name regardless of its location. But it won't change the information in the **Saved Path**.

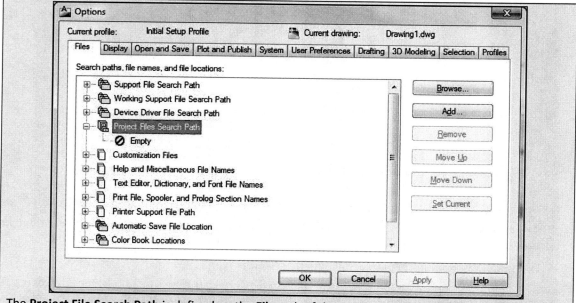

The **Project File Search Path** is defined on the **Files** tab of the Options dialog box (above). Pick the "+" beside the **Project File Search Path** listing to see the folders AutoCAD will search. To add a folder, select the **Project File Search Path** listing and pick the **Add** button. AutoCAD will add a listing, and you must type in the desired name.

> Then pick the plus beside the new project name, double click on the **Empty** slot, and select the folder you wish to add to the search path.
> *Warning: Don't make changes in the Options dialog box without first consulting the CAD guru on your project.*

> You'll find another useful, if less obvious, tool within the External References palette. When you select a reference in the **File References** section, AutoCAD will highlight that referenced file in the drawing to help you find it. Alternately, if you select a referenced file in the drawing, AutoCAD will also select it in the **File References** section of the External References palette.

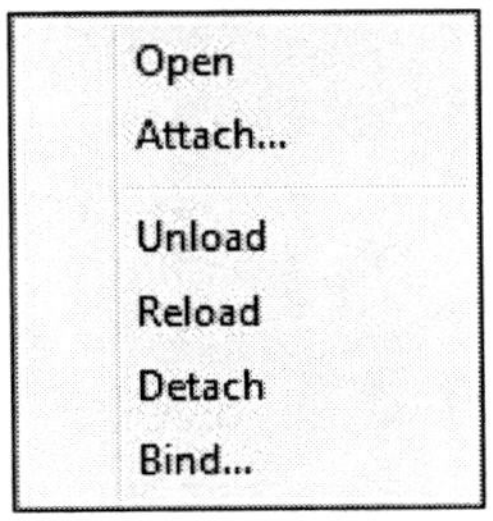

- AutoCAD may provide different cursor menus within the **File Reference** frame, depending upon where you right click 🖱. Right clicking in an open area produces a menu that includes the same tools called by the buttons we've already discussed. Right clicking on a referenced file, however, calls a menu (Figure 27.005) that proves useful in managing your references.

  Figure 27.005

  o The **Open** option opens the selected reference. Drawings open in a new AutoCAD window where you can edit them (we'll look at another way to edit referenced drawings later in this lesson); DWFs open in Design Review, PDFs in the default Adobe viewer on your computer. You can't **Open** or edit a DGN.
  o The **Attach** option reopens the appropriate Attach dialog box. We've already discussed these.
  o **Unload** a drawing to remove a reference that you may want to use later. Unloading doesn't permanently remove the reference from the drawing, but it does remove it from display.
  o Use the **Reload** option to update a reference without having to reopen your primary drawing, or to display an unloaded drawing.
  o Use the **Detach** option to remove selected references. (Note: You can't detach a nested reference. You must detach its primary drawing.)
  o Use the **Bind** option only on a referenced drawing; it isn't available for PDFs, DWFs, or DGNs. It will permanently attach a reference drawing to a primary drawing. (This isn't the same as the *XBind* command, which we'll discuss in Section 27.1.3, p.642.) AutoCAD presents the Bind Xrefs dialog box (Figure 27.006).

Figure 27.006

> Before you bind a referenced drawing, you should understand something about dependent symbols. *Dependent symbols* are the layers, blocks, text styles, dimension styles, and linetypes that are part of a referenced drawing. To be available in the primary drawing, these symbols require that the referenced drawing be attached. PDFs and DWFs do not include dependent symbols and, although they will reflect the freeze/thaw/on/off settings of the layer on which they're inserted, they won't reflect color or linetype settings.

As referenced information, *dependent symbols* are identified by the name of the referenced drawing followed by a bar and the symbol's name (i.e., **board|slots** is the **slots** layer on the referenced *board* drawing). The two options in the Bind Xrefs dialog box determine how Xref dependent symbols will be treated.

- **Bind**ing a referenced drawing causes AutoCAD to rename the dependent symbols but retain the referenced drawing's name in the new name. Thus, the **board|slots** layer will become **board0slots**. This approach means that you'll always be able to trace where the symbol originated. [If the **board0slots** already exists, the number between the dollar signs increases (**board1slots**, **board2slots**, etc.).]

- **Insert**ing a referenced drawing changes the referenced drawing into a block with the name of the referenced drawing becoming the name of the block. Dependent symbols become a part of the drawing but drop the name of the referenced drawing (i.e., **board|slots** becomes **slots**).
- The **Details** frame of the External Reference palette (Figure 27.002, p.632) can provide details about the reference selected in the **File References** frame. Most of the details are informational, although some can be modified (such as the **Attach/Overlay Type** of referenced drawing or the **Found At** location of the reference). The **Found At** box indicates where the selected drawing was found when AutoCAD loaded. This is where AutoCAD actually found the drawing and may not be the same as the saved path. Use the **Browse** button ••• (to the right of the selected box) to select a different copy of the reference.

  Alternately, you can have the **Details** frame provide a preview of the selected reference by using the **Preview** toggle on the right end of the frame's title bar.

Whew! That was a lot of material to cover! Let's try our hands at referencing.

Here's the scenario for this lesson:

> *Your employer is developing a new product – a kit designed to help children learn about electricity. The kit will contain a circuit board where the child can attach a battery pack, switches, lights, resistors, and a galvanometer (kids love those big scientific words). The kit must be capable of creating any of several different wiring layouts.*
>
> *Your job is to lay out some wiring diagrams of things the child can build using the objects listed. You have the initial board and component designs from the engineer, but some of these objects may change as the project develops. You don't have time to wait for engineering (what a surprise!), so you'll begin your layouts now using Xrefs so you can easily modify the component designs as they develop.*

> In the current release of AutoCAD, referenced items appear faded. To adjust this fading, go to the ribbon's **Insert** tab, **Reference** subpanel and change **the Xref fading** slider to *zero* (lower numbers mean less transparency). Alternately, you can set the **XDwgFadeCtl** system variable to *zero*.

Do This: 27.1.1A	Working with Xrefs

I. Start a new drawing using the *Start* template found in the C:\Steps\Lesson27 folder.
II. Set the Xref display fade control to zero (refer to the previous insert).
III. Save the drawing as *MyBoard* in the C:\Steps\Lesson27 folder.
IV. Follow these steps.

27.1.1A: WORKING WITH XREFS

1. Open the External Reference palette.
   Command: *xr*

2. Pick the **Attach DWF** option (under the **Attach** flyouts).

3. AutoCAD presents the Select DWF File dialog box. Select the *title block-Model.dwf* file in the C:\Steps\Lesson27 folder. (Double click on the file name or pick the **Open** button [Open] to continue.)

27.1.1A: WORKING WITH XREFS

4. AutoCAD presents the Attach DWF Underlay dialog box (Figure 27.004, p.634). Be sure the selections match those shown in the following figure.
   - Use a **Relative path**.
   - Remove the checks from the boxes in the **Insertion** and **Scale** frames.

   Pick the **OK** button to continue.

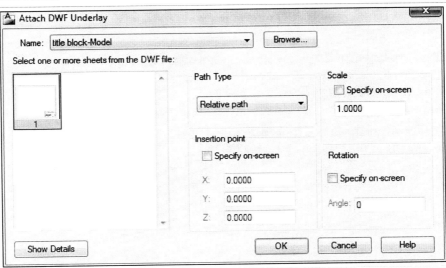

5. Draw the board shown.
   - The board is on the **board** layer.
   - The rectangle's corners are at coordinates *1,1* and *8.5,9*.
   - The fillets are ½".
   - The slots are on the **slots** layer.
   - The slots are 3/16"dia., ½" apart, and begin at coordinate *1.5,1.5*.
   - There are 15 rows and 14 columns of slots.

6. Save the drawing and close it.
   Command: *qsave*
   Command: *close*

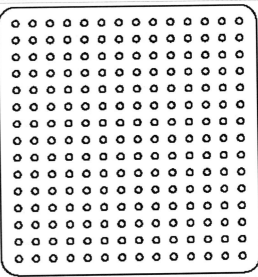

7. Start a new drawing using the *Start* template found in the C:\Steps\Lesson27 folder. Save the drawing as *MyCircuit1* in the C:\Steps\Lesson27 folder.

8. Repeat Steps 1 to 4 using the **Attach DWG** button, but this time, reference the *MyBoard* you created earlier in this lesson. (If you have a problem with this one, you can attach the *board* file instead.) Be sure to use an **Attachment Reference Type** and a **Relative** path.

9. Save the drawing.
   Command: *qsave*

27.1.1A: WORKING WITH XREFS

10. Check the **Layer** control box. Notice that AutoCAD has added the layers **board|board**, and **board|slots**. (The other layers were part of the template.) These are the layer included in the referenced drawing. Note the limits of referenced layers – you can turn them off, freeze them, lock them, or plot them; but they can't be made current. Nor can you draw on these layers, as they aren't actually part of the primary drawing.

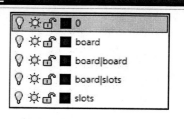

11. Using the **Attach** button on the ribbon, we'll attach the battery and switch drawings. (You'll notice that this procedure is identical to the one we used to attach the *MyBoard* file except that we skip the External Reference palette.)

   Command: *attach*

12. AutoCAD presents the Select Reference File dialog box. Select the *.dwg* file type, and then the *battery* file in the C:\Steps\Lesson27 folder.

13. Insert the file at coordinates *2.5,4* as shown. (Be sure to use a **Relative** path.)

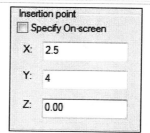

14. Repeat Steps 11 through 13 to insert the *switch* file at coordinates 2.5,5.5. Insert this file at 90° (see the figure).
Your board looks like the figure at right.

15. Save the drawing but don't exit.

   Command: *qsave*

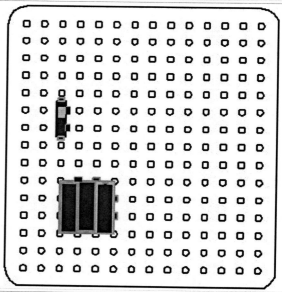

16. Select the battery. Notice that AutoCAD highlights the battery in the External References palette and opens the ribbon's **External Reference** tab.

17. Clear the selection.

18. In the External References palette, select the switch. Notice that AutoCAD highlights the switch in the drawing. (This works best if the palette is not docked and you can see both the listing in the palette and the switch in the drawing.)

27.1.1A: WORKING WITH XREFS

19. Clear the selection [Esc].

Notice the two pairs of wire connections on both the switch and the battery. Our engineers haven't decided where to put the single required pair, so they've created two for now. We have a fair idea which ones they'll eventually use, but we don't want to change the actual component drawings yet. We can, however, alter the reference to show only those connections we want to show.

To remove part of a reference, use the *Clip* command. The sequence looks like this:

Where to Find It:	
Command Line:	*Clip*
Ribbon (Tab/Panel):	Insert – Reference – **Clip**
Menu:	Modify – Clip – **Xref**
Toolbar:	Reference – **Clip Xref**

    Command: *clip*
    Select Object to clip: *[Select the reference(s) to clip.]*
    Enter clipping option [ON/OFF/Clipdepth/Delete/generate Polyline/New boundary] <New>: *[Tell AutoCAD what you want to do – accept the default to clip a reference.]*
    Outside mode - Objects outside boundary will be hidden.
    Specify clipping boundary or select invert option:
    [Select polyline/Polygonal/Rectangular/Invert clip] <Rectangular>: *[Tell AutoCAD what type of clipping boundary you wish to use.]*
    Specify first corner:
    Specify opposite corner:

> The *Clip* command replaces several reference-specific commands: XClip, DWFClip, and DGNClip. All perform essentially the same function and the consolidation really simplifies clipping.

- The first line of options presents several choices:
  - The **ON/OFF** options determine whether AutoCAD will present only those portions of the referenced drawing visible through the clipping window (**ON**) or the entire referenced drawing (**OFF**).
  - **Clipdepth** allows you to set the front/back clipping planes of three-dimensional objects. A two-dimensional boundary must exist before you can define front or back clip points. Identify the clip points by coordinate or distance from the two-dimensional clipping boundary.
  - Use **Delete** to remove a boundary.
  - **Generate Polyline** will automatically draw a polyline along the clipping boundary. Again, a boundary must exist before AutoCAD can generate the polyline.
  - **New boundary** (the default) permits you to identify a new clipping boundary.
- The next line lets you know that AutoCAD will (by default) hide those parts of the referenced drawing that lie outside the clipping boundary. You can change this using the **Invert clip** option on the next line.
- The next line of options (**[Select polyline/Polygonal/Rectangular/Invert clip] <Rectangular>:**) offers some choices for how to create the new boundary.
  - Use the **Select polyline** option to use an existing polyline to define the boundary. The polyline doesn't have to be closed, but it must not intersect itself. (To create a "round" clipping boundary, use a polygon with a large number of sides.)
  - Use the **Polygonal** option to create a multisided or nonlinear shape for a boundary (much like using a poly-window to create a selection set).
  - Use the **Rectangular** option (the default) to use a window to define the clipping boundary.

- Finally, use the **Invert clip** option to switch the viewing mode to hide those parts of the referenced drawing that lie *inside* the boundary.

> AutoCAD provides a system variable to allow you to see a boundary (frame) even when it isn't a polyline. The system variable for Xrefs is **Frame** and AutoCAD provides a simple toggle on the ribbon (**Insert** tab, **Reference** panel). A setting of **0** (the default - ▢) means that frames will be hidden. A setting of **1** ▢ means that frames will both appear and be plotted. A setting of **2** ▢ means that frames will appear, but *not* be plotted. The last setting – **3** ▢ – means that your settings are different for the various file types that have been referenced.
> The **Frame** system variable overrides the older **XClipFrame**, **DWFFrame**, **PDFFrame**, and **DGNFrame** system variables.

> You can edit the clipping frame using grips (very convenient!) ... *and* you can even invert the clipping area (again, using grips) to see everything *outside* rather than *inside* the frame!
> Will wonders never cease?!

We'll remove the extra wire connectors from the views of our battery and switch.

Do This: 27.1.2A	Clipping Xrefs

I. Be sure you're still in the *MyCircuit1* file. If not, please open it now.
II. This exercise is easier if you zoom in around the battery and switch.
III. Follow these steps.

27.1.2A: CLIPPING XREFS

1. Enter the *Clip* command.

    Command: *clip*

2. Select the battery.

    Select Object to clip:

3. Accept the default **New boundary** option [New boundary].

    **Enter clipping option
    [ON/OFF/Clipdepth/Delete/generate Polyline/New boundary] <New>:** *[ENTER]*

4. Tell AutoCAD that you'll create a **Rectangular** [Rectangular] clipping boundary (the default).

    **Specify clipping boundary or select invert option:
    [Select polyline/Polygonal/Rectangular/Invert clip] <Rectangular>:** *[ENTER]*

5. Place the first corner at the lower right corner of the battery (use OSNAPs).

    Specify first corner: _endp of

6. ... and the opposite corner northward and westward of the battery.

    Specify opposite corner:

AutoCAD clips the reference, and the battery now looks like the figure at right. (Note that you haven't actually modified the battery, just how much of it you can see through the reference window.)

7. Let's use **the Invert clip** option for the switch. Repeat steps 1 through 3, but this time, select the switch.

641

27.1.2A: CLIPPING XREFS

8. Select the **Invert clip** option [Invert clip].
   Specify clipping boundary or select invert option:
   [Select polyline/Polygonal/Rectangular/Invert clip] <Rectangular>: *I*

9. Now select the **Rectangular** option [Rectangular].
   Specify clipping boundary or select invert option:
   [Select polyline/Polygonal/Rectangular/Invert clip] <Rectangular>: *R*

10. Put your rectangular window around the side switch connections. Your board looks like the figure at right.

11. Have you noticed that the title block affects your ability to zoom in and out on the board with ease? Let's remove it temporarily while we work on our board.
    Open the External Reference palette.
    Command: *xr*

12. Notice (see the following figure) that all the referenced drawings are now listed – even the nested title block drawing (remember, it was part of the *Board* drawing).
    Right click on **board|title block** and select **Unload** [Unload] from the cursor menu.

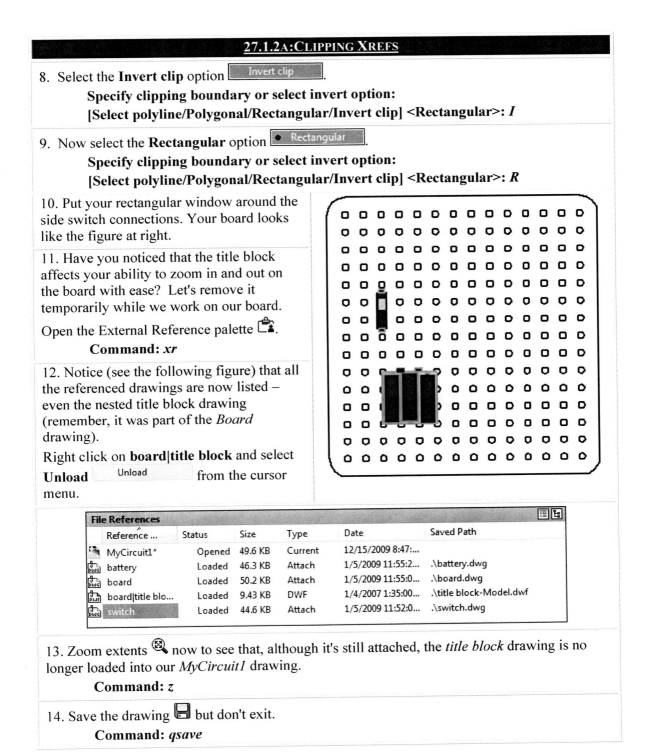

13. Zoom extents now to see that, although it's still attached, the *title block* drawing is no longer loaded into our *MyCircuit1* drawing.
    Command: *z*

14. Save the drawing but don't exit.
    Command: *qsave*

27.1.3 Xrefs and Dependent Symbols

AutoCAD allows two methods of permanently attaching referenced drawing data to the primary file – Binding and Xbinding. Use the **Bind** option on the External Reference palette's cursor menu to permanently attach, or bind, an entire referenced drawing to the primary drawing. We discussed this in Section 27.1.1 (p.632) and will demonstrate it in Section 27.4 (p.650). But you can also attach

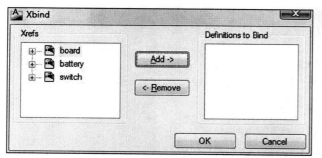

Where to Find It:	
Command Line:	*XBind*
Hotkey(s):	*xb*
Menu:	Modify – Object – External Reference – **Bind**
Toolbar:	Reference – **Xbind**

Figure 27.007

parts (specific dependent symbols) of the referenced drawing individually to the primary drawing. The command for this is *Xbind*, and it presents the Xbind dialog box (Figure 27.007).

The Xbind dialog box is one of AutoCAD's easiest. Simply select the dependent symbol you wish to bind in the Xrefs frame, pick the **Add** button, and **OK** your changes!

Let's try one. The battery drawing has a **wire** layer and a **times** (or **calibri**) text style; we'll add these to our primary file.

Do This: 27.1.3A	Xbinding Dependent Symbols

I. Be sure you're still in the *MyCircuit1* file. If not, please open it now.
II. Follow these steps.

27.1.3A: XBINDING DEPENDENT SYMBOLS

1. Enter the *Xbind* command.
   Command: *xb*

2. AutoCAD presents the Xbind dialog box (Figure 27.007). Pick the "+" beside the battery reference. AutoCAD presents a list of dependent symbol categories. Categories with symbols available for binding have a "+" beside them.

3. Pick the "+" beside the **Layer** category, and then select the **battery|wire** layer.

4. Pick the **Add** button. Notice that the **battery|wire** layer disappears from the **Xrefs** frame and appears in the **Definitions to Bind** frame (following figure).

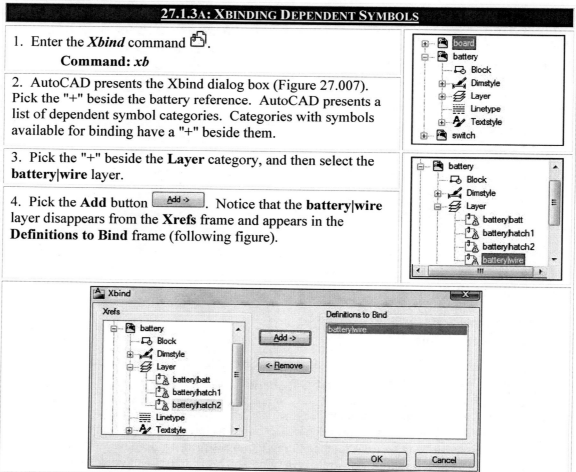

643

27.1.3A: XBINDING DEPENDENT SYMBOLS

5. Pick the **OK** button [OK] to complete the command. If necessary, reconcile the new layer.

6. Check the **Layer** control box. Notice that the **battery|wire** layer has been replaced with **battery0wire**. You can now use this layer as you would any other layer in the primary drawing.

7. Repeat Steps 1 to 5 to bind the **battery|times** (or **battery - v|calibri**) text style to the primary drawing. [Check your success by looking in the **Style Name** control box in the Text Style dialog box to be sure you have the **battery0times** (or **battery - v0calibri**)style.

8. Save the drawing 💾 but don't exit.
   Command: *qsave*

27.1.4 Unloading, Reloading, and Overlaying Xrefs

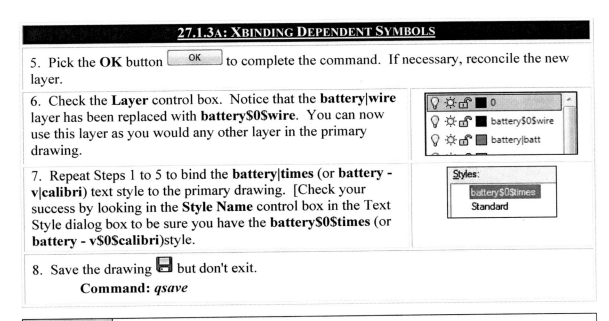

In Exercise 27.1.2A (p.641), we unloaded the *title block* reference to make our work easier. All we had to do was to select the reference to unload (*title block*) in the External Reference palette, and then pick the **Unload** option on the cursor menu. Unloading a reference

Figure 27.008

can speed work by reducing regeneration time. And as we saw, unloading a reference can remove unnecessary, distracting, and even obstructing objects from the display. But eventually, you'll probably have to reload the unloaded reference.

Reloading an unloaded reference is as easy as unloading it. Simply select the drawing to reload and pick the **Reload** option on the same cursor menu. We'll reload the title block in our next exercise.

But first, let's consider the **Attach** versus **Overlay** types of reference. Consider the diagram in Figure 27.008, p.644.

As you can see, the **Overlay** type of reference is effective only when used on a nested reference. In other words, overlaying a reference into the primary drawing won't affect the nested reference's visibility. To hide the nested reference, it must be overlain into the drawing being referenced by the primary drawing. We'll see this in our next exercise.

> AutoCAD drawings reference PDF, DWF, and DGN files through "underlaying." To hide an underlain reference, freeze the layer on which it was attached.

This is what we'll do: We'll save our drawing and insert it as a reference into a new drawing. The new drawing will be the **Primary** drawing in our schematic (Figure 27.008), and *MyCircuit1* will be **Reference #2**. Since *MyBoard* (or *board*) is referenced by MyCircuit1 (using the **Attach** type of reference), it will become a nested reference (**Reference #1**). In the exercise, we'll change the reference type between *MyCircuit1* and *MyBoard* and see how that affects our primary drawing. Let's begin.

Do This: 27.1.4A	Overlaying References

I. Be sure you're still in the *MyCircuit1* file. If not, please open it now.
II. Reload the title block (repeat Steps 11 – 12 in Exercise 27.1.2A, p.642, but pick **Reload** in Step 12). Save and close *MyCircuit1*.
III. Start a new drawing using the *Start* template in C:\Steps\Lesson27. Save it as *MyCircuit1a*.
IV. Follow these steps.

27.1.4A: OVERLAYING REFERENCES

1. Open the External Reference palette.

   Command: *xr*

2. Reference the *MyCircuit1* drawing in the C:\Steps\Lesson27 folder. (If *MyCircuit1* isn't available, use *Circuit1*.) Use the **Attachment Reference Type** and a **Relative path**. Reconcile the layers as needed.

3. Save the drawing.

   Command: *save*

4. Leave *MyCircuit1a* open, but open *MyCircuit1* (or *Circuit1*), too.

   Command: *open*

5. In the External Reference palette of the *MyCircuit1* (*Circuit1*) drawing, select *board* (or *MyBoard*). Then change the **Reference Type** in the **Details** frame as shown in the following figure.

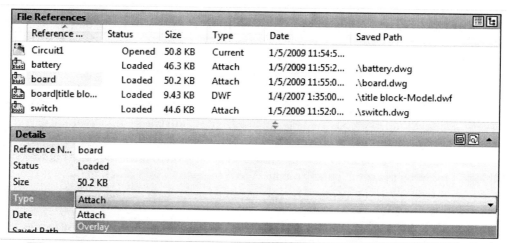

6. Save *MyCircuit1* (or *Circuit1*) ...

   Command: *save*

... and return to *MyCircuit1a*. (You can use the Window drop down menu and select the drawing you wish to be current or use Quick View Drawings.)

27.1.4A: OVERLAYING REFERENCES

7. AutoCAD returns to the *MyCircuit1a* drawing and presents a bubble message telling you that a referenced drawing has changed. Pick on the link to reload the changed drawing.

Notice that the *board* reference disappears.

8. Repeat Steps 4 through 7, but this time set the **Reference Type** back to **Attach**. You'll notice that the *board* reference returns to the *MyCircuit1a* drawing.

9. Save and close any open drawings.

Command: *qsave*

27.2 Editing Xrefs

Although AutoCAD has made it quite simple to edit a referenced drawing, you should do so with caution. Remember that others may have referenced the same drawing and any changes made to it will reflect in their drawings as well.

AutoCAD's reference editing tools provide the ability for us to edit a referenced drawing without having to leave the primary drawing. Even better is the fact that they're so easy to use.

Three commands make up the reference editing tools.

- *RefEdit* begins the editing session. It prompts you to select the referenced objects to edit, like this:

    Command: *RefEdit*

    Select reference: *[Select the referenced drawing.]*
    [Here, AutoCAD presents the Reference Edit dialog box (Figure 27.009) and the Edit Reference panel on the ribbon. Complete the options and pick the OK button to continue. AutoCAD then closes the dialog box and makes the referenced drawing available for editing.]
    REFCLOSE or one of the close buttons on the ribbon's **Edit Reference** panel will end the editing session. *[AutoCAD lets you know how to end the editing session.]*

Where to Find It:	
Command Line:	*Refedit*
Ribbon (Tab/Panel):	Insert – Reference (subpanel) – **Edit Reference**
Menu:	Tools – Xref and Block In-place Editing – **Edit Reference In-place**
Toolbar:	Refedit – **Edit Reference In-place**

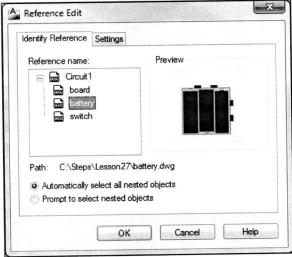

Figure 27.009

- The **Identity Reference** tab of the Reference Edit dialog box (Figure 27.009) provides a list box showing the name of the selected reference and a **Preview** frame where you can see the referenced drawing. It also provides two options:
    - **Automatically select all nested objects** controls whether or not AutoCAD will automatically include nested references in the editing session.
    - A bullet next to **Prompt to select nested objects** lets AutoCAD know that you'd rather individually select nested objects in your editing session.
- The **Settings** tab of the Reference Edit dialog box (Figure 27.010) provides some options that may make your editing session run more smoothly.

    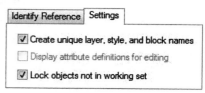

    Figure 27.010 (not actual size)

    - A check next to **Create unique layer, style, and block names** tells AutoCAD to use the $#$ procedure for dealing with symbols much as it does when binding an Xref. If unchecked, layer, style, and block names remain the same as in the referenced drawing.
    - Selecting **Display attribute definitions for editing** means that the attribute definitions of blocks within the referenced drawing will be available for your editing.
    - It's best to leave the check next to **Lock objects not in working set**. This prevents you from accidentally editing something in the primary drawing when you're editing your referenced drawing.

Where to Find It:	
Command Line:	*RefSet*
Ribbon (Tab/Panel):	*Edit Reference – Edit Reference – **Refset Add** or **Refset Remove**
Menu:	Tools – Xref and Block In-place Editing – **Add to Working Set** or **Remove from Working Set**
Toolbar:	Refedit – **Add to Working Set** or **Remove from Working Set**

Where to Find It:	
Command Line:	*RefClose*
Ribbon (Tab/Panel):	*Edit Reference – Edit Reference – **Refclose Don't Save** or **Refclose Save**
Menu:	Tools – Xref and Block In-place Editing – **Refclose Don't Save** or **Refclose Save**
Toolbar:	Refedit – **Close Reference** or **Save Reverence Edits**

- Use *RefSet* to add or remove referenced objects from the editing session. It prompts:

    Command: *RefSet*

    Transfer objects between the RefEdit working set and host drawing...

    Enter an option [Add/Remove] <Add>: *[Enter to add or type R to remove objects from the set.]*

    Select objects: *[Select the object(s) to add or remove.]*

- *RefClose* ends the editing session. It prompts

    Command: *RefClose*

    Enter option [Save/Discard reference changes] <Save>: *[Enter to save the changes or type D to discard the changes.]*

---

* available when you edit a referenced drawing

AutoCAD presents a warning box (Figure 27.011) telling you what you're about to do. Pick **OK** to continue or **Cancel** to prevent saving.

Let's try *RefEdit*.

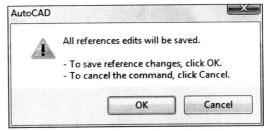

Figure 27.011

*Our engineer has determined where the wire connections should be on our battery and switch. We could open those drawings for the changes, but we're in the MyCircuit1 file now and can make the changes here using **RefEdit**.*

Do This: 27.2A	Editing Referenced Drawings

I. Open the *MyCircuit1* file.
II. Follow these steps.

27.2A: EDITING REFERENCED DRAWINGS

1. First, let's remove the clipping planes. Enter the *Clip* command.
   Command: *clip*

2. Select the battery.
   Select objects:

3. Tell AutoCAD you want to **Delete** the clipping planes.
   **Enter clipping option
   [ON/OFF/Clipdepth/Delete/generate Polyline/New boundary] <New>:** *d*

4. Repeat Steps 1 through 3 for the switch.
AutoCAD presents both references in their entirety.

5. Now we begin the editing session. Enter the *RefEdit* command.
   Command: *refedit*

6. Select the battery.
   Select reference:

7. AutoCAD presents the Reference Edit dialog box (Figure 27.009, p.646). We'll use the default settings, so pick the **OK** button.

8. Erase the wiring connections on the right side of the battery. (There are actually three objects – a hatching and two rectangles.)
   Command: *e*

9. Complete the *RefEdit* command with the *RefClose - Save* command.
   Command: *refclose*

10. AutoCAD presents a warning box (Figure 27.011, p.648). Pick the **OK** button to complete the procedure.

11. Repeat Steps 5 through 10 to remove the right set of wiring connections on the switch.

12. Save and close the drawing.
    Command: *qsave*

27.3 Using Our Drawing as a Reference

Now that we've created a drawing with circuit board, battery, and switch, we can begin to create our wiring diagrams. We'll plan on three layouts using the battery and switch in these locations, so we'll use the *MyCircuit1* file as a reference over which we'll create our layouts.

Let's begin.

Do This: 27.3A	Using Xrefs to Create Three Wiring Diagrams

I. Start a new drawing using the *Start* template in the C:\Steps\Lesson27 folder. Save the new drawing as *MyCircuit1b* in the C:\Steps\Lesson27 folder.
II. Follow these steps.

27.3A: USING XREFS TO CREATE THREE WIRING DIAGRAMS

1. Begin by attaching the *MyCircuit1* (or *Circuit1* if *MyCircuit1* isn't available) drawing as an Xref. Attach as an **Attachment Reference Type** with a **Relative path**. Accept the other defaults.
   Command: *attach*

2. Reconcile the layers as needed.

3. Xbind the **battery|wire** layer and the **battery|times (battery|calibri)** text style to your new drawing. (Use the procedure detailed in Exercise 27.1.3A, p.643.)
   Command: *xb*

4. Rename the new layer to **Wire** and the new text style to **Times** (or Calibri).

5. Save the drawing.
   Command: *qsave*

6. Reference the *lamp* file in the C:\Steps\Lesson\Steps\Lesson27 folder and place it at coordinates **6.5,6**.
   Command: *attach*

7. Reconcile the layers as needed.

8. Using splines, draw the wire shown in the figure at right. (Be sure to use the **Wire** layer.) Add the text on an appropriate layer. (Text size is 3/16" and uses the **Times** or **Calibri** style.)

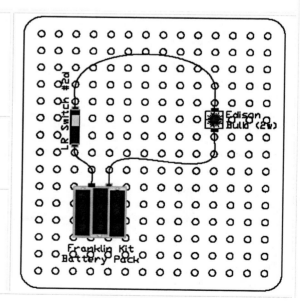

27.3A: USING XREFS TO CREATE THREE WIRING DIAGRAMS

9. Insert the block file *title info* at the lower left corner of the title block. (Place it on its own layer, and fill in the data as appropriate for your situation as shown.)

10. Save 💾 and close the drawing.
    Command: *qsave*

Sam Houston State University				
Wiring Project #1 Simple Light & Switch Diagram				
Size:	Project Number:	Drawing Number:		Rev:
B	10372	B14314		0
Scale: NTS	Drawn By: T. Sykes		Sheet 1	of 1

11. Repeat Steps 1 through 8 (as needed) to create the drawings shown in the following figures. Adjust the title blocks to read *Wiring Project #2, Galvanometer Diagram* (drawing number) *B-14319a*, and *Wiring Project #3, Resistor Diagram* (drawing number *B-14319b*. (Appropriate reference drawings have been provided in the C:\Steps\Lesson27 folder.)

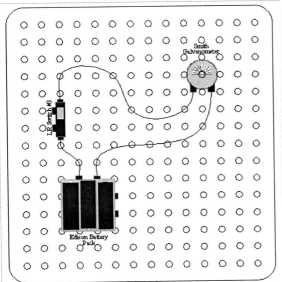 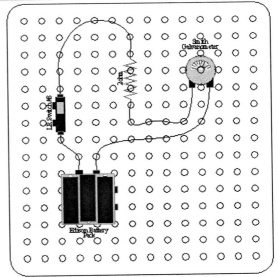

12. Save the drawings as *MyCircuit1c* and *MyCircuit1d* in the C:\Steps\Lesson27 folder.

Congratulations! You've completed three diagrams according to your original job requirements! You'll have an opportunity to create other diagrams at the end of this lesson. First, however, let's take a look at some special requirements for sending your referenced drawings to your client.

27.4 Binding an Xref to Your Drawing

The biggest problem that arises from the use of Xrefs lies in the location of the referenced file and how AutoCAD keeps track of that location.

By default, AutoCAD will look for referenced files in the **Saved Path** first, and then in folders listed in the **Project Files Search Path** (p.635). Again, by default, the **Project Files Search Path** is empty. Therefore, you'll need to add the location (path) of your references to the **Project Files Search Path** or rely on the **Saved Path**. Of course, you could put all the drawings in the same folder, but that's not very organized.

A problem develops, however, when you decide to send your drawing to the client (or somewhere else). First, you must remember to send all the reference files along with the primary file. But even when you do this, the client may not (indeed, probably won't) place the files in a folder with the same name you were using. The result will be that the primary drawing won't be able to locate the references (your client gets frustrated and you look for another job).

You could solve the problem by fully documenting necessary project/folder/path information and sending the documents along with the files (much as eTransmit does). But this means that the client must read instructions about how to view your work – probably not a strong selling point for giving you more business.

You might use the eTransmit tool and/or sheet sets. Indeed, that's probably the best approach, but there is another way. AutoCAD allows you to bind all the references to the primary drawing as blocks. Then you simply send the one file, and the client reads it with any AutoCAD program or viewer.

Binding references, however, dramatically increases the size of the drawing. So you need to wait until the project is complete before doing it. It might even be a good idea to archive the project before binding in case you need to do additional work on the original(s).

As we saw in our study of the External Reference palette (Section 27.1.1, p.632), binding isn't difficult. Simply select the references to bind and pick the **Bind** button on the cursor menu. AutoCAD will ask if you wish to **Bind** or **Insert**. (See the discussion of these options in Section 27.1.1.)

Let's see what happens when we bind our references.

> Although you can bind referenced drawings, you cannot bind PDFs, DWFs, or DGNs. Keep this in mind when creating your referenced drawings. It might be a good idea to insert these files without a path – then keep them in the same location as the drawing you're creating.

Do This: 27.4A	Binding References

I. Open the *MyCircuit1b* file in the C:\Steps\Lesson27 folder. (If this file isn't available, you can open *Circuit1b* instead.)

II. Follow these steps.

27.4A: BINDING REFERENCES

1. Before we bind our references, let's see how large our drawing file is. Enter the ***dwgprops*** command.

    Command: *dwgprops*

2. AutoCAD presents the [Drawing Name] Properties dialog box.

    Note the size of the drawing (it's on the **General** tab), and then exit the dialog box.

    | Type: | AutoCAD Drawing |
    | Location: | C:\Steps\Lesson27\ |
    | Size: | 53.20KB (54,513 bytes) |

3. Open the External Reference palette.

    Command: *xr*

4. Select the *MyCircuit1* (or *Circuit1*) drawing and then pick **Bind** on the cursor menu.

5. AutoCAD presents the Bind Xrefs dialog box. Remember, if you use the **Bind Bind Type**, dependent symbols adopt this format: **[ref name]$[#]$[layer]** (as in **battery0hatch1**). If you use **Insert**, dependent symbols drop the reference drawing name and $[#]$ from their name.

    Put a bullet next to **Insert** and pick the **OK** button.

27.4A: BINDING REFERENCES

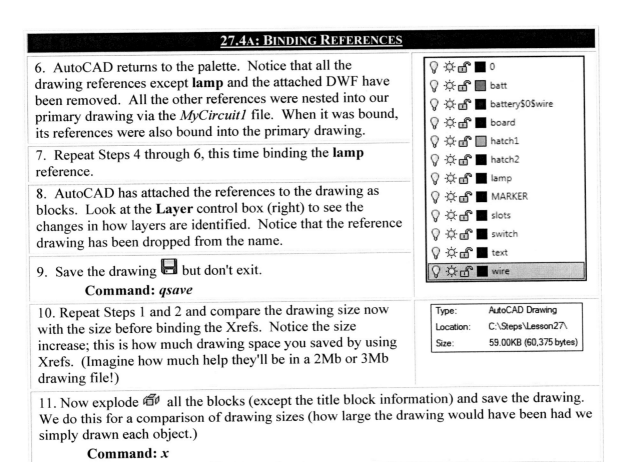

6. AutoCAD returns to the palette. Notice that all the drawing references except **lamp** and the attached DWF have been removed. All the other references were nested into our primary drawing via the *MyCircuit1* file. When it was bound, its references were also bound into the primary drawing.

7. Repeat Steps 4 through 6, this time binding the **lamp** reference.

8. AutoCAD has attached the references to the drawing as blocks. Look at the **Layer** control box (right) to see the changes in how layers are identified. Notice that the reference drawing has been dropped from the name.

9. Save the drawing 💾 but don't exit.
   Command: *qsave*

10. Repeat Steps 1 and 2 and compare the drawing size now with the size before binding the Xrefs. Notice the size increase; this is how much drawing space you saved by using Xrefs. (Imagine how much help they'll be in a 2Mb or 3Mb drawing file!)

11. Now explode 💥 all the blocks (except the title block information) and save the drawing. We do this for a comparison of drawing sizes (how large the drawing would have been had we simply drawn each object.)
    Command: *x*

12. Repeat Steps 9 and 10. Notice how large the drawing file is now!

13. Close the drawing. (Save the changes.)
    Command: *close*

You can now send the drawing to your client without fear of losing something in the transmittal!

27.5 Extra Steps

Asking is only half the battle. Listening to the answer reveals the path to success.
Anonymous

So, what do you think of External References? When used properly, they can be tremendous timesavers. But something else is required. Let me tell you a story.

When I was working for one of the big petrochemical companies in Houston many years ago, a friend of mine (another guru) was assigned to a new project. He was to set up the CAD system for the project.

For some wild reason (wild reasoning isn't uncommon in petrochem), he decided to buck the norm and set up the project the way AutoCAD was designed to work – using things like attributes to track materials and Xrefs to save time. Jim worked diligently for weeks setting up and starting the project.

Then the company fired him. It seemed the project's lead knew nothing about Xrefs and very little about attributes. All he saw was weeks gone with only a few drawings created (remember that proper setup of an AutoCAD project takes time in the beginning but saves time in the end).

The project lead hired a beginning CAD operator, had him teach himself about Xrefs and attributes, and then spent more weeks disassembling, binding, exploding, and so forth all of the drawings and setups which Jim had created.

What is the moral of this story? Jim forgot one crucial fact of CAD operations. That fact involves communication. Many supervisors are simply not aware of AutoCAD's potential or proper use (yet). Jim didn't explain to the boss what he was doing, nor did the boss bother to learn Jim's "new" system. He replaced it with something he understood (more expensive and time consuming, but also more comfortable for him).

Your Extra Steps exercise is this: Go to your supervisor (or contact a supervisor in the industry in which you hope to work) and set up an interview. Ask if the company uses Xrefs and/or attributes to track materials. Then ask why or why not. If they're using Xrefs, try to convince them to stop (explain that this is an assignment and don't get too combative) and listen closely to their arguments. If they're not using Xrefs, try to convince them to do so. Again, listen closely to their arguments.

27.6 What Have We Learned?

Items covered in this lesson include:

- *The Xref Manager*
- *Dependent Symbols*
- *Manipulating Xrefs – unloading, reloading, overlaying, and binding*
- *Editing Xrefs*

- *Commands:*
  - *Xref*
  - *Attach*
  - *XAttach*
  - *Xbind*
  - *Clip*
  - *Frame*
  - *RefEdit*
  - *RefClose*
  - *RefSet*
  - *Frame*
  - *XClipFrame*
  - *DWFFrame*
  - *PDFFrame*
  - *DGNFrame*
  - *XDwfFadeCtl*

While only a fraction of the people to whom I've spoken about Xrefs actually use them, the number seems to be growing. Xrefs rank well behind blocks in popularity, but they can be almost as useful in their own way. Take some time to get comfortable with them in the exercises at the end of the lesson. Then talk it over with your employer and make an educated decision about how you'll proceed.

The objects of our next lesson – raster images and OLE – closely resemble Xrefs. There we'll discuss referencing other forms of graphics into our drawings. The procedures and dialog boxes resemble those of Xrefs, so the lesson should go fairly smoothly.

But first, let's try a few more Xrefs.

27.7 Exercises

1. Using Xrefs whenever possible, create the drawing in the following figure.
    1.1. Use the following references (included in the C:\Steps\Lesson27 folder):
        1.1.1. *Title block*
        1.1.2. *Battery*
        1.1.3. *Lamp*
        1.1.4. *Galvanometer*
        1.1.5. *MyBoard* (you created this in Exercise 27.1.1A, p.637) or *Board*
        1.1.6. *title info* (insert this as a block to use the attributes)
    1.2. Create layers as needed (you'll need a wire and a text layer)
    1.3. Save the drawing as *MyCircuit1e* in the C:\Steps\Lesson27 folder.

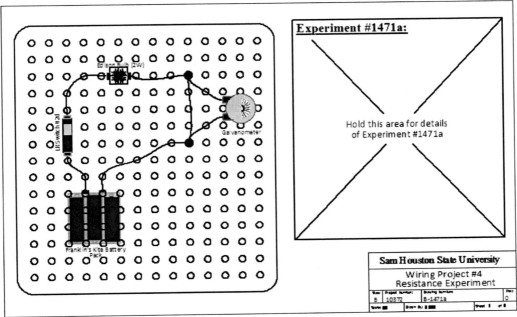

MyCircuit1e.dwg

2. Create the drawings in the following two figures, using Xrefs whenever possible.
    2.1. Use the following references (included in the C:\Steps\Lesson27 folder):
        2.1.1. *Title block-a*
        2.1.2. Other drawings you might wish to create
        2.1.3. *title info* (insert this as a block to use the attributes)
    2.2. Create layers as needed (you'll need a pipe and a text layer)
    2.3. Save the drawings as *Pump Config 1* and *Pump Config 2* in the C:\Steps\Lesson27 folder.

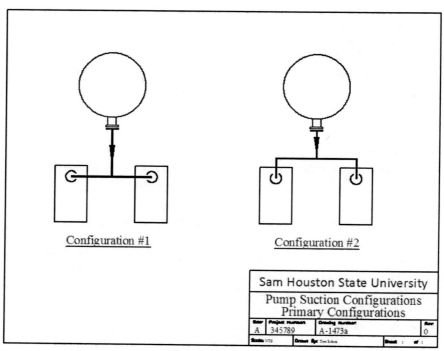
Pump Config 1.dwg

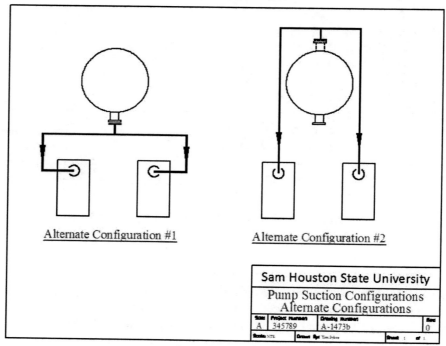
Pump Config 2.dwg

27.8 **For Web-Based Review Questions and Additional Exercises, visit: http://foragerpub.com/AcadFiles/2011/2011.htm**

Lesson 28

Following this lesson, you will:

- ✓ Know how to use graphic images from other applications in AutoCAD drawings
- ✓ Know how to use AutoCAD drawings in other applications
- ✓ Be familiar with raster and vector images
- ✓ Know how to use Object Linking and Embedding to share information with other applications

Other Application Files and AutoCAD

I can think of several reasons why you might want to use other application files in an AutoCAD drawing. How about including a photograph of a completed project on a mechanical drawing? I've used them to show completed pieces of furniture I designed. Or perhaps you might want to include a corporate logo in the title block of your drawing. If you can find a nice, colorful scan of the logo, your employer might be pleased to find it on the design! Architects and plant designers might show an outline of the proposed new building or unit against an aerial photograph of the area.

You can probably think of many uses for other applications' files in your drawings. Thank goodness that the thinkers and doers at AutoCAD also considered the possibilities.

28.1 Two Types of Graphics

Two categories of graphic files are associated with AutoCAD: *Vector Images* and *Raster Images*.

- Vector Images

  By default, AutoCAD creates vector images. In fact, you've been drawing vector images since you began using AutoCAD.

  Vector images locate geometry by definition and coordinate. For example, a line has two defined endpoints. AutoCAD treats it as a single object – a line – by definition.

  Vector images hold their definition regardless of the resolution of the screen or drawing view. Since they only have to define coordinates and objects, their size is quite small compared to a raster image of the same objects.

  Types of files found in this category include:
  - *DWG* – You're already familiar with how AutoCAD shares its drawing files within the AutoCAD program (blocks and Xrefs).
  - *DWF* – The Drawing Web Format is a 2D vector file designed for publishing AutoCAD drawings on the web.
  - *3DS* – This is another Autodesk file. This one works with 3D Studio.
  - *DXF* – ASCII version of the image file.
  - *SAT* – The ACIS file format.
  - *EPS* – Adobe graphic file. Useful for sharing information with Adobe's publishing applications.
  - *WMF* – Windows MetaFiles were created by Microsoft for use in its graphics programs. You can use them to share information with most Microsoft publishing software. (Note: A WMF file actually contains both vector and raster imaging.)

- Raster Images

  Most non-CAD programs use raster images. These images locate geometry by screen pixel coordinate. In other words, objects aren't defined as circles, lines, and so forth, but are the result of the color definition of a series of pixels.

  This method of graphics creation generally requires more memory and tends to lose image quality (sharpness) as the viewer gets close to the image.

  The more popular types of files in this category include:
  - *GIF* – An ideal Internet file because of its small size.
  - *JPG* – Also good for Internet use, the JPG (pronounced "jay'-peg") file doesn't have the GIF's color limitations, making it perfect for photographs. However, the JPG file can become quite large.
  - *TIF* – Microsoft had a hand in developing this file type, so its acceptability in various graphics programs is almost universal.

- *BMP* – Microsoft's original bitmap, this file type also is almost universal in acceptability. Virtually all IBM-type computers (all those with the Windows operating system) use BMPs in the Windows' Paint program.
- *PCT* – Use this Apple graphics file for easy exchange with the Apple-type computer.

Other and lesser-known file types include *RLE, DIB, RST, GP4, MIL, CAL, CG4, FLC, FLI, BIL, IG4, IGS, PCX, PCT, PNG, RLC, AND TGA.* (Note: most scanners create raster images.)

Since we're already familiar with the important vector images, let's look at raster images.

28.2 Working with Raster Images: The Image Manager

For working with raster images, AutoCAD uses the External Reference palette you studied in our last lesson (so this one should be easy!). You can access it with the *Image* command or using the methods you learned in Lesson 27.

You'll find very little difference in the way you manage images and the way you manage Xrefs.

28.2.1 Attaching, Detaching, Loading, and Unloading Image Files

AutoCAD provides two methods for opening the Attach Image dialog box (Figure 28.001): the *Attach* command (use the ribbon's button: Insert – Reference – **Attach**) or the External Reference palette Image Attach button (see the WTFI table). You can also use the External Reference palette to detach, load, or unload image files just as you did Xrefs. The *Attach* and *ImageAttach* commands both begin with a standard Windows Select File dialog box, then proceed to the Attach Image dialog box in Figure 28.001. Notice the similarity

	Where to Find It:
Command Line:	*ImageAttach*
Hotkey(s):	*iat*
Ribbon (Tab/Panel):	Insert – Reference – **Attach**
Menu:	Insert – **Raster Image Reference**
Toolbar:	Reference – **Attach Image**

between the Attach Image dialog box and the External Reference dialog box we discussed in our last lesson (Figure 27.003, p.633). They're almost identical and the function of each button/frame is the same.

There are, however, two notable exceptions; the **Scale** frame of the Image dialog box has only one input box. The reason for this is simple: while you could scale a reference drawing file along the XYZ axes, a raster image

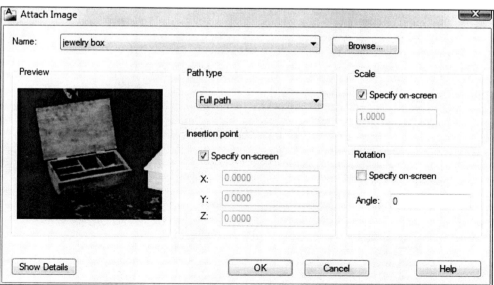

Figure 28.001

has no XY or Z-axis along which to scale. Therefore, AutoCAD restricts us to a uniform scale for raster images. Additionally, images are underlain, as were PDFs, DWFs, and DGNs, and you have no **Geographic Data** option.

Let's reference some images into a file.

Do This: 28.2.1A	Working with Images

I. Open the *cutting table28* file in the C:\Steps\Lesson28 folder. The drawing looks like the figure at right.
II. Thaw the **MARKER** layer.
III. Hide frames (Lesson 27, p.641). (We'll discuss image framing after this exercise, but for now, it's best to have frames turned off.)
IV. Follow these steps.

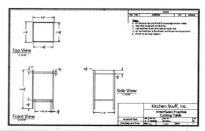

28.2.1A: WORKING WITH IMAGES

1. We'll attach a logo in the title block, so pick the **Attach** button on the ribbon's **Insert** tab, **Reference** panel.

2. AutoCAD presents a Windows Select… File dialog box. Select the *logo.jpg* file in the C:\Steps\Lesson28 folder.

Pick the **Open** button to continue.

3. AutoCAD presents the Image dialog box (refer to the following figure).

- Be sure there is a check in the **Specify on-screen** check box in the **Insertion point** frame. Clear the check boxes in the other frames.
- Enter *0.2* into the **Scale** frame's text box.
- Use a **Relative path**.

Pick the **OK** button to continue.

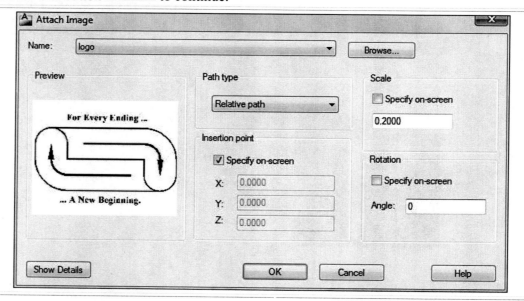

28.2.1A: WORKING WITH IMAGES

4. Insert the image at the node in the title block (just left of the words "AutoCAD Text").

5. Now let's use the *ImageAttach* command to insert a photograph of the cutting table.
   Command: *iat*

6. Select the *cutting table.jpg* file, and then pick the **Open** button [Open] to continue.

7. This time enter a scale of **.75**.
Pick the **OK** button to continue and insert the image at the node in the middle of the screen (~6,5).
Your drawing looks like the following figure. (Did you notice that the *ImageAttach* approach was the same as the *Attach* approach?)

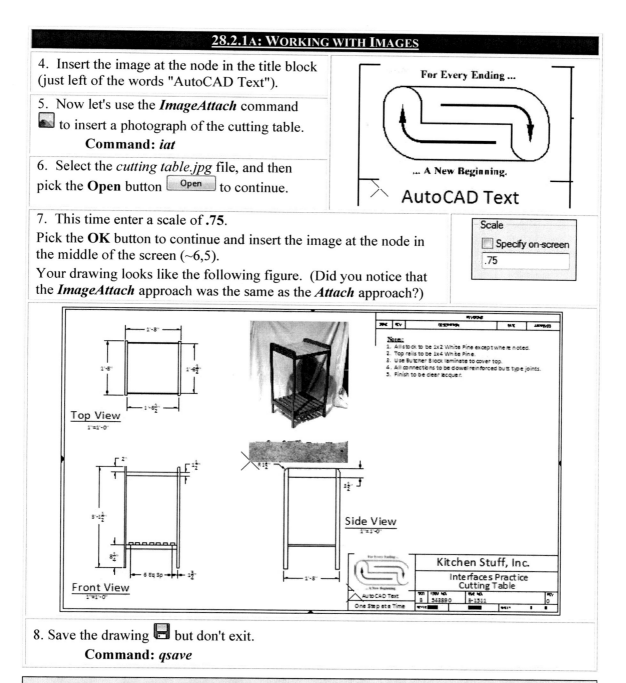

8. Save the drawing but don't exit.
   Command: *qsave*

Images attach on the current layer. You can freeze them, turn them off, lock them, and so forth just like any other object.

Over the course of this exercise, did you try to adjust either of the images we referenced? Did you notice that you couldn't select either one? If not, try to erase one of them now.

Unlike Xrefs, which allow you to use OSNAPs and other modifying tools, images have no vector geometry to select. But AutoCAD doesn't leave you with something you can't modify. You can select the image's frame. Simply pick the **Display and plot frames** or **Display and don't plot frames** options we discussed for Xrefs in Lesson 27 (p.641) to access the frame.

AutoCAD includes an **ImageFrame** system variable that can control the frame just for images – independent of the **Frame** system variable. This can be useful if you want to show the image frames but not the Xref frames. See the WTFI table (right) for access information.	Where to Find It:	
	Command Line:	*ImageFrame*
	Menu:	Modify – Object – Image – Frame
	Toolbar:	Reference – **Image Frame**

In our next exercise, we'll use the frame to adjust our view of the photograph.

28.2.2 Clipping Image Files

Clipping an image is no different from clipping an Xref. The command sequence is identical (begin with either the *Clip* or the *ImageClip* command):

Where to Find It:	
Command Line:	*Clip* or *ImageClip*
Hotkey(s):	*icl*
Ribbon (Tab/Panel):	Insert – Reference – **Clip**
Menu:	Modify – Clip – **Image**
Toolbar:	Reference – **Clip Image**

      Command: *clip or imageclip*
      Select Object to clip: *[Select the image you wish to clip.]*
      Enter image clipping option [ON/OFF/Delete/New boundary] <New>: *[Tell AutoCAD what you want to do.]*
      Outside mode - Objects outside boundary will be hidden.
      Specify clipping boundary or select invert option:
      [Select polyline/Polygonal/Rectangular/Invert clip] <Rectangular>: *[Tell AutoCAD what type of clipping boundary you wish to use.]*
      Specify first corner point: Specify opposite corner point: *[Identify the corners of the rectangle or the vertices of the polygon.]*

The *ImageClip* command automatically activates the **ImageFrame** system variable so that you can select an image, the *Clip* command does not.

Let's clip our photograph.

Do This: 28.2.2A	Clipping Images

I. Be sure you're still in the *cutting table28* file in the C:\Steps\Lesson28 folder. If not, please open it now.
II. Follow these steps.

28.2.2A: CLIPPING IMAGES

1. Enter the *ImageClip* command.
   Command: *icl*

2. (Notice that AutoCAD activates the **ImageFrame** system variable.) Select the frame around the cutting table photograph.
   Select image to clip:

3. Hit ENTER twice to accept the **New** and **Rectangular** boundary defaults.
   Enter image clipping option [ON/OFF/Delete/New boundary] <New>: *[ENTER]*
   Outside mode - Objects outside boundary will be hidden.
   Specify clipping boundary or select invert option:
   [Select polyline/Polygonal/Rectangular/Invert clip] <Rectangular>: *[ENTER]*

28.2.2A: CLIPPING IMAGES

4. Place a rectangle around the image of the cutting table as shown in the following figure (left). AutoCAD clips the photograph.
   Specify first corner point:
   Specify opposite corner point:
The results look like the right figure.

5. Use the ribbon toggle to **Display...frames** 🖻 or 🖾.

6. This time, use the *Clip* command to trim the logo image so it doesn't overlap the border.

7. **Hide frames** 🗙.

8. Save the drawing 💾 but don't exit.
   Command: *qsave*

Did you prefer the *ImageClip* approach or the *Clip* approach?

28.2.3 Working with Image Files

AutoCAD provides four tools for use with graphic images that aren't available (nor would they be useful) for Xrefs. These include: *ImageAdjust*, *ImageQuality*, *Transparency*, and *DrawOrder*.

- The *ImageAdjust* command calls the Image Adjust dialog box (Figure 28.002, p.663). Here you can adjust the brightness, contrast, or fading effect of the image (how well the image blends with the background color of the drawing). The dynamic preview image shows your modifications before you accept them with the **OK** button.

Where to Find It:	
Command Line:	*ImageAdjust*
Hotkey(s):	*iad*
Menu:	Modify – Object – Image – Adjust
Toolbar:	Reference – **Adjust Image**

Like *ImageClip*, *ImageAdjust* automatically activates the **ImageFrame** system variable.

- **ImageQuality** is a system variable that affects only the display of your graphic images. It presents the following prompt:
   Command: *imagequality*
   Enter image quality setting [High/Draft]

Where to Find It:	
Command Line:	*ImageQuality*
Menu:	Modify – Object – Image – Quality
Toolbar:	Reference –**Image Quality**

<High>:

High quality presents the best possible image in the display, but it may cause some delay in the initial display or redraw/regen time. Draft quality isn't quite as pretty, but it allows for much faster displays. AutoCAD will always use **High** quality when plotting despite this setting.

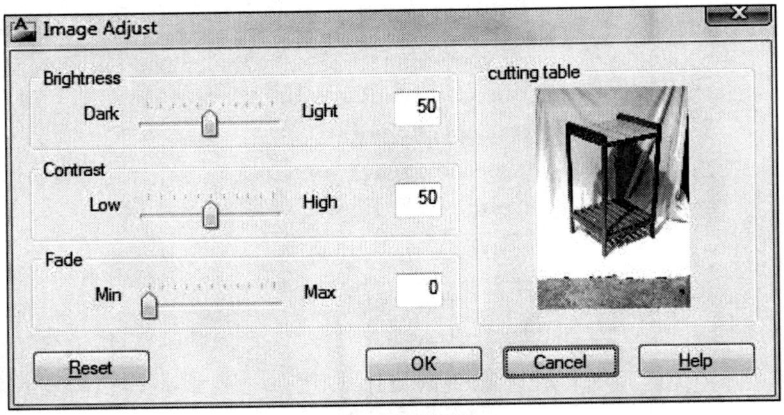

- The *Transparency* command works only on certain types of graphics that have a transparency property.

   You can set transparency on a per-image basis.

- [We'll discuss *DrawOrder* after the next exercise.]

Let's experiment with these commands before looking at our last tool.

Figure 28.002

	Where to Find It:
Command Line:	*Transparency*
Menu:	Modify – Object – Image – **Transparency**
Toolbar:	Reference – **Image Transparency**

Do This: 28.2.3A	Image Quality

I. Be sure you're still in the *cutting table28* file in the C:\Steps\Lesson28 folder. If not, please open it now.

II. Follow these steps.

28.2.3A: IMAGE QUALITY

1. We'll begin by adjusting the cutting table photograph. Zoom 🔍 in a bit closer to it.

    Command: *z*

2. Enter the *ImageAdjust* command .

    Command: *iad*

3. Notice that *ImageAdjust* activates the **ImageFrame** system variable. Select the frame around the photograph.

    Select image(s):
    Select image(s): *[ENTER]*

4. AutoCAD presents the Image Adjust dialog box (Figure 28.002, p.663). Work with the three settings until you're pleased with the preview image. (Use the **Reset** button in the lower left corner if you need to return to the original settings.)

    Pick the **OK** button when you're satisfied.

5. Now we'll set the *ImageQuality* to speed our display time. Enter the command .

    Command: *imagequality*

6. Tell AutoCAD to use a **Draft** quality. Notice the change in the photograph. Remember that the *ImageQuality* setting affects only the display.

28.2.3A: IMAGE QUALITY

7. Zoom all.

   Command: *z*

8. Add the *Product* text, freeze the **MARKER** layer, and save the drawing.

   Command: *qsave*

It now looks like the following figure.

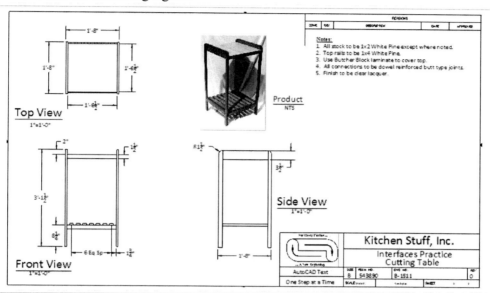

- Our last image modification tool is called *DrawOrder*. By default, AutoCAD displays the last object drawn or inserted atop (or over) previously drawn objects. This may cause certain objects to "hide" behind other objects. *DrawOrder* allows you to control the order in which objects are displayed. In other words, use *DrawOrder* to control which object appears on top. It presents the following sequence:

Where to Find It:	
Command Line:	*DrawOrder*
Hotkey(s):	*dr*
Ribbon (Tab/Panel):	Home – Modify – Draw Order (flyout) – **[Option]**
Menu:	Tools – Draw Order – **[Option]**
Toolbar:	Draw Order – **[Option]**

   Command: *draworder*

   Select objects: *[Select an object to modify.]*

   Select objects: *[Select more objects or hit ENTER to confirm the set.]*

   Enter object ordering option [Above objects/Under objects/Front/Back] <Back>: *[Tell AutoCAD how to reposition the image/object.]*

The options are best explained in an exercise, so let's begin.

Do This: 28.2.3B	**Controlling Drawing Order**

I. Open the *DrawOrd* file in the C:\Steps\Lesson28 folder. The drawing looks like the figure at right. (The drawing has two polylines and an image.)

II. Follow these steps.

28.2.3B: CONTROLLING DRAWING ORDER

1. Enter the *DrawOrder* command.
   Command: *dr*
2. Select the red horizontal polyline.
   Select objects:
   Select objects: *[ENTER]*
3. Move it to the front **Front**.
   Enter object ordering option [Above objects/Under objects/Front/Back] <Back>: *f*
   Notice the change in the display.

I created this file to give you a chance to play with the various options of the *DrawOrder* command. Take a few minutes to experiment with each of the options. (Note: You can also use the buttons on the Draw Order toolbar or in the ribbon or menu flyouts.)

28.3 Exporting Image Files

It seems that there are almost as many ways to export image files as there are types of image files to export. Luckily, AutoCAD has also provided a single command that covers the spectrum. That command is *Export*, and it works with a Windows standard Save-type dialog box.

	Where to Find It:
Command Line:	*Export*
Hotkey(s):	*exp*
Application Menu:	Export – [format]
Menu:	File – **Export**

The Export Data dialog box works as easily as the Save dialog box. Simply select the type of file you wish in the **Save as type** control box, and then pick the **Save** button.

Let's give it a try.

Do This: 28.3A	Exporting Image Files

I. Be sure you're still in the *DrawOrd* file in the C:\Steps\Lesson28 folder. If not, please open it now.
II. Follow these steps.

28.3A: EXPORTING IMAGE FILES

1. Enter the *Export* command.
   Command: *export*

2. AutoCAD presents the Export Data dialog box. Pick the down arrow next to the Files of type control box. Notice the file types available for exporting.
   Select **Bitmap (*.bmp)**.

3. Pick the **Save** button **Save** to return to the command line.

4. Hit ENTER to select all the objects in the drawing.
   Select objects or <all objects and viewports>: *[ENTER]*

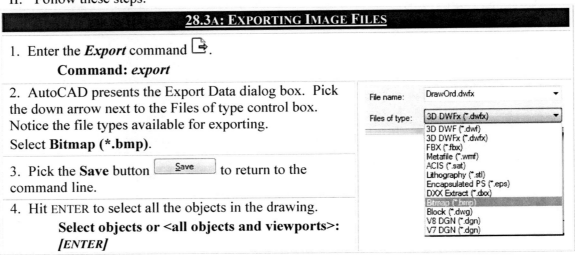

665

28.3A: EXPORTING IMAGE FILES

5. Use Windows' Paint (or any other graphics program) to view the *DrawOrd.bmp* file you just created in the C:\Steps\Lesson28 folder.

Try repeating the last exercise using the *Cutting table28* file. Does it work?*

28.4 Working with Linked Objects – Object Linking and Embedding (OLE)

Let's begin with the obvious: Exactly what is OLE? You've probably heard the term since you began using computers, but what does it mean?

OLE stands for *Object Linking and Embedding*. It refers to the method of creating a compound document – a document that requires more than one application (program) to create. In other words, you might use OLE to show a bill of materials on an AutoCAD drawing by linking directly to the Excel spreadsheet or MS Access database file where the information is stored. (These are Microsoft examples; other programs can be used as well.)

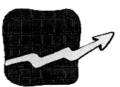

> There are actually three *OLE* terms. The first – *OLE Object* – I explain below. The second – *OLE Server* – refers to the source program that originated the object. The third – *OLE Container* – refers to the program in which you wish to link or embed the object.

The *Object* part of OLE refers to something that you'll add to your drawing (or other Windows program file). The object is actually a file, part of a file, or graphic image that originated in another program.

The *Linking and Embedding* part of OLE refers to how you wish to attach the OLE Object from the OLE Server to the OLE Container document. A *linked* document maintains its ties with the server file. Any changes made to it will reflect in the original linked document. An *embedded* document has no ties with the server file. Changes made to it are unique within your compound document. Both linked and embedded documents require the OLE Server application (program) for editing.

28.4.1 Inserting Other Application Data into AutoCAD Drawings

Microsoft and AutoCAD each provide a method for inserting an OLE Object into a document/drawing. The Microsoft method utilizes the clipboard and an AutoCAD dialog box. It gives greater control over exactly what will be inserted but requires that the server file be opened in the server application. The AutoCAD method utilizes two dialog boxes and doesn't require that the server application be opened, but in the past, it didn't offer as much control over what would be inserted.

Where to Find It:	
Command Line:	*InsertOBJ*
Hotkey(s):	*io*
Ribbon (Tab/Panel):	Insert – Data – **OLE Object**
Menu:	Insert – **OLE Object**
Toolbar:	Insert – **OLE Object**

We'll see each of these methods in our exercise. But first, let's take a look at the AutoCAD dialog boxes.

AutoCAD presents the Insert Object dialog box (Figure 28.003, p.667) when you enter the *InsertObj* command. Here you find two options for creating the image you wish to insert, and a **Display As Icon** check box.

---

*In earlier releases, **Export** had trouble with viewports. But progress has eliminated the difficulties and you should now get a good bitmap.

- When you place a check in the **Display As Icon** box, AutoCAD displays the server application's icon in the drawing instead of the file itself. Double clicking on the icon calls the object file for editing. This can speed display time.

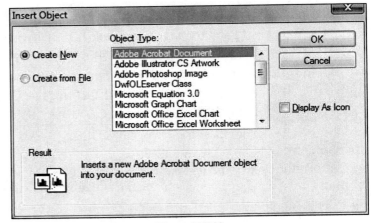

Figure 28.003

- A bullet in the **Create New** option causes the **Object Type** list box to display a list of all the programs on the computer available for OLE communication (Figure 28.003). Use this approach when an OLE Object doesn't exist and you want to create one. AutoCAD automatically embeds objects created using this method.

- A bullet in the **Create from File** option will cause AutoCAD to switch the **Object Type** list box with the **File** text box, **Browse** button, and **Link** check box shown in Figure 28.004.

Figure 28.004

You'll use the **File** text box to enter the name and location of the file you wish to insert. Alternately, you can pick the **Browse** button to use a Windows standard Open File dialog box to select the file you wish to insert.

A check in the **Link** box will link the object to your drawing rather than embedding it.

Let's insert an Excel spreadsheet into our cutting table drawing. (Note: This exercise requires the use of MS Excel – Office 97 release or better. If you don't have this software on your computer, read through the exercise, but don't try to do it.)

OLE Objects are hidden in a rendered drawing. (We'll discuss rendering in the 3D text.)

You can edit OLE properties using the Properties palette.

Do This: **28.4.1A**	**Inserting OLE Objects**

I. Reopen the *cutting table28* file in the C:\Steps\Lesson28 folder. If you haven't completed work on that drawing, open the *cutting table-done* file in the same folder.

We'll insert the OLE Object (an Excel spreadsheet) twice to see both methods of insertion and both types of insertion.
- Using the AutoCAD method, we'll embed the object.
- Using the Microsoft method, we'll link the object.

II. Zoom extents.
III. Follow these steps.

28.4.1A: INSERTING OLE OBJECTS

1. Enter the *InsertObj* command.
    Command: *io*

28.4.1A: INSERTING OLE OBJECTS

2. On the Insert Object dialog box, pick the **Create from File** option. AutoCAD presents the dialog box you saw in Figures 28.003 (p.667) and 28.004 (p.667).

3. Pick the **Browse** button.

4. AutoCAD presents the Browse dialog box (a Windows Select File dialog box). Select the *Cutting List.xlsx* file (an Excel spreadsheet) in the C:\Steps\Lesson28 folder. Pick the **Open** button to continue.

5. AutoCAD returns to the Insert Object dialog box. We'll embed the spreadsheet object into our drawing, so be sure the **Link** check box is clear.

6. Pick the **OK** button to complete the command. AutoCAD places the OLE Object in the upper left corner of the screen.

7. Select the OLE object, and then open the Properties palette.

8. In the **Geometry** section of the Properties palette, change the width of the object to **5.5**. Exit the Properties palette.

Geometry	
Position X	-2.9926
Position Y	6.7565
Position Z	0.0000
Width	5.5000
Height	3.1401
Scale width ...	77.1095
Scale height...	77.2816
Lock aspect	Yes

9. Move the object to the open area on the right side of the drawing.
   Command: *m*

10. Open the *Cutting List.xls* file in Excel.

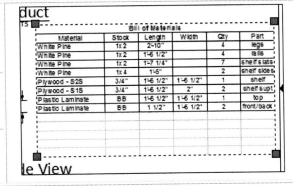

11. Return to AutoCAD. (You can pick the **AutoCAD** button on the taskbar or hold down the ALT key while tabbing to AutoCAD.)

12. Double click within the OLE Object to edit it. Notice that the object opens in Excel for editing (following figure). But notice also that the name of the file being edited is *Worksheet in cutting table*.
Change the value of cell A3 to read **Red Oak**. (Be sure to hit ENTER to finish the procedure.)

	A	B	C	D	E	F
1	Bill of Materials					
2	Material	Stock	Length	Width	Qty	Part
3	Red Oak	1x2	2'-10"		4	legs
4	White Pine	1x2	1'-6 1/2"		4	rails
5	White Pine	1x2	1'-7 1/4"		7	shelf slats
6	White Pine	1x4	1'-5"		2	shelf sides
7	Plywood - S2S	3/4"	1'-6 1/2"	1'-6 1/2"	1	shelf
8	Plywood - S1S	3/4"	1'-6 1/2"	2"	2	shelf supt
9	Plastic Laminate	BB	1'-6 1/2"	1'-6 1/2"	1	top
10	Plastic Laminate	BB	1 1/2"	1'-6 1/2"	2	front/back

28.4.1A: INSERTING OLE OBJECTS

13. Repeat Step 11. Notice that the value of the material in the first row has changed. (You may need to regen the drawing to see the change.) This is how you'll edit embedded OLE Objects.

Material	Stock	Length	Width	Qty	Part
Red Oak	1x2	2'-10"		4	legs
White Pine	1x2	1'-6 1/2"		4	rails
White Pine	1x2	1'-7 1/4"		7	shelf slats
White Pine	1x4	1'-5"		2	shelf sides
Plywood - S2S	3/4"	1'-6 1/2"	1'-6 1/2"	1	shelf
Plywood - S1S	3/4"	1'-6 1/2"	2"	2	shelf supt
Plastic Laminate	BB	1'-6 1/2"	1'-6 1/2"	1	top
Plastic Laminate	BB	1 1/2"	1'-6 1/2"	2	front/back

14. Return to Excel and close the *Worksheet in cutting table* file. Excel then presents the original *Cutting List. xlsx* file. Notice that it hasn't changed. Remember that the OLE Object was *embedded* into the OLE Container, not linked to the original file.

15. Leave Excel running, but return to AutoCAD and undo all the changes (until the OLE Object is no longer in the file).
    Command: *u*

16. Now we'll use the Microsoft method for inserting an OLE Object. Return to the spreadsheet and highlight the data cells (A1:F10) as shown in the following figure.

Cutting List.xlsx

	A	B	C	D	E	F
1	Bill of Materials					
2	Material	Stock	Length	Width	Qty	Part
3	White Pine	1x2	2'-10"		4	legs
4	White Pine	1x2	1'-6 1/2"		4	rails
5	White Pine	1x2	1'-7 1/4"		7	shelf slats
6	White Pine	1x4	1'-5"		2	shelf sides
7	Plywood - S2S	3/4"	1'-6 1/2"	1'-6 1/2"	1	shelf
8	Plywood - S1S	3/4"	1'-6 1/2"	2"	2	shelf supt
9	Plastic Laminate	BB	1'-6 1/2"	1'-6 1/2"	1	top
10	Plastic Laminate	BB	1 1/2"	1'-6 1/2"	2	front/back

17. Hold down the CTRL key on the keyboard while typing *C* (Windows method for copying something to the clipboard).

18. Return to AutoCAD.

19. On AutoCAD's Edit pull-down menu, select **Past Special** Paste Special... . (Notice that there is also a **Paste** option. This one will embed the clipboard material into the document/drawing.) Alternately, you can enter the *Pastespec* command at the command prompt.

20. AutoCAD presents the Paste Special dialog box (right). By default, it'll embed the objects. Notice the formats available for embedding objects.

Pick the **Paste Link** option. Notice that, when linked, the image must be either in the format of the OLE Server application or as **AutoCAD Entities**. If you select **AutoCAD Entities**, the data will appear as a linked AutoCAD table. (We did this in Lesson 19, p.443.)

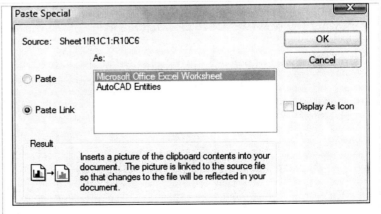

28.4.1A: INSERTING OLE OBJECTS

21. Select the **Excel** option and pick the **OK** button  to link the clipboard object (the Excel spreadsheet) into your AutoCAD drawing.

22. AutoCAD asks where to place the object. Put it under the **Product** callout (where we put in previously).

23. Select the OLE object and open the Properties palette.
    Command: *props*

24. Resize the object as you did in Step 8, p.668.

25. Exit the Properties palette.

26. Return to Excel and exit the program.

27. Now let's see what happens when we edit a linked object. Double click within the OLE Object. AutoCAD opens the OLE Server application (Excel) with the appropriate file. (Notice that the name of the file is *Cutting List. xlsx* – our original file, not an embedded copy.)

28. To make our OLE Object consistent with our drawing, change the font to **Times New Roman** or **Calibri** (in Excel).

29. Now change the **White Pine** materials to **Ponderosa Pine** (see the following figure).

	A	B	C	D	E	F
1	Bill of Materials					
2	Material	Stock	Length	Width	Qty	Part
3	Ponderosa Pine	1x2	2'-10"		4	legs
4	Ponderosa Pine	1x2	1'-6 1/2"		4	rails
5	Ponderosa Pine	1x2	1'-7 1/4"		7	shelf slats
6	Ponderosa Pine	1x4	1'-5"		2	shelf sides
7	Plywood - S2S	3/4"	1'-6 1/2"	1'-6 1/2"	1	shelf
8	Plywood - S1S	3/4"	1'-6 1/2"	2"	2	shelf supt
9	Plastic Laminate	BB	1'-6 1/2"	1'-6 1/2"	1	top
10	Plastic Laminate	BB	1 1/2"	1'-6 1/2"	2	front/back

30. Save the changes in Excel and exit the program.

31. Return to AutoCAD and notice the changes. These changes have been incorporated both in the original document and in the AutoCAD link.

32. Save the drawing.
    Command: *qsave*

It now looks like the following figure.

28.4.1A: INSERTING OLE OBJECTS

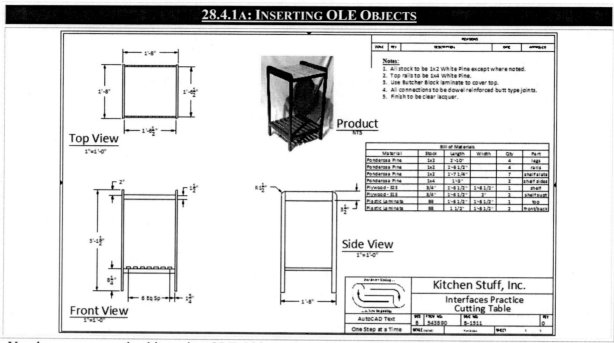

You have now practiced inserting OLE Objects. You've seen the difference between embedding and linking, but you haven't seen some of the OLE editing tools AutoCAD provides. Let's take a look at these now.

| 28.4.2 | **Modifying OLE Objects** |

For many reasons, I prefer OLE to imaging via the External Reference palette. I particularly enjoy the versatility offered by being able to edit an OLE Object using its source application. I'm no longer restricted to AutoCAD's commands and procedures. (Although an impressive program, AutoCAD can't do all things.) But thanks to some brilliant programmers at both Autodesk and Microsoft, I can use all of the programs and applications on my computer in a holistic approach to providing my client with a more complete product.

AutoCAD does, however, provide some additional tools for controlling the OLE image once it becomes part of our drawing. Let's look at these now.

- **Olehide**. You can use any of four settings for the **Olehide** system variable to control the display of an OLE Object both on the screen and when plotting. A setting of
    - **0** (the default) makes all OLE Objects visible and printable.
    - **1** makes only OLE Objects in Paper Space visible and printable.
    - **2** makes only OLE Objects in Model Space visible and printable.
    - **3** hides all OLE Objects and none are printable.

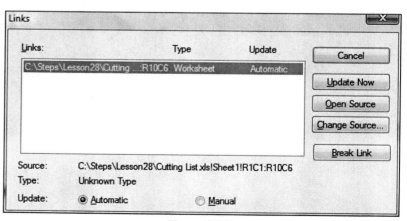

Figure 28.005

- **Olelinks**. The *Olelinks* command presents the Links dialog box (Figure 28.005). Here you find several options.
  - The list box shows all the linked OLE Objects in the drawing. Select the one with which you want to work.
  - The **Update Now** button will update the OLE Object by reading the source file and making any necessary adjustments.
  - **Open Source** opens the OLE Object in its source application.
  - **Change Source** provides an opportunity to change the source of the link.
  - **Break Link** converts the selected OLE Object into a picture of the object. Use caution when doing this because you can't modify the picture. It's a handy way, however, to send the drawing to a client who might not have the OLE Server application. Remember: *The OLE Server application must be loaded on the computer for an OLE Object to be visible!*
  - At the bottom of the dialog box, you'll find two radio buttons. Selecting **Automatic** means that AutoCAD will update the links as they change. Selecting **Manual** means that you must update the links yourself (using the **Update Now** button) whenever you want them updated.
- **Olescale**. *Olescale* calls an OLE Text Size dialog box (Figure 28.006) where you can adjust the size and font of the OLE text. (You have to select an OLE object *before* entering the *Olescale* command.)

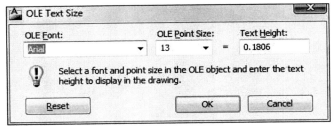

Figure 28.006

- **Olestartup**. This system variable controls whether or not the OLE Server application loads when you plot the drawing. A setting of **0** (the default) won't load the application. A setting of **1** will load the application. Loading the application often improves the quality of a plot.
- **Olequality**. This system variable controls the plot quality of the OLE objects. The possible settings and their meanings follow:
  - **0** produces a monochrome plot. This is the lowest quality but is appropriate for embedded spreadsheets.
  - **1** (the default) produces low graphics (text) quality. Use this for text documents (like Word or WordPad). (I prefer a setting of **1** for spreadsheets as well, since a spreadsheet is mostly text.)
  - **2** produces high graphics quality. It works well for charts, bitmaps, or simple artwork.
  - **3** tells AutoCAD to automatically select the quality based on the type of object being plotted.

28.4.3 AutoCAD Data in Other Applications

The methods and procedures for using an AutoCAD drawing in another application (such as Word or Excel) don't differ from what you've already seen. The only limitation involves Paper Space and Model Space views. Remember that you can't select objects in Paper Space if Model Space is active (and vice versa). This presents a problem when copying objects to the clipboard. But let's try an exercise copying our cutting table drawing into a WordPad document.

Do This: 28.4.3A Using AutoCAD Drawings in Other Applications

I. Be sure you're still in the *cutting table28* (or *cutting table28 – done*) file in the C:\Steps\Lesson28 folder. If not, please open it now.
II. Follow these steps.

28.4.3A: USING AUTOCAD DRAWINGS IN OTHER APPLICATIONS

1. Using **Quick View Layouts**, open the model.

2. Without entering a command, select all of the objects on the screen.

3. Hold down the CTRL key and type *C*. This places all of the selected objects on the Windows clipboard.

   Command: ^c

4. Open WordPad. Follow this path from the **Start** button on Windows' desktop:

   Start – All Programs – Accessories – WordPad

5. In WordPad, pick the **Paste Special** option on the Edit pull-down menu.

6. WordPad presents the Paste Special dialog box (following figure).

   Highlight **AutoCAD Drawing** in the list box, and make sure there is a bullet next to **Paste**. Then pick the **OK** button.

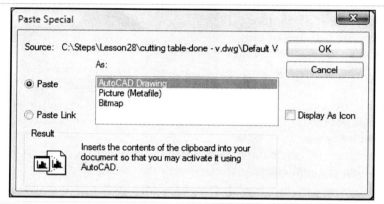

7. Return to AutoCAD and close the cutting table drawing. (Don't save changes, but leave the application running.)

8. Back in WordPad, double click on the drawing image. Notice that AutoCAD launches with *Drawing in Document* on the title bar. You've embedded an AutoCAD drawing into a WordPad document!

9. Close both applications without saving.

Of course, linking an AutoCAD drawing into another application's document is as easy!

28.5 Extra Steps

Return to the *MyCircuit* series of drawings (*1b* through *1e*) you created in Lesson 27. Create a bill of materials in Excel to include all of the block objects in *MyCircuit1a*. Be sure to save the spreadsheet outside of AutoCAD (save as *MyMats.xls* in the C:\Steps\Lesson28 folder). Then insert the spreadsheet into each of the drawings as an OLE Object, changing the list as necessary to reflect the materials in the drawing. Will you embed or link the spreadsheet?[*]

---

[*] You'll need to embed the OLE Object so that your changes are specific to the individual drawings.

28.6 What Have We Learned?

Items covered in this lesson include:
- *Differences between raster and vector imaging*
- *The Image Manager*
- *Exporting image files*
- *Object Linking and Embedding*
- *Using Windows' clipboard in AutoCAD*
- *AutoCAD Commands:*
    - *Image*
    - *ImageAttach*
    - *ImageFrame*
    - *ImageClip*
    - *ImageAdjust*
    - *ImageQuality*
    - *Transparency*
    - *DrawOrder*
    - *Export*
    - *InsertObj*
    - *Olehide*
    - *Olelinks*
    - *Olequality*
    - *Olescale*
    - *Olestartup*
    - *Clip*

In this lesson you learned how to use material from several types of applications in your drawing – and even how to use your drawings in other applications! These steps toward publication-quality drawings will serve you well in almost any industry.

28.7 Exercises

1. Open the *Wine Rack* file in the C:\Steps\Lesson28 folder. Create the layout shown in the following figure. Use the following to guide your setup:
    1.1. Set up to plot on a B-size (17x11) sheet of paper.
    1.2. Use the *ANSI B Title Block.dwg* file in the C:\Steps\Lesson28 folder as your border/title block.
    1.3. Use the *title info28* file to insert data into the title block.
    1.4. The photograph is *winerack.tif* and can be found in the C:\Steps\Lesson28 folder.
    1.5. Embed the BOM using the *Winerack.xls* Excel file.
    1.6. The logo in the title block is *logo.jpg* and can be found in the C:\Steps\Lesson28 folder.
    1.7. Complete the layout and dimension the viewports as indicated.
    1.8. Save the drawing as *MyWineRack* in the C:\Steps\Lesson28 folder.

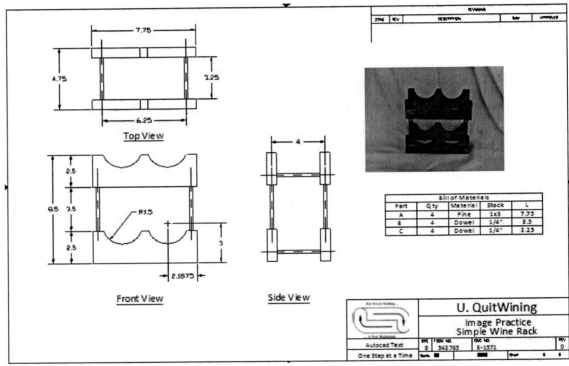

Wine Rack

2. Open the *Jewelry Box* file in the C:\Steps\Lesson28 folder. Create the layout shown. Use the following to guide your setup:
   2.1. Set up to plot on a B-size (17x11) sheet of paper.
   2.2. Use the *ANSI B Title Block.dwg* file in the C:\Steps\Lesson28 folder as your border/title block.
   2.3. Use the *title info28* file to insert data into the title block.
   2.4. The photograph is *jewelery box.tif* and can be found in the C:\Steps\Lesson28 folder.
   2.5. Embed the BOM using the *JBox.xls* Excel file.
   2.6. The logo in the title block is *logo-mini.jpg* and can be found in the C:\Steps\Lesson28 folder.
   2.7. Complete the layout and dimension the viewports as indicated.
   2.8. Save the drawing as *MyJewelryBox* in the C:\Steps\Lesson28 folder.

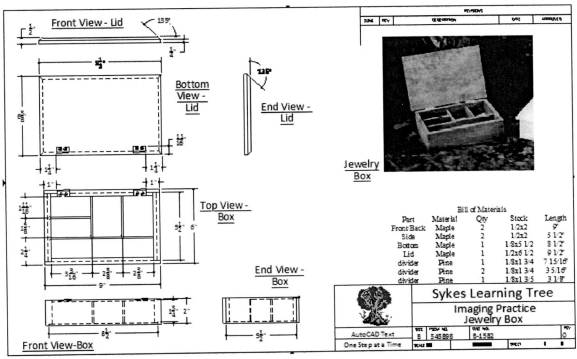

Jewelry Box

28.8 **For Web-Based Review Questions and Additional Exercises, visit: http://foragerpub.com/AcadFiles/2011/2011.htm**

To continue, please pick up a copy of *3D AutoCAD 2011: One Step at a Time*.

3D AutoCAD 2011: One Step at a Time

Contents

Chapter 1: "Z" Basics
Chapter 2: More of Z Basics
Chapter 3: Wireframes and Surface Modeling
Chapter 4: Predefined Meshes
Chapter 5: Complex Mesh & Surface Models
Chapter 6: Z-Space Editing
Chapter 7: Mesh & Surface Editing
Chapter 8: Solid Modeling Creation Tools – The Basics
Chapter 9: Composite Solids
Chapter 10: Editing 3D Solids
Chapter 11: Three-Dimensional Blocks and Three-Dimensional Plotting Tools
Chapter 12: Presentation Tools
Chapter 13: Animation Tools

Where to from here?

Appendix – A: Initial Setup

When you installed AutoCAD, you had an opportunity to create an *initial setup* – that is, the default workspace AutoCAD uses. In this text, you see how to use the 2D Drafting and Annotation workspace (which contains just about everything you'll need for general AutoCAD functionality). You can, however, have AutoCAD install the tools most often required for several specific industries: Architecture, Civil Engineering, Electrical Engineering, Manufacturing, MEP, Structural Engineering, and General Design and Documentation.

If you neglected to go through this setup when you installed AutoCAD, or if you wish to change the initial setup, follow these steps.

APPENDIX A: INITIAL SETUP

1. Open the Options dialog box. (Caution: the Options dialog box is *not* a place for experimentation!)

    Command: *op*

2. Pick the **User Preferences** tab.

3. On the User Preferences tab, pick the **Initial Setup** button.

    AutoCAD begins the Initial Setup wizard.

4. On the first page of the wizard, select the industry which most closely describes your work. For this text, select **Other**.

    ○ Other (General Design and Documentation)

5. Pick the **Next** button to continue.

6. AutoCAD will create a default set of ribbon panels depending upon your selection in Step 4, but on the second page, you can add additional task-based tools that you may require. These include tools for: **3D Modeling**, **Photorealistic Rendering**, **Review and Markup**, and **Sheet Sets**. Select these now. (For this text, select them all.)

7. Pick the **Next** button to continue.

8. When AutoCAD opens, it starts a new file with a default template. On the last page of our wizard, you can tell AutoCAD to use its default template, one of your own, or a default template based on your industry and unit requirements. For this text, **Use AutoCAD 2011's default drawing template file**.

9. Pick the **Finish** button to complete the setup.

10. Close the Options dialog box.

AutoCAD now begins with an Initial Setup Workspace. Use the procedures detailed in Lesson 1 to change the workspace should you desire.

Appendix – B: Drawing Scales

SCALE (= 1')	SCALE FACTOR	\multicolumn{5}{c}{DIMENSIONS OF DRAWING WHEN FINAL PLOT SIZE IS:}				
		8½"x11"	11"x17"	17"x22"	22"x34"	24"x36"
1/16"	192	136'x176'	176'x272'	272'x352'	352'x544'	384'x576'
3/32"	128	90'8x117'4	117'4x181'4	181'4x234'8	234'8x362'8	256'x384'
1/8"	96	68'x88'	88'x136'	136'x176'	176'x272'	192'x288'
3/16"	64	45'4x58'8	58'8x90'8	90'8x117'4	117'4x181'4	128'x192'
¼"	48	34'x44'	44'x68'	68'x88'	88'x136'	96'x144'
3/8"	32	22'8x29'4	29'4x45'4	45'4x58'8	58'8x90'8	64'x96'
½"	24	17'x22'	22'x34'	34'x44'	44'x68'	48'x72'
¾"	16	11'4x14'8	14'8x22'8	22'8x29'4	29'4x45'4	32'x48'
1"	12	8'x6'11	11'x17'	17'x22'	22'x34'	24'x36'
1½"	8	5'8x7'4	7'4x11'4	11'4x14'8	14'8x22'8	16'x24'
3"	4	34"x44"	3'8x5'8	8'x6'11	7'4x11'4	8'x12'
(1" =)						
10'	120	85'x110'	110'x170'	170'x220'	220'x340'	240'x360'
20'	240	170'x220'	220'x340'	340'x440'	440'x680'	480'x720'
25'	300	212'6x275'	275'x425'	425'x550'	550'x850'	600'x900'
30'	360	255'x330'	330'x510'	510'x660'	660'x1020'	720'x1080'
40'	480	340'x440'	440'x680'	680'x880'	880'x1360'	960'x1440'
50'	600	425'x550'	550'x850'	850'x1100'	1100'x1700'	1200'x1800'
60'	720	510'x660'	660'x1020'	1020'x1320'	1320'x2040'	1440'x2160'
80'	960	680'x880'	880'x1360'	1360'x1760'	1760'x2720'	1920'x2880'
100'	1200	850'x1100'	1100'x1700'	1700'x2200'	2200'x3400'	2400'x3600'
200'	2400	1700'x2200'	2200'x3400'	3400'x4400'	4400'x6800'	4800'x7200'

Appendix – C: Function Keys and Their Uses

Key	Function	Key	Function	Key	Function
F1	AutoCAD's Help	F5	Isoplane toggle	F9	Snap toggle
F2	Toggle between graphics & text screen	F6	Toggle coordinate indicator on or off	F10	Polar Tracking Toggle
F3	OSNAP settings	F7	Grid toggle	F11	Object Tracking Toggle
F4	Tablet (Digitizer) toggle	F8	Ortho toggle	F12	Dynamic Input Toggle

Appendix – D: MText Keystrokes

	THESE KEYS:	DO THIS:
P O S I T I O N	← → ↑ ↓	Move the cursor through the text one space at a time
	Ctrl + ← Ctrl + →	Move the cursor through the text one word at a time
	Home	Move the cursor to the beginning of the line
	End	Move the cursor to the end of the line
	Page Up Page Down	Move the cursor through the document up to 28 lines at a time
	Ctrl + Home	Move the cursor to the beginning of the document
	Ctrl + End	Move the cursor to the end of the document
	Ctrl + Page Up Ctrl + Page Down	Move the cursor to the top or bottom of the screen
S E L E C T I O N	Ctrl + A	Selects all the text in the mtext object
	Shift + ← Shift + →	Selects / deselects text one character at a time
	Shift + ↑ Shift + ↓	Selects / deselects text one line at a time
	Ctrl + Shift + ← Ctrl + Shift + →	Selects / deselects text one word at a time
A C T I O N	Delete	Deletes the character to the right of the cursor
	Backspace	Deletes the character to the left of the cursor
	Ctrl + Backspace	Deletes the word to the left of the cursor
	Esc	Leaves the text editor without saving the changes
	Ctrl + C	Copy the selected text to the Windows clipboard
	Ctrl + V	Paste text from the Windows clipboard into the text editor
	Ctrl + X	Remove the selected text and place it on the Windows clipboard
	Ctrl + Z	Undo the last edit
	Enter	Starts a new paragraph

Appendix – E: Dimension Variables

Dimension Variable	Default Setting	Explanation
DIMADEC	-1	Angular decimal places (set from 0-8); a setting of –1 tells AutoCAD to use the DIMDEC setting
DIMALT	Off	Use of alternate units
DIMALTD	2	Decimal places in alternate units
DIMALTF	25.4	Scale factor in alternate units
DIMALTTD	2	Decimal places in tolerances of alternate units
DIMALTTZ	0	Suppression of zeros in alternate units (0 or 1)
DIMALTU	2	Units format for alternate units (except angular)
DIMALTZ	0	Suppression of zeros in tolerance values
DIMAPOST	""	Identifies a text suffix / prefix to be used with alternate dim
DIMASO	on	Associative dimensioning (on or off)
DIMASZ	0.18	Arrowhead size
DIMAUNIT	0	Format for angular dimensions
DIMBLK	""	Specify a block to place on the dimension line rather than an arrowhead
DIMBLK1 DIMBLK2	"" ""	Specifies blocks for both ends of the dimension line (if the DIMSAH variable is **on**)
DIMCEN	0.09	Controls what, if any, center marks are placed in circles and arcs (Centerlines are drawn if the number is less than 0; Center marks are drawn if the number is above 0; no center marks/lines are drawn if value equals 0)
DIMCLRD	0	Color for dimension lines, arrowheads, and leaders
DIMCLRE	0	Color for extension lines
DIMCLRT	0	Color for dimension text
DIMDEC	4	Primary units decimal places
DIMDLE	0.0	Extension of dimension line beyond extension line
DIMDLI	0.38	Spacing of baseline dimension lines
DIMEXE	0.18	Extension of extension line beyond dimension line
DIMEXO	0.0625	Distance from origin to beginning of extension line
DIMFIT	3	This controls whether or not a dimension line will be placed between the extension lines if there is enough space. Settings are: 0 – arrows and text; 1 – just text; 2 – just arrows; 3 – best fit (puts text and arrows between the extension lines as space is available); 4 – leader (when no space is available for the text, it is placed aside and connected to the dimension line with a leader line); 5 – no leader (same as 4 but will not use a leader line).
DIMGAP	0.09	Space around the text when it is placed inside the dimension line or between bottom of text and dimension line when text is placed above the line
DIMJUST	0	Horizontal position of dimension text. Possible settings include: 0 – centered between extension lines; 1 – next to the first extension line; 2 – next to the second extension line; 3 – next to and aligned with the first extension line; 4 – next to and aligned with the second extension line

Dimension Variable	Default Setting	Explanation
DIMLFAC	1.0	Global scale factor for dimensioning
DIMLIM	off	Places dimension limits as default dimension text
DIMPOST	""	Identifies a text prefix or suffix for the dimension text
DIMRND	0.00	Rounds the dimension distances to this value
DIMSAH	off	Determines whether or not blocks will be used instead or arrowheads
DIMSCALE	1.0	Sets an overall scale factor for all dimension variables requiring size settings
DIMSD1 DIMSD2	off off	Suppression of the first and second dimension line
DIMSE1 DIMSE2	off off	Suppression of the first and second extension line
DIMSHO	on	This actually controls whether or not the dimension will be redefined as the dimension is dragged
DIMSOXD	off	Controls whether or not dimension lines will be drawn outside the extension lines
DIMSTYLE	"Standard"	Current dimension style
DIMTAD	0	Vertical position of text. Possible settings are: 0 – centered inside the dimension line; 1 – place text above the dimension line; 2 – on the side of the dimension line farthest from the defining point; 3 – Japanese standard
DIMTDEC	4	Decimal places in tolerance value
DIMTFAC	1.0	Scale factor the text height of tolerance values
DIMTIH	on	Whether (**on**) or not (**off**) text inside the dimension lines will be horizontal
DIMTIX	off	Force text between the extension lines
DIMTM	0.0	Sets the minimum tolerance limits when either DIMLIM or DIMTOL is **on**
DIMTOFL	off	Force a dimension line between extension lines
DIMTOH	on	Position of text outside dimension lines (**on** forces horizontal text)
DIMTOL	off	Add tolerances to dimension text
DIMTOLJ	1	Vertical justification of tolerance values (0 – bottom; 1 – middle; 2 – top)
DIMTP	0.0	Upper tolerance limit
DIMTSZ	0.0	Size of ticks used instead of arrowheads
DIMTVP	0.0	Vertical position of text above / below dimension line
DIMTXSTY	"Standard"	Text style used for the dimension text
DIMTXT	0.18	Height of dimension text

Dimension Variable	Default Setting	Explanation
DIMTZIN	0	Suppression of zeros in tolerance values. Possible values: 0 – suppresses zero feet and zero inches; 1 – includes zero feet and inches; 2 – includes zero feet but suppresses zero inches; 3 – includes zero inches but suppresses zero feet; 4 – suppresses leading zeros in decimals; 8 – suppresses trailing zeros in decimals; 12 – suppresses both leading and trailing zeros in decimals.
DIMUNIT	2	Dimension units for everything but angular dimensions. Possible settings are: 1 – Scientific; 2 – Decimal; 3 – Engineering; 4 – Architectural (stacked); 5 – Fractional (stacked); 6 – Architectural; 7 – Fractional; 8 – windows settings
DIMUPT	off	If **on**, the cursor controls placement of both dimension and text; if **off**, the cursor controls placement only of the dimension
DIMZIN	0	Suppression of zeros in the primary units. . Possible values: 0 – suppresses zero feet and zero inches; 1 – includes zero feet and inches; 2 – includes zero feet but suppresses zero inches; 3 – includes zero inches but suppresses zero feet; 4 – suppresses leading zeros in decimals; 8 – suppresses trailing zeros in decimals; 12 – suppresses both leading and trailing zeros in decimals.

Appendix – F: Hotkeys

Command	Hotkey
ADCenter	ADC
Align	AL
Area	AA
Array	AR
Attdisp	ATT
Attedit	ATE
Block	B, -B
Break	BR
Chamfer	CHA
Circle	C
color	COL
Copy	CO, CP
DDEdit	ED
Dimaligned	DAL
Dimangular	DAN
Dimbaseline	DBA
Dimcontinue	DCO
Dimdiameter	DDI
Dimedit	DED
Dimlinear	DLI
Dimordinate	DOR
Dimradius	DRA
Dist	DI
Divide	DIV
Donut	DO
Dsettings	DS
Ellipse	EL
Erase	E
Explode	X
Extend	EX
Fillet	F
Filter	FI
Grid	F7
Grips	GR
Group	G
Hatch	H
HatchEdit	HE
Insert	I
Layer	LA, -LA
Leader	LE
Lengthen	LEN
Line	L
Linetype	LT, -LT

Command	Hotkey
List	LI
LTScale	LTS
Lweight	LW
MatchProp	MA
Measure	ME
Mirror	MI
MLine	ML
Move	M
MText	MT, T
Offset	O
Open	^O
Ortho	F8
OSNAP	OS, -OS
Pan	P, -P
Pedit	PE
Pline	PL
Point	PO
Polygon	POL
Properties	PROPS
Purge	PU
QSave	^S
Rectangle	REC
Redraw	R
Regen	Re
Rotate	RO
Save	^S
Scale	SC
Snap	SN
Solid	SO
Spell	SP
Spline	SPL
Splinedit	SPE
Stretch	S
Style	ST
Text	DT
Trim	TR
Units	UN
View	V
WBlock	W, -W
Xline	XL
Zoom	Z

Appendix – G: Actions & Parameters Chart for Dynamic Blocks

Parameters	Actions								
	Move	Stretch	Scale	Array	Polar Stretch	Rotation	Flip	Lookup	No Action
Point	■	■							
Linear	■	■	■	■					
Polar	■	■	■	■	■				
XY	■	■	■	■					
Rotation						■			
Alignment									■
Flip							■		
Visibility									■
Lookup								■	
Base Point									■

Index

Absolute Coordinates, 47
Action Recorder, 287, 288, 289, 291
    ActRecord, 288
    ActStop, 288
ActRecord, 288
ActStop, 288
ADCenter, 169, 685
AddSelected, 180
Align, 180, 181, 182, 685
Angbase, 207
AnnoAllVisible, 586
Arc, 109, 112, 113
Area, 685
Array, 184, 185, 186, 685, 686
AttDef, 501, 502
AttDia, 502, 509
Attdisp, 685
ATTipe, 502
AttReq, 509
AUI, 3
AutoCAD Design Center, 169
AutoConstrain, 230, 231, 232
Autodesk Seek, 170
AutoLISP, 104, 105
BAttMan, 513, 516, 518
Block, 469, 470, 471, 473, 474, 475, 476, 477, 478, 491, 495, 685
    Insert, 470, 471, 473, 474, 475, 476
Block Attributes
    AttDef, 501, 502
    AttDia, 502, 509
    ATTipe, 502
    AttReq, 509
    BAttMan, 513, 516, 518
    DataExtraction, 520
    Insert, 502, 514, 516, 520
Block Editing
    Authoring Palettes, 477, 478
    BAction, 478, 484
    BConstruction, 478
    BEdit, 476, 477
    BParameter, 478, 479
    BSave, 477
    BTable, 477
    BTestBlock, 477
    BVMode, 491
    DelConstraint, 478
Break, 195, 196, 685
BTable, 507
Calculator, 266, 267, 268, 269
Cartesian Coordinate System, 47
CConstraintForm, 394
Chamfer, 196, 197, 198, 685
Circle, 35, 40, 685
Clip, 661, 662
Close, 9, 11, 19, 26
Closeall, 19
Color, 127, 128, 129, 130, 131, 132, 133, 142, 145, 682
columns, 337, 346
CommandLine, 13
CommandLineHide, 13
ConstraintBar, 227

ConstraintSettings, 230, 397
Copy, 178, 179, 184, 190, 681, 685
CopyToLayer, 165
Cut, 190
DataExtraction, 520
DDEdit, 100, 685
DDPType, 246
Delete, 45
Delobj, 211
Dependent Symbols, 642
Design Center, 464
design intent, 225, 394
Design Review, 538, 539, 540, 542
DGNFrame, 641
dictionary, 348
Dimaligned, 685
Dimangular, 685
Dimbaseline, 685
DimConstraint, 394
Dimcontinue, 685
Dimdiameter, 685
Dimedit, 685
Dimension
    Annotative, 378, 379, 384, 390
    Dimaligned, 362, 370, 372
    Dimangular, 355
    Dimarc, 358, 359
    DimAssoc, 352
    Dimbaseline, 363
    Dimbreak, 352, 368, 371
    Dimcontinue, 360, 363
    Dimdiameter, 357
    DimDisAssociate, 352
    DimInspect, 373, 374
    Dimjogged, 357
    Dimlinear, 353, 354, 355, 360, 362
    Dimordinate, 364, 365
    Dimradius, 357
    DimReAssociate, 352
    DimRegen, 352
    DimSpace, 363
    DimTEdit, 368, 369
    Isometric, 372
    QDim, 366, 367
    terminology, 351
Dimension Style
    Angular, 377, 378, 386
    Diameter, 377, 378
    -dimstyle, 400
    Leaders, 377
    Linear, 377, 378, 381, 384, 394
    Ordinal, 377
    Radial, 377
    Tolerances, 377
Dimension Styles, 377
Dimensional Constraint
    aligned, 394
    angular, 394
    diameter, 394
    linear, 394
    radial, 394
Dimensional Constraints, 394, 397